ENGINEERING PHYSICS

ENGINEERING PHYSICS

By

DIPAK CHANDRA GHOSH

Emeritus Professor, Jadavpur University
M.Sc (Gold Medalist), Ph. D
Professor, Deptt. of Physics, Jadavpur University
Coordinator of Nuclear and Particle
Physics Research Center, Jadavpur University
Formerly Dean, Faculty of Science, Head,
Deptt. of Physics, Jadavpur University
Recipient of Sir C.V. Raman Award (U.G.C)

NRIPESH CHANDRA GHOSH

Retd. Professor of Physics
M.Sc, Ph. D
Formerly Reader, Deptt. of Physics,
Jadavpur University
At present Professor of Physics,
RCC - Institute of Information Technology, Kolkata

PRABIR KUMAR HALDAR

M.Sc, Ph. D
Assistant Professor of Physics, Dinhata College, Cooch Behar
Formerly Lecturer in Physics, SIT, Siliguri
Recipient of the DST Project Under SERC FAST TRACK
Scheme for Young Scientist & U.G.C. Project

UNIVERSITY SCIENCE PRESS
(An Imprint of Laxmi Publications Pvt. Ltd.)
An ISO 9001:2008 Company
BENGALURU • CHENNAI • COCHIN • GUWAHATI • HYDERABAD
JALANDHAR • KOLKATA • LUCKNOW • MUMBAI • RANCHI • NEW DELHI
BOSTON (USA) • ACCRA (GHANA) • NAIROBI (KENYA)

ENGINEERING PHYSICS

Printed and bound in India
Typeset at Shubham Composer, Delhi
First Edition: 2008, Second Edition: 2016
ISBN 978-81-318-0366-0

Branches

✆	Bengaluru	080-26 75 69 30	
✆	Chennai	044-24 34 47 26,	24 35 95 07
✆	Cochin	0484-237 70 04,	405 13 03
✆	Guwahati	0361-254 36 69,	251 38 81
✆	Hyderabad	040-27 55 53 83,	27 55 53 93
✆	Jalandhar	0181-222 12 72	
✆	Kolkata	033-22 27 43 84	
✆	Lucknow	0522-220 99 16	
✆	Mumbai	022-24 91 54 15,	24 92 78 69
✆	Ranchi	0651-220 44 64	

PUBLISHED IN INDIA BY

UNIVERSITY SCIENCE PRESS
(An Imprint of Laxmi Publications Pvt.Ltd.)
An ISO 9001:2008 Company
113, GOLDEN HOUSE, DARYAGANJ,
NEW DELHI - 110002, INDIA
Telephone : 91-11-4353 2500, 4353 2501
Fax : 91-11-2325 2572, 4353 2528
www.laxmipublications.com info@laxmipublications.com

C—11832/016/04
Printed at : N.K. BOOK BINDER

Contents

FOREWORD

It gives me immense pleasure in writing a few lines about the book **Engineering Physics** written by my colleagues Prof. Dipak Ghosh, Dr. Nripesh Chandra Ghosh and my student Dr. Prabir Kumar Haldar. There was need of a good comprehensive book for undergraduate engineering students all over the country—the authors have done a commendable job venturing to write this book. The exposition is very lucid and the presentation has been carefully formatted. I am confident that this book will be interesting and useful for young mind.

Prof. Ashoke Nath Bose
Emeritus Professor of Physics
Former Vice Chancellor
Jadavpur University, Kolkata
India

Preface to the Second Edition

It is a matter of extreme pleasure to us that the first edition of our book **Engineering Physics** has been widely appreciated by the students and teachers community all over India. However, the constructive suggestions of our well wishers have prompted us to bring out the second edition of the book. This edition includes several valuable addition of topics like Elasticity, surface Tension, Viscosity, Heat and Thermodynamics, to accommodate the syllabi of the most of undergraduate Engineering curriculum. Exhaustive numerical problems covering all topics have been incorporated which will be of extreme useful for practice.

We hope this edition will also receive wide acceptance by students and teachers all over the country. We take the opportunity to thank all teachers who suggested several points for the benefit of the students.

Suggestions for the improvement of this book will be gratefully acknowledged.

—Authors

PREFACE TO THE FIRST EDITION

Our long experience in teaching physics to engineering students of the Jadavpur University has revealed that there are very scanty number of textbooks in engineering physics, which are comprehensive. This book is a single volume text which covers all the topics outlined in syllabi of engineering physics for graduate level students of most of the Indian Universities of Technology and IITs. This book will be of great help to the graduate level students of physics of all the Indian Universities.

To familiarize the students with the basic principles of physics and their implementation we have included numerous solved examples. We have also given a number of MCQs and examples to be solved by the students. We have included in-focus section at the end of each chapter to apprise the students about the most important aspects in it. We have also included a table indicating the connection between basic scientific input and corresponding technology. This will motivate the students and will create in them an urge to aspire for greater things.

We have been enriched by the suggestions from many of our students and research workers who are now teachers in Universities and Degree Colleges. Mentioning of the individual names in particular will be rather too long to be enlisted within such a short span of space. However, special mention ought to be made of the contributions by Dr. Argha Deb, Reader, Dept. of Physics, Jadavpur University, Sri Arindam Mandal and Sri Sunil Maiti, RCC-IIT, Kolkata, in developing a few chapters by adding and editing the materials. We are much too indebted to them for their earnest and prudent participation.

We also appreciate gratefully the untiring help rendered by Kanchan Kumar Patra and Rosalima Sengupta during the preparation of the manuscript.

We have every hope that the students will find this book very much useful to them.

Suggestions from teachers and students will be of great value to the authors in improving this textbook.

—Authors

Isaac Newton

1 Chapter

Classical Mechanics

1.1 INTRODUCTION

Mechanics is the branch of physical science which treats of the simplest form of motion of matter–the mechanical motion. Such motion is made up of the changes in the relative positions of bodies (or their parts) in space in the course of time.

Bodies are macroscopic systems consisting of a very large number of molecules or atoms, so that the sizes of these systems are many times larger than the intermolecular distances.

Classical or Newtonian Mechanics deals with the motions of bodies travelling at velocities that are very much less than that of light in vacuum (which is $3 \times 10^{+10}$ cm/sec). The investigation of the motion of bodies travelling with velocities very close to that of the velocity of light in vacuum is taken up in Relativistic Mechanics which is based on the theory of relativity. Specific features of the motion of microparticles are dealt within Quantum Mechanics. Microparticles are those particles whose rest mass (*i.e.*, the mass of the particle measured in the co-ordinate system with respect to which the particle is stationary) is very close to or smaller than the rest mass of atoms.

Classical mechanics possesses three principal branches—(*a*) statics (*b*) kinematics and (*c*) dynamics.

Statics deals with the laws of the composition of forces and with the conditions of equilibrium of solid, liquid and gaseous bodies.

Kinematics is the study of the mechanical motion of bodies without regard for the interaction between bodies causing this motion.

Dynamics treats of the influence of the interaction between bodies on their mechanical motion.

Thus, kinematics deals with all possible motions of material systems without reference to the agency which causes the motion. Dynamics deals with the motions to the forces associated with it along with the properties of moving bodies. As such it introduces the concept of forces and other related problems of motion. Statics, however deals with the system of forces which actually give no motion to the system. The net effect of the forces should be zero.

Newton's laws of motion is the basis of Mechanics. Alternative development of the same by Lagrange, Hamilton and others goes by the name Classical Mechanics.

Newtonian mechanics deals with vectors such as force, acceleration, momentum etc. The Newtonian laws of motion are valid for **inertial frames** in which a particle subjected to no external force moves along a straight line with constant velocity, An inertial frame, therefore, is unaccelerated.

The position and time recorded together constitute an event represented by (x, y, z, t) and the so-used reference system is the space-time reference system.

To describe the motion of a particle in space knowledge of a co-ordinate system is required the simplest one being cartesian.

Systems of co-ordinates most frequently used in mechanics are :

(*i*) right handed rectangular cartesian.
(*ii*) cylindrical and
(*iii*) spherical polar.

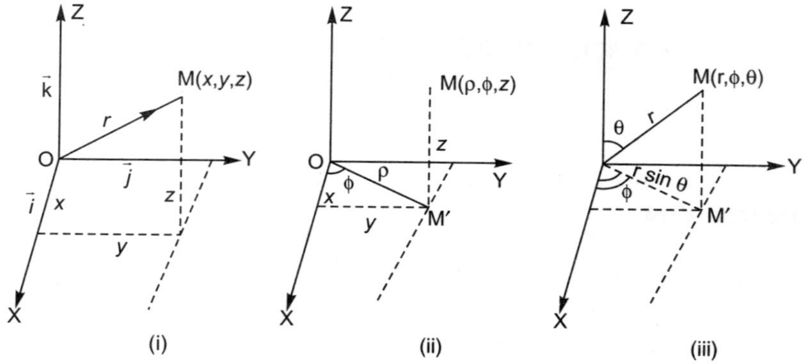

Figure 1.1

$\hat{i}, \hat{j}, \hat{k}$ are unit vectors along X, Y, Z respectively

(*i*) $\vec{r} = \hat{i}\,x + \hat{j}\,y + \hat{k}\,z$

(*ii*) $\rho = \sqrt{x^2 + y^2}$ and $x = \rho \cos \phi$,

$\phi = \tan^{-1} \dfrac{y}{x}$ $\quad y = \rho \sin \phi$,

$z = z$.

(*iii*) $x = r \sin \theta \cos \phi$

$y = r \sin \theta \sin \phi$

$z = r \cos \theta$

Frame of Reference is a set of co-ordinate axes by which the position of any object may be specified as it changes with time.

1.1.1 Newton's Laws of Translational Motion

Newtonian mechanics is based principally on the three natural laws of motion as stated below.

1.1.1.1 Newton's First Law

A body at rest remains at rest, and a body in motion continues to move with a constant velocity in a straight line, unless, in either case, it is acted upon by an unbalanced force or a non-zero resultant force.

Discussion : The first part of this law states that it takes an unbalanced force or a non-zero resultant force to change the state of rest of a body. Thus a body at rest has forces acting upon it but the resultant of these forces is zero. As an example one can consider a book lying on a table and is at rest simply because the forces acting on the book are balanced. These forces are gravitational downwards force on the book and equal and upward opposite force of reaction so that the resultant vertical force is zero.

The second part says that a body which is set in motion and left to itself will maintain a motion of translation with constant velocity unless there is an unbalanced or non-zero resultant force acting on the body. The presence of an unbalanced force may change the magnitude or the direction of the velocity (or both). It should be remembered that a change in direction of velocity also is a type of acceleration.

The first law gives a qualitative relationship between the action of an unbalanced force and the resulting acceleration.

1.1.1.2 Newton's Second Law

The time rate of change of the linear momentum of a body is equal to the resultant force acting upon the body and has the same direction as the resultant force.

Discussion : Let $\sum \vec{F}$ represent the resultant force acting on a body whose mass is m and whose velocity at any time t is v. Then by Newton's second law

$$\sum \vec{F} = \frac{d}{dt}(m\vec{v}) = \frac{d\vec{p}}{dt}$$

where $p = mv$ is the linear momentum.

$$\therefore \qquad \sum \vec{F} = m\frac{d\vec{v}}{dt} + \vec{v}\frac{dm}{dt} \quad \text{provided mass } m \text{ is a variable one.}$$

Taking the mass as constant and considering $\frac{dv}{dt}$ to be acceleration $\vec{a}$ we have,

$$\sum \vec{F} = m\vec{a}.$$

Hence one can say that the acceleration of a body is directly proportional to the resultant force acting on the body and is in the same direction as the resultant force. The equation states that it requires an unbalanced force to impart an acceleration to a body and if the resultant force is zero there is no acceleration, the body being either at rest or in uniform motion in a straight line. Further the larger the value of m, the smaller is the acceleration $\vec{a}$. This means that under the action of a *constant* resultant force, mass and acceleration are inversely proportional to each other.

1.1.1.3 Newton's Third Law

To every action (force) there is an equal and opposite reaction (force).

Discussion : If there is an interaction between two bodies A and B, the force exerted by A on B is equal in magnitude but opposite in direction to the force exerted by B on A.

It is to be noted that forces always occur in pairs as action and reaction forces and these never act on the same body but always on different bodies. In case of a bat striking a ball, during the impact the bat exerts a force on the ball and the ball exerts an equal and opposite force on the bat. Action and reaction may also act on different parts of the same body.

Whenever two bodies interact the force $\overrightarrow{F_{12}}$ on the second body (say, 2) due to the first (say, 1) is equal and opposite to the force $\overrightarrow{F_{21}}$ on the first (1) due to the second (2)

$$\overrightarrow{F_{12}} = -\overrightarrow{F_{21}}$$

The above relation possess the inherent requirement that both the forces are to be measured at the same time.

The first two laws hold only when observed in unaccelerated reference frames. The fundamental law of classical mechanics is the Newton's second law of motion.

For a body of constant mass $\vec{F} = m\vec{a}$, $\vec{a}$ being the acceleration. Sometimes m is not constant. As for example, m is not constant for a satellite that is ejecting fuel. For any body moving in the region of relativistic velocities (*i.e.*, close to the velocity of light) the mass m is not constant. Hence, we define an *inertial frame* as one which is unaccelerated. Thus, $\vec{F} = \frac{d\vec{p}}{dt}$ and $\vec{F} = m\vec{a}$ define the *true force* acting on the body. To get a valid law in the given simple forms we must *refer the acceleration to a reference system which is unaccelerated*. Such a reference system is called *Inertial or Galilean*. In an *accelerated (non-inertial)* reference system $\vec{F}$ does not equal $m\vec{a}$ if $\vec{a}$ is observed from a *non-inertial* system.

1.1.2 Inertial System of Reference

It is necessary to have an *unaccelerated reference system* with respect to which we can measure accelerations. Such a system assumes that we must have some independent way of knowing that there is no force upon it. We know that the forces between two bodies must fall off fairly rapidly as the distance between them increases (inverse square law). A typical star is at least 10^{18} cm from its nearest neighbour and is expected to have only a small acceleration (negligible). Hence fixed stars may define a convenient unaccelerated co-ordinate system to a good approximation. So, *all reference frames in uniform motion of translation relative to fixed stars are the inertial frames*. For many purposes, *the earth is a fairly good approximation to an inertial frame*.

1.1.3 Pseudo Forces or Fictitious Forces

To an observer in an accelerated frame of reference there appears to operate a force in a direction opposite to that of the acceleration the magnitude of the force so caused is equal to the product of the mass of the body and the acceleration of the frame of reference. This force is called Pseudo force or fictitious force.

In an inertial frame $\vec{F} = m\overrightarrow{a_I}$

where $\overrightarrow{a_I}$: acceleration in inertial frame

If $\vec{a}$ is the acceleration of a body in a non-inertial frame then, $\overrightarrow{a_I} = \vec{a} + \overrightarrow{a_0}$ giving

$$\vec{F} = m\left(\vec{a} + \overrightarrow{a_0}\right)$$

For experiments in a non-inertial frame one has to include $\vec{a_0}$ in the force equation.

Thus, $$\vec{F} + \vec{F_0} = m\vec{a}$$

where $F_0 = -m\vec{a_0}$ is the fictitious force.

Example

(*i*) A man in an elevator moves up with an acceleration $\vec{a_0}$ and so he is in a non-inertial frame of reference. His apparent weight will be given by $m\left(\vec{g} + \vec{a_0}\right) = \vec{F}$; $m\vec{a_0}$ is the pseudo force.

(*ii*) Consider a point of mass m at rest in a non-inertial frame so that in this frame $\vec{a} = 0$. If now this non-inertial frame rotates uniformly about an axis with respect to an inertial frame then the acceleration of the point-mass will be

$\vec{a_0} = -\omega^2\vec{r}$; $\vec{r}$: position vector of the rotating mass.

The mass might be kept at rest by a string so as to make $\vec{a_0} = 0$ in the non-inertial frame whereby,

$$\vec{F} = -\vec{F_0} = m\vec{a_0} = -m\omega^2\vec{r}$$

The fictitious force F_0 in the above is the centrifugal force and is directed away from the axis of rotation.

(*iii*) Suppose the force applied to a mass m by a stretched spring in the x-direction is $F_x = -Kx$, K being a constant. Consider a non-inertial frame with the acceleration $\vec{a_0} = b\vec{x}$ in the x-direction. If the spring is at rest in this non-inertial frame then its acceleration $\vec{a}$ in this frame is zero, and

$$\vec{F} = m\left(\vec{a} + \vec{a_0}\right) \text{ reduces to } F_x = ma_0$$

and $$\vec{F} + \vec{F_0} = m\vec{a} \text{ reduces to } F_x + (F_0)_x = 0$$

i.e., $$ma_0 = -(F_0)_x$$

Now, $$\vec{F} = -K\vec{x} \quad \therefore \quad -Kx + (F_0)_x = 0 \text{ or } -Kx - ma_0 = 0$$

or $$x = -\frac{ma_0}{K}$$

Hence, the displacement is proportional to the acceleration a_0 of the non-inertial frame.

1.1.4 Frictional Force

In order to ascertain the state of motion of a body we have to consider the combined effects of all the forces that act upon the body. Whenever an object moves or tends to move while in contact with another object or medium there is always a frictional force, which opposes the motion or the tendency of motion. The three kinds of friction are—*sliding (static or kinetic), rolling* and *viscous*. Sliding and rolling frictions apply to the *relative motion* of one solid body in contact with another solid body. Force of friction between a pair of surfaces is *independent of the area of contact of their surfaces*.

The laws of friction are :

(*i*) The frictional force (static or kinetic) is proportional to the normal force pressing the surfaces together. Thus,
$f_s = \mu_s F_N$ and $f_k = \mu_k F_N$
where f_s : static frictional force, μ_s : coefficient of static friction, f_k : kinetic frictional force, μ_k : coefficient of kinetic friction and F_N : force normal to the contact surface of the two objects.

(*ii*) Static friction is greater than kinetic friction. The maximum static friction known as limiting friction.

(*iii*) The kinetic frictional force is independent of the relative velocity of the sliding surfaces.

1.1.5 Motion in Two Dimensions (Motion in a plane)

The position of a particle moving in a plane has to be specified by two co-ordinates such as x and y. This particular type of motion of the particle is termed "motion in a plane", the plane being defined by X- and Y- axes.

(*i*) **Motion Under Gravity in a Vertical Circle :** A body of mass "m" is tied to a string of length "l" whose other end is fixed at O. The body now made to move in a vertical circle by imparting a large initial velocity u.

The resultant force acting at the position A on the object is

$$T_1 - mg \text{ (Figure 1.2)}$$

$$\therefore \quad T_1 - mg = \frac{mu^2}{l} \text{ or } T_1 = m\left(g + \frac{u^2}{l}\right)$$

Resultant force acting on the object at B is

$$T_2 + mg = \frac{mu'^2}{l} \text{ where } u' \text{ : velocity of the body at B}$$

$$\therefore \quad T_2 = m\left(\frac{u'^2}{l} - g\right)$$

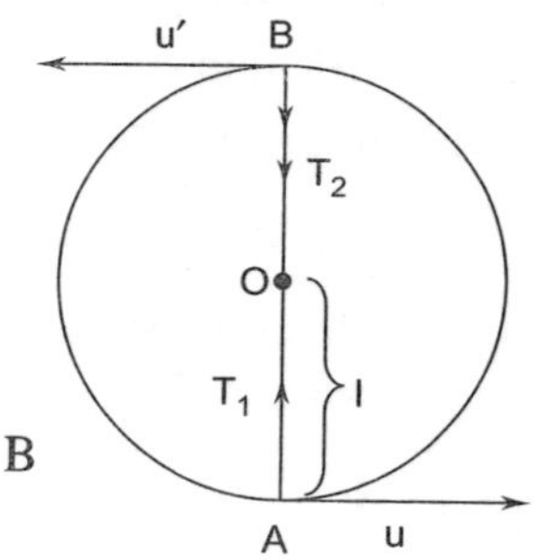

Figure 1.2

If T_2 is positive the string will remain tight and from the previous equation T_1 is positive. The condition for the object to complete a circle is that the tension in the string should always be positive. For just completing the vertical circle T_2 must be zero, when $u' = u_c$, u_c being known as **critical speed**.

$$\therefore \quad 0 = \frac{mu_c^2}{l} - mg$$

or $\quad u_c = \sqrt{lg}$ gives the critical velocity which will keep the body rotating in a vertical circle of radius l. The speed at A is given by

$$u^2 = u_c^2 + 4lg \text{ or } u = \sqrt{5lg}\,.$$

(*ii*) **Motion of a Projectile :**

Here, $\quad u_x = u \cos \theta$

$u_y = u \sin \theta$ where θ : angle of projection with the horizontal

R : horizontal range of the projectile

T : total time of flight of the projectile

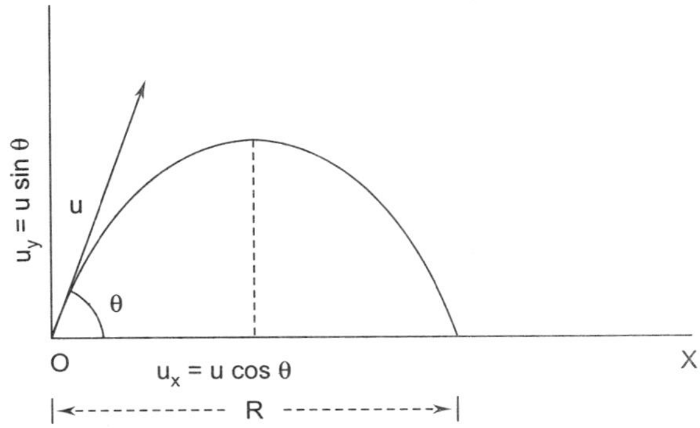

Figure 1.3

(*a*) The trajectory of the projectile is a parabola (Figure 1.3), the equation being

$$y = x \tan\theta - \frac{1}{2}\frac{g}{u^2 \cos^2\theta}\,.x^2$$

(*b*) The total time of flight $T = \dfrac{2u \sin\theta}{g}$

(*c*) The time of ascent = the time of descent $= \dfrac{T}{2} = \dfrac{u \sin\theta}{g}$

(*d*) Maximum height the projectile can reach is $h = \dfrac{u^2 \sin^2\theta}{2g}$

(*e*) The horizontal distance (range) $R = \dfrac{u^2 \sin 2\theta}{g}$

The maximum range $R_{max} = \dfrac{u^2}{g}$ (when $\theta = 45°$)

1.2 CONSERVATION LAWS (FOR PARTICLES IN MOTION)

1. Conservation of Linear Momentum : *When the vector sum of the external forces acting on a system is zero, the total linear momentum of the system remains constant.*

According to Newton's 2nd law,

$$\vec{F} = \frac{d\vec{p}}{dt} = \frac{d}{dt}\left(m\vec{v}\right) = 0 \text{ i.e., } \vec{p} = \text{constant.}$$

where the linear momentum of the particle $= m\vec{v}$.

If the system contains n number of particles, then

$$\vec{p} = \sum_{i=1}^{n} p_i = \sum_{i=1}^{n} m_i v_i$$

From Newton's 2nd and 3rd law, we get

$$F^{ext} = \frac{d\vec{p}}{dt}$$

So, the rate of change of linear momentum of a system of particles is equal to the net external force acting on the system.

If $$F^{ext} = 0, \frac{d\vec{p}}{dt} = 0 \quad i.e., \vec{p} = \text{constant}.$$

If the external force acting on a system of particles is zero, the net linear momentum of the system remains constant.

2. Conservation of Angular Momentum : *When the net external torque acting on a system is zero, the total angular momentum of the system is constant (conserved).*

$$\vec{\tau}\ (\text{torque}) = \frac{d}{dt}\vec{J} = \frac{d}{dt}\left(\vec{r} \times \vec{p}\right) \qquad \left(\because\ \vec{J} \times 2 = \vec{r} \times \vec{p} : \text{Angular Momentum}\right)$$

For a system of n particle we can write

$$\vec{J} = \sum_{i=1}^{n} \vec{J} = \sum_{i=1}^{n} \vec{r_i} \times \vec{p_i}$$

$$\therefore \quad \frac{d\vec{J}}{dt} = \sum_{i=1}^{n} \vec{r_i} \times \frac{d\vec{p_i}}{dt} + \sum_{i=1}^{n} \frac{d\vec{r_i}}{dt} \times \vec{p_i}$$

$$= \sum_{i=1}^{n} \vec{r_i} \times \vec{F_i} \qquad \left[\because \frac{d\vec{r_i}}{dt} = \vec{v_i}, \vec{p_i} = m\vec{v_i} \quad \therefore \vec{v_i} \times m\vec{v_i} = 0\right]$$

where $\vec{F_i} = \frac{d\vec{p_i}}{dt}$ = net force acting on the ith particle.

Though, the internal forces occur in equal and opposite pairs, the net internal force acting on the system of n particles is zero.

$$\therefore \quad \vec{\tau}^{ext}\ (\text{Torque}) = \frac{d\vec{J}}{dt} = \sum_{i=1}^{n} \vec{r_i} \times \vec{F_i}^{ext}$$

If $\vec{\tau}^{ext} = 0$, $\vec{J}$ = constant

Therefore, if external torque acting on a system of particles is zero, the angular momentum of the system remains constant.

3. Conservation of Energy : *Energy can neither be created nor destroyed but can only be converted from one form to another.*

Or

The total energy of an isolated system always remains constant.

For conservative forces (*i.e.*, forces derivable from a scalar potential function V)

$$F = -\Delta V \text{ } i.e., \text{ } F = -\frac{dV}{dr}$$

or
$$F.\,dr = -\,dV$$

Under this condition,

Total Energy (T.E.) = K.E. + P.E. (energy conservation)

If a particle of mass *m* moves from position 1 to position 2 under the action of a conservative force then its work done is,

$$W_{12} = \int_1^2 F\,dr = \int_1^2 \frac{dp}{dt}\,.\,dr$$

Now,
$$p = m\dot{r} \text{ and } dr = \dot{r}\,dt$$

∴
$$W_{12} = \int_1^2 \frac{d}{dt}(m\dot{r})\,\dot{r}\,dt = \int_1^2 \frac{d}{dt}\left(\frac{1}{2}m\dot{r}^2\right)dt = T_2 - T_1$$

Again,
$$W_{12} = \int_1^2 F\,dr = -\int_1^2 \Delta V\,.\,dr = -\int_1^2 \frac{dV}{dr}\,.\,dr$$

$$= V_1 - V_2 \text{ (for a conservative system).}$$

Hence,
$$W_{12} = T_2 - T_1 = V_1 - V_2$$

and so,
$$T_1 + V_1 = T_2 + V_2 = \text{constant.}$$

Therefore, the total energy of a particle is conserved.

1.2.1 Elastic and Inelastic Collisions

Two bodies may collide and give rise to an impulsive force in several ways :

(*a*) **Direct Impact :** When the centers of mass of the bodies approach each other along their common line of motion it is *direct impact*.

(*b*) **Oblique Impact :** When the colliding bodies do not move on the common line of motion it is termed as *oblique impact*.

(*c*) **Central Impact :** When the force of impact is along the line that joins the centers of mass of bodies, it is called *central impact*. Otherwise, the impact is *eccentric*.
We shall concentrate on direct central impact.

The time of impact can be divided into two periods of time :

(*i*) A short period of time known as the *period of deformation* commences at the instant the bodies come in contact, when they begin to compress or deform one another at the surfaces of contact. During this compression period the force between the surfaces increases and reaches its largest value at the instant the deformation stops, at which instant the centers of mass of the bodies are moving with the same speed.

(*ii*) A period succeeding the first known as the *period of restitution* is present for bodies that are partially or entirely elastic. During this period the mass-centers begin separating and the force between the surfaces decreases, reaching a value zero at the instant of separation, when the deformed surfaces have returned to their initial form.

When two or more bodies collide, linear momentum is always conserved but the total kinetic energy before impact may or may not be equal to the total kinetic energy after impact depending on the type of collision.

A perfectly elastic collision is one in which the kinetic energy is conserved. In this case, the bodies separate after impact with velocities such that the total kinetic energy remain constant.

If the collision is such that the bodies separate after collision but the kinetic energy is not conserved due to some of the initial kinetic energy going into dissipative forms or causing permanent displacement of surface layer, the collision is termed *inelastic*.

If the collision is of the type in which the bodies *do not separate after impact* but stick together, it is a *perfectly inelastic collision*.

(*i*) **Perfectly elastic collision**

A head-on collision is viewed as below.

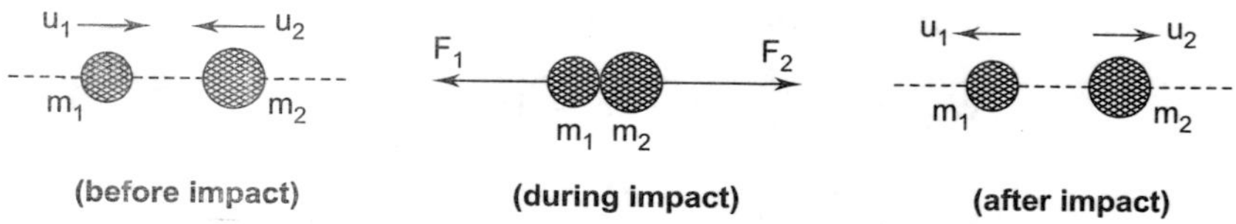

F_1 is the impulsive force by m_2 on m_1 and F_2 is the impulsive force exerted by m_1 on m_2 the duration of impact is instantaneous the force are: impulsive)

Thus, $$F_1 = -F_2 \quad \text{(by Newton's third law)}$$

$\therefore$ $$\text{impulse of } F_1 = -\text{ impulse of } F_2$$

or $$(m_1 v_1 - m_1 u_1) = -(m_1 v_2 - m_2 u_2)$$

or $$(m_1 v_1 + m_2 v_2) = (m_1 u_1 + m_2 u_2) \qquad \ldots(i)$$

The above equations shows that the total linear momentum of the system remains unchanged.

Hence, for a completely isolated system the total linear momentum remains constant.

If the **impact is perfectly elastic** then the kinetic energy has to be conserved. So,

$$\frac{1}{2} m_1 u_1^2 + \frac{1}{2} m_2 u_2^2 = \frac{1}{2} m_1 v_1^2 + \frac{1}{2} m_2 v_2^2$$

or $$m_1 (u_1 - v_1)(u_1 + v_1) = m_2 (u_2 - v_2)(u_2 + v_2) \qquad \ldots(ii)$$

From, conservation of total linear momentum,

$$m_1 (u_1 - v_1) = m_2 (v_2 - u_2) \qquad \ldots(iii)$$

Dividing (*ii*) by (*iii*) we get,

$$u_1 + v_1 = u_2 + v_2$$

or $$(u_1 - u_2) = -(v_1 - v_2) \qquad \ldots(iv)$$

NOTE (Velocity of approach (before collision) = – (Velocity of separation) (after collision)]

Eliminating v_2 between (*iv*) and (*i*),

$$(m_1 + m_2) v_1 = m_1 u_1 - m_2 u_1 + 2m_2 u_2$$

or
$$v_1 = \frac{(m_1 - m_2)\,u_1 + 2m_2\,u_2}{(m_1 + m_2)} = \frac{(m_1 - m_2)}{m_1 + m_2}\,u_1 + \frac{2m_2}{m_1 + m_2}\,u_2 \qquad \ldots(v)$$

In a similar way we get,

$$v_2 = \frac{(m_2 - m_1)\,u_2 + 2m_1\,u_1}{(m_1 + m_2)} = \left(\frac{2m_1}{m_1 + m_2}\right)u_1 + \left(\frac{m_2 - m_1}{m_2 + m_1}\right)u_2 \qquad \ldots(vi)$$

The following cases be considered now:

(A) If $m_1 = m_2$ then from equations (v) and (iv) we have

$v_1 = u_2$ and $v_2 = u_1$ $\rightarrow$ which shows

that in an one dimensional elastic collisions of two bodies of equal masses there is only an exchange of velocities amongst the bodies as a result of collision. So the first body moving initially with a velocity u_1 **to collide** with the second body **approaching it** with an initial velocity u_2 will **after collision** move with velocity u_2 while the second will proceed with a final velocity u_1.

(B) If $u_2 = 0$ is the **second mass** m_2 **before impact** be **at rest** then from equations (v) and (iv).

$$v_1 = \frac{(m_1 - m_2)}{(m_1 + m_2)} \cdot u_1 \quad \text{and} \quad v_2 = \frac{2m_1}{(m_1 + m_2)}\,u_1$$

(a) Now, consider $m_1 = m_2$, then $v_1 = 0$ and $v_2 = u_1$

So, in this case both the momentum and the kinetic energy of the first body is completely transferred to the second body on collision and the first body gets stopped.

(b) In this case consider $m_2 >> m_1$ *i.e.*, m_1 is very very small in mass compared to m_2 and is, therefore, negligible (taking $u_2 = 0$) $\therefore$ $v_1 \approx -u_1$ and $v_2 \approx 0$

So, in the case of a very very light body impinging with a **very heavy body at rest** the velocity of the light body is **approximately reserved** and the second (or heavy) body remains **approximately at rest.**

(c) If, however, $m_2 << m_1$ *i.e.*, a heavy body collides with a very very light body which is **at rest.** (taking $u_2 = 0$) The velocity of the heavy body remains **practically** the same but the **light body moves** along the direction motion of the **heavy body before collision** with approximately **twice the velocity of the heavy body (before collision)**.

Maximum energy transfer in a head-on collision (elastic)

Consider a ball of mass m_1 with velocity u_1 to collide with a ball of mass m_2 **at rest**.

Let the initial kinetic energy of the ball of mass m_1 be K_i and the final kinetic energy of the same ball be K_f, the final velocity being v_1.

The **fractional decrease** in the kinetic energy of the ball with mass m_1 is,

$$\frac{K_i - K_f}{K_i} = \frac{u_1^2 - v_1^2}{u_1^2} = \left(1 - \frac{v_1^2}{u_1^2}\right)$$

$$\therefore \qquad \frac{K_i - K_f}{K_i} = 1 - \frac{\left(\frac{m_1 - m_2}{m_1 + m_2}\right) u_1^2}{u_1^2} \qquad \because \quad v_1 = \left[\left(\frac{m_1 - m_2}{m_1 + m_2}\right) u_1\right] \text{ when } u_2 = 0$$

$$= 1 - \left[\frac{(m_1 + m_2)^2 - 4m_1 m_2}{(m_1 + m_2)^2}\right] = 1 - 1 + \frac{4m_1 m_2}{(m_1 + m_2)^2}$$

If $\qquad m_1 = m$ and $m_2 = nm$ (where n : integer)

then $$1 - \frac{K_f}{K_i} = \frac{4nm^2}{(m + nm)^2} = \frac{4n}{(1+n)^2}$$

For maximum transfer of KE, there shall be **no decrease** in the initial kinetic energy of the ball of mass m that is colliding with the ball of mass nm **at rest.** So $\frac{K_f}{K_i}$ must be zero.

Putting $\qquad \frac{K_f}{K_i} = 0$ in above,

$$1 = \frac{4n}{1+n^2} \qquad \text{or} \qquad 1 + n^2 = 4n$$

or $$(1 - n)^2 = 0 \qquad \therefore \qquad n = 1$$

So when the mass ratio (nm/m) of the two masses is 1, the **whole** of the kinetic energy of the **moving ball** is transferred to the ball **at rest before collision.**

(ii) **Perfectly inelastic collision**

If two bodies **do not separate after impact** but move as a **combined single unit** with common velocity v, then the conservation of linear momentum becomes

$$m_1 u_1 + m_2 u_2 = (m_1 + m_2)\, v$$

or $$v = \frac{m_1 u_1 + m_2 u_2}{m_1 + m_2}$$

The loss of kinetic energy due to collision can be derived as

$$\frac{1}{2} m_1 u_1^2 + \frac{1}{2} m_2 u_2^2 - \frac{1}{2}(m_1 + m_2)\, v^2.$$

where, the kinetic Energy (K.E.) before collision is $\left(\frac{1}{2} m_1 u_1^2 + \frac{1}{2} m_2 u_2^2\right)$ and the kinetic energy after collision is $\frac{1}{2}(m_1 + m_2)\, v^2$.

Hence, the loss of K.E. $= \frac{1}{2} m_1 u_1^2 + \frac{1}{2} m_2 u_2^2 - \frac{1}{2}\left\{(m_1 + m_2) \cdot \left(\frac{m_1 u_1 + m_2 u_2}{m_1 + m_2}\right)^2\right\}$

$$= \frac{1}{2}\left[m_1 u_1^2 + m_2 u_2^2 - \left(\frac{m_1^2 u_1^2 + m_2^2 u_2^2 + 2m_1 m_2 u_1 u_2}{m_1 + m_2}\right)\right]$$

$$= \frac{1}{2}\left[\frac{m_1^2 u_1^2 + m_1 m_2 u_2^2 + m_1 m_2 u_1^2 + m_2^2 u_2^2 - m_1^2 u_1^2 - m_2^2 u_2^2 - 2m_1 m_2 u_1 u_2}{m_1 + m_2}\right]$$

$$= \frac{m_1 m_2}{2(m_1 + m_2)}\left(u_1^2 + u_2^2 - 2u_1 u_2\right)$$

or loss of K.E. due to **perfectly inelastic** collision

$$= \frac{m_1 m_2}{2(m_1 + m_2)}\left[u_1 - u_2\right]^2$$

This shows that due to perfectly inelastic collision there will be a positive loss of K.E.

(iii) **Imperfectly elastic collision**

In between the perfectly elastic collision and the perfectly inelastic collision all other collisions are **imperfectly elastic collisions** or **partially elastic collisions.**

It was observed by Newton that in case of collisions of any type between two bodies,

the velocity of separation = $e \times$ (velocity of approach)

where, e is constant known as **coefficient of restitution.**

This e has magnitude $0 < e < 1$ *i.e.*, its value lies between 0 and 1.

$$e = \frac{v_2 - v_1}{u_1 - u_2} = \frac{-(v_1 - v_2)}{(u_1 - u_2)} = -\frac{\text{velocity of separation}}{\text{velocity of approach}}$$

or $$\frac{(v_1 - v_2)}{(u_1 - u_2)} = -e \quad \text{where} \quad e : \text{positive number}$$

The coefficient of restitution indicates the degree to which two bodies in impact approach perfect elasticity or perfect inelasticity. For $e = 1$ we have **perfectly elastic** collision, the kinetic energy being conserved. For $e = 0$, $v_1 = -v_2$ we have **perfectly inelastic collision.**

For change of kinetic energy in an **imperfectly elastic collision**, we proceed as follows —

$$m_1 u_1 + m_2 u_2 = m_1 v_1 + m_2 v_2 \qquad \text{(conservation of linear momentum)}$$

$$\text{Decrease in K.E.} = \Delta E = \left(\frac{1}{2} m_1 u_1^2 + \frac{1}{2} m_2 u_2^2\right) - \left(\frac{1}{2} m_1 v_1^2 - \frac{1}{2} m_2 v_2^2\right)$$

or $$2\Delta E = m_1 u_1^2 + m_2 u_2^2 - m_1 v_1^2 - m_2 v_2^2$$

Now, $$e = -\frac{(v_1 - v_2)}{(u_1 - u_2)} \quad \text{or,} \quad (v_1 - v_2) = -e\,(u_1 - u_2) \qquad \ldots(i)$$

and $$m_1 u_1 + m_2 u_2 = m_1 v_1 + m_2 v_2 \qquad \ldots(ii)$$

multiplying equation (*i*) by m_2 and adding to equation (*ii*) we have,

$$m_2 v_1 - m_2 v_2 + m_1 v_1 + m_2 v_2 = m_1 u_1 + m_2 u_2 - m_2 e (u_1 - u_2)$$

or $$v_1 (m_1 + m_2) = [(m_1 u_1 + m_2 u_2) - m_2 e (u_1 - u_2)]$$

$\therefore$ $$v_1 = \frac{1}{m_1 + m_2}\left[\left(m_1 u_1 + m_2 u_2\right) - m_2 e \left(u_1 - u_2\right)\right] \qquad \ldots(iii)$$

In a similar way multiplying (*i*) by m_1 and subtracting from equation (*ii*)

$$v_2 = \frac{1}{m_1 + m_2}\left[\left(m_1 u_1 + m_2 u_2\right) + m_1 e \left(u_1 - u_2\right)\right] \qquad \ldots(d)$$

Using the values of v_1 and v_2 from (*iii*) and (*iv*) in the equation for 2ΔE we get,

$$2\Delta E = m_1 u_1^2 + m_2 u_2^2 - m_1 \left[\frac{1}{m_1 + m_2}\left\{\left(m_1 u_1 + m_2 u_2\right) - m_2 e \left(u_1 - u_2\right)\right\}\right]^2$$

$$= m_2 \left[\frac{1}{m_1 + m_2}\left\{\left(m_1 u_1 + m_2 u_2\right) + m_1 e \left(u_1 - u_2\right)\right\}\right]^2$$

or $$2\Delta E = \left(\frac{m_1 m_2}{m_1 + m_2}\right) \cdot \left(e^2 - 1\right)\left(u_1 - u_2\right)^2$$

or Loss in Energy $$= \Delta E = \frac{1}{2}\left(\frac{m_1 m_2}{m_1 + m_2}\right)\left(e^2 - 1\right)\left(u_1 - u_2\right)^2$$

1.2.2 Conservation of Angular Momentum

The most general sort of motion can be considered to be a combination of translational motion and rotational motion. In pure translation all parts of the body have the same velocity or accelaration. In pure rotational motion each particle of a rigid body, remaining at a **fixed distance** from the axis of rotation, moves in a circle of a radius equal to the distance of the particle from the axis of rotation.

If θ be the angle formed by an arc of length **S** at the center of a circle of radius *r*, then when θ remains constant **S** ∝ *r*. On the otherhand if *r* is kept constant **S** ∝ θ. Taking both the variations together the combined variation gives,

$$s = \theta \,.\, r \qquad \text{(Variational constant = 1)}$$

or $$\theta = \frac{s}{r} \,:$$ s and r are measured in same unit

Thus the angle may be measured by the ratio of the arc length to its radius. This angle is in "radians" rather than in 'degrees'.

The angular velocity 'ω' (or uniform angular speed) is the angular displacement (θ) per unit time,

$$\omega = \frac{d\theta}{dt}$$

The angular acceleration is defined as the rate of angular velocity. If ω_0 be the initial angular velocity and ω the final angular velocity after a time interval t then the angular acceleration,

$$\alpha = \frac{\omega - \omega_o}{t} = \frac{d\omega}{dt} \text{ at the limit } t \to 0$$

(*a*) **Kinematic equations for translation and rotation**

Translation (*a* : acceleration)	*Rotation* (α : angular acceleration)	*Relation*
$v = v_0 + at$	$\omega = \omega_0 + \alpha t$	$s = \theta . r$
$s = v_0 t + \frac{1}{2} at^2$	$\theta = \omega_0 t + \frac{1}{2} \alpha t^2$	$v = \omega r$
$v^2 = v_0^2 + 2as$	$\omega^2 = \omega_0^2 + 2\alpha\theta$	$a = \alpha r$
$v_{av} = (v + v_0)/2$	$\omega_{av} = (\omega + \omega_0)/2$	

(*b*) **Newton's Second Law for Rotation**

Let a body be acted upon by some force F the body being at a distance r from a point (say, O). This force can be resolved into two components—tangential (F_t) and radial (F_r) *i.e.*, along the radius and perpendicular to the radius (*i.e.*, along the tangent at that point) of the rotational motion caused by the force F. The tangential components F_t produces the rotational force (*i.e.*, Torque). We shall thus have the total torque

$$\tau = F_t \times r = F_{1t} \times r_1 + F_{2t} \times r_2 + F_{3t} \times r_3 + \ldots\ldots$$

$$= m_1 \alpha_{1t} \times r_1 + m_2 \alpha_{2t} \times r_2 + m_3 \alpha_{3t} \times r_3 + \ldots\ldots$$

m_1, m_2, m_3 etc. are the mass points constituting the body positioned at distances r_1, r_2, r_3 etc. from O respectively

Now, $\alpha_{1t} \times r_1 = \alpha \times r_1, \alpha_{2t} \times r_2 = \alpha \times r_2,$

$\alpha_{3t} \times r_3 = \alpha \times r_3$ etc where α is the common **angular acceleration** for all the particles of the body.

Hence, $$\tau = \left(m_1 r_1^2 + m_2 r_2^2 + m_3 r_3^2 + \ldots\ldots\right) \times \alpha = \vec{F}_t \times \vec{r}$$

The expression $$\left(m_1 r_1^2 + m_2 r_2^2 + m_3 r_3^2 + \ldots\ldots\right) = \sum_{i=1}^{n} m_i r_i^2 = I$$

I is known as the Moment of Inertia (or Rotational Inertia) and the resultant torque is $\tau = \mathbf{I}\alpha$

From the equation, $\tau = I\alpha$, it can be stated that a resultant torque acting on a rigid body causes an angular acceleration which is directly proportional to the torque and inversely to the moment of inertia of the body **about the axis of rotation** $\left(\alpha = \frac{\tau}{I}\right)$. In **rotational motion**, therefore, 'I' replaces '*m*' and 'α' replaces '*a*' of the Newton's second law of **translational** motion, the rotational force being represented by τ.

Angular Impulse = **Torque** × **time interval** = Change in angular momentum. Moment of Inertia of a body is that property which needs an unbalanced torque to give the body an angular accelaration. Although moment of inertia in rotation corresponds to mass in translation, the mass of a body is always the **same** while its moment of inertia **depends upon the axis of rotation.**

Another characteristic of moment of inertia is that it is a **scalar quantity**.

Radius of Gyration: This is a fictitious radius which is **distance R from any given axis** at which the total mass may be **imagined concentrated** without altering the moment of inertia of the body about that axis.

At a distance *k* from the axis of rotation

$$I = Mk^2 \left(M = \sum_{i=1}^{n} m_i\right)$$

and

$$k = \sqrt{\frac{I}{m}} \left[I = \sum_{i=1}^{n} m_i\, r_i^2 \text{ Actual M.I.}\right]$$

(*c*) **Angular Momentum** of a rotating body **about an axis** is the product of the moment of inertia about this axis and the angular velocity: L (angular momentum) = Iω

which is analogous to the linear momentum $p = mv$ for translational motion.

In a similar way the rotational kinetic energy is $\frac{1}{2}I\omega^2$ (in analogy with translational kinetic energy $\frac{1}{2}mv^2$). One must note that in **purely rotational** motion moment of inertia plays the role of mass (the constant of inertia) as in the case of only translational motion.

Angular momentum of a **single particle** also is $mr^2\omega$ when it is rotating keeping a **constant distance** *r* with the axis about which it is rotating with angular velocity ω.

So, $\quad I = mr^2\omega = mvr \qquad [\because \; v \text{ (the tangential velocity)} = \omega r].$

In fact, moment of inertia of the particle (or the body) about the axis mentioned, which is a **vector perpendicular** to the plane of rotation of the particle along the rotational axis.

(*d*) **Principle of conservation of Momentum**

In general, Iω = constant, which means – when there is no external unbalanced angular force (*i.e.*, torque) acting there is **no change** in the angular momentum of a rotating body.

$$I_1 \omega_1 = I_2 \omega_2$$

The above relation shows that if, while a body is rotating, the **distribution of mass** about the axis of rotation changes so that'I' changes, then 'ω' must change in inverse proportion.

Example: When a person standing at the center of a practically frictionless turn-table is given an angular velocity about the vertical axis of the platform, he can control his speed of rotation by the following methods — (*a*) by increasing his moment of inertia (I) by extending his arms outwards when 'ω' gets reduced and (*b*) by pulling his arms in whereby his moment of inertia (I) gets decreased causing increase in ω.

Units of Angular Impulse: (m.nt. sec) – MKS; (*m*-dyne. sec) – CGS

Units of Angular Momentum – (kg. m^2/sec) in MKS and (gm. cm^2/sec) in CGS.

Solved Problems

Problem 1. *At an instant, two trains start from two stations A and B with velocities v_1 and v_2 respectively. After meeting on the way at O the train from A reaches B in time t_1 and the train from B reaches A in time t_2 simultaneously.*

Show that : $$\frac{v_1}{v_2} = \sqrt{\frac{t_1}{t_2}}$$

Solution.

From the problem,

$$OA = v_2 t_2, \; OB = v_1 t_1$$

As they traverse OA and OB during the same time

$$t = \frac{AO}{v_1} = \frac{BO}{v_2} \text{ or } \frac{v_2 t_2}{v_1} = \frac{v_1 t_1}{v_2} \quad \therefore \quad \frac{v}{v_2} = \sqrt{\frac{t_1}{t_2}}.$$

Problem 2. *A particle is moving along a straight line. The relation between time (t) and distance x is $t = \alpha x^2 + \beta x$ where α and β are positive constants. Obtain the retardation of the particle.*

Solution. Given, $$t = \alpha x^2 + \beta x \quad \text{or} \quad \frac{dt}{dx} = 2\alpha x + \beta$$

$$\therefore \quad v = \frac{dx}{dt} = \frac{1}{2\alpha x + \beta} = (2\alpha x + \beta)^{-1}$$

Now, $$vt = x \text{ or } \frac{dv}{dx}.t = 1 \text{ (considering } t \text{ to be constant)}$$

$$\therefore \quad \frac{dv}{dx} = \frac{1}{t}$$

$$f\text{(acceleration)} = \frac{v}{t}$$

$$= v.\frac{dv}{dx}$$

$$= v \cdot \frac{d}{dx}\left\{(2\alpha x + \beta)^{-1}\right\} = v \cdot \left[-(2\alpha x + \beta)^{-2} \cdot 2\alpha\right]$$

$$= v \cdot [-(2\alpha x + \beta)^{-2} \cdot 2\alpha] = v \cdot (-2\alpha v^2)$$

$$\therefore \quad f = -v \cdot 2\alpha \cdot v^2 = -2\alpha v^3 \text{ (retardation).}$$

Problem 3. *A smooth plane of length L and height h is inclined at an angle θ with the horizontal. A body at the top starts from rest and slides down the inclined surface. How much time is taken by the body to reach the bottom ?*

Solution. Acceleration along the plane = $g \sin \theta$

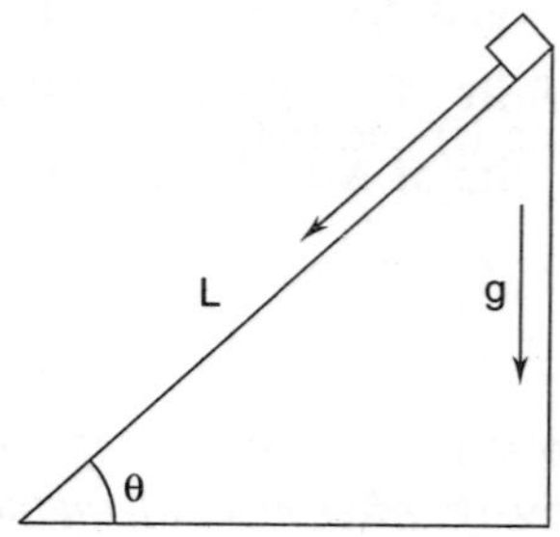

Now, $$s = ut + \frac{1}{2} ft^2$$

In this case, $$L = \frac{1}{2} (g \sin \theta)\, t^2$$

or $$t = \sqrt{\frac{2L}{g \sin \theta}}$$

Again, $$L \sin \theta = h$$

or $$L = \frac{h}{\sin \theta}$$

$$\therefore \quad t = \frac{1}{\sin \theta}\sqrt{\frac{2h}{g}}.$$

Problem 4. *A particle is projected up along an inclined plane of slope 45° with the horizontal. If the coefficient of friction be 0.5, what will be its retardation ?*

Solution. Total retarding force = gravitational force + frictional force

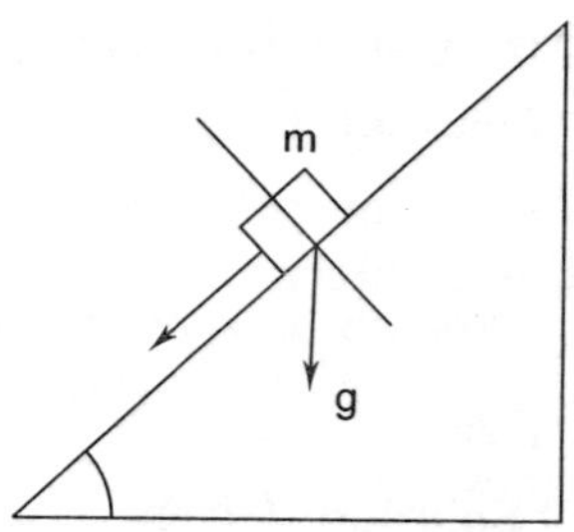

$$= mg \sin \theta + \mu\, mg \cos \theta$$

∴ retardation $$f = g (\sin \theta + \mu \cos \theta)$$

or $$f = g\left(\frac{1}{\sqrt{2}} + 0.5 \times \frac{1}{\sqrt{2}}\right) \quad (\because \theta = 45°)$$

$$= \frac{1.5}{\sqrt{2}} g.$$

Problem 5. *Two bodies of masses, m_1 = 2.0 kg and m_2 = 1.0 kg are in contact on a frictionless table. A horizontal force, F = 3.0 newton is applied to m_1. Find the force of contact between m_1 and m_2. Again find the force of contact if F is applied to m_2.*

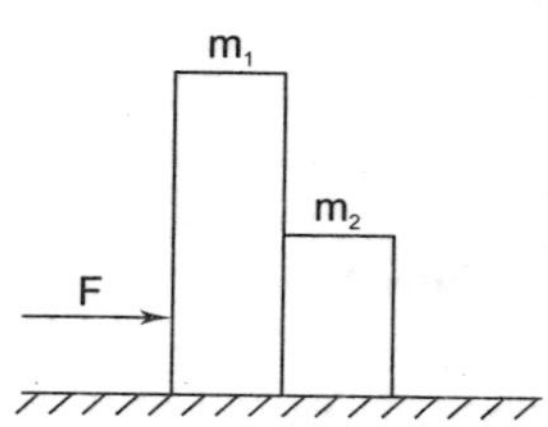

Solution. Let F′ be the force of contact between m_1 and m_2.

Suppose m_1 block exerts a force F′ on m_2, m_2 also exerts force F′ on m_1

According to the law of action and reaction,

The resultant force on m_1 is F – F′ and that on m_2 is F′.

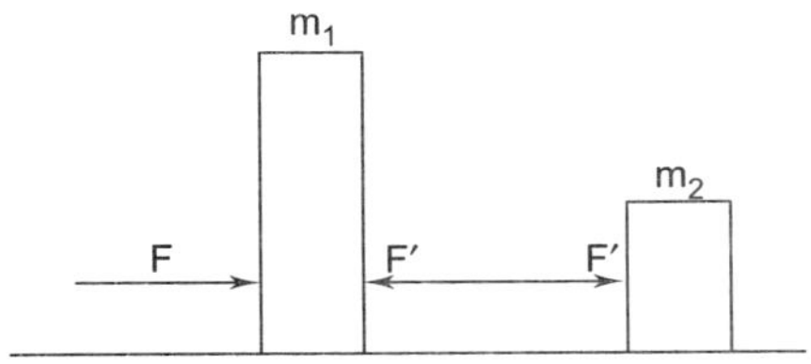

From the Newton's second law, we have

$$F - F' = m_1 a \quad \ldots(i)$$

and $$F' = m_2 a \quad \ldots(ii)$$

Solving (*i*) and (*ii*), we get

or, $$F = (m_1 + m_2) a$$

or, $$F' = \frac{m_2}{m_1 + m_2} . F$$

Here $$m_1 = 2.00 \text{ kg}, m_2 = 1.0 \text{ kg}, F = 3 \text{ newton}$$

∴ $$F' = \frac{1}{2+1} . 3 = 1.0 \text{ newton.}$$

Again if F is applied to m_2, then

$$F' = \frac{2}{2+1} . 3 = 2.0 \text{ newton}$$

Problem 6. *Two balls A and B having different but unknown masses, collide A is initially at rest while B has a speed v. After collision B has a speed $\frac{v}{2}$ and moves at right angle to its original motion. Find the direction in which ball A moves after collision. Can you determine the speed of A from the given information ?*

Solution. Let v' be the speed of A after the collision.

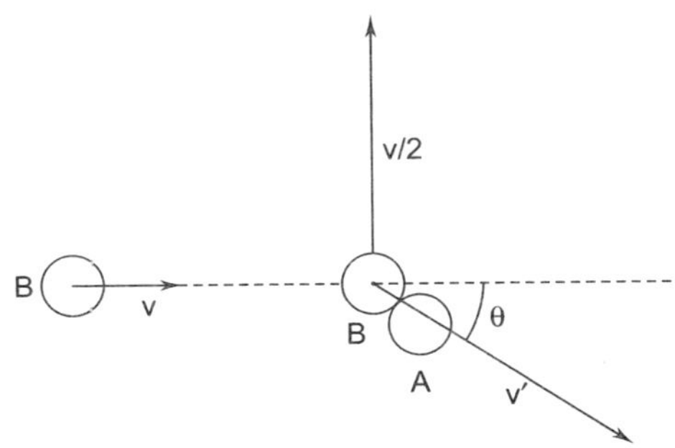

Applying the conservation of momentum along the original direction of motion,

$$m_B v = m_A v' \cos\theta \quad \ldots(i)$$

Again, by the conservation of momentum at right angles to the original motion,

$$0 = m_B \frac{v}{2} - m_A v' \sin\theta$$

or, $$m_B \frac{v}{2} = m_A v' \sin\theta \quad \ldots(ii)$$

From equations (*i*) and (*ii*), we get

$$\frac{1}{2} = \tan\theta$$

∴ $$\theta = \tan^{-1}\left(\frac{1}{2}\right) = 27°.$$

Thus, the ball A moves at (90° + 27°) or 117° from the original direction of B.

We cannot determine the speed of A after the collision unless we know m_A and m_B or their ratio.

Problem 7. *A ball of mass m moving with a velocity v along X-axis, strikes another ball of mass 2m kept at rest. The first ball comes to rest after collision and the other breakes into two equal pieces. One of the pieces starts moving along Y-axis with a speed v_1. What will be the velocity of other piece.*

Solution. Here the total momentum of the balls before collision is mv along X-axis. After the collision, momentum of the 1st ball = 0, momentum of the 1st piece = mv_1 along the Y-axis and momentum of the 2nd piece = mv_2 and momentum of the 2nd piece = mv_2 along its direction of motion where v_2 is the speed of second piece.

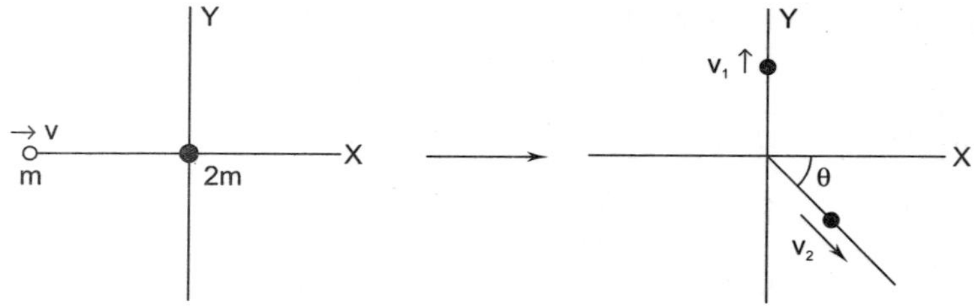

Applying the conservation of momentum along X-axis

$$mv_2 \cos\theta = mv \qquad \ldots(i)$$

and along Y-axis

$$mv_2 \sin\theta = mv_1 \qquad \ldots(ii)$$

From equations (*i*) and (*ii*), we get

$$\therefore \quad m^2 v_2^2 = m^2 (v^2 + v_1^2)$$

or

$$v_2 = \sqrt{v_1^2 + v_2^2} \text{ and } \tan\theta = \frac{v_2}{v_1}.$$

Problem 8. *A block slides down an inclined plane of slope angle θ with constant velocity. It is then projected up the same plane with an initial speed v_0. How far up the incline will it move before coming to rest ? Will it slide down again.* [Lucknow University]

Solution. Here net force along the plane is $mg \sin\theta - f_K$, where f_K is the frictional force, which is zero because the block slides down with constant velocity.

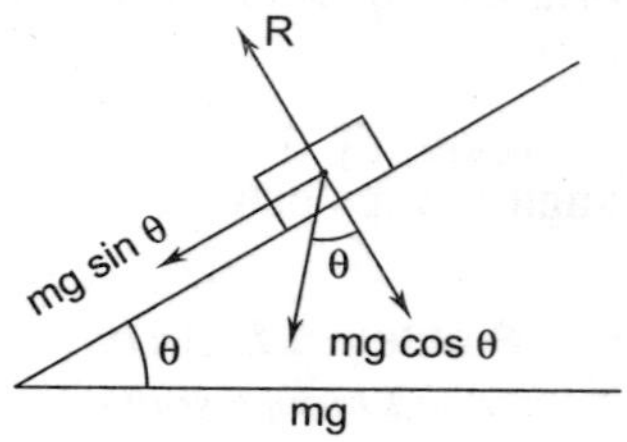

$$\therefore \quad mg \sin\theta - f_K = 0$$

or,

$$f_K = mg \sin\theta$$

When the block is projected up (Here f_K will be down the plane), thus the net force down the plane becomes

$$= mg \sin\theta + f_K$$

$$= mg \sin\theta + mg \sin\theta = 2mg \sin\theta$$

Hence the acceleration down the plane would be

$$a = \frac{2mg \sin\theta}{m} = 2g \sin\theta.$$

Now, using the relation, $V^2 = v_0^2 + 2ax$

Here $V = 0$,

or,

$$-x = \frac{v_0^2}{+2a} = +\frac{v_0^2}{4g \sin\theta}$$

Hence, the distance covered up the plane

$$-x = \frac{v_0^2}{4g \sin \theta}.$$

As soon as the block stops, the frictional force becomes static, which is larger than the kinetic. Hence, the block will not slide down again.

Problem 9. *A wheel mounted on a shaft of negligible moment of inertia is rotating at 500 r.p.m. An exactly similar second wheel, initially at rest, is suddenly coupled to the same shaft. Compute the angular velocity of the combination.*

Solution. Let ω be the final angular velocity of the combination then, we have

Initial angular momentum = Final angular momentum

$$I_1 \omega_1 + I_2 \omega_2 = (I_1 + I_2)\omega$$

Here $I_1 = I_2 = I$; $\omega_1 = 500$ r.p.m. and $\omega_2 = 0$

$\therefore$ $I \times 500 = 2I\omega$

or, $\omega = 250$ r.p.m.

Problem 10. *A 1200-kg car is moving east at 30.0 m/sec. and collides with a 3600-kg truck moving at 20.0 m/sec. in a direction 60° north of east. The vehicle interlock and move off together. Find their common velocity.*

Solution. We know that the $m_1 \vec{u_1} + m_2 \vec{u_2} = (m_1 + m_2)\vec{v}$.

Let x be eastward and y northward. For x-component

$$(1200 \text{ kg})(30.0 \text{ m/s}) + (3600 \text{ kg})(20.0 \text{ m/s})(\cos 60°) = (4800 \text{ kg})\, v_x$$

$\therefore$ $v_x = 15$ m/sec.

For y-component, $0 + (3600 \text{ kg})(20.0 \text{ m/sec.})(\sin 60°) = (4800 \text{ kg})\, v_y$.

or, $v_y = 13.0$ m/sec.

$\therefore$ $v = \sqrt{v_x^2 + v_y^2} = 19.8$ m/sec.

and $\tan \theta = \dfrac{v_y}{v_x}$ yields $\theta =$ **40.9°** north of east.

Problem 11. *An object at rest explodes into three pieces of equal mass. One moves east at 20 m/sec ; a second moves south-east at 30 m/sec. What is the velocity of the third piece ?*

Solution. Let x be east and y will be north.

Then, $m\vec{v_1} + m\vec{v_2} + m\vec{v_3} = 0$, yields in x direction,

$(20 \text{ m/s}) + (30 \text{ m/s}) \cos 45° + v_3 \cos \theta = 0$

$\therefore$ $v_3 \cos \theta = -41.2$ m/s ...(*i*)

In y-direction,

$0 - (30 \text{ m/s}) \sin 45° + v_3 \sin \theta = 0$

$\therefore$ $v_3 \sin \theta = 21.3$ m/s. ...(*ii*)

From equations (*i*) and (*ii*), we get

$\tan\theta = -0.516$ and $\theta = 153°$ or $27°$ north of west. Solving for v_3, we get

$$v_3 = 46.9 \text{ m/sec.}$$

1.3 MECHANICS OF A SYSTEM OF PARTICLES (CONSISTING OF TWO OR MORE PARTICLES)

Center of Mass

When a system contains many particles, its translatory motion can be represented by the center of mass of the system.

Consider a system containing *n* particles of masses $m_1, m_2, \ldots, m_n$ having position vectors from an arbitrary origin O as $r_1, r_2, \ldots, r_n$ respectively then the position vector of the center of mass of the system is given by,

$$R = \frac{m_1 r_1 + m_2 r_2 + \ldots + m_n r_n}{m_1 + m_2 + \ldots + m_n}$$

$$= \frac{\sum m_i r_i}{\sum m} = \frac{\sum m_i r_i}{M}$$

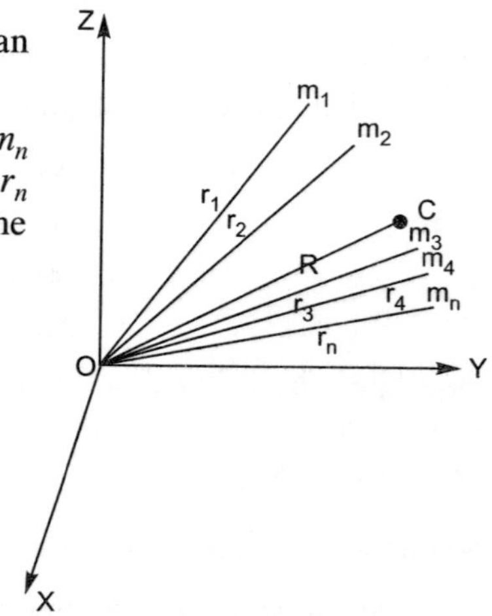

Figure 1.4

In cartesian co-ordinates,

$$\text{if } r_i = (x_i, y_i, z_i) \text{ and } R = (\bar{x}, \bar{y}, \bar{z})$$

$$\bar{x} = \frac{\sum m_i x_i}{M}, \; \bar{y} = \frac{\sum m_i y_i}{M}, \text{ and } \bar{z} = \frac{\sum m_i z_i}{M}$$

NOTE The position of center of mass is independent of the origin.

For mechanical systems consisting of two or more particles two types of forces become active :

(*i*) External Forces not belonging to the system act upon the particles of the system.
(*ii*) Internal forces arising due to interaction between the particles of the system are active.

1.4 LIMITATIONS OF NEWTONIAN MECHANICS

In any dynamical problem two essential steps are involved :

(*i*) differential equations of motion are to be set up
(*ii*) those equations are to be solved.

In setting up of the differential equations of motion a particular co-ordinate system is chosen to simplify the problem.

Newtonian equations of motion do not maintain their simple form in all the co-ordinate systems.

Example :

Vector Form	*Cartesian*	*Cylindrical*	*Spherical Polar*
(Independent	$F_x = m\ddot{x}$	$F_\rho \neq m\ddot{\rho}$	$F_r \neq m\ddot{r}$
co-ordinate	$F_y = m\ddot{y}$	$F_\phi \neq m\ddot{\phi}$	$F_\theta \neq m\ddot{\theta}$

system) $F_z = m\ddot{z}$ $F_z = m\ddot{z}$ $F_\phi \neq m\ddot{\phi}$

$F = m\ddot{r}$

as $x = r \sin\theta \cos\phi$

$y = r \sin\theta \sin\phi$

and $z = r\cos\theta$

Thus, Newtonian method does not help.

Lagrangian formulation and vector equations are both independent of co-ordinate systems but Lagrangian formulation is of very general nature. Lagrange's equations of motion are of great importance as they can be written in any system of co-ordinates.

In an inertial frame (*i.e.*, a frame in which a body will not be accelerated in the absence of interacting bodies in the environment, in short, where $\sum F = 0$,

$$F = ma_j$$

$\begin{cases} F & : \text{applied force} \\ a_j & : \text{acceleration as observed in inertial frame} \end{cases}$

In non-inertial frame (such as rotating earth)

$F = m(a + a_0)$ where, a_0 : acceleration of the non-inertial frame relative to an inertial frame

$\therefore$ $a_1 = a + a_0$

a : acceleration of the body in the non-inertial frame.

Hence, $F - ma_0 = ma$

or $F + F_0 = ma$ (where, $F_0 = -ma_0$)

1.5 CONSTRAINTS

A material system is said to be free if all of its constituent particles or bodies can occupy arbitrary positions in space or have arbitrary velocities. Otherwise, it is called a *constrained system.*

Constraints (mechanical) are restrictions imposed on the position or motion of the material system being considered in space.

The constraints can be classified as :

(*a*) *Holonomic and Non-holonomic Constraints*

Let $r_1, r_2, r_3, \ldots, r_n$ be the position co-ordinates of the particles of a system and t denotes the time, then if the conditions of all the constraints be expressed as equations of the form $f(r_1, r_2, r_3, \ldots, r_n, t) = 0$ then the constraints will be holonomic. The function should not contain derivatives of the co-ordinates of the points in the system or can be reduced to above form by integration otherwise the constraints are non-holonomic.

(*b*) *Unilateral and Bilateral Constraints*

A constraint is said to be unilateral if the restrictions it imposes are expressed by an inequality of the form

$$f(x_1, y_1, z_1, \ldots, x_i, y_i, z_i, \ldots, \dot{x}_i, \dot{y}_i, \dot{z}_i \ldots t) \geq 0$$

where, x_i, y_i, z_i, etc., and $\dot{x}_i, \dot{y}_i, \dot{z}_i$ are the positions and the velocities of the particles of the system in motion and t, the time.

When the above constraint relations are not in the form of inequalities but are in the form of equations they are known as bilateral.

$$f(.., x_i, y_i, z_i, ..., \dot{x}_i, \dot{y}_i, \dot{z}_i ... t) = 0$$

(*c*) *Scleronomic and Rheonomic Constraints*

Constraints are called stationary or Scleronomous if the corresponding constraint equations do not explicitly contain time. They are also called constraints independent of time. Otherwise, they are called time-dependent or Rheonomous constraints.

$x^2 + z^2 = l^2 =$ constant, is the equation of constraint for a simple pendulum which is scleronomic.

The constraint equation for pendulum with variable length of string —

$$|r(t)|^2 = l^2(t) \text{ is Rheonomic.}$$

(*d*) *Conservative and Dissipative Constraints*

Constraint forces do not perform any work during constrained motion in a conservative set up.

In this case, the total mechanical energy of the system is conserved while performing the constrained motion.

When the total mechanical energy is not conserved and the constraint forces do some work the performance is under dissipative constraint.

$$x^2 + z^2 = l^2 = \text{constant (conservative)}$$

and
$$|r(t)|^2 = l^2(t) \text{ ... (dissipative).}$$

Therefore, A constraints is :

(*i*) **Either**-*SCLERONOMIC* : constraint relations do not explicitly depend on time.
or-*RHEONOMIC* : constraint relations depend explicitly on time.

(*ii*) **Either** - *HOLONOMIC* (means integrable) : constraint relations are or can be made independent of velocities $(\dot{x}, \dot{y}, \dot{z})$
or - *NON-HOLONOMIC* : constraint relations are not holonomic (*i.e.*, dependent on velocities.)

(*iii*) **Either** - *CONSERVATIVE* : total mechanical energy of the system is conserved while performing the constrained motion constraint forces do not do any work.
or - *DISSIPATIVE* : constraint forces do work and the total mechanical energy is not conserved.

(*iv*) **Either** - *BILATERAL* : at any point on the constraint surface both the forward and the backward motions are possible. Constraint relations are not in the form of inequalities but in the form of equations.
or - *UNILATERAL* : at some points no forward motion is possible. Constraint relations are expressed in the form of inequalities.

Examples of Constraints

1. **Rigid Body :** In a rigid body the distance between any two particles remain constant. This constancy is maintained by interatomic forces acting between them. The magnitude of these forces is not known but their effect viz. to maintain the particle-separation constant under any circumstances is known. The problem is how to use this information. The equations of constraint are

$$(|r_i = r_k|)^2 = \text{constant (for all } i \text{ and } k)$$

where, r_i and r_k are the position vectors of ith and k^{th} particle.

The constraint is conservative, scleronomic, holonomic and bilateral.

2. **Deformable Bodies :** Deformable bodies are those whose shape can change. Suppose that the deformation of the body is changing with time according to a certain prescribed function of time. Then the motion of such a body is constrained by the equation.

$$|r_i - r_k| = f(\varepsilon), \quad r_i, r_k : \text{position vectors.}$$

In this case, one can show that, these constraint relations cannot give the work, $\Delta W = 0$ (*i.e.*, $\Delta W \neq 0$)

This is a case of rheonomic, holonomic, bilateral and dissipative constraint.

3. **Simple Pendulum :** In a simple pendulum unlike the weight "*mg*" of the bob, we do not know the tension "T" in the string to begin with. We, however, know its effect viz., to maintain the length "*l*" of the pendulum at a constant value. Here, T is the force of constraint and the equation of constraint is,

$$x^2 + z^2 = l^2 = \text{constant},$$

where the co-ordinate of the center of the bob is (x, z) with respect to the origin at the center of suspension.

NOTE the motion is restricted to the $x - z$ plane, not by constraint, but by choice *i.e.*, the initial conditions.

This is a scleronomic, holonomic, bilateral, conservative constraint.

4. **A Body Sliding Down a Smooth Inclined Plane Under Gravity :**

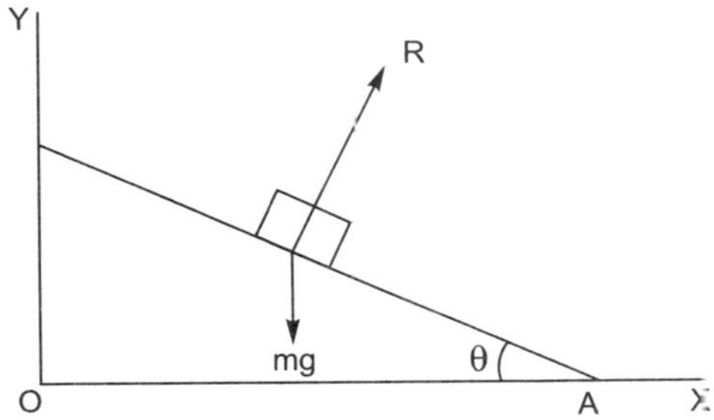

Figure 1.5

In this case, the normal reaction R (not, in general, to be regarded as known force) is the force of constraint. The constraint equation is of the form,

$$y - x \tan\theta = 0 \left(\because \; \frac{y}{x} = \tan\theta \right)$$

where x and y are measured from A.

This is a case of scleronomic, holonomic, bilateral and conservative constraint.

5. **Rolling Without Slipping :** Consider a spherical ball or a cylinder rolling on a plane without slipping. Assume the surfaces in contact are rough. The starting point is C (where the body makes its 1st contact with the planes.

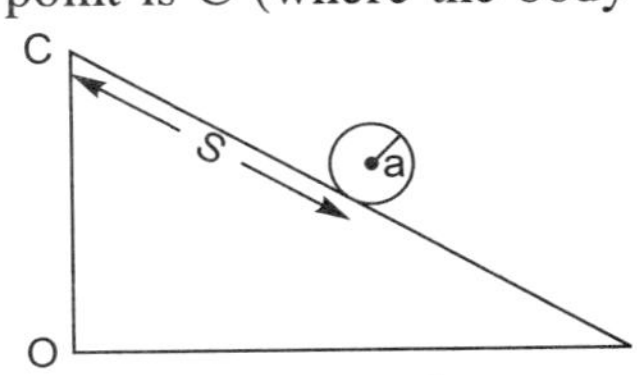

Figure 1.6

For rolling without slipping the equation of constraint is,

$$ds = a d\theta \text{ or } \dot{s} = a\dot{\theta} \quad a : \text{radius of the body}$$
$$\theta : \text{angle of rotation}$$

On integration, $s - a\theta = \text{constant}$

For a cylinder, the constraint is holonomic but for a sphere the constraint is non-holonomic as in this case the constraint equation is non-integrable.

6. **A Particle on the Surface of a Solid Sphere :** The particle is restricted by the constraints so that it can move only on the surface or in the region outside the sphere. It must obey the equation $r \geq a$ (a : radius of the sphere).
   This is a case of scleronomic, non-holonomic unilateral constraint.
7. **A Bead Sliding on a Rotating Wire :** Consider the bead B to slide on the wire rotating with an angular velocity ω about an axis perpendicular to the XY plane. The equation of constraint is given by,

$$y - x \tan \omega t = 0.$$

The constraint is rheonomic, holonomic, bilateral.

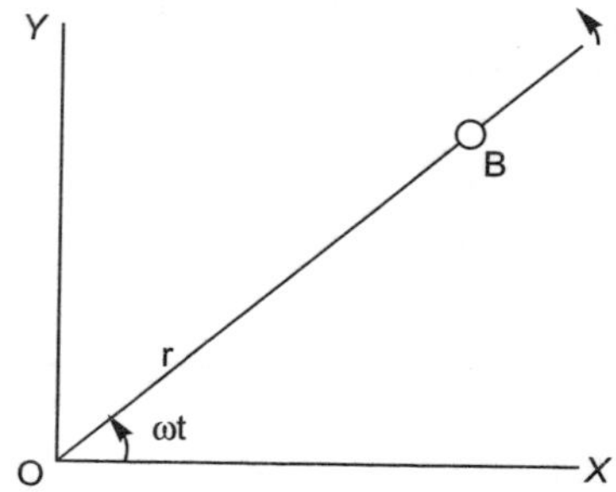

Figure 1.7

1.6 DEGREES OF FREEDOM

The number of independent ways in which a mechanical system can move without violating any constraint which may be imposed, is called the degrees of freedom of the system. It is indicated by the least possible number of co-ordinates to describe the system completely.

The **position** of a **system** consisting of N particles is **determined by** 3N number of **independent co-ordinates (or parameters)**, that is 3-cartesian co-ordinates for each particle.

Now, if the distribution of points **be fixed in some way** then the **number** of degrees of freedom may be less than 3N.

Example : Consider two points (x_1, y_1, z_1) and (x_2, y_2, z_2) are *constrained* by *rigid non-deformable* coupling. Then upon the six co-ordinates $x_1, y_1, z_1, x_2, y_2, z_2$ of these points is *imposed the condition,*

$$(x_2 - x_1)^2 + (y_2 - y_1)^2 + (z_2 - z_1)^2 = C_{12}^2 \qquad \ldots(i)$$

where, C_{12} : fixed distance between the points = constant.

The condition equation shows that the cartesian co-ordination are not independent ; *a relation exists* between them. Only *five* of the *six values* $(x_1, y_1, z_1, x_2, y_2, z_2)$ are now independent. Hence the *system of two particles* have *only* FIVE degrees of freedom.

For *three particles* fixed in space on the *vertices of a triangle the third particle* must satisfy the *TWO* equations.

$$(x_3 - x_1)^2 + (y_3 - y_1)^2 + (z_3 - z_1)^2 = C_{31}^2 \qquad \ldots(ii)$$

$$(x_3 - x_2)^2 + (y_3 - y_2)^2 + (z_3 - z_2)^2 = C_{32}^2 \qquad \ldots(iii)$$

Thus, the NINE co-ordinates of the vertices of the rigid triangle are defined by the three equations (*i*), (*ii*) and (*iii*). Hence out of nine co-ordinates only SIX are independent. The triangle has SIX degrees of freedom.

In a system of N particles subject to "*k*" independent constraints of the form, $f_i\,(x_1, y_1, z_1, x_2, y_2, z_2, \ldots, x_n, y_n, z_n\, t) = c_i$ the number of degrees of freedom = (3N – *k*).

The specified functions are of 3N co-ordinates and possibly of *t* (time) if the constraints depend explicitly on *t*.

It can be concluded that "*imposing constraint is a way of reducing the number of equations of motion and make these equal to that of degrees of freedom*".

1.7 GENERALISED CO-ORDINATES

The elimination of dependent co-ordinates can be effected by introducing a set of (3N – *k*) independent variables known as Generalised Co-ordinates. These are designated by "*q*" with numerical subscripts ; $q_1, q_2, q_3, \ldots, q_n$ to represent a set of "*n*" generalised co-ordinates. They are :

(*i*) general, in the sense that they need not be lengths, angles or positional co-ordinates in conventional orthogonal (*i.e.*, mutually perpendicular) system.

(*ii*) in number, just equal to the number of degrees of freedom of the system.

(*iii*) independent of each other.

The generalised co-ordinates need not belong to any particle in a group of three (as in the case of cartesian co-ordinates). All sorts of quantities may be used to serve as generalised co-ordinates. Number of independent variables being minimised, constraints are automatically obeyed. So, one should not introduce the forces of constraints in the treatments with generalised co-ordinates. When we switch over to describe a specific problem, the symbols $q_1, q_2, q_3, \ldots, q_n$ correspond to co-ordinates that we choose to describe the motion. As for example, for cartesian co-ordinates, or polar co-ordinates,

$$q_1 = x \qquad q_1 = r\left(=\sqrt{x^2+y^2}\right)$$

or

$$q_2 = y \qquad q_2 = \theta\left(=\tan^{-1}\frac{y}{x}\right)$$

for spherical polar co-ordinates (when the problem involves spherical symmetry)

$$q_1 = r\left(=\sqrt{x^2+y^2+z^2}\right)$$

$$q_2 = \theta\left(=\cot^{-1}\frac{z}{\sqrt{x^2+y^2}}\right)$$

$$q_3 = \phi\left(=\tan^{-1}\frac{y}{x}\right)$$

In case of a co-ordinate system moving uniformly with velocity *v* in *x* - direction,

$$q_1 = x - \dot{x}t \qquad \dot{x} = vt = \text{constant}$$

$$q_2 = y$$

$$q_3 = z$$

NOTE For holonomic systems the number of generalised co-ordinates is exactly equal to the number of degrees of freedom whereas for non-holonomic systems it is larger than the number of degrees of freedom.

Generalised Displacement

Consider a small displacement of an N-particle system defined by changes δr_i in cartesian co-ordinates r_i (i = 1, 2, 3, ..., N) with time t held fixed. If one considers an arbitrary virtual displacement δr_i, then since r_i are functions of generalised co-ordinates given by the equation,

$$r_i = r_i(q_1, q_2, \ldots, q_{3N}, t)$$

We have,

$$\delta r_i = \sum_{j=1}^{3N} \frac{\partial r_i}{\partial q_j} . \delta q_j \quad (\because \delta t = 0) \qquad \ldots(1.1)$$

At the beginning we have chosen r_i to represent the 3N co-ordinates ($x_1, y_1, z_1, \ldots, z_n$) for our convenience of symbol where each r_i is equivalent to three components x_1, y_1, z_1 and so on. The δq_j are called generalised displacements or virtual arbitrary displacements. If q_j is an angle co-ordinate then δq_j is an angular uisplacement.

Generalised Velocity

The generalised velocities of a system are the total time derivatives of the generalised co-ordinates of the system. Thus,

$$\dot{q}_i = \frac{dq_i}{dt} \quad (i = 1, 2, \ldots., k)$$

For an unconstrained system.

$$r_i = r_i(q_1, q_2, \ldots q_{3N}, t)$$

Then,

$$\dot{r}_i = \sum_{j=1}^{3N} \frac{\partial r_i}{\partial q_j} \dot{q}_j + \frac{\partial r_i}{\partial t} \qquad \ldots(1.2)$$

If the N-system contains k constraints then the number of generalised co-ordinates is 3N – k = f (say) and in that case,

$$\dot{r}_i = \sum_{j=1}^{f} \frac{\partial r_i}{\partial q_j} \dot{q}_j + \frac{\partial r_i}{\partial t} \; [= v_i] \qquad \ldots(1.3)$$

NOTE If the generalised co-ordinate q_j involves both cartesian and angle co-ordinates, the general velocity associated with a cartesian co-ordinate x is just $\dot{x}$ while the generalised velocity associated with an angular co-ordinate θ is $\frac{d\theta}{dt}$ (or $\dot{\theta}$). If the generalised co-ordinate has the dimensions of momentum the generalised velocity will have the dimensions of force and so on.

$\frac{dr_i}{dt}$ appears only when there are constraints varying with time.

For solving non-holonomic problems one uses the fact that "the virtual work done by forces of constraint is zero.

1.8 PRINCIPLE OF VIRTUAL WORK

Principle of virtual work is a form of differential variational principle of mechanics which expresses the difference for the actual motion from those permitted by the constraints in each given instant of time.

If the system of particles goes from one configuration (*i.e.*, description in a co-ordinate space) to another, each particle of the system is imagined to be displaced by an infinitesimal vector δr_i called the **virtual displacement** to distinguish the same from dr_i which occurs in a time interval dt during which the forces and the constraints may be changing. In order that a system be in equilibrium the resultant force acting on each particle must be zero *i.e.*, $F_i = 0$, where F_i is the force acting on the ith particle of the system. Then for the system $\sum_i F_i \cdot \delta r_i$ = virtual work = 0.

If constraints are present then, $F_i = F_i^{(a)} + f_i$ where, $F_i^{(a)}$: actual (or apparent) force and f_i : force of constraint on i^{th} particle.

Hence, $$\sum_i \left(F_i^{(a)} + f_i\right) \cdot \delta r_i = 0$$

or, $$\sum_i F_i^{(a)} \cdot \delta r_i + \sum_i f_i \cdot \delta r_i = 0$$

Restricting to the systems where the virtual work done by the forces of constraint is zero *i.e.*, $\sum_i f_i \cdot \delta r_i = 0$ the above equation reduces to $\sum_i F_i^{(a)} \cdot \delta r_i = 0$.

This is the condition for equilibrium of a system. Thus, "A system is in equilibrium only if the total virtual work of the actual (or applied) forces is ZERO".

1.9 D'ALEMBERT'S PRINCIPLE

This principle is particularly useful in cases of motion with constraints, where application of **Newton's** equations may often be difficult.

In a system of particles, the total force F_i on the ith particle may be split up into two parts :

$$F_i = F_i^a + F_i^c$$

where F_i^c : force of constraints.

Forces other than reactions of constraints are called "*applied forces*" denoted by F_i^a.

The equation of motion according to Newton's 2nd law is,

$$F_i = F_i^a + F_i^c = \dot{p}_i$$

and the work done in a virtual displacement of the system is,

$$\sum_i F_i^a \, \delta r_i + \sum_i F_i^c \cdot \delta r_i = \sum_i \dot{p}_i \cdot \delta r_i$$

[Virtual displacement is denoted by δr_i rather than dr_i]

Virtual work done by forces of constraint being zero,

we write, $\sum_i \left(\mathrm{F}_i^a - \dot{p}_i\right) . \delta r_i = 0$

This equation is called D'Alembert's principle.

Here, the forces of constraint do not appear and a single equation is obtained instead of a set of simple equation for each individual particle in the Newtonian scheme. In this equation the coefficients of δr_i cannot be set equal to zero, as all δr_i are not completely independent. The above equation must be transformed into an expression involving virtual displacements δq_i of generalised co-ordinates which are then independent of each other (for holonomic constraints) so that the coefficients of δq_i can be set separately equal to zero – which gives the Lagrange's Equations.

Application of D'Alembert's Principle

Atwood's Machine : m_1, m_2 are two unequal masses connected by an inextensible string passing over a smooth pulley P. To determine the acceleration of the masses under gravity.

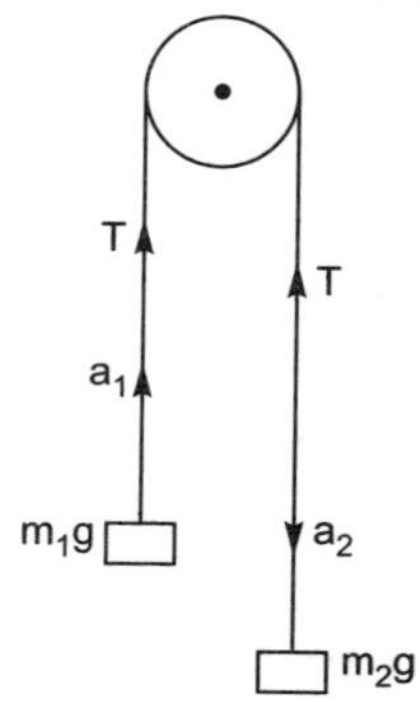

Figure 1.8

Here, T (the tension in the string) and the reaction of the pulley are the forces of constraint. The two applied forces are $m_1 g$ and $m_2 g$. Let a_1 and a_2 be the accelerations of m_1 and m_2 respectively then according to D'Alembert's principle,

$$(m_1 g - m_1 a_1) . \delta r_1 + (m_2 g - m_2 a_2)\, \delta r_2 = 0$$

The virtual displacements, in this case, are taken positive in the downward direction. The length of the string being constant,

$$\delta r_1 = - \delta r_2 \text{ and } a_1 = - a_2$$

$\therefore$ $$(m_1 g - m_1 a_1) . \delta r_1 - (m_2 g + m_2 a_1)\, \delta r_1 = 0$$

$\Rightarrow$ $$(m_1 g - m_1 a_1) = (m_2 g + m_2 a_1)$$

$\therefore$ $$|a_1| = |a_2| = \frac{m_1 - m_2}{m_1 + m_2} \cdot g$$

1.10 GENERALISED FORCE : LAGRANGE'S EQUATION

In terms of generalised co-ordinates the virtual work of the applied force F_i^a (say F_i) becomes,

$$\delta W = \sum_i F_i . \delta r_i = \sum_j \left(\sum_i F_i . \frac{\partial r_i}{\partial q_j} \right) \delta q_j = \sum_j Q_j . \delta q_j$$

$$\because \quad \delta r_i = \sum_j \frac{\partial r_i}{\partial q_j} . \delta q_j$$ (as there is no passage of time in virtual displacement)

where $$Q_j = \sum_i F_i . \frac{\partial r_i}{\partial q_j}$$

From the equation (as given above)

$$\delta W = \sum_j Q_j . \delta q_j \qquad \text{...(1.4)}$$

It appears that Q_j behaves in the same way as F_i in $\delta W \left(= \sum_i F_i . \delta r_i \right)$ *i.e.*, Q_j plays the role similar to F_i. So, Q_j is called the components of generalised force.

Now, as q's need not have the dimensions of length Q's also need not have the dimensions of force, however, $Q_j.\delta q_j$ must have the dimensions of work.

Again one can write, $\sum_i \dot{p}_i . \delta r_i = \sum_i m_i \ddot{r}_i . \delta r_i$

$$= \sum_j \left(\sum_i m_1 \ddot{r}_i . \frac{\partial r_i}{\partial q_j} \right) \delta q_j \qquad \text{...(1.5)}$$

But $$\sum_i m_i \ddot{r}_i . \frac{\partial r_i}{\partial q_j} = \sum_i \left\{ \frac{d}{dt} \left(m_i \dot{r}_i . \frac{\partial r_i}{\partial q_j} \right) - m_i \dot{r}_i \frac{d}{dt} \left(\frac{\partial r_i}{\partial q_j} \right) \right\}$$

Which on solving becomes,

$$\sum_i m_i \ddot{r}_i . \frac{\partial r_i}{\partial q_j} = \frac{d}{dt} \left\{ \frac{\partial}{\partial \dot{q}_i} \left(\sum_i \frac{1}{2} m_i v_i^2 \right) \right\} - \frac{\partial}{\partial q_j} \left(\sum_i \frac{1}{2} m_i v_i^2 \right)$$

$$= \frac{d}{dt} \frac{\partial T}{\partial \dot{q}_j} - \frac{\partial T}{\partial q_j} \qquad \text{...(1.6)} \left\{ \because \sum_i \frac{1}{2} m_i v_i^2 = T \text{ (K.E.)} \right\}$$

Here T is the Kinetic Energy of the system of particles.

With the help of equations (1.4), (1.5) and (1.6) D'Alembert's principle becomes,

$$\sum_j \left[Q_j - \left\{ \frac{d}{dt} \left(\frac{\partial T}{\partial \dot{q}_j} \right) - \frac{\partial T}{\partial q_j} \right\} \right] \delta q_j = 0$$

Whereform, $$Q_j = \frac{d}{dt} \left(\frac{\partial T}{\partial \dot{q}_j} \right) - \frac{\partial T}{\partial q_j} \qquad \text{...(1.7)}$$

where $j = 1, 2, \ldots, n$

There are n such equations in all.

For Conservative System, forces are derivable from scalar potential function V of position only.

Then,

$$V = V(r_1, r_2, ..., r_n) = V(q_1, q_2, ..., q_n)$$

$$\therefore \quad dV = \sum_i \frac{\partial V}{\partial x_i} dx_i + \frac{\partial V}{\partial y_i} dy_i + \frac{\partial V}{\partial z_i} dz_i = \sum_i \nabla_i V . dr_i$$

$$\therefore \quad \frac{\partial V}{\partial q_j} = \sum_i \nabla_i V . \frac{\partial r_i}{\partial q_j}$$

Therefore, the generalised force is,

$$Q_j = \sum_i F_i \frac{\partial r_i}{\partial q_j} = -\sum_i \nabla_i V \frac{\partial r_i}{\partial q_j} = -\frac{\partial V_i}{\partial q_j} \quad ...(1.8)$$

Using equations (1.7) and (1.8) one can obtain,

$$\frac{d}{dt}\left(\frac{\partial T}{\partial \dot{q}_j}\right) - \frac{\partial (T - V)}{\partial q_j} = 0 \quad ...(1.9)$$

The potential V being a function of position only and not of generalised velocities the above equation (1.9) can be written as,

$$\frac{d}{dt}\left(\frac{\partial L}{\partial \dot{q}_j}\right) - \frac{\partial L}{\partial q_j} = 0 \quad ...(1.10)$$

where $j = 1, 2, ..., n$

and $L = (T - V)$

$L = (T - V)$ is known as the Lagrangian of the system.

When the system is not conservative (*i.e.*, say, depends on the generalised velocities $\dot{q}$), the generalised force Q_j associated with q_j is defined by,

$$Q_j = \frac{\partial U}{\partial q_j} + \frac{d}{dt}\left(\frac{\partial U}{\partial \dot{q}_j}\right) \quad ...(1.11)$$

$$(i.e., L = T - U)$$

where, $U(q, \dot{q})$ is a function of generalised velocity $\dot{q}$ in addition to the position co-ordinates q.

The U may be called velocity dependent potential or generalised potential.

The set of n equations as determined by equation (1.7) are valid in the case of conservative as well as non-conservative forces. These are called LAGRANGE'S EQUATIONS.

However, this name is used mostly for equations of the systems whenever conservative forces are acting. The equations are then of the form given by equation (1.10) and are known as LAGRANGE'S EQUATIONS of motion for a conservative holonomic system.

NOTE Lagrange's equations for a system is equivalent to Newton's equation of motion and so it is not a new theory, it is only an alternative and equivalent formulation.

In Newtonian approach the forces are due to external agencies and they represent the cause of the acceleration produced in the body. The resulting acceleration is the effect of the force.

In Lagrangian approach, we consider the kinetic and potential energies of the system. This is the difference between the two formulations. The kinetic and the potential energies are scalar functions and do not vary under co-ordinate transformations. Thus one can always find a set of generalized co-ordinates that makes the problem simpler to solve.

Another difference is that in some problem it may not be possible to know all the forces acting on the system, but the expression for the kinetic and potential energy may be given. Under these situations, Lagrange's equations are very much useful.

It is always easy to deal with scalar quantities than with vector quantities such as force, momentum etc., and there lies the importance of Lagrange's Equations.

1.11 CYCLIC OR IGNORABLE CO-ORDINATES

The Lagrangian L is written as a function of q_j and $\dot{q}_j$. If any of the co-ordinates (say) q_k is absent in the expression for L then $\frac{\partial L}{\partial q_k} = 0$

The equation of motion corresponding to q_k becomes

$$\frac{d}{dt}\left(\frac{\partial L}{\partial \dot{q}_k}\right) = 0 \quad \text{or} \quad \frac{\partial L}{\partial \dot{q}_k} = \text{constant}$$

Thus, whenever q_k does not occur explicitly in L, the corresponding linear momentum p_k is constant of the motion. Such a co-ordinate q_k (called the first integral of motion) is said to be ignorable (or cyclic co-ordinate). When the generalised translational co-ordinate q_k is cyclic (*i.e.*, absent in the Lagrangian function), the function L is unchanged under this translation and the equation of motion remains the same.

Example : Consider the motion of two bodies under the action of internal force. As no external force is acting on the system, the translation of the whole system will not affect the internal forces or the velocities. Thus T (the KE), V (the PE) and hence the Lagrangian L (*i.e.*, T – V) are unaffected. If R (X, Y, Z) represents the position vector of the centre of mass of the system, then

$$\frac{\partial L}{\partial X} = \frac{\partial L}{\partial Y} = \frac{\partial L}{\partial Z} = 0 \text{ and hence, } M\dot{R} \text{ (the linear momentum of c.m.) = constant (} i.e.\text{, conserved).}$$

Superiority of Lagrangian Approach over Newtonian Approach

1. Lagrange's equations of motion are invariant in form with respect to co-ordinate transformation.
2. They put emphasis on energy and work (which are scalar quantities) rather than on force (vector quantities).
3. They are expressed in generalised co-ordinates which do not have to refer directly to the position or displacements of actual particles.
4. As stated in (1) above, Lagrange's equations of motion neither change their form under point transformation nor do they change if we add to the Lagrangian a total time derivative of any function F (q, t).
5. It turns out that every symmetry (with respect to arbitrary translation or rotation or passage of time) in the Lagrangian corresponds to a conservation law. Hence, symmetries in Lagrangian have very important consequences.

1.12 HAMILTON'S EQUATION OF MOTION AND HAMILTONIAN

The space for the description of the motion of a single particle must be three-dimensional. In case of two particles, six-dimensional space is required. That is, the problem of two-particles forming a system can be considered as a problem of a single particle in a six-dimensional space. Similarly, for a N-particle system, the problem can be treated as the motion of a single particle moving in 3-N dimensional space known as CONFIGURATION SPACE. The single particle is then a SYSTEM POINT. In case there are "k" equations of constraints the (3N-k) dimensional space will be the CONFIGURATION SPACE.

In Lagrangian formulation the independent variables are the generalised co-ordinates q_i and the time t and generalised velocities $\dot{q}_j$ being derivatives of the generalised co-ordinates are dependent variables.

The view point of Hamiltonian method is to remove this dependence and to introduce a new independent variable the generalised momentum p_j.

Lagrangian is a function of generalised co-ordinate, generalised velocity and time *i.e.*,

$$L = L(q_j, \dot{q}_j, t)$$

Hamiltonian is a function of generalised co-ordinate, generalised momentum and time *i.e.*,

$$H = H(q_j, p_j, t)$$

It is derived from : $H = \sum_j p_j \dot{q}_j - L(q_j, \dot{q}_j, t)$ and $\frac{\partial L}{\partial \dot{q}_j} = p_j =$ generalised momentum.

Hamilton's Equations of Motion can be derived using Lagrangian function L. Those are :

$$\frac{\partial H}{\partial q_j} = -\dot{p}_j \text{ and } \frac{\partial H}{\partial p_j} = \dot{q}_j$$

Further,
$$-\frac{\partial L}{\partial t} = \frac{\partial H}{\partial t}$$

Hamilton's equations of motion are also known as Hamilton's Canonical Equations of motion.

Physical Significance of the Hamiltonian

(*i*) If the Lagrangian of the system does not contain time explicitly then the Hamiltonian is a constant of motion.

(*ii*) Hamiltonian of a conservative system is equal to its total energy *i.e.*, H = T + V (for a conservative system) for a time independent co-ordinate transformation.

$$T = K.E. \; ; \; V = P.E.$$

Important steps involved in Hamiltonian Form :

(*a*) Find L = T – V

(*b*) Find H from
$$H = \sum_j p_j \dot{q}_j - L$$

(*c*) Express $H = H(q_j, p_j, t)$

(*d*) Write down Hamilton's canonical equations.

Applications of the Hamiltonian Formulation

A particle of mass m moves under a two dimension potential $V(x, y) = -(ax^2 + by^2)$ where a, b are constants. Obtain the Hamiltonian of the particle.

In the problem, P.E. ; $V = -(ax^2 + by^2)$

$$\text{K.E. ; } T = \frac{1}{2} m \left(\dot{x}^2 + \dot{y}^2\right)$$

The Lagrangian is, $L = T - V$

$$= \frac{1}{2} m \left(\dot{x}^2 + \dot{y}^2\right) + (ax^2 + by^2)$$

Hamiltonian,

$$H = \sum_j p_j \cdot \dot{q}_j - L$$

$$= p_x \dot{x} + p_y \dot{y} - L$$

$$\left\{ \begin{array}{l} \text{Here } q_j \to x, y \\ \dot{q}_j \to \dot{x}, \dot{y} \\ p_j \to p_x, p_y \end{array} \right.$$

or

$$H = (m\dot{x} \cdot \dot{x} + m\dot{y} \cdot \dot{y}) - \frac{1}{2} m \left[\left(\dot{x}^2 + \dot{y}^2\right) + \left(ax^2 + by^2\right)\right]$$

or

$$H = m\dot{x}^2 + m\dot{y}^2 - \frac{1}{2} m\dot{x}^2 - \frac{1}{2} m\dot{y}^2 - ax^2 - by^2$$

$$= \frac{1}{2}\left(m\dot{x}^2 + m\dot{y}^2\right) - \left(ax^2 + by^2\right)$$

$$\text{Also} \quad m\dot{x} = p_x = \frac{\partial L}{\partial \dot{x}}$$

$$m\dot{y} = p_y = \frac{\partial L}{\partial \dot{y}} \quad \left\{ \begin{array}{l} \because p_j / m = j \\ \therefore \dot{x} = \dfrac{p_x}{m} \text{ and } \dot{y} = \dfrac{p_y}{m} \end{array} \right\}$$

$$\therefore \quad H = \frac{1}{2}\left(\frac{p_x^2 + p_y^2}{m}\right) - \left(ax^2 + by^2\right)$$

Hence, the Hamiltonian of the system is,

$$H = \frac{p_x^2 + p_y^2}{2m} - \left(ax^2 + by^2\right)$$

NOTE *Phase Space.* While using the Lagrangian formulation for describing the motion of a system with N degrees of freedom the system is represented in an N-dimensional co-ordinate space called **configuration space.** Any point in this space describes all the constituents of the system and the point is known as **system point.**

In Hamiltonian formulation N co-ordinates q_j and N momenta p_j are taken as independent variables. If only N momenta are used as axes in N-dimensional space, we get momentum space. Any point in the momentum space will describe the state of motion of the whole system. To describe a function like H (p_j, q_j) both the co-ordinate and momentum spaces are needed which is 2N-dimensional space (PHASE SPACE).

IN FOCUS

■ System of Co-ordinates

(*i*) Right handed rectangular cartesian

$$\vec{r} = \hat{i}x + \hat{j}y + \hat{k}z$$

(*ii*) Cylindrical

$$\rho = \sqrt{x^2 + y^2} \qquad x = \rho\cos\phi$$

$$\phi = \tan^{-1}\frac{y}{x} \quad \text{and} \quad y = \rho\sin\phi$$

$$z = z$$

(*iii*) Spherical Polar

$$x = r\sin\theta\cos\phi$$
$$y = r\sin\theta\sin\phi$$
$$z = r\cos\theta$$

■ Conservation of Linear Momentum

$$F^{ext} = \frac{d\vec{P}}{dt} \quad \text{where } \vec{P} = \sum_{i=1}^{n} P_i = \sum_{i=1}^{n} m_i v_i$$

■ Conservation of Angular Momentum

$$\tau^{ext}\text{ (torque)} = \frac{dJ}{dt} = \sum_{i=1}^{n} r_i \times F_i^{ext}$$

■ Conservation of Energy

For conservative force $F = -\dfrac{dv}{dr}$

Total energy = K.E. + P.E.

■ Center of Mass

$$\bar{x} = \frac{\Sigma m_i x_i}{M}, \bar{y} = \frac{\Sigma m_i y_i}{M}, \text{ and } \bar{z} = \frac{\Sigma m_i z_i}{M}$$

Constraints

Scleronomic : Constant relation do not explicitly depend on time.

Rheonomic : Constant relation depend explicitly on time.

Holonomic : Constant relation can be made independent of velocities.

Non-Holonomic : Constant relation are dependent on velocities.

Conservative : Total mechanical energy is conserved while performing the motion constraint forces do not do any work.

Dissipative : Constant force do work and total mechanical energy is not conserved.

Bilateral : Constraint relations are not in the form of inequalities but in the form of equations.

Unilateral : Constraint relations are expressed in the form of inequalities.

- **Degree of Freedom:** The number of independent ways in which a mechanical system can move without violating any constraint which may be imposed is called the degree of freedom of the system.
- **Generalised Co-ordinates:** Generalised co-ordinates are designated by "q" with numerical subscripts;
  - **In cartesian co-ordinate**

$$q_1 = x \qquad\qquad q_1 = r\left(= \sqrt{x^2 + y^2}\right)$$

or

$$q_1 = y \qquad\qquad q_2 = \theta\left(= \tan^{-1}\frac{y}{x}\right)$$

  - **For spherical polar co-ordinates**

$$q_1 = r\left(= \sqrt{x^2 + y^2 + z^2}\right)$$

$$q_2 = \theta\left(= \cot^{-1}\frac{z}{\sqrt{x^2 + y^2}}\right)$$

$$q_3 = \phi\left(= \tan^{-1}\frac{y}{x}\right)$$

- **Generalised Displacement:** A small displacement

$$\delta r_i = \sum_{j=1}^{3N}\frac{\partial r_i}{\partial q_j}.\delta q_j$$

- **Generalised Velocity**

$$v_i = \dot{r}_i = \sum_{j=1}^{f}\frac{\partial r_i}{\partial q_j}\dot{q}_j + \frac{\partial r_i}{\partial t}$$

- **Principle of Virtual Work:** A system is in equilibrium only if the total virtual work of the actual forces is zero

$$\sum_i F_i^{(a)}.\delta r_i = 0$$

where $F_i^{(a)}$: actual (or apparent) force

- **D'Alembert's Principle**

$$\Sigma\left(F_i^{(a)} - \dot{p}_i\right).\delta r_i = 0$$

■ **Generalised Force: Lagrange's Equation**

$$\frac{\partial}{\partial t}\left(\frac{\partial L}{\partial \dot{q}_j}\right) - \frac{\partial L}{\partial \dot{q}_j} = 0 \qquad \text{where } j = 1, 2, \ldots n \text{ and } L = (T - V)$$

L is known as the Lagrangian of the system.

The generalised force Q_j associated with q_j is defined by

$$Q_j = \frac{\partial U}{\partial q_j} + \frac{d}{dt}\left(\frac{\partial U}{\partial \dot{q}_j}\right)$$

where U $(q, \dot{q})$ is a function of generalised velocity $\dot{q}$.

■ **Hamiltonian's Equation of Motion:** Hamiltonian's is a function of generalised co-ordinate, generalised momentum and time *i.e.*, $H = H(q_j, p_j, t)$

It is derived from : $H = \sum_j p_j\dot{q}_j - L(q_j, \dot{q}_j, t)$

and $\dfrac{\partial L}{\partial \dot{q}_j} = p_j$ = generalised momentum.

Solved Problems

Problem 1. *For a mass m attached to a spring of force constant k at any instant, the kinetic energy is given by $\frac{1}{2}mv^2$ and potential energy by $\frac{1}{2}kx^2$, where v is a instantaneous velocity and x is the displacement. Set up the Lagrangian of system and show that the dynamical equation of motion of mass is given by $a = -\omega^2 x$, where a is the acceleration and $\omega^2 = \frac{k}{m}$.*

Solution. Using Lagrange's equation

The kinetic energy $\quad T = \frac{1}{2}mv^2 = \frac{1}{2}m\dot{x}^2$

Potential energy $\quad V = \frac{1}{2}kx^2 \qquad$ [Here x is the generalized co-ordinate]

∴ Lagrangian $\quad L = T - V = \frac{1}{2}m\dot{x}^2 - \frac{1}{2}kx^2$

The Lagrange's equation in co-ordinate x is

$$\frac{d}{dt}\left(\frac{\partial L}{\partial \dot{x}}\right) - \frac{\partial L}{\partial x} = 0 \qquad \ldots(i)$$

From above $\quad \dfrac{\partial L}{\partial \dot{x}} = m\dot{x}, \; \dfrac{\partial L}{\partial x} = -kx$

Substituting these values in equation (*i*), we get

$$\frac{d}{dt}(m\dot{x}) - (-kx) = 0,$$

or
$$m\ddot{x} + kx = 0$$

$$\ddot{x} + \frac{k}{m}x = 0 \qquad \ldots(ii)$$

As
$$\ddot{x} = \frac{d^2x}{dt} = \text{acceleration} = -\omega^2 x,$$

∴
$$\ddot{x} + \frac{k}{m}x = 0$$

$$\ddot{x} = -\frac{k}{m}x = -\omega^2 x$$

where $\omega^2 = \frac{k}{m}$,

Using Hamilton's Equation of Motion

The Hamiltonian is $H = \sum p_i \dot{q}_i - L$

$$= p_x \dot{x} - L \qquad \ldots(iii)$$

Since
$$p_x = \frac{\partial L}{\partial \dot{x}} = -m\dot{x}$$

$$H = m\dot{x}\dot{x} - \left[\frac{1}{2}m\dot{x}^2 - \frac{1}{2}kx^2\right]$$

$$= \frac{1}{2}m\dot{x}^2 + \frac{1}{2}kx^2 \qquad \ldots(iv)$$

= Kinetic energy (T) + Potential energy (V)

$$= \frac{1}{2}m\left(\frac{p_x}{m}\right)^2 + \frac{1}{2}kx^2$$

[where $p_x = \frac{\partial L}{\partial \dot{x}} = m\dot{x}$ or, $\dot{x} = \frac{p_x}{m}$]

Then,
$$\frac{\partial H}{\partial p_x} = \frac{p_x}{m}$$

and
$$\frac{\partial H}{\partial x} = kx \qquad \ldots(v)$$

Thus Hamilton's equations of motion for this system will be

$$\dot{x} = \frac{\partial H}{\partial p_x} = \frac{p_x}{m} \qquad \ldots(vi)$$

$$\dot{p}_x = -\frac{\partial H}{\partial x} = -kx \qquad \ldots(vii)$$

Putting the value of p_x from equation (*vi*) into equation (*vii*) we find

$$m\ddot{x} = -kx$$

or $$\ddot{x} + \frac{k}{m} x = 0 \qquad \ldots(viii)$$

Which is the equation of motion of the system as obtained from Lagrange's equation.

Problem 2. *Derive the Lagrangian and Hamilton's equations of motion for a particle falling freely under the the influence of gravity.* [W.B.U.T].

Solution. Consider a particle of mass m falls freely under the influence of gravity from the height h from ground.

After releasing the particle at A, it covers some distance x and reaches at B.

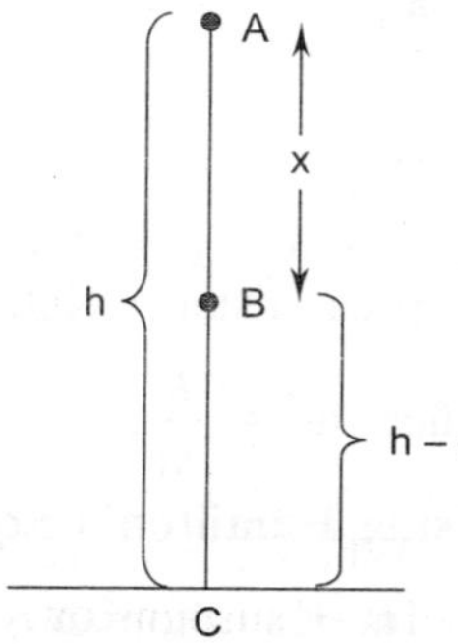

Now the potential energy at B

$$V = mg\,(h - x) \text{ and kinetic energy}$$

$$T = \frac{1}{2} m\dot{x}^2$$

Hence the Lagrangian is $L = T - V$

$$= \frac{1}{2} m\dot{x}^2 - mg\,(h - x)$$

$$\frac{2L}{\partial \dot{x}} = m\dot{x}, \frac{\partial L}{\partial x} = mg$$

So, the Lagrangian equation of motion,

$$\frac{d}{d\tau}\left(\frac{\partial L}{\partial \dot{x}}\right) - \frac{\partial L}{\partial x} = 0$$

or $$m\ddot{x} - mg = 0, \text{ or } \ddot{x} - g = 0 \ldots(.) \qquad [\because \; m \neq 0]$$

Using Hamilton's equation:

The Hamiltonian is $$H = \sum p_i\,\dot{q}_i - L = p_x\,\dot{x} - L$$

Here $$p_x = \frac{\partial L}{\partial \dot{x}} = m\dot{x} \text{ or } \dot{x} = \frac{p_x}{m}$$

or $$H = \frac{p_x^2}{m} - \left\{\frac{1}{2} m \frac{p_x^2}{m^2} - mg\,(h - x)\right\} = \frac{p_x^2}{2m} + mg\,(h - x)$$

$$= T + V$$

Thus Hamilton's equation of motion for this system will be $\dot{p}_x = -\dfrac{\partial H}{\partial x} = mg$ and $\dot{x} = \dfrac{\partial H}{\partial p_x} = \dfrac{p_x}{m}$

on $p_x = m\dot{x}$ or, $\dot{p}_x = m\ddot{x} = mg$ or, $\ddot{x} - g = 0$

Problem 3. *Derive the equation of motion for a simple pendulum of length l and mass of bob m.*

[W.B.U.T]

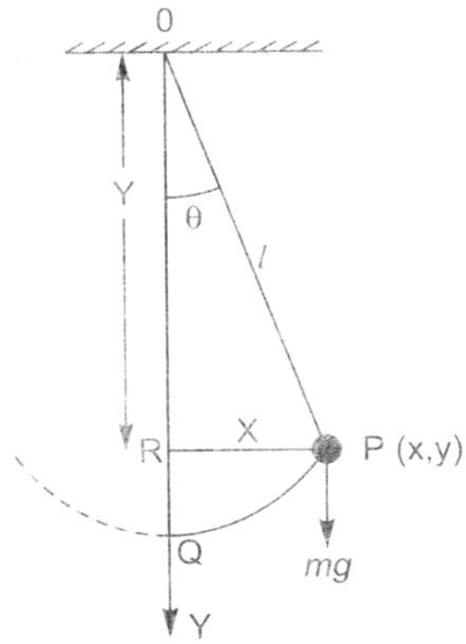

Solution. Let the angle θ between the rest position and the deflected position is chosen as generalized co-ordinate.

$\therefore$ Here, kinetic energy is $T = \frac{1}{2} m (l\dot{\theta})^2$

$$= \frac{1}{2} ml^2 \dot{\theta}^2$$

In coming from position P to Q, the mass has fallen freely through a vertical distance RQ. The potential energy is

$$V = mg\,(OQ - OR) = mg\,(l - l\cos\theta)$$
$$= mgl\,(1 - \cos\theta)$$

Where the reference level or zero level of potential energy has been taken at a distance l below the point of suspension.

Using Lagrange's Equation

The Lagrangian is L = T – V

$$= \frac{1}{2} ml^2 \dot{\theta}^2 - mgl\,(1 - \cos\theta) \qquad \ldots(i)$$

So that $$\frac{\partial L}{\partial \dot{\theta}} = ml^2 \dot{\theta} \text{ and } \frac{\partial L}{\partial \theta} = -mgl \sin\theta$$

Putting in Lagrange's equation

$$\frac{d}{dt}\left(\frac{\partial L}{\partial \dot{\theta}}\right) - \frac{\partial L}{\partial \theta} = 0,$$

$$ml^2 \ddot{\theta} + mgl \sin\theta = 0$$

$$\ddot{\theta} + \frac{g}{l} \sin\theta = 0$$

If the amplitude of the motion is small enough, $\sin\theta \approx \theta$,

$$\therefore \quad \ddot{\theta} + \frac{g}{l}\theta = 0 \qquad \ldots(ii)$$

which is equation for S.H.M. with period $T = 2\pi\sqrt{\frac{l}{g}}$.

Using Hamilton's equations

The Hamiltonian is $H = \sum p_i \dot{q}_i - L$

$$= p_\theta \dot{\theta} - L \qquad \ldots(iii)$$

where $p_\theta = \frac{\partial L}{\partial \dot{\theta}} = ml^2 \dot{\theta}$

$$\therefore \quad H = ml^2 \dot{\theta}^2 - \left\{\frac{1}{2} ml^2 \dot{\theta}^2 - mgl\,(1 - \cos\theta)\right\}$$

$$= \frac{1}{2} ml^2 \dot{\theta}^2 + mgl\,(1 - \cos\theta) = T + V$$

$$= \frac{1}{2} ml^2 \left(\frac{p_\theta}{ml^2}\right)^2 + mgl\,(1 - \cos\theta) \qquad \ldots(iv)$$

Thus, $$\frac{\partial H}{\partial p_\theta} = \frac{p_\theta}{ml^2}, \quad \frac{\partial H}{\partial \theta} = mgl \sin\theta$$

Thus Hamilton's equations of motion for this system will be

$$\dot{\theta} = \frac{\partial H}{\partial p_\theta} = \frac{p_\theta}{ml^2} \qquad \ldots(v)$$

$$\dot{p}_\theta = -\frac{\partial H}{\partial \theta} = -mgl \sin\theta \qquad \ldots(vi)$$

Putting the value of p_θ from equation (*v*) into equation (*vi*), we get

$$ml^2 \ddot{\theta} = -mgl \sin\theta$$

$$\ddot{\theta} + \frac{g}{l}\theta = 0 \qquad \ldots(vii)$$

Problem 4. *Obtain the equation of motion of a system of two particles of unequal masses connected by an inextensible string passing over a small smooth pulley [Atwood's Machine].*

[Agra University].

Solution. Let two masses M_1 and M_2 be connected by a light inextensible string l passing over a small smooth pulley. It is found that the heavier particle (M_2) decends while the lighter (M_1) ascends and the system is moving with a constant acceleration.

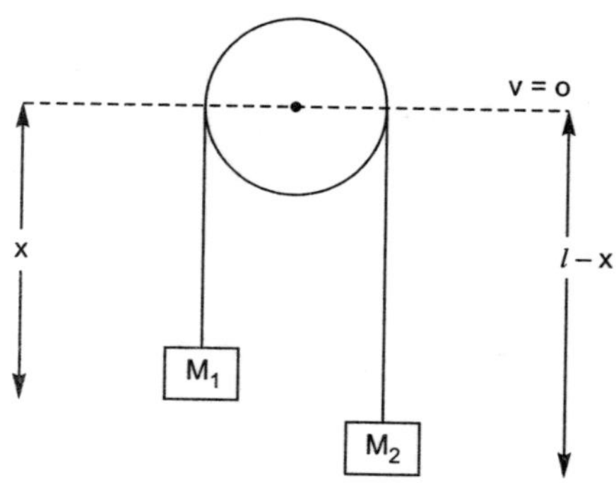

There is only one independent co-ordinate x.

The potential energy of the system is $V = -M_1 g x - M_2 g\,(l - x)$

The kinetic energy of the system $T = \frac{1}{2}(M_1 + M_2)\,\dot{x}^2$

Using Lagrange's Equation

The Lagrangian is L = T – V

$$= \frac{1}{2}(M_1 + M_2)\dot{x}^2 + M_1 g x + M_2 g(l - x) \quad \ldots(i)$$

$$\therefore \quad \frac{\partial L}{\partial \dot{x}} = (M_1 + M_2)\,\dot{x}\,,\ \frac{\partial L}{\partial x} = M_1 g + M_2 g(-1)$$

∴ The Lagrange's equation in terms of generalised co-ordinate x is

$$\frac{d}{dt}\left(\frac{\partial L}{\partial \dot{x}}\right) - \frac{\partial L}{\partial x} = 0$$

or, $(M_1 + M_2)\,\ddot{x} - (M_1 - M_2)\,g = 0$

which yields acceleration $\ddot{x} = \dfrac{(M_1 - M_2)}{M_1 + M_2}\,g \quad \ldots(ii)$

Using Hamilton's Equation

The Hamiltonian is $H = \sum p_i\,\dot{q}_i - L$

$$= p_x\,\dot{x} - L \quad \ldots(iii)$$

where $p_x = \dfrac{\partial L}{\partial \dot{x}} = (M_1 + M_2)\,\dot{x}$

$$\therefore \quad H = (M_1 + M_2)\,\dot{x}^2 - \left\{\frac{1}{2}(M_1 + M_2)\dot{x}^2 + M_1\,gx + M_2\,g(l - x)\right\}$$

$$= \frac{1}{2}(M_1 + M_2)\dot{x}^2 - M_1\,gx - M_2\,g(l - x)$$

$$= T + V$$

$$= \frac{1}{2}(M_1 + M_2)\left\{\frac{p_x}{M_1 + M_2}\right\}^2 - M_1 g\,x - M_2\,g(l - x) \quad \ldots(iv)$$

Thus $\dot{x} = \dfrac{\partial H}{\partial p_x} = \dfrac{p_x}{(M_1 + M_2)},\ \dfrac{\partial H}{\partial x} = -(M_1 - M_2)\,g$

Thus, Hamilton's equation of motion for the system will be

$$\dot{x} = \frac{\partial H}{\partial p_x} = \frac{p_x}{M_1 + M_2} \quad \ldots(v)$$

$$\dot{p}_x = -\frac{\partial H}{\partial x} = +(M_1 - M_2)\,g \quad \ldots(vi)$$

Putting the value of p_x from equation (*v*) into equation (*vi*),

we get

$$(M_1 + M_2)\ \ddot{x} = +\ (M_1 - M_2)\ g$$

$$\ddot{x} = \frac{M_1 - M_2}{M_1 + M_2}\ g$$

Problem 5. *An electrical circuit contains an inductance L and a capacitance C. Find the Lagrangian equation of motion when the charge of the condenser is q.* [G.N.D. Univ.]

Solution. Suppose an electrical circuit consists of an inductance, capacitance C and a condenser. The Lagrangian for this circuit (L_E) can be expressed as

$$L_E = T_M - V_E$$

where T_M is the magnetic energy of the electrical circuit analogous to the kinetic energy of the mechanical system and V_E the electrical energy of the electrical circuit analogous to the potential energy of the mechanical system.

Here $$T_M = \frac{1}{2}\ Li^2 = \frac{1}{2}\ L\dot{q}^2$$

and Electrical energy, $$V_E = \frac{1}{2}\frac{q^2}{C}$$

or $$L_E = \frac{1}{2}\ L\dot{q}^2 - \frac{1}{2}\frac{q^2}{C}.$$

Hence, the Lagrangian equation of motion for such a circuit can be expressed as

$$\frac{d}{dt}\left(\frac{\partial L_E}{\partial \dot{q}_J}\right) - \frac{\partial L_E}{\partial q_J} = Q_J$$

where Q_J = Generalised force due to friction while the conservative force are included in L_E. If there is no frictional or dissipative force. Then $Q_J = 0$, So the equation of motion becomes

$$\frac{d}{dt}\left(\frac{\partial L_E}{\partial \dot{q}}\right) - \frac{\partial L_E}{\partial q} = 0$$

or $$\frac{d}{dt}(L\dot{q}) - \left(\frac{-q}{C}\right) = 0$$

or $$L\ddot{q} + \frac{q}{C} = 0$$

This is the required equation of motion.

Problem 6. *A bead is sliding on a uniformly rotating wire in a force free space. Find the resulting motion.* [Agra University]

Solution. Here the constraint is time dependent. We have $x = r \cos \omega t$, $y = r \sin \omega t$

Here, ω is the angular frequency and $\theta = \omega t$

The kinetic energy (T) of the bead is given by

$$T = \frac{1}{2} m\left(\dot{x}^2 + \dot{y}^2\right) = \frac{1}{2} m\,\dot{r}^2 + \frac{1}{2} m\,r^2\,\dot{\theta}^2 = \frac{1}{2} m\,\dot{r}^2 + \frac{1}{2} mr^2\,\omega^2$$

Using Lagrange's Equation

The Lagrange's equation in terms of r is given by

$$\frac{d}{dt}\left(\frac{\partial T}{\partial \dot{r}}\right) - \frac{\partial T}{\partial r} = Q_r$$

In this case under consideration $Q_r = 0$, since the bead is sliding in a force-free space.

Therefore the Lagrange's equation becomes

$$\frac{d}{dt}\left(\frac{\partial T}{\partial \dot{r}}\right) - \frac{\partial T}{\partial r} = 0$$

or
$$\frac{d}{dt}(m\,\dot{r}) - mr\omega^2 = 0$$

$$\ddot{r} = r\omega^2$$

which represents the centripetal acceleration due to which the bead moves outward.

Problem 7. *Show that if a given co-ordinate is cyclic in Lagrangian it will also be cyclic in Hamiltonian.* [Pune Univ.]

Solution. Hamiltonian is a function of q, p, and t. It can be expressed as $H = H(q_J, p_J, t)$

or
$$dH = \sum_j \frac{\partial H}{\partial q_j} dq_j + \sum \frac{\partial H}{\partial P_j} dp_j + \sum \frac{\partial H}{\partial t} dt \qquad \ldots(i)$$

Again, Hamiltonian (H) can be expressed as

$$H = \sum p_j\,\dot{q}_j - L$$

or
$$dH = \sum_j \left(P_j\,d\dot{q}_j + \dot{q}_j\,dP_j\right) - dL \qquad \ldots(ii)$$

But, Lagrangian (L) can be expressed in terms of q, $\dot{q}$ and t.

So
$$L = L\left(q_j, \dot{q}_j, t\right)$$

or
$$dL = \sum_j \frac{\partial L}{\partial q_j} \cdot dq_j + \sum_j \frac{\partial L}{\partial \dot{q}_j} \cdot d\dot{q}_j + \frac{\partial L}{\partial t} dt \qquad \ldots(iii)$$

Using equations (*ii*) and (*iii*), we have.

$$dH = \sum_j \left(P_j\,d\dot{q}_j + \dot{q}_j\,dP_j\right) - \left\{\sum_j \frac{\partial L}{\partial q_j} dq_j + \sum_j \frac{\partial L}{\partial \dot{q}_j} d\dot{q}_j + \frac{\partial L}{\partial t} dt\right\} \ldots(iv)$$

But we know,

$$\frac{\partial L}{\partial \dot{q}_j} = \mathrm{P}_j \text{ and } \frac{\partial L}{\partial q_j} = \dot{\mathrm{P}}_j$$

Now equation (*iv*) becomes.

$$dH = \sum \dot{q}_j \, dp_j - \sum \dot{p}_j \, dq_j - \frac{\partial L}{\partial t} dt \qquad \ldots(v)$$

Now comparing equations (*i*) and (*v*) we get

$$\dot{q}_j = \frac{\partial H}{\partial p_j} \quad \text{and} \quad p_j = -\frac{\partial H}{\partial q_j} \qquad \ldots(vi)$$

If q is a cyclic co-ordinate in Lagrangian, Then

$$\dot{p}_j = \frac{\partial L}{\partial q_j} = 0, \text{ or } p_j = \text{count},$$

Now from equation (*vi*) we get, $\dot{p}_j = -\dfrac{\partial H}{\partial q_j} = 0$

or $$\frac{\partial H}{\partial q_j} = 0$$

i.e., q_j does not appear in H

So q is also cyclic co-ordinate in Hamiltonian.

Problem 8. *To find the Lagrange's equation of motion for an electrical circuit comprising of an inductance L and a capacitance C. The condenser is charged to q coulombs and the current flowing in the circuit is i amperes.*

Solution. The magnetic energy of the circuit $T_m = \frac{1}{2} Li^2 = \frac{1}{2} L\dot{q}^2 \left[i = \frac{dq}{dt} = \dot{q}\right]$

and electrical energy $$V_E = \frac{1}{2}\frac{q^2}{C}$$

Therefore, Lagrangian can be represented as

$$L_E = \frac{1}{2} L\dot{q}^2 - \frac{1}{2}\frac{q^2}{C}$$

Here q is the generalized co-ordinate, the equation of motion will be

$$\frac{d}{dt}\left(\frac{\partial L_E}{\partial \dot{q}}\right) - \frac{\partial L_E}{\partial q} = 0$$

$$L\ddot{q} + \frac{q}{C} = 0$$

This is the equation of motion.

Problem 9. *Use Lagrange's equation to find the equation of motion of a compound pendulum which oscillates in a vertical plane about a fixed horizontal axis. Hence find the period of oscillation of the compound pendulum.* [Agra University]

Solution. Let us consider G to be the centre of mass of the compound pendulum and O be its intersection with axis of rotation. Also let θ be the instantaneous angle which OG makes with the vertical axis. OA and I be the moment of inertia of the body about the axis of rotation.

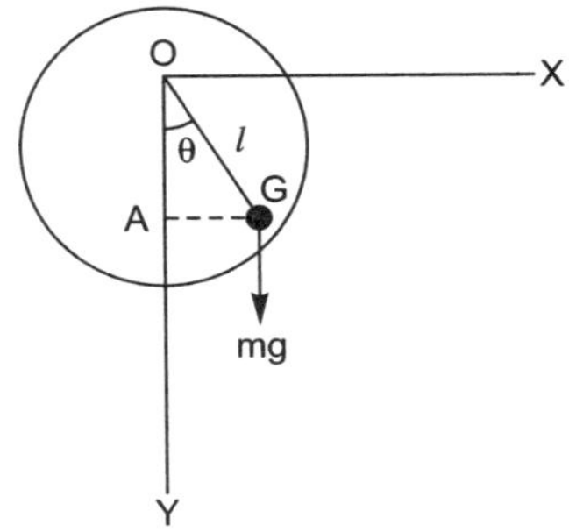

Kinetic energy of the system,

$$T = \frac{1}{2} I\dot{\theta}^2$$

where $\left(\dot{\theta} = \frac{d\theta}{dt}\right)$

The potential energy of the system,

$$V = -mg\,(OA) = -mgl\cos\theta$$

∴ The Lagrangian,

$$L = T - V$$

$$= \frac{1}{2} I\dot{\theta}^2 + mgl\cos\theta$$

∴ Now, the Lagrange's equation of motion is

$$\frac{d}{dt}\left(\frac{\partial L}{\partial \dot{\theta}}\right) - \frac{\partial L}{\partial \theta} = 0$$

$$I\ddot{\theta} + mgl\sin\theta = 0$$

If the amplitude of oscillation is small, then the equation of motion takes the form

$$\ddot{\theta} + \frac{mgl}{I}.\theta = 0.$$

Hence, the period of oscillation of the compound pendulum,

$$T = \frac{2\pi}{\omega} = 2\pi . \frac{1}{\sqrt{\frac{mgl}{I}}}$$

$$= 2\pi\sqrt{\frac{I}{mgl}}$$

If k is the radius of gyration of the body about an axis through the centre of gravity G then

$$I = mk^2 + ml^2$$

∴

$$T = 2\pi\sqrt{\frac{m\left(k^2 + l^2\right)}{mgl}} = 2\pi\sqrt{\frac{k^2 + l^2}{gl}}$$

Problem 10. *A bead slides on a wire in the shape of a cycloid described by the equation*

$$x = a\,(1 - \cos\theta)$$

$$y = a\,(1 + \cos\theta)$$

where

$$0 \le \theta \le 2\pi$$

Find the Lagrangian function and the equation of motion. Neglect the friction between the bead and the wire. [Pune Univ.]

Solution.

(*a*) Let p (x, y) be the position of the bead of mass m at any instant of time t. The path of the bead is a cycloid and is represented by

$$x = a\,(\theta - \sin\theta) \text{ and } y = a\,(1 + \cos\theta)$$

Hence, the kinetic energy of the bead will be

$$\mathrm{T} = \frac{1}{2} m \left(\dot{x}^2 + \dot{y}^2\right) = \frac{1}{2} m \left[\left\{a\left(\dot{\theta} - \cos\theta\,\dot{\theta}\right)\right\}^2 + a\left\{-a\sin\theta\,\dot{\theta}\right\}^2\right]$$

$$= \frac{1}{2} m\,a^2 \left[\dot{\theta}^2 + \left(\cos^2\theta + \sin^2\theta\right)\dot{\theta}^2 - 2\cos\theta\,\dot{\theta}^2\right]$$

$$= m\,a^2\,\dot{\theta}^2\,(1 - \cos\theta)$$

and the potential energy of the bead, considering X-axis as the reference level will be

$$\mathrm{V} = mgy = mg\,a\,(1 + \cos\theta)$$

Thus, the Lagrangian is

$$\mathrm{L} = \mathrm{T} - \mathrm{V} = m\,a^2\,\dot{\theta}^2\,(1 - \cos\theta) - mga\,(1 + \cos\theta)$$

Therefore, the Lagrangian equation of motion is

$$\frac{d}{dt}\left(\frac{\partial \mathrm{L}}{\partial \dot{\theta}}\right) - \frac{\partial \mathrm{L}}{\partial \theta} = 0$$

or $$\frac{d}{dt}\left\{2m\,a^2\,\dot{\theta}\,(1 - \cos\theta)\right\} - mg\,a\sin\theta - m\,a^2\,\dot{\theta}^2\sin\theta = 0$$

or $$2m\,a^2\,\ddot{\theta}\,(1 - \cos\theta) + 2m\,a^2\,\dot{\theta}\sin\theta\,\dot{\theta} - m\,a^2\,\dot{\theta}^2\sin\theta - mg\,a\sin\theta = 0$$

or $$\ddot{\theta}\,(1 - \cos\theta) + \frac{1}{2}\dot{\theta}^2\sin\theta - \frac{g}{2a}\sin\theta = 0$$

Multiple Choice Questions

1. If a co-ordinate corresponding to a rotation is cyclic, rotation of the system about the given axis remains invariant, then the following quantity is conserved.
   (*a*) Linear momentum (*b*) Kinetic energy (*c*) Angular momentum.
2. If a generalised co-ordinate has the dimension of momentum, the generalised velocity will have the dimension of
   (*a*) Force (*b*) Acceleration (*c*) Velocity.
3. A physical system is invariant under rotation about a fixed axis. Then the following quantity is conserved.
   (*a*) Total angular momentum.
   (*b*) Angular momentum along the axis of rotation
   (*c*) Linear momentum along the axis of rotation.
4. The product of any generalised momentum and the associated (or conjugate) co-ordinate must have the dimension of
   (*a*) Linear momentum (*b*) Angular momentum (*c*) Energy.

5. Whatever dimension a generalized co-ordinate has, the product of generalized force and generalized displacement must have the dimension of
(*a*) Work (*b*) Torque (*c*) Force.

6. The Lagrange's equation for a linear harmonic oscillator is
(*a*) $L = \frac{1}{2} m \ddot{x} - \frac{1}{2} kx^2$ (*b*) $L = \frac{1}{2} m \ddot{x} + \frac{1}{2} kx^2$ (*c*) $L = \frac{1}{2} m \dot{x}^2 + \frac{1}{2} kx^2$

7. Equation of motion for a bead sliding on a uniformly rotating wire in a force free space is
(*a*) $\ddot{r} = r\omega^2$ (*b*) $\ddot{\theta} + \frac{g}{l}\theta = 0$ (*c*) $\ddot{\theta} + \frac{mgl\,\theta}{I} = 0$

8. The path followed by a particle in sliding from one point to another in the absence of friction in the shortest time is a
(*a*) Sphere (*b*) Cycloid (*c*) Sigmoid.

9. Hamilton's canonical equations of motion are
(*a*) $\dot{q}_i = \frac{\partial H}{\partial p_i}$ and $\dot{p}_i = -\frac{\partial H}{\partial q_i}$
(*b*) $q_i = \frac{\partial H}{\partial \dot{p}_i}$ and $p_i = \frac{\partial H}{\partial \dot{q}_i}$
(*c*) $q_i = \frac{\partial H}{\partial \dot{p}_i}$ and $p_i = \frac{\partial H}{\partial \dot{q}_i}$

10. If a co-ordinate is cyclic, the Hamiltonian would reduce the number of variables in new formulation by
(*a*) Three (*b*) Two (*c*) One.

11. If the Lagrange's equation of motion are second order equations, then the degrees of freedom for this are
(*a*) 2^{n-1} (*b*) 2^n (*c*) 2^{n+1}.

12. Degree of freedom of hydrogen molecule is
(*a*) 1 (*b*) 2 (*c*) 5.

13. Degree of freedom of a hydrogen molecule are
(*a*) 1 (*b*) 2 (*c*) 5 [*W.B.U.T*]

14. Hamiltonian is defined as
(*a*) H = T − V (*b*) H = T + V (*c*) H = 2T + L [*W.B.U.T*]

15. A system consists of 3 point masses. If the mutual distances between the masses do not vary with time. Then the degree of freedom such a system are
(*a*) 0 (*b*) 6 (*c*) 9.

Answers

1. (*c*)	2. (*a*)	3. (*b*)	4. (*b*)	5. (*a*)	6. (*a*)
7. (*a*)	8. (*b*)	9. (*a*)	10. (*b*)	11. (*c*)	12. (*c*).
13. (*c*).	14. (*b*)	15. (*b*)			

Review Questions

1. What are constraints ? How do they affect the motion of a mechanical system ? Take specific examples to explain the force of constraint. **[Hints : 1.4]**
2. What are generalised co-ordinates ? What is the advantage of using them ? **[Hints : 1.6]**

3. What do you mean by constraint motion ? Distinguish clearly between :
   (*a*) Holonomic and Non-holonomic constraints.
   (*b*) Scleronomic and Rheonomic constraints. **[Hints :** 1.4]
4. State and explain D′ Alembert's principle and derive the Lagrange's equation from it. **[Hints :** 1.8 and 1.9]
5. Define cyclic co-ordinates. **[Hints :** 1.10]
6. Define Lagrange's equation in the form *[WBUT]*

$$\frac{d}{dt}\left(\frac{\partial T}{\partial \dot{q}_k}\right) - \frac{\partial T}{\partial q_k} = Q_k$$

   where the symbols have their usual meaning. **[Hints :** 1.9.]
7. Define Hamiltonian of a system. When is it equal to the total energy of the system ? **[Hints :** 1.11] *[WBUT]*
8. Write a short note on 'Generalized potential in classical mechanics'. *[WBUT]*
9. Write down the Lagrange's equation for a particle moving freely in one dimension. *[WBUT]*
10. A particle is falling vertically under the influence of gravity. Obtain the Hamiltonian and the equations of motion corresponding to that Hamiltonian. *[WBUT]*
11. Obtain the equation of motion of a system of two particles of unequal masses connected by an inextensible string passing over a small smooth pulley. *[WBUT]*
12. Derive the equation of motion for a simple pendulum of length l and mass of bob m. *[WBUT]*
13. Write down the Lagrangian for a system of two particles of masses m_1 and m_2 connected by an inextensible string passing through a hole in a horizontal table such that mass m_2 hangs vertically and m_1 moves on the smooth top of the table. Hence obtain the equation of motion for the system and solve the same.
14. Apply Lagrange's equation to determine the motion of a bead sliding on a uniformly rotating wire in a force free space.
15. What are constraints? Explain various types of constraints. What do you mean by (*i*) holonomic and non-holonomic constraints (*ii*) Scleronomic and Rheonomic constraints? Explain with examples. *[M.D. Univ., Pune Univ., G.N.D. Univ.]*
16. Define generalised co-ordinates and obtain expression for generalised displacement and generalised velocity. What is real and virtual displacement? *[M.D. Univ., Pune Univ.]*
17. Explain the principle of virtual work and D′ Alembert's principle. *[Pune Univ.]*

Unsolved Problems

1. Three blocks of masses $m_1 = 10$ kg, $m_2 = 20$ kg and $m_3 = 30$ kg are connected as shown on a smooth horizontal table and pulled to the right with a force $T_3 = 60$ newton. Find the tensions T_1 and T_2. *[Lucknow University]* **[Ans.** $T_1 = 10$ N; $T_2 = 30$ N]
2. A ball moving at a speed of 2.2 m/s strikes an identical stationary ball. After collision one ball moves at 1.1 m/s at 60° angle with the original line of motion. Find the velocity of other ball. **[Ans.** 1.9 m/sec.]
3. Find the fractional decrease in the kinetic energy of neutron when it collides elastically head-on with a (*i*) Lead nucleus ; (*ii*) hydrogen nucleus, initially at rest. The ratio of nuclear mass to neutron mass is 206 for lead and 1 for hydrogen. **[Ans.** (*i*) 2% ; (*ii*) 100%]
4. A body with a mass of 0.80 Kg is on a plane inclined at 30° to the horizontal. What force must be applied on the body so that it slides with an accelerations of 0.10 m/s^2 ; (*a*) uphill ; (*b*) downhill ? The coefficient of sliding friction with the plane is 0.30. *[Lucknow University]*
   [Ans. (*a*) 6.04 N ; (*b*) – 1.80 N]

5. An ice skater spins with arms outstretched at 1.9 rev/sec. Her moment of inertia at this time is 1.33 kg.m^2. She pulls in her arms to increase her rate of spin. If her moment of inertia is 0.48 kg. m^2 after she pulls in her arms, what is her new rate of rotation ? **[Ans.** 6 rad.]
6. A 7500 kg truck travelling at 5 m/s east collides with a 1500 kg car moving at 20 m/s in a direction 30° south of west. After collision, the two vehicles remain tangled together with what speed and in what direction does the wreckage begin to move ? **[Ans.** v = 2.1 m/s, θ = 53°]
7. A piece of ice slides down a 45° inclined plane in twice the time it takes to slide down a frictionless 45° inclined plane. What is the co-efficient of kinetic friction between ice and inclined plane ? **[Ans.** 0.75]
8. Two masses, one n times heavier than the other are dropped from the same height. How do their moment compare just before they hit the ground ? If these masses have same kinetic energy then how do their moment compare ? **[Ans.** $n : 1$; $\sqrt{n} : 1$]
9. A 5 kg object with a speed of 30 m/s, strikes a steel plate at an angle of 45° and rebounds at the same speed and same angle. What is the change in magnitude and direction of the linear momentum of the object. **[Ans.** 212 kg-m/s, perpendicular and away from the plate]
10. A massless string pulls a mass of 50 kg upward against gravity. The string would break if subjected to a tension greater than 600 newtons. What is the maximum acceleration with which the mass can be moved upward ? **[Ans.** 2.2 m/s^2]
11. A body is projected with a speed u m/s at an angle α with the horizontal, the kinetic energy at the highest point is $\frac{3}{4}$ th of the initial kinetic energy. What is the value of α ? **[Ans.** α = 30°]

Maxwell

2

Chapter

Electricity and Magnetism

2.1 BASIC CONCEPTS OF VECTOR

A **vector** is a quantity having both magnitude and direction, such as displacement, velocity, force etc.

Graphically a vector is represented by the arrow OB (figure 2.1) defining the direction, the magnitude of the vector being indicated by the length of the arrow.

Figure 2.1

Analytically a vector is represented by a letter with an arrow over it, as $\vec{A}$ in figure 2.1 and its magnitude is denoted by $|\vec{A}|$ or A.

Scalar : A scalar is a quantity having magnitude but no direction, *e.g.,* mass, length, time, temperature etc. Scalars are indicated by letters in ordinary type as in elementary algebra.

Unit Vector : A unit vector is a vector having unit magnitude. If $\vec{A}$ is a vector with magnitude $A \neq 0$, then $\vec{A}$ / A is a unit vector having the same direction as $\vec{A}$.

2.1.1 Scalar and Vector Products

(*a*) Scalar Product or Dot Product : The dot or scalar product of two vectors $\vec{A}$ and $\vec{B}$, denoted by $\vec{A}\cdot\vec{B}$ (read $\vec{A}$ dot $\vec{B}$), is defined as the product of the magnitudes of $\vec{A}$ and $\vec{B}$ and the cosine of the angle θ between them. In symbols, $\vec{A}\cdot\vec{B} = AB\cos\theta,\ 0 \le \theta \le \pi$.

NOTE $\vec{A}\cdot\vec{B}$ is a scalar and not a vector.

The following laws are valid :

1. $\vec{A}\,.\,\vec{B} = \vec{B}\,.\,\vec{A}$ (Commutative law for Dot products)

2. $\vec{A}\,.\left(\vec{B}+\vec{C}\right) = \vec{A}\,.\,\vec{B}+\vec{A}\,.\,\vec{C}$ (Distributive law)

3. $m\left(\vec{A}\,.\,\vec{B}\right) = \left(m\vec{A}\right).\,B = \vec{A}\,.\,\left(m\vec{B}\right) = \left(\vec{A}\,.\,\vec{B}\right)m$, where m is a scalar

4. $\hat{i}\,.\,\hat{i} = \hat{j}\,.\,\hat{j} = \hat{k}\,.\,\hat{k} = 1$ and $\hat{i}\,.\,\hat{j} = \hat{j}\,.\,\hat{k} = \hat{k}\,.\,\hat{i} = 0$, where $\hat{i}$, $\hat{j}$ and $\hat{k}$ are unit vectors along X, Y and Z axes.

5. If $\vec{A} = A_1\,\hat{i} + A_2\,\hat{j} + A_3\,\hat{k}$ and $\vec{B} = B_1\,\hat{i} + B_2\,\hat{j} + B_3\,\hat{k}$

$$\vec{A}\,.\,\vec{B} = A_1B_1 + A_2B_2 + A_3B_3$$

$$\vec{A}\,.\,\vec{A} = A^2 = A_1^2 + A_2^2 + A_3^2$$

$$\vec{B}\,.\,\vec{B} = B^2 = B_1^2 + B_2^2 + B_3^2$$

6. If $\vec{A}\,.\,\vec{B} = 0$ and $\vec{A}$ and $\vec{B}$ are not null vectors, then $\vec{A}$ and $\vec{B}$ are perpendicular.

Null Vector : If $\vec{A}$ and $\vec{B}$ are identical in both magnitudes and directions then $\vec{A} - \vec{B} = 0$. This zero is a zero-vector or null-vector.

***(b)* Cross or Vector Products :** The cross or vector product of $\vec{A}$ and $\vec{B}$ is a vector $\vec{C} = \vec{A}\times\vec{B}$ (read $\vec{A}$ cross $\vec{B}$). The magnitude of $\vec{A}\times\vec{B}$ is defined as the product of the magnitudes of $\vec{A}$ and $\vec{B}$ and the sine of the angle θ between them.

The direction of the vector $\vec{C} = \vec{A}\times\vec{B}$ is perpendicular to the plane of $\vec{A}$ and $\vec{B}$ such that $\vec{A}$, $\vec{B}$ and $\vec{C}$ form a right-handed system. In symbols,

$$\vec{A}\times\vec{B} = AB\sin\theta\;\hat{n},\quad 0 \le \theta \le \pi$$

where $\hat{n}$ is a unit vector indicating the direction of $\vec{A}\times\vec{B}$.

If $\vec{A} = \vec{B}$, $\vec{A}$ is parallel to $\vec{B}$, then $\sin\theta = 0$ and we define $\vec{A}\times\vec{B} = 0$

The following laws are valid :

1. $\vec{A}\times\vec{B} = -\,\vec{B}\times\vec{A}$ (Commutative law for cross product fails)

2. $\vec{A}\times\left(\vec{B}+\vec{C}\right) = \vec{A}\times\vec{B}+\vec{A}\times\vec{C}$ (Distributive law)

3. $m\left(\vec{A}\times\vec{B}\right) = \left(m\vec{A}\right)\times\vec{B} = \vec{A}\times\left(m\vec{B}\right) = \left(\vec{A}\times\vec{B}\right)m$, where m is a scalar

4. $\hat{i}\times\hat{i} = \hat{j}\times\hat{j} = \hat{k}\times\hat{k} = 0, \quad \hat{i}\times\hat{j} = \hat{k},\ \hat{j}\times\hat{k} = \hat{i},\ \hat{k}\times\hat{i} = \hat{j}$

5. If $\vec{A} = A_1\hat{i} + A_2\hat{j} + A_3\hat{k}$ and $\vec{B} = B_1\hat{i} + B_2\hat{j} + B_3\hat{k}$

then
$$\vec{A}\times\vec{B} = \begin{vmatrix} \hat{i} & \hat{j} & \hat{k} \\ A_1 & A_2 & A_3 \\ B_1 & B_2 & B_3 \end{vmatrix}$$

6. The magnitude of $\vec{A}\times\vec{B}$ is the same as the area of a parallelogram with sides $\vec{A}$ and $\vec{B}$.

7. If $\vec{A}\times\vec{B} = 0$, and $\vec{A}$ and $\vec{B}$ are not null vectors, then $\vec{A}$ and $\vec{B}$ are parallel.

2.1.2 Areal Vector

The vector which represents the area of a parallelogram caused by means of vector product of two vectors $\vec{A}$ and $\vec{B}$ acting at an angle θ between them, is known as **areal vector.** This vector is always perpendicular to the plane containing the two vectors as stated.

If two vectors $\vec{A}$ and $\vec{B}$ creates the cross product $\vec{C} = \vec{A}\times\vec{B} = \hat{n}\,AB\sin\theta$, where $AB\sin\theta$ represents the area of the parallelogram created by $\vec{A}$ and $\vec{B}$, then $\vec{C}$ is an areal vector as shown in figure 2.2.

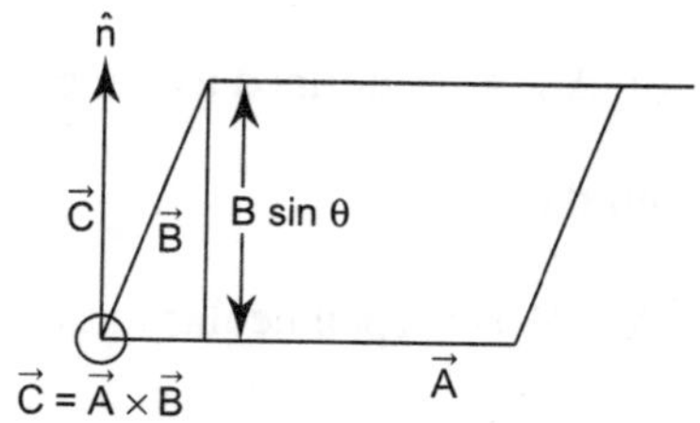

Figure 2.2

The direction of areal vector $\vec{C}$ is perpendicular to the plane of the paper obeying the right hand screw rule as shown in figure.

2.1.3 Concepts of Vector and Scalar Fields

A **field** is a region in space in which a function is defined at all points.

(*a*) **Scalar field :** The region in which a scalar quantity is continuous and is defined by a single value at every point of position variable is called scalar field.

If ϕ is a scalar quantity of position variable $\vec{r}$ with co-ordinate (x, y, z), then the scalar field

$$\phi(\vec{r}) = \phi(x, y, z) = x^2y - z^2 \text{ (say)}$$

(*b*) **Vector field :** The region in which a vector quantity is continuous and is defined by a single value at every point of position variable is called vector field.

If $\vec{V}$ is a vector function of position variable $\vec{r}$ with co-ordinate (x, y, z), then the vector field $\vec{V} = \vec{V}(x, y, z)$.

Scalar and Vector Point Functions Related to Field

(*a*) **Scalar Field :** If to each point (x, y, z) of a region R in space there corresponds a number or scalar $\phi n(x, y, z)$, then n is called a scalar function of position or scalar point function and we say that a scalar field n has been defined in R.

A scalar field which is independent of time is called a **stationary or steady state scalar field.**

Examples : 1. $n(x, y, z) = x^3y - z^2$ defines a scalar field.

2. The temperature at any point within or on the earth's surface at a certain time defines a scalar field.

(*b*) **Vector Field :** If to each point (x, y, z) of a region R in space there corresponds a vector $\vec{A}(x, y, z)$ then $\vec{A}$ is called a vector function of position or vector point function and we say that a vector field $\vec{A}$ has been defined in R.

A vector field which is independent of time is called stationary or steady state vector field.

Examples : 1. $\vec{A} = xy\hat{i} + y^2z\hat{j} + z^3x\hat{k}$ is a vector field.

2. If the velocity at any point (x, y, z) within a moving fluid is known at a certain time, it is said to be a vector field.

2.1.4 Derivative of Vectors

2.1.4.1 *Ordinary Derivative of Vectors*

Let $\vec{y}(x)$ be a vector depending on a single scalar variable x.

Then $$\frac{\Delta\vec{y}}{\Delta x} = \frac{\vec{y}(x+\Delta x) - \vec{y}(x)}{\Delta x}$$

where Δx denotes an increment in x (figure 2.3)

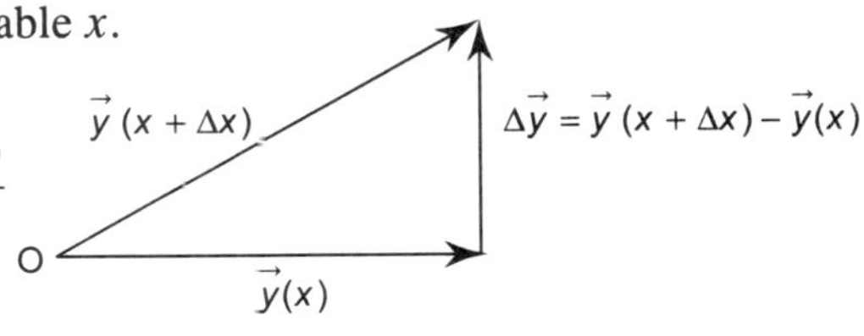

Figure 2.3

The ordinary derivative of the vector $\vec{y}(x)$ with respect to the vector x is given by :

$$\frac{d\vec{y}}{dx} = \lim_{\Delta x \to 0} \frac{\Delta\vec{y}}{\Delta x} = \lim_{\Delta x \to 0} \frac{\vec{y}(x+\Delta x) - \vec{y}(x)}{\Delta x}$$

if limit exists.

Since $\frac{d\vec{y}}{dx}$ is itself a vector depending on x, we can consider its derivative with respect to x. If this derivative exists it is denoted by $\frac{d^2\vec{y}}{dx^2}$. In this manner higher order derivatives are described.

Differentiation Formulae

If $\vec{A}$, $\vec{B}$ and $\vec{C}$ are differentiable vector functions of a scalar x, and n is a differentiable scalar function of x, then

1. $\frac{d}{dx}\left(\vec{A}+\vec{B}\right) = \frac{d\vec{A}}{dx}+\frac{d\vec{B}}{dx}$

2. $\frac{d}{dx}\left(\vec{A}.\vec{B}\right) = \vec{A}.\frac{d\vec{B}}{dx}+\frac{d\vec{A}}{dx}.\vec{B}$

3. $\frac{d}{dx}\left(\vec{A}\times\vec{B}\right) = \vec{A}\times\frac{d\vec{B}}{dx}+\frac{d\vec{A}}{dx}\times\vec{B}$

4. $\frac{d}{dx}\left(n\vec{A}\right) = n\frac{d\vec{A}}{dx}+\frac{dn}{dx}\vec{A}$

5. $\frac{d}{dx}\left(\vec{A}.\vec{B}\times\vec{C}\right) = \vec{A}.\vec{B}\times\frac{d\vec{C}}{dx}+\vec{A}.\frac{d\vec{B}}{dx}\times\vec{C}+\frac{d\vec{A}}{dx}.\vec{B}\times\vec{C}$

6. $\frac{d}{dx}\left\{\vec{A}\times\left(\vec{B}\times\vec{C}\right)\right\} = \vec{A}\times\left(\vec{B}\times\frac{d\vec{C}}{dx}\right)+\vec{A}\times\left(\frac{d\vec{B}}{dx}\times\vec{C}\right)+\frac{d\vec{A}}{dx}\times\left(\vec{B}\times\vec{C}\right)$

2.1.4.2 Partial Derivative of Vectors

If $\vec{A}$ is a vector depending on more than one scalar variables say x, y, z, then we write $\vec{A} = \vec{A}$ (x, y, z). The partial derivative of $\vec{A}$ with respect to x is defined as

$$\frac{\partial\vec{A}}{\partial x} = \lim_{\Delta x\to 0}\frac{\vec{A}(x+\Delta x, y, z)-\vec{A}(x, y, z)}{\Delta x}$$

if this limit exists

Similarly,

$$\frac{\partial\vec{A}}{\partial y} = \lim_{\Delta x\to 0}\frac{\vec{A}(x, y+\Delta y, z)-\vec{A}(x, y, z)}{\Delta y}$$

$$\frac{\partial\vec{A}}{\partial z} = \lim_{\Delta x\to 0}\frac{\vec{A}(x, y, z+\Delta z)-\vec{A}(x, y, z)}{\Delta z}$$

Rules for partial differentiation of vectors are similar to those used in elementary calculus for scalar functions. Thus if $\vec{A}$ and $\vec{B}$ are functions of x, y, z, then for example,

1. $\dfrac{\partial}{\partial x}\left(\vec{A} . \vec{B}\right) = \vec{A} . \dfrac{\partial \vec{B}}{\partial x} + \dfrac{\partial \vec{A}}{\partial x} . \vec{B}$

2. $\dfrac{\partial}{\partial x}\left(\vec{A} \times \vec{B}\right) = \vec{A} \times \dfrac{\partial \vec{B}}{\partial x} + \dfrac{\partial \vec{A}}{\partial x} \times \vec{B}$

3. $\dfrac{\partial^2}{\partial y . \partial x}\left(\vec{A} . \vec{B}\right) = \dfrac{\partial}{\partial y}\left\{\dfrac{\partial}{\partial x}\left(\vec{A} . \vec{B}\right)\right\} = \dfrac{\partial}{\partial y}\left\{\vec{A} . \dfrac{\partial \vec{B}}{\partial x} + \dfrac{\partial \vec{A}}{\partial x} . \vec{B}\right\}.$

The Vector Differential Operator DEL

Operator is a mathematical tool which acts on some function and produce or generate a new function.

The vector differential operator del, written $\vec{\nabla}$, is defined by

$$\vec{\nabla} = \frac{\partial}{\partial x}\hat{i} + \frac{\partial}{\partial y}\hat{j} + \frac{\partial}{\partial z}\hat{k}$$

This vector operator possesses properties analogous to those of ordinary vectors. The operator $\vec{\nabla}$ is also known as Nabla.

2.1.5 Directional Derivative of Vector Point Function

The rate of change of a scalar quantity with respect to position co-ordinate in a certain direction is called the directional derivative.

For finding the directional derivative of a scalar quantity ϕ in a direction of a vector $\vec{A}$, we have to use the following steps.

Step 1. Take the gradient of the scalar quantity of which we have to find out the directional derivative *i.e.*, $\vec{\nabla}\phi$

Step 2. Find out the unit vector of the vector along which we have to find out the directional derivative *i.e.*, $\hat{n} = \dfrac{\vec{A}}{|\vec{A}|}$

Step 3. Take the dot product between grad ϕ and unit vector $\hat{n}$ which gives the directional derivative. *i.e.*, the Directional derivative $= \vec{\nabla}\phi . \hat{n}$

2.1.6 Gradient of Scalar Field

Let $n(x, y, z)$ be defined and differentiable at each point (x, y, z) in a certain region of space (*i.e.*, n defines a differentiable scalar field). Then the gradient of n, written $\vec{\nabla}n$ or gradient n, is defined by

$$\vec{\nabla}n = \left(\frac{\partial}{\partial x}\hat{i} + \frac{\partial}{\partial y}\hat{j} + \frac{\partial}{\partial z}\hat{k}\right)n = \frac{\partial n}{\partial x}\hat{i} + \frac{\partial n}{\partial y}\hat{j} + \frac{\partial n}{\partial z}\hat{k}$$

NOTE $\vec{\nabla}n$ defines a vector field, whose magnitude at any point is equal to the maximum rate of increase of n at that point and whose direction is along the normal to the level surface at that point.

The component of $\vec{\nabla}n$ in the direction of a unit vector $\hat{n}$ is given by $\vec{\nabla}n \cdot \hat{n}$ and is called the directional derivative of n in the direction $\hat{n}$.

Physically, this is the rate of change of n at (x,y,z) in the direction $\hat{n}$.

Example 1. *If $\phi = 2xz^4 - x^2y$, find $\vec{\nabla}\phi$ and $\left|\vec{\nabla}\phi\right|$ at the point (2, – 2, – 1).*

Solution:

$$\phi = 2xz^4 - x^2y$$

$$\vec{\nabla}\phi = \frac{\partial}{\partial x}\left(2xz^4 - x^2y\right)i + \frac{\partial}{\partial y}\left(2xz^4 - x^2y\right)j + \frac{\partial}{\partial z}\left(2xz^4 - x^2y\right)k.$$

$$\left.\vec{\nabla}\phi\right|_{(2,-2,-1)} = \left(2z^4 - 2xy\right)i + \left(-x^2\right)j + 8xz^3\,k.$$

$$= \left\{2\cdot(-1)^4 - 2\cdot 2(-2)\right\}i + \left(-2^2\right)j + 8\cdot 2\cdot(-1)^3\,k$$

$$= (2+8)\,i - 4j - 16k$$

$$= 10i - 4j - 16k$$

$$\left|\vec{\nabla}\phi\right| = \sqrt{(10)^2 + (-4)^2 + (-16)^2}$$

$$= \sqrt{100 + 16 + 256}$$

$$= \sqrt{372} = 2\sqrt{93}$$

Directional Derivative of Scalar Field

If $\phi = x^3 - yz$ is a scalar field then we take the gradient of ϕ *i.e.*, $\vec{\nabla}\phi$. After that operate the dot product of $\vec{\nabla}\phi$ along the direction having unit vector $\hat{n}$, *i.e.*, $\vec{\nabla}\phi \cdot \hat{n}$ gives, the directional derivative of scalar field.

2.1.7 Line Integral, Potential Energy and Force

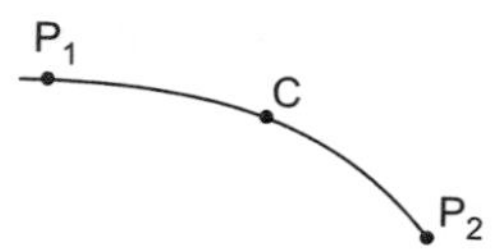

Let $\vec{r}(u) = x\hat{i} + y\hat{j} + z\hat{k}$ where $\vec{r}$ is the position vector of (x, y, z) defining a curve C joining points P_1 and P_2, where $u = u_1$ and $u = u_2$ respectively.

We assume that C is composed of a finite number of points for each of which $\vec{r}(u)$ has continuous derivative. Let $\vec{A}(x, y, z) = A_1\hat{i} + A_2\hat{j} + A_3\hat{k}$ be a vector function of position defined and continuous along C.

Then the integral of the tangential component of $\vec{A}$ along C from P_1 to P_2, written as

$$\int_{P_1}^{P_2} \vec{A} \cdot \vec{dr} = \int_C \vec{A} \cdot \vec{dr} = \int A_1 dx + A_2 dy + A_3 dz$$

is an example of a line integral.

In general, any integral which is to be evaluated along a curve is called a **line integral**. Such integral can be defined in terms of limits of sums as are the integrals of elementary calculus.

NOTE If $\vec{A} = \vec{\nabla} \phi$ everywhere in a region R of space, defined by $a_1 \le x \le a_2$, $b_1 \le y \le b_2$, $c_1 \le z \le c_2$, where ϕ (x, y, z) is single valued and has continuous derivatives in R, then

1. $\int_{P_1}^{P_2} \vec{A} \cdot dr$ is independent of the path C in R joining P_1 and P_2.
2. $\oint_C \vec{A} \cdot dr = 0$ around any closed curve C in R.

In such case $\vec{A}$ is called conservative vector field or 'non-curl' or 'lamellar' field and ϕ is called scalar potential.

It is used to find out the work-done in a field against the force and also to find out the type of force and field.

2.2 VECTOR FIELD AND FLUX

If a 'vector' physical quantity is revealed from point to point in a region of space by a continuous function $\vec{V}$ (x, y, z), the region is a vector field and the function $\vec{V}$ is 'vector point function'. Magnetic and electric field intensity in space are examples of vector field. It can be completely mapped out by curved lines, known as flux lines, the tangent at any point of a line giving the direction of $\vec{V}$ at that point. The magnitude of $\vec{V}$ is given by the number of flux lines crossing unit area perpendicular to their direction drawn at that point. Obviously flux lines cannot intersect.

2.2.1 Divergence of Vector Field

The divergence of a vector field at any point is defined as the amount of flux per unit volume diverging from the point. So the amount of flux is essentially a scalar.

Also note that $\vec{\nabla} \cdot \vec{V} \neq \vec{V} \cdot \vec{\nabla}$

A vector field which satisfies the condition $\vec{\nabla} \cdot \vec{A} = 0$, is called 'solenoidal'. In case of fluid, it is incompressible.

2.2.2 Representation of Divergence in Cartesian Coordinate

Let $\vec{V}$ $(x, y, z) = V_1 \hat{i} + V_2 \hat{j} + V_3 \hat{k}$ be defined and differentiable at each point in a certain region of space *i.e.*, $\vec{V}$ defines a differentiable vector field.

Then the divergence of $\vec{V}$, written $\vec{\nabla} \cdot \vec{V}$ or div $\vec{V}$, is defined as, the dot product of $\vec{\nabla}$ with $\vec{V}$.

Thus,
$$\vec{\nabla}.\vec{V} = \left(\frac{\partial}{\partial x}\hat{i} + \frac{\partial}{\partial y}\hat{j} + \frac{\partial}{\partial z}\hat{k}\right).\left(V_1\hat{i} + V_2\hat{j} + V_3\hat{k}\right)$$
$$= \frac{\partial V_1}{\partial x} + \frac{\partial V_2}{\partial y} + \frac{\partial V_3}{\partial z}$$

Example 1. *Prove that the vector* $\vec{A} = 3y^4z^2\hat{j} + 4x^3z^2\hat{j} - 3x^2y^2\hat{k}$ *is solenoidal.*

Solution.

$$\text{div}\,\vec{A} = \vec{\nabla}\left(3y^4z\hat{i} + 4x^3z^2\,\hat{j} - 3x^2y^2\,\hat{k}\right)$$
$$= \frac{\partial}{\partial x}\left(3y^4z\right) + \frac{\partial}{\partial y}\left(4x^3z^2\right) + \frac{\partial}{\partial z}\left(-3x^2y^2\right). = 0 + 0 + 0 = 0$$

As the div $\vec{A}$ is zero, hence the vector $\vec{A}$ is solenoidal.

Example 2. *Show that* $\vec{A} = \left(2x^2 + 8xy^2z\right)i + \left(3x^3y - 3xy\right)i - \left(4y^2z^2 + 2x^3z\right)k$ *is not a solenoidal but* $B = xyz^2\,\vec{A}$ *is solenoidal.*

Solution.

$$\vec{A} = \left(2x^2 + 8xy^2z\right)\hat{i} + \left(3x^3y - 3xy\right)\hat{j} - \left(4y^2z^2 + 2x^3z\right)\hat{k}.$$
$$\text{div}\,\vec{A} = \vec{\nabla}\cdot\vec{A} = \frac{\partial}{\partial x}\left(2x^2 + 8xy^2z\right) + \frac{\partial}{\partial y}\left(3x^3y - 3xy\right) + \frac{\partial}{\partial z}\left(-4y^2z^2 - 2x^3z\right)$$

$\therefore$
$$= 4x + 8y^2z + 3x^3 - 3x - 8y^2z - 2x^3 = x + x^3$$

$\therefore$ $\vec{A}$ is not solenoidal.

$\therefore$
$$\vec{B} = xyz^2\,\vec{A} = xyz^2\left(2x^2 + 8xy^2z\right)i + xyz^2\left(3x^3y - 3xy\right)j - xyz^2\left(4y^2z^2 + 2x^3z\right)k.$$
$$= \left(2x^3yz^2 + 8x^2y^3z^3\right)i + \left(3x^4y^2z^2 - 3x^2y^2z^2\right)j - \left(4xy^3z^4 + 2x^4yz^3\right)k.$$
$$\text{div}\,\vec{B} = \vec{\nabla}\cdot\vec{B} = \frac{\partial}{\partial x}\left(2x^3yz^2 + 8x^2y^3z^3\right) + \frac{\partial}{\partial y}\left(3x^4y^2z^2 - 3x^2y^2z^2\right)$$
$$+ \frac{\partial}{\partial z}\left(-4xy^2z^4 - 2x^4yz^3\right).$$
$$= 6x^2yz^2 + 16xy^3z^3 + 6x^4yz^2 - 6x^2yz^2 - 16xy^3z^3 - 6x^4yz^2$$
$$= 0$$

$\therefore$ $\vec{B}$ vector is solenoidal.

Example 3. *If $\vec{A} = 2x^2 i - 3yzj + yz^2 k$ and $\phi = 2z - x^3y$, find $\vec{A}\cdot\vec{\nabla}\phi$ and $\vec{A}\times\vec{\nabla}\phi$ at the point (1, – 1, 1).*

Solution.

$$\vec{\nabla}\phi = \frac{\partial}{\partial x}\left(2x - x^3y\right) i + \frac{\partial}{\partial y}\left(2z - x^3y\right) j + \frac{\partial}{\partial z}\left(2z - x^3y\right) k.$$

$$= -3x^2yi + \left(-x^3\right) j + 2k$$

$$\vec{A} = 2x^2i - 3yzj + xz^2k$$

$$\vec{A}\times\vec{\nabla}\phi = \begin{vmatrix} i & j & k \\ 2x^2 & -3yz & xz^2 \\ -3x^2y & -x^3 & 2 \end{vmatrix}$$

$$\vec{A}\cdot\vec{\nabla}\phi = \left(-3x^2yi - x^3j + 2k\right)$$

$$= \left(2x^2i - 3yzj + xz^2k\right)\cdot\left(-3x^2yi - x^3j + 2k\right)$$

$$= -6x^4y + 3x^3yz + 2xz^2$$

$$\vec{A}\cdot\vec{\nabla}\phi\Big|_{(1,-1,1)} = -6\cdot1\cdot(-1) + 3\cdot1\cdot(-1)1 + 2\cdot1\cdot1$$

$$= 6 - 3 + 2 = 5.$$

or, $$\vec{A}\times\vec{\nabla}\phi\Big|_{(1,-1,1)} = i\left\{-6yz + x^4z^2\right\} - j\left\{4x^2 + 3x^3yz^2\right\} + k\left\{-2x^5 - 9x^2y^2z\right\}.$$

$$i\left\{-6\cdot(-1)\cdot + 1\cdot1\right\} - j\left\{4\cdot1 + 3\cdot1(-1)\cdot1\right\} + k\left\{-2\cdot1 - 9\cdot1\cdot(-1)^2\cdot1\right\}^2$$

$$= i\,(6+1) - j\,(4-3) + k\,(-2-9) = 7i - j - 11k.$$

Example 4. *Find* $\vec{\nabla}\left|\vec{r}\right|^3$

Solution. Let $\vec{r} = xi + yj + zk$. $\qquad r = \left|\vec{r}\right| = \sqrt{x^2 + y^2 + z^2}$

$$\therefore \qquad \vec{\nabla}\left|\vec{r}\right|^3 = \vec{\nabla}\cdot\left(\sqrt{x^2 + y^2 + z^2}\right)^3 = \vec{\nabla}\cdot\left(x^2 + y^2 + z^2\right)^{3/2}$$

$$= \frac{\partial}{\partial x}\left(x^2 + y^2 + z^2\right)^{3/2}\cdot\hat{i} + \frac{\partial}{\partial y}\left(x^2 + y^2 + z^2\right)^{3/2}\cdot\hat{j} + \frac{\partial}{\partial z}\left(z^2 + y^2 + z^2\right)^{3/2}\cdot\hat{k}$$

$$= \frac{3}{2}\cdot\left(x^2 + y^2 + z^2\right)^{1/2}\cdot 2x\hat{i} + \frac{3}{2}\left(x^2 + y^2 + z^2\right)^{1/2}\cdot 2y\cdot\hat{j} + \frac{3}{2}\left(x^2 + y^2 + z^2\right)^{1/2}\cdot 2z\cdot\hat{k}$$

$$= 3\cdot\left(x\hat{i} + y\hat{j} + z\hat{k}\right)\left(x^2 + y^2 + z^2\right)^{1/2} = 3\vec{r}\cdot\left|\vec{r}\right|.$$

2.2.3 Electrostatic Field

The *electrostatic field* at a point is defined as the limit of the ratio between the force on a test charge placed at the point to the magnitude of the test charge as the magnitude of the test charge approaches zero.

If $\vec{F}$ is the force experienced by the test charge q placed at a distance $\vec{r}$ the electric field $\vec{E}$ at a distance $\vec{r}$ will be

$$\vec{E}(\vec{r}) = \lim_{q \to 0} \frac{\vec{F}}{q}$$

The electric field in S.I. units is measured in Newton coulomb^{-1} or volt meter^{-1} (in C.G.S. dyne per esu of charge).

Thus the dimension of E are :

$$E = \frac{\text{volt}}{\text{m}} = \frac{\text{Joule}}{\text{Coulomb - m}} = \frac{\text{kg ms}^{-2}}{\text{As}} = \text{kg ms}^{-3}\ \text{A}^{-1} = \text{MLT}^{-3}\ \text{I}^{-1}.$$

Electrostatic Potential : The electrostatic potential at a point is the work done in bringing a unit test charge from infinity to that point. This work is stored in the test charge as potential energy.

The electrostatic potential therefore gives the potential energy per unit charge, the potential energy at infinity being zero.

Note that the electrostatic potential $\phi(\vec{r})$ is a function of $\vec{r}$ only and is independent of the path along which the test charge is moved from infinity to $\vec{r}$ (figure 2.4).

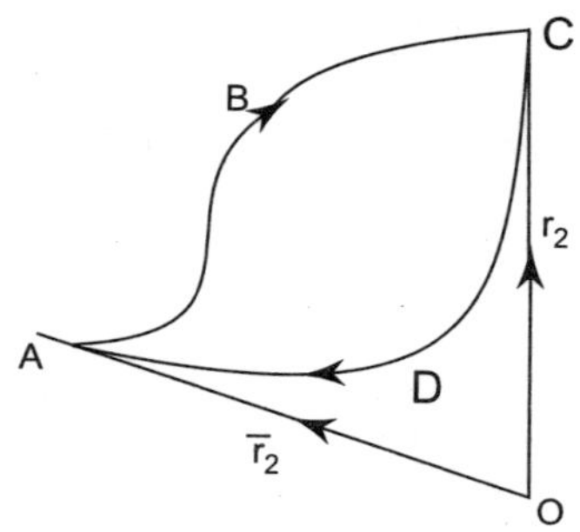

Figure 2.4

In SI units, the potential is measured in Joule per Coulomb or Volt. In C.G.S. – ergs per esu of charge or stat volt.

The dimension of electrostatic potential is $ML^2T^{-3}I^{-1}$.

Note that the potential along closed path is zero *i.e.*, the net work-done is zero. Therefore electrostatic force is conservative force.

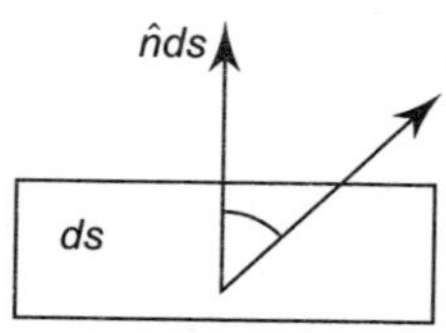

Figure 2.5

Electrostatic Flux : The product of the normal component of the electric field at the surface element *ds* with the surface area is known as the flux of the electric field through *ds*, *i.e.*, Flux $d\phi_E = E ds \cos\theta$ where θ is the angle between the direction of electric field $\vec{E}$ and positive normal to the surface area *ds* (figure 2.5).

$$d\phi_E = \vec{E}\,.\,\vec{ds}$$

In vector form, $$d\phi_E = \vec{E}\,.\,\hat{n}\,.ds$$

Thus the electric flux through a vector area $\vec{ds}$ is given by the dot product of electric field vector $\vec{E}$ and area vector $\vec{ds}$.

Unit of flux in SI units is given by Newton/Coulomb [NC^{-1}].

2.2.4 Divergence of Electrostatic Field

We know that the electric field or intensity is the '–ve' gradiant of electric potential.

i.e., $$\vec{E} = -\vec{\nabla}\phi$$

where $\phi = \frac{q}{r}$, $$\vec{E} = \frac{q}{r^3}\vec{r}$$

∴ The divergence of electrostatic field is

$$\vec{\nabla}\,.\,\vec{E} = \frac{q}{r^3}\vec{\nabla}\,.\,\vec{r} = \frac{3q}{r^3}$$, so it is a scalar quantity.

2.2.5 Gauss's Law of Electrostatics

Gauss's theorem states that in S.I. Unit, the total electric flux in free space (or vacuum) through a closed surface is equal to $1/\varepsilon_0$ times the charges q enclosed by the surface or,

$$\phi_E = \oiint \vec{E}\,.\,\vec{ds} = \frac{q}{\varepsilon_0}$$

where $\vec{E}$ is the electric field at the center of the area element ds on the closed surface and ε_0 is the free space permittivity.

Consider a charge q situated at 0, the origin of the rectangular co-ordinate system (figure 2.6). Let S be the Gaussian surface around the charge. Now consider a small area $\vec{ds}$ at a vector distance $\vec{r}$ from the charge q.

Therefore, the electric flux through the area $\vec{ds}$ is given by

$$d\phi_E = \vec{E}\,.\,\vec{ds}$$ where $\vec{E}$ is the electric field vector at $\vec{r}$.

Therefore, total flux over the (closed) Gaussian surface due to charge q inside it in free space is given by

$$\phi_E = \int d\phi_E = \oiint \vec{E}\,.\,\vec{ds}$$

Now the electric intensity $\vec{E}$ at a point on the elementary surface $\vec{ds}$ at a position vector $\vec{r}$ due to the charge q is given by

$$\vec{E} = \frac{1}{4\pi\varepsilon_0}\frac{q}{r^2}\hat{r}$$

where $\hat{r}$ is the unit vector in the direction of electric intensity $\vec{E}$.

Also $\vec{ds} = ds\,\hat{n}$, where $\hat{n}$ is the unit vector in the direction of positive normal to the surface $\vec{ds}$.

$$\phi_E = \frac{1}{4\pi\varepsilon_0}\oiint \frac{q}{r^2}\hat{r}\,.\hat{n}\,ds$$

$$= \frac{1}{4\pi\varepsilon_0}\oiint q\frac{ds\cos\theta}{r^2} \quad [\because \hat{r}\,.\hat{n} = \cos\theta]$$

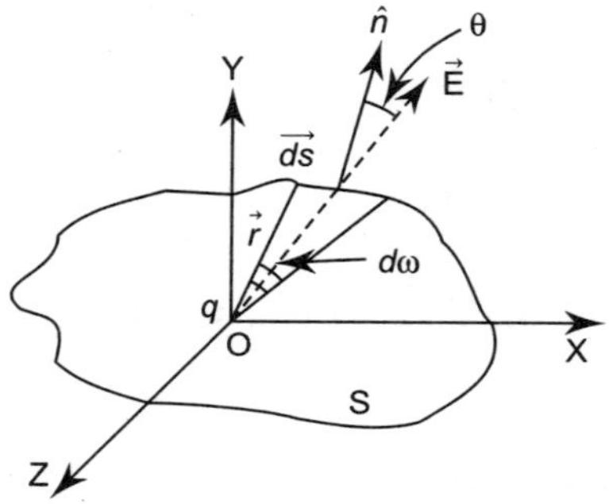

Figure 2.6

Now, $\frac{ds\cos\theta}{r^2} = d\omega$, the solid angle subtended by the elementary area $\vec{ds}$ at q.

$\therefore$ $$\phi_E = \frac{1}{4\pi\varepsilon_0}\oiint q\,d\omega = \frac{q}{4\pi\varepsilon_0}\oiint d\omega$$

As, $$\oiint d\omega = 4\pi$$

$\therefore$ $$\phi_E = \frac{q}{\varepsilon_0}$$

Therefore, the statement $\oiint \vec{E}\,.\vec{ds} = \frac{q}{\varepsilon_0}$ is known as integral form of Gauss's law.

Application of Gauss's theorem to calculate the electric field due to an infinitely long straight charged wire (or infinite line charge).

Consider an infinitely long straight wire having a constant line charge density ρ. The electric field due to electric charge will be radially outwards due to symmetry.

Let us draw a cylinder of height h and radius r co-axial with the line charges and closed at both ends by plain caps A and B normal to the axis to represent the Gaussian surface (figure 2.7). The closed surface consists of three parts A, B and C. Now, we consider a very small area $\vec{ds}$ on the surface A, then the area vector $\vec{ds}$ acts along the outward drawn normal and is at right angles to the electric field vector $\vec{E}$.

For the surface A : $\iint_A \vec{E} \cdot \vec{ds} = 0$

Similarly for the surface B : $\iint_B \vec{E} \cdot \vec{ds} = 0$

For the surface C the area vector $\vec{ds}$ and the electric field vector $\vec{E}$ are parallel and act in the same direction.

$\iint_C \vec{E} \cdot \vec{ds} = \iint_C E\, ds$ where E and ds are the magnitudes of $\vec{E}$ and $\vec{ds}$ respectively.

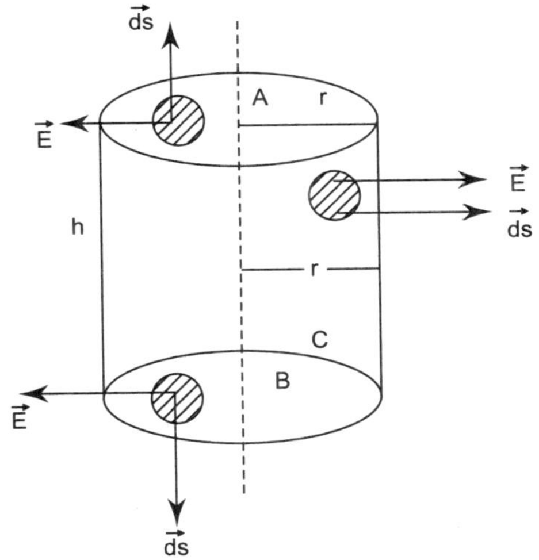

Figure 2.7

As all point on the curved surface of the cylinder C are equidistant from the axis, the electric field is the same at every point. Hence, E = constant.

$$\therefore \qquad \iint_C E\, ds = E \iint_C ds = 2\pi r h E$$

where $2\pi rh$ is the area of the curved surface C of height h and radius r.

According to Gauss's theorem, the electric flux over the closed surface is

$$\phi = \oiint \vec{E} \cdot \vec{ds} = \int_A \vec{E} \cdot \vec{ds} + \int_B \vec{E} \cdot \vec{ds} + \int_C \vec{E} \cdot \vec{ds}$$

$$= \frac{q}{\varepsilon_0} \qquad \text{[where } q = \text{total charge within the Gaussian surface]}$$

$$\Rightarrow \qquad \oiint \vec{E} \cdot \vec{ds} = \frac{q}{\varepsilon_0} = \frac{\rho h}{\varepsilon_0} \qquad \rho : \text{Line charge density}$$

$$\Rightarrow \qquad 2\pi r h E = \frac{\rho h}{\varepsilon_0}$$

$$\Rightarrow \qquad E = \frac{1}{2\pi r\varepsilon_0} \cdot \rho$$

$$= \frac{1}{2\pi\varepsilon_0}\left(\frac{\rho}{r}\right)$$

The direction of $\vec{E}$ is perpendicular to the charged wire at every point.

2.2.6 Gauss's Divergence Theorem

It states that if V is the volume bounded by a closed surface S and $\vec{A}$ is a vector function of position with continuous derivative, then

$$\iiint_V \vec{\nabla} \cdot \vec{A}\, dv = \iint_S \vec{A} \cdot \hat{n}\, ds \quad \text{where } \hat{n} \text{ is the positive normal to } ds.$$

The surface integral of the normal component of a vector $\vec{A}$ taken over a closed surface is equal to the integral of the divergence of $\vec{A}$ taken over the volume enclosed by the surface.

Deduction of the Differential form of Gauss's Law

Gauss's law states that

$$\oiint \vec{E} \cdot \vec{ds} = q/\varepsilon_0$$

Now applying the Gauss's divergence theorem of vector calculus, we have for electrostatic field

$$\oiint \vec{E} \cdot \vec{ds} = \iiint_V \vec{\nabla} \cdot \vec{E}\, dv$$

∴ Gauss's law then can be written as

$$\int_V \vec{\nabla} \cdot \vec{E}\, dv = q/\varepsilon_0 \qquad \ldots(2.1)$$

If ρ is the charge density at a point within the volume V, then

$$q = \iiint_V \rho\, dv$$

$$\therefore \qquad \int_V \vec{\nabla} \cdot \vec{E}\, dv = \int_V \frac{\rho\, dv}{\varepsilon_0}$$

$$\Rightarrow \qquad \vec{\nabla} \cdot \vec{E} = \rho/\varepsilon_0$$

This is the differential form of Gauss's law.

To show $\vec{E} = -\vec{\nabla}V$

The electric potential at a point A in an electric field $\vec{E}$ is defined as the work done in bringing a unit positive charge from infinity to that point against the electric forces.

$$\therefore \quad V = -\int_{\infty}^{A} \vec{E}\,.\,\vec{dr}$$

or
$$dV = -\vec{E}\,.\,\vec{dr} \qquad \ldots(2.2)$$

Now if the co-ordinate of the point A is x, y, z, then V (x, y, z) is a function of these co-ordinates.

$$\therefore \quad dV = \frac{\partial V}{\partial x}dx + \frac{\partial V}{\partial y}dy + \frac{\partial V}{\partial z}dz$$

$$= \left(\hat{i}\frac{\partial V}{\partial x} + \hat{j}\frac{\partial V}{\partial y} + \hat{k}\frac{\partial V}{\partial z}\right).\left(\hat{i}\,dx + \hat{j}\,dy + \hat{k}\,dz\right) = \vec{\nabla}V\,.\,\vec{dr} \ldots(2.3)$$

∴ From equations (2.2) and (2.3), we have

$$-\vec{E}\,.\,\vec{dr} = \vec{\nabla}V\,.\,\vec{dr}$$

$$\Rightarrow \quad \vec{E} = -\vec{\nabla}V = -\text{grad } V$$

Significance of (–) ve sign :

The (–) ve sign implies that the increase in electric intensity $\vec{E}$ points in the direction of decreasing V.

2.2.7 Laplace's Equation

When the enclosed charge density is zero *i.e.*, the point of observation is in empty space, $\rho = 0$, we get

$$\nabla^2 V = 0$$

or
$$\frac{\partial^2 V}{\partial x^2} + \frac{\partial^2 V}{\partial y^2} + \frac{\partial^2 V}{\partial z^2} = 0$$

This is Laplace's equation.

2.2.8 Poisson's Equation

The differential form of Gauss's law states

$$\vec{\nabla}.\vec{E} = \rho/\varepsilon_0$$

where $\vec{E}$ is the electric field intensity and ρ is the volume charge density.

Again
$$\vec{E} = -\vec{\nabla}V$$

$$\therefore \quad \vec{\nabla}.\left(-\vec{\nabla}V\right) = \rho/\varepsilon_0 \Rightarrow \nabla^2 V = -\rho/\varepsilon_0$$

$$\Rightarrow \quad \left(\frac{\partial^2 V}{\partial x^2} + \frac{\partial^2 V}{\partial y^2} + \frac{\partial^2 V}{\partial z^2}\right) = -\rho/\varepsilon_0$$

This is the Poisson's equation which holds good for each point in space.

Poisson's and Laplace's Equation

These equations are the fundamental differential equations, satisfied by the potential V. We have seen that the electric field $\vec{E}$ is the negative of the potential V *i.e.*,

$$\vec{E} = -\vec{\nabla}V$$

Also, the electric field satisfies the differential form of Gauss's law, $-\vec{\nabla}.\vec{E} = \rho/\varepsilon_0$ where ρ = charge density and ε_0 = permittivity of free space.

$$\vec{\nabla}.\vec{\nabla}V = -\rho/\varepsilon_0$$

The operator $\vec{\nabla}.\vec{\nabla}$ is called the Laplacian operator ∇^2 (delsquared), which is a linear scalar operator.

Now $$\nabla^2 V = -\rho/\varepsilon_0$$

This equation is known as Poisson's equation.

The operator ∇^2 has the following forms in the three co-ordinate systems generally used :

(*i*) **Rectangular Cartesian Co-ordinates :**

$$\nabla^2 \equiv \frac{\partial^2}{\partial x^2} + \frac{\partial^2}{\partial y^2} + \frac{\partial^2}{\partial z^2}$$

(*ii*) **Spherical Polar Co-ordinates :**

$$\nabla^2 \equiv \frac{1}{r^2}\frac{\partial}{\partial r}\left(r^2\frac{\partial}{\partial r}\right) + \frac{1}{r^2 \sin\theta}\frac{\partial}{\partial\theta}\left(\sin\theta\frac{\partial}{\partial\theta}\right) + \frac{1}{r^2\sin^2\theta}\frac{\partial^2}{\partial\phi^2}$$

(*iii*) **Cylindrical Co-ordinates :**

$$\nabla^2 \equiv \frac{1}{r}\frac{\partial}{\partial r}\left(r\frac{\partial}{\partial r}\right) + \frac{1}{r^2}\frac{\partial^2}{\partial\theta^2} + \frac{\partial^2}{\partial z^2}$$

Thus, the Poisson's equation is a partial differential equation which may be solved for V if the functional dependence of ρ and the boundary conditions are known. The choice of the co-ordinate system depends on the symmetry of the problem.

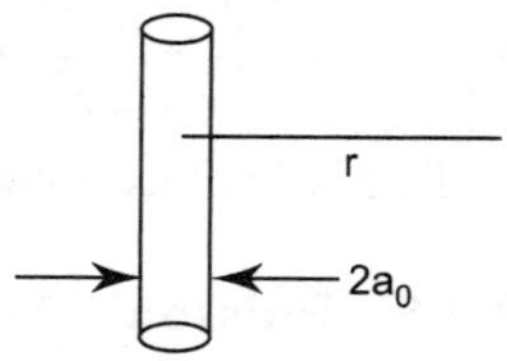

Figure 2.8

If $\rho = 0$ in some region of space, then in that region Poisson's equation reduces to

$$\nabla^2 V = 0$$

which is known as Laplace's equation.

In problems involving conductors, ρ is usually zero in the space between the conductors, hence we can find out the solution of Laplace's equation in this region satisfying the boundary condition at the surface of the conductors.

Solution of Laplaces Equation in three Co-ordinate Systems :

Case 1. *Solve Laplace's equation to find out the potential at a distance r from the axis of an infinitely long conducting cylinder of radius a_0 charged with a surface charge density σ. Take the potential of the cylinder to be zero.*

Solution. By symmetry, the electrostatic potential ϕ will depend on r only (figure 2.8). In the region exterior to the conducting cylinder, Laplace's equation is satisfied *i.e.*, $\nabla^2\phi = 0$. Since ϕ is a function of r only *i.e.*, the equation reduces to $\frac{1}{r}\frac{d}{dr}\left(r \cdot \frac{d\phi}{dr}\right) = 0$

$\Rightarrow \quad \phi = al_n r + b$, where a and b are constants.

We have to find out the value of a and b applying boundary conditions. The boundary condition is

$$\phi = 0 \quad \text{at } r = a_0$$

$$\therefore \quad b = -a \ln a_0$$

We know that at the surface of a conductor the electric field $E = \sigma/\varepsilon_0$ *i.e.*,

$$-\left(\frac{d\phi}{dr}\right)_{r = a_0} = \sigma/\varepsilon_0$$

Thus $\quad -a/a_0 = \sigma/\varepsilon_0 \quad \text{or} \quad a = -\frac{a_0\sigma}{\varepsilon_0}$

Therefore, the total solution is $\phi = \frac{-a_0\sigma}{\varepsilon_0}\ln\left(\frac{r}{a_0}\right)$ is a function of a single variable. Laplace's equation reduces to ordinary differential equation when the potential is a function of a single variable.

In the case where ϕ is a function of the single rectangular co-ordinate x, Laplace's equation becomes $\frac{d^2\phi}{dx^2} = 0$, the solution is $\phi = ax + b$, where a and b are constants to be determined from boundary conditions.

Case 2. *Two concentric spherical shells of radii a_1 and a_2 ($a_2 > a_1$), are charged to potentials ϕ_1 and ϕ_2, respectively. Determine the electric potential and field in the region between the shells. Also determine the charge on the inner shell.*

Solution. Due to spherical symmetry, let us consider spherical polar co-ordinates with the origin at the centre of the shell. So, potential ϕ in the region between the shells will be independent of θ and ϕ from symmetry considerations. Therefore, Laplace's equation reduces to

$$\frac{1}{r^2}\frac{d}{dr}\left(r^2\frac{d\phi}{dr}\right) = 0$$

$\Rightarrow \quad \phi = \frac{a}{r} + b$, where a and b are constants. From the boundary condition,

$$\phi = \phi_1 \quad \text{at } r = a_1 \quad \text{gives } \phi_1 = \frac{a}{a_1} + b \qquad \ldots(i)$$

and that

$$\phi = \phi_2 \quad \text{at } r = a_2 \quad \text{yields } \phi_2 = \frac{a}{a_2} + b \qquad \ldots(ii)$$

Solving equations (*i*) and (*ii*) for a and b we get

$$a = \frac{a_1 \cdot a_2}{(a_2 - a_1)}(\phi_1 - \phi_2) \text{ and } b = \frac{a_2\phi_2 - a_1\phi_1}{a_2 - a_1}$$

Therefore,

$$\phi = \frac{a_1 \cdot a_2}{(a_2 - a_1)} \cdot \frac{(\phi_1 - \phi_2)}{r} + \frac{a_2\phi_2 - a_1\phi_1}{a_2 - a_1}$$

The electric field is $$\vec{E} = -\hat{r}\,\frac{d\phi}{dr} = \frac{a_1 \cdot a_2}{(a_2 - a_1)} \cdot \frac{(\phi_1 - \phi_2)\,\hat{r}}{r^2}$$

The electric field just outside the inner shell is

$$\overrightarrow{E_{a_1}} = \frac{a_2}{a_2 - a_1} \cdot \frac{\phi_1 - \phi_2}{a_1}\,\hat{r}$$

Therefore, the charge density of the inner shell is

$$\vec{\sigma} = \varepsilon_0\,\overrightarrow{E_{a_1}} \cdot \hat{r} = \frac{\varepsilon_0\,a_2}{a_1}\left(\frac{\phi_1 - \phi_2}{a_2 - a_1}\right)$$

So, the total charge on the inner shell is

$$Q_1 = 4\pi a_1^2\,\vec{\sigma} = 4\pi\,\varepsilon_0\,a_1\,a_2\,\frac{(\phi_1 - \phi_2)}{(a_2 - a_1)}.$$

2.2.9 Equation of Continuity

Consider a small area element $\vec{ds}$ of a closed surface S. If P is a point on the area element $\vec{ds}$ and $\vec{J}$ the current density vector at P, then the charge flowing out of the area $\vec{ds}$ per second.

$$dI = \vec{J} \cdot \vec{ds}$$

The total charge crossing the closed surface S outward per unit time

$$I = \oiint_s \vec{J} \cdot \vec{ds}$$

If ρ is the volume density of charge at any point, then the total charge within the volume V enclosed by the surface S

$$q = \iiint_V \rho\,dv$$

Since the current is flowing outward the charge within the enclosed surface is decreasing with time. The time rate of decrease of charge is given by

$$\frac{dq}{dt} = I = -\frac{\partial}{\partial t}\iiint_V \rho\,dv$$

According to the principle of conservation of charge.

Total charge crossing the closed surface S per second = Rate of decrease of charge in the enclosed volume.

$$\oiint_S \vec{J} \cdot \vec{ds} = -\frac{\partial}{\partial t}\iiint_V \rho\,dv = -\iiint_V \frac{\partial \rho}{\partial t}\,dv \qquad \text{...(2.4)}$$

But according to the Gauss's divergence theorem

$$\oiint_S \vec{J} \cdot \vec{ds} = \iiint_V (\vec{\nabla} \cdot \vec{J})\, dv \qquad \ldots(2.5)$$

$\therefore$ From equations (2.4) and (2.5) $\Rightarrow \iiint_V \vec{\nabla} \cdot \vec{J}\, dv = -\iiint_V \frac{\partial \rho}{\partial t}\, dv$

$$\Rightarrow \qquad \vec{\nabla} \cdot \vec{J} = -\frac{\partial \rho}{\partial t} \Rightarrow \vec{\nabla} \cdot \vec{J} + \frac{\partial \rho}{\partial t} = 0$$

This is known as equation of continuity. It is also known as charge conservation equation in space.

2.3 THE CURL OF A VECTOR FIELD

The curl of a (non lamellar) vector field at any point is defined as the maximum line integral of the vector computed per unit area along the boundary of an infinitesimal area at that point. It is essentially a vector. In case of hydrodynamics, curl is sensed as rotation of a fluid.

2.3.1 Representation of Curl in Cartesian Co-ordinate

If $\vec{A}$ (x, y, z) is a differentiable vector field then curl or rotation of $\vec{A}$, is defined by the cross product of $\vec{A}$ with the vector differential operator del $(\vec{\nabla})$ *i.e.*,

$$\text{Curl A} = \vec{\nabla} \times \vec{A} = \left(\frac{\partial}{\partial x}\hat{i} + \frac{\partial}{\partial y}\hat{j} + \frac{\partial}{\partial z}\hat{k}\right) \times \left(A_1\hat{i} + A_2\hat{j} + A_3\hat{k}\right)$$

$$= \begin{vmatrix} \hat{i} & \hat{j} & \hat{k} \\ \frac{\partial}{\partial x} & \frac{\partial}{\partial y} & \frac{\partial}{\partial z} \\ A_1 & A_2 & A_3 \end{vmatrix}$$

2.3.2 Stoke's Theorem

The line integral of the tangential component of a vector $\vec{A}$ taken around a simple closed curve C is equal to the surface integral of the normal component of the curl of $\vec{A}$ taken over any surface S having C as its boundary.

i.e.,
$$\oint_C \vec{A} \cdot \vec{dr} = \iint_S \left(\vec{\nabla} \times \vec{A}\right) \cdot \hat{n}\, ds = \iint_s \left(\vec{\nabla} \times \vec{A}\right) \cdot \vec{ds}$$

2.3.3 Potential Fields

A region in space in which the potential is continuous and is defined by a single value at every point of the position variable is called **potential field.**

2.3.4 Application of the Curl of Vector Field

Consider a vector field $\vec{A}$, the magnitude and direction of which is a function of position co-ordinates at a point, then the vector product of the differential operator $\vec{\nabla}$ and vector $\vec{A}$ is a vector function of position co-ordinates x, y, z and is known as Curl of $\vec{A}$

$$\text{Curl } \vec{A} = \vec{\nabla} \times \vec{A} = \begin{vmatrix} \hat{i} & \hat{j} & \hat{k} \\ \frac{\partial}{\partial x} & \frac{\partial}{\partial y} & \frac{\partial}{\partial z} \\ A_x & A_y & A_z \end{vmatrix}$$

Physical Significance : Consider a small area $\overrightarrow{\Delta s}$ enclosed by a closed path which forms a boundary. Let $\vec{A}$ be a vector field defined everywhere in the region and $\vec{dl}$ a vector representing a small element of the path, its direction being that of the tangent to the path at the element $\vec{dl}$. If θ is the angle between the vector $\vec{A}$ at any point on the curve and the direction of element, then

$$\vec{A} \,.\, \vec{dl} = A\, dl \cos\theta.$$

The line integral of the vector field $\vec{A}$ for the closed path

$$= \oint A\, dl \cos\theta = \oint \vec{A} \,.\, \vec{dl}$$

where the symbol $\oint$ = integration over the entire closed path.

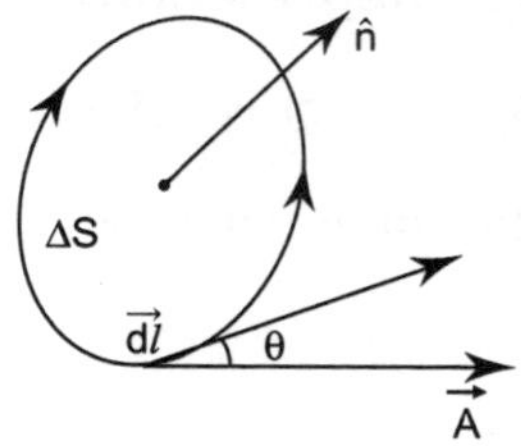

Figure 2.9

∴ Line integral per unit area

$$= \frac{1}{\Delta s} \oint \vec{A} \,.\, \vec{dl}$$

Now consider a unit vector $\hat{n}$ in a direction perpendicular to the area Δs, the (+)ve direction of $\hat{n}$ bring related to the direction of line integral by right handed screw rule.

The curl of the vector field $\vec{A}$ is defined as the vector, the magnitude of whose component in the direction of the unit vector $\hat{n}$ is given by the line integral of the vector $\vec{A}$ for the closed path per unit area enclosed by it when the area becomes very small.

$$\therefore \quad \text{Curl } \vec{A} \,.\, \hat{n} = \lim_{\Delta S \to 0} \frac{1}{\Delta S} \oint \vec{A} \,.\, \vec{dl}$$

As $\Delta s \to 0$, let ΔS be denoted by ds

$$\text{Curl } \vec{A} \,.\, \hat{n}\, ds = \oint \vec{A} \,.\, \vec{dl}$$

$$\therefore \quad \text{Curl } \vec{A} \,.\, \vec{ds} = \oint \vec{A} \,.\, \vec{dl}$$

or $$\text{Curl } \vec{A} = \frac{1}{\vec{ds}} \cdot \oint \vec{A} \cdot \vec{dl}$$

Irrotational field (Conservative field) : A vector field whose curl vanishes (= 0) is called an irrotational field and conservative field also.

Example : Electric field.

Rotational field : A vector field whose curl is not equal to zero is called a rotational vector field.

Example : Magnetic field.

NOTE *Circulation* : The line integral of a vector function $\vec{A}$ along a closed path is called circulation.

∴ Circulation = $\oint \vec{A} \cdot \vec{dl}$ where $\vec{dl}$ is an elementary vector length along the closed path and $\oint$ denotes integral over a closed path.

2.3.5 Magnetic Field

Magnetic field is a field in which a moving charge experiences a velocity dependent force given by

$$\vec{F_m} = q\vec{v} \times \vec{B}$$

Unit : The magnitude of the force on a charge q moving with a velocity $\vec{v}$ in a direction perpendicular to the magnetic field $\vec{B}$ is given by

$$F = q\,v\,B$$

$$B = \frac{F}{qv}.$$

when F = 1 Newton, q = 1 coulomb, v = 1 meter/sec. ∴ B = 1

The S.I. unit of magnetic field is Tesla or Weber/mH^2

Tesla : The magnetic field has strength of 1 Tesla if it exerts a force of 1 newton on a charge of 1 coulomb moving with velocity 1 metre per second in a direction perpendicular to the field.

The C.G.S. unit of magnetic field is Gauss.

i.e., When F = 1 dyne, q = 1 e.m.u., v = 1 cm/sec.

Relation :

$$\text{Tesla} = \frac{\text{Newton}}{\text{Coulomb} \times \text{meter sec}^{-1}}$$

$$\text{Gauss} = \frac{\text{Dyne}}{\text{e.m.u.} \times \text{cm sec}^{-1}}$$

$$\frac{\text{Tesla}}{\text{Gauss}} = \frac{\text{Newton/dyne}}{(\text{Coulomb/e.m.u.})\,(\text{meter/cm})}$$

$$= \frac{10^5}{\frac{1}{10} \times 100} = 10^4$$

$\therefore$ 1 Tesla = 10^4 Gauss.

Magnetic Flux : Through a surface is defined as the total number of lines of magnetic induction passing through the surface.

Consider an elementary surface area $\overrightarrow{dA}$ of the surface area $\vec{A}$. Let $\hat{n}$ be a unit vector normal to the area element $\overrightarrow{dA}$, hence $\hat{n}$ represents the direction of the area vector $\overrightarrow{dA}$ (figure 2.10)

Hence, the number of lines of magnetic induction over the area $\overrightarrow{dA}$ *i.e.*, magnetic flux over area $\overrightarrow{dA}$ is given by

$$d\phi_B = \vec{B} . \overrightarrow{dA} = \vec{B} . \hat{n}\, dA$$

$\therefore$ Magnetic flux over the whole surface A

$$= \phi_B = \int_A \vec{B} . \hat{n}\, dA$$

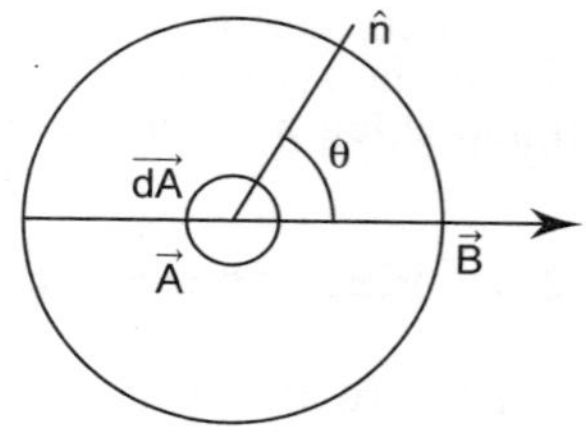

Figure 2.10

2.3.6 Ampere's Circuital Law

It states that "the line integral of a magnetic induction field vector along any closed path is equal to μ_0 times the net current enclosing the path".

If a magnetic field $\vec{B}$ encloses a current I flowing through a straight conducting wire in a direction perpendicular to the plane of the closed path and $\vec{dl}$ be a vector element of the path, then

$\oint \vec{B} . \vec{dl} = \mu_0 I$ where $\oint \vec{B} . \vec{dl}$ is the line integral of the magnetic field for the closed path.

Proof : The magnetic induction field $\vec{B}$ due to an infinitely long straight conductor carrying a current I at a point lying at a perpendicular distance r from it, is given by

$$B = \frac{\mu_0}{4\pi} \frac{2I}{r}.$$

Now we draw a circle with the point where the wire leaves the paper as centre and with radius r then the circumference of the circle is of constant induction.

A small element dl of the circular line of constant magnetic induction is closen, which is tangential to the circle so that the vectors $\vec{B}$ and $\vec{dl}$ are parallel.

$$\therefore \quad \vec{B} . \vec{dl} = B\, dl \cos\theta = B\, dl$$

$\therefore$ Value of line integral for the whole closed path is

$$\oint \vec{B} . \vec{dl} = \oint B\, dl = B\, 2\pi r.$$

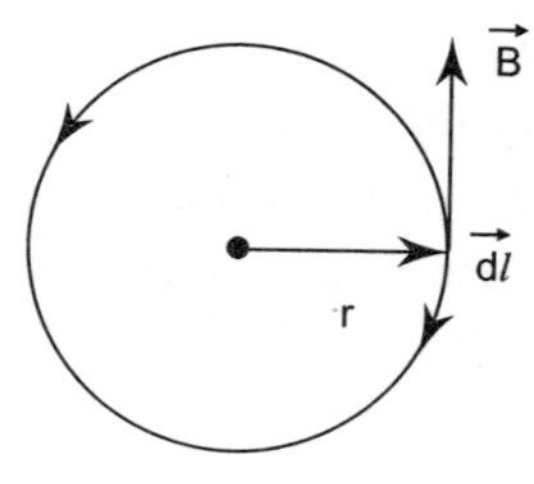

Figure 2.11

Since B is constant, putting the value of B we have

$$\oint \vec{B} . \vec{dl} = \frac{\mu_0}{4\pi} . \frac{2I}{r} \times 2\pi r = \mu_0 I$$

NOTE The line integral of the magnetic field is independent of the path. This means that the value of line integral of the magnetic field along all the circular paths of any radius would be the same and equal to $\mu_0 I$.

(*i*) **Magnetic field due to current in a straight conductor of infinite length :**

Consider an infinitely long wire along the Y-axis through which a current I is flowing. The lines of magnetic induction will consist of concentric circles with the wire as the centre. Let $\vec{B}$ be the magnetic induction field at a point P at a perpendicular distance *r* from the wire. (Figure 2.12)

According to Ampere's law, the line integral of the magnetic field for a closed path.

$$\oint \vec{B} \cdot \vec{dl} = \mu_0 I$$

From the symmetry, the magnitude of B is the same at all points lying at a distance *r* from the wire *i.e.*, on the circle of radius *r*.

$$\therefore \quad \oint \vec{B} \cdot \vec{dl} = B\, 2\pi r = 2\pi\, Br$$

$$\therefore \quad 2\pi\, Br = \mu_0 I$$

$$B = \frac{\mu_0}{2\pi} \frac{I}{r}$$

or
$$B = \frac{\mu_0}{4\pi} \times \frac{2I}{r} \qquad \ldots(2.6)$$

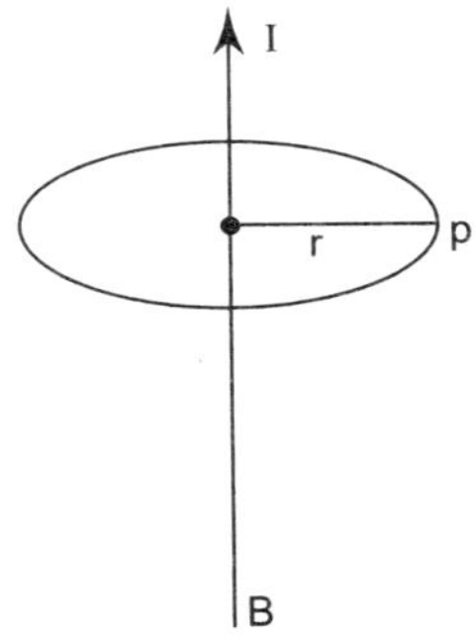

Figure 2.12

(*ii*) **Magnetic field in the case of a long coaxial cable :** Let us consider a coaxial cable containing a central cylindrical conductor of radius *a* and a coaxial outer cylindrical conductor of radius *b* (figure 2.13). The two conductors carry the same current I in opposite directions. By symmetry, the magnetic field at a distance *r* from the center of the conductors will be a function of *r* only, where $a < r < b$. Also, $\vec{B}$ must be tangent everywhere to a circle of radius *r* and concentric with the conductors. We can show as in case (*i*), that

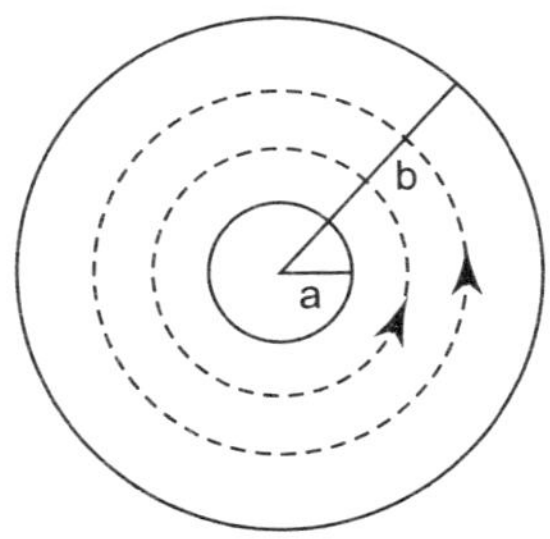

Figure 2.13

$$B = \frac{\mu_0 I}{2\pi r}, \quad a < r < b \qquad \ldots(2.7)$$

If $r > b$, $B = 0$, Thus the total current enclosed by the circle of radius *r* is $(I - I) = 0$.

(*iii*) **Force between two infinitely long parallel current carrying conductors :**

If I_1 and I_2 be the currents in the two wires N and M spaced at a distance *d* apart in free space. The magnetic field produced by the current I_2 in M at any point in N is given by

$$\therefore \quad B = \frac{\mu_0\, I_2}{2\pi d} \qquad \ldots(2.8)$$

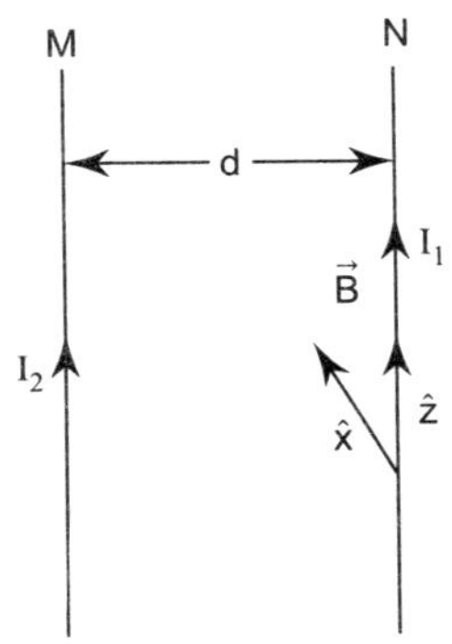

Figure 2.14

In this the case radius of the wire is smaller than d

The direction of B is at right angles to the plane of the wires as shown in figure 2.14. It is given by

$$\overrightarrow{F_0} = I_1\, \hat{z} \times \vec{B} \qquad \text{...(2.9)}$$

where $\hat{z}$ is a unit vector in the direction of current I_1.

Substituting for B from equation (2.8) we obtain

$$\overrightarrow{F_0} = \frac{\mu_0\, I_1 \,.\, I_2}{2\pi d}\,(\hat{z} \times \hat{x}) \qquad \text{...(2.10)}$$

where $\hat{x}$ is a unit vector in the direction of $\vec{B}$. The force F_0 is directed towards the wire M, if I_1 and I_2 flow in the same direction. On the other hand, F_0 is directed away from the wire M, if I_1 and I_2 flow in the opposite direction. Thus wires carrying parallel current attract, while wires carrying anti-parallel current repel each other. The magnitude of the force per meter of the wires is determined from

$$F_0 = \frac{\mu_0\, I_1 \,.\, I_2}{2\pi d} \qquad \text{...(2.11)}$$

(*iv*) **Magnetic field of a toroid.** The configuration in which we have closely spaced windings of a wire wound around a ring is called a toroid (figure 2.15). Suppose a current I flow through the windings. Consider a circular path C_1 inside the ring and concentric with it. The symmetry shows that the magnetic field is constant everywhere along the path C_1 and is tangential to it. If the current direction is as shown in figure 2.15, the direction of the line of constant B is counter clockwise. The total current enclosed by the path C_1 is NI, where N is the total number of turns on the toroid.

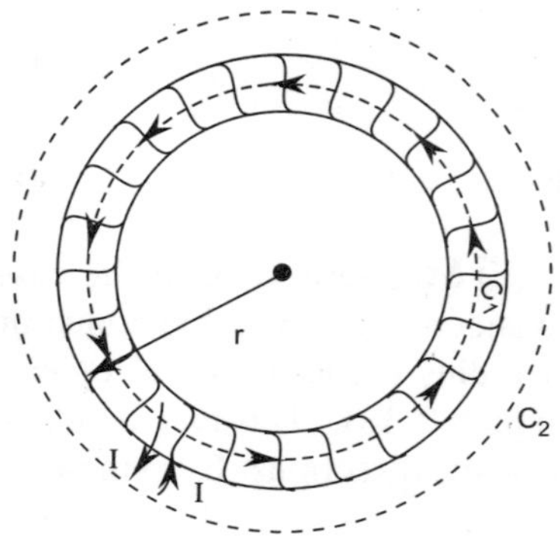

Figure 2.15

Applying Ampere's circuital law, for the path C_1, we obtain

$$\oint_{C_1} \vec{B} \,.\, \vec{dl} = \mu_0\, NI \quad \text{or} \quad B\,(2\pi r) = \mu_0\, NI$$

where r is the radius of the circular path C_1. Thus,

$$B = \frac{\mu_0\, NI}{2\pi r} \qquad \text{...(2.12)}$$

If $r \simeq R$, and we put $n = \dfrac{N}{2\pi R}$, then the equation reduces to

$$B = \mu_0\, nI \qquad \text{...(2.13)}$$

where n = number of turns per unit length.

2.3.7 Curl of Magnetic Field

Ampere's line integral theorem states that the line integral of the magnetic field $\vec{B}$ along a closed curve is equal to μ_0 times the total current enclosed by the curve.

$$\oint \vec{B} \cdot \vec{dl} = \mu_0 I \qquad \text{...(2.14)}$$

If $\vec{J}$ is the current density which varies from place to place but is constant in time, then

$$I = \iint_S \vec{J} \cdot \vec{ds}$$

where $\vec{ds}$ is a small vector element of surface area and the whole surface is bounded by a closed path C.

∴ From equation (2.14), we have

$$\oint \vec{B} \cdot \vec{dl} = \mu_0 \iint_S \vec{J} \cdot \vec{ds} \qquad \text{...(2.15)}$$

But according to Stoke's theorem

$$\oint \vec{B} \cdot \vec{dl} = \iint_S \text{Curl } \vec{B} \cdot \vec{ds} = \iint_S \left(\vec{\nabla} \times \vec{B}\right) \cdot \vec{ds}$$

∴ Then equation (2.15) reduces to

$$\iint_S \left(\vec{\nabla} \times \vec{B}\right) \cdot \vec{ds} = \mu_0 \iint_S \vec{J} \cdot \vec{ds}$$

or

$$\vec{\nabla} \times \vec{B} = \mu_0 \vec{J} \qquad \text{...(2.16)}$$

The equation $\vec{\nabla} \times \vec{B} = \mu_0 \vec{J}$ is known as **Ampere's law in differential form.** It is a point equation and gives the relation between magnetic field at a point and the current density at the same point in space.

2.3.8 Divergence of Magnetic Field $\left(\vec{B}\right)$

According to the definition of the divergence of vector field

$$\text{Div } \vec{B} = \frac{\text{Flux of } \vec{B} \text{ through a surface enclosing volume } v}{\text{Volume } v \text{ in the limit } v \to 0}$$

$$= \lim_{v \to 0} \frac{1}{v} \iint_S \vec{B} \cdot \vec{ds}$$

In the case of an electric field, the lines of electric force originate from a positive charge and terminate on a negative charge. But magnetic lines of force have no such source or sink but are continuous and join back on themselves. If therefore, we consider a closed surface S in a magnetic field, we always have as many lines of the magnetic field entering into the surface as leave it. Thus the magnetic flux of $\vec{B}$ over any surface is zero.

or

$$\iint \vec{B} \cdot \vec{ds} = 0$$

⇒

$$\vec{\nabla} \cdot \vec{B} = \text{Div } \vec{B} = 0$$

In other words, $\vec{\nabla} \cdot \vec{B} = 0$ whether the point lies outside or inside the current loop. Thus the divergence of magnetic field vector is zero everywhere.

Isolated magnetic pole does not exist : The relation magnetic flux $\phi_B = \oiint \vec{B} \cdot \vec{ds} = 0$ ...(2.17)

through any closed surface is known as Gauss's law in magnetism. Gauss's law in electrostatics in S.I. units states that

$$\varepsilon_0 \oiint \vec{E} \cdot \vec{ds} = q \qquad ...(2.18)$$

Comparing equations (2.17) and (2.18) we find that in magnetism there is no counter part to the free charge q in electricity which in other words means, that isolated magnetic pole do not exist. This is the significance of $\oiint \vec{B} \cdot \vec{ds} = 0$.

2.3.9 Concepts of Magnetic Scalar Potential and Vector Potential

(*a*) **Magnetic Scalar Potential :** In case of electric field, we know $\vec{E} = -\vec{\nabla}\phi$ where $\vec{E}$ is the electric field intensity and ϕ is the electric potential, which is a scalar quantity. Electric field is a conservative field and hence $\vec{\nabla} \times \vec{E} = 0$

or
$$\vec{\nabla} \times \left(-\vec{\nabla}\phi\right) = 0 \qquad ...(2.19)$$

Again Curl $\vec{B} = \vec{\nabla} \times \vec{B} = 0$ only in the special case when the line integral $\oint \vec{B} \cdot \vec{dl} = 0$ *i.e.*, when the line integral does not enclose a current.

If the line integral encloses a current of current density $\vec{J}$ then $\vec{\nabla} \times \vec{B} = \mu_0 \vec{J}$, is the differential form of ampere's law.

For a current free space (J = 0) $\vec{\nabla} \times \vec{B} = 0$ ...(2.20)

Comparing equations (2.19) and (2.20)

We can write $\vec{B} = -\vec{\nabla}\phi_m$ for a current free space, where ϕ_m is called magnetic scalar potential. It exists only for a current free space (J = 0).

(*b*) **Magnetic Vector Potential :** We know that $\vec{\nabla} \cdot \vec{B} = 0$. As the divergence of the curl of a vector vanishes, we can write the magnetic induction in the form

$$\vec{B} = \vec{\nabla} \times \vec{A},$$

$$\vec{\nabla} \cdot \vec{B} = \vec{\nabla} \cdot \left(\vec{\nabla} \times \vec{A}\right) = 0$$

where the vector field $\vec{A}$ is called the magnetic vector potential.

Magnetic vector potential is defined as a curl of which is equal to $\vec{B}$, the magnetic induction field.

as $\vec{\nabla} \times \vec{B} = \mu_0 \vec{J}$, we have

$$\vec{\nabla} \times \left(\vec{\nabla} \times \vec{A}\right) = \mu_0 \vec{J}$$

$$\vec{\nabla}\left(\vec{\nabla} . \vec{A}\right) - \nabla^2 \vec{A} = \mu_0 \vec{J}$$

and choosing $\vec{\nabla} . \vec{A} = 0$

we get $\nabla^2 \vec{A} = -\mu_0 \vec{J}$

The equations for the cartesian components of $\vec{A}$ are

$$\nabla^2 A_x = -\mu_0 J_x,$$

$$\nabla^2 A_y = -\mu_0 J_y$$

and $\nabla^2 A_z = -\mu_0 J_z$

Each of these component equations has the form of Poisson's equation ; $\nabla^2 \phi = -\left(\frac{\rho}{\varepsilon_0}\right)$ in electrostatic which has the solution

$$\phi\left(\vec{r}\right) = \frac{1}{4\pi\varepsilon_0}\int \frac{\rho\left(\vec{r_0}\right) dV_0}{\left(\vec{r} - \vec{r_0}\right)}$$

2.4 ELECTROMAGNETIC INDUCTION

The phenomenon of electromagnetic induction is the development of an induced electromotive force ε in a conducting circuit placed in a varying magnetic field.

2.4.1 Faraday's Laws of Electromagnetic Induction

Faraday noticed experimentally that whenever the magnetic flux linked with an electric circuit changed, a current was induced in the circuit - a phenomenon known as electromagnetic induction. The results of Faraday led to the development of two useful laws.

1. **Neumann's Law :** The induced electromotive force (e.m.f.) in a circuit is equal to the time rate of change of the magnetic flux linked with the circuit.
2. **Lenz's Law :** The direction of the induced e.m.f. or the current is such that it will oppose the change of flux producing it,

If ϕ is the flux linked with a circuit at time t, then $\frac{d\phi}{dt}$ gives the time rate of variation of flux. The combination of the two laws of electromagnetic induction gives

$$\varepsilon = -\frac{d\phi}{dt} \qquad \text{...(2.21)}$$

where ε is the induced e.m.f.

This equation is often referred to as the integral form of Faraday's law of electromagnetic induction and is a fundamental equation of electromagnetism. The '– ve' sign indicates that the e.m.f. ε opposes the changes of flux. Hence the induced e.m.f. is often referred to as back e.m.f. If R is the resistance of the circuit, then the induced current is

$$i = \frac{\varepsilon}{R} = -\frac{1}{R}\frac{d\phi}{dt}.$$

In S.I. units, ε is expressed in volt, R in ohm, i in ampere, ϕ in weber and t in second.

Integral and Differential Form of Faraday's Law

Consider a circuit C placed in a magnetic field $\vec{B}$. If S is the surface bounded by the current circuit, then the magnetic flux ϕ linked with the circuit is given by

$$\phi = \iint_S \vec{B} \cdot \vec{ds}$$

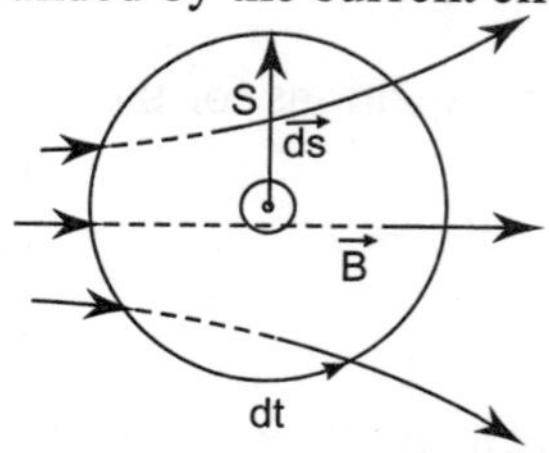

Figure 2.16

If the magnetic field changes and $\frac{\partial B}{\partial t}$ is the time rate of the change of the field then,

$$\frac{d\phi}{dt} = \iint_S \frac{\partial \vec{B}}{\partial t} \cdot \vec{ds} \qquad \left[\because \vec{B} = \vec{B}(r, t)\right]$$

∴ The e.m.f. induced in the circuit due to the changing magnetic flux is

$$e = -\frac{d\phi}{dt} = -\iint_S \frac{\partial \vec{B}}{\partial t} \cdot \vec{ds} \qquad \ldots(2.22)$$

This is **integral form** of Faraday's law.

Now e.m.f. is defined as the work done in taking a unit positive charge completely round the circuit. *i.e.*, it is the line integral of the electric field $\vec{E}$ over the circuit, given by

$$e = \oint \vec{E} \cdot \vec{dl} \qquad \ldots(2.23)$$

Comparing equations (2.22) and (2.23), we have

$$\oint \vec{E} \cdot \vec{dl} = -\iint_S \frac{\partial \vec{B}}{\partial t} \cdot \vec{ds}$$

But according to Stoke's Law,

$$\iint_S \vec{\nabla} \times \vec{E} \cdot \vec{ds} = \oint \vec{E} \cdot \vec{dl}$$

$$\therefore \quad \iint_S \vec{\nabla} \times \vec{E} \cdot \vec{ds} = -\iint_S \frac{\partial \vec{B}}{\partial t} \cdot \vec{ds}$$

$$\therefore \quad \vec{\nabla} \times \vec{E} = -\frac{\partial \vec{B}}{\partial t} \quad \text{or} \quad \text{Curl } \vec{E} = -\frac{\partial \vec{B}}{\partial t}.$$

This is the **differential form** of Faraday's Law. It indicates that the changing magnetic field produces an electric field.

2.4.2 Concepts of Displacement of Current

Differential form of Ampere's circuital law becomes (in general, if material is electrically conducting and is magnetized),

$$\vec{\nabla} \times \vec{B} = \mu_0 \left(\vec{J} + \overrightarrow{J_m}\right) \qquad \ldots(2.24)$$

where $\vec{J}$ is the free current density and $\overrightarrow{J_m}$ is the bound or the magnetization current density.

Since $$\overrightarrow{J_m} = \vec{\nabla} \times \vec{M}$$

we can write $$\vec{\nabla} \times \left(\vec{B} - \mu_0 \vec{M}\right) = \mu_0 \vec{J}$$

or, $$\vec{\nabla} \times \vec{H} = \vec{J} \quad \text{where } \vec{H} = \frac{\vec{B}}{\mu_0} - \vec{M} \qquad \ldots(2.25)$$

The quantity $\vec{H}$ is called the magnetic intensity.

From Stoke's theorem we obtain from equation (2.25)

$$\int_S \left(\vec{\nabla} \times \vec{H}\right) . \hat{n}\, ds = \oint \vec{H} . \vec{dl} = \int_S J . \hat{n}\, ds$$

or $$\oint \vec{H} . \vec{dl} = I \qquad \ldots(2.26)$$

That is, the line integral of the tangential component of magnetic intensity around a closed path C equals the transport current enclosed by the path C. This is the new form of Ampere's circuital law.

In free space $\vec{M} = 0$, $\vec{B} = \mu_0 \vec{H}$.

From equation (2.26), we can write

$$\oint \vec{H} . \vec{dl} = \int_s J . \hat{n}\, ds$$

Let us apply equation (2.25) to the circuit consisting of a parallel-plate capacitor in figure 2.17. Suppose that the capacitor is charged by a current I. Consider a closed path C and the surface S_1 yields

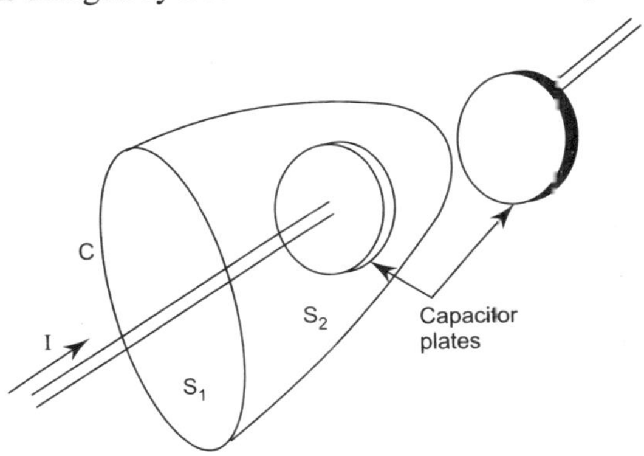

Figure 2.17

$$\oint_C \vec{H} \cdot \vec{dl} = \int_S \vec{J} \cdot \hat{n}\, ds = I \qquad \ldots(2.27)$$

The same law for the closed path C and the surface S_2 gives

$$\oint \vec{H} \cdot \vec{dl} = \int_{S_2} \vec{J} \cdot \hat{n}\, ds = 0 \qquad \ldots(2.28)$$

The current density being zero on every point on S_2. Clearly, equations (2.27) and (2.28) are contradictory. Equation (2.27) is correct since it does not involve the capacitor that has been introduced. On the otherhand equation (2.28) must be modified with the inclusion of the effect of the capacitor.

Since the divergence of any curl is zero, we have from equation (2.25)

$$\vec{\nabla} \cdot \left(\vec{\nabla} \times \vec{H}\right) = \vec{\nabla} \cdot \vec{J} = 0$$

This implies that $\vec{\nabla} \cdot \vec{J} = 0$, which is contradictory to the continuity equation. As electric charge is accumulated on the plate of the capacitor contained within the volume enclosed by the surface S_1 and S_2, the continuity equation is

$$\vec{\nabla} \cdot \vec{J} + \frac{\partial \rho}{\partial t} = 0 \qquad \ldots(2.29)$$

where ρ is the charge density on the capacitor plate and is changing with time. Therefore, some quantity must be added to the R.H.S., of equation (2.25) that will be consistent with equation (2.29).

To find the quantity, the electric displacement $\vec{D}$ is related to charge density ρ by

$$\vec{\nabla} \cdot \vec{D} = \rho \qquad \ldots(2.30)$$

Using equations (2.29) and (2.30), we obtain

$$\vec{\nabla} \cdot \vec{J} + \frac{\partial}{\partial t}\left(\vec{\nabla} \cdot \vec{D}\right) = \vec{\nabla} \cdot \left(J + \frac{\partial \vec{D}}{\partial t}\right) = 0$$

If we add $\frac{\partial \vec{D}}{\partial t}$ to the R.H.S. of equation (2.25), then its divergence will satisfy equation (2.29), so that the discrepancy is removed.

With $\frac{\partial \vec{D}}{\partial t}$ included, we have the generalized Ampere's law

$$\vec{\nabla} \times \vec{H} = \vec{J} + \frac{\partial \vec{D}}{\partial t}$$

The quantity $\frac{\partial \vec{D}}{\partial t}$ was first introduced by Maxwell, and is called the displacement current density.

2.4.3 Maxwell's Equations

The following four equations are known as Maxwell's equations concerning electromagnetic theory,

1. $$\vec{\nabla}\cdot\vec{D} = \rho \qquad ...(2.31)$$

2. $$\vec{\nabla}\cdot\vec{B} = 0 \qquad ...(2.32)$$

3. $$\vec{\nabla}\times\vec{E} = -\frac{\partial\vec{B}}{\partial t} \qquad ...(2.33)$$

4. $$\vec{\nabla}\times\vec{H} = \vec{J} + \frac{\partial\vec{D}}{\partial t} \qquad ...(2.34)$$

2.4.4 Maxwell's Equations and its Physical Significance

1. $\vec{\nabla}\cdot\vec{D} = \rho$ or *i.e.*, $\vec{\nabla}\cdot\vec{E} = \dfrac{\rho}{\varepsilon_0}$ [where $\vec{D} = \varepsilon_0\vec{E}$]

This steady state equation represents the differential form of Gauss s law in electrostatics.

It has following significance :

(*i*) This steady state equation is independent of time.
(*ii*) According to this equation, the flux line of electric force depends upon the charge density ρ.
(*iii*) It also signifies that ρ the charge density is scalar quantity. The positive charge acts as a source and negative charge as a sink for electric lines of force.

2. $\vec{\nabla}\cdot\vec{B} = 0$

This steady state equation gives the differential form of Gauss's law in magnetism. The physical signficance of this equation are

(*i*) This steady state equation is independent of time.
(*ii*) According to this equation 'Isolated magnetic poles do not exist'.
(*iii*) This equation signifies that equal number of lines of magnetic force enter and leave a given volume.
(*iv*) Lines of magnetic force are endless *i.e.*, there is no source or sink for lines of magnetic force.

3. $\vec{\nabla}\times\vec{E} = -\dfrac{\partial\vec{B}}{\partial t}$

This time varying equation represents the differential form of Faraday's law of electromagnetic induction.

It has following significance :

(*i*) It is a time dependent equation varying with time.
(*ii*) It gives the relation between the electric field vector $\vec{E}$ and the magnetic field vector $\vec{B}$.
(*iii*) Electric field is produced by changing magnetic flux.
(*iv*) It is a statement of Faraday's and Lenz's law of electromagnetic induction.

4. $\vec{\nabla}\times\vec{B} = \mu_0\left[\vec{J} + \varepsilon_0\dfrac{\partial\vec{E}}{\partial t}\right] = \mu_0\left[\vec{J} + \dfrac{\partial\vec{D}}{\partial t}\right]$

This equation represents the differential form of Ampere's law for steady as well as varying currents.

It has following significance :

(*i*) It is a time dependent equation and relates the space variation of $\vec{B}$ with time variation of $\vec{D}$ or $\vec{E}$.

(*ii*) It relates the magnetic field vector $\vec{B}$ with displacement vector $\vec{D}$ and current density vector $\vec{J}$.

(*iii*) It is a statement of Ampere's law.

(*iv*) Conduction current $\left(\vec{J}\right)$ as well as displacement current generates magnetic field $\left(\vec{B}\right)$ jointly as well as separately.

2.4.5 Maxwell's Wave Equation and its Solution

We shall see that as a consequence of Maxwell's equations the electric and magnetic fields can travel through the space in the form of waves. Such waves are called electromagnetic waves.

We assume that medium is linear *i.e.*, permittivity ε, the permeability μ and the electrical conductivity σ_c are constants. The wave equation for the magnetic intensity $\vec{H}$ is obtained by taking the curl of equation (2.34)

$$\vec{\nabla} \times \left(\vec{\nabla} \times \vec{H}\right) = \vec{\nabla} \times \vec{J} + \vec{\nabla} \times \frac{\partial \vec{D}}{\partial t} \qquad \ldots(2.35)$$

Since the conduction current density is $\vec{J} = \sigma_c \vec{E}$ and the electric displacement $\vec{D} = \varepsilon \vec{E}$.

We get from equation (2.35)

$$\therefore \qquad \vec{\nabla} \times \left(\vec{\nabla} \times \vec{H}\right) = \sigma_c\, \vec{\nabla} \times \vec{E} + \varepsilon \frac{\partial}{\partial t}\left(\vec{\nabla} \times \vec{E}\right) \qquad \ldots(2.36)$$

where the space and time derivatives being independent, have been interchanged. Substituting for $\vec{\nabla} \times \vec{E}$ from (2.36) and noting that $\vec{B} = \mu \vec{H}$, we have

$$\vec{\nabla} \times \left(\vec{\nabla} \times \vec{H}\right) = -\mu\, \sigma_c \frac{\partial \vec{H}}{\partial t} - \mu\varepsilon \frac{\partial^2 \vec{H}}{\partial t^2}$$

Using the vector identity,

$$\vec{\nabla} \times \left(\vec{\nabla} \times \vec{A}\right) = \vec{\nabla}\left(\vec{\nabla} . \vec{A}\right) - \nabla^2 \vec{A}$$

and identifying $\vec{A}$ with $\vec{H}$, we obtain from equation (2.36)

$$\vec{\nabla}\left(\vec{\nabla} . \vec{H}\right) - \nabla^2 \vec{H} = -\sigma_c \mu \frac{\partial \vec{H}}{\partial t} - \varepsilon \mu \frac{\partial^2 \vec{H}}{\partial t^2} \qquad \ldots(2.37)$$

Now $\vec{\nabla}.\vec{H} = \frac{1}{\mu}\vec{\nabla}.\vec{B} = 0$ from equation (2.36). Hence, equation (2.37) yields

$$\nabla^2 \vec{H} - \varepsilon\mu \frac{\partial^2 \vec{H}}{\partial t^2} - \sigma_c \mu \frac{\partial \vec{H}}{\partial t} = 0 \qquad \text{...(2.38)}$$

which is the desired wave equation for $\vec{H}$.

The wave equation for $\vec{E}$ is derived by taking the curl of equation (2.32)

$$\vec{\nabla}\times\left(\vec{\nabla}\times\vec{E}\right) = -\vec{\nabla}\times\frac{\partial\vec{B}}{\partial t} \qquad \text{...(2.39)}$$

Noting that $\vec{B} = \mu\vec{H}$, using equation (2.34) and putting $\vec{J} = \sigma_c\vec{E}$ and $\vec{D} = \varepsilon\vec{E}$, we obtain from equation (2.39)

$$\vec{\nabla}\times\left(\vec{\nabla}\times\vec{E}\right) = -\mu\sigma_c\frac{\partial\vec{E}}{\partial t} - \varepsilon\mu\frac{\partial^2\vec{E}}{\partial t^2}$$

and

$$\nabla\left(\vec{\nabla}.\vec{E}\right) - \nabla^2\vec{E} = -\sigma_c\mu\frac{\partial\vec{E}}{\partial t} - \varepsilon\mu\frac{\partial^2\vec{E}}{\partial t^2} \qquad \text{...(2.40)}$$

If the medium contains no free charge, $\rho = 0$, so that $\vec{\nabla}.\vec{E} = \left(\frac{1}{\varepsilon}\right)\vec{\nabla}.\vec{D} = 0$ from equation (2.31). Hence, equation (2.40) reduces to

$$\nabla^2\vec{E} - \varepsilon\mu\frac{\partial^2\vec{E}}{\partial t^2} - \sigma_c\mu\frac{\partial\vec{E}}{\partial t} = 0 \qquad \text{...(2.41)}$$

which is the wave equation for $\vec{E}$, note that $\vec{H}$ and $\vec{E}$ satisfy the same wave equation.

Here the conductivity σ_c vanishes. Hence the wave equations (2.38) and (2.40) reduces to

$$\nabla^2\vec{H} = \varepsilon\mu\frac{\partial^2\vec{H}}{\partial t^2} \qquad \text{...(2.42)}$$

and

$$\nabla^2\vec{E} = \varepsilon\mu\frac{\partial^2\vec{E}}{\partial t^2} \qquad \text{...(2.43)}$$

Again, we know that the wave equation is given

$$\frac{\partial^2 y}{\partial x^2} = \frac{1}{c^2}\frac{\partial y^2}{\partial t^2} \qquad \text{...(2.44)}$$

where c = velocity of light

Comparing equations (2.42) and (2.43) with (2.44), we obtain

$$\frac{1}{c^2} = \mu\varepsilon$$

$$c = \frac{1}{\sqrt{\mu\varepsilon}}$$

In free space $\mu = \mu_o = 4\pi \times 10^{-7}$ kg m/Coulomb^2

and $\varepsilon_0 = 8.86 \times 10^{-12}$ farad/m

$\therefore$ $c = 3 \times 10^8$ m/sec

IN FOCUS

- The vector differential operator del is defined as

$$\vec{\nabla} = \frac{\partial}{\partial x}\hat{i} + \frac{\partial}{\partial y}\hat{j} + \frac{\partial}{\partial z}\hat{k}$$

The gradient of scalar field n

$$\vec{\nabla}n = \frac{\partial n}{\partial x}\hat{i} + \frac{\partial n}{\partial y}\hat{j} + \frac{\partial n}{\partial z}\hat{k}$$

where n is a differentiable scalar field.

- The divergence of vector field

$$\vec{\nabla}.\vec{V} = \frac{\partial V_1}{\partial x} + \frac{\partial V_2}{\partial y} + \frac{\partial V_3}{\partial z}$$

where $\vec{V} = \hat{i}V_1 + \hat{j}V_2 + \hat{k}V_3$

The divergence of a vector field at any point is defined as the amount of flux per unit volume diverging from the point. The amount of flux is essentially a scalar.

If $\vec{\nabla}.\vec{V} = 0$, the vector field is called 'solenoidal' incase of fluid, it is incompressible.

- **Gauss law of Electrostatics :** In S.I unit, the total electric flux (ϕ_E) in free space (or vacuum) through a closed electric surface is equal to $\frac{1}{\varepsilon_0}$ times the charges (q) enclosed by the surface

$$\phi_E = \oint\!\!\oint \vec{E}.\vec{ds} = \frac{q}{\varepsilon_0}$$

It is also known as integral form of Gauss's law. The differential form of Gauss's law :

$$\vec{\nabla}.\vec{E} = \frac{\rho}{\varepsilon_0}$$

- **Laplace's equation :** $\nabla^2 V = 0$

or $$\frac{\partial^2 V}{\partial x^2} + \frac{\partial^2 V}{\partial y^2} + \frac{\partial^2 V}{\partial z^2} = 0$$

- **Poisson's equation :** $\nabla^2 V = -\frac{\rho}{\varepsilon_0}$

- **The equation of continuity :** $\vec{\nabla}.\vec{j} + \frac{\partial \rho}{\partial t} = 0$

- **The curl of a vector field :** $\vec{\nabla} \times \vec{V} = \begin{vmatrix} \hat{i} & \hat{j} & \hat{k} \\ \frac{\partial}{\partial x} & \frac{\partial}{\partial y} & \frac{\partial}{\partial z} \\ V_1 & V_2 & V_3 \end{vmatrix}$

where = $\vec{V} = V_1\hat{i} + V_2\hat{j} + V_3\hat{k}$

The curl of a vector field at any point is defined as the maximum line integral of the vector computed per unit area along the boundary of an infinitesimal area at that point. It is essentially a vector. In case of hydrodynamics, it is sensed as rotation of a fluid. **If $\vec{\nabla} \times \vec{V} = 0$ is called Irrotational** or **conservative field.**

- **Stoke's Theorem :** $\oint_C \vec{A}.\vec{dr} = \iint_S \left(\vec{\nabla} \times \vec{A}\right).\vec{ds}$

- **Ampere's Circuital law :** The line integral of a magnetic induction field vector $\left(\vec{B}\right)$ along any closed path is equal to μ_0 times the net current (I) enclosing the path,

$$\oint \vec{B}.\vec{dl} = \mu_0 I$$

- **Ampere's law in differential form :** $\vec{\nabla} \times \vec{B} = \mu_0 \vec{j}$

where $\vec{j}$ is the current density.

- **Integral form of Faraday's law :** $e = \frac{d\phi}{dt} = -\iint_S \frac{\partial \vec{B}}{\partial t}.\vec{ds}$

where e is induced e.m.f and ϕ is the magnetic flux linked with the circuit.

- **Differential form of Faraday's Law :**

$$\vec{\nabla} \times \vec{E} = -\frac{\partial \vec{B}}{\partial t} \text{ or curl } \vec{E} = -\frac{\partial \vec{B}}{\partial t}$$

$\vec{E}$ = electric field and $\vec{B}$ = magnetic field

- **The generalized Ampere's law :** $\vec{\nabla} \times \vec{H} = \vec{j} + \frac{\partial \vec{D}}{\partial t}$

where $\frac{\partial \vec{D}}{\partial t}$ is called the displacement current density.

- **Maxwell's equations :**

1. $\vec{\nabla}.\vec{D} = \rho$ or, $\vec{\nabla}.\vec{E} = \frac{\rho}{\varepsilon_0}$ is differential form of Gauss's law in electrostatics.
2. $\vec{\nabla}.\vec{B} = 0$ is differential form of Gauss's law in magnetism. It also implies that the Isolated magnetic poles do not exist.
3. $\vec{\nabla} \times \vec{E} = -\frac{\partial \vec{B}}{\partial t}$ is differential form of Faraday's law in electromagnetic Induction.
4. $\vec{\nabla} \times \vec{E} = \mu_0 \left[\vec{J} + \frac{\partial D}{\partial t}\right]$ represents the differential form of Ampere's law for steady as well as varying currents.

- **Wave Equations :** $\nabla^2\vec{H} = \varepsilon_0\mu_0 \frac{\partial^2 \vec{H}}{\partial t^2}$

$$\nabla^2\vec{E} = \varepsilon_0\mu_0 \frac{\partial^2 \vec{E}}{\partial t^2}$$

Solved Problems

Problem 1. *The force $\vec{F}$ on an electron of charge – e, moving with a velocity $\vec{V}$ in the presence of an electric field $\vec{E}$ and magnetic field $\vec{B}$ is given by $\vec{F} = -e\left[\vec{E} + \vec{V} \times \vec{B}\right]$*

If $\vec{B} = \hat{k}\, B$, write down the components of $\vec{F}$ along the axes of x, y and z.

Solution. Suppose $\vec{E}$, $\vec{V}$ and $\vec{B}$ in terms of their components are

$$\vec{E} = E_x \hat{i} + E_y \hat{j} + E_z \hat{k},$$

$$\vec{V} = \hat{i}\, V_x + \hat{j}\, V_y + \hat{k}\, V_z$$

and $$\vec{B} = \hat{k}\, B.$$

Then, we have $$\vec{V} \times \vec{B} = \begin{vmatrix} \hat{i} & \hat{j} & \hat{k} \\ V_x & V_y & V_z \\ 0 & 0 & B \end{vmatrix} = \hat{i}\, V_y B - \hat{j}\, V_x B$$

The expression $$\vec{F} = -e\left[\left(E_x \hat{i} + E_y \hat{j} + E_z \hat{k}\right) + \left(\hat{i}\, V_y B - \hat{j}\, V_x B\right)\right]$$

$$= -e\left[\hat{i}\left(E_x + V_y B\right) + \hat{j}\left(E_y - V_x B\right) + \hat{k}\, E_z\right]$$

Hence the components of $\vec{F}$ are

$$-e\,(E_x + V_y B),\ -e(E_y - V_x B),\ -eE_z.$$

Problem 2. *If $\vec{P} = \hat{i} + 3\hat{j} + 5\hat{k}$; $\vec{Q} = -\hat{i} + 3\hat{j} - 4\hat{k}$ and $\vec{R} = -\hat{i} - 2\hat{j} + \hat{k}$, find $\left(\vec{P} \times \vec{Q}\right) \cdot \vec{R}$ and $\left(\vec{P} \times \vec{Q}\right) \times \vec{R}$.*

Solution. Here, $$\vec{P} \times \vec{Q} = \begin{vmatrix} \hat{i} & \hat{j} & \hat{k} \\ 1 & 3 & 5 \\ -1 & 3 & -4 \end{vmatrix} = \hat{i}\,(-12 - 15) - \hat{j}\,(-4 + 5) + \hat{k}\,(3 + 3)$$

$$= -27\hat{i} - \hat{j} + 6\hat{k}$$

∴ $$\left(\vec{P} \times \vec{Q}\right) \cdot \vec{R} = (-27\hat{i} - \hat{j} + 6\hat{k}) \cdot (-\hat{i} - 2\hat{j} + \hat{k})$$

$$= 27 + 2 + 6 = 35.$$

Now $$\left(\vec{P} \times \vec{Q}\right) \times \vec{R} = \begin{vmatrix} \hat{i} & \hat{j} & \hat{k} \\ -27 & -1 & 6 \\ -1 & -2 & 1 \end{vmatrix}$$

$$= \hat{i}\ [-1 + 12] - \hat{j}\ [-27 + 6] + \hat{k}\ [54 - 1]$$

$$= 11\hat{i} + 21\hat{j} + 53\hat{k}\,.$$

Problem 3. *Prove the following relation in the vector triple product:*

$$\vec{A} \times \left(\vec{B} \times \vec{C}\right) = \vec{B}\left(\vec{A}\,.\,\vec{C}\right) - \vec{C}\left(\vec{A}\,.\,\vec{B}\right)$$

Solution. We have $\vec{A} \times \left(\vec{B} \times \vec{C}\right)$

$$= \left(A_x \hat{i} + A_y \hat{j} + A_z \hat{k}\right) \times \begin{vmatrix} \hat{i} & \hat{j} & \hat{k} \\ B_x & B_y & B_z \\ C_x & C_y & C_z \end{vmatrix}$$

$$= \left(A_x \hat{i} + A_y \hat{j} + A_z \hat{k}\right) \times \left[\hat{i}\left(B_yC_z - C_yB_z\right) - \hat{j}\ (B_xC_z - B_zC_x) + \hat{k}\left(B_xC_y - B_yC_x\right)\right]$$

$$= \begin{vmatrix} \hat{i} & \hat{j} & \hat{k} \\ A_x & A_y & A_z \\ \left(B_yC_z - C_yB_z\right) & -(B_xC_z - B_zC_x) & \left(B_xC_y - B_yC_x\right) \end{vmatrix}$$

$$= \hat{i}\ B_x\,(A_yC_y + A_zC_z) - \hat{i}\ C_x\,(A_yB_y + A_zB_z) + \hat{j}\ B_y\,(A_xC_x + A_zC_z) - \hat{j}\ C_y\,(A_xB_x + A_zB_z) +$$
$$\hat{k}\ B_z\,(A_xC_x + A_yC_y) - \hat{k}\ C_z\,(A_xB_x + A_yB_y)$$

$$= \hat{i}\ B_x\,(A_yC_y + A_zC_z + A_xC_x) - \hat{i}\ C_x\,(A_yB_y + A_zB_z + A_yB_y) - \hat{j}\ B_y\,(A_xC_x + A_zC_z + A_yC_y) -$$
$$\hat{j}\ C_z\,(A_xB_x + A_yB_y + A_zB_z) + \hat{k}\ B_z\,(A_xC_x + A_yC_y + A_zC_z) - \hat{k}\ C_z\,(A_xB_x + A_yB_y + A_zBz)$$

$$= \left(\hat{i}\,B_x + \hat{j}\,B_y + \hat{k}\,B_z\right)(A_xC_x + A_yC_y + A_zC_z) - \left(\hat{i}\,C_x + \hat{j}\,C_y + \hat{k}\,C_z\right)(A_xB_x + A_yB_y + A_zB_z)$$

$$= \vec{B}\left(\vec{A}\,.\,\vec{C}\right) - \vec{C}\left(\vec{A}\,.\,\vec{B}\right)$$

It may also be expressed in the determinant form as

$$\vec{A} \times \left(\vec{B} \times \vec{C}\right) = \begin{vmatrix} \vec{B} & \vec{C} \\ \left(\vec{A}\,.\,\vec{B}\right) & \left(\vec{A}\,.\,\vec{C}\right) \end{vmatrix}$$

Problem 4. *Prove that* $\left(\vec{A} \times \vec{B}\right) \times \vec{C} = \vec{B}\left(\vec{A}\,.\,\vec{C}\right) - \vec{A}\left(\vec{B}\,.\,\vec{C}\right)$

Given $\vec{A} \times \left(\vec{B} \times \vec{C}\right) = \vec{B}\left(\vec{A}\,.\,\vec{C}\right) - \vec{C}\left(\vec{A}\,.\,\vec{B}\right)$

Solution. $\vec{A}\times\left(\vec{B}\times\vec{C}\right) = \begin{vmatrix} \vec{B} & \vec{C} \\ (\vec{A}.\vec{B}) & (\vec{A}.\vec{C}) \end{vmatrix}$

Here $\left(\vec{A}\times\vec{B}\right)\times\vec{C} = -\vec{C}\times\left(\vec{A}\times\vec{B}\right)$

$$= -\begin{vmatrix} \vec{A} & \vec{B} \\ (\vec{C}.\vec{A}) & (\vec{C}.\vec{B}) \end{vmatrix}$$

$$= -\left[\vec{A}\left(\vec{C}.\vec{B}\right) - \vec{B}\left(\vec{C}.\vec{A}\right)\right]$$

$$= \vec{B}\left(\vec{C}.\vec{A}\right) - \vec{A}\left(\vec{C}.\vec{B}\right)$$

$$= \vec{B}\left(\vec{A}.\vec{C}\right) - \vec{A}\left(\vec{B}.\vec{C}\right)$$ **(Proved)**

Problem 5. *Prove :* $\vec{A}\times\left(\vec{B}\times\vec{C}\right)+\vec{B}\times\left(\vec{C}\times\vec{A}\right)+\vec{C}\times\left(\vec{A}\times\vec{B}\right) = 0$ *symbols have usual meaning.*

[WBUT]

Solution. Here $\vec{A}\times\left(\vec{B}\times\vec{C}\right)+\vec{B}\times\left(\vec{C}+\vec{A}\right)+\vec{C}\times\left(\vec{A}\times\vec{B}\right)$

$$= \begin{vmatrix} \vec{B} & \vec{C} \\ (\vec{A}.\vec{B}) & (\vec{A}.\vec{C}) \end{vmatrix} + \begin{vmatrix} \vec{C} & \vec{A} \\ (\vec{B}.\vec{C}) & (\vec{B}.\vec{A}) \end{vmatrix} + \begin{vmatrix} \vec{A} & \vec{B} \\ (\vec{C}.\vec{A}) & (\vec{C}.\vec{B}) \end{vmatrix}$$

$$= \vec{B}\left(\vec{A}.\vec{C}\right)-\vec{C}\left(\vec{A}.\vec{B}\right)+\vec{C}\left(\vec{A}.\vec{B}\right)-\vec{A}\left(\vec{B}.\vec{C}\right)+\vec{A}\left(\vec{C}.\vec{B}\right)-\vec{B}\left(\vec{A}.\vec{C}\right)$$

$= 0$ **(Proved)**.

Problem 6. *Prove* $\left(\vec{A}\times\vec{B}\right).\left(\vec{C}\times\vec{D}\right) = \left(\vec{A}.\vec{C}\right)\left(\vec{B}.\vec{D}\right)-\left(\vec{A}.\vec{D}\right)\left(\vec{B}.\vec{C}\right)$ [Agra University]

Solution. Let $\left(\vec{A}\times\vec{B}\right) = \vec{X}$,

then we may write

$$\vec{X}.\left(\vec{C}\times\vec{D}\right) = \left(\vec{X}\times\vec{C}\right).\vec{D}$$

$$= \left\{\left(\vec{A}\times\vec{B}\right)\times\vec{C}\right\}.\vec{D} \qquad \left[\text{Putting } \vec{X} = \vec{A}\times\vec{B}\right]$$

$$= -\left\{\vec{C}\times\left(\vec{A}\times\vec{B}\right)\right\}.\vec{D}$$

$$= - \begin{vmatrix} \vec{A} & \vec{B} \\ (\vec{C}.\vec{A}) & (\vec{C}.\vec{B}) \end{vmatrix} . \vec{D}$$

$$= \left\{\vec{B}\left(\vec{C}.\vec{A}\right) - \vec{A}\left(\vec{C}.\vec{B}\right)\right\}.\vec{D}$$

$$= \left(\vec{A}.\vec{C}\right)\left(\vec{B}.\vec{D}\right) - \left(\vec{A}.\vec{D}\right)\left(\vec{C}.\vec{B}\right) \textbf{ (Proved)}$$

Problem 7. *Prove that* $\left(\vec{A}\times\vec{B}\right)\times\left(\vec{C}\times\vec{D}\right) = \vec{C}\left(\vec{A}.\vec{B}\times\vec{D}\right) - \vec{D}\left(\vec{A}.\vec{B}\times\vec{C}\right)$ [Agra University]

Solution. Let $\vec{A}\times\vec{B} = \vec{P}$

$$\therefore \quad \vec{P}\times\left(\vec{C}\times\vec{D}\right) = \begin{vmatrix} \vec{C} & \vec{D} \\ (\vec{P}.\vec{C}) & (\vec{P}.\vec{D}) \end{vmatrix} = \vec{C}\left(\vec{P}.\vec{D}\right) - \vec{D}\left(\vec{P}.\vec{C}\right)$$

$$= \vec{C}\left(\vec{A}\times\vec{B}.\vec{D}\right) - \vec{D}\left(\vec{A}\times\vec{B}.\vec{C}\right)$$

$$= \vec{C}\left(\vec{A}.\vec{B}\times\vec{D}\right) - \vec{D}\left(\vec{A}.\vec{B}\times\vec{C}\right)$$

Problem 8. *Prove that* $\left(\vec{A}\times\vec{B}\right).\left(\vec{B}\times\vec{C}\right)\times\left(\vec{C}\times\vec{A}\right) = \left(\vec{A}.\vec{B}\times\vec{C}\right)^2$

Solution.

$$\text{L.H.S.} = \left(\vec{A}\times\vec{B}\right).\left(\vec{B}\times\vec{C}\right)\times\left(\vec{C}\times\vec{A}\right)$$

$$= \left(\vec{A}\times\vec{B}\right).\left\{\vec{X}\times\left(\vec{C}\times\vec{A}\right)\right\} \qquad \left[\because \vec{X} = \vec{B}\times\vec{C}\right]$$

$$= \left(\vec{A}\times\vec{B}\right).\begin{vmatrix} \vec{C} & \vec{A} \\ (\vec{X}.\vec{C}) & (\vec{X}.\vec{A}) \end{vmatrix}$$

$$= \left(\vec{A}\times\vec{B}\right).\left\{\vec{C}\left(\vec{X}.\vec{A}\right) - \vec{A}\left(\vec{X}.\vec{C}\right)\right\}$$

$$= \left(\vec{A}\times\vec{B}\right).\left\{\vec{C}\left\{\left(\vec{B}\times\vec{C}\right).\vec{A}\right\} - \vec{A}\left(\vec{E}\times\vec{C}.\vec{C}\right)\right\}$$

$$= \left\{\left(\vec{A}\times\vec{B}\right).\vec{C}\right\}\left\{\vec{A}.\left(\vec{B}\times\vec{C}\right)\right\} - 0 \qquad \left[\vec{X} = \vec{B}\times\vec{C}\right]$$

$$= \left(\vec{A}.\vec{B}\times\vec{C}\right)\left(\vec{A}.\vec{B}\times\vec{C}\right) \qquad \text{[Dot Product is Commutative]}$$

$$= \left(\vec{A}.\vec{B}\times\vec{C}\right)^2 \textbf{ (Proved)}$$

Problem 9. *Prove that* $\frac{d}{dt}\left[\vec{V}\cdot\frac{d\vec{V}}{dt}\times\frac{d^2\vec{V}}{dt^2}\right] = \vec{V}\cdot\frac{d\vec{V}}{dt}\times\frac{d^3\vec{V}}{dt^3}$

Solution.

$$\text{L.H.S.} = \frac{d}{dt}\left(\vec{V}\cdot\frac{d\vec{V}}{dt}\times\frac{d^2\vec{V}}{dt^2}\right)$$

$$= \vec{V}\cdot\frac{d\vec{V}}{dt}\times\frac{d^3\vec{V}}{dt^3} + \vec{V}\cdot\frac{d^2\vec{V}}{dt^2}\times\frac{d^2\vec{V}}{dt^2} + \frac{d\vec{V}}{dt}\cdot\frac{d\vec{V}}{dt}\times\frac{d^2\vec{V}}{dt^2}$$

$$= \vec{V}\cdot\frac{d\vec{V}}{dt}\times\frac{d^3\vec{V}}{dt^3} + 0 + 0 \qquad \left[\because \begin{array}{l}\vec{A}\cdot\vec{B}\times\vec{B} = 0\\ \vec{A}\cdot\vec{A}\times\vec{B} = 0\end{array}\right]$$

= R.H.S. **(Proved)**

where $\vec{A} = \frac{d\vec{V}}{dt}$

$\vec{B} = \frac{d^2\vec{V}}{dt^2}$

Problem 10. *Prove that* $\vec{A}\times\frac{d^2\vec{B}}{dt^2} - \frac{d^2\vec{A}}{dt^2}\times\vec{B} = \frac{d}{dt}\cdot\left(\vec{A}\times\frac{d\vec{B}}{dt} - \frac{d\vec{A}}{dt}\times\vec{B}\right)$

Solution.

$$\text{R.H.S.} = \frac{d}{dt}\left(\vec{A}\times\frac{d\vec{B}}{dt} - \frac{d\vec{A}}{dt}\times\vec{B}\right)$$

$$= \left\{\vec{A}\times\frac{d^2\vec{B}}{dt^2} + \frac{d\vec{A}}{dt}\times\frac{d\vec{B}}{dt}\right\} - \left\{\frac{d\vec{A}}{dt}\times\frac{d\vec{B}}{dt} + \frac{d^2\vec{A}}{dt^2}\times\vec{B}\right\}$$

$$= \vec{A}\times\frac{d^2\vec{B}}{dt^2} - \frac{d^2\vec{A}}{dt^2}\times\vec{B} \text{ (Proved)}$$

Problem 11. *Prove that* $\vec{A}\cdot\frac{d\vec{A}}{dt} = A\,\frac{dA}{dt}$

Solution. We know that, $\frac{d}{dt}\left(\vec{A}\cdot\vec{A}\right) = \vec{A}\cdot\frac{d\vec{A}}{dt} + \frac{d\vec{A}}{dt}\cdot\vec{A}$

$$= 2\vec{A}\cdot\frac{d\vec{A}}{dt} \qquad \left[\because \vec{X}\cdot\vec{Y} = \vec{Y}\cdot\vec{X}\right]$$

and $$\frac{d}{dt}\,(A^2) = 2A\,\frac{dA}{dt}.$$

So we can write

$$\frac{d}{dt}\left(\vec{A}\,.\,\vec{A}\right) = \frac{d}{dt}\,(A^2)$$

or,
$$2\vec{A}\,.\,\frac{d\vec{A}}{dt} = 2A\,\frac{dA}{dt}$$

∴
$$\vec{A}\,.\,\frac{d\vec{A}}{dt} = A\,\frac{dA}{dt} \quad \textbf{(Proved)}$$

Problem 12. *Find the workdone W in moving a particle in a force field given by*

$$\vec{F} = 3xy\,\hat{i} - 5z\,\hat{j} + 10x\,\hat{k}$$

along the curve
$$x = t^2 + 1,\quad y = 2t^2,\quad z = t^3$$

from t = 1 to t = 2.

Solution. Workdone $= \int_C \vec{F}\,.\,\vec{dr} = \int_C \left(3xy\,\hat{i} - 5z\,\hat{j} + 10x\,\hat{k}\right)\left(\hat{i}\,dx + \hat{j}\,dy + \hat{k}\,dz\right)$

$$= \int_C (3xy\,dx - 5z\,dy + 10x\,dz)$$

$$= \int_1^2 \left\{3\left(t^2+1\right)2t^2\,.\,(2t\,dt) - 5t^3\,(4t\,dt) + 10\left(t^2+1\right)\left(3t^2\,dt\right)\right\}$$

$$\left[\begin{array}{lllllll} \because & x & = & t^2+1 & \text{or} & dx & = 2t\,dt, \\ & y & = & 2t^2 & \text{or} & dy & = 4t\,dt, \\ \text{and} & z & = & t^3 & \text{or} & dz & = 3t^2\,dt \end{array}\right]$$

$$= \int_1^2 \left(12t^5 + 10t^4 + 30t^2 + 12t^3\right) dt$$

$$= \left[12\,\frac{t^6}{6} + 10\,\frac{t^5}{5} + 30\,\frac{t^3}{3} + 12\,\frac{t^4}{4}\right]_1^2$$

$$= [(128 + 64 + 80 + 48) - (2 + 2 + 10 + 3)]$$

$$= 320 - 17 = 303.$$

Problem 13. *Evaluate* $\int_C \vec{F}\,.\,\vec{dr}$ *where* $\vec{F} = xy\,\hat{i} + (x^2 + y^2)\,\hat{j}$ *and C is the arc of the curve,* $y = x^2 - 4$ *from (2, 0) to (4, 12) in the x-y plane.*

Solution. $\int_C \vec{F}\,.\,\vec{dr} = \int_C \left\{xy\,\hat{i} + \left(x^2+y^2\right)\hat{j}\right\}\left\{dx\,\hat{i} + dy\,\hat{j}\right\}$

$$= \int_2^4 \left\{xydx + (x^2 + y^2)\,dy\right\}$$

$$= \int_2^4 \left[x\left(x^2-4\right)dx + \left\{x^2 + \left(x^2-4\right)^2\right\}(2x\,dx)\right] \quad \left[\begin{array}{lll} \because\ y & = & x^2 - 4 \\ \therefore\ dy & = & 2x\,dx \end{array}\right]$$

$$= \int_2^4 \left\{x^3 - 4x + 2x^3 + 2x^5 - 16x^3 + 32x\right\} dx$$

$$= \int_2^4 \left\{2x^5 - 13x^3 + 28x\right\} dx$$

$$= \left[2 \cdot \frac{x^6}{6} - 13 \cdot \frac{x^4}{4} + 28 \cdot \frac{x^2}{2}\right]_2^4$$

$$= 732.$$

Problem 14. *If $\phi(x, y, z) = 3x^2y - y^3z^2$, find the value of grad ϕ at the point (1, – 2, – 1).* [WBUT]

Solution. We know that grad $\phi = \vec{\nabla}\phi = \left(\hat{i}\frac{\partial}{\partial x} + \hat{j}\frac{\partial}{\partial y} + \hat{k}\frac{\partial}{\partial z}\right)\phi$

$$= \hat{i}\frac{\partial \phi}{\partial x} + \hat{j}\frac{\partial \phi}{\partial y} + \hat{k}\frac{\partial \phi}{\partial z}$$

$$= \hat{i}\frac{\partial}{\partial x}\left(3x^2y - y^3z^2\right) + \hat{j}\frac{\partial}{\partial y}\left(3x^2y - y^3z^2\right) + \hat{k}\frac{\partial}{\partial z}\left(3x^2y - y^3z^2\right)$$

$$= \hat{i}\, 6xy + \hat{j}\,(3x^2 - 3y^2z^2) + \hat{k}\,(-2y^3z)$$

$$\vec{\nabla}\phi\Big|_{(1,-2,-1)} = -12\hat{i} - 9\hat{j} - 16\hat{k}$$

Problem 15. *Equipotential surface in some problems are spherical in shape : $V(x, y, z) = C(x^2 + y^2 + z^2)$. Show that the force is radial i.e., along the radius vector.*

Solution. We know that the force $\vec{F} = -\vec{\nabla}V$

i.e., The force at a point is given by the negative potential gradient at that point.

$$\vec{F} = -\vec{\nabla}V = -\frac{\partial V}{\partial x}\hat{i} - \frac{\partial V}{\partial y}\hat{j} - \frac{\partial V}{\partial z}\hat{k}$$

Here $V = C(x^2 + y^2 + z^2)$

$$\vec{F} = -2C\left(x\hat{i} + y\hat{j} + z\hat{k}\right)$$

$$= -2C\vec{r} \text{ [where } \vec{r} = \hat{i}x + \hat{j}y + \hat{k}z = \text{radius vector]}$$

$\therefore$ $\vec{F} \propto -\vec{r}$ C is some constant.

This indicates that the force is radial.

Problem 16. *The electric field at any point (x, y, z) is given by $\vec{E} = 2x\hat{i} + 2y\hat{j} + 2z\hat{k}$ Derive an equation for the equipotential surface. What is such surface called?*

Solution. We know that the electric field at any point

$$\vec{E} = -\vec{\nabla}V = -\left(\frac{\partial V}{\partial x}\hat{i} + \frac{\partial V}{\partial y}\hat{j} + \frac{\partial V}{\partial z}\hat{k}\right) \quad \ldots(i)$$

We have $\vec{E} = 2x\hat{i} + 2y\hat{j} + 2z\hat{k}$ ...(*ii*)

Comparing equations (*i*) and (*ii*), we get

$$\frac{\partial V}{\partial x} = -2x,\ \frac{\partial V}{\partial y} = -2y,\ \frac{\partial V}{\partial z} = -2z$$

By a Theorem of partial differentiation, we have

$$dV = \frac{\partial V}{\partial x}dx + \frac{\partial V}{\partial y}dy + \frac{\partial V}{\partial z}dz$$

By Integration,

$$V = \int \frac{\partial V}{\partial x}dx + \int \frac{\partial V}{\partial y}dy + \int \frac{\partial V}{\partial z}dz$$

$$= \int(-2x)\,dx + \int(-2y)\,dy + \int(-2z)\,dz$$

$$= -(x^2 + y^2 + z^2) + \text{constant.}$$

This is a spherical surface.

Problem 17. *If $\vec{r}$ is the position vector of a point, then evaluate (i) grad r ; (ii) grad $\left(\frac{1}{r}\right)$; and (iii) grad (log r)*

Solution. We know that

$$\vec{r} = \hat{i}x + \hat{j}y + \hat{k}z$$

$$\therefore \quad r = \sqrt{x^2 + y^2 + z^2} = \left(x^2 + y^2 + z^2\right)^{\frac{1}{2}}$$

(*i*)

$$\text{grad } r = \vec{\nabla}r = \hat{i}\frac{\partial}{\partial x}\left(x^2 + y^2 + z^2\right)^{\frac{1}{2}} + \hat{j}\frac{\partial}{\partial y}\left(x^2 + y^2 + z^2\right)^{\frac{1}{2}} + \hat{k}\frac{\partial}{\partial z}\left(x^2 + y^2 + z^2\right)^{\frac{1}{2}}$$

$$= \frac{1}{2}\frac{2x}{\left(x^2 + y^2 + z^2\right)^{\frac{1}{2}}}\hat{i} + \frac{1}{2}\frac{2y}{\left(x^2 + y^2 + z^2\right)^{\frac{1}{2}}}\hat{j} + \frac{1}{2}\frac{2z}{\left(x^2 + y^2 + z^2\right)^{\frac{1}{2}}}\hat{k}$$

$$= \frac{x\hat{i} + y\hat{j} + z\hat{k}}{\left(x^2 + y^2 + z^2\right)^{\frac{1}{2}}} = \frac{\vec{r}}{r}$$

(*ii*)

$$\text{grad}\left(\frac{1}{r}\right) = \vec{\nabla}\left(\frac{1}{r}\right) = \left(\hat{i}\frac{\partial}{\partial x} + \hat{j}\frac{\partial}{\partial y} + \hat{k}\frac{\partial}{\partial z}\right)\left(x^2 + y^2 + z^2\right)^{-\frac{1}{2}}$$

$$= \hat{i}\,\frac{\partial}{\partial x}\left(x^2+y^2+z^2\right)^{-\frac{1}{2}} + \hat{j}\,\frac{\partial}{\partial y}\left(x^2+y^2+z^2\right)^{-\frac{1}{2}} + \hat{k}\,\frac{\partial}{\partial z}\left(x^2+y^2+z^2\right)^{-\frac{1}{2}}$$

$$= -\frac{1}{2}\cdot\frac{2x}{\left(x^2+y^2+z^2\right)^{\frac{3}{2}}}\,\hat{i} - \frac{2y}{2\left(x^2+y^2+z^2\right)^{\frac{3}{2}}}\,\hat{j} - \frac{2z}{2\left(x^2+y^2+z^2\right)^{\frac{3}{2}}}\,\hat{k}$$

$$= -\frac{\left(\hat{i}x+\hat{j}y+\hat{k}z\right)}{\left(x^2+y^2+z^2\right)^{\frac{3}{2}}} = -\frac{\vec{r}}{r^3}$$

(*iii*) grad (log *r*) $= \vec{\nabla}\left(\log\sqrt{x^2+y^2+z^2}\right) = \frac{1}{2}\vec{\nabla}\left\{\log\left(x^2+y^2+z^2\right)\right\}$

$$= \frac{1}{2}\left\{\hat{i}\,\frac{\partial}{\partial x}\log\left(x^2+y^2+z^2\right) + \hat{j}\,\frac{\partial}{\partial y}\log\left(x^2+y^2+z^2\right) + \hat{k}\,\frac{\partial}{\partial z}\log\left(x^2+y^2+z^2\right)\right\}$$

$$= \frac{1}{2}\cdot\left\{\hat{i}\cdot\frac{2x}{\left(x^2+y^2+z^2\right)} + \hat{j}\,\frac{2y}{\left(x^2+y^2+z^2\right)} + \hat{k}\,\frac{2z}{\left(x^2+y^2+z^2\right)}\right\}$$

$$= \frac{\hat{i}x+\hat{j}y+\hat{k}z}{\left(x^2+y^2+z^2\right)}$$

$$= \frac{\vec{r}}{r^2} = \frac{\hat{r}}{r} \qquad \left[\hat{r} = \frac{\vec{r}}{r}\right]$$

Problem 18. *Evaluate* $\vec{\nabla}\times\left(\frac{\vec{r}}{r^2}\right)$.

Solution. Let $\vec{r} = x\hat{i}+y\hat{j}+z\hat{k},\ r = |\vec{r}| = \sqrt{x^2+y^2+z^2}$

$$\vec{\nabla}\times\left(\frac{x\hat{i}+y\hat{j}+z\hat{k}}{x^2+y^2+z^2}\right) = \begin{vmatrix} \hat{i} & \hat{j} & \hat{k} \\ \frac{\partial}{\partial x} & \frac{\partial}{\partial y} & \frac{\partial}{\partial z} \\ \frac{x}{x^2+y^2+z^2} & \frac{y}{x^2+y^2+z^2} & \frac{z}{x^2+y^2+z^2} \end{vmatrix}$$

$$= \hat{i}\left\{\frac{\partial}{\partial y}\left(\frac{z}{x^2+y^2+z^2}\right) - \frac{\partial}{\partial z}\left(\frac{y}{x^2+y^2+z^2}\right)\right\}$$

$$- \hat{j}\left\{\frac{\partial}{\partial x}\left(\frac{z}{x^2+y^2+z^2}\right) - \frac{\partial}{\partial z}\left(\frac{x}{x^2+y^2+z^2}\right)\right\}$$

$$+ \hat{k}\left\{\frac{\partial}{\partial x}\left(\frac{y}{x^2+y^2+z^2}\right) - \frac{\partial}{\partial y}\left(\frac{x}{x^2+y^2+z^2}\right)\right\}$$

$$= \hat{i}\left\{-z\cdot\left(x^2+y^2+z^2\right)^{-2}\cdot 2y + y\cdot\left(x^2+y^2+z^2\right)^{-2}\cdot 2z\right\}$$

$$-\hat{j}\left\{-z\cdot\left(x^2+y^2+z^2\right)^{-2}\cdot 2x + x\cdot\left(x^2+y^2+z^2\right)^{-2}\cdot 2z\right\}$$

$$+\hat{k}\left\{-y\left(x^2+y^2+z^2\right)^{-2}\cdot 2x + x\cdot\left(x^2+y^2+z^2\right)^{-2}\cdot 2y\right\}$$

$$= \mathrm{i}\cdot 0 - j\cdot 0 + k\cdot 0 = 0.$$

Problem 19. *Prove curl (ϕ grad ϕ) = 0*

Solution. Curl (ϕ grad ϕ) $= \vec{\nabla}\cdot\left(\phi\vec{\nabla}\phi\right)$

$$= \vec{\nabla}\times\left\{\phi\left(\frac{\partial\phi}{\partial x}\hat{i}+\frac{\partial\phi}{\partial y}\hat{j}+\frac{\partial\phi}{\partial z}\hat{k}\right)\right\}$$

$$= \vec{\nabla}\times\left\{\phi\frac{\partial\phi}{\partial x}\hat{i}+\phi\frac{\partial\phi}{\partial y}\hat{j}+\phi\frac{\partial\phi}{\partial z}\hat{k}\right\}$$

$$= \begin{vmatrix} \hat{i} & \hat{j} & \hat{k} \\ \frac{\partial}{\partial x} & \frac{\partial}{\partial y} & \frac{\partial}{\partial z} \\ \phi\frac{\partial\phi}{\partial x} & \phi\frac{\partial\phi}{\partial y} & \phi\frac{\partial\phi}{\partial z} \end{vmatrix}$$

$$= \hat{i}\left\{\frac{\partial}{\partial y}\left(\phi\frac{\partial\phi}{\partial z}\right)-\frac{\partial}{\partial z}\left(\phi\frac{\partial\phi}{\partial y}\right)\right\}-\hat{j}\left\{\frac{\partial}{\partial x}\left(\phi\frac{\partial\phi}{\partial z}\right)-\frac{\partial}{\partial z}\left(\phi\frac{\partial\phi}{\partial x}\right)\right\}$$

$$+\hat{k}\left\{\frac{\partial}{\partial x}\left(\phi\frac{\partial\phi}{\partial y}\right)-\frac{\partial}{\partial y}\left(\phi\frac{\partial\phi}{\partial x}\right)\right\}$$

$$= \hat{i}\left\{\phi\frac{\partial^2\phi}{\partial y\cdot\partial z}+\frac{\partial^2\phi}{\partial y\,\partial z}-\phi\frac{\partial^2\phi}{\partial z\cdot\partial y}-\frac{\partial^2\phi}{\partial z\,\partial y}\right\}$$

$$-\hat{j}\left\{\phi\frac{\partial^2\phi}{\partial x\cdot\partial z}+\frac{\partial^2\phi}{\partial x\,\partial z}-\phi\frac{\partial^2\phi}{\partial z\cdot\partial x}-\frac{\partial^2\phi}{\partial z\,\partial x}\right\}$$

$$+\hat{k}\left\{\phi\frac{\partial^2\phi}{\partial x\cdot\partial y}+\frac{\partial^2\phi}{\partial x\,\partial y}-\phi\frac{\partial\phi}{\partial y\cdot\partial x}-\frac{\partial^2\phi}{\partial y\,\partial x}\right\}$$

$$= \hat{i}\cdot 0 - \hat{j}\cdot 0 + \hat{k}\cdot 0 = 0.$$

Problem 20. *If $\vec{A} = x^2z\hat{i} + yz^2\hat{j} - 3xy\hat{k}$, $\vec{B} = y^2\hat{i} - yz\hat{j} + 2x\hat{k}$ and $\phi = 2x^2 + yz$, find (a) $\vec{A}\cdot\left(\vec{\nabla}\phi\right)$, (b) $\left(\vec{A}\cdot\vec{\nabla}\right)\cdot\phi$, (c) $\left(\vec{A}\cdot\vec{\nabla}\right)\cdot\phi$, (d) $\vec{B}\cdot\left(\vec{A}\cdot\vec{\nabla}\right)$, and (e) $\left(\vec{\nabla}\cdot\vec{A}\right)\cdot\vec{B}$.*

$$\vec{\mathrm{A}} = x^2z\hat{i} + yz^2\hat{j} - 3xy\hat{k},\ \vec{\mathrm{B}} = y^2\hat{i} - yz\hat{j} + 2x\hat{k}.$$

$$\phi = 2x^2 + yz.$$

Solution.

(a) $$\vec{A}\cdot\left(\vec{\nabla}\phi\right) = \vec{A}\left\{\frac{\partial}{\partial x}\left(2x^2+yz\right)\hat{i}+\frac{\partial}{\partial y}\left(2x^2+yz\right)\hat{j}+\frac{\partial}{\partial z}\left(2x^2+yz\right)\hat{k}\right\}$$

$$= \vec{A}\cdot\left(4x\hat{i}+z\hat{j}+y\hat{k}\right)=\left(x^2z\hat{i}+yz^3\hat{j}-3xy\hat{k}\right)\cdot\left(4x\hat{i}+z\hat{j}+y\hat{k}\right)$$

$$= 4xz^3+yz^4-3xy^2.$$

(b) $$\left(\vec{A}\cdot\vec{\nabla}\right)\phi = \left\{\left(x^2z\hat{i}+yz^3\hat{j}-3xy\hat{k}\right)\cdot\left(\frac{\partial}{\partial x}\hat{i}+\frac{\partial}{\partial y}\hat{j}+\frac{\partial}{\partial z}\hat{k}\right)\right\}\phi$$

$$= \left(x^2z\frac{\partial}{\partial x}+yz^3\frac{\partial}{\partial y}-3xy\frac{\partial}{\partial z}\right)\phi = x^2z\frac{\partial}{\partial x}\left(2x^2+yz\right)+ yz^3\frac{\partial}{\partial y}\left(2x^2+yz\right) - 3xy\frac{\partial}{\partial z}\left(2x^2+yz\right)$$

$$= x^2z\cdot 4x+yz^3\cdot z-3xy\cdot y$$

$$= 4x^3z+yz^4-3xy^2 \textbf{ Ans.}$$

(c) $$\left(\vec{A}\cdot\vec{\nabla}\right)\cdot\vec{B} = \left(x^2z\frac{\partial}{\partial x}+yz^3\cdot\frac{\partial}{\partial y}-3xy\frac{\partial}{\partial z}\right)\cdot\left(y^2\hat{i}-yz\hat{j}+2x\hat{k}\right)$$

$$= yz^2\frac{\partial}{\partial y}y^2\hat{i}+\left\{yz^3\frac{\partial}{\partial y}(-yz)-3xy\frac{\partial}{\partial z}(-yz)\right\}\hat{j}+x^2z\frac{\partial}{\partial x}(2x)\hat{k}$$

$$= yz^2\cdot 2y\hat{i}+\left\{yz^3\cdot(-z)-3xy(-yz)\right\}\hat{j}+x^2z\cdot 2\hat{k}$$

$$= 2y^2z^2\hat{i}+\left(3xy^3z-yz^4\right)\hat{j}+2x^2\cdot z\hat{k}.$$

(d) $$\vec{B}\cdot\left(\vec{A}\cdot\vec{\nabla}\right) = \vec{B}\cdot\left\{\left(x^2z\hat{i}+yz^3\hat{j}-3xy\hat{k}\right)\cdot\left(\frac{\partial}{\partial x}\hat{i}+\frac{\partial}{\partial y}\hat{j}+\frac{\partial}{\partial z}\hat{k}\right)\right\}$$

$$= \vec{B}\cdot\left(x^2z\frac{\partial}{\partial x}+yz^3\frac{\partial}{\partial y}-3xy\frac{\partial}{\partial z}\right)$$

$$= \left(y^2\hat{i}-yz\hat{j}+2x\hat{k}\right)\cdot\left(x^2z\frac{\partial}{\partial x}+yz^3\frac{\partial}{\partial y}-3xy\frac{\partial}{\partial z}\right)$$

$$= \left(x^2y^2z\hat{i}-x^2y\cdot z^2\hat{j}+2x^3z\hat{k}\right)\frac{\partial}{\partial x}+\left(y^3z^3\hat{i}-y^2z^4\hat{j}+2xyz^3\hat{k}\right)\frac{\partial}{\partial y}$$

$$+\left(-3xy^3\hat{i}+3xy^2z\hat{j}-6x^2y\hat{k}\right)\frac{\partial}{\partial z}.$$

(e) $$\left(\vec{\nabla}\cdot\vec{A}\right)\vec{B} = \left\{\frac{\partial}{\partial x}\left(x^2z\right)+\frac{\partial}{\partial y}\left(yz^3\right)+\frac{\partial}{\partial z}(-3xy)\right\}\cdot\vec{B}$$

$$= \left(2xz+z^3\right)\left(y^2\hat{i}-yz\hat{j}+2x\hat{k}\right)$$

$$= \left(2xy^2z+y^2z^3\right)\hat{i}-\left(2xyz^2+yz^4\right)\hat{j}+\left(4x^2z+2xz^3\right)\hat{k}.$$

Problem 21. *Find curl* $\left(\vec{r}\, f(r)\right)$, *where f (r) is differentiable curl* $\left(\vec{r}\, f(r)\right) = \vec{\nabla}\times\left(\vec{r}\, f(r)\right)$

$$= \vec{\nabla}\times\left(xf(\hat{r})\,i + yf(r)\,\hat{j} + zf(r)\,\hat{k}\right)$$

Solution.

$$= \begin{vmatrix} \hat{i} & \hat{j} & \hat{k} \\ \frac{\partial}{\partial x} & \frac{\partial}{\partial y} & \frac{\partial}{\partial z} \\ xf(r) & yf(r) & zf(r) \end{vmatrix}$$

$$= \hat{i}\left\{\frac{\partial}{\partial y} zf(r) - \frac{\partial}{\partial z} yf(r)\right\} - \hat{j}\left\{\frac{\partial}{\partial x} zf(r) - \frac{\partial}{\partial z} xf(r)\right\} + \hat{k}\left\{\frac{\partial}{\partial x} yf(r) - \frac{\partial}{\partial y} xf(r)\right\}.$$

$$= \hat{i}\left(z\frac{\partial f}{\partial y} - y\frac{\partial f}{\partial z}\right) - \hat{j}\left(z\frac{\partial f}{\partial x} - x\cdot\frac{\partial f}{\partial z}\right) + \hat{k}\left(y\frac{\partial f}{\partial x} - x\frac{\partial f}{\partial y}\cdot\right)$$

But,

$$\frac{\partial f}{\partial x} = \left(\frac{\partial f}{\partial r}\right)\left(\frac{\partial r}{\partial x}\right) = \frac{\partial f}{\partial r}\frac{\partial}{\partial x}\left(\sqrt{x^2+y^2+z^2}\right) = \frac{f'(r)\,2x}{2\sqrt{x^2+y^2+z^2}}$$

$$= \frac{f'x}{r}. \text{ Similarly } \frac{\partial f}{\partial y} = \frac{f'y}{r}, \frac{\partial f}{\partial z} = \frac{f'z}{r}.$$

Then the result

$$= \left(z\frac{f'y}{r} - y\frac{f'z}{r}\right)\hat{i} - \hat{j}\left(z\frac{f'x}{r} - x\frac{f'z}{r}\right) + \hat{k}\left(y\frac{f'x}{r} - x\frac{f'y}{r}\right) = 0.$$

Problem 22. *Show that if* $\phi(x, y, z)$ *is any solution of Laplace's equation, then* $\vec{\nabla}\phi$ *is a vector which is both solenoidal and irrotational.*

Solution. By hypothesis, ϕ satisfies Laplace's equation $\vec{\nabla}^2\phi = 0$, *i.e.*, $\vec{\nabla}\cdot\left(\vec{\nabla}\phi\right) = 0$, then $\nabla\phi$ is solenoidal.

Again, $\vec{\nabla}\times\left(\vec{\nabla}\phi\right) = 0$ so, that $\vec{\nabla}\phi$ is also irrotational.

Problem 23. *(a) A vector* $\vec{V}$ *is called irrotational if curl* $\vec{V} = 0$. *Find constants a, b, c so that* $V = (x + 2y + az)\,\hat{i} + (bx - 3y - z)\,\hat{j} + (4x + cy + 2z)\,\hat{k}$ *is irrotational.*

(b) Show that $\vec{V}$ *can be expressed as the gradient of a scalar function.*

Solution (*a*) Curl

$$= \vec{V} = \vec{\nabla}\times\vec{V} = \begin{vmatrix} \hat{i} & \hat{j} & \hat{k} \\ \frac{\partial}{\partial x} & \frac{\partial}{\partial y} & \frac{\partial}{\partial z} \\ x+2y+az & bx-3y-z & 4x+cy+2z \end{vmatrix} = i\left\{\frac{\partial}{\partial y}(4x+cy+2z) - \frac{\partial}{\partial z}(bx-3y-z)\right\}$$

$$-\hat{j}\left\{\frac{\partial}{\partial x}(4x+cy+2z) - \frac{\partial}{\partial z}(x+2y+az)\right\} + \hat{k}\left\{\frac{\partial}{\partial x}(bx-3y-z) - \frac{\partial}{\partial y}(x+2y+az)\right\}$$

$$= (c+1)\,\hat{i} + (a-4)\,\hat{j} + (b-z)\,\hat{k}$$

This equals zero when $a = 4$, $b = 2$, $c = -1$ and

$$V = (x + 2y + 4z)\,\hat{i} + (2x - 3y - z)\,\hat{j} + (4x - y + 2z)\,\hat{k}.$$

(*b*) Assume $\vec{V} = \vec{\nabla}\phi = \frac{\partial \phi}{\partial x}\hat{i} + \frac{\partial \phi}{\partial y}\hat{j} + \frac{\partial \phi}{\partial z}\hat{k}$.

Then,
$$\frac{\partial \phi}{\partial x} = x + 2y + 4z \quad \ldots(i)$$

$$\frac{\partial \phi}{\partial y} = 2x - 3y - z \quad \ldots(ii)$$

$$\frac{\partial \phi}{\partial z} = 4x - y + 2z \quad \ldots(iii)$$

Integrating (*i*) partially with respect to x, keeping y and z constant.

$$\phi = \frac{x^2}{2} + 2xy + 4xz + f(y, z) \quad \ldots(iv)$$

where $f(y, z)$ is an arbitrary function of y and z. Similarly from (*ii*) and (*iii*),

$$\phi = 2xy - \frac{3y^2}{2} - yz + g(x, z) \quad \ldots(v)$$

$$\phi = 4xz - yz + z^2 + h(x, y) \quad \ldots(vi)$$

Comparison of (*iv*), (*v*) and (*vi*) shows that there will be a common value of ϕ if we choose

$$f(y, z) = -\frac{3y^2}{2} + z^2,\; g(x, z) = \frac{x^2}{2} + z^2,\; h(x, y) = \frac{x^2}{2} - \frac{3y^2}{2}.$$

So, that
$$\phi = \frac{x^2}{2} - \frac{3y^2}{2} + z^2 + 2xy + 4xz - yz.$$

Note that we can also add any constant to ϕ. In general, if $\vec{\nabla} \times \vec{V} = 0$, then we can find ϕ. So, that $\vec{V} = \vec{\nabla}\phi$. A vector field $\vec{V}$ which can be derived from a scalar field ϕ so that $\vec{V} = \vec{\nabla}\phi$ is called a conservative vector field and ϕ is called the scalar potential. Note that conversely if $\vec{V} = \vec{\nabla}\phi$, then $\vec{\nabla} \times V = 0$.

Problem 24. *If the electric potential at any point due to a charge q is given by $V = \frac{q}{r}$, then show that the electric intensity $\vec{E}$ is given by $\vec{E} = \frac{q}{r^3}\vec{r}$.*

Solution. We know that the electric intensity is the negative gradient of electric potential, *i.e.*,

$$\vec{E} = -\vec{\nabla}V = -\vec{\nabla}\left(\frac{q}{r}\right)$$

$$= -q\,\vec{\nabla}\left(\frac{1}{r}\right) \qquad \left[\begin{array}{lll}\text{where} & r = & (x^2 + y^2 + z^2)^{\frac{1}{2}} \\ \text{and} & \vec{r} = & \hat{i}x + \hat{j}y + \hat{k}z\end{array}\right]$$

But we get from the problem {17 (*ii*)}

$$\vec{\nabla}\left(\frac{1}{r}\right) = -\frac{\vec{r}}{r^3}$$

$$\therefore \qquad \vec{E} = \frac{q}{r^3}\vec{r} \textbf{ (Proved)}$$

Problem 25. *Prove that $\vec{\nabla} r^n = nr^{n-2}\vec{r}$ where $\vec{r}$ is the position vector of any point.*

Solution.

$$\vec{\nabla} r^n = \hat{i}\frac{\partial}{\partial x}\left(x^2+y^2+z^2\right)^{\frac{n}{2}} + \hat{j}\frac{\partial}{\partial y}\left(x^2+y^2+z^2\right)^{\frac{n}{2}} + \hat{k}\frac{\partial}{\partial z}\left(x^2+y^2+z^2\right)^{\frac{n}{2}}$$

$$= \frac{n}{2}\left(x^2+y^2+z^2\right)^{\frac{n}{2}-1}.\left(2x\hat{i}\right) + \frac{n}{2}\left(x^2+y^2+z^2\right)^{\frac{n}{2}-1}\left(2y\hat{j}\right) + \frac{n}{2}\left(x^2+y^2+z^2\right)^{\frac{n}{2}-1}\left(2z\hat{k}\right)$$

$$= n\left(x^2+y^2+z^2\right)^{\frac{n}{2}-1}.\left(\hat{i}x+\hat{j}y+\hat{k}z\right)$$

$$= n.\left\{r^2\right\}^{\frac{n}{2}-1}.\vec{r} \qquad [\because r^2 = (x^2+y^2+z^2)]$$

$$= n\, r^{n-2}\,\vec{r}$$ **(Proved).**

Problem 26. *The electric potential in space is given by $V = 3x + 4y - 7z$. Obtain the expression for potential gradient and electric intensity.* [H.P. University]

Solution. Given,

$$V = 3x + 4y - 7z$$

$$\text{Potential gradient} = \vec{\nabla}V = \hat{i}\frac{\partial V}{\partial x} + \hat{j}\frac{\partial V}{\partial y} + \hat{k}\frac{\partial V}{\partial z}$$

or,

$$\vec{\nabla}V = \hat{i}\frac{\partial}{\partial x}(3x+4y-7z) + \hat{j}\frac{\partial}{\partial y}(3x+4y-7z) + \hat{k}\frac{\partial}{\partial z}(3x+4y-7z)$$

$$= 3\hat{i} + 4\hat{j} - 7\hat{k}$$

$$\therefore \quad \text{Electric intensity } \vec{E} = -\vec{\nabla}V = -\left(3\hat{i}+4\hat{j}-7\hat{k}\right)$$

Problem 27. *(a) Electric potential at a point is given by $V = x^2y + 2z$. What are the components of electric field at that point? Find the electric field at a point (2, 1, 2).* [G.N.D. University]

(b) Calculate the force acting on the charge 2×10^{-10} coulomb located at the point (2, 0, – 3) in the field region in terms of $\hat{i}, \hat{j}, \hat{k}$, when the electric potential at any point in the region is given by $V(x, y, z) = (20 + 6x^2 - 5xy + 4y^2 + 3z^2)$. [Lucknow University]

Given $$V = x^2y + 2z$$

Solution. We know, $\vec{E} = -\vec{\nabla}V$

or,

$$E_x\hat{i} + E_y\hat{j} + E_z\hat{k} = -\left[\hat{i}\frac{\partial V}{\partial x} + \hat{j}\frac{\partial V}{\partial y} + \hat{k}\frac{\partial V}{\partial z}\right]$$

or,

$$\left.\begin{aligned} E_x &= -\frac{\partial V}{\partial x} = -2xy,\\ E_y &= -\frac{\partial V}{\partial y} = -x^2,\\ E_z &= -\frac{\partial V}{\partial z} = -2 \end{aligned}\right\}$$

These are the components of electric field at that point.

At the point (2, 1, 2). The electric field is

$$\vec{E} = -\hat{i}\,2xy - \hat{j}\,x^2 - 2\hat{k}$$

$$\vec{E}\Big|_{2,1,2} = -\hat{i}\,2\times2\times1 - \hat{j}\,2\times2 - 2\,\hat{k}$$

$$= -4\hat{i} - 4\hat{j} - 2\hat{k}$$

(*b*) Here

$$\vec{E} = -\vec{\nabla}V = -\hat{i}\frac{\partial V}{\partial x} - \hat{j}\frac{\partial V}{\partial y} - \hat{k}\frac{\partial V}{\partial z}$$

$$\frac{\partial V}{\partial x} = \frac{\partial}{\partial x}(20 + 6x^2 - 5xy + 4y^2 + 3z^2) = 12x - 5y$$

$$\frac{\partial V}{\partial y} = \frac{\partial}{\partial y}(20 + 6x^2 - 5xy + 4y^2 + 3z^2) = 8y - 5x$$

and

$$\frac{\partial V}{\partial z} = \frac{\partial}{\partial z}(20 + 6x^2 - 5xy + 4y^2 + 3z^2) = 6z$$

$$\vec{E} = -\hat{i}\,(12x - 5y) - \hat{j}\,(8y - 5x) - \hat{k}\,(6z)$$

At the point (2, 0, – 3), the electric field

$$\vec{E} = -\left[24\hat{i} - 10\hat{j} - 18\hat{k}\right]$$

∴ The force acting on the charge 2×10^{-10} coulomb is

$$\vec{F} = e\,\vec{E} = 2\times10^{-10}\left(-24\hat{i} + 10\hat{j} + 18\hat{k}\right)$$

Problem 28. *(a) The electric potential in space is given by (3x + 4y – z). Show that the electric field intensity is uniform everywhere.* [H.P. University]

(b) If the electric potential is constant in a certain region of space, what inference can be drawn about the electric field ? [Lucknow University]

Solution. (*a*) Given that,

Electric potential $V = 3x + 4y - z$

∴ Electric field intensity $\vec{E} = -\vec{\nabla}V$

$$= -\left\{\hat{i}\frac{\partial V}{\partial x} + \hat{j}\frac{\partial V}{\partial y} + \hat{k}\frac{\partial V}{\partial z}\right\}$$

$$= -\left\{\hat{i}\,3 + \hat{j}\,4 + \hat{k}\,(-1)\right\}$$

$$= -3\hat{i} - 4\hat{j} + \hat{k}$$

Here, the electric field does not depend upon the space co-ordinates (*x*, *y*, *z*). So it is constant and uniform everywhere.

(*b*) The electric field would be zero when the electric potential is constant.

$$\because \qquad \vec{E} = -\vec{\nabla}V$$

If V = constant = C

$$\vec{E} = -\vec{\nabla}C = 0$$

Problem 29. *If $\vec{r}$ is the position vector of a point then show that*

(*i*) $div\,\vec{r} = \vec{\nabla}.\vec{r} = 3$

(*ii*) $div\left(\frac{\vec{r}}{r^3}\right) = \vec{\nabla}.\left(\frac{\vec{r}}{r^3}\right)$

(*iii*) $div\left(r^n\,\vec{r}\right) = (3+n)r^n$

Solution. We know that the position vector $\vec{r}$ is written

as
$$\vec{r} = \hat{i}x + \hat{j}y + \hat{k}z$$
$$r^2 = x^2 + y^2 + z^2$$

(*i*)
$$\text{div}\,\vec{r} = \vec{\nabla}.\vec{r} = \left(\hat{i}\frac{\partial}{\partial x} + \hat{j}\frac{\partial}{\partial y} + \hat{k}\frac{\partial}{\partial z}\right).\left(\hat{i}x + \hat{j}y + \hat{k}z\right)$$

$$= \frac{\partial x}{\partial x} + \frac{\partial y}{\partial y} + \frac{\partial z}{\partial z} = 1 + 1 + 1 = 3. \textbf{ (Proved)}.$$

(*ii*)
$$\text{div}\left(\frac{\vec{r}}{r^3}\right) = \vec{\nabla}.\left(\frac{\vec{r}}{r^3}\right)$$

$$= \vec{\nabla}.\left\{\hat{i}\frac{x}{\left(x^2+y^2+z^2\right)^{\frac{3}{2}}} + \hat{j}\frac{y}{\left(x^2+y^2+z^2\right)^{\frac{3}{2}}} + \hat{k}\frac{z}{\left(x^2+y^2+z^2\right)^{\frac{3}{2}}}\right\}$$

$$= \frac{\partial}{\partial x}.\left\{\frac{x}{\left(x^2+y^2+z^2\right)^{\frac{3}{2}}}\right\} + \frac{\partial}{\partial y}\left\{\frac{y}{\left(x^2+y^2+z^2\right)^{\frac{3}{2}}}\right\} + \frac{\partial}{\partial z}\left\{\frac{z}{\left(x^2+y^2+z^2\right)^{\frac{3}{2}}}\right\}$$

Now
$$\frac{\partial}{\partial x}\left\{\frac{x}{\left(x^2+y^2+z^2\right)^{\frac{3}{2}}}\right\} = \left(x^2+y^2+z^2\right)^{-\frac{3}{2}} + x.\left(-\frac{3}{2}\right)\left(x^2+y^2+z^2\right)^{-\frac{5}{2}}.(2x)$$

$$= \left(x^2+y^2+z^2\right)^{-\frac{5}{2}}\left\{x^2+y^2+z^2-3x^2\right\}$$

$$= \frac{y^2+z^2-2x^2}{\left(x^2+y^2+z^2\right)^{\frac{5}{2}}}$$

Similarly,

$$\frac{\partial}{\partial y}\left\{\frac{y}{\left(x^2+y^2+z^2\right)^{\frac{3}{2}}}\right\} = \frac{x^2+z^2-2y^2}{\left(x^2+y^2+z^2\right)^{\frac{5}{2}}}$$

and $$\frac{\partial}{\partial z}\left\{\frac{z}{\left(x^2+y^2+z^2\right)^{\frac{3}{2}}}\right\} = \frac{x^2+y^2-2z^2}{\left(x^2+y^2+z^2\right)^{\frac{5}{2}}}$$

$$\therefore \quad \text{div}\left(\frac{\vec{r}}{r^3}\right) = \vec{\nabla}\cdot\left(\frac{\vec{r}}{r^3}\right)$$

$$= \frac{1}{\left(x^2+y^2+z^2\right)^{\frac{5}{2}}}\left\{y^2+z^2-2x^2+x^2+z^2-2y^2+x^2+y^2-2z^2\right\}$$

$= 0$ **(Proved)**

(iii) $$\text{div}\left(r^n\vec{r}\right) = \vec{\nabla}\cdot\left\{r^n\left(\hat{i}x+\hat{j}y+\hat{k}z\right)\right\}$$

$$= \frac{\partial}{\partial x}\left(r^n x\right)+\frac{\partial}{\partial y}\left(r^n y\right)+\frac{\partial}{\partial z}\left(r^n z\right)$$

$$\therefore \quad \frac{\partial}{\partial x}\left(r^n x\right) = \left(x^2+y^2+z^2\right)^{\frac{n}{2}}+x\cdot\frac{n}{2}\cdot\left(x^2+y^2+z^2\right)^{\frac{n}{2}-1}\cdot 2x$$

$$\left[\because r^n = \left(x^2+y^2+z^2\right)^{\frac{n}{2}}\right]$$

$$= r^n + nx^2 \cdot r^{n-2}$$

Similarly, $$\frac{\partial}{\partial y}\left(r^n y\right) = r^n + ny^2 \cdot r^{n-2}$$

and $$\frac{\partial}{\partial z}\left(r^n z\right) = r^n + nz^2 \cdot r^{n-2}$$

$\therefore$ Adding we get

$$\text{div}\left(r^n\vec{r}\right) = r^n + nx^2 . r^{n-2} + r^n + ny^2 . r^{n-2} + r^n + nz^2 . r^{n-2}$$

$$= 3r^n + nr^{n-2}\left(x^2+y^2+z^2\right)$$

$$= 3r^n + nr^{n-2} \cdot r^2$$

$$= (3 + n)\, r^n \textbf{ (Proved)}$$

Problem 30. *(a) From the differential form of Gauss's law develop the Poisson's and Laplace's equations.* [G.N.D. University, Pune University, H.P. University, K. University]

(b) What is the value of $\nabla^2\phi$ at a point in empty space? What is the significance of Laplace's equation ? [Pune University]

Solution. Poisson's Equation :

(*a*) The differential form of Gauss's law states

$$\vec{\nabla} \cdot \vec{E} = \frac{\rho}{\varepsilon_0} \text{ (In S.I. Unit)} \qquad \ldots(i)$$

where ρ is the volume density of charge and $\vec{E}$ is the electric field intensity.

But we know that $\vec{E} = -\vec{\nabla}\phi$,

where ϕ is the electric potential.

$$\therefore \quad \vec{\nabla} \cdot \left(-\vec{\nabla}\phi\right) = \frac{\rho}{\varepsilon_0}$$

or,
$$-\nabla^2\phi = \frac{\rho}{\varepsilon_0}$$

or,
$$\nabla^2\phi = -\frac{\rho}{\varepsilon_0} \qquad \ldots(ii)$$

or,
$$\left(\frac{\partial^2\phi}{\partial x^2} + \frac{\partial^2\phi}{\partial y^2} + \frac{\partial^2\phi}{\partial z^2}\right) = -\frac{\rho}{\varepsilon_0}$$

This is the poisson's equation for each point in space.

Laplace's Equation :

When the enclosed charge density is zero *i.e.*, $\rho = 0$ or we can say for the observation is in empty space ($\rho = 0$), we get from equation (*ii*)

$$\nabla^2\phi = 0$$

or,
$$\frac{\partial^2\phi}{\partial x^2} + \frac{\partial^2\phi}{\partial y^2} + \frac{\partial^2\phi}{\partial z^2} = 0 \qquad \ldots(iii)$$

This equation is known as Laplace's equation.

(*b*) The value $\nabla^2\phi$ in empty space is zero because in empty space $\rho = 0$.

$$\therefore \quad \nabla^2\phi = 0.$$

Significance : The very important significance of such function is "The average value of a function ϕ (x, y, z) satisfying Laplace's equation over any spherical surface is equal to its value at the center of the spherical surface".

Problem 31. *Show that the potential function $V = V_0 (x^2 - 2y^2 + z^2)$ satisfy Laplace's equation where V_0 is a constant.* [WBUT]

Solution. We know from the Laplace's equation

$$\nabla^2 V = 0$$

or, $$\frac{\partial^2 V}{\partial x^2} + \frac{\partial^2 V}{\partial y^2} + \frac{\partial^2 V}{\partial z^2} = 0, \text{ where V is the potential function.}$$

Here, given potential function,

$$V = V_0 (x^2 - 2y^2 + z^2)$$

$$\therefore \quad \frac{\partial V}{\partial x} = V_0 (2x)$$

or, $$\frac{\partial^2 V}{\partial x^2} = 2V_0,$$

$$\frac{\partial V}{\partial y} = -4y\, V_0, \quad \text{or} \quad \frac{\partial^2 V}{\partial y^2} = -4V_0$$

and, $$\frac{\partial V}{\partial z} = +2z\, V_0, \quad \text{or,} \quad \frac{\partial^2 V}{\partial z^2} = 2V_0$$

$$\therefore \quad \frac{\partial^2 V}{\partial x^2} + \frac{\partial^2 V}{\partial y^2} + \frac{\partial^2 V}{\partial z^2} = 2V_0 - 4V_0 + 2V_0$$

$$= 0.$$

Here the potential function V satisfies the Laplace's equation.

Problem 32. *Write down Laplace's equation and show if the functions*

$$f(xy) = x^2 - y^2 \text{ and } g(xy) = x^2 + y^2$$

satisfy the two dimensional Laplace's equation. [H.P. University]

Solution. From Laplace's equation

$$\nabla^2 \phi = 0$$

or, $$\frac{\partial^2 \phi}{\partial x^2} + \frac{\partial^2 \phi}{\partial y^2} + \frac{\partial^2 \phi}{\partial z^2} = 0$$

In two dimensions

$$\frac{\partial^2 \phi}{\partial x^2} + \frac{\partial^2 \phi}{\partial y^2} = 0$$

(*i*) For $\phi = f(xy) = x^2 - y^2$

$$\frac{\partial \phi}{\partial x} = 2x \quad \text{or,} \quad \frac{\partial^2 \phi}{\partial x^2} = 2,$$

and $$\frac{\partial \phi}{\partial y} = -2y \quad \text{or,} \quad \frac{\partial^2 \phi}{\partial y^2} = -2$$

$$\therefore \qquad \frac{\partial^2 \phi}{\partial x^2} + \frac{\partial^2 \phi}{\partial y^2} = 2 - 2 = 0.$$

Hence, $f(xy) = x^2 - y^2$ satisfies two dimensional Laplace's equation.

(*ii*) Again for $\phi = g(xy) = x^2 + y^2$

$$\therefore \qquad \frac{\partial \phi}{\partial x} = 2x \quad \text{or} \quad \frac{\partial^2 \phi}{\partial x^2} = 2$$

and
$$\frac{\partial \phi}{\partial y} = 2y \quad \text{or} \quad \frac{\partial^2 \phi}{\partial y^2} = 2,$$

$$\therefore \qquad \frac{\partial^2 \phi}{\partial x^2} + \frac{\partial^2 \phi}{\partial y^2} = 2 + 2 = 4 \neq 0,$$

$$\therefore \qquad \nabla^2 \phi \neq 0$$

The function $g(xy) = x^2 + y^2$ does not satisfy two dimensional Laplace's equation.

Example 33. *What is charge density in a region of space where electrostatic potential is given by* $V = a - b(x^2 + y^2) - c \log(x^2 + y^2)$? [G.N.D. University]

Solution. We know from poisson's equation

$$\nabla^2 V = -\frac{\rho}{\varepsilon_0} \quad \text{where } \rho \text{ is the charge density}$$

or,
$$\frac{\partial V}{\partial x^2} + \frac{\partial^2 V}{\partial y^2} + \frac{\partial^2 V}{\partial z^2} = -\frac{\rho}{\varepsilon_0} \qquad \ldots(i)$$

Now,
$$\frac{\partial V}{\partial x} = \frac{\partial}{\partial x}\left[a - b\left(x^2 + y^2\right) - c \log\left(x^2 + y^2\right)\right] = -2bx - \frac{2cx}{x^2 + y^2}$$

$$\frac{\partial^2 V}{\partial x^2} = -2b - \frac{2c\left\{x^2 + y^2 - x \,.\, 2x\right\}}{\left(x^2 + y^2\right)^2} = -2b - 2c\frac{\left(y^2 - x^2\right)}{\left(x^2 + y^2\right)^2}$$

and
$$\frac{\partial V}{\partial y} = -2by - \frac{2cy}{x^2 + y^2}, \quad \frac{\partial^2 V}{\partial y^2} = -2b - 2c\frac{\left(x^2 - y^2\right)}{\left(x^2 + y^2\right)^2}$$

$$\frac{\partial V}{\partial z} = 0 \quad \text{or,} \quad \frac{\partial^2 V}{\partial z^2} = 0.$$

Putting these values in equation (*i*), we get

$$-2b - 2c\frac{\left(y^2 - x^2\right)}{\left(x^2 + y^2\right)^2} - 2b - 2c\frac{\left(x^2 - y^2\right)}{\left(x^2 + y^2\right)} = -\frac{\rho}{\varepsilon_0}$$

or,
$$-4b - 2c \,.\, \frac{\left\{y^2 - x^2 + x^2 - y^2\right\}}{\left(x^2 + y^2\right)^2} = -\frac{\rho}{\varepsilon_0}$$

or, $$\frac{\rho}{\varepsilon_0} = 4b$$

$\therefore$ $$\rho = 4\varepsilon_0 \, b. \text{ (Ans.)}$$

Problem 34. *(a) Prove that div grad* $r^m = \nabla^2 r^m = m(m+1)\, r^{m-2}$

[Kanpur University, Gorakhpur University]

(b) Hence, evaluate $\nabla^2 r$ *where* $\vec{r}$ *is the position vector* $\left(r = \sqrt{x^2+y^2+z^2}\right)$.

[Agra University]

Solution. (*a*) div. grad $(r^m) = \nabla^2 r^m = \left(\frac{\partial^2}{\partial x^2} + \frac{\partial^2}{\partial y^2} + \frac{\partial^2}{\partial z^2}\right)\left(x^2+y^2+z^2\right)^{\frac{m}{2}}$

$$\therefore \quad \frac{\partial}{\partial x}\left(x^2+y^2+z^2\right)^{\frac{m}{2}} = mx\left(x^2+y^2+z^2\right)^{\frac{m}{2}-1}$$

and $$\frac{\partial^2}{\partial x^2}\left(x^2+y^2+z^2\right)^{\frac{m}{2}} = m\left(x^2+y^2+z^2\right)^{\frac{m}{2}-1} + mx.\left(\frac{m}{2}-1\right)\left(x^2+y^2+z^2\right)^{\frac{m}{2}-2}.2x$$

$$= m\left(x^2+y^2+z^2\right)^{\frac{m}{2}-2}\left[x^2+y^2+z^2+(m-2)\,x^2\right]$$

$$= m\, r^{m-4}\,[(m-1)\,x^2 + y^2 + z^2] \qquad \left[\because \left(x^2+y^2+z^2\right)^{\frac{m-4}{2}} = r^{m-4}\right]$$

Similarly, $$\frac{\partial^2}{\partial y^2}\left(x^2+y^2+z^2\right)^{\frac{m}{2}}$$

$$= m\, r^{m-4}\,[x^2 + (m-1)\,y^2 + z^2]$$

and $$\frac{\partial^2}{\partial z^2}\left(x^2+y^2+z^2\right)^{\frac{m}{2}} = m\, r^{m-4}\,[x^2 + y^2 + (m-1)\,z^2]$$

Putting these values, we get

$$\text{div. grad } (r^m) = m\, r^{m-4}\,[(m-1)\,x^2 + y^2 + z^2 + x^2 + (m-1)\,y^2 + z^2 + x^2 + y^2 + (m-1)\,z^2]$$

$$= m\,(m+1)\, r^{m-4}\,[x^2 + y^2 + z^2)] \ldots$$

$$= m\,(m+1)\, r^{m-4} \,.\, r^2 = m\,(m+1)\, r^{m-2} \text{ (Proved)}$$

(*b*) $\because$ $$\nabla^2 r^m = m\,(m+1)\, r^{m-2}$$

Putting $m = 1$, we get

$$\nabla^2 r = 1\,.\,(1+1)\, r^{1-2}$$

$$= \frac{2}{r}$$

Problem 35. *The electric field at any point in a given space is directed radially outward along the line joining the point to a fixed point (r = 0) and is given by ar + br². Evaluate the charge enclosed in a sphere of radius R with the fixed point (r = 0) as center.* [Gorakhpur University]

Solution. Given that, the magnitude of the radial electric field is

$$E = ar + br^2$$

Its value at the surface of the sphere of radius R is therefore

$$E = aR + bR^2.$$

The surface integral of the normal component of electric field $(\vec{E})$ over the closed surface S of the sphere is

$$\iint_S \vec{E} \cdot \overrightarrow{dS} = \iint_S E\, dS = \iint_S (aR + bR^2)\, dS$$

$$= (aR + bR^2) \iint_S dS \qquad [\because \vec{E} \text{ and } \overrightarrow{dS} \text{ are in same direction}]$$

$$= (a + bR)\, 4\pi R^3 \qquad \left[\because \iint_S dS = 4\pi R^2\right]$$

But we know from Gauss's law of electrostatics, $\iint_S \vec{E} \cdot \overrightarrow{dS} = \frac{q}{\varepsilon_0}$ [q = charge enclosed by the surface]

or, $$(a + bR)\, 4\pi R^3 = \frac{q}{\varepsilon_0}$$

or, $$q = 4\pi\varepsilon_0 R^3 (a + bR).$$ **(Ans.)**

Problem 36. *The electric field $\vec{E}$ in a certain space is given by*

$$E_x = (ax + by + c),\ E_y = 0,\ E_z = 0$$

Use Gauss's theorem, to evaluate the charge enclosed in a cube with one of its sides of length L along the x-axis, the two faces of the cube are perpendicular to the X axis lying in the planes x = l and x = l + L.

Solution. According to the Gauss's theorem of divergence,

$$\iint_S \vec{E} \cdot \overrightarrow{dS} = \iiint_V \text{div}\, \vec{E}\, dV$$

or, $$\iint_S \vec{E} \cdot \overrightarrow{dS} = \iiint_V \text{div}\, \vec{E}\, dV$$

$$= \iiint_V \left(\frac{\partial E_x}{\partial x} + \frac{\partial E_y}{\partial y} + \frac{\partial E_z}{\partial z} \right) dV$$

$$= \iiint_V a\, dV = aV = a L^3 \qquad [\text{Volume of the cube } V = L^3]$$

From Gauss's law in electrostatics,

we have

$$\iint_S \vec{E}\,.\,\overrightarrow{dS} = \frac{q}{\varepsilon_0}$$

or
$$a\,L^3 = \frac{q}{\varepsilon_0}$$

$$q = \varepsilon_0\, a\, L^3 \text{ (Ans.)}$$

Problem 37. *If $\vec{E} = 3\hat{i} + 4\hat{j} + 8\hat{k}$. Calculate the electric flux through a surface area 100 units lying in X-Y plane.* [G.N.D. University, Pbi University, Pune University]

Solution. Given $\vec{E} = 3\hat{i} + 4\hat{j} + 8\hat{k}$, thus field is uniform.

As the surface area is in X-Y plane, the area vector is directed in the z direction.

∴ Area vector $\vec{S} = 100\,\hat{k}$ [$\hat{k}$ is the unit vector along the $+\,z$ axis]

For free space, electric flux $\phi_E = \iint_S \vec{E}\,.\,\overrightarrow{dS} = \vec{E}\,.\,\vec{S}$ as the field is uniform.

∴
$$\phi_E = \vec{E}\,.\,\vec{S} = \left(3\hat{i} + 4\hat{j} + 8\hat{k}\right).\ 100\hat{k}$$

$$= 800 \text{ Nm}^2\text{ C}^{-1} \left[\hat{i}\,.\,\hat{i} = 1, \hat{i}\,.\,\hat{j} = \hat{j}\,.\,\hat{k} = 0\right].$$

Problem 38. *In a region of space, the electric field is given by $\vec{E} = 8\hat{i} + 4\hat{j} + 3\hat{k}$. Calculate the electric flux through a surface of area 100 unit in X-Y plane.* [WBUT]

Solution. Given that $\vec{E} = 8\hat{i} + 4\hat{j} + 3\hat{k}$

As the surface area is in X-Y plane, the area vector is directed in the Z direction.

For free space: Area vector $\vec{S} = 100\,\hat{k}$

∴ electric flux $\phi_E = \vec{E}\,.\,\vec{S} = \left(8\hat{i} + 4\hat{j} + 3\hat{k}\right)\left(100\,\hat{k}\right)$

$$= 300 \text{ Nm}^2\text{ C}^{-1}.$$

Problem 39. *In a region an electric field is $\vec{E} = 2\hat{i} + 3\hat{j} + 4\hat{k}$. Calculate the electric flux through the surface $\vec{S} = 10\hat{i}$* [H.P. University]

Solution. Given $\vec{E} = 2\hat{i} + 3\hat{j} + 4\hat{k}$ *i.e.*, the field is uniform

and $\vec{S} = 10\hat{i}$

∴ electric flux $\phi_E = \vec{E}\,.\,\vec{S} = \left(2\hat{i} + 3\hat{j} + 4\hat{k}\right).\ 10\hat{i}$

$$= 20 \text{ Nm}^2\text{ C}^{-1} \text{ (Ans.)}$$

Problem 40. *If* $\vec{F} = y\hat{i} + (x^2 + y^2)\hat{j} + (yz + zx)\hat{k}$, *then evaluate Curl* $\vec{F}$. [Agra University]

Solution. Curl $\vec{F} = \vec{\nabla} \times \vec{F}$

$$= \left(\hat{i}\frac{\partial}{\partial x} + \hat{j}\frac{\partial}{\partial y} + \hat{k}\frac{\partial}{\partial z}\right) \cdot \left\{y\hat{i} + (x^2 + y^2)\hat{j} + (yz + zx)\hat{k}\right\}$$

$$= \begin{vmatrix} \hat{i} & \hat{j} & \hat{k} \\ \frac{\partial}{\partial x} & \frac{\partial}{\partial y} & \frac{\partial}{\partial z} \\ y & (x^2 + y^2) & (yz + zx) \end{vmatrix}$$

$$= \hat{i}\left\{\frac{\partial}{\partial y}(yz + zx) - \frac{\partial}{\partial z}(x^2 + y^2)\right\} - \hat{j}\left\{\frac{\partial}{\partial x}(yz + zx) - \frac{\partial}{\partial z}y\right\} + \hat{k}\left\{\frac{\partial}{\partial x}(x^2 + y^2) - \frac{\partial y}{\partial y}\right\}$$

$$= \hat{i}\{z + 0 - 0\} - \hat{j}(z) + \hat{k}\{2x - 1\}$$

$$= \hat{i}z - \hat{j}z + \hat{k}(2x - 1) \text{ **(Ans.)**}$$

Problem 41. *(a) What is an irrotational field? Give an example.*

[H. P. University, G. N. D. University]

(b) If $\vec{F} = (2xy + z^3)\hat{i} + x^2\hat{j} + 3xz^2\hat{k}$. *Show that* $\vec{F}$ *is an irrotational vector.* [WBUT]

Solution. (*a*) When the curl of a vector field vanishes it is called an irrotational field.

If $\vec{F}$ represents an irrotational field, then Curl $\vec{F} = 0$

i.e., $\vec{\nabla} \times \vec{F} = 0,$

electric field is an example of irrotational field.

When $\vec{\nabla} \times F \neq 0,$

then the field is called a rotational field. Magnetic field $(\vec{B})$ is an example of rotational field.

(*b*) $$\vec{\nabla} \times \vec{F} = \begin{vmatrix} \hat{i} & \hat{j} & \hat{k} \\ \frac{\partial}{\partial x} & \frac{\partial}{\partial y} & \frac{\partial}{\partial z} \\ (2xy + z^3) & x^2 & 3xz^2 \end{vmatrix}$$

$$= \hat{i}\left(\frac{\partial}{\partial y}3xz^2 - \frac{\partial}{\partial z}x^2\right) - \hat{j}\left(\frac{\partial}{\partial x}3xz^2 - \frac{\partial}{\partial z}(2xy + z^3)\right) + \hat{k}\left(\frac{\partial x^2}{\partial x} - \frac{\partial}{\partial y}(2xy + z^3)\right)$$

$$= \hat{i}(0) - \hat{j}(3z^2 - 3z^2) + \hat{k}(2x - 2x)$$

$$= 0,$$

So, $\vec{F}$ is an irrotational field.

Problem 42. *Consider a vector field $\vec{A} = x^2\hat{i} + y^2\hat{j} + z^2\hat{k}$, (i) Is the field solenoidal (ii) Is the field irrotational ?* [Pbi. University]

Solution. (*i*) If the divergence of the vector field vanishes, the field is solenoidal.

$$\vec{\nabla}.\vec{A} = \left(\hat{i}\frac{\partial}{\partial x} + \hat{j}\frac{\partial}{\partial y} + \hat{k}\frac{\partial}{\partial z}\right).\left(x^2\hat{i} + y^2\hat{j} + z^2\hat{k}\right)$$

$$= \frac{\partial x^2}{\partial x} + \frac{\partial y^2}{\partial y} + \frac{\partial z^2}{\partial z} = 2x + 2y + 2z = 2\,(x + y + z)$$

$$\because \qquad \vec{\nabla}.\vec{A} \neq 0,$$

the field is not solenoidal.

(*ii*)
$$\vec{\nabla}\times\vec{A} = \begin{vmatrix} \hat{i} & \hat{j} & \hat{k} \\ \frac{\partial}{\partial x} & \frac{\partial}{\partial y} & \frac{\partial}{\partial z} \\ x^2 & y^2 & z^2 \end{vmatrix} = 0$$

$$\therefore \qquad \vec{\nabla}\times\vec{A} = 0,$$

the field is irrotational.

Problem 43. *A vector $\vec{r} = \hat{i}x + \hat{j}y + \hat{k}z$.*

Show that $\oint \vec{r}.\vec{dr} = 0$

Solution. According to Stoke's law,

$$\oint \vec{A}.\vec{dl} = \iint_S \left(\vec{\nabla}\times\vec{A}\right).\vec{dS}$$

Here $\qquad \vec{A} = \vec{r}$ and $\vec{dl} = \vec{dr}$

$$\therefore \qquad \oint \vec{A}.\vec{dl} = \oint \vec{r}.\vec{dr} = \iint_S \left(\vec{\nabla}\times\vec{r}\right).\vec{dS} \qquad \text{[Applying Stoke's law]}$$

$$\therefore \qquad \vec{\nabla}\times\vec{r} = \begin{vmatrix} \hat{i} & \hat{j} & \hat{k} \\ \frac{\partial}{\partial x} & \frac{\partial}{\partial y} & \frac{\partial}{\partial z} \\ x & y & z \end{vmatrix} = 0$$

$$\therefore \qquad \oint \vec{r}.\vec{dr} = 0.$$

Problem 44. *A long wire carries a current of 5 mA. Find the line integral of $\vec{B}$ around the path (of radius 10 cm) enclosing the wire.* $\frac{\mu_0}{4\pi} = 10^{-7}$ *S.I. Units.*

Solution. Here $\quad I = 5 \text{ mA} = 5 \times 10^{-3}$ Amp.

$$\mu_0 = 4\pi \times 10^{-7} \text{ S.I. unit.}$$

We know from Ampere's circuital law, "the line integral of a magnetic field along any closed path is equal to μ_0 times the net current enclosing the path".

i.e.,
$$\oint \vec{B} \cdot \vec{dl} = \mu_0 I = 4\pi \times 10^{-7} \times 5 \times 10^{-3}$$
$$= 62.9 \times 10^{-10} \text{ Tesla meter.}$$

Problem 45. *(a) Calculate the magnitude of the magnetic field due to a long thin wire carrying current of 15 Amp at a distance of 1 cm from the wire, given $\mu_0 = 4\pi \times 10^{-7}$ N/A^2.*

[H.P. University]

(b) Calculate the magnetic field at a distance of 5 m from an infinite straight conductor carrying current of 100 A. [H.P. University]

Solution. (*a*) Current I = 15 Amp. Distance x = 1 cm = 0.01 m

∴ Magnitude of magnetic field $\left|\vec{B}\right| = \frac{\mu_0}{4\pi} \cdot \frac{2I}{x}$

$$= \frac{4\pi \times 10^{-7}}{4\pi} \times \frac{2 \times 15}{0.01}$$
$$= 3 \times 10^{-4} \text{ Tesla}$$

(*b*) Here, given that x = 5 m, and I = 100 Amp.

∴ Magnitude of the magnetic field

$$\left|\vec{B}\right| = \frac{\mu_0}{4\pi} \cdot \frac{2I}{x}$$
$$= \frac{4\pi \times 10^{-7}}{4\pi} \times \frac{2 \times 100}{5}$$
$$= 4 \times 10^{-6} \text{ Tesla.}$$

Problem 46. *Two straight wires are kept in air 2 m apart carrying currents of 80 A and 30 A in the same direction. Calculate the force between them and specify its nature.* [H.P. University]

Solution. Force per meter of the wire $= \frac{\mu_0}{4\pi} \cdot \frac{2I_1I_2}{x}$

Given, the distance between the wire = 2 *m*

Current in one wire $\quad I_1 = 30$ Amps

Current in the other wire $I_2 = 40$ Amps

$$= \frac{10^{-7} \times 2 \times 30 \times 40}{2} \quad \left[\frac{\mu_0}{4\pi} = 10^{-7} \text{ S.I.}\right]$$

$$= 1.2 \times 10^{-4} \text{ N.}$$

Here, the currents are in the same direction, so the two wires attract each other.

Problem 47. *(a) If at any position, the vector potential $\vec{A} = 5\,(x^2 + y^2 + z^2)\,\hat{i}$, evaluate the magnetic field at that position.*

Solution. Given, the vector potential $\vec{A} = 5\,(x^2 + y^2 + z^2)\,\hat{i}$

$\therefore$ Magnetic field $\vec{B} = \text{curl } \vec{A} = \vec{\nabla} \times \vec{A}$

$$= \begin{vmatrix} \hat{i} & \hat{j} & \hat{k} \\ \frac{\partial}{\partial x} & \frac{\partial}{\partial y} & \frac{\partial}{\partial z} \\ 5\left(x^2 + y^2 + z^2\right) & 0 & 0 \end{vmatrix}$$

$$= \hat{j}\,\frac{\partial}{\partial z}\left[5\left(x^2 + y^2 + z^2\right)\right] - \hat{k}\,\frac{\partial}{\partial y}\left[5\left(x^2 + y^2 + z^2\right)\right]$$

$$= \hat{j} \times 10z - \hat{k}\,10y = 10\left(z\hat{j} - \hat{k}y\right) \textbf{ (Ans.)}$$

Multiple Choice Questions

1. If $\hat{n}$ is the unit vector in the direction of $\vec{A}$, then

   (a) $\hat{n} = \dfrac{\vec{A}}{|\vec{A}|}$ (b) $\hat{n} = \vec{A}\,|\vec{A}|$ (c) $\hat{n} = \dfrac{|\vec{A}|}{|\vec{A}|}$

2. The direction of $\vec{\nabla}\phi$ is always

   (a) Perpendicular to the surface $\phi = a$ constant
   (b) Parallel to the surface $\phi = a$ constant
   (c) Perpendicular or parallel depending upon the shape of the surface.

3. If a vector field can be expressed as the gradient of a scalar, then the vector field is called

   (a) Solenoidal (b) Lamellar (c) Irrotational.

4. Two vectors $\vec{A}$ and $\vec{B}$ are parallel, when

   (a) $\vec{A} \times \vec{B} = 0$ (b) $\vec{A} \cdot \vec{B} = 0$ (c) $\vec{A} \cdot \vec{B} = 1$

5. The vector field is solenoidal if

   (a) $\vec{\nabla} \times \vec{A} = 0$ (b) $\vec{\nabla} \cdot \vec{A} = 0$ (c) $\vec{A} = 0$

6. The value of $\vec{\nabla} \cdot \vec{r}$ is

   (a) 0 (b) 1 (c) 3.

7. The vector field is irrotational if

(*a*) $\vec{\nabla} . \vec{A} = 0$ (*b*) $\vec{\nabla} \times \vec{A} = 0$ (*c*) $\vec{\nabla} . \vec{A} = 1$

8. A vector field A is conservative if

(*a*) $\vec{A} = \vec{\nabla} \times \phi$ (*b*) $\vec{A} = \vec{\nabla}\phi$ (*c*) $\vec{A} = \vec{\nabla}\phi$

9. The divergence of curl of a vector is always

(*a*) 1 (*b*) 0 (*c*) $\frac{1}{2}$

10. The Z-component of the curl of the field $E_x = ky$, $E_y = kx$, $E_z = 0$ is
(*a*) $2k$ (*b*) $-2k$ (*c*) 0

11. The line integral of the vector field around a closed path can always be written as

(*a*) $\iint_S (\vec{\nabla} \times \vec{A}) . \vec{dS}$ (*b*) $\oiint (\vec{\nabla} \times \vec{A}) . \vec{dS}$ (*c*) $\iint_S (\vec{\nabla} . \vec{A})\, dS$

12. For a conservative field

(*a*) $\oint \vec{E} . \vec{dl} = El$ (*b*) $\oint \vec{E} . \vec{dl} = 0$ (*c*) $\oint \vec{E} . \vec{dl} = 1$

13. The relation between electric field and potential is

(*a*) $\vec{E} = \vec{\nabla}V$ (*b*) $\vec{E} = -\vec{\nabla}V$ (*c*) $\vec{E} = \vec{\nabla^2}V$

14. In a free space poisson's equation is

(*a*) $\nabla^2 V = \frac{\rho}{\varepsilon_0}$ (*b*) $\nabla^2 V = 0$ (*c*) $\nabla^2 V = -\frac{\rho}{\varepsilon_0}$.

15. The potential is constant in a certain region of space. The value of $\vec{E}$ in that region is
(*a*) Zero (*b*) Infinite
(*c*) Depends on the location of the point.

16. The displacement current through a circuit is given by

(*a*) $\varepsilon_0 \frac{\partial \vec{E}}{\partial t}$ (*b*) $\mu_0 \varepsilon_0 \frac{\partial \vec{E}}{\partial t}$ (*c*) $\varepsilon_0 \frac{\partial \phi}{\partial t}$

17. An electromagnetic field satisfies
(*a*) Gauss's law (*b*) Faraday's law and Ampere's law
(*c*) All of them.

18. An isolated magnetic pole does not exist. This is equivalent to saying

(*a*) $\vec{\nabla} \times \vec{A} = 0$, where A is any vector potential

(*b*) $\vec{\nabla} . \vec{B} = 0$, where B is magnetic field intensity

(*c*) $\vec{\nabla} \times \vec{B} = \mu_0 \vec{J}$, where $\vec{J}$ is current density.

19. The differential form of Faraday's law of electromagnetic induction is

(*a*) $\vec{\nabla} \times \vec{B} = -\frac{\partial \vec{E}}{\partial t}$ (*b*) $\vec{\nabla} \times \vec{B} = \mu_0 \vec{J}$ (*c*) $\vec{\nabla} \times \vec{E} = -\frac{\partial \vec{B}}{\partial t}$

20. The direction of induced e.m.f. in a circuit is given by
(*a*) Faraday's law (*b*) Lenz's law (*c*) None of these.

21. The displacement current arises due to
(*a*) Positive charges only
(*b*) Negative chages only
(*c*) Time varying electric field.

22. A vector $\vec{A}$ points vertically upward and $\vec{B}$ points towards north. The vector product $\vec{A} \times \vec{B}$ is
(*a*) Along west (*b*) Along east (*c*) Zero.

23. Let $\vec{C} = \vec{A} + \vec{B}$
(*a*) C is never equal to A + B
(*b*) C is always equal to A + B
(*c*) It is possible to have $|\vec{C}| < |\vec{A}|$ and $|\vec{C}| < |\vec{B}|$

24. If the line integral of $\vec{A}$ over a closed path $\left(\oint \vec{A} \cdot \vec{dr}\right)$ is zero, the field is
(*a*) Solenoidal (*b*) Conservative (*c*) Lamellar.

25. Two parallel wires carrying currents flowing in opposite directions will
(*a*) attract each other (*b*) repel each other (*c*) neither attract nor repel.

26. The electric field produced by a uniformly moving charge is
(*a*) conservative (*b*) non-conservative
(*c*) anything depending upon velocity.

27. Which quantity is invariant under lorentz transformation ?
(*a*) electric field intensity (*b*) charge (*c*) magnetic induction

28. The angle between $\hat{i}$ and $(2\hat{i} + \hat{j})$ is
(*a*) $\cos^{-1} \frac{2}{5}$ (*b*) $\cos^{-1} \frac{2}{\sqrt{5}}$ (*c*) $\cos^{-1} \frac{2}{3}$

29. In free space poisson's equation is
(*a*) $\nabla^2 V = 0$ (*b*) $\nabla^2 V = \frac{\varepsilon_0}{\rho}$ (*c*) $\nabla^2 V = \infty$

30. If $\vec{B} = \vec{\nabla} \times \vec{A}$, $\vec{B}$ and $\vec{A}$ are any vector, then
(*a*) $\vec{\nabla} \cdot \vec{B} = 0$ (*b*) $\vec{\nabla} \cdot \vec{B} = 1$ (*c*) $\vec{\nabla} \cdot \vec{B} = -1$.

31. Maxwell's electromagnetic wave equations in terms of electric field vector in free space is
(*a*) $\nabla^2 \vec{E} = -\varepsilon_0 \mu_0 \frac{\partial^2 \vec{E}}{\partial t^2}$ (*b*) $\nabla^2 E = \mu_0 \varepsilon_0 \frac{\partial^2 E}{\partial t^2}$ (*c*) $\nabla^2 \vec{E} = \mu_0 \varepsilon_0 \frac{\partial^2 \vec{E}}{\partial t^2}$

Answers

1. (*a*)	**2.** (*a*)	**3.** (*b*)	**4.** (*a*)	**5.** (*b*)	**6.** (c)
7. (*b*)	**8.** (*c*)	**9.** (*b*)	**10.** (*c*)	**11.** (*a*)	**12.** (*b*)
13. (*b*)	**14.** (*c*)	**15.** (*a*)	**16.** (*a*)	**17.** (*c*)	**18.** (*b*)
19. (*c*)	**20.** (*b*)	**21.** (*c*)	**22.** (*a*)	**23.** (*c*)	**24.** (*c*)
25. (*b*)	**26.** (*b*)	**27.** (*b*)	**28.** (*b*)	**29.** (*a*)	**30.** (*a*)
31. (*c*).					

Review Questions

1. Comment on the superposition principle in relation to Coulomb's law. [*Burdwan Univ.*]
2. Find an expression for the electrostatic scalar potential at a distance r due to a point charge placed at the origin. [*Burdwan Univ.*]
3. State and explain Gauss law in electrostatics. Obtain its differentiation form. [*Calcutta Univ. W.B.U.T.*]
4. State Gauss's law in electrostatics and extend same to Poisson's equations. When does it reduce to Laplace's equation ? [*Calcutta Univ.*]
5. Derive an expression for the capacitance of a cylindrical capacitor applying Gauss's law. [*W.B.U.T.*]
6. State and prove Poisson's equation in electrostatics. [*Calcutta Univ.*]
7. What is Laplace's equation ? What are their importances in electrostatics ?
8. State Biot-Savart's law. Using Biot-Savart's law, calculate the field at the centre of a circular current carrying coil. [*W.B.U.T. Univ.*]
9. Starting from the expression of $\vec{B} = \frac{\mu_0 I}{4\pi}$. Show that (*i*) $\vec{\nabla} \times \vec{B} = \mu_0 \vec{J}$ and (*ii*) $\vec{\nabla} \cdot \vec{B} = 0$ expression (*ii*)? [*Calcutta University*]
10. Write down Faraday's law of electromagnetic induction in integral form. Comment on the negative sign in the equation. [*B. Univ.*]
11. Write down Maxwell's equations for an electromagnetics field mentioning the law (*s*) represented by each of these equations. [*W.B.U.T., Calcutta Univ.*]
12. Write down Maxwell's electromagnetic field equation in conductors. [*Calcutta Univ., Burdwan Univ*]

Unsolved Problems

1. Find a unit vector perpendicular to $\hat{i} + 2\hat{k}$ and $\hat{i} + \hat{j} - \hat{k}$. Find also the area of a triangle with these two vectors at adjacent sides. $\left[\textbf{Ans.} \pm \frac{1}{\sqrt{14}}\left(-2\hat{i} + 3\hat{j} + \hat{k}\right), \frac{1}{2}\sqrt{14}\right]$ [*Gorakhpur*]
2. Prove that $\hat{i} \times (\vec{v} \times \hat{i}) + \hat{j} \times (\vec{v} \times \hat{j}) + \hat{k} + (\vec{v} \times \hat{k}) = 2\vec{v}$. [*Rohil khand*]
3. Prove that $(\vec{A} \times \vec{B}) \cdot (\vec{C} \times \vec{D}) = (\vec{A} \cdot \vec{C})(\vec{B} \cdot \vec{D}) - (\vec{A} \cdot \vec{D})(\vec{B} \cdot \vec{C})$.
4. Show that the vectors $\hat{i} - 2\hat{j} + 3\hat{k}, -2\hat{i} + 3\hat{j} - 4\hat{k}$ and $\hat{i} - \hat{j} + 2\hat{k}$ are coplaner. [*Bundelkhand*]
5. Show that $\frac{d}{dt}\left(\vec{r} \times \frac{d\vec{r}}{dt}\right) = \vec{r} \times \frac{d^2\vec{r}}{dt^2}$.
6. If $\phi(x, y, z) = 3x^2y - y^3z^2$, find the value of $\vec{\nabla}\phi$ at the point (1, - 2, - 1). $\left[\textbf{Ans.} -12\hat{i} - 9\hat{j} - 16\hat{k}\right]$ [*WBUT, Ravishanker*]
7. In an electric field the electric potential is given by $V(x, y, z) = \left(2x^2 + 4y^2 + 3z^2\right)^{-\frac{1}{2}}$, find the electric field at a point (1, 1, 1). [*WBUT*]

8. If $\phi = x^2 - y^2 + 2z$, find $\vec{\nabla} \cdot \left(\vec{\nabla}\phi\right)$. [*WBUT*]

9. Show that $\vec{B}\ (x, y, z) = 2xyz\hat{i} + \left(x^2z + 2y\right)\hat{j} + x^2y\hat{k}$ is irrotational. [*WBUT*]

10. Find the components of the vector $\vec{A} = 3\hat{i} - \hat{j} - \hat{k}$ along and perpendicular to the vector $\vec{B} = \hat{i} - \hat{j}$. [The components of $\vec{A}$ along $\vec{B}$ $(A\cos\theta)\ \hat{b} = \dfrac{(AB\cos\theta)}{B}\hat{b} = \dfrac{(\vec{A}\cdot\vec{B})\hat{b}}{B} = \dfrac{(\vec{A}\cdot\vec{B})\vec{B}}{B^2}$ the sum of the components must be equal to the vector $\vec{A}$ therefore the component of $\vec{A}$ perpendicular to $\vec{B}$ is $\vec{A}$ – (component of $\vec{A}$ along $\vec{B}$) $= \left(3\hat{i} - \hat{j} - \hat{k}\right) - \left(2\hat{i} - 2\hat{j}\right) = \hat{i} + \hat{j} - \hat{k}$]

11. Equipotential surface in some problems are spherical in shape : $V(x, y, z) = C(x^2 + y^2 + z^2)$ show that the force is radial is along the radius vector.

[**Hints:** $\vec{F}\alpha - \vec{r}$] [Meerut]

12. The electric field at any point (x, y, z) b given by $\vec{E} = 2x\hat{i} + 2y\hat{j} + 2z\hat{k}$. Derive an equation for the equipotential surface. What is such a surface called ?

[**Ans.** $-(x^2 + y^2 + z^2)$ + constant, spherical surface] [*Gorakhpur*]

13. Show that $\vec{\nabla}\phi$ is a vector perpendicular to the surface $\phi(x, y, z) = e$, where C is a constant.

14. Find $\vec{\nabla}\phi$ if (*a*) $\phi = \ln|r|$ (*b*) $\phi = \dfrac{1}{r}$ [**Ans.** (*a*) $\dfrac{\vec{r}}{r^2}$, (*b*) $-\dfrac{\vec{r}}{r^3}$] [*Meerut, Agra, Rajasthan*]

15. Prove that $\vec{\nabla}r^n = nr^{n-2}\vec{r}$ [*Meerut, Awadh, Agra*]

16. If $\vec{A} = i + 2j - 3k$ and $\vec{B} = 3i - j + 2k$, show that the vectors $\vec{A} + \vec{B}$ and $\vec{A} - \vec{B}$ are perpendicular vectors.

17. Show that $(\vec{w} \times \vec{r})^2 = \vec{w}\left[\vec{r} \times (\vec{w} \times \vec{r})\right]$

18. Verify the relation $\left(\vec{A} \times \vec{B}\right)\cdot\left(\vec{C} \times \vec{D}\right) = \left(\vec{A}\cdot\vec{C}\right)\left(\vec{B}\cdot\vec{D}\right) - \left(\vec{A}\cdot\vec{D}\right) - \left(\vec{A}\cdot\vec{D}\right)\left(\vec{B}\cdot\vec{C}\right)$.

19. If r be the position vector, prove the following :

(*a*) $\vec{\nabla}\cdot\dfrac{\vec{r}}{r^3} = 0$ (*b*) $\vec{\nabla}\cdot\vec{r} = 3$ (*c*) $\vec{\nabla}\cdot r^n = nr^{n-2}\vec{r}$ (*d*) $\vec{\nabla}^2 r^n = n(n+1)r^{n-2}$.

20. If $v(t)$ is a time-dependent vector show that

$$\vec{v}\cdot\frac{d\vec{v}}{dt} = v\frac{dv}{dt},$$

where v is the magnitude of the vector $\vec{v}$.

21. Prove the identity $\left(\vec{A}\cdot\vec{\nabla}\right)\vec{A} = \dfrac{1}{2}\vec{\nabla}\vec{A}^2 - \vec{A} \times \left(\vec{\nabla} \times \vec{A}\right)$

22. Prove the identity $\vec{\nabla} \times \left(\vec{\nabla} \times \vec{A}\right) = \vec{\nabla}\left(\vec{\nabla}\cdot\vec{A}\right) - \vec{\nabla}^2\vec{A}$

23. A force $\vec{F}$ is given by $\vec{F} = 2xy\,i + x^2\,j + Axz^2\,k$. For that value of A is the force $\vec{F}$ conservative?

24. Verify the divergence theorem for a vector $\vec{A}$ given by

$$\vec{A} = i\frac{x}{r} + j\frac{y}{r} + k\frac{z}{r} = \frac{\vec{r}}{r}.$$

where, i, j and k are the unit vectors along x, y and z respectively, the region of integration being a sphere of radius r with the centre at the origin.

25. Use poisson's equation to find the charge density in a region where the potential energy function is

$$v = a - b\left(x^2 + y^2\right) - c \log \sqrt{x^2 + y^2}$$ [**Ans.** $\rho = 4\varepsilon_0\left(b + \frac{e}{x^2 + y^2}\right)$]

a, b, c being constants.

26. Explain whether the field given by

$$\vec{E} = a\left(xy^2 i + y^3 j\right)$$

are conservative. Here i and j are unit vectors and 'a' is a constant.

27. Use Laplaces equation to show that the function $v = 2x^2 + 7y - 2z^2$ represents the potential function in a charge free region.

[**Hints :** $\nabla^2 v = 4 + 0 - 4 = 0$. Hence $\rho(x, y, z) = 0$]

Robert Hooke

2A

Chapter

Elasticity

2A.1 INTRODUCTION

A body remains in natural equilibrium under the internal molecular forces of attraction whose magnitude depends upon the spacing between the molecules. When external forces are applied to a body, the distance between different parts of the body changes and the body is deformed. The body so deformed has a natural tendency to regain its primary state by an internal resistive force developed with the body to oppose the external applied force. When the external forces are removed, the new internal forces bring the body to its normal state. This property of the body is known as elasticity. So we can say "the property, by virtue of which a body tends to recover its original configuration (shape and size) on the removal of the deforming forces, is called elasticity". The bodies which show such a property are called perfectly elastic bodies. If the body does not show any tendency to regain its original configuration is known as plastic bodies.

2A.2 STRESS AND STRAIN

When external forces acts on a body, internal forces or restoring forces opposing the former are developed at each section of the body. The restoring force per unit area called into play inside the body is called the stress.

If there is no permanent change produced in the body, the restoring force is equal and opposite to the applied force.

If F is restoring force acting normally on the area of cross-section 'A'

$$\text{Stress} = \frac{F}{A}.$$

Its units are Newton/metre^2 (in M.K.S.) or dyne/cm^2 (in C.G.S.)

Strain

When a deforming force is acted upon a body, its configuration (length, volume or shape) suffers a change. Relative change in configuration due to the application of deforming forces is called strain. There are three types of strain :

(*i*) **Longitudinal Strain**

It is ratio between change in length to its original length.

Consider a wire of length L is stretched by a length. So longitudinal strain $= \frac{l}{L}$.

It has no unit, since it is pure ratio.

(*ii*) **Volume Strain**

It is the ratio between change in volume to its original volume. If v is the change in volume produce by the force in the initial volume V,

Volumetric strain $= \frac{v}{V}$.

It has no unit.

(*iii*) **Shear Strain**

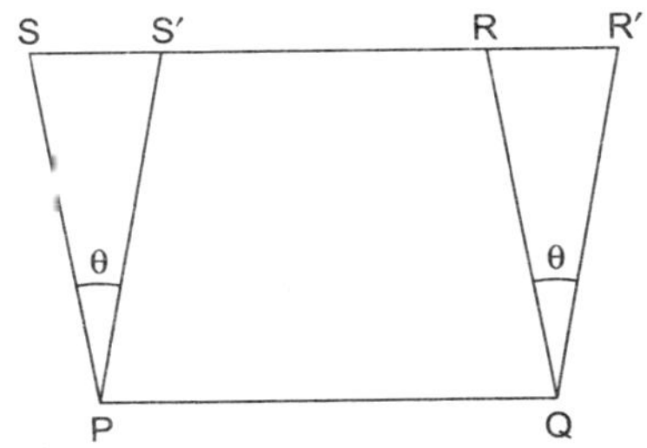

Consider PQRS be cross-sectional view of a cube having its F lower face PQ fixed. A tangential force F is applied to the upper face and it gets distorted to the position PQR′S′, then

$$\text{Shearing Strain} = \frac{\text{Displacement of the place}}{\text{Distance from the fixed place}}$$

$$= \frac{SS'}{SP} = \tan\theta = \theta \qquad \text{[If } \theta \text{ is small]}$$

So shearing strain can be defined as the ratio between displacement in one place to its distance from the fixed plane, it has no unit.

2A.3 HOOKE'S LAW

It states that, "Within the elastic limit, the stress applied to a body is directly proportional to the strain developed". That is Stress ∝ Strain

or, $$\frac{\text{Stress}}{\text{Strain}} = \text{Constant.}$$

The constant of proportionality is known as modulus of elasticity of the body and depends upon the material of the body.

Corresponding to the three types of strain, there are three types of co-efficient of elasticity.

(*i*) **Young's Modulus of Elasticity (Y)**

It is defined as the ratio between longitudinal stress to the longitudinal strain. It is denoted by 'Y'

$$Y = \frac{\text{longitudinal stress}}{\text{longitudinal strain}}.$$

If L is the length of a wire of radius 'r' and an increase in length l is caused by a force F then

$$Y = \frac{F/\pi r^2}{l/L} = \frac{FL}{\pi r^2 l}$$

It has the unit of stress, since strain does not possess any unit. In S.I. system Newton/metre^2 and C.G.S. system. Dyne/cm^2.

(*ii*) **Bulk modulus of Elasticity (K)**

It is defined as the ratio between normal stress to the volumetric strain. It is denoted by 'K'.

$$K = \frac{\text{Normal stress}}{\text{volumetric strain}}.$$

Consider a sphere of volume 'V'. On applying deforming forces from all sides it gets reduced in size. Now,

$$K = \frac{F/\pi r^2}{v/V} = \frac{P}{v/V} = \frac{PV}{v}.$$

where P is the pressure and v is the change in volume.

(*iii*) **Modulus of rigidity (η)**

It is defined as the ratio between tangential stress to the shearing strain. It is denoted by 'η'.

$$\eta = \frac{\text{Tangential stress}}{\text{Shearing strain}}$$

Consider a cube of face area 'A' having lower face fixed and a tangential force 'F' applied at the upper face and produces a shear strain θ, then modulus of rigidity, $\eta = \frac{F/A}{\theta}$.

2A.4 POISSON'S RATIO

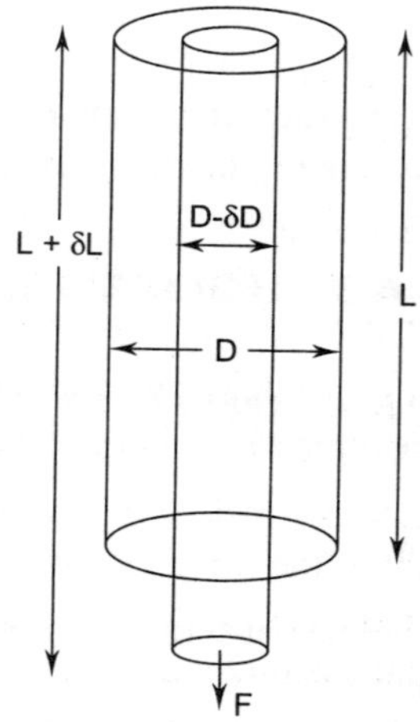

When a wire is stretched, it elongates but at the same time its diameter contracts. In general, when a body extends in a particular direction due to longitudinal stress, it contracts in perpendicular direction. Thus, longitudinal strain is attended always with a lateral strain. Within the elastic limit, lateral strain is directly proportional to longitudinal strain. So the ratio of lateral strain to the longitudinal strain is defined as Poisson's ratio and denoted by 'σ'.

Consider a wire of length L, diameter D is increases to its length L + δL and the diameter decreases to D – δD, then

$$\sigma = \frac{\text{Lateral strain}}{\text{Longitudinal strain}} = \frac{\delta D/D}{\delta L/L} = \frac{\delta D \cdot L}{D \cdot \delta L}$$

It has no unit and it will be shown in next that σ should lie between $+\frac{1}{2}$ and –1.

2A.5 A SHEAR IS EQUIVALENT TO A COMPRESSION AND AN EXTENSION AT RIGHT ANGLES TO EACH OTHER

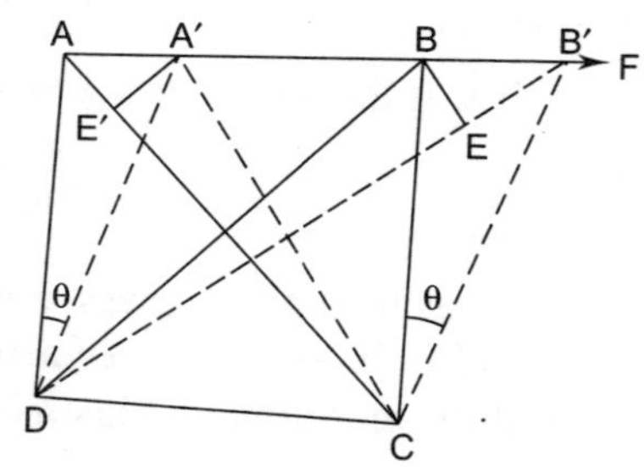

Let ABCD be the section of a cube fixed at the lower surface. Let a tangential force F be applied to its upper surface, so that it is sheared through a small angle θ and takes the shape A′B′CD. In this process DB is extended to DB′ while AC is compressed to A′C.

Let Arc BE and A′E′ be drawn with centres D and C, and the radii DB and CA′ respectively. These may be taken as perpendiculars to DB′ and CA respectively, since θ is small.

From the diagram, DB = DE and DC = DB cos 45°

$$= \frac{DB}{\sqrt{2}} \text{ or } DB = \sqrt{2}\,DC = \sqrt{2}l$$

[l is the length of each side of the cube]

Now $\quad EB' = BB' \cos < EB'B = BB' \cos 45°$

[<EBB′ = 45°]

$$= \frac{BB'}{\sqrt{2}}.$$

But $\quad BB' = BC \cdot \theta = l\theta$

$$EB' = \frac{l\theta}{\sqrt{2}}.$$

Now linear elongation of DB is EB′. Let α be the linear extensional strain

$$\alpha = \frac{EB'}{DE} = \frac{EB'}{DB} = \frac{\frac{l\theta}{\sqrt{2}}}{\sqrt{2}l} = \frac{\theta}{2}$$

or $\quad \theta = 2\alpha.$

Thus θ is the twice the linear extensional strain or fractional elongation. Similarly, the diagonal AC contracts by AE′, and the linear contractional strain or fractional contraction is found to be equal to θ/2. Thus a simple shear is produced as a result of equal amount of fractional linear elongation and fractional linear contraction at right angles to each other, either (extension or contraction) being inclined at an right angle of 45° to the direction of the shearing displacement.

2A.6 INTER RELATION AMONG THE ELASTIC CONSTANT

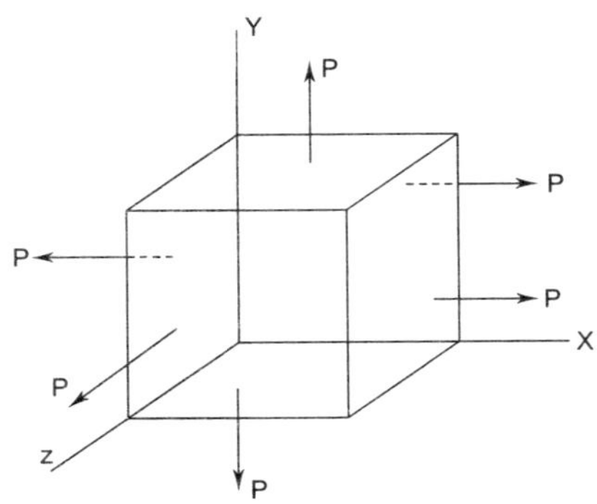

Let us consider a cube whose sides, each of length l, are parallel to the three co-ordinate axes x, y and z. Let a uniform normal stress P be applied over its surface, Then each side of the cube is under an extension stress P. The stress P acting parallel to the x-axis will produce extension parallel to the x-axis, and compressions parallel to y and z axes.

These will be as follows :

Extension parallel to x-axis =

longitudinal strain × initial length

$$= \frac{\text{Stress}}{\text{Young's modulus}} \times l = \frac{P}{Y} \times l$$

Compression parallel to y axis

= Lateral strain × initial length

= Poisson's ratio × Longitudinal strain × Initial length

$$= \sigma \times \frac{P}{Y} \times l \quad \left[\sigma = \frac{\text{lateral strain}}{\text{longitudinal strain}}\right]$$

Compression parallel (||) to z-axis

$$= \sigma \times \frac{P}{Y} \times l$$

Similarly the stress P acting parallel to *y*-axis produces extension parallel to *y*-axis

$$= \frac{P}{Y} \times l$$

Compression parallel to *x*-axis $= \sigma \frac{P}{Y} l$

and Compression parallel to *z*-axis $= \sigma \times \frac{P}{Y} \times l$

Also, the stress P acting parallel to z-axis produces extension parallel to *z*-axis

$$= \frac{P}{Y} \times l$$

Compression parallel to *x*-axis $= \sigma \times \frac{P}{Y} \times l$

and Compression parallel to *y*-axis $= \sigma \times \frac{P}{Y} \times l$

Hence net extension parallel to *x*-axis

$$= \frac{P}{Y} l - \sigma \frac{P}{Y} l - \sigma \frac{P}{Y} l$$

$$= (1 - 2\sigma) \frac{P}{Y} l$$

Net extension parallel to *y*-axis $= (1 - 2\sigma) \frac{P}{Y} l \equiv \left[-\frac{\sigma P}{Y} l + \frac{P}{Y} l - \sigma \frac{P}{Y} l \right]$

and net extension parallel to *z*-axis

$$= - \sigma \frac{P}{Y} l - \sigma \frac{P}{Y} l + \frac{P}{Y} l = \frac{P}{Y} (1 - 2\sigma) l$$

Thus each side of the cube increase from l to $l + \frac{P}{Y} (1 - 2\sigma) l$

The new volume is therefore

$$= \left[l + \frac{P}{Y} (1 - 2\sigma) l \right]^3$$

$$= l^3 \left[1 + \frac{P}{Y} (1 - 2\sigma) \right]^3$$

$$= l^3 \left[1 + \frac{3P}{Y} (1 - 2\sigma) \right] \quad \text{[Higher order terms are neglected]}$$

Since the initial volume was l^3, the change in volume

$$= l^3\left[1+\frac{3P}{Y}(1-2\sigma)\right]-l^3=\frac{3Pl^3}{Y}(1-2\sigma)$$

$$\text{Hence, volume strain} = \frac{\text{Change in volume}}{\text{Initial volume}} = \frac{\frac{3P}{Y}(1-2\sigma)\,l^3}{l^3}$$

$$= \frac{3P}{Y}(1-2\sigma)\,.$$

Now, considering the cube as a whole, its surface being under a uniform normal stress P undergoes the volume strain $\frac{3P}{Y}(1-2\sigma)$. Its bulk modulus is therefore,

$$k = \frac{\text{Normal stress}}{\text{Volume strain}} = \frac{P}{\frac{3P}{Y}(1-2\sigma)} = \frac{Y}{3(1-2\sigma)}$$

$$\boxed{Y = 3k(1-2\sigma)} \qquad \text{...(2A.1)}$$

Let us consider a cube of side l, upon which an extensional stress P parallel to x-axis, and a compressional stress P parallel to y-axis have been applied. There is no stress along z-axis

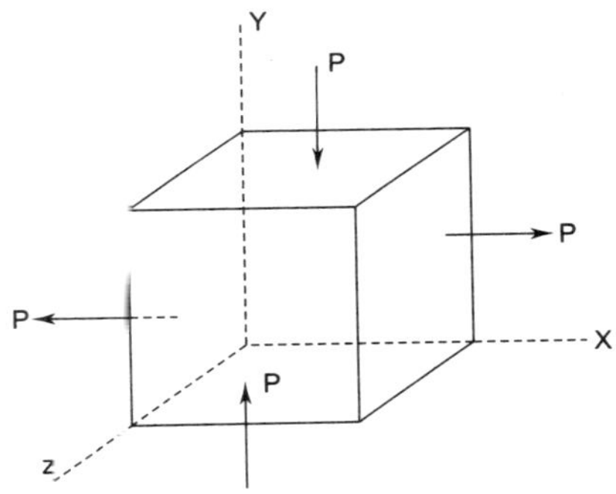

The extensional stress P produces extension along x-axis

$$= \frac{P}{Y}\times l$$

$$\text{Compression along } y\text{-axis} = \sigma\times\frac{P}{Y}\times l$$

and $$\text{compression along z-axis} = \sigma\times\frac{P}{Y}\times l$$

The compression stress P produces

$$\text{Compression along } y\text{-axis} = \frac{P}{Y}\times l$$

$$\text{extension along } x\text{-axis} = \sigma\times\frac{P}{Y}\times l$$

and $$\text{extension along } z\text{-axis} = \sigma\times\frac{P}{Y}\times l$$

$$\text{Thus, net extension along } x\text{-axis} = \frac{P}{Y}l+\sigma\frac{P}{Y}l = l(1+\sigma)\frac{P}{Y}$$

and $$\text{net compression along } y\text{-axis} = \sigma\frac{P}{Y}l+\frac{P}{Y}l=\frac{P}{Y}(1+\sigma)\,l\,.$$

There are equal extension and compression along the z-axis. Hence, there is no change along z-axis.

Since each side of the cube is of initial length l, the extensional and compressional strains along x- and y-axes are each equal to

$$\frac{P}{Y}\frac{(1+\sigma)\,l}{l} = \frac{P}{Y}(1+\sigma).$$

Now simultaneous equal extensional and compressional strains at right angles to each other, are equivalent to a shear θ (say)

$$\therefore \quad \frac{P}{Y}(1+\sigma)+\frac{P}{Y}(1+\sigma) = \theta$$

$$\frac{P}{\theta} = \frac{Y}{2\,(1+\sigma)}.$$

Further, the extensional stress P parallel to the x-axis, and compessional stress P parallel to the y-axis will be equivalent to a shearing stress P. Hence, the modulus of rigidity of the material of the cube is

$$\eta = \frac{\text{shearing stress}}{\text{shear}} = \frac{P}{\theta}.$$

Putting the value of $\frac{P}{\theta}$ from above, we have

$$\eta = \frac{Y}{2\,(1+6)}$$

$$\boxed{Y = 2\eta\,(1+\sigma)} \qquad \ldots(2A.2)$$

$$\frac{Y}{2\eta} = 1+\sigma,$$

or

$$\sigma = \frac{Y}{2\eta} - 1.$$

Let us put this value of σ in (2A.1), then

$$Y = 3k\left[1-2\cdot\left(\frac{Y}{2\eta}-1\right)\right] = 3k\left(3-\frac{Y}{\eta}\right)$$

or

$$Y = qk - \frac{3kY}{\eta}$$

$$\therefore \quad \boxed{Y = \frac{9\eta k}{\eta+3k}} \qquad \ldots(2A.3)$$

Equations (2A.1) and (2A.2) give

$$Y = 3k\,(1-2\sigma) = 2\eta\,(1+\sigma)$$

or,

$$\boxed{\sigma = \frac{3k-2\eta}{6k+2\eta}} \qquad \ldots(2A.4)$$

Equations (2A.1) – (2A.4) are the various relations among the elastic constants.

2A.7 THEORETICAL LIMITING VALUES OF POISSON'S RATIO

The relation between Poisson's ratio σ to bulk modulus k and modulus of rigidity η is

$$3k(1 - 2\sigma) = 2\eta(1 + \sigma)$$

If σ is positive, R.H.S. is positive. The L.H.S. must also be positive.

This is so when $\quad 2\sigma < 1$

or, $\quad \sigma < \frac{1}{2}$.

or, Now if is σ negative the L.H.S. is positive and hence R.H.S. must also be positive. This is so when,

$$1 + \sigma > 0$$

or, $\quad \sigma > -1$.

Theoretically σ must lie between -1 and $\frac{1}{2}$. σ is however, never negative. Hence, in practice, σ lies between 0 and $\frac{1}{2}$.

For an Incompressible Material

Whose volume remain unchanged by changing pressure, we have

$$k = \alpha. \qquad \left[\because \; k = \frac{\frac{P}{\delta v}}{v} = \frac{P}{0}\right]$$

An infinite value of bulk modulus means that the body is incompressible.

Putting $\quad k = \alpha$.

$$\frac{Y}{3k} = 1 - 2\sigma, \text{ we get}$$

$$1 - 2\sigma = 0, \sigma = \frac{1}{2}.$$

2A.8 CANTILEVER

Expression for the depression of a uniform bar held horizontally at one end and loaded at the other

Beam Clamped at one end and loaded at the other (cantilever) – **Let us consider a thin,** uniform and light beam of length l camped horizontally at one end A (Fig. 2A.1), when it is loaded with a weight W at the free end B, the end B is depressed downward compare to A so that the beam undergoes bending. (The dotted boundary in Fig. 2A.1 represents the position of the beam before loading). Such a system is called a 'cantilever'.

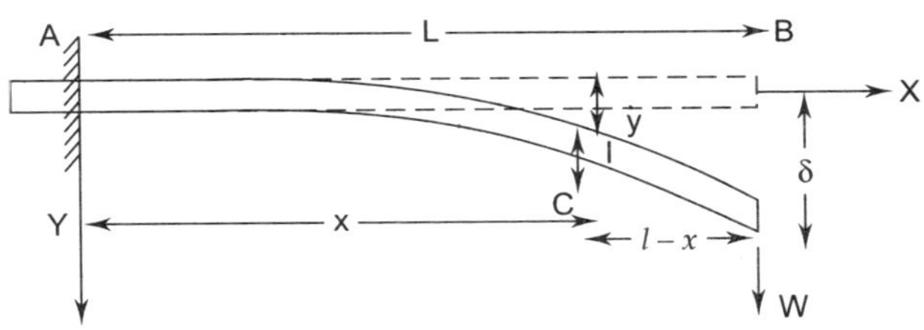

Figure 2A.1

The beam is light, the whole depression may be taken as due to the load W.

Let us take a section of the beam at C, distant x from A and consider the equilibrium of the part CB. Since the beam is fixed at A, the load W at B exists an external torque tending to rotate the beam clockwise. Its magnitude is W $(l - x)$. The torque is balanced by an anticlockwise restoring torque formed by the internal forces arise as a result of elastic reaction against the extension of filaments above the neutral surface and compressed below the neutral surface. The magnitude of the restoring torque is $\frac{YI}{\rho}$, where Y is the Young's modulus of the material of the beam, I is the geometrical moment of Inertia of the section C about the neutral surface and ρ is the radius of curvature of the bent beam at C. At equilibrium, therefore,

$$W(l - x) = \frac{YI}{\rho}$$

$$\rho = \frac{YI}{W(l - x)} \qquad \ldots(2A.5)$$

This expression shows that the radius of the curvature at a point of the beam varies inversely as $(l - x)$, the distance of the point from loaded end. Thus, in this case the beam is not bent in circular arc.

Now, let y be the depression at point C. Taking the end A as origin, Let us draw X, Y axes as shown in figure. Then (x, y) are the co-ordinates of the point C, and the radius of curvature at this point is given by

$$\rho = \frac{\left[1 + \left(\frac{dy}{dx}\right)^2\right]^{\frac{3}{2}}}{\frac{d^2y}{dx^2}},$$

where $\frac{dy}{dx}$ is the slope of the tangent at a point (x, y). If the depression be within elastic limit, the slope will be small. Therefore, $\left(\frac{dy}{dx}\right)^2$ will be negligible compared to 1, and we can put,

$$\rho = \frac{1}{\frac{d^2y}{dx^2}}$$

Substituting for ρ in (2A.5), we get

$$\frac{1}{\frac{d^2y}{dx^2}} = \frac{YI}{W(l - x)}$$

or

$$\frac{d^2y}{dx^2} = \frac{W(l - x)}{YI}$$

Integrating this, we get

$$\frac{dy}{dx} = \frac{W}{YI}\left(lx - \frac{x^2}{2}\right) + A_1$$

where A_1 is constant of integration. At the clamped end A of the beam the tangent is horizontal, *i.e.*, at $x = 0$, we have $\frac{dy}{dx} = 0$, putting this value in last expression we find $A_1 = 0$

$$\therefore \quad \frac{dy}{dx} = \frac{W}{YI}\left(lx - \frac{x^2}{2}\right)$$

Integrating again, we get

$$y = \frac{W}{YI}\left(l\frac{x^2}{2} - \frac{x^3}{6}\right) + A_2$$

where A_2 is again a constant of integration. Again at $x = 0$, we have $y = 0$, so that $A_2 = 0$

$$y = \frac{W}{YI}\left(l\frac{x^2}{2} - \frac{x^3}{6}\right).$$

This is the expression for the depression at a distance x from the fixed end.

At B, we have $x = l$, and the depression y is a maximum. Let it be equal to δ. Then, substituting l and δ for x and y respectively in the above expression, we have

$$\delta = \frac{W}{YI}\left(\frac{l^3}{2} - \frac{l^3}{6}\right) = \frac{Wl^3}{3YI} \qquad \ldots(2A.6)$$

If the beam is of rectangular cross-section, $I = \frac{bd^3}{12}$, where b is the breadth and d is the thickness of the beam, Therefore from equation (2A.6), we get

$$\delta = \frac{4Wl^3}{Ybd^3} \qquad \ldots(2A.7)$$

If the beam is circular cross-section of radius r, then

$I = \frac{\pi r^4}{4}$, Then from (2A.6), we get

$$\delta = \frac{4Wl^3}{3Y\pi r^4} \qquad \ldots(2A.8)$$

Uniform cantilever is more likely to break at fixed end

The external load W is applied at the free end of the cantilever exerts a bending torque at each section of the cantilever. Its value at a distance x from the fixed end is W $(l - x)$, where l is the length of the cantilever. This is maximum for $x = 0$, that is, at the fixed end. Hence, the cantilever is more likely to break near the fixed end.

Expression for the Amount of Depression for a beam Loaded at the Middle and Supported at the ends

Beam loaded at the middle point:

Let AB (Fig. 2A.2) be a beam of rectangular cross-section resting symmetrically on two horizontal knife-edges K_1 and K_2 and loaded at its middle-point D by a weight W. The reaction at each Knife-edge is $\frac{W}{2}$, acting vertically upward. The beam bends symmetrically with respect to its middle point D at which the depression is maximum. Let l be the distance between K_1 and K_2.

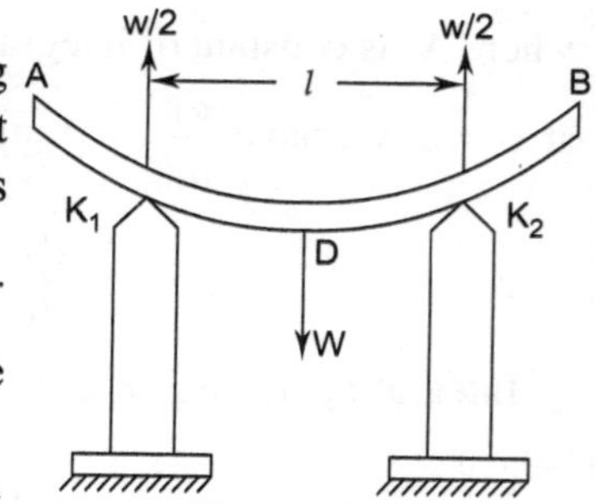

Figure 2A.2

On account of symmetrical lending on two sides of D, a tangent drawn at D will be horizontal. Hence, each half of the beam, DA and DB, may be considered as a cantilever of length $l/2$ fixed at one end D and loaded by an upward load $\frac{W}{2}$ at the other. Therefore, we have simply to determine the elevation of B (say) above D (Fig. 2A.3).

Let us take a section at C, at a distance x from D. Let us now consider the equilibrium of the part CB. Since beam is fixed at D, the load $\frac{W}{2}$ at B exerts a torque on CB tending to rotate it anti-clockwise. Its magnitude is $\frac{W}{2}\left(\frac{l}{2}-x\right)$. This torque is balanced by a clockwise restoring torque supplied by the internal forces exerted by the part DC over the section C. These forces arises due to elastic reaction of the beam against the extension of filaments on one side of the neutral surface and compression on the other. The magnitude of the restoring torque is $\frac{YI}{\rho}$.

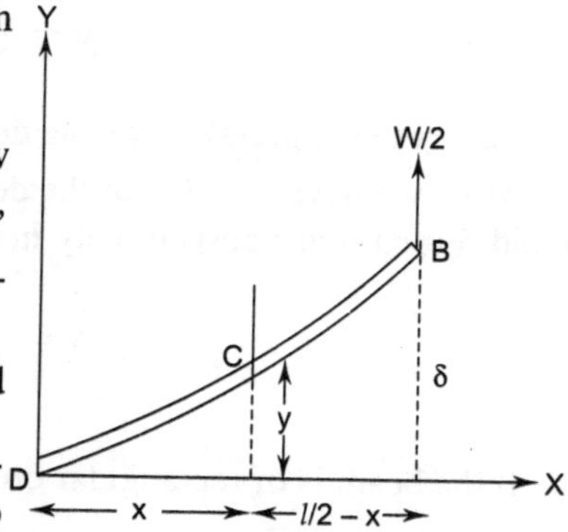

Figure 2A.3

At equilibrium, therefore

$$\frac{W}{2}\left(\frac{l}{2}-x\right) = \frac{YI}{\rho}$$

$$\rho = \frac{2YI}{W\left(\frac{l}{2}-x\right)} \qquad \text{...(2A.9)}$$

Let us choose the coordinate axes X, Y as shown, the origin O being at D. Let y be the elevation at C. The co-ordinates of C are (x, y). Therefore, if the elevation is small, the radius of curvature at C is given by

$$\rho = \frac{1}{d^2y/dx^2}$$

Substituting this in (2A.9), we get

$$\frac{1}{d^2y/dx^2} = \frac{2YI}{W\left(\frac{l}{2}-x\right)} \quad \text{or} \quad \frac{d^2y}{dx^2} = \frac{W}{2YI}\left(\frac{l}{2}-x\right).$$

Integrating this, we get

$$\frac{dy}{dx} = \frac{W}{2YI}\left(\frac{l}{2}x - \frac{x^2}{2}\right) + A_1$$

where A_1 = constant. At the fixed point D, the tangent is horizontal, *i.e.*, at $x = 0$, we have $\frac{dy}{dx} = 0$. Substituting these values in the last expression, we find that $A_1 = 0$.

$$\therefore \qquad \frac{dy}{dx} = \frac{W}{2YI}\left(\frac{l}{2}x - \frac{x^2}{2}\right)$$

Integrating again, we get

$$y = \frac{W}{2YI}\left(\frac{lx^2}{4} - \frac{x^3}{6}\right) + A_2$$

where A_2 is again a constant of integration. Again at D we have $x = 0$ and $y = 0$, so that $A_2 = 0$.

$$y = \frac{W}{2YI}\left(\frac{lx^2}{4} - \frac{x^3}{6}\right) \qquad \text{...(2A.10)}$$

At B we have $x = \frac{l}{2}$, and the elevation y is maximum. Let be equal to δ. Then substituting in (2A.10) δ for y and $\frac{l}{2}$ for x, we have

$$\delta = \frac{W}{2YI}\left(\frac{l^3}{16} - \frac{l^3}{48}\right) = \frac{Wl^3}{48\,YI}.$$

If b and d be the width and thickness of the beam respectively, we have,

$$I = \frac{bd^3}{12}. \text{ Therefore, } \delta = \frac{Wl^3}{4Y\,bd^3}.$$

This is the same as the depression at the middle point D of the beam for the cantilever.

Example 1. *The rectangular cross-section of long rod has sides in the ratio 1 : 2. Calculate the ratio of the depressions under the same load when*

(i) the smaller side is vertical,

(ii) the longer side is vertical. [Jlwaji & Vikram University]

Solution. We know that the depression at the free end of a cantilever of rectangular cross-section under a load W is

$$\delta = \frac{4Wl^3}{Ybd^3}.$$

where l is length, b is width and d is thickness of the cantilever-rod. Consider a and $2a$ be the sides (ratio 1 : 2) of the cross-section.

(*i*) When the smaller side is vertical, $b = 2a$, $d = a$.

$$\delta_1 = \frac{1}{2}\,\frac{4Wl^3}{Ya^4}.$$

(*ii*) When the longer side is vertical $b = a$, $d = 2a$.

$$\delta_2 = \frac{4Wl^3}{Y(a)(2a)^3} = \frac{1}{8}\frac{4Wl^3}{Ya^4}$$

$$\frac{\delta_1}{\delta_2} = \frac{1/2}{1/8} = \frac{4}{1}$$

$$\delta_1 : \delta_2 = 4 : 1.$$

Example 2. *For the same cross-sectional areas, show that the beam of a square cross-section is stiffer than one of circular cross-section of the same length and material. Show also that for a given load the depressions are in the ratio 3 : π* [Rajasthan, Kanpur, Gorakhpur, Ranchi University]

Solution. The depression δ at the loaded end of a beam whose other end is clamped horizontally is given by

$$\delta = \frac{Wl^3}{3YI}.$$

where W is the load, l the length of the beam and I the geometrical moment of inertia of first sections. For a rectangular beam, we know

$$I = \frac{bd^3}{12}.$$

where b and d are the breadth and depth of its section. If the beam has a square section, $b = d$, then

$$I = \frac{b^4}{12}.$$

I $r = \delta_1$ be the depression for this beam,

then,
$$\delta_1 = \frac{Wl^3}{3Y\left(\frac{b^4}{12}\right)} = \frac{4Wl^3}{Yb^4}.$$

For a beam of circular cross-section of radius r,

$$I = \frac{\pi r^4}{4}.$$

Therefore, if δ_2 be the depression for this beam, we have

$$\delta_2 = \frac{Wl^3}{3Y\left(\frac{\pi r^4}{4}\right)} = \frac{4Wl^3}{3Y\pi r^4}$$

(*i*) ÷ (*ii*) ⇒

$$\frac{\delta_1}{\delta_2} = \frac{3\pi r^4}{b^4}.$$

But the cross-sectional areas of the two beams are equal so, $b^2 = \pi r^2$

$$\therefore \qquad \frac{\delta_1}{\delta_2} = \frac{3\pi r^2}{(\pi r^2)^2} = \frac{3}{\pi},$$

which is less than 1

$$\therefore \qquad \delta_1 < \delta_2.$$

Here, the depression at the loaded end of the beam of square cross-section is smaller than that of the beam of circular cross section. Hence, the square cross-section is stiffer than the circular cross-section.

Example 3. *Calculate the length of the wire that will break under its own weight when suspended vertically. Given breaking stress = 9.8 × 10⁸ Nm⁻², density of wire = 10⁴ kg m⁻³ and g = 9.8 ms⁻².*

[Bangalore Univ., Kanpur Univ.]

Answer. Given that the wire hangs vertically. So its weight mg act as a longitudinal force.

Consider L be the maximum length of the wire and which can hang without breaking.

$$\text{Weight of wire mg} = La\,\rho\,g \text{ Newton.}$$

According to the problem,

$$\text{Breaking loud} = \text{Breaking stress} \times a$$

So,
$$La\,\rho\,g = 9.8 \times 10^8 \text{ N m}^{-2} \times a$$

or,
$$L = \frac{9.8 \times 10^8}{9.8 \times 10^4} = 10^4 \text{ meter.}$$

Minimum length for breaking stresses

$$= 10^4 \text{ meter.}$$

Example 4. *Find the work done in stretching a wire of 1 sq. mm cross-section Young's modulus 2×10^{11} N/m² and 2 m long through 0.1 mm.* [Gharwal University]

Answer. We know that the workdone in a stretching wire is

$$W = \frac{1}{2}\frac{Y\alpha l^2}{L}$$

Here,
Y = Young's modulus of the wire = 2×10^{11} N/m²
α = Area of cross-section = 1 sq. mm = 10^{-6} m²
l = 0.1 mm = 0.0001 meter, and length of the wire, L = 2 m

or,
$$W = \frac{1}{2} \times \frac{2 \times 10^{11} \times 10^{-6} \times (0.0001)^2}{2\,\text{m}} = 5 \times 10^{-4} \text{ Joule.}$$

Example 5. *(a) Calculate the Poisson's ratio for silver. Given Young's modulus for silver is 7.25 $\times 10^{10}$ N/m² and bulk modulus is 11×10^{10} N/m².*

(b) The volume of a solid does not vary with pressure. Find Poisson's ratio for the solid.

[Meerut University, Gharwal University]

Answer. (*a*) Given that, $Y = 7.25 \times 10^{10}$ N/m², $K = 11 \times 10^{10}$ N/m²

Now
$$Y = 3K(1 - 2\sigma)$$

$$\therefore \qquad 1 - 2\sigma = \frac{Y}{3K} \quad \text{or} \quad 2\sigma = 1 - \frac{Y}{3K}$$

$$\therefore \qquad \sigma = \frac{1}{2}\left[1 - \frac{Y}{3K}\right] = \frac{1}{2}\left[1 - \frac{7.25 \times 10^{10}}{3 \times 11 \times 10^{10}}\right] = 0.39.$$

(*b*) Again, we know $\quad Y = 3K\,(1 - 2\sigma) \quad$ or $\quad 1 - 2\sigma = \dfrac{Y}{3K},$

where Young's modulus Y, bulk modulus K, and Poisson's ratio σ

Now bulk modulus $\quad K = \dfrac{\text{Applied pressure}}{\text{Volumetric strain}} = \dfrac{P}{\dfrac{v}{V}}.$

As the volume of the solid does not vary with pressure, the volumetric strain is zero

or $$\frac{v}{V} = 0.$$

Therefore, $K \propto$ (infinity)

and $$\frac{Y}{3K} = 0$$

Hence, $$1 - 2\sigma = 0 \quad \text{or,} \quad 2\sigma = 1$$

$$\therefore \qquad \sigma = \frac{1}{2} = 0.5.$$

2A.9 TORSION OF A RIGIDITY

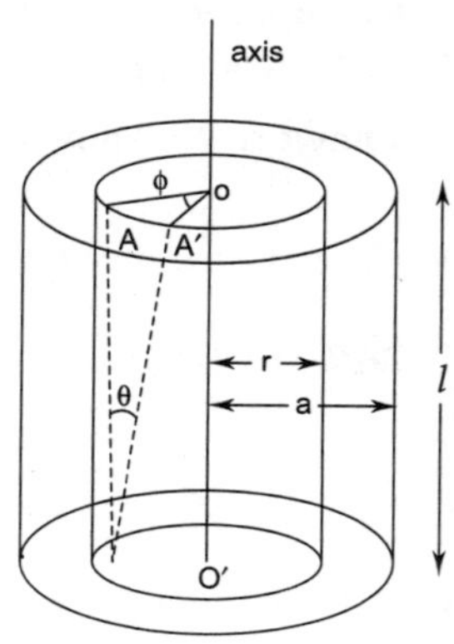

We consider a cylinder of length l and radius a. Let the lower end of the cylinder be clamped and at the upper end a torque is applied. We now consider an imaginary co-axial cylinder of radius r. A straight line is considered over the surface of this imaginary cylinder parallel to its axis OO′. After twisting the line gets inclined to the original line. The angle between the two lines measures the angle of shear. AA′ represents the amount of turn of an element originally at A on the surface of the cylinder.

$$\therefore \qquad l\theta = AA'.$$

As r is the distance of the element from the axis of the cylinder and ϕ the angle of tiwst, we have

$$r\phi = AA' = l\theta$$

$$\therefore \qquad \theta = \frac{r\phi}{l}.$$

Now due to elastic property of the material the relative turning of the co-axial disc gives rise to restoring torque which tends to prevent the twisting of the cylinder.

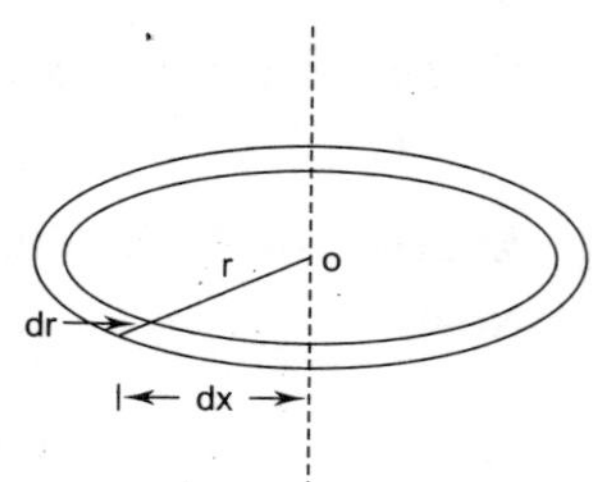

To calculate the restoring torque we take into consideration one such shell of radius r and thickness dr. On the face of this shell at the position A we imagine an element of length dx and breadth dr.

Over this element, Let F be the tangential force that comes to play due to relative turning.

$$\therefore \qquad \text{Tangential stress} = \frac{F}{dr\, dx}$$

$$\therefore \qquad \eta = \frac{F/dr\, dx}{\theta} = \frac{F/dr\, dx}{r\phi/l} = \frac{Fl}{r\phi\, dr\, dx}$$

where η is the rigidity modulus of the material of the cylinder

$$\therefore \qquad F = \frac{\eta\phi r}{l}\, dr\, dx$$

The moment of this force about the axis of the cylinder is given by

$$F \cdot r = \frac{\eta\phi}{l} r^2\, dr\, dx\,.$$

Therefore, the restoring torque over the entire surface of the annulus is given by

$$\delta\Gamma = \frac{\eta\phi}{l} r^2\, dr \sum dx$$

$$= \frac{2\pi\eta\phi r^3}{l}\, dr \qquad \left[\because \sum dx = 2\pi r\right]$$

Therefore, the total restoring torque that gives to play during the twisting of the entire cylinder is

$$\Gamma = \frac{2\pi\eta\phi}{l}\int_0^a r^3\, dr = \frac{\pi\eta a^4}{2l}\phi\,.$$

The expressions for torque remains unchanged, if we consider that the upper end is fixed and at the lower end a torque is applied. The restoring torque Γ is equal in magnitude to the applied torque. $\frac{\Gamma}{\phi}$ is called torsional rigidity τ of the cylinder and is given by

$$\tau = \frac{\pi\eta a^4}{2l}\,.$$

If the cylinder is hollow of outer and inner radii r_1 and r_2 respectively, Then

$$\Gamma = \frac{2\pi\eta\phi}{l}\int_{r_2}^{r_1} r^3\, dr = \frac{\pi\eta\phi}{2l}\left(r_1^4 - r_2^4\right)$$

The torsional rigidity in this case is

$$\tau = \frac{\pi\eta\left(r_1^4 - r_2^4\right)}{2l}\,.$$

Example 1. *Two cylindrical shaft have the same length and mass and are made of the same material. One is solid, while the other, which is hollow, has an external radius twice the internal radius. Compare their torsional rigidities.*

Solution. The mass of the solid cylinder of radius r, length l and density ρ

$$= \pi r^2 \, l \, \rho.$$

The mass of the hollow cylinder of same length and density having external radius $2r_1$ and internal radius r_1

$$= \pi\left[(2r_1)^2 - r_1^2\right] = 3\pi r_1^2 \, l \, \rho$$

$\because$ The masses of the two cylindrical shafts are same

$$\pi r^2 \, l \, \rho = 3\pi r_1^2 \, l \, \rho$$

$$r^2 = 3r_1^2 .$$

Now, the torsional rigidity of the solid cylindrical shaft is given by

$$\tau_1 = \frac{\pi \eta r^4}{2l} .$$

For the hollow cylindrical shaft the torsional rigidity is given by

$$\tau_2 = \pi\eta\left[\frac{(2r_1)^4 - r_1^4}{2l}\right] = \frac{15\pi\eta r_1^4}{2l}$$

$$\therefore \quad \frac{\tau_1}{\tau_2} = \frac{r^4}{15r_1^4} = \frac{9}{15} = \frac{3}{5} \qquad \left[\because \frac{r^4}{r_1^4} = 9\right]$$

2A.10 HOLLOW SHAFT IS STRONGER THAN A SOLID ONE OF THE SAME MASS

Let us consider a solid cylinder shaft of length l, radius r and of a material whose co-efficient of rigidity is η. The torque necessary to twist it through an angle of ϕ radian is

$$\tau = \frac{\pi \eta r^4 \phi}{2l} = \frac{\pi \eta r^2 \left(r^2 \phi\right)}{2l} \qquad \text{...(2A.11)}$$

If the shaft be hollow with internal and external radii r_1 and r_2 respectively, the torque necessary to twist it through the same angle of ϕ radian will be

$$\tau' = \frac{\pi\eta\left(r_2^4 - r_1^4\right)\phi}{2l} = \frac{\pi\eta\left(r_2^2 - r_1^2\right)\left(r_2^2 + r_1^2\right)\phi}{2l} \qquad \text{...(2A.12)}$$

The masses (Volume $\times$ density) and lengths of the hollow and solid shafts are equal.

Thus, $\quad \pi\left(r_2^2 - r_1^2\right) l \rho = \pi r^2 l \rho$

where ρ is the density of the material of the shaft

$$\therefore \qquad r_2^2 - r_1^2 = r^2.$$

Substituting, r^2 for $r_2^2 - r_1^2$ in (*ii*) and comparing it with (*i*) we get

$$\frac{\tau'}{\tau} = \frac{r_2^2 + r_1^2}{r^2} > 1$$

$$\tau' > \tau.$$

i.e., the torque necessary to twist a hollow cylinder is greater than the torque necessary to twist a solid cylinder of the same mass length and material through the same angle. Hence, a greater amount of energy will be stored in the hollow cylindrical shaft when twisted. For the same reason, the shaft in the form of a hollow cylinder will be stronger than a solid shaft.

Example 1. *A cylinder of radius r and length l is recast into a pipe of same length. If the pipe possesses torsional rigidity which is 19 times greater than that of original cylinder, Calculate the inner radius of the pipe.* [Bundelkhand Univ.]

Solution. Consider, c and c' be the torsional rigidities of the solid and hollow cylinders respectively.

Then, $$c = \frac{\pi \eta r^4}{2l} \quad \text{and} \quad c' = \frac{\pi \eta \left(r_2^4 - r_1^4\right)}{2l}$$

where r is the radius of the solid cylinder, and r_1 and r_2 the internal and external radii of the hollow cylinder.

Here, $$\frac{c'}{c} = 19,$$

Then $$r_2^4 - r_1^4 = 19r^4$$

or, $$\left(r_2^2 + r_1^2\right)\left(r_2^2 - r_1^2\right) = 19r^4 \qquad \ldots(i)$$

or, since the length and mass remain unchanged, we have

$$\pi\left(r_2^2 - r_1^2\right) l\, \rho = \pi r^2\, l\, \rho$$

$$r_2^2 - r_1^2 = r^2 \qquad \ldots(ii)$$

or, Making this substitution in equation (*i*), we get

$$r_2^2 + r_1^2 = 19r^2 \qquad \ldots(iii)$$

Subtracting (*ii*) from (*iii*) and solving we get

$$r_1 = 3r.$$

Example 2. *What couple must be applied to a wire one metre long, 1 mm in diameter in order to twist one end of it, through 90°, the other end remaining fixed. Rigidity of material of the wire is 2.8 × 10^{10} N-m^2.* [Burdwan Univ.]

Solution. Given, $l = 1$ m radius, $a = 0.5$ mm $= 0.5 \times 10^{-3}$ m

$$\eta = 2.8 \times 10^{10} \text{ N-m}^2, \theta = 90° = \frac{\pi}{2} \text{ radian.}$$

Couple per unit angular twist

$$c = \frac{\pi \eta a^4}{2l}$$

∴ Couple for angular twist,

$$c\theta = \frac{\eta \pi a^4}{2l} \times \frac{\pi}{2} = \frac{\eta \pi^2 a^4}{4l}$$

$$= \frac{2.8 \times 10^{10} \times \pi^2 \times (0.5 \times 10^{-3})^4}{4 \times 1}$$

$$= 43.\backslash 20 \times 10^{-4} \text{ N-m.}$$

2A.11 STRAIN ENERGY

When a body is strained internal forces due to elastic property of the body are called into play and hence work is done upon the body against the internal forces. This work is stored up within body as potential energy which is called strain energy. We consider here the following cases :

(a) Stretching of a Wire

We consider a wire of length L stretched by an amount l by a stretching force F. Now, for a small increase in length dl of the wire, the work done will be $F \cdot dl$, Hence, the work done for stretching the wire by a length l is given by

$$W = \int_0^l F \cdot dl = \int_0^l \frac{Y \alpha l}{L} dl = \frac{1}{2} \frac{Y \alpha l^2}{L}$$

$$= \frac{1}{2} Fl = \frac{1}{2} \times \text{Stretching force} \times \text{amount stretched}$$

∴ Work done per unit volume

$$= \frac{1}{2} \left(\frac{F}{\alpha}\right)\left(\frac{l}{L}\right)$$

$$= \frac{1}{2} \times \text{Stress} \times \text{Strain.}$$

(b) Twisting of a rod of Circular Cross-section

We know from the theory of torsion of a cylinder that the torsional couple is given by

$$C = \frac{\pi \eta a^4}{2l} \phi$$

∴ The work done and hence the energy stored up in twisting the cylinder by $d\phi$ is

$$dV = C\,d\phi = \frac{\pi \eta a^4}{2l}\,\phi\,d\phi$$

∴ The total work done or the energy stored up in twisting the cylinder by ϕ is

$$V = \frac{\pi \eta a^4}{2l}\int_0^{\phi_0} \phi\,d\phi = \frac{\pi \eta a^4}{4l}\,\phi_0^2$$

If C_0 be the torque when ϕ_0 is the angle of twist, we have

$$C_0 = \frac{\pi \eta a^4}{2l}\,\phi_0$$

∴ $$V = \frac{1}{2} C_0\,\phi_0 = \frac{C_0^2\,l}{\eta \pi a^4}.$$

IN FOCUS

- **Young's Modulus of elasticity** $Y = \dfrac{FL}{\pi r^2 l}$

  where L is the length of wire of radius r and an increase in length l is caused by a force F.

- **Bulk modulus of elasticity** $K = \dfrac{PV}{v}$

  where P and V is the pressure and volume and v is the change in volume

- **Modulus of rigidity** $\eta = \dfrac{F/A}{\theta}$

  where θ is the shear strain.

- **Poisson's ratio** $\sigma = \dfrac{\delta D}{D}\cdot\dfrac{l}{\delta l}$

  where L and D is the length and diameter of the wire respectively.
- Angle of shear is the twice the linear extensional strain.
- Relation among the Elastic constants:

  $$Y = 3k\,(1 - 2\sigma); \quad Y = 2\eta\,(1 + \sigma); \quad \sigma = \frac{3k - 2\eta}{6k + 2\eta}$$

- Depression (δ) of a uniform bar of length l held horizontally at one end and loaded W at the other end

  $$\delta = \frac{Wl^3}{3YI}$$

  where I is geometrical moment of Inertia.

  - For the beam of rectangular Gross-section $I = \dfrac{bd^3}{12}$, where b is the breadth and d is the thickness of the beam.
  - beam of circular cross-section of radius 'r'; $I = \dfrac{\pi r^4}{4}$.

- Depression (δ) for a beam loaded at the middle and supported at the ends

$$\delta = \frac{Wl^3}{48\ YI}$$

- Torsional rigidity of the cylinder $\zeta = \dfrac{\pi\eta\, a^4}{2l}$.

  where η is the rigidity modulus.
- The work done per unit volume for stretching the wire

$$W = \frac{1}{2} \times \text{stress} \times \text{strain}$$

Solved Problems

Problem 1. *Young's modulus Y for brass is 9×10^{10} nt/m². Determine the energy stored in a brass rod of 2 cm² cross-section, 0.01 meter in length if compressed by a force of 10 Newton.*

[Lucknow University]

Solution. If we know that the work done (energy stored) in compressing the rod is given by

$$W = \frac{1}{2}\ \text{Volume} \times \text{stress} \times \text{strain}$$

Here, volume of the rod = length × cross section

$$= 0.01 \text{ m} \times (2 \times 10^{-4} \text{ m}^2)$$

$$= 2 \times 10^{-6} \text{ m}^3$$

Again, stress = compression force/cross-section

$$= \frac{10\ nt}{\left(2 \times 10^{-4}\ \text{m}^2\right)} = 5 \times 10^4\ \text{nt/m}^2.$$

and

$$\text{strain} = \frac{\text{Stress}}{\text{Y}} = \frac{5 \times 10^4}{9 \times 10^{10}} = \frac{5}{9} \times 10^{-6}$$

$\therefore$

$$W = \frac{1}{2} \times \left(2 \times 10^{-6}\ \text{m}^3\right) \times \left(5 \times 10^4\ nt/\text{m}^2\right) \times \left(\frac{5}{9} \times 10^{-6}\right)$$

$$= 2.78 \times 10^{-8} \text{ Joule.}$$

Problem 2. *A material has Poisson's ratio 0.20. If a uniform rod of it suffers longitudinal strain 4.0×10^{-3}, deduce the percentage change in its volume.*

Solution. Poisson's ratio,

$$\sigma = \frac{\text{lateral strain}\left(\dfrac{\Delta r}{r}\right)}{\text{longitudinal strain}\left(\dfrac{\Delta l}{l}\right)}$$

Here,

$$\frac{\Delta l}{l} = 4.0 \times 10^{-3}$$

and

$$\frac{\Delta r}{r} = 6 \times \frac{\Delta l}{l} = 0.20 \times \left(4.0 \times 10^{-3}\right) = 0.80 \times 10^{-3}$$

Let v be the initial volume, and V + ΔV be the final volume.

The $$V = \pi r^2 l \qquad \ldots(i)$$

and $$V + \Delta V = \pi(r - \Delta r)^2 (l + \Delta l) \qquad [\because \ \pi r^2 l = v] \ \ldots(ii)$$

$$= \pi(r^2 - 2r\,\Delta r + \Delta^2 r^2)(l + \Delta l)$$

$$\therefore \quad \Delta V = \pi r^2 \Delta l - 2\pi r l\,\Delta r$$

$$\frac{\Delta V}{V} = \frac{\Delta l}{l} - \frac{2\Delta r}{r}$$

Substituting the above values :

$$\frac{\Delta V}{V} = 4.0 \times 10^{-3} - 2\,(0.80 \times 10^{-3}) = 2.4 \times 10^{-3}$$

∴ percentage change in volume

$$\frac{\Delta V}{V} \times 100 = (2.4 \times 10^{-3}) \times 100$$

$$= 0.24.$$

Problem 3. *An iron wire of length 1 meter and radius. 5 mm elongates by 0.32 mm when stretched by a force of 5 kg-weight, and twists through 0.4 radian when equal and opposite torques of 3×10^4 dyne-cm are applied at its end. Calculate elastic constants for iron.*

[Jiwaji, Ravishankar University]

Solution. We know that the Young's modulus of the material is given by,

$$Y = \frac{\text{Stress}}{\text{Strain}} = \frac{F/\pi r^2}{x/l} = \frac{Fl}{\pi r^2 x}$$

Given $F = 5000 \times 981$ dyne, $r = 0.5$ mm $= 0.05$ cm $x = 0.32$ mm $= 0.032$ cm and $l = 1$ meter $= 100$ cm.

$$\therefore \quad Y = \frac{5000 \times 981 \times 100}{3.14 \times (0.05)^2 \times 0.032} = 19.5 \times 10^{11} \text{ dyne/cm}^2.$$

Now, the torque necessary to produce a twist of ϕ radian in the wire is given by

$$\tau = \frac{\pi \eta r^4}{2l}\phi$$

$$\eta = \frac{\tau\,(2l)}{\pi r^4 \phi}$$

$$\therefore \quad \eta = \frac{3 \times 10^4 \times 200}{3.14 \times (0.05)^4 \times 0.4}$$

$$= 7.64 \times 10^{11} \text{ dyne/cm}^2.$$

Again, Poisson's ratio

$$\sigma = \frac{Y}{2\eta} - 1 = \frac{19.5 \times 10^{11}}{2 \times 7.64 \times 10^{11}} - 1 = 0.276.$$

Problem 4. *A square bar of length 1 m and cross-section 1cm^2 is clamped horizontally at one end and a weight of 1 kg m is applied at the other end. Neglecting weight of the bar calculate the depression of the loaded end. Given Y = 9.78 × 10^{10} N/m^2 and g = 9.78 m/sec^2*

[M.D. Univ., Nagpur Univ.]

Solution. Given, l = 1 m. As it is a square bar $d = b = 1$ cm $= 10^{-2}$ m

$$W = 1 \text{ kg} = 9.78 \text{ N} ; Y = 9.78 \times 10^{10} \text{ Nm}^{-2}$$

∴ Depression of the loaded end

$$y = \frac{4Wl^3}{Ybd^3} \quad \text{for a rectangular bar.}$$

$$= \frac{4Wl^3}{Yb^4} \quad \text{for a square bar}$$

$$= \frac{4 \times 9.78 \times 1}{9.78 \times 10^{10} \times 10^{-8}} = 4 \times 10^{-2} \text{m}$$

$$= 4 \text{ cm.}$$

Problem 5. *Compare the loads required to produce equal depressions for two beams made of the same material and having the same length and weight with only difference that one has circular ecross-section while the cross-section of the other is square.* [Agra Univ.]

Solution. Here,

$$\text{Mass of square bar} = a^2 \, l\rho$$

$$\text{Mass of circular bar} = \pi r^2 l\rho$$

where l is the length of each bar, ρ its density, r the radius of circular bar and a each side of the face of square bar, then

If as the mass of the bars are equal

then $$\pi r^2 l\rho = a^2 l\rho$$

or, $$\pi r^2 = a^2 \qquad \ldots(i)$$

If I_1 is the geometrical M.I. of the square bar and I_2 that of the circular bar, then

Depression for square bar $$y = \frac{W_1}{YI_1} \cdot \frac{l^3}{3} \qquad \ldots(ii)$$

and Depression for circular bar

$$y = \frac{W_2}{YI_2} \cdot \frac{l^3}{3} \qquad \ldots(iii)$$

From equations (*ii*) and (*iii*), we get

$$\frac{W_1}{I_1} = \frac{W_2}{I_2}$$

or, $$\frac{W_1}{W_2} = \frac{I_1}{I_2}$$

Now $\qquad I_1 = \dfrac{a^4}{12}$ and $I_2 = \dfrac{\pi r^4}{4}$

$\therefore \qquad \dfrac{W_1}{W_2} = \dfrac{a^4}{12} \times \dfrac{4}{\pi r^4} = \dfrac{a^4}{3\pi r^4} = \dfrac{\pi^2 r^4}{3\pi r^4} = \dfrac{\pi}{3} = 1.05$

Problem 6. *A wire of radius 1 mm and length 2 meter is twisted through 90°. Calculate the angle of shear at the surface at the axis of the wire and at a point mid-way. If the modulus of rigidity of material is 5.0 × 10¹¹ dyne/cm², what is the torsional couple ?* [Agra University]

Solution. Consider l be the length of the wire, r the radius and ϕ the twist at the free end. Then the angle of the shear at a radial distance x from the axis of the wire is

$$\theta = \frac{x\phi}{l}$$

At the axis of the wire, $\quad x = 0$, so that $\theta = 0$.

At the surface of the wire,

$$x = r,$$

so that $$\theta = \frac{r\phi}{l}.$$

Given $\qquad r = 1$ mm $= 0.01$ cm $l = 200$ cm and $\phi = 90°$

$\therefore \qquad \theta = \dfrac{1 \times 90°}{200} = 0.045°.$

At a point mid way, $\qquad x = \dfrac{r}{2}$

$\therefore \qquad \theta = \dfrac{1}{2} \times 0.045° = 0.0225°$

The torsional couple is

$$\tau = \frac{\eta \pi r^4}{2l} \phi$$

For $\qquad \phi = 90° = \dfrac{\pi}{2}$ radian,

$$\tau = \frac{\pi^2 \eta r^4}{4l} = \frac{(3.14)^2 \times (5.0 \times 10^{11}) \times (0.1)^4}{4 \times 200}$$

$$= 6.2 \times 10^5 \text{ dyne cm.}$$

Problem 7. *Two solid cylinders of the same material having lengths l and 2l, and radii r and 2r, are joined coaxially. Under a couple applied between the free ends. The shorter cylinder shows a twist of 30°. Calculate the twist of the longer cylinder.* [Kanpur University]

Solution. Let τ be the couple, and let it produce a twist ϕ in the shorter cylinder and twist ϕ' in the larger cylinder.

Then, $\qquad \tau = \dfrac{\pi \eta r^4}{2l} \phi = \dfrac{\pi \eta (2r)^4}{2(u)} \phi'$

Then, $\qquad \phi' = \dfrac{\phi}{8} = \dfrac{30°}{8} = 3.75°.$

As the two cylinders have the same mass, Then

$$\pi a^2 \, l\rho = \pi 3r_1^2 l\rho$$

$$\therefore \qquad a^2 = 3r_1^2$$

Couple per unit angular twist (or torsional rigidity) of solid cylinder

$$c = \frac{\pi\eta a^4}{2l} = \frac{\pi\eta}{2l} 9r_1^4$$

Couple per unit angular twist (or torsinal rigidity) of hollow cylinder

$$c' = \frac{\pi\eta}{2l}\left(r_2^4 - r_1^4\right) = \frac{\pi\eta}{2l}\left[(2r_1)^4 - r_1^4\right] = \frac{\pi\eta}{2l} 15r_1^4$$

$$\therefore \qquad \frac{c'}{c} = \frac{\frac{\pi\eta}{2l} 15r_1^4}{\frac{\pi\eta}{2l} 9r_1^4} = \frac{5}{3}$$

Therefore, the hollow cylinder has a greater torsional rigidity, it is stronger than the solid cylinder of the same mass, length and material.

Problem 8. (*a*) *A solid cylinder of radius 5 cm is converted into a hollow cylinder of same mass and length and external radius 7 cm. If the restoring couple per unit radian twist in original cylinder is c, deduce the same for the new hollow cylinder.* [Kanpur University, Agra University]

(b) Two cylinders have the same length and mass are made of same material, one is solid while the other which is hollow has an external radius twice the internal radius. Compare the torsional rigidities of the two cylinders. Show that second cylinder is stronger. [Calcutta University]

Solution. (*a*) If r is the radius of solid cylinder and r_2 and r_1 the outer and inner radius of the hollow cylinder of the same mass and length as that of solid cylinder, then

$$\pi r^2 l\,\rho = \pi\left(r_2^2 - r_1^2\right) l\,\rho .$$

or,

$$r^2 = r_2^2 - r_1^2 .$$

Here, $\qquad r = 5$ cm and $r_2 = 7$ cm

$$\therefore \qquad r_1^2 = r_2^2 - r^2 = 49 - 25 = 24 \text{ cm}^2.$$

If c' is the restoring couple for the hollow cylinder,

then

$$\frac{c'}{c} = \frac{r_2^2 + r_1^2}{r^2} = \frac{49 + 25}{24} = 2.92$$

$$\therefore \qquad c' = 2.92\, c.$$

(*b*) Length of each cylinder $= l$; Density of material of each cylinder $= \rho$.

Radius of solid cylinder $= a$.

Outer radius of hollow cylinder

$= r_2$, inner radius of hollow cylinder $= r_1$.

Given $\qquad r_2 = 2r_1$.

Mass of the solid cylinder $= \pi a^2 \, l \, \rho$

Mass of the hollow cylinder $= \pi\left(r_2^2 - r_1^2\right) l\,\rho = \pi\, 3r_1^2\, l\,\rho$.

Multiple Choice Questions

1. Young's modulus of a wire of length L and radius r is Y N/m^2. If length is reduced to L/2 and radius $r/2$, its Young's modulus will be—
   (*a*) Y/2 (*b*) Y (*c*) 2Y (*d*) 4Y
2. Young's modulus of air is—
   (*a*) infinity (*b*) more than 1 but not infinite
   (*c*) less than one but not zero (*d*) zero
3. Energy per unit volume of a strained wire is given by—
   (*a*) $\frac{1}{2}$ load × extension (*b*) $\frac{1}{2}$ extension × stress
   (*c*) $\frac{1}{2}$ stress × strain (*d*) $\frac{1}{2}$ strain × load.
4. Bulk modulus of perfectly rigid body—
   (*a*) zero (*b*) 1 (*c*) infinity (*d*) none of these
5. How does Young's modulus vary with increase of temperature?
   (*a*) decreases (*b*) increases
   (*c*) remains constant (*d*) first increases and then decreases.
6. A wire is stretched to double of its length. The strain is—
   (*a*) 2 (*b*) 1 (*c*) zero (*d*) 0.5
7. For obtaining appreciable extension, the wire should be—
   (*a*) short & thin (*b*) long & thin (*c*) short & thick (*d*) long & thick
8. A metallic rod of Young's modulus 2×10^{11} N/m^2 undergoes a strain of 0.05%. The energy stored per unit volume of the rod will be
   (*a*) 0.5×10^4 J/m^3 (*b*) 2.5×10^4 J/m^3
   (*c*) 0.5×10^8 J/m^3 (*d*) 2.5×10^8 J/m^3
9. A metallic rod of length L, area of cross-section A, Young's modulus Y has coefficient of linear expansion α. If the rod is heated through a temperature T, the energy stored per unit volume will be
   (*a*) $\frac{1}{2}$ Y α T (*b*) $\frac{1}{2}$ Y α^2 T^2 (*c*) $\frac{1}{2}$ Y α T (*d*) $\frac{1}{2}$ Y α^2 T
10. Force required to stretch a steel wire of 1 sqcm cross-section to 1.1 times its length would be (Y = 2×10^{11} N/m^2)
    (*a*) 2×10^6 N (*b*) 3×10^6 N (*c*) 2×10^{-6} N (*d*) 4.2×10^{-7} N
11. In the following data L is the initial length of the wire, r is the radius, Under a given load, in which case the extension will be minimum?
    (*a*) L = 40 cm, r = 0.4 mm (*b*) L = 50 cm, r = 0.5 mm
    (*c*) L = 60 cm r = 0.6 mm (*d*) L = 70 cm, r = 0.7 mm

Answers

1. (*b*)	**2.** (*d*)	**3.** (*c*)	**4.** (*c*)	**5.** (*a*)	
6. (*b*)	**7.** (*b*)	**8.** (*b*)	**9.** (*b*)	**10.** (*a*)	**11.** (*d*)

Review Questions

1. Explain the terms stress and strain. Define Young's modulus Y, the bulk moduluas κ, rigidity modulas η and poisson's ratio σ. Write dimensions of σ. [*Madurai, Nagpur, Meerut, Calcutta Univ.*]

2. (*a*) Show that shear is equivalent to an elongation strain and compression strain at right angles to each other and each is half of shearing angles? *[Jadavpur, Calcutta, Kanpur, Ranchi Univ.]*
   (*b*) Prove that a shearing stress is equivalent to a linear tensile stress and an equal compression stress mutually at right angles. *[Vidya Sagar, Kanpur Univ.]*
3. If Y, κ and σ represent Young's modulus, Bulk modulus and Poisson's ratio respectively, then prove that $k = \dfrac{Y}{3(1-2\sigma)}$ *[Calcutta, Jadavpur, Meerut Univ.]*
4. (*a*) What is the difference between angle of twist and angle of shear? Deduce an expression for the couple required to twist a uniform solid cylinder by an angle. *[Kerala Univ., Meerut Univ. Gharwal Univ. Nagpur Univ.]*
   (*b*) What is the value of the couple for a hollow cylinder of inner radius r_1 and outer radius r_2? *[Kanpur Univ.]*
5. Explain why a hollow cylinder is stronger than solid cylinder of same length, mass and material. *[Calcutta Univ.]*
6. What is a cantilever? Derive an expression for the depression of the loaded end of a cantilever of (*i*) circular cross-section and (*ii*) rectangular cross-section of negligible weight. *[Ranchi Univ. Meerut, Kanpur, Calcutta Univ.]*

Unsolved Problems

1. Calcualte Y for rubber if a rubber tube 0.4 meter long, whose external and internal diameters are 0.01 meter and 0.004 meter respectively, extends 0.0006 meter when stretched by a force of 5 Kgm-wt. (*Rajasthan Univ.*) [**Ans.** 4.9×10^8 nt/m^2.]
2. A wire of length 1 meter and diameter 2 mm is clamped at one of its ends. Calculate the couple required to twist the other end by 45° ($\eta = 5 \times 10^{11}$ dyne/cm^2). (*Gorakhpur Univ.*) [**Ans.** 6.2×10^5 dyne-cm.]
3. The solid cylindrical rods of same material having lengths/and 1 and 2/and radii 2*r* and *r* respectively are soldered coaxially. The free end is twisted through 11°, the other end being clamped. Calculate the twist of the thicker rod. (*Kanpur ; Rajasthan Univ.*) $\left[\textbf{Ans.} \left(\dfrac{1^\circ}{3}\right)\right]$
4. The restoring couple per unit twist in a solid cylinder of radius 5.0 cm is 1×10^8 dyne-cm. Find the restoring couple per unit twist in a hollow cylinder of the same material, mass and length but of internal radius 12.0 cm. *[Agra Univ.]* [**Ans.** 1.25×10^9 dyne-cm.]
5. A 0.5 kg solid cylinder is suspended by a thin wire in such a way that its axis remains vertical. It the period of torsional oscillations is 2.5 sec, what will be the period if the cylinder is replaced by a 0.4 kg solid sphere of the same radius? (*Agra Univ.*) [**Ans.** (2.0 sec)]
6. A cantilever of length *l* and uniform cross-section shows a depression δ at the loaded end. What will be the depressions at distances *l*/4, *l*/2 and 3*l*/4 from the fixed end $\left[\dfrac{11}{128}\delta, \dfrac{5}{16}\delta, \dfrac{81}{128}\delta\right]$ *[Agra Univ.]*

2B

Chapter

Surface Tension

2B.1 INTRODUCTION

The free surface of a liquid has a natural tendency to contract to a smallest possible area as if it is in a state of tension like a stretched membrane. Liquid drop if small in size, acquire spherical shape due to surface tension as sphere has the minimum surface area but large drop flatten out due to its weight. The phenomenon of surface tension are closely related to the following terms:

(*i*) **Cohesive Forces:** It is the force of attraction between two molecules of similar nature. The force varies inversely as the eighth power of the distance between the two molecules. It is very strong in case of solids, weak in the case of liquids and weakest in case of gases.

(*ii*) **Adhesive Forces:** It is the force of attraction between two molecules of different substances. Due to this force water wets a glass and cement adheres to brick.

(*iii*) **Molecular Range:** Every two molecules of matter, attract each other. The force of attraction decreases with an increase in distance. The maximum distance between two molecules so that the force of attraction remains effective is called molecular range. It is of the order, 10^{-7} cm.

(*iv*) **Surface Film:** It is defined as the portion of liquid on the surface and caught between two parallel planes situated at molecular range apart.

2B.2 SURFACE TENSION

Consider a line PQ is drawn in any direction in the liquid surface, (figure 2B.1), the surface on either side of this line exerts a pulling force on the surface on the other side in order to decrease the surface area. This force lies in the plane of the surface and is at right angles to the line PQ. The magnitude of this force per unit length of PQ is considered as the measure of the surface tension of the liquid. If F is the force acting along 'l' cm of this line, the force of surface tension 'T' is defined as $T = \frac{F}{l}$.

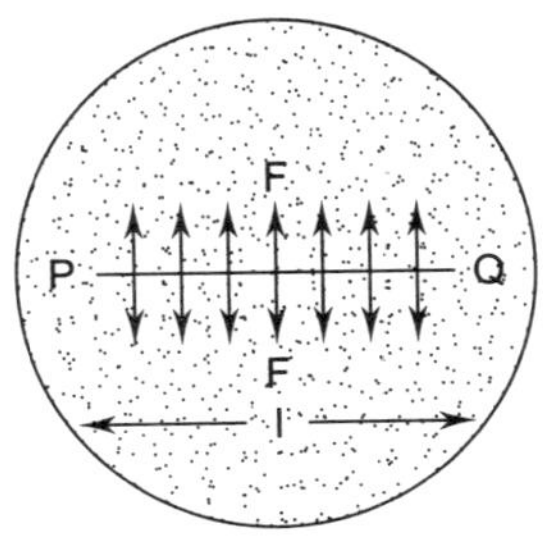

Figure 2B.1

If $\quad l = 1, \quad$ then $\quad T = F.$

i.e., the surface tension of a liquid is defined as the amount of tangential force acting per unit length on either side of an imaginary line drawn over the liquid surface. Its unit in S.I. system Newton/metre and in C.G.S. System dyne/cm.

2B.3 MOLECULAR THEORY OF SURFACE TENSION

Let XY be the free surface of a liquid contained in a beaker. The molecules of the liquid attract each other lay force of cohesion. The maximum distance upto which the force of cohesion act is of the order of 10^{-7} cm. It is called their molecular range.

A sphere with molecule as centre and radius equal to molecular range is called sphere of influence. Any other molecule entering this sphere will be at a distance less than molecular range and hence will be attracted by the molecule.

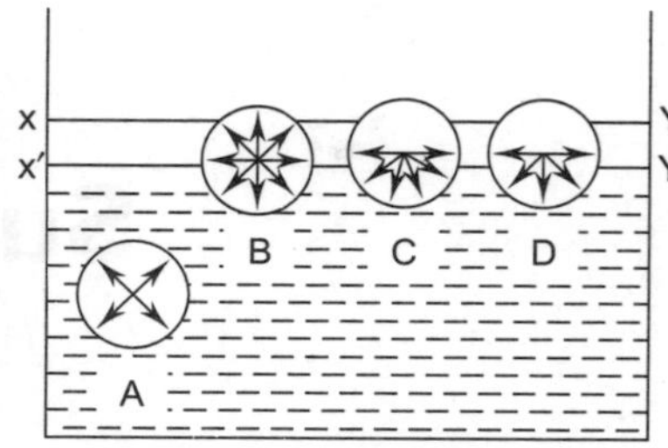

Figure 2B.2

A sphere of influence 'A' lies inside the liquid, (figure 2B.2), the molecule is attracted equally in all direction as shown. So the resultant force on 'A' is zero. This condition prevails for the molecule 'B'.

The sphere of influence 'C' lies partly outside the liquid and this part contains a few molecules of the liquid only than that of the lower half. Thus, the molecule 'C' will experience a higher attractive force in the downward direction than in the upward. Hence, there is a resultant downward force acting on it.

The sphere of influence of 'D' is exactly half outside the liquid and half inside it. So it will get a resultant downward force perpendicular to the free surface of the liquid and the force will be maximum.

Consider another plane X′Y′ lying below the surface XY and parallel to it while the distance between them is equal to the molecular range. The amount liquid caught between XY and X′Y′ is called surface film. If the molecule is raised up within the liquid to the surface film, the work has to be done against the downward cohesive force. This work done is stored in the molecule and is known as potential energy of the molecule. Hence, the potential energy of the molecules lying within the surface film is greater than the potential energy of the molecule lying below.

Hence, every system in equilibrium tries to have minimum potential energy. If the surface area is minimum, the potential energy is minimum. For this reason, the free surface of a liquid always tends to have a spherical shape (minimum surface area).

2B.4 SURFACE ENERGY

We know that the force of surface tension acts in such a way that the area of the surface tension should be minimum. If we extend the area of the surface we shall have to do it against the forces of surface tension. The necessary mechanical workdone on the molecules gets stored in the new surface as a potential energy in the new surface formed. Further when the surface is increased, its temperature tends to fall. Therefore, heat flows into it from the surroundings to keep its temperature constant and is added to its energy.

This total energy per unit area of the surface is called the 'intrinsic surface energy'.

2B.5 RELATION BETWEEN SURFACE TENSION AND SURFACE ENERGY

Consider a wire frame PQRS in which the wire PQ can slide without frictions (figure 2B.3). The frame is dipped into soap solution and taken out when a thin film of soap solution will be formed in the space PQRS. Due to the surface tension, a forec will act normally at every point of PQ towards SR. If the film PQ is pulled by a small distance δx to the position P′Q′ keeping the temperature constant, then

$$\text{workdone} = 2T \times \delta x \times l = 2T\ \delta x\ l$$

Here the factor 2 appears because, there are two surfaces of the soap film.

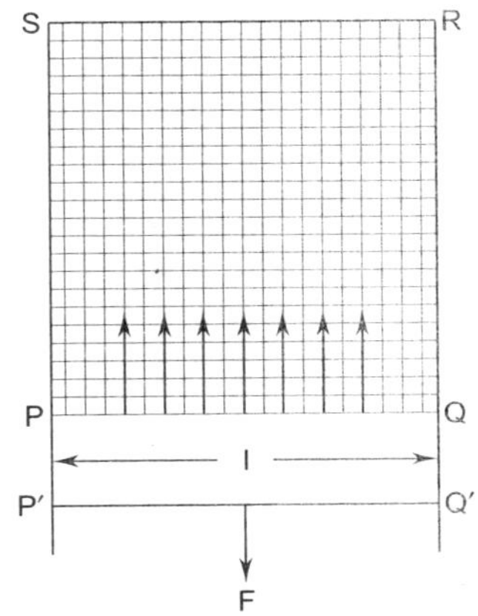

Figure 2B.3

$$\text{Increase in area} = 2\delta x\ l$$

∴ Workdone for increase in unit area (W)

$$= \frac{2Tl\,\delta x}{2l\,\delta x} = T.$$

Thus, the surface tension is numerically equal to the mechanical workdone for unit increase in surface area, *i.e.*, the surface energy per unit area of a surface is numerically equal to the surface tension.

Case (*i*):

Expansion under isothermal condition: As the surface expands more molecules from within the film are pulled on the surface. Energy required to do so is derived from the film, as a result which gets cooled. To maintain constant temperature, we have to supply heat from outside.

Let H be the heat supplied per unit area, Then total energy required

$$E \cdot 2l\ \delta x = T \cdot 2l\ \delta x + H \cdot 2l\ \delta x$$

where 'E' is the surface energy,

$$\therefore \qquad E = T + H.$$

Case (*ii*):

Expansion under adiabatic condition: Here, the temperature of the system does not remain constant.

In this case no heat can be supplied from outside.

$$H \cdot 2l \cdot \delta x = 0$$

$$\therefore \qquad E = T.$$

Under adiabatic conditions, surface tension is numerically equal to the surface energy.

2B.6 ANGLE OF CONTACT

If a plate of glass is dipped in a liquid, the surface of liquid near the region of contact becomes curved shaped.

If a tangent is drawn to the curved liquid surface at the point of contact with the solid, then the angle (θ) which the tangent makes with the solid surface inside liquid is known as the angle of contact (figure 2B.4). Angle of contact may have value between 0° to 180°. For ordinary water and glass it is 8° and for silver and water it is 90°. In the case of Mercury and glass, the angle of contact is 140°.

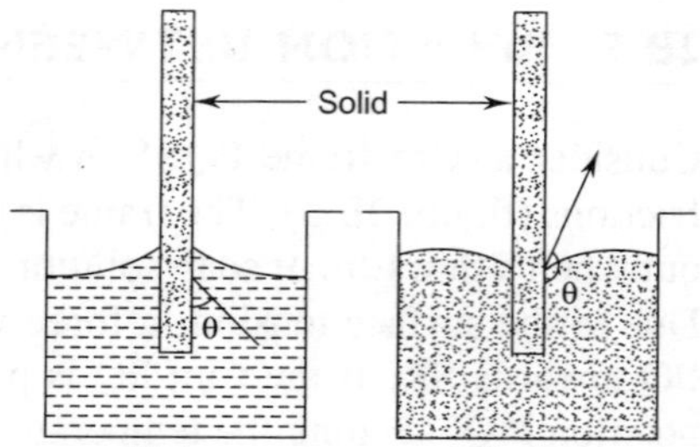

Figure 2B.4

Angle of contact depends on:

(*a*) nature of solid and liquid.
(*b*) medium exists above the free surface of the liquid.

The angle of contact also changes in presence of impurity in the liquid.

2B.7 RELATION BETWEEN SURFACE TENSION, RADIUS OF CURVATURE AND EXCESS PRESSURE ON A CURVED SURFACE

If the pressure on both sides of a liquid suface is equal to the surface is flat without any curvature. If there is a convex curved liquid surface at rest, the inward pressure on it due to surface tension is balanced by an equal pressure acting outward. In order to find the excess pressure, consider a small rectangular element PQRS of a curved liquid surface which is convex on the upper side (figure 2B.5). Consider the adjacent PQ and QR right angles to each other and also consider, PQ = x and QR = y. If the normals to the surface drawn at P and Q intersect at O_1 and normals at Q and R intersect at O_2.

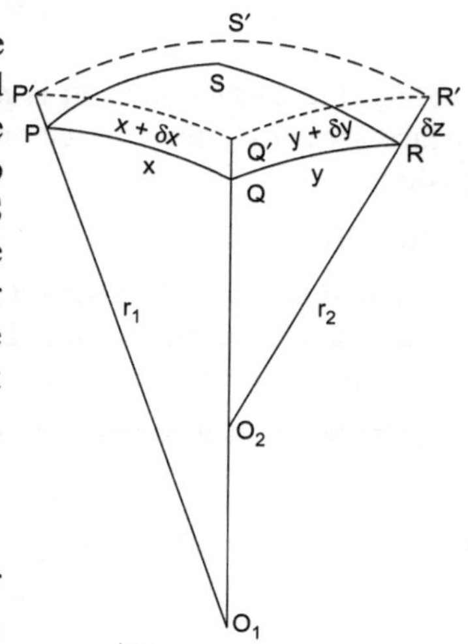

Figure 2B.5

Let $$PO_1 = QO_1 = r_1$$

and $$QO_2 = RO_2 = r_2.$$

Here r_1 and r_2 are known as the maximum and minimum radii of curvatures respectively. They are called principal radii of curvature of the surface. If the excess pressure on the concave side is P when the liquid at rest, then the outward thrust on the element PQRS = P_{xy}. If the liquid surface is moved outward parallel to itself through a small distance δz and occupied the position P′Q′R′S′, keeping the temperature constant.

Let $$P'Q' = x + \delta x \text{ and } Q'R' = y + \delta y.$$

The mechanical workdone by the excess pressure P is

$$W = \text{Force} \times \text{distance}$$
$$= \text{Pressure} \times \text{area} \times \text{distance}$$
$$= P_{xy}\,\delta z \qquad \text{...(2B.1)}$$

The increase in the area of the surface is [Area P′Q′R′S′ – Area PQRS]

$$= [(x + \delta x)(y + \delta y) - xy]$$
$$= x\,\delta y + y\,\delta x \qquad \text{...(2B.2)}$$

Here $\delta x\,\delta y$ is neglected, because δx and δy, both are small quantity.

Here, $\Delta P'Q'O_1$ and ΔPQO_1 are similar triangle.

$$\frac{P'Q'}{PQ} = \frac{P'O_1}{PO_1}$$

i.e.,
$$\frac{x+\delta x}{x} = \frac{r_1+\delta z}{r_1} \quad \text{or} \quad \frac{\delta x}{x} = \frac{\delta z}{r_1}$$

or,
$$\delta x = x \cdot \frac{\delta z}{r_1} \qquad \text{...(2B.3)}$$

Similarly, from $\Delta Q'R'O_2$ and ΔQRO_2,

$$\delta y = y\,\frac{\delta z}{r_2} \qquad \text{...(2B.4)}$$

Putting these values of δx and δy in equation (2B.2), the increase in the surface area

$$= \frac{xy\,\delta z}{r_2} + \frac{xy\,\delta z}{r_1}$$

$$= xy\,\delta z\left(\frac{1}{r_1}+\frac{1}{r_2}\right).$$

Hence, the increase in free surface energy

= Increase in surface area × Surface Tension

$$= xy\,\delta z\left(\frac{1}{r_1}+\frac{1}{r_2}\right)T \qquad \text{...(2B.5)}$$

This increase in energy is the result of mechanical workdone by the excess pressure P and is numerically equal to it. Hence equating (2B.1) and (2B.3), we get

$$Pxy\,\delta z = xy\,\delta z\left(\frac{1}{r_1}+\frac{1}{r_2}\right)T$$

or,
$$P = T\left(\frac{1}{r_1}+\frac{1}{r_2}\right) \qquad \text{...(2B.6)}$$

If we consider the liquid film instead of liquid surface, the expression will be

$$P = 2T\left(\frac{1}{r_1}+\frac{1}{r_2}\right) \qquad \text{...(2B.7)}$$

because a film has two surfaces.

Special Cases:

(*a*) **Spherical Surface:** In case of spherical surface

$$r_1 = r_2 = r$$

and we have
$$P = T\left(\frac{1}{r_1}+\frac{1}{r_2}\right)$$

$$= \frac{2T}{r}.$$

If instead of spherical surface, we have a spherical film, lika a 'soap bubble', then

$$P = \frac{4T}{r} \qquad \text{[Soap bubble has two surfaces]}$$

(*b*) **Cylindical Surfaces:** In this case, one radius of curvature is infinite.

$$\therefore \qquad P = T\left(\frac{1}{r} + \frac{1}{\alpha}\right) = \frac{T}{r}$$

For cylindrical film, $P = \frac{2T}{r}$.

2B.8 CAPILLARITY

Capillare means as fine as hair, capillar tube means a tube having very fine uniform diameter. Capillarity is a phenomenon by virtue of which liquid rises or depresses along a capillary tube partly immersed in the liquid and it is different from that outside it.

If a capillary tube is dipped in a liquid, the liquids having acute angle of contact with solids rise into it like water while those having obtuse angle get depressed in it like mercury (figure 2B.6). It is due to the effect of surface tension.

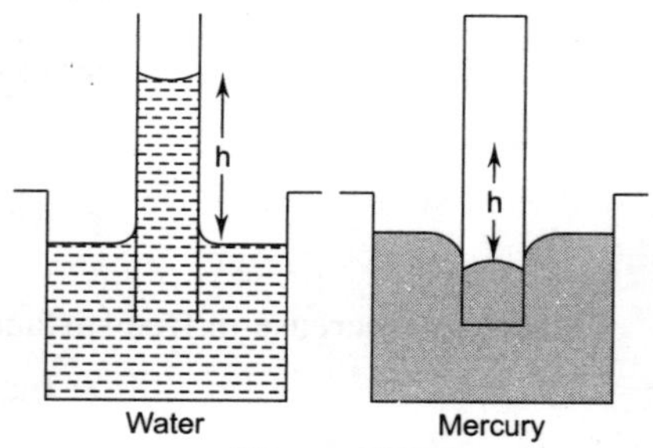

Figure 2B.6

2B.9 RISE OF LIQUID IN A CAPILLARY TUBE

Consider a glass capillary tube of circular bore of radius r is dipped vertically in a liquid. Liquid rises up along the tube and h be the height of the liquid column upto the bottom of the meniscus (figure 2B.7). Consider

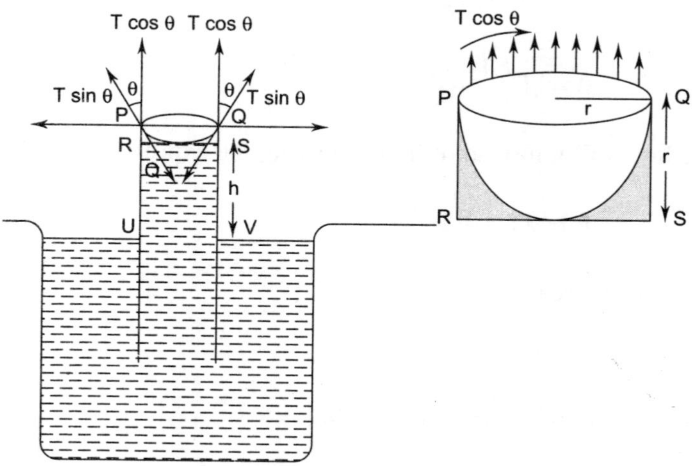

Figure 2B.7

R and S lying on the surface of meniscus in diametrically opposite positions. Due to the surface tension, water surface tends to contract. While doing so it pulls the glass molecules at P and Q tangentially. Molecules of the glass exert an equal and opposite reaction force T and this force T drags water up in the tube. Water keeps on rising up into tube till the total upward force due to surface tension is equal to the downward weight of water inside the tube.

The reaction force T can be resolved into two rectangular components:

(*a*) T cos θ acting vertically upwards.
(*b*) T sin θ acting horizontally outward.

Considering the whole meniscus, the horizontal components are cancelled out, do not make any contribution to its rise. But the vertical components are added up.

∴ Total force in upward direction

$$\begin{aligned} F &= \sum T \cos\theta \\ &= \text{Circumference of the circle of contact} \times \text{Force per unit length} \\ &= 2\pi r \times T\cos\theta \\ &= 2\pi r\, T\cos\theta \end{aligned}$$

Here, r is the radius of the meniscus.

At the equilibrium of the liquid column, upward force becomes equal to the weight of the lifted liquid.

Now weight of the lifted liquid,

$$W = V\rho g$$

where ρ is the density of the liquid.

∴ V = volume of liquid in cylinder RSUV + volume of water below the meniscus and above the tangent RS

= $\pi r^2 h$ + [volume of the cylinder PQRS – volume of the hemisphere r of radius r]

$$= \pi r^2 h + \left[\pi r^2 r - \frac{2}{3}\pi r^3\right]$$

$$= \pi r^2\left(h + \frac{r}{3}\right).$$

So according to the conditions,

$$2\pi r\, T\cos\theta = \pi r^2\left(h + \frac{r}{3}\right)\rho g$$

$$T = \frac{r\left(h + \frac{r}{3}\right)\rho g}{2\cos\theta}.$$

In case of fine capillary tube $\frac{'r'}{3}$ is so small compared to h and can be neglected.

$$T = \frac{rh\rho g}{2\cos\theta}.$$

Here T, ρ, g, θ are constant for a particular liquid.

$$rh = \text{constant}$$

$$h \propto \frac{1}{r}.$$

Hence, the height of liquid risen in a capillary tube varies inversely as the radius of the tube. This equation is known as Jurin's law.

IN FOCUS

- **Surface tension (T)** $= \dfrac{\text{Force (F)}}{\text{length } (l)}$
- Relation between surface tension (T), Radius of curvature (r) and excess pressure (P) on a curve surface

$$P = 2T\left(\frac{1}{r_1} + \frac{1}{r_2}\right)$$

  - For liquid film $P = 2T\left(\dfrac{1}{r_1} + \dfrac{1}{r_2}\right)$
  - For spherical surface $P = T\left(\dfrac{1}{r_1} + \dfrac{1}{r_2}\right) = \dfrac{2T}{r}$ $[\because r_1 = r_2 = r]$
  - For spherical Film $(P) = \dfrac{4T}{r}$
  - For cylindrical surface $(P) = T\left(\dfrac{1}{r} + \dfrac{1}{\alpha}\right) = \dfrac{T}{r}.$
  - For cylindrical Film $(P) = \dfrac{2T}{r}$

Problem 1. *What amount of energy will be liberated if 1000 droplets of water each of 10^{-8} m in diameter coalesce to form a one large spherical drop ? Surface tension of water = 75×10^{-3} Nm^{-1}.*

[Meerut University]

Solution. Given, Radius of each droplet $= 0.5 \times 10^{-8}$ m

So, volume of 1000 droplets

$$= \frac{4}{3}\pi\left(0.5 \times 10^{-8}\right)^3 \times 1000.$$

If R is the radius of bigger drop formed, then

$$\frac{4}{3}\pi R^3 = \frac{4}{3}\pi\left(0.5 \times 10^{-8}\right)^3 \times 1000$$

$$\therefore \quad R = 10 \times 0.5 \times 10^{-8} = 5 \times 10^{-8} \text{ m}$$

Surface area of big drop $= 4\pi\,(5 \times 10^{-8})^2\ m^2$

Surface area of 1000 droplets $= 1000 \times 4\pi\left(0.5 \times 10^{-8}\right)^2\ m^2 = 10 \times 4\pi\left(5 \times 10^{-8}\right)^2\ m$

$$\text{Decrease in area} = 10 \times 4\pi\left(5 \times 10^{-8}\right)^2 - 4\pi\left(5 \times 10^{-8}\right)^2$$

$$= 4\pi \times 9 \times \left(5 \times 10^{-8}\right)^2\ m^2 .$$

$$\therefore \quad \text{Energy liberated} = 4\pi \times 9 \times \left(5 \times 10^{-8}\right)^2 \times 75 \times 10^{-3} = 2.12 \times 10^{-14}\ \text{Joule.}$$

Problem 2. *Calculate the excess pressure inside a soap bubble of radius 3×10^{-3} m. Surface tension of soap solution is $20 \times 10^{-3}\ Nm^{-1}$. Also calculate surface energy.*

[Calcutta University, Bangalore University]

Solution. Radius of soap bubble

$$= 3 \times 10^{-3}\ m,\ \text{surface Tension } T = 20 \times 10^{-3}\ Nm^{-1}.$$

$$\text{Excess pressure,} \qquad p = \frac{4T}{r} = \frac{4 \times 20 \times 10^{-3}}{3 \times 10^{-3}} = 26.67\ Nm^{-2}.$$

$$\text{Surface energy} = \text{surface tension} \times \text{area of surface}$$
$$= 4\pi r^2\, T = 4\pi \times 9 \times 10^{-6} \times 20 \times 10^{-3}$$
$$= 22.62 \times 10^{-7}\ \text{Joule.}$$

Problem 3. *A soap bubble is slowly enlarged from a radius of 0.01 m to 0.1 m. Calculate the work done in the process. Surface tension of soap solution is $26 \times 10^{-3}\ Nm^{-1}$.* [Bangalore University]

Solution. As the radius of soap bubble is increased its surface area increases. Hence, workdone in the process = surface tension × increase in surface area

$$= 26 \times 10^{-3} \times 4\pi\,[(0.1)^2 - (0.01)^2]$$

[Given $r_1 = 0.01$ m ; $r_2 = 0.1$ m]

$$= 3.235 \times 10^{-3}\ J.$$

Problem 4. *Calculate the workdone against surface tension in blowing a soap bubble from a radius of 10 cm to 20 cm, if the surface tension of soap solution is 25 dynes/cm.* [Meerut University]

Solution. Original area of both the surfaces of the bubble

$$= 2 \times 4\pi\,(10\ cm)^2 = 800\ \pi\ cm^2$$

Final area of the surfaces of the bubble

$$= 2 \times 4\pi\,(20\ cm)^2$$
$$= 320\ \pi\ cm^2$$

∴ Increase in surface area

$$= 3200\ \pi - 800\ \pi = 2400\ \pi\ cm^2.$$

Hence, the workdone against surface tension is

$$W = \text{Surface tension} \times \text{increase in area}$$
$$= 25\ \text{dynes/cm} \times (2400 \times 3.14\ cm^2)$$
$$= 1.88 \times 10^5\ \text{erg.}$$

Problem 5. *Two drops of a liquid, each of radius r, coalesce to form a larger drop. Derive an expression for the rise in temperature.* [Gorakhpur, Awadh University]

Solution. When two or more small drops of a liquid coalesce to form a single large drop, the surface area decreases, and hence heat, resulting in the rise in temperature of the drop.

Consider R be the radius of the larger drop formed. Its volume is the same as that of the two smaller drops, each of radius r. Thus,

$$\frac{4}{3}\pi R^3 = 2 \times \frac{4}{3}\pi r^3$$

$$R = (2)^{1/3} r.$$

Hence, the decrease in surface area

$$= \text{Surface area of 2 small drops} - \text{Surface area of larger drop}$$

$$= 2 \times 4\pi r^2 - 4\pi RL$$

$$= 4\pi\,(2r^2 - (2)^{2/3} r^2) \qquad [\text{Again } R = 2^{1/3} r]$$

$$= 8\pi r^2 \{1 - (2)^{-1/3}\}$$

If W be the energy released, we have

$$W = \text{Surface tension} \times \text{decrease in surface area}$$

$$= T \times 8\pi r^2 \{(1) - (2)^{-1/3}\} \qquad \ldots(i)$$

This energy is spent in increasing the temperature of the liquid. Suppose θ be the rise in temperature. If ρ be the density and S the specific heat of the liquid, the heat H absorbed by the drop is

$$H = \text{mass} \times \text{specific heat} \times \text{temp-rise.}$$

$$= \frac{4}{3}\pi R^3 \rho \times s \times \theta$$

$$= \frac{4}{3}\pi\left(2r^3\right)\rho s\theta \qquad \ldots(ii)$$

From (*i*) and (*ii*), we get

$$W = JH$$

$$T \times 8\pi r^2 \{1 - (2)^{-1/3}\} = J \times \frac{4}{3}\pi\left(2r^3\right)\rho s\theta$$

$$\therefore \qquad \theta = \frac{3T}{rJ\rho s}\left\{1 - (2)^{-\frac{1}{3}}\right\}.$$

Problem 6. *If a number of little droplets of water, each of radius r coalesce to form a single drop of radius R, show that the rise in temperature will be given by* $\frac{3s}{J}\left(\frac{1}{r} - \frac{1}{R}\right)$, *where s is the surface tension of water and J is the mechanical equivalent of heat.* [Vikram University, Awadh University]

Solution. When a number of droplets coalesce to form a single drop, the surface area decreases and hence the surface energy also decreases. Consider *n* droplets coalesce. Then,

$$\text{Surface area of } n \text{ droplets} = n \times 4\pi r^2$$

Surface area of the combined drop $= 4\pi R^2$

$\therefore$ Decrease in surface area $= n \times 4\pi r^2 - 4\pi R^2$

$$= 4\pi\,(nr^2 - R^2)$$

If W be the energy released, we have

$$W = \text{decrease in surface area} \times \text{surface tension}$$

$$= 4\pi\,(nr^2 - R^2) \times S \qquad \ldots(i)$$

This energy is spent in increasing the temperature of the water drop. The heat H absorbed by the drop is

$$H = \text{mass} \times \text{specific heat} \times \text{temperature rise}$$

$$= \frac{4}{3}\pi R^3 \times 1 \times 1 \times t$$

$$= \frac{4}{3}\pi R^3 t \qquad \ldots(ii)$$

Considering Joule's relation W = JH, we have

$$4\pi\,(nr^2 - R^2) \times S = J \times \frac{4}{3}\pi R^3 t\,.$$

or,
$$t = \frac{3s}{J}\left(\frac{nr^2 - R^2}{R^3}\right) = \frac{3s}{J}\left(\frac{nr^2}{R^3} - \frac{1}{R}\right) \qquad \ldots(iii)$$

Since the total volume of n droplets is equal to that of the combined drop, we have

$$n \times \frac{4}{3}\pi r^3 = \frac{4}{3}\pi R^3$$

or,
$$nr^3 = R^3$$

or,
$$\frac{nr^2}{R^3} = \frac{1}{r}\,.$$

Putting this value of $\frac{nr^2}{R^3}$ in equation (*iii*), we get

$$t = \frac{3s}{J}\left(\frac{1}{r} - \frac{1}{R}\right).$$

Problem 7. *Obtain an expression for the work necessary to break up a liquid drop of radius R into n equal smaller drops.* [Gorakhpur University]

Solution. When a liquid drop is broken into a number of smaller drops, the surface area increases. Hence, work is to be done against the surface tension. To calculate the increase in surface consider 'r' be the radius of each of the 'n' smaller drops. Since the total volume of all the 'n' smaller drops is the same as the volume of the original drop of radius 'R'.

We have,

$$n \times \frac{4}{3}\pi r^3 = \frac{4}{3}\pi R^3$$

$$\therefore \qquad r = \frac{R}{(n)^{\frac{1}{3}}}.$$

Thus, increase in surface area

$$= n \times 4\pi r^2 - 4\pi R^2$$

$$= 4\pi\,(nr^2 - R^2)$$

$$= 4\pi \left\{ n \frac{R^2}{(n)^{\frac{2}{3}}} - R^2 \right\}$$

$$= 4\pi R^2\, \{(n)^{1/3} - 1\}$$

The necessary work is, therefore,

$$W = \text{Surface tension} \times \text{increase in surface area}$$

$$= T \times 4\pi R^2\, \{(n)^{1/3} - 1\}$$

$$= 4\pi R^2 T\, \{(n)^{1/3} - 1\}.$$

Problem 8. *A small hollow sphere which has a small hole in it is immersed in water to a depth of 0.4 m before any water penetrates into it. If the surface tension of water is 73×10^{-3} Nm^{-1}, find the radius of hole.* [Buradwan University]

Solution. Given, depth of water $h = 0.4$ m, density of water $= 10^3$ kg m^{-3}.

Excess pressure of water at a depth h metre

$$\therefore \qquad h\,dg = 0.4 \times 10^3 \times 9.8 = 3.92 \times 10^3 \text{ N m}^{-2}.$$

If r is the radius of the hole, then before water enters into the hollow sphere, an air bubble of radius r will escape. Thus, excess pressure inside the bubble

$$p = \frac{2T}{r} = \frac{2 \times 73 \times 10^{-3}}{r} \text{ N m}^{-2}.$$

In equilibrium condition,

$$h\,dg = \frac{2T}{r}$$

$$r = \frac{2T}{hdg} = \frac{2 \times 73 \times 10^{-3}}{3.92 \times 10^3} = 37.25 \times 10^{-6} \text{ m}.$$

Problem 9. *The two arms of a U-tube have diameters 2 mm and 1 mm. The tube is partly filled with water and is held with its arms vertical. Find the difference in the levels of water in the two limbs if the surface tension of water is 70×10^{-3} N m^{-1}.* [Calcutta University]

Solution. We know that the rise of liquid in a capillary tube is given by the relation $\dfrac{2T\cos\theta}{r} = h\rho g$.

where T is the surface tension of the liquid, θ the angle of contact, r the radius of the tube, h the height of the liquid column, ρ the density of the liquid and g the acceleration due to gravity.

Consider a U-tube, the two limbs of which have different radius say r_1 and r_2, then

Rise of liquid in the limb of the radius $h_1 = \dfrac{2T\cos\theta}{r_1 \rho g}$

∴ Difference in the height of liquid column in the two limbs of the U-tube :

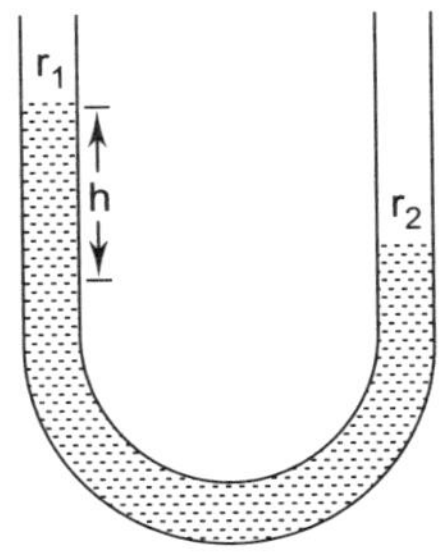

$$h = \frac{2T\cos\theta}{\rho g}\left[\frac{1}{r_1} - \frac{1}{r_2}\right]$$

Now $= r_1 = 1 \text{ mm} = 1 \times 10^{-3} \text{ m}, r_2 = 2 \text{ mm} = 2 \times 10^{-3} \text{ m}$

$T = 70 \times 10^{-3} \text{ N m}^{-1}, \theta = 0, \rho = 10^3 \text{ kg m}^{-3}, g = 9.8 \text{ m s}^{-2}$

$$\therefore \quad h = \frac{2 \times 70 \times 10^{-3}}{10^3 \times 9.8}\left[\frac{1}{1 \times 10^{-3}} - \frac{1}{2 \times 10^{-3}}\right]$$

$$= 7.1 \times 10^{-3} \text{ m} = 7.1 \text{ mm}.$$

Problem 10. Two spherical soap bubbles A and B, of radii 3 and 5 cm coalesce so as to have a portion of their surface in common. Calculate the radius of curvature of this common surface.

Solution. Suppose r be the radius of curvature of the common surface of the soap bubbles P and Q.

The excess pressure in P above atmospheric is $\dfrac{4T}{3}$, while that in Q is $\dfrac{4T}{5}$. Thus, there is an excess pressure on the P-side of the common surface above that on the Q-side of $\dfrac{4T}{3} - \dfrac{4T}{5}$.

But this excess pressure must be equal to $\left(\dfrac{4T}{r}\right)$. Hence,

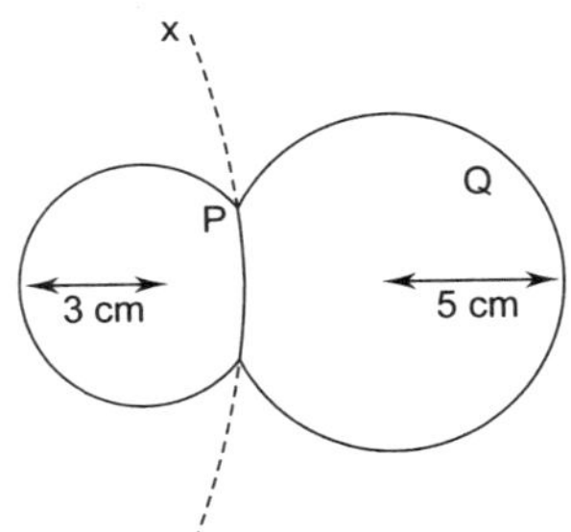

$$\frac{4T}{r} = \frac{4T}{3} - \frac{4T}{5}$$

or,
$$\frac{1}{r} = \frac{1}{3} - \frac{1}{5}$$

$$\therefore \quad r = 7.5 \text{ cm}.$$

Problem 11. *Two spherical soap bubbles coalesec to form a bigger bubble without any leakage of air. If V is the change in volume of the contained air and S the change in total surface area, show that S PV + 4ST = 0, where T is the surface tension of the soap solution and P is the atmospheric pressure.* [Allahabad, Gorakhpur, Awadh University]

Solution. Consider r_1 and r_2 be the radii of the two soap bubbles before they coalesce and R the radius of the combined bubble. In each case the pressure outside the combined bubble, we have

$$p_1 = P + \frac{4T}{r_1}, \; p_2 = P + \frac{4T}{r_2} \quad \text{and} \quad p = P + \frac{4T}{R}.$$

If V_1, V_2 and V be the corresponding volumes, we have

$$V_1 = \frac{4}{3}\pi r_1^3, \; V_2 = \frac{4}{3}\pi r_2^3 \quad \text{and} \quad V = \frac{4}{3}\pi R^3.$$

If the temperature kept constant, we have, from Boyle's law,

$$p_1 V_1 + p_2 V_2 = pV$$

i.e.,
$$\left(p+\frac{4T}{r_1}\right)\left(\frac{4}{3}\pi r_1^3\right)+\left(p+\frac{4T}{r_2}\right)\left(\frac{4}{3}\pi r_2^3\right) = \left(p+\frac{4T}{R}\right)\left(\frac{4}{3}\pi R^2\right)$$

or,
$$p\left(\frac{4}{3}\pi r_1^3+\frac{4}{3}\pi r_2^3-\frac{4}{3}\pi R^3\right)+\frac{4T}{3}\left(4\pi r_1^2+4\pi r_2^2-4\pi R^2\right)=0$$

But
$$\frac{4}{3}\pi r_1^3+\frac{4}{3}\pi r_2^3-\frac{4}{3}\pi R^3 = \text{V (change in volume)}$$

and
$$4\pi r_1^2+4\pi r_2^2-4\pi R^2 = \text{S (change in surface)}$$

Hence, the last expression gives

$$pV+\frac{4T}{3}S=0$$

or $3pV + 4ST = 0$ **(proved).**

Problem 12. *Two soap bubbles of radii r_1 and r_2 coalesce to form a single bubble of radius R. If the external pressure is P, prove that the surface tension T of the soap solution is given by* $T = \dfrac{P\left(R^3-r_1^3-r_2^3\right)}{4\left(r_1^2+r_2^2-R^2\right)}$ [Calcutta University, Gorakhpur University, Delhi University]

Solution. Consider P_1 and P_2 be the pressures inside the bubbles before they coalesce, and p the pressure inside the bigger bubble formed. Then, we have

$$p_1 = P+\frac{4T}{r_1}$$

$$p_2 = P+\frac{4T}{r_2}$$

and
$$p = P+\frac{4T}{R}$$

If v_1, v_2 and V be the corresponding volumes, we have

$$v_1=\frac{4}{3}\pi r_1^3,\; v_2=\frac{4}{3}\pi r_2^3 \text{ and } V=\frac{4}{3}\pi R^3.$$

If the temperature unchanged, Boyle's law gives

$$p_1 v_1 + p_2 v_2 = PV$$

or,
$$\left(P+\frac{4T}{r_1}\right)\left(\frac{4}{3}\pi r_1^3\right)+\left(P+\frac{4T}{r_2}\right)\left(\frac{4}{3}\pi r_2^3\right)=\left(P+\frac{4T}{R}\right)\left(\frac{4}{3}\pi R^3\right)$$

or,
$$\left(P+\frac{4T}{r_1}\right)r_1^3+\left(P+\frac{4T}{r_2}\right)r_2^3=\left(P+\frac{4T}{R}\right)R^3$$

or,
$$4T\left(r_1^2+r_2^2-R^2\right) = P\left(R^3-r_1^3-r_2^3\right)$$

∴
$$T=\frac{P\left(R^3-r_1^3-r_2^3\right)}{4\left(r_1^2+r_2^2-R^2\right)}.$$

Multiple Choice Questions

1. While olive oil spreads over water surface, water does not spread over surface of olive oil. It is because,
   (*a*) surface tension of water is much greater than that of olive oil.
   (*b*) water is denser than olive oil.
   (*c*) surface tension of water is less than that of olive oil.
   (*d*) olive oil is more viscous than water.
2. A vertical glass capillary tube, open at both ends, contains some mercury, which of the following shapes is likely to be taken by the mercury in the tube

(a) (b) (c) (d)

3. Surface tension of water at 20° is 72.9 dyne/cm. The surface energy (E) of the water surface (assume that the increase in surface area occurs under isothermal condition) is then,
   (*a*) E = 72.9 erg/cm^2 (*b*) E = 36.45 erg/cm^2
   (*c*) E = 145.8 erg/cm^2 (*d*) impossible to predict.
   [Numerically S.T. = S.E.]
4. The shape of the liquid meniscus in a capillary tube depends on the magnitudes of the force of cohesion (F_1) and the force of adhesion (F_2). The meniscus is,
   (*a*) Concave when $F_1 > \frac{F_2}{\sqrt{2}}$ (*b*) Convex when $F_2 > \frac{F_2}{\sqrt{2}}$
   (*c*) Convex when $F_1 < \frac{F_2}{\sqrt{2}}$ (*d*) Convex when $\frac{F_1}{\sqrt{2}} > F_2$.
5. Kerosene oil rises up in the wick of a lamp because of
   (*a*) diffusion (*b*) osmosis
   (*c*) surface tension (*d*) the gravitational pull of the wick.
6. Which of the following graphs represents the variation of the height of the liquid column (*h*) in a capillary tube with its radius (*r*) ?

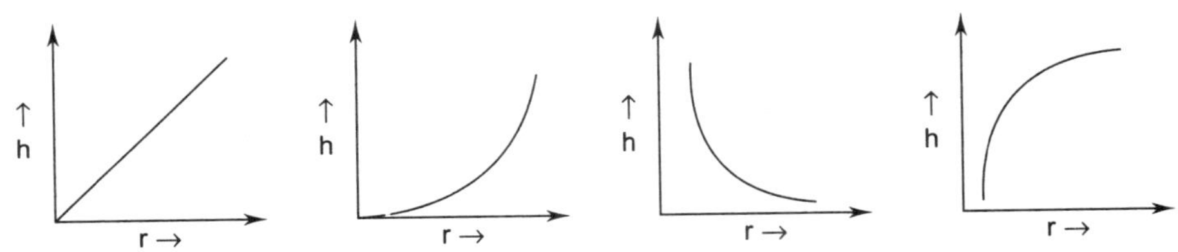

(a) (b) (c) (d)

$\left[r \, \alpha \, \frac{1}{h}\right]$

7. Which one of the graphs represents the variation of surface tension with temperature over a small range of temperature for water?

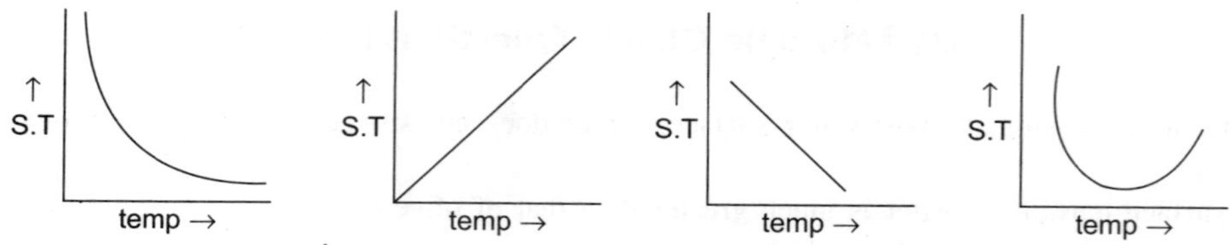

(a) (b) (c) (d)

Review Questions

1. (*a*) Explain what do you understand by surface tension of a liquid.
   [*Bhopal, Gharwal, Nagpur, Calcutta University*]
   (*b*) Give an account of molecular theory of surface tension.
   [*Calcutta, Burdwan, Nagpur, Bhopal University*]
2. (*a*) Explain what do you understand by the angle of contact in the case of a liquid.
   [*Meerut, Calcutta University*]
   (*b*) Account for the curvature of the surface of a liquid in the neighbourhood of a solid surface.
3. (*a*) Explain the term surface energy. Derive the relation between surface tension and surface energy.
   [*Gharwal University, Meerut University, Indore University*]
   (*b*) Calculate the workdone in spraying a drop of mercury of 1 mm radius into one million identical drops all of the same size. Surface tension of mercury is 550×10^{-3} N m.
4. Show that the excess pressure acting on the curved surface of a curved membrane is given by $p = 2T\left(\frac{1}{r_1}+\frac{1}{r_2}\right)$, where r_1 and r_2 are the radii of curvature and T the surface tension of the membrane.
   [*Calcutta, Jadavpur, Bhopal University*]
5. (*a*) Derive an expression for the height h through which a liquid of surface tension T will rise in a capillary tube of radius r. Explain clearly from where the energy comes when the liquid rises against gravity in the capillary tube. [*Calcutta, Vidyasagar, Burdwan University*]
   (*b*) What will happen if the length of capillary tube is smaller than *h* ?

Unsolved Problems

1. What amount of energy will be liberated if 1000 droplets of water each of 10^{-8} m in diameter coalesce to form a one large spherical drop? Surface tension of water = 75×10^{-3} N/m [*Meerut University*]
   [**Ans.** 2.1×10^{-14} Joule]
2. Calculate the workdone in blowing a soap bubble of radius 10 cm and surface tension 30 dynes per cm.
   [*Gharwal University*]
   [**Ans.** 7.5×10^{-3} Joule]
3. Find the amount of workdone in blowing a soap bubble of surface tension 30 dyne/cm from 10 cm radius to 15 cm radius.
   [*Lucknow University*]
   [**Ans.** 9.4×10^4 erg]
4. Calculate the amount of energy in breaking a water drop of radius 0.4 cm into 125 smaller drops. (Surface tension of water = 72 dynes/cm)
   [*Guwahati University*]
   [**Ans.** 579 ergs]
5. The pressure of air in a soap bubble of 7 mm diameter is 8 mm of water above the atmospheric pressure. Calculate the surface tension of the soap solution (g = 980 cm/sec^2).
   [*Meerut, Vikram University*]
   [**Ans.** 68.6 dyne/cm]

Viscosity

2C.1 INTRODUCTION

Liquids and gases together are termed as fluids as there are some analogus in behaviour between them. Fluid cannot permanent by withstand any shearing stress. However, gases are compressible but liquids are not the same way. When a liquid is at rest, do not show any kind of rigidity. But when ever a liquid flows, there is a relative motion between the adjacent layers of the liquid, internal tangential forces act between two adjacent layers opposing this relative motion arise.

If we consider any particular layer of the liquid we find that the layer immediately above it is moving faster than the layer immediately below it.

The property of a liquid owing to which it opposes the relative motion between its different layers is called viscosity. Their magnitude differing from one liquid to another.

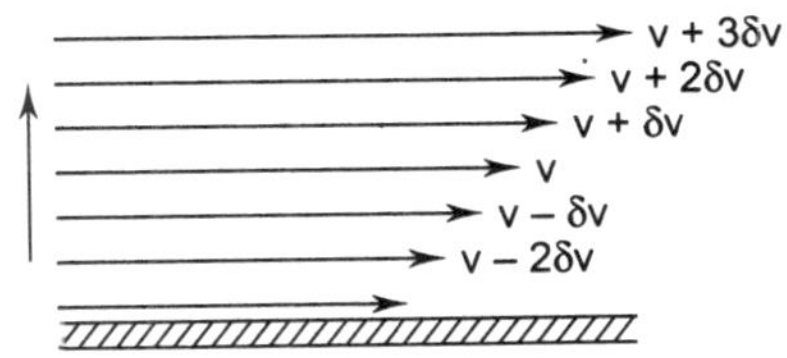

Figure 2C.1

2C.2 FLOW OF LIQUIDS

The liquids flow by two different ways:

(a) Streamlined flow

When a liquid flows such that each particle passing a certain point follows exactly the same path as the preceding particles which passed the same point, the flow is said to be streamlined and the path is called a streamline.

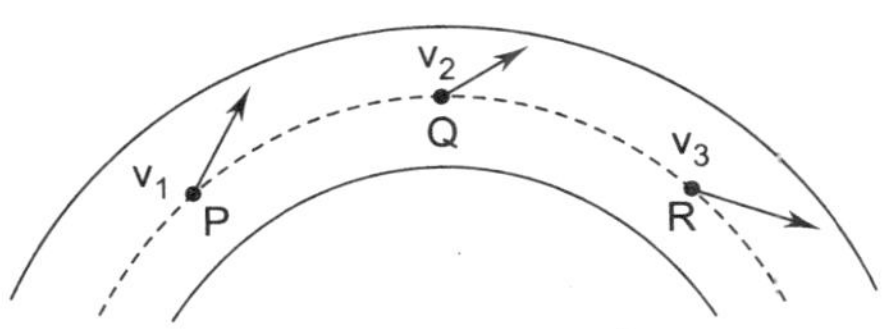

Figure 2C.2

Suppose a liquid flow through a tube having the velocities V_1 at P, V_2 at Q and V_3 at R which is shown in Fig. 2C.2. In this tube the velocity of a particle will vary along its line of flow, but at any fixed point the velocity of the particle will vary along its line of flow, but at any fixed point the velocity of the particle, whichever is there, is always the same.

In case of uniform cross-sectional tubes, all the streamlines are parallel to the axis of the tube (Fig. 2C.3). In a streamlined flow, the external energy maintaining the flow is used mainly overcoming the viscous drag between the different layers of the liquid. The streamline flow is maintained when the velocity is below a certain limit known as critical velocity.

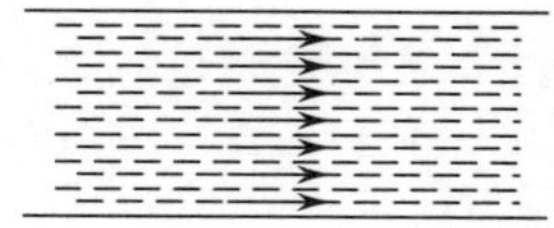

Figure 2C.3

(b) Turbulent flow

When the velocity of the particles of the liquid passing through any given point varies with time both in magnitude and direction, is called turbulent flow. Most of the external energy maintaining the flow is used in setting up eddy currents in the liquid as shown in Fig. 2C.4.

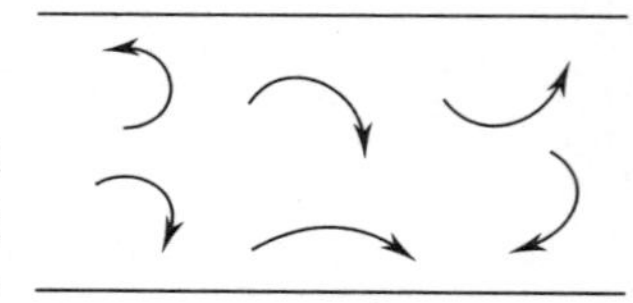

Figure 2C.4

2C.3 VELOCITY GRADIENT

Consider a liquid is flowing along OX. Where the liquid layer in contact with the fixed surface is stationary and velocity of each layer increased with its distance from the fixed surface.

Let us consider two parallel infinitesimally close plane PQ and RS distant y and $y + dy$ respectively from OX. If v and $v + dv$ is the velocities in PQ and RS respectively, then $\frac{dv}{dy}$ is the rate of change velocity with distance perpendicular to the direction of flow. It is termed as velocity gradient at points in the plane PQ. Here the dimension of velocity gradient $\frac{dv}{dy}$

$$= \frac{[LT^{-1}]}{[L]} = [T^{-1}]$$

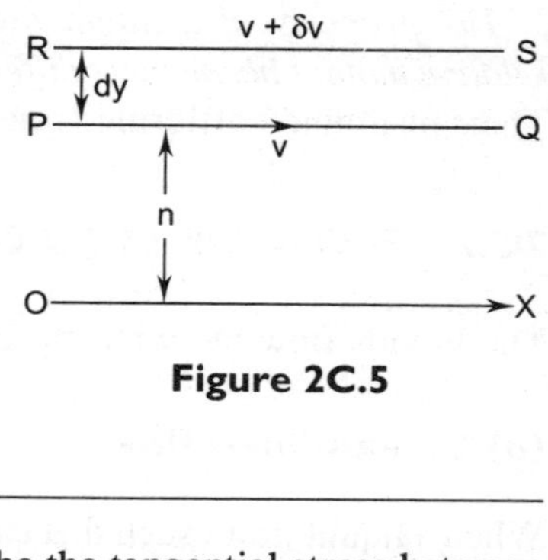

Figure 2C.5

2C.4 CO-EFFICIENT OF VISCOSITY

Consider a liquid flow in a streamline over a fixed surface OX. If P be the tangential stress between two layers of a liquid which are moving relative to each other, then according to Newton's law of viscous resistance

$$P \propto \frac{dv}{dy} \text{ (The velocity gradient)}$$

Where dv is the change of velocity between the two consecutive close layers separated by a distance dy

$$\therefore \qquad P = \eta \frac{dv}{dy}$$

Where η is the constant of proportionality known as co-efficient of viscosity of the liquid. If $\frac{dv}{dy} = 1$ then $P = \eta$.

Thus, the co-efficient of viscosity of a liquid is the tangential stress per unit velocity gradient.

Dimension:

$$\eta = \frac{P}{\frac{dv}{dy}} = \frac{F/A}{\frac{dv}{dy}} \qquad \left[\begin{aligned} P &= \text{Force per unit area} \\ &= \frac{F}{A} \end{aligned}\right]$$

$$= \frac{\left[MLT^{-2}\right]}{\left[L^2\right]} \frac{[L]}{\left[LT^{-1}\right]}$$

$$= \left[ML^{-1}\,T^{-1}\right]$$

In the C.G.S system the unit of co-efficient of viscosity is gm/(cm – sec). It is also known as 'poise' to honour Poiseuille. Thus, 1 poise means that a viscous force of 1 dyne acts per cm^2 to maintain a velocity difference of 1 cm/sec between two layers 1 cm apart.

2C.5 CRITICAL VELOCITY AND REYNOLDS NUMBER

Osborne Reynolds experimentally showed that if the average velocity [the average velocity of flow of a liquid is the total volume of the liquid flowing per second per unit cross-section area of the tube) of flow of a given liquid is below a certain value, the motion becomes streamline, but if it exceeds the value of motion is turbulent. The limiting velocity is called "critical velocity" (V_C) of the liquid.

Reynolds established that the critical velocity v_c of a liquid is

(*i*) Proportional to the co-efficient of viscosity; *i.e.*, $v_c \propto \eta$

(*ii*) Inversely proportional to the density of the liquid; $v_c \propto \frac{1}{\rho}$

(*iii*) Inversely proportional to the radius of the tube; $v_c \propto \frac{1}{r}$

Combining we have, $v_c \propto \frac{\eta}{\rho r}$ or $v_c = \frac{K\eta}{\rho r}$

The constant of proportionality K is known as Reynolds number and its value is approximately 1000 for narrow tubes. The significance of the Reynolds number K is that its value determines the nature of flow of a liquid through a tube.

2C.6 POISEUILLE'S EQUATION FOR FLOW OF LIQUID THROUGH A HORIZONTAL NARROW TUBE

Poiseuille derived an equation for the volume of liquid flowing per second through a narrow horizontal tube of uniform bore under stream line motion. He made the following assumptions:

(*i*) The flow of liquid is steady and streamline.
(*ii*) There is no radial flow of liquid and pressure is constant over every cross-section of the tube.
(*iii*) The tube is horizontal, so that gravity does not affect the flow.
(*iv*) Layer of liquid in contact with the wall of the tube is stationary and velocity increases from the wall towards the axis of the tube.

Let us consider a horizontal tube of length l and radius 'a' and a liquid is flowing through the tube under a constant pressure difference 'P' at its two ends.

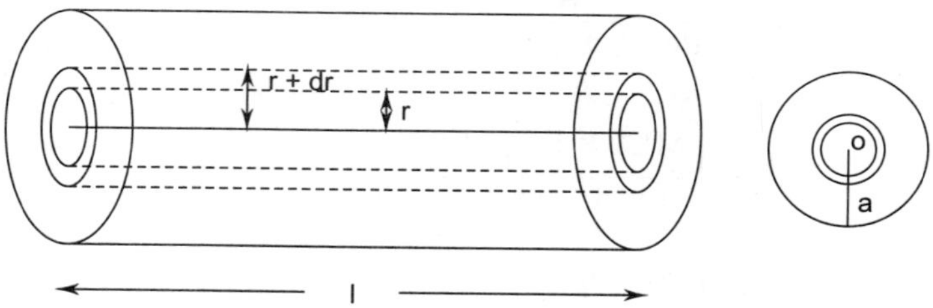

Figure 2C.6

The layer of liquid in contact with the wall of tube at being rest, the velocity of this layer is. We also consider in the steady state, the velocity of a cylindrical liquid layer at a distance r from the axis of the tube is v. So the velocity gradient perpendicular to the direction of flow is $-\frac{dv}{dr}$ at the every point of the liquid layer. The negative sign indicates as r increases v decreases slowly moving liquid outside the cylinder exerts a backward viscous force on the cylinder. According to Newton's equation,

The viscous force = co-efficient of viscosity × surface area of the layer × velocity gradient

$$\therefore \qquad F = -\eta A \frac{dv}{dr} \qquad \ldots(2C.1)$$

Here $\quad A$ = surface area of the liquid = $2\pi rl$

and $\quad \eta$ = coefficient of viscosity.

On the other hand, the flow of liquid is due to pressure difference across the two ends of the tube. Therefore, the force tending to accelerate the liquid cylinder = Pressure difference between its ends × Area of cross-section

$$= P\pi r^2 \qquad \ldots(2C.2)$$

If there is no acceleration of the liquid, these forces must balance, so that

$$P\pi r^2 = -2\pi rl\eta \frac{dv}{dr}$$

or

$$\frac{dv}{dr} = -\frac{P}{2\eta l} r$$

Integrating we get,

$$v = -\frac{P}{2\eta l}\int r\,dr = -\frac{Pr^2}{4\eta l} + C.$$

where C is the integrating constant and value is determined from the boundary conditions.

As the layer of the liquid contact with the tube, the velocity $v = 0$, so $v = 0$ at $r = a$. So we get

$$0 = -\frac{Pa^2}{4\eta l} + C \quad \text{or} \quad C = \frac{Pa^2}{4\eta l}$$

$$\therefore \qquad v = \frac{P}{4\eta l}\left(a^2 - r^2\right) \qquad \text{...(2C.3)}$$

Thus, the velocity of liquid on the axis is

$$v = \frac{Pa^2}{4\eta l}, \qquad [\because \quad r = 0] \qquad \text{...(2C.4)}$$

This is the equation for parabola (Fig. 2C.7). It increases from zero at the wall $x = r$ of the tube to a maximum at the centre $x = 0$.

We now assume a thin cylindrical shell of radius r and $r + dr$. The volume of the liquid flowing per second through it is given by

$$dV = \text{Velocity} \times \text{cross-sectional area of the shell}$$

$$= v \times 2\pi r dr$$

$$= \frac{P\left(a^2 - r^2\right) 2\pi\, r\, dr}{4\eta l}$$

$$= \frac{P\pi}{2\eta l}\int_0^a \left(a^2 - r^2\right) r\, dr$$

or

$$V = \frac{P\pi}{2\eta l}\left[a^2 \cdot \frac{r^2}{2} - \frac{r^4}{4}\right]_0^{a^0} = \frac{\pi P a^4}{8\eta l}$$

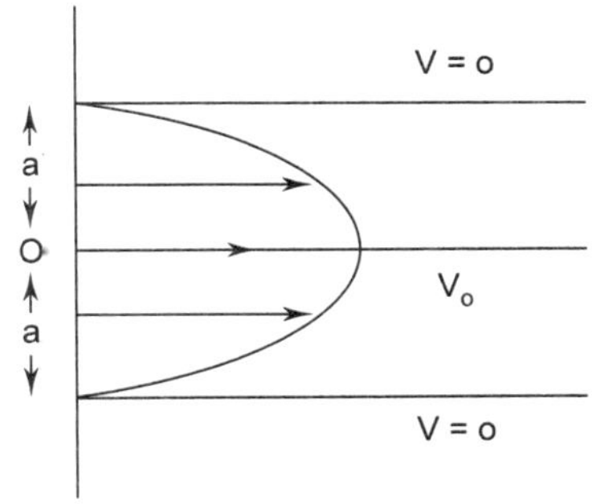

Figure 2C.7

This is known as Poiseuille's formula for the streamline flow of liquid through a narrow horizontal tube.

Limitations and Corrections

(*i*) It is assumed that the flow of liquid is steady throughout the length of the tube. In actual practice, the flow of the liquid is slightly accelerated at the entrance and becomes steady only after traversing a small length of the tube. It may be reduced by using a long tube and effective length of the tube should be taken as $(l + 1.64a)$ where a is the radius of the tube.

(*ii*) The equation is derived on the basis that the applied pressure difference is totally used for overcoming viscous resistance. But in reality a part of the applied pressure is used to give kinetic energy to the liquid. To move allowance for this the effective pressure difference should be taken equal to $\left(P - \frac{V^2\rho}{\pi^2 a^4}\right)$, where ρ is the density of the liquid.

(*iii*) Poiseuille's formula hold for narrow tube only but in case of tube of wide bore the flow will tend to become turbulent.

After proper corrections, the Poiseuille's equation can be expressed as

$$\eta = \frac{P\pi r^4}{8V\,(1 + 164a)} - \frac{V\rho}{8\pi\,(l + 1.64a)}$$

2C.7 STOKE'S LAW

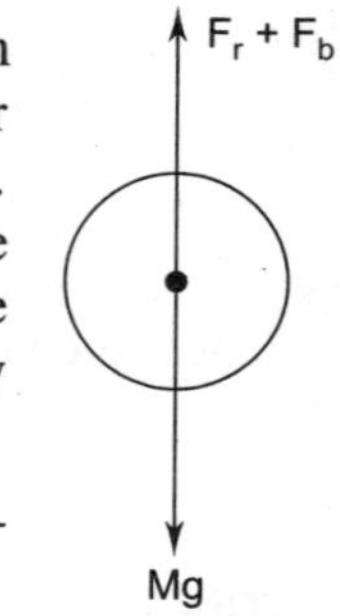

When a body falls under gravity through viscous medium, the layer of the fluid in contact with the body moves with the velocity of the body, and velocities of the other layers gradually decrease producing a relative motion between layers of the fluids. This is opposed by the viscous force in the fluid. This viscous force increases with the velocity of the body until the viscous force acting upwards plus upward thrust of the fluid on the body become equal to the weight of the body acting downwards. The body then falls with a constant velocity which is known as terminal velocity.

Consider F_v be the upward viscous force on the body and F_b be the upward buoyancy force on the body, then we can write

$$F_v + F_b = Mg \qquad \ldots(2C.5)$$

Stokes derived that when a small sphere of radius r is moving slowly with a constant velocity v through a perfectly homogeneous viscous fluid (viscosity-η) of infinite extension, the viscous force experienced by the Sphere is $F_v = 6\pi\eta rv$

This is called Stoke's law.

$$\boxed{F_v = 6\pi\eta rv} \qquad \ldots(2C.6)$$

Now the buoyancy force

$$F_b = \text{Weight of the fluid displaced}$$

$$F_v = \text{Volume of the body density of the fluid } (\sigma) \times g$$

$$= \frac{4}{3}\pi r^3 \rho g \qquad \ldots(2C.7)$$

If the body falls with terminal velocity then from equating (2C.5) we get

$$6\pi\eta rv + \frac{4}{3}\pi r^3 \sigma g = Mg$$

$$= \frac{4}{3}\pi r^3 \sigma g$$

or,

$$6\pi\eta rv = \frac{4}{3}\pi r^3 \rho g - \frac{4}{3}\pi r^3 \sigma g \qquad (\sigma = \text{density of the material})$$

$$= \frac{4}{3}\pi r^3 g\,(\rho - \sigma)$$

or,

$$v = \frac{2}{9}\frac{r^2 g\,(\rho - \sigma)}{\eta}$$

2C.8 EFFECTS OF TEMPERATURE AND PRESSURE ON VISCOSITY

Temperature

The viscosity of liquids decreases markedly with rise in temperature.

Example: *The viscosity of water decreases from 0.0101 poise at 20°C to 0.0047 poise at 60°C.*

NOTE No definite relation has been established between viscosity and temperature.

One of the relation has the form.

$$\ln \eta_T = \alpha + \frac{\beta}{T}$$

where α, β are constant and T is the absolute temperature. η_T is the viscosity at temperature TK.

However, in case of gases the viscosity increases with rise in temperature.

For small ranges of temperature the variation is according to the following Sutherland's formula

$$\eta = \frac{\eta_o\, kT^{1/2}}{1 + \frac{S}{T}}$$

where S is known on Sutherland's, constant

Pressure

Practically the viscosity of a gas is independent of pressure. The viscosity of gas rapidly decreases with decrease of pressure at very low pressure.

On the other hand at very high pressure the viscosity of a gas increases with increase of pressure. The viscosity of liquids in general increases with increase of pressure.

2C.9 EQUATION OF CONTINUITY

It is a special case of law of conservation of mass. It illustrates that the rate of flow of incompressible moving fluid is same throughout a pipe if the flow of liquid is streamline.

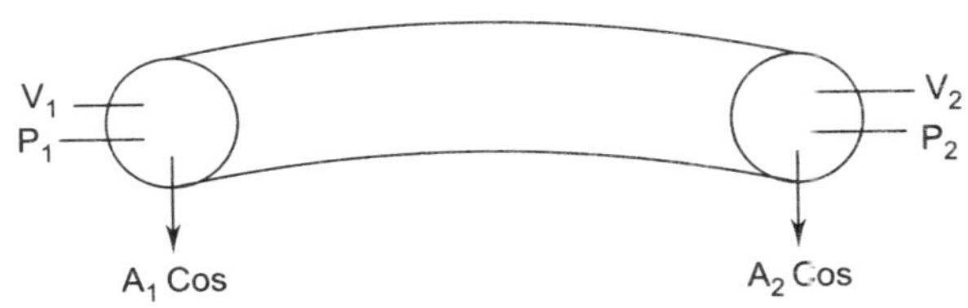

Figure 2C.8

Let us consider a fluid is flowing steadily through a pipe, entering through non-uniform pipe at the edge P_1 with velocity V_1 and leaving P_2 with velocity V_2 (fig. 2C.8). Area of cross section of P_1 and P_2 are A_1 and A_2 respectively. Here we assume that the fluid is an incompressible liquid.

Hence, the mass of the liquid flowing in unit time through the section P_1 is

$$M_1 = A_1 V_1 \rho \qquad (\rho = \text{the density of the fluid})$$

and again through the section P_2 is

$$M_2 = A_2 V_2 \rho$$

If no liquid is absorbed in the region between P_1 and P_2 then we may write

$$M_1 = M_2$$

or

$$A_1 V_1 = A_2 V_2$$

This is the equation of continuity *i.e.*, when an incompressible, non-viscous fluid flows steadily through a tube of non-uniform cross-section then the product of area of cross section and velocity of flow is same at every point in the tube.

2C.10 BERNOULLI'S THEOREM

It states that "For a steady flow of one incompressible, non-viscous fluid the sum of kinetic energy, potential energy and pressure energy per unit volume of a fluid is constant along any section of flow".

Proof:

Let us consider an incompressible, non-viscous fluid flowing steadily through a non-uniform pipe as shown in Fig. 2C.9. Let P_1, A_1, V_1 and h_1 are the pressure, area of cross-section velocity and height from the horizontal reference level at a cross-section X. The corresponding quantities in the right horizontal part of the tube are P_2, A_2, V_2 and h_2 at cross-section Y.

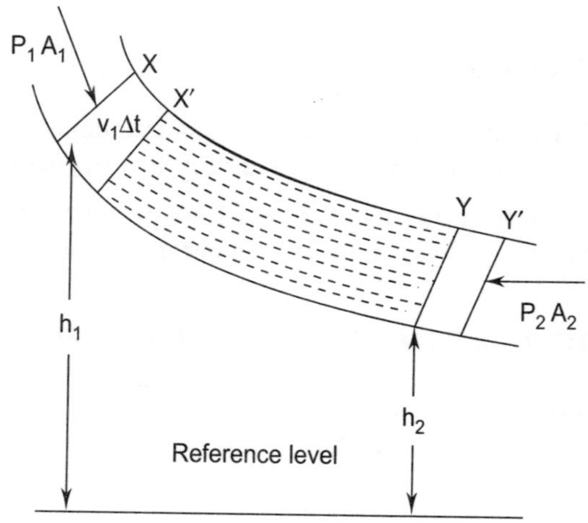

Figure 2C.9

Let us consider the flow of fluid included between X and Y. Suppose, in a very small time interval Δt, the left end of the fluid XY advances from X to X′ through a displacement $v_1 \Delta t$ parallel to the force $P_1 A_1$.

Therefore, work done on the fluid XY be the force $P_1 A_1$ is

$$W_1 = \text{Force} \times \text{displacement} = P_1 A_1 v_1 \Delta t$$

In the same time interval the right end of the fluid XY flows from Y to Y′ through a distance $v_2 \Delta t$ against the force $P_2 A_2$. Therefore, the work done by the liquid against the force $P_2 A_2$ is

$$W_2 = P_2 A_2 v_2 \Delta t$$

Thus, the resultant work done by the pressure force on the fluid XY is

$$W = W_1 - W_2 = P_1 \frac{\Delta m_1}{\rho_1} - P_2 \frac{\Delta m_2}{\rho_2}$$

where $\Delta m_1 = A_1 v_1 \Delta t \rho_1$ [Mass of the liquid crossed × at Δt time]

and $\Delta m_2 = A_2 v_2 \Delta t \rho_2$ [Mass of the liquid crossed Y]

For an Incompressible fluid

$\Delta m_1 = \Delta m_2$ [The density of the liquid at X and Y are same *i.e.*, $\rho_1 = \rho_2$]

Thus net work done on the fluid XY.

$$W = \frac{\Delta m}{\rho}(p_1 - p_2)$$

When the liquid from XX′ advances to YY′, its potential energy decreases. The loss of potential energy

$$\Delta E_P = \Delta m\, gh_1 - \Delta mgh_2$$
$$= \Delta mg\,(h_1 - h_2)$$

Similarly, the increase in kinetic energy (K.E.) of the fluid as it displaced from XY to X′Y′ is equal to the kinetic energy of the fluid which appears YY′ minus the K.E. which disappears from XX′. Therefore, increase in kinetic Energy of the fluid

$$\Delta E_k = \frac{1}{2}\Delta m\left(v_2^2 - v_1^2\right)$$

Again, from the principle of the conservation of energy, we have

Work done on the liquid = gain of K.E. of the fluid – loss of P.E. of the fluid

or, $$\frac{\Delta m}{\rho}(P_1 - P_2) = \frac{1}{2}\Delta m\left(v_2^2 - v_1^2\right) - \Delta mg\left(h_1 - h_2\right)$$

or, $$P_1 + \frac{1}{2}\rho v_1^2 + \rho gh_1 = P_2 + \frac{1}{2}\rho v_2^2 + \rho gh_2$$

Thus $$P + \frac{1}{2}\rho v_d^2 + \rho gh = \text{constant}$$

This is the Mathematical form of Bernoulli's equation for steady, non-viscous, incompressible flow.

or, $$\frac{P}{\rho g} + \frac{v^2}{2g} + h = \text{constant.}$$

$$\frac{P}{\rho g} = \text{pressure. head,}\quad \frac{v^2}{2g} = \text{velocity head}$$

and $$h = \text{elevation head}$$

So, Pressure head + velocity head + elevation head = constant

IN FOCUS

- **Coefficient of viscosity (η)** → Poiseuille ($ML^{-1}\,T^{-1}$)
- **Critical velocity (v_c)** $= \dfrac{K\eta}{\rho r}$

  Where K → Reynolds number

  δ → Density of the liquid

  r → radius of the Tube

  η → coefficient of viscosity
- Poiseuille's formula for the stream line flow of liquid through a narrow horizontal tube

$$v = \frac{\pi \rho a^4}{\delta \eta l}$$

P → pressure difference between its ends.

a → radius of the tube

l → length of the tube

- **Viscous force from Stoke's law** $Fv = 6\,\pi\eta r v$
- **The equation of continuity** $A_1 V_1 = A_2 V_2$
- **Effect of Temperature on viscosity**
  → The viscosity of liquids decreases markedly with rise in temperature.
  → In case of gases viscosity increases with rise in temperature.
- The mathematical form of Bernoulli's equation for steady, non-viscous, incompressible flow

$$\text{Pressure head}\left(\frac{P}{\rho g}\right) + \text{velocity head}\left(\frac{v^2}{2g}\right) + \text{elevation head } (h) = \text{constant}$$

Solved Problems

Problem 1. *In the Poiseuille experiment the following observations were made. Volume of water collected in 5 minutes = 40 c.c.; Head of water = 0.4 m; length of capillary tube = 0.602 m and radius of capillary tube = 0.52×10^{-3} m. Calculate the co-efficient of viscosity of water.*

[Bang. Univ.]

Solution. Volume of water collected per second

$$V = \frac{40}{5 \times 60}\,\text{cm}^3 = \frac{40}{50 \times 60} \times 10^{-6}\,\text{m}^3 = \frac{2}{15} \times 10^{-6}\,\text{m}^3$$

Head of water $h = 0.4$ m

∴ Difference of pressure $P = h\rho g = 0.4 \times 10^3 \times 9.8\ \text{N m}^{-2} = 3.92 \times 10^3\ \text{N m}^{-2}$

Length of capillary tube $l = 0.602$ m

Radius of the capillary tube

$$a = 0.52 \times 10^3\ \text{m}$$

Now co-efficient of viscosity

$$\eta = \frac{\pi p a^4}{8lV}$$

$$= 1.4 \times 10^{-3}\ \text{N m}^{-2}$$

Problem 2. *A plate of metal 100 sq. cm in area rests on a layer of caster oil 2 mm thick whose co-efficient of viscosity is 15.5 poise. Calculate the horizontal force required to move the plate with a speed of 0.03 m s^{-1}.*

[Bang. Univ.]

Solution. Given,

$$A = 100\ \text{sq. cm} = 10^{-2}\ \text{m}^2$$

$$v = 0.03\ \text{m s}^{-1}$$

$$r = 2\ \text{mm} = 0.2 \times 10^{-2}\ \text{m}$$

$$\eta = 15.5\ \text{poise} = 1.55\ \text{dec - poise.}$$

Horizontal viscous force $F = -\eta A \dfrac{v}{r} = \dfrac{1.55 \times 10^{-2} \times 0.03}{0.2 \times 10^{-2}}$

$= -0.2325$ N

External force required = 2.325 N

Problem 3. *Water is conveyed through a horizontal tube of length 4 cm and diameter 8 cm at the rate of 2 litres/s. Assuming only viscous resistance, calculate the pressure difference required to maintain the flow. Coefficient of viscosity of water is 0.01 cgs unit.* [Rohilkhand Univ.]

Solution. We know,

$$\eta = \frac{\pi P r^4}{8l \cdot V} \quad \text{or,} \quad P = \frac{8l \cdot V \cdot \eta}{\pi r^4}$$

Given that l = 4 km = 4×10^5 cm, V = 20 litres/s = 20×10^3 cm³/s, r = 4 cm and η = 0.01

$$\therefore \quad P = \frac{8 \times 4 \times 10^5 \times 20 \times 10^3 \times 0.01}{3.14 \times (4)^4} = 0.796 \times 10^6 \text{ dynes/cm}^2$$

Problem 4. *Water flows through a horizontal tube of length 20 cm and internal radius 0.081 cm under the constant pressure head of water 20 cm high. In 2 minutes 864 cc of water flows through the tube. Calculate the coefficient of viscosity of water.*

Solution. We know, coefficient of viscosity,

$$\eta = \frac{\pi P r^4}{8Vl} = \frac{\pi h \rho g r^4}{8Vl}$$

Here, h = 20 cm, ρ = 1 gram/cc, g = 980 cm/s², l = 20 cm and r = 0.081 cm

Volume of water flows through the tube per second,

$$V = \frac{864}{60 \times 12} \, cc/\text{second} = 1.2 \, cc/\text{second}$$

$$\therefore \quad \eta = \frac{22}{7} \cdot \frac{20 \times 980 \times (0.081)^4}{8 \times 1.2 \times 20} = 0.01381 \text{ poise.}$$

Problem 5. *Find the terminal velocity of an oil-drop of density 0.95/cm² and radius 10^{-4} cm falling through air of density 0.0013 g/cm³, if the viscosity of air is 181×10^{-6} cgs unit, g = 980 cm/s².*

Solution. We can write $v = \dfrac{2}{9} \cdot \dfrac{r^2 (\rho - \sigma) g}{\eta}$

Here $r = 10^{-4}$ cm, ρ = 0.95 g/cm³, σ = 0.0013 g/cm³ and $\eta = 181 \times 10^{-6}$

$$\therefore \quad v_r = \frac{2}{9} \cdot \frac{(10^{-4})^2 (0.95 - 0.0013) \times 980}{181 \times 10^{-6}} = \frac{2}{9} \cdot \frac{10^{-8} \times 0.9487 \times 980}{181 \times 10^{-6}}$$

$= 0.0114$ cm/s.

Problem 6. *Two drops of water of the same size are falling through air with terminal velocity of 1 m s^{-1}. If the two drops combine to form a single drop what would be the new terminal velocity? Explain the formula used.*

Solution. We know that the terminal velocity, $V_r = \frac{2}{9} \cdot \frac{r^2 (\rho - \sigma) g}{\eta}$ for single drop and pair of drops ρ, σ, g and η are constants.

Thus we have $V_r \propto r^2$.

Let r = radius of each of the five small drops

and R = radius of the combined drop. We have

$$2r^3 = R^3 \quad \text{or,} \quad R = (2)^{1/3} r.$$

Hence

$$\frac{(v_r)_1}{(v_r)_2} = \frac{r^2}{R^2} = \frac{r^2}{(2)^{2/3} r^2} = \frac{1}{2^{2/3}}$$

[Where $(v_r)_1$ = terminal velocity of small drops and $(v_r)_2$ = terminal velocity of large drop].

Therefore, $(V_r)_2 = (V_r)_1 \times (2)^{2/3} = 1 \times (2)^{2/3}$ m–s^{-1} = $2^{2/3}$ m s^{-1}

Problem 7. *Three capillaries of lengths 8L, 0.2L, and 2L and radii r, 0.2r and 0.5r respectively are connected in series. If the total pressure across the system in an experiment p, deduce the pressure across the shortest capillary.* [Kanpur, Agra, Allahabad Univ.]

Solution. Consider, P_1, P_2, P_3 be the pressure across the first, second, third capillaries respectively. The rate of flow must be the same throughout the system.

Thus,

$$Q = \frac{\pi P_1 r^1}{8\eta (8L)} = \frac{\pi P_2 (0.2r)^4}{8\eta (.02L)} = \frac{\pi P_3 (.05r)^4}{8\eta (2L)}$$

or

$$\frac{P_1}{8} = \frac{P_2}{125} = \frac{P_3}{32}$$

$$P_1 = \frac{8}{125} P_2 \quad \text{and} \quad P_3 = \frac{32}{125} P_2$$

$$P_1 + P_2 + P_3 = P$$

$$\frac{165}{125} P_2 = P$$

$$P_2 = \frac{125}{165} P,$$

Problem 8. *Water flow in a streamline through a horizontal pipe of cross-sectional radius a from a vertical tank of cross-section. Find the time when the height of the level in the tank becomes half.*

[Allahabad Univ.]

Solution. Consider the height of liquid in the tank above the horizontal tube at any instant, the pressure-difference across the ends of the tube at that instant is $P = \rho g h$, where ρ is the density of the liquid. If the height of liquid falls through dh in a small time-interval dt, then the volume of the liquid flowed through the tube in time dt is A (dh), where A is the area of cross-section of the tank. Therefore, the rate of flow of the liquid at the instant t is

$$= -\frac{A\,(dh)}{dt}$$

The '–' sign indicates that the, decreases with increase in t. But from Poiseuille's formula, the rate of flow

$$= \frac{\pi P a^4}{8 l \eta} = \frac{\pi\,(\rho g h)\,a^4}{8 l \eta}$$

$$= -\frac{A\,(dh)}{dt} = \frac{\pi\,(\rho g h)\,a^4}{8 l \eta}$$

or, $$dt = -\frac{8 l \eta\, A\, dh}{\eta\, \rho g a^4\, h}$$

Suppose H is the initial height of the liquid and t be the time required by the level to fall to H/2. Then, integrating the last expression between proper limits, we have,

$$\int_0^t dt = -\frac{8 l \eta\, A}{\pi \rho g a^4} \int_H^{H/2} \frac{dh}{h}$$

$$t = -\frac{8 l \eta\, A}{\pi \rho g a^4}\left[\log_e \frac{H}{2} - \log_e H\right] = \frac{8 l \eta\, A}{\pi \rho g a^4} \log_e 2$$

∴ This is required equation.

Problem 9. *A tube of radius R, length L is connected in series with another of radius R/2 and length L/4. If the pressure across the two tubes taken together is P, deduce the pressure across the tubes separately.*

Solution. Consider P_1 and P_2 be the pressure across the first and second tubes respectively. Then,

$$P = P_1 + P_2$$

The rate of flow of a liquid (viscosity n) throughout the system is the same. Thus.

$$\frac{\pi P_1 R^4}{8 \eta L} = \frac{\pi P_2 \left(\frac{R}{2}\right)^2}{8 n \left(\frac{L}{4}\right)}$$

$$P_1 = \frac{P_2}{4}$$

$$P_1 = \frac{P}{5} \quad \text{and} \quad P_2 = \frac{4P}{5}$$

Problem 10. *Two drops of water of the same size are falling through air with terminal velocity 1 m s^{-1}. If the two drops combine to form a single drop, calculate the terminal velocity.*

Solution. Let r be the radius of each drop and v_1 that of the combined drop, then

$$\frac{4}{3}\pi r_1^3 = 2 \times \frac{4}{3} r^3$$

$$r_1 = 2^{1/3} r$$

If v is the terminal velocity of each drop and v_1 that of the combined drop then according to Stoke's law

$$v = \frac{2r^2(\rho - d)g}{9\eta}$$

$$v_1 = \frac{2r_1^2(\rho - d)g}{9\eta}$$

$$\frac{v_1}{v} = \frac{r_1^2}{r^2} = \frac{2^{2/3} r^2}{r^2} = 2^{2/3} = 1.588 \text{ m/sec.}$$

Multiple Choice Questions

1. The dimensions of viscosity are
   (*a*) $M^{-1}L^{-1}T^{-2}$ (*b*) $ML^{-1}T^{-2}$ (*c*) $ML^{-1}T^{-1}$ (*d*) $ML^{1}T^{-1}$

   [Hints: $F = \eta A \frac{dv}{dr}$]

2. The viscous force is
   (*a*) nuclear (*b*) gravitational (*c*) electrical (*d*) electromagnetic
3. Reynold's number (k) has the dimension:
   (*a*) $M^0L^{-1}T^{-2}$ (*b*) $M^1L^0T^{-2}$ (*c*) $ML^{-1}T^{-1}$ (*d*) $M^0L^0T^0$

   [Hint: $k = \frac{v_c \rho r}{\eta}$]

4. A small spherical ball is dropped in a viscous medium of very large extent. The speed (v) of the ball versus time graph is properly represented by

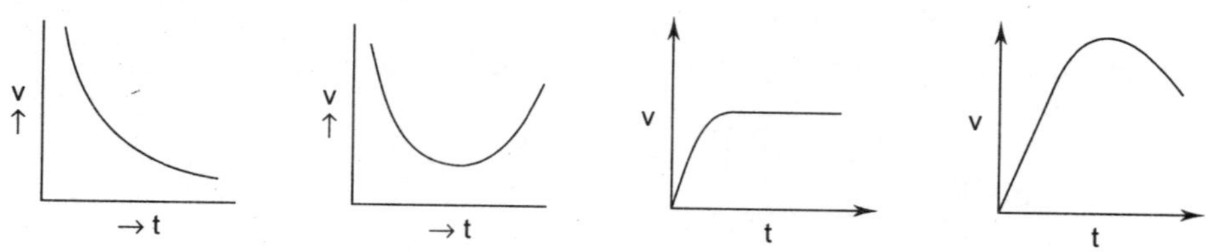

(a) (b) (c) (d)

5. A tiny spherical metal ball is dropped in a viscous liquid of infinite extent. The velocity of fall (v) of the ball as a function of distances (s) travelled by it along the liquid, is best represented by

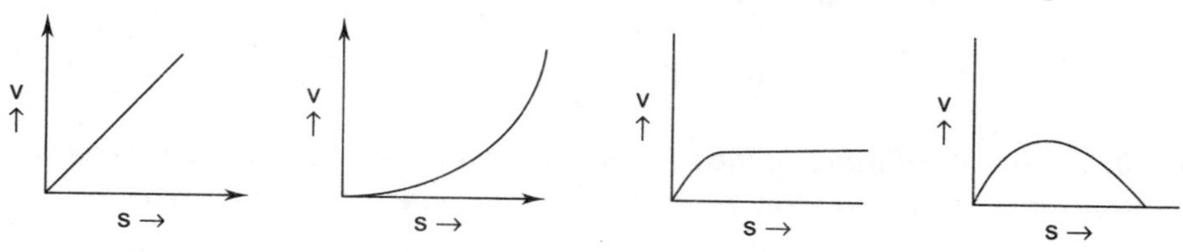

(a) (b) (c) (d)

6. Equation of continuity for compressible fluid is given by

(*a*) $\alpha_1 v_1 = \alpha_2 v_2$ (*b*) $\alpha_1 v_1 \rho_1 = \alpha_2 v_2 \rho_2$ (*c*) $\alpha_1 v_2 \rho_1 = \alpha_2 v_1 \rho_2$ (*d*) $\alpha_1 v_1 \rho_2 = \alpha_2 v_2 \rho_1$

[α_1, v & ρ stand for area of cross-section, velocity of flow and density of the fluid respectively]

7. Two tiny balls of the same metal, one having a radius thrice the other, are dropped in a long jar filled with a liquid. The terminal velocity of the larger ball as compared to that of the smaller one will be

(*a*) three times (*b*) six times (*c*) nine times (*d*) same

[Hint: $v = \frac{2g}{9} a^2 \frac{(\rho - \sigma)}{\eta}$; $v \propto a^2$]

8. A liquid flows steadily through a narrow horizontal cylindrical pipe having radii 2R &R at two points A and B respectively. If the velocity of flow at A is V, then the flow-velocity at B will be—

(*a*) 2V (*b*) $\frac{V}{2}$ (*c*) 4V (*d*) $\frac{V}{4}$

[Hint: $\alpha_A V_A = \alpha_B V_B$]

Answers

1. (*c*) **2.** (*b*) **3.** (*d*) **4.** (*c*) **5.** (*c*)
6. (*b*) **7.** (*c*) **8.** (*c*)

Review Questions

1. State and prove Bernoulli's theorem for a liquid along a streamline *(Osm. U., Nag. U. Indore U. Burd. U. Cal. U.)*

2. Explain the term viscosity and coefficient of viscosity. Give the dimension of coefficient of viscosity *(Bhopal U, Nag. V, Gharwal U. Bang. U, A.U)*

3. Give with necessary theory Poiseuille's method of determining the coefficient of viscosity of a liquid. State clearly the assumptions made.

4. Distinguish between streamline and turbulent flow of a liquid, discuss briefly the idea of critical velocity and explain the significance of the Reynold's number. *(Meerut U., Osm U., Nag. U., Bang. U., J.P. U., Cal. U., V.S.U, Burd U., P.U)*

5. How does the viscosity vary with temperature and pressure? *(Meerut U., Gharwal U., Cal. U.)*

6. (*a*) Why does a body attain terminal velocity in falling through a viscous liquid? *(Cal. U)*

(*b*) Derive Stoke's formula for the velocity of a small sphere falling through a viscous liquid using the method of dimensions. *(Guwahati U, V.S.U., Meerut U.)*

7. Calculate the mass of water flowing in 10 minute through a tube 0.1 cm in diameter, 40 cm long if there is a constant pressure head of 20 cm of water. The co-efficient of viscosity of water is 0.0089 C.G.S. units. *(Kan. U, AU)*

8. Two drops of water of the same size are falling through air with terminal velocity 1 m s^{-1}. If the two drops combine to form a single drop, calculate the terminal velocity. *(Vidya Sagar University)*

9. A steel ball of radius 2×10^{-3} m falls in a vertical column of castor oil. The co-efficient of viscosity of castor oil is 0.7 N m^{-2} and its density 0.98×10^3 kg m^{-3}. The density of steel is $7.8 \times$ kg m^{-3} and g = 9.8 m s^{-2}. Find its terminal velocity. *(Bang. U.)*

James Clerk Maxwell

2D Chapter

Heat & Thermodynamics

2D.1 INTRODUCTION

Kinetic Theory of Gases

Description of the characteristic behaviour of gases has been provided by the kinetic theory of gases. The theory assumes that a gas is composed of individual molecules which are: (1) on an average so far apart in relation to their small size that there are no interacting forces between them except during collision, (2) in continuous rapid straight line random motion and strike each other and walls of the container with perfectly elastic impact. Further (3) between two successive impacts the molecules travel in straight line with uniform velocity. The average distance through which they move between successive collisions is known as the Mean Free Path (MFP). Also (4) the number of collisions per unit volume of the gas remains the same.

According to the kinetic theory, gas molecules impinging on the walls of the container rebounds back with direct elastic impacts, thereby a change in momentum takes place on the wall of the container. This change in momentum per unit area of the container walls per unit time causes pressure on the walls of the container which is the pressure of the enclosed gas.

Ideal Gas

An ideal gas is thought of as containing point molecules (*i.e.*, molecules without dimension). Thus a point-molecule does not occupy any space. Consider a point molecule in the container to move with velocity v. If the enclosure is a rectangular parallelopiped with sides a, b, c then its volume is abc. The components of velocity 'v' along sides a, b and c be denoted by v_a, v_b and v_c respectively. Considering the face A (of area bc) the change in momentum component mv_a after elastic impact at the face A is $mv_a - (-mv_a) = 2mv_a$. The time elapsed for the same molecule to move through the distance 'a' after its first collision the other face A′ (parallel to A) and reach the plane A to have its second impact with A is $t' = a/v_a$. The molecule after its impact at A travels back

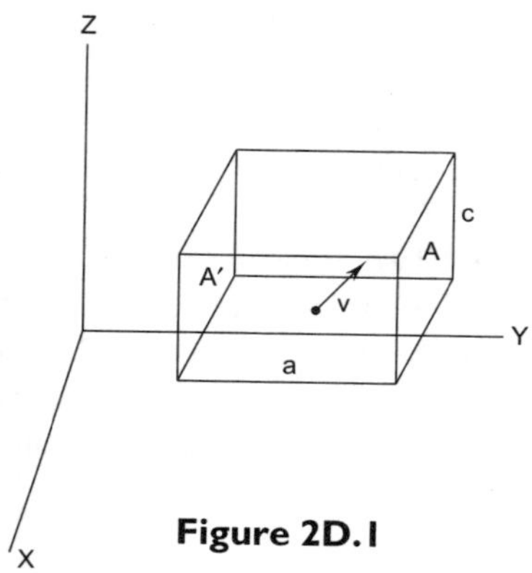

Figure 2D.1

to A′ for its third impact with A′. Thus between the first and the second impact at A′ only one collision takes place at the plane A. Hence the time elapsed in having one collision at face A in between two successive collisions at the face A′ is $t = t' + t' = \frac{2a}{v_a}$. Therefore, the number of collisions per second on face A $= \frac{v_a}{2a}$. The rate of change of momentum

$$= 2mv_a \times \frac{v_a}{2a} = \frac{mv_a^2}{a}$$

where, m = mass of a molecule of the gas.

The total rate at which momentum is imparted to face A

$$= \frac{mv_{a1}^2}{a} + \frac{mv_{a2}^2}{a} + \cdots\cdots + \frac{mv_{aN}^2}{a}$$

where v_{a1}, v_{a2} etc., stand for velocity components of 1st, 2nd ... Nth molecules in the direction parallel to 'a'.

$\therefore$ Force along 'a', $F_a = \frac{m}{a}\left(v_{a1}^2 + v_{a2}^2 + \cdots\cdots + v_{aN}^2\right)$

and p_a (pressure on face A) $= \frac{m}{\vec{a} \times \vec{b} \cdot \vec{c}}\left(v_{a1}^2 + v_{a2}^2 + \cdots\cdots + v_{aN}^2\right)$

Similarly, $p_b = \frac{m}{\vec{b} \times \vec{a} \cdot \vec{c}}\left(v_{b1}^2 + v_{b2}^2 + \cdots\cdots v_{bN}^2\right)$

and $p_c = \frac{m}{\vec{c} \times \vec{b} \cdot \vec{a}}\left(v_{c1}^2 + v_{c2}^2 + \cdots\cdots + v_{cN}^2\right)$

The pressure in a fluid being the same in all directions

$$p_a = p_b = p_c = p \text{ (say)}$$

$$\therefore \quad p = \frac{1}{3}(p_a + p_b + p_c)$$

$$= \left[\frac{1}{3}\frac{m}{abc}\left(v_{a1}^2 + v_{a2}^2 + \cdots\cdots + v_{aN}^2\right) + \left(v_{b1}^2 + v_{b2}^2 + \cdots\cdots + v_{bN}^2\right) + \left(v_{c1}^2 + v_{c2}^2 + \cdots\cdots + v_{cN}^2\right)\right]$$

$$= \frac{1}{3}\frac{mN}{V}\left[\frac{v_1^2 + v_2^2 + \cdots\cdots + v_N^2}{N}\right],$$

$$= \frac{1}{3}\frac{m}{V} \cdot V_{rms}^2$$

v_{rms} is the root mean square of the gas molecules.

where $v_1^2 = \left(v_{a1}^2 + v_{b1}^2 + v_{c1}^2\right)$

$v_2^2 = \left(v_{a2}^2 + v_{b2}^2 + v_{c2}^2\right)$ and so on.

N is the total number of molecules of the gas.

$$V_{rms} = \sqrt{\bar{v}^2} = \sqrt{\frac{v_1^2 + v_2^2 + \cdots\cdots + v_N^2}{N}}.$$

Now, $$p = \frac{1}{3}\frac{M}{V}v_{rms}^2 = \frac{2}{3}\frac{1}{V}\left(\frac{1}{2}Mv_{rms}^2\right) = \frac{2}{3}\frac{E}{V}$$

E : translatory KE of the gas molecules.

or $$pV = \frac{2}{3}E.$$

From experiments,

$$pV_m = RT$$

V_m : molar gas volume
R : Universal gas constant

For V volume of gas,

$$V_m = \frac{V}{n}$$

R = 8.31 J/mole/K

so, $$pV = n\,RT$$

n : number of moles in V

For 1 mole of the gas we have

$$p\,V_m = RT$$
$$= N_A\,kT$$

where N_A = Avogadro's Number

$\therefore$ $$p = \frac{N_A}{V_m}\cdot kT$$

$N_A = 6.023 \times 10^{23}$/mole

$$= N\,kT$$

k = Boltzmann constant $= \dfrac{R}{N_A}$

$k = 1.38 \times 10^{-23}$ J/K.

$$\left(\frac{N_A}{V_m}\right) = N : \text{number of molecules/unit volume.}$$

2D.2 REAL GAS

In dealing with ideal gas the space available for each molecule to roam about is the entire volume of the enclosure of the gas. In case of real gas the volume of the container is partially occupied by each of the dimensional molecule of the real gas as such this part is inaccessible for all others. The equation $pV = RT$ needs corrected as $p\,(V - b) = RT$ where 'b' is the limiting volume which a real gas would occupy at infinitely great pressure [*i.e.*, $(V - b) = 0$ at $p = \infty$].

The magnitude of b has been determined as,

$$b = \frac{16}{3}\pi r^3 N_A,$$

r : radius of a molecule.

Further, a molecule near the wall of the container would be subjected to the action of a resultant force exercised by the other molecules and this force is directed into the gas. Owing to this, the pressure on the wall of the container will become smaller by an amount Δp so that the form

$p = \dfrac{RT}{(V - b)}$ would become

$$p = \frac{RT}{(V-b)} + \Delta p \quad \text{or} \quad p + \Delta p = \frac{RT}{(V-b)}$$

The expression for Δp has been obtained as $\Delta p = \frac{a}{V^2}$, a : constant.

The equation of state for real gas, thus, is of the form

$$\left(p + \frac{a}{V^2}\right)(V-b) = RT. \text{ (for one mole of the gas)}$$

This is the Van der Waal's equation for real gases with a and b as Van der Waal's constant. The values of the constants 'a' and 'b' are temperature dependent.

At present the state of a real gas is described by the equation

$$pV = RT + \frac{B}{V} + \frac{C}{V^2} + \cdots\cdots.$$

The coefficients B, C etc., are called Virial Coefficients and are to be evaluated experimentally. They depend on the temperature.

2D.3 DEGREES OF FREEDOM

Degrees of Freedom of a material system is the number of possible independent motions of the system. Thus, the degrees of freedom is the number of independent co-ordinates required to specify the dynamic position of a particle or a system.

2D.4 THE LAW OF EQUIPARTITION OF ENERGY

The Law of Equipartition of energy states that the energy of a gas is on the average distributed equally among the various degrees of freedom of the system. The value **per degree of freedom** of the gas system is $\frac{1}{2}kT$.

The average kinetic energy of the **translational** motion per molecule is $\frac{1}{2}mv_{rms}^2 = \frac{3}{2}kT$. For a monatomic molecule the average K.E. is $E = \frac{1}{2}mv_x^2 + \frac{1}{2}mv_y^2 + \frac{1}{2}mv_z^2$. The number of degrees of freedom for a monatomic molecule is **three**. Hence for each degree of freedom the K.E. is $\frac{1}{3} \times \frac{3}{2}kT = \frac{1}{2}kT$. For a diatomic molecule the number of degrees of freedom is five (three translational and two rotational) and so its average K.E. is $\frac{5}{2}kT$ (being $\frac{1}{2}kT$ for each degree of freedom).

2D.5 MAXWELL'S SPEED (OR VELOCITY) DISTRIBUTION LAW

The kinetic theory of gases is based on the idea that the gas molecules are moving randomly within the container, with speeds ranging between zero and infinity. Thus, the molecules possess varied speeds within this range. The variation in speeds goes by the name - distribution of speed (or velocity). It is therefore necessary to obtain a functional form of this speed distribution of molecules of the gas enclosed.

For the deduction of the distribution law the molecules of the gas are to be in state of equilibrium after making elastic impacts amongst them and the walls of the container. In this case the molecules being point molecules, they do not have impact among them. The gas molecules are assumed to obey the following conditions :

(*a*) There are large number of molecules with complete isotropy. This means that the number of molecules in any region of space within the container is the same and their velocities in any direction are the same.
(*b*) The velocity components along any three coordinate axis are independent of each other.
(*c*) The chance of the velocity of a molecule to lie between certain limits is dependent only on the velocity and its limits and nothing else.

Consider a molecule with velocity vector c (Fig. 2D.2). The components of c along the three cartesian axes OX, OY, OZ be u, v and w respectively

$$\therefore \qquad u^2 + v^2 + w^2 = c^2.$$

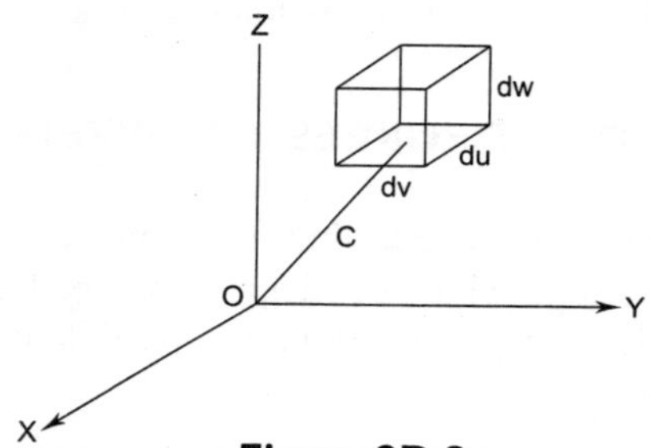

Figure 2D.2

According to the assumption (*c*), the chance that the velocity component along OX ends between u and $u + du$ is $f(u)\,du$ and similary for the components along OY and OZ are respectively $f(v)\,dv$ and $f(w)\,dw$. By assumption (*b*) the chance that the velocity vector c ends simultaneously between u and $u + du$, v and $v + dv$, w and $w + dw$ is given by

$$f(u)\,f(v)\,f(w)\,du\,dv\,dw$$

The above expression can be written as,

$$\text{F}(c)\,du\,dv\,dw$$

or alternatively $\phi(c^2)\,du\,dv\,dw$

$$\phi(c^2) = f(u)\,f(v)\,f(w)$$

$$\text{or} \qquad \phi(u^2 + v^2 + w^2) = f(u)\,f(v)\,f(w).$$

NOTE Independent probabilities (or chances) can be combined as product for obtaining the resultant probability.

If N be the number of gas molecules per unit volume of the container and $d\text{N}_c$ be the number within the velocity range c and $c + dc$, then $d\text{N}_c = \text{N}\,f(u)\,f(v)\,f(w)\,du\,dv\,dw$.

For a specified value of c (*i.e.*, fixed value), $\phi(c^2)$ is constant.

$$\therefore \qquad d[\phi(c^2)] = 0 = d[f(u)\,f(v)\,f(w)]$$

$$\text{Thus,} \qquad f'(u)\,f(v)\,f(w)\,du + f(u)\,f'(v)\,f(w)\,dv + f(u)\,f(v)\,f'(w)\,dw = 0$$

Dividing by $f(u)\,f(v)\,f(w)$, we get

$$\frac{f'(u)\,du}{f(u)} + \frac{f'(v)\,dv}{f(v)} + \frac{f'(w)\,dw}{f(w)} = 0 \qquad \ldots(2D.1)$$

Again as c is constant and $c^2 = u^2 + v^2 + w^2$, we get by differentiating $O = u\,du + v\,dv + w\,dw$ ...(2D.2)

Multiplying (2D.2) by a constant λ we get

$$0 = \lambda u\,du + \lambda v\,dv + \lambda w\,dw \qquad \ldots(2D.3)$$

Adding equations (2D.1) and (2D.3)

$$\left(\frac{f'(u)}{f(u)} + \lambda u\right)du + \left(\frac{f'(v)}{f(v)} + \lambda v\right)dv + \left(\frac{f'(w)}{f(w)} + \lambda w\right)dw = 0 \qquad \ldots(2D.4)$$

In the equation (2D.4) the term being additive and du, dv, dw are arbitrarily chosen

$$\frac{f'(u)}{f(u)} + \lambda u = 0\ ; \quad \frac{f'(v)}{f(v)} + \lambda v = 0 \quad \text{and} \quad \frac{f'(w)}{f(w)} + \lambda w = 0.$$

Thus, $\dfrac{f'(u)\,du}{f(u)} = -\lambda u$ or $f(u) = A e^{-\lambda u^2/2}$ (on integration)

and similarly,

$$f(v) = A e^{-\frac{\lambda v^2}{2}} \quad \text{and} \quad f(w) = A e^{-\frac{\lambda w^2}{2}}$$

Hence, $f(u)\,f(v)\,f(w)\,du\,dv\,dw = A^3 e^{-\frac{\lambda}{2}\left(u^2+v^2+w^2\right)}\,du\,dv\,dw$.

On evaluating, $A = \left(\dfrac{m}{2\pi k T}\right)^{\frac{1}{2}}$ where m : mass of the molecule

and $\lambda = \dfrac{m}{kT}$ k : Boltzmann constant

T : Temperature of the gas in absolute scale.

$$\therefore f(u)\,f(v)\,f(w)\,du\,dv\,dw = \left(\frac{m}{2\pi kT}\right)^{\frac{3}{2}} e^{-\frac{mc^2}{2kT}}\,du\,dv\,dw\,.$$

Now $du\,dv\,dw$ when transformed to spherical polar coordinates becomes $dc \times c\,d\theta \times c \sin\theta\,d\phi$

($\because$ $d\theta$, $d\phi$ being very small).

$$\therefore \qquad dN_c = N \cdot \left(\frac{m}{2\pi kT}\right)^{\frac{3}{2}} \cdot e^{-\frac{m}{2kT}c^2} \cdot c^2 \sin\theta\,d\theta\,d\phi\,dc$$

Integrating the expression for dN_c over all values of θ from 0 to π and for all values of ϕ from 0 to 2π, we get

$$dN_c = N \cdot \left(\frac{m}{2\pi kT}\right)^{\frac{3}{2}} e^{-\frac{m}{2kT}c^2} \cdot 4\pi c^2 \, dc \qquad \ldots(2D.5)$$

$4\pi c^2\, dc$: The volume of the space between the spherical shells of radius c and $c + dc$

dN_c thus expresses the number of molecules having velocity lying between c and $c + dc$.

The expression (2D.5) is the Maxwell's velocity (or speed) Distribution law.

The probability that a molecule has its velocity lying between c and $c + dc$ at a particular temperature is given by $\frac{dN_c}{N} = \left(\frac{m}{2\pi kT}\right)^{\frac{3}{2}} e^{-\frac{m}{2kT}c^2} \cdot 4\pi c^2\, dc$

$$= f(c)\, dc$$

where $$f(c) = 4\pi\left(\frac{m}{2\pi kT}\right)^{\frac{3}{2}} \cdot e^{-\frac{m}{2kT}c^2} \cdot c^2$$

If the function $f(c)$ be plotted against c (fig. 2D.3) the distribution curve for the molecular velocity is obtained. With an increase in the temperature of the gas the maximum of the distribution curve shifts towards the higher velocities and its magnitude is decreased.

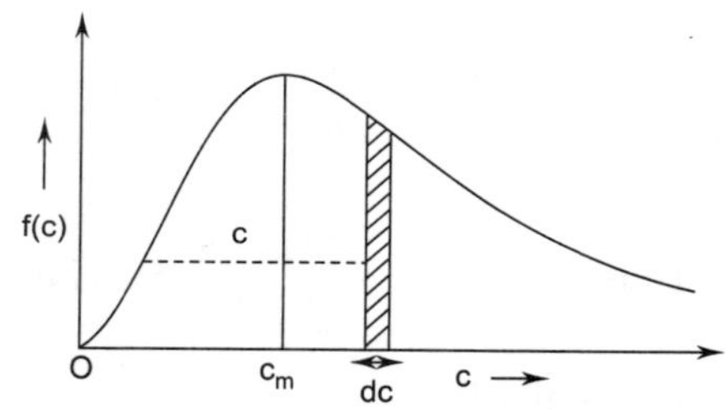

Figure 2D.3

Distinction is made between the following velocities that characterize a gas at a given temperature.

(*i*) Most probable velocity (c_m) corresponds to the maximum of function $f(c)$. This is – $c_m = \sqrt{\frac{2kT}{m}}$.

(*ii*) The Mean (or Average) velocity $(\bar{c})$ of molecular motion is given by $(\bar{c}) = \sqrt{\frac{8kT}{\pi m}}$.

(*iii*) The Root Mean Square (RMS) velocity of molecular motion is given by $c = \sqrt{\frac{3kT}{m}}$

$$\therefore \quad c_m : \bar{c} : c = \sqrt{2} : \sqrt{\frac{8}{\pi}} : \sqrt{3} = 1.41 : 1.60 : 1.73.$$

2D.6 MEAN FREE PATH (M.F.P.)

Mean Free Path (M.F.P.) of molecules is the average distance travelled by the molecules between two successive collisions. The mean free path is usually denoted by λ. The expression for the mean free path (according to Maxwell is) $\lambda = \frac{1}{\sqrt{2}\pi\sigma^2 N}$, where σ is the diameter of the molecule.

IN FOCUS

- **The pressure in a fluid** $\mathbf{P} = \frac{1}{3}\frac{m}{V} \cdot V^2rms$
- **The Vander wall equation of state,**

$$\left(P + a/v^2\right)(v - b) = RT$$

- **Law of equipartition of energy:** The energy of a gas is on the average distributed equally among the various degrees of the system. The value per degree of freedom of the gas system is $\frac{1}{2}$ KT.
- **The Maxwell velocity distribution law**

$$f(c) = 4\pi\left(\frac{m}{2\pi\,KT}\right)^{3/2} e\,\frac{-mc^2}{2\not{\pi}\,KT} \cdot c^2$$

→ **Most probable velocity** $c_m = \sqrt{\frac{2KT}{m}}$

→ **Mean velocity** $\bar{c} = \sqrt{\frac{8KT}{\pi m}}$

→ **The Root mean square velocity (R.M.S)** $c = \sqrt{\frac{3KT}{m}}$

Solved Problems

Problem 1. *If the root mean square speed of molecules of hydrogen is 1.84 km s^{-1}, calculate root mean square speed of oxygen at N.T.P. The molecular weight of hydrogen and oxygen are 2 and 32 respectively.*

Solution. Let c_1 and c_2 be the root mean square speed of hydrogen and oxygen respectively, then

$$c_1 = 1.84 \text{ km } s^{-1} = 1.84 \times 10^3 \text{ m s}^{-1}$$

$$\frac{c_2}{c_1} = \sqrt{\frac{\rho_1}{\rho_2}} = \sqrt{\frac{\text{Molecular weight of hydrogen}}{\text{Molecular weight of oxygen}}} = \sqrt{\frac{2}{32}} = \frac{1}{4}$$

$$c_2 = \frac{c_1}{4} = \frac{1.84 \times 10^3}{4} = 460 \text{ m s}^{-1}.$$

Problem 2. *(a) Calculate the r.m.s. velocity of a molecule of mercury vapour at 300 K. Gas constant R = 8.31 S.I. units. Molecular weight of mercury – weight of mercury = 221×10^{-3} kg.*

(b) The mean kinetic energy of a molecule of hydrogen gas at 0°C is 5.62×10^{-23} J and the molar gas constant is 8.31 JK^{-1}. Calculate Avogadro's number.

Solution. (*a*) Root mean square velocity $c = \sqrt{\frac{3RT}{M}}$

$$R = 8.31 \text{ S.I. units, } T = 300\,k \quad \text{and} \quad M = 221 \times 10^{-3} \text{ kg}$$

$$c = \sqrt{\frac{3 \times 8.31 \times 300}{221 \times 10^{-3}}}$$
$$= 1.84 \times 10^2 \text{ m s}^{-1}.$$

(*b*) Kinetic energy per molecule $= \frac{3}{2} kT = \frac{3}{2}\frac{R}{N} T$

where k is Boltzmann gas constant, R gas constant per gm molecule, N the Avogadro's number and T the absolute temperature.

$$N = \frac{3RT}{2 \times \text{K.E. per molecule}}$$
$$= \frac{3 \times 8.31 \times 273}{2 \times 5.62 \times 10^{-21}} = 6.05 \times 10^{23}$$

Problem 3. *At what temperature, pressure remaining unchanged will the velocity of hydrogen be double of its value at N.T.P.*

Solution. $c^2 = \frac{3R}{M} T \quad \therefore \quad c \propto \sqrt{T}$

Let the velocity at T_0 K be c and at TK be $2c$, then

$$\frac{2c}{c} = \sqrt{\frac{T}{T_0}}$$

or $$\frac{T}{T_0} = 4$$

or $$T = 4T_0 = 4 \times 273 = 1092 \text{ K}$$
$$= (1092 - 273) = 819°\text{C}.$$

Problem 4. *Calculate the root mean square velocity of air molecules at N.T.P., density of air = 1.293 kg/m³.*

Solution. We have, $C_{RMS} = \sqrt{\frac{3P}{\rho}}$

Here, $$P = 0.76 \times 13.6 \times 10^3 \times 9.8 \text{ N/m}^2 = 1.013 \times 10^5 \text{ N/m}^2.$$
$$\rho = 1.293 \text{ kg/m}^3.$$

So, $$C_{RMS} = \sqrt{\frac{3 \times 1.013 \times 10^5}{1.293}} = 4.85 \times 10^2 \text{ m/s}.$$

Problem 5. *The average kinetic energy of a molecule of hydrogen at 0° is 5.64×10^{-21} Joule/K and the g-molecular gas constant (R_0) = 8.32 Joules/K. Calculate Avogadro's number (N_a).*

Solution. The average kinetic energy of a molecule $= \frac{3}{2}\frac{R_0}{N_a} T$.

Hence, $$5.64 \times 10^{-21} = \frac{3}{2} \cdot \frac{8.32}{N_a} \times 273 \quad [0°\text{C} = 273 \text{ K}]$$

or, $$N_a = \frac{3 \times 8.32 \times 273}{2 \times 5.64 \times 10^{-21}} = 6.04 \times 10^{23}.$$

Problem 6. *2 moles of hydrogen gas at 30°C are mixed with one mole of helium gas at 60°C. Find the temperature of the mixture.*

Solution. The average K.E. of a mole of a gas, $E = \frac{3}{2} R_0 T$,

where R_0 = molar gas constant.

If K and N_A be the Boltzmann's constant and Avogadro's number respectively, then $R_0 = KN_A$.

$$\therefore \text{ K.E. of a mole of a gas} = E = \frac{3}{2} KN_A T$$

For 2 moles of hydrogen at 30°C

$$(E)H_2 = \frac{3}{2} \times K \times 2N_A \times (30 + 273) = \frac{3}{2} K \times 2N_A \times 303$$

For 1 mole of helium at 60°C

$$(E)_{He} = \frac{3}{2} K \times N_A \times (60 + 273) = \frac{3}{2} \times K \times N_A \times 333$$

If the two gases are mixed we have $3N_A$ molecules in the mixture. Let the temperature of mixture be TK. For the mixture, we have

$$E = \frac{3}{2} \times 3N_A K \times T$$

As the energy is conserved,

$$\frac{3}{2} \times 3N_A K \times T = \frac{3}{2} K \times 2N_A \times 303 + \frac{3}{2} KN_A \times 333$$

or, $$3 \times T = 2 \times 303 + 333 = 939$$

$$T = 313 \text{ K}$$

Hence, $$t = 313 - 273 = 40°\text{C}.$$

Multiple Choice Questions

1. At very low temperature the co-efficient of viscosity of a gas —
   (*a*) decreases with decrease of pressure
   (*b*) increases with increase of pressure
   (*c*) is independent of pressure.
2. In diffusion the transport of which of the following occurs —
   (*a*) momentum (*b*) energy (*c*) mass.
3. The kinetic energy per unit volume of a perfect gas is equal to —
   (*a*) $\frac{2}{3}$ P (*b*) $\frac{3}{2}$ P (*c*) $\frac{P}{3}$ where P is the pressure.

4. Molecules of an ideal gas have –
   (*a*) Only kinetic energy
   (*b*) Only potential energy
   (*c*) Both kinetic and potential energy.

Answers

1. (*c*) 2. (*c*) 3. (*b*) 4. (*a*)

Review Questions

1. Derive an expression for the pressure of an ideal gas on the basis of kinetic theory. Prove that the pressure of a gas is equal to $\frac{2}{3}$ of the translational kinetic energy of the molecules in a unit volume. [*Guwahati University, Vidya Sagar. Univ., Burdwan Univ., Lucknow Univ.*]
2. What is the law of equitation energy?
3. Find an expression for the energy of a gas or kinetic energy of gases. What is the interpretation of temperature of this theory?
4. Write down Maxwell's equation for the distribution of molecular velocity and explain it. [*Vidya Sagar University*]

Unsolved Problems

1. Calculate the R.H.S. speed of carbon dioxide molecule at 77°C. Given molecular weight of CO_2 = 44 gm.
2. Calculate the molecular kinetic theory of 1 gm of hydrogen gas at 50°C, given molecular weight of hydrogen is 2 and R = 8.3 joule per gram. [**Ans.** 2×10^3 joule]
3. The average kinetic energy of a molecule of helium at 27°C is 6.17×10^{-18} joule. Calculate the Avogadro's number, given molar gas constant R = 8.3 joule. [*Calcutta University*] [**Ans.** 6.053×10^{23}]
4. The density of oxygen at N.T.P. is 1.423 kg . m^3. Calculate the value of R.M.S. velocity, given g = 9.8 m/sec^2. [*N.B. University*] [**Ans.** 4.61×10^2 m/sec]

2D.7 THERMODYNAMICS

Therodynamics is mainly concerned with the transformation of heat to mechanical work and vice versa. The laws of thermodynamics are postulates based on experiments. The **FIRST LAW** relating to mechanical and heat energies is the law of conservation of energy. It is expressed as

$$dQ = dU + dW$$

when a system passes over from state 1 to state 2

$$\int_1^2 dU = \int_1^2 dQ - \int_1^2 dW$$

dU = change in internal energy of the system
dW = mechanical work done
dQ = Heat supplied

It is very important to note that the quantity of heat supplied, like the work done by the system (or on it), depends on **how the system passed** from state 1 to state 2. The change in the internal energy dU is independent of path followed and depends only on **the initial and final states**. The internal energy is called **a function of state,** whereas Q and W are **not** functions of state, Q and W are functions of the **process** of the change of state. In short, Q and W depend on the path followed but U does not.

In case of a **cyclic process** in which a system returns to its initial state after passing through a number of intermediate states $\oint dU = 0$ but this **does not** at all mean Q and W are also equal to **zero.**

So, in a cyclic process $\oint dQ = \oint dW$.

When heat energy is transferred from a body at higher temperature to one with a lower temperature without the performance of any body moving during the transfer (*i.e.*,) without any work being done) there will be an increase in internal energy. This transfer of heat will by itself stop when the temperatures get equalised. But if heat is transferred to a body that does work (by expansion) then,

$$dW = dQ - dU.$$

Maximum work is done in an isothermal process when the **internal energy does not change** so that,

$$dW = dQ.$$

To achieve this the body performing work after receiving heat from the souce must return to its initial state so as to start the same process again. This is known as cyclic process. Here, A*a*B*b* A depicts a cyclic process (or cycle). The work done during a cycle is equal to the area enclosed by A*a*B*b*A.

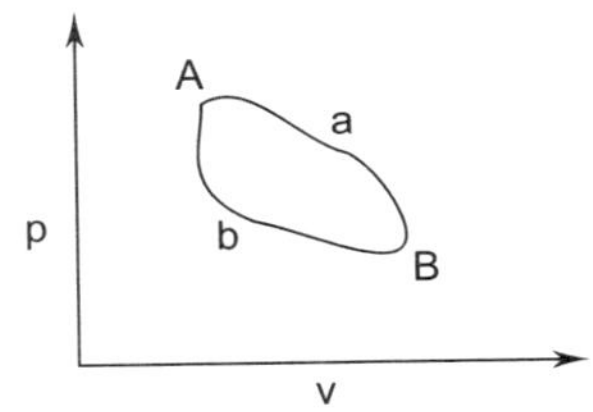

Figure 2D.4

It is observed that to convert heat into mechanical work there is a necessity of a "working body" which will receive heat from a "source", do certain amount of work and discharge the residual heat to a "sink" (or cold reservoir). In a cyclic process the working body after having expanded and done work at the expense of the heat received along the curve A to B (in Fig. A*a*B) must return to its initial state A. It must, therefore, be compressed along the curve from B to A (in Fig. B*b*A). The lower curve B*b*A corresponds to lower temperature, whereas the higher curve A*a*B corresponds to higher temperature. For operation of a cyclic engine we therefore need a **heat sink** (or cooler). The work done during a cycle is equal to the area enveloped by this closed curve. Work done depends on the path.

2D.8 THE CARNOT'S CYCLE

It is an idealised engine which is never realisable in practice. Carnot's engine possess maximum possible efficiency which cannot be achieved by any other machines that are realised in practice. But Carnot's engine provides the best guide for the construction of an actual heat engine and its improvement. The main features of a Carnot engine are :

(*i*) It operates **in a cyclic process** between two constant temperatures, one a hotter body called the **source** and the other a lower temperature body called the **sink**.

(*ii*) It receives heat **from the source** which is at **constant** temperature.

(*iii*) It performs some mechanical work in **each cycle**, the correspoding energy being **a part** of the heat drawn from the source.

(*iv*) It discharges the **balance of heat** after the performance of the work to the sink.

The operation of the engine is schematically depicted in figure 2D.5 (*a*), (*b*), (*c*) and (*d*).

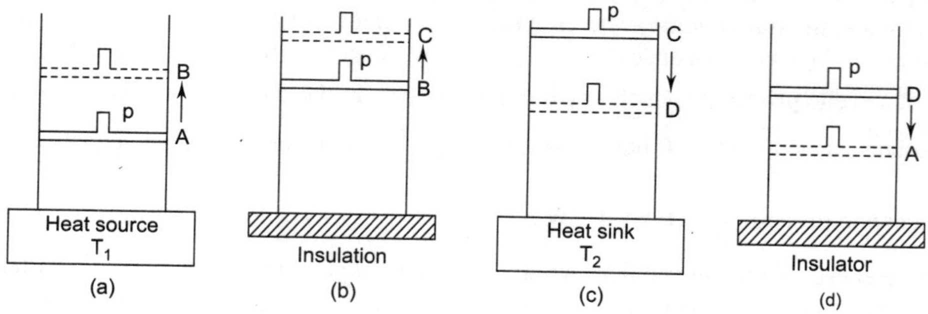

Figure 2D.5

Working Steps in Carnot's Engine:

1. **Isothermal Expansion [Figure 2D.5 (*a*)]:** The cylinder with the working substance is placed in thermal contact with the heat source so that the working substance (gas enclosed by the cylinder) expands isothermally at a constant temperature T_1. The volume increases from V_1 to V_2. The amount of heat absorbed being, Q_1.
2. **Adiabatic Expansion [Figure 2D.5 (*b*)]:** The cylinder is placed on heat insulating platform and the gas is further expanded by putting up the piston. The volume of gas increases from V_2 to V_3 with a consequent fall in temperature from T_1 to T_2.

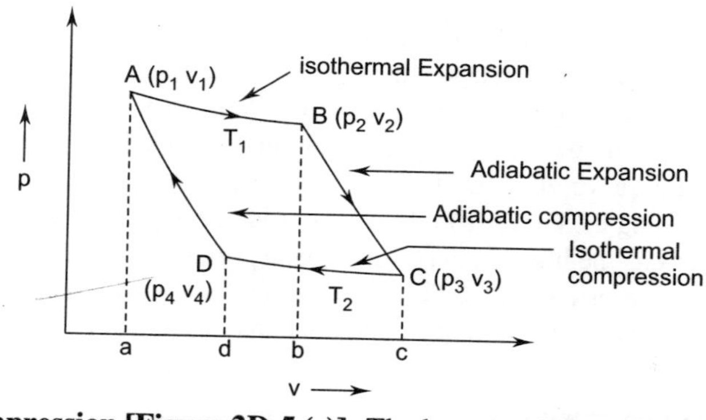

3. **Isothermal Compression [Figure 2D.5 (*c*)]:** The bottom of the cylinder is now placed in perfect contact with the heat sink at temperature T_2. The gas is compressed from C to D until it attains a volume V_4 corresponding to the state point D. The pressure of the gas is now p_4. During the isothermal compression the rejected by the gas (working substance) to the sink be Q_2.
4. **Adibatic Compression [Figure 2D.5 (*d*)]:** The cylinder is again placed on the insulator and then further compressed to reach the initial starting state point A. The volume of the gas decreases further from V_4 to V_1 and its temperature rises from T_2 to T_1. The gas thus goes through a cyclic operation each part of which is considered reversible.

$$\text{Thermal Efficiency, } \eta = \frac{\text{Workdone}}{\text{Heat supplied}} = \frac{W}{Q_1} = \frac{Q_1 - Q_2}{Q_1}$$

or

$$\eta = 1 - \frac{Q_2}{Q_1}$$

If the working substance is a **perfect gas** in a **Carnot's engine,** then $\frac{Q_2}{Q_1} = \frac{T_2}{T_1}$. Hence, $\eta = \left(1 - \frac{T_2}{T_1}\right)$, which means that the efficiency (η) of a Carnot's engine depends on the ratio of the absolute temperature of the heat-sink and the heat source.

The **Carnot Theorem** states that the thermodynamic efficiency of a reversible Carnot cycle does not depend upon the working substance and is a function of only the absolute temperatures of the heat source (T_1) and heat sink (T_2).

2D.9 ZEROTH LAW OF THERMODYNAMICS

Zeroth Law of Thermodynamics is a statement for **thermal equilibrium.** It states that - 'Two systems in thermal equilibrium with a third are in thermal equilibrium with each other.

2D.10 SECOND LAW OF THERMODYNAMICS

The first law accounts for the transfer and the transformation of energy associated with thermodynamic changes in a system. It does not, however, impose any restriction on **the direction** in which the process may proceed. The second law makes it apparent that heat cannot pass **of itself from a colder body to a hotter** body. It thereby suggests indirectly that there is **no such restriction** in its flow **from a hotter body** to a colder body as long as the colder body remains colder compared to the hotter one. It further suggests that while heat may so flow work may be derived out of it. The statement of the second law can be made in the following ways :-

(*a*) **Clausius's Statement:** It is impossible for a self acting machine, unaided by any external agency, to transfer heat from a body at a lower temperature to a body at a higher temperature.

(*b*) **Kelvin's Statement:** It is impossible by any inanimate material agency to derive any mechanical work from any portion of matter by cooling it to a temperature colder than the coldest of its surroundings.

Both the statements of the second law contain specific references to an engine or device operating in a cycle. The enunciations consider two bodies from the hotter of which heat may pass of itself to the colder one. Kelvin's statement further suggests that when heat will so pass there is a possibility of transforming it into work. Carnot's engine demonstrates that only a part of the heat absorbed can be converted to work, the remaining portion has to be rejected to sink. The mechanical work is equivalent to the stated difference of heat between the source and the sink.

Entropy

The second law of thermodynamics can be stated as quantative relation with the concept of entropy. To introduce this concept, we consider an infinitesimal isothermal expansion of an ideal gas. The expansion of gas increases its disorder because the molecule have greater randomness of position after the expansion than before.

Suppose we add heat dQ and the gas expand just enough to keep the temperature constant. In this case internal energy, unaltered because the internal energy depends on temperature only. Hence, from the 1st law, the work dW done by the gas is equal to the amount of heat supplied (dQ).

$$dQ = dW = P\,dV = \frac{nRT}{V}\,dV$$

$$\text{Again } \frac{dV}{V} = \frac{dQ}{nRT}.$$

Therefore, the fractional change of volume $\left(\frac{dV}{V}\right)$ is a measure of the increase in disorder and it is proportional to the quantity $\frac{dQ}{T}$. So we can define the infinitesimal change of entropy dS during an infinitesimal reversible process at absolute temperature T as

$$dS = \frac{dQ}{T}.$$

Entropy is a state of gas (working substance) just like pressure, volume, internal energy etc. The introduction of the concept of entropy has been focussed to the change of thermal state of the working substance which undergoes a thermodynamical change in the system. In a given equilibrium state the system has a definite value of entropy.

For the process consisting of a continuous series of equilibrium states, $dS = \frac{dQ}{T}$, dS being the change in entropy due to the infinitesimal gain of heat dQ by the system at a fixed absolute temperature T of the source of heat supply.

For all cycles $\oint \frac{dQ}{T} \leq 0$ $\qquad$ $\oint$: integral extended over a cycle.

For only reversible cycles

$$\oint \frac{dQ}{T} = 0$$

Entropy is related to the disorder of the system. Higher entropy of the system means greater disorder.

NOTE A transformation is *reversible* only when any work done by the system in direct process is compensated by the equal work done on the system in the corresponding reverse process. This process of transformation is an *idealised* one in which all the changes in the system take place infinitely slowly (known as Quasi-static process). Further there should be *no dissipation* of energy during the process. A *reversible cyclic* process is the one in which every operation in the cycle is reversible. [See Figures 2D.6 (*a*) and (*b*)].

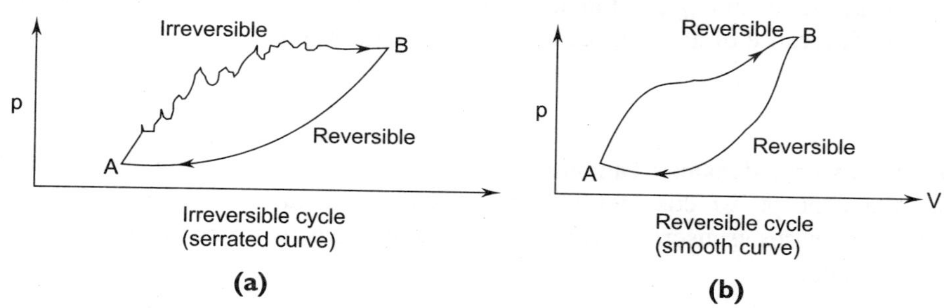

Figure 2D.6

2D.11 THIRD LAW OF THERMODYNAMICS

Third Law of Thermodynamics (Nernst's heat theorem) states that "the entropy of every system *at absolute zero* can always be taken equal to *zero*."

The third law can also be expressed as "it is impossible by any procedure, no matter how idealised, to reduce any system to the absolute zero temperature in a *finite* number of operation".

IN FOCUS

- **1st law of Thermodynamics** $dQ = du + dw$
- **Efficiency of the Carnot's Engine**

$$\eta = 1 - \frac{Q_2}{Q_1} = 1 - \frac{T_2}{T_1}$$

$du \rightarrow$ change in internal energy of the system

$dw \rightarrow$ Mechanical workdone

$dQ \rightarrow$ Heat supplied.

- **Zeroth law of thermodynamics:** "Two systems in thermal equilibrium with a third are in thermal equilibrium with each other."
- **Second law of Thermodynamics:**
  (*a*) **Clausius statement:** It is impossible for a self acting device, unaided by any external agency to transfer heat from a body to at lower temperature to a body at higher temperature
  (*b*) **Kelvin's statement:** It is impossible by any inanimate material agency to derive any mechanical work from any portion of matter by cooling it to a temperature colder then the coldest of its surroundings.
- **Entropy:** $dS = \dfrac{dQ}{T}$

  dS being the change in entropy due to the infinitesimal gain of heat dQ by the system at a fixed absolute temperature T of the source of heat supply.
- **Third law of Thermodynamics:** It is impossible by any procedure, no matter how idealized, to reduce any system to the absolute zero temperature in a finite number of operation.

Problem 1. *1 gm mol of a monatomic perfect gas* $\left(\gamma = \frac{5}{3}\right)$ *at 27°C is adiabatically compressed a reversible process from an initial pressure of 1 atm to a final pressure of 50 atm. Calculate the resulting difference of temperature.*

Solution. We have $P_1 = 1$ atm, $P_2 = 50$ atm, $T_1 = 273 + 27 = 300$ K, $T_2 = ?$

In a reversible adibatic process, we have

$$\left(\frac{P_2}{P_1}\right)^{\gamma-1} = \left(\frac{T_2}{T_1}\right)^{\gamma}$$

$$(50)^{\frac{5}{3}-1} = \left(\frac{T_2}{300}\right)^{\frac{5}{3}}$$

$$\frac{2}{3}\log 50 = \frac{5}{3}(\log T_2 - \log 300)$$

$$\frac{2}{3} \times 1.6990 = \frac{5}{3}\log T_2 - \frac{5}{3} \times 2.4771$$

$$\frac{5}{3}\log T_2 = 5.2611$$

$\therefore$ $$T_2 = 1434 \text{ K}.$$

$\therefore$ Difference in temperature

$$= 1434 - 300 = 1134 \text{ K} \quad \text{or} \quad 1134°\text{C}.$$

Problem 2. *(a) One mole of gas, assumed to be perfect, at 0°C is heated Isobarically till its volume is doubled, find the amount of heat absorbed. Given : C_v = 20.9 J/mole/K, R = 8.3 J/mol/K.*

Solution. Since the gas is heated isobarically, we obtain from Charle's law,

$$\frac{V_1}{V_2} = \frac{T_1}{T_2}$$

Here, by the problem, $\dfrac{V_2}{V_1} = 2$ and $T_1 = 0°\text{C} = 273 \text{ K}$

$$T_2 = T_1 \frac{V_2}{V_1} = 273 \times 2 = 546 \text{ K}.$$

Heat required to raise the temperature of 1 mole of the gas by (546 – 273) or 273 K is

$$dQ = C_p\, dT$$
$$= (C_v + R)\, dT$$
$$= (20.9 + 8.3) \times 273$$
$$= 29.2 \times 273$$
$$= 79.7 \times 10^2 \text{ J}.$$

Problem 3. *A certain mass of a gas at N.T.P. is expanded to three times its volume under adiabatic conditions. Calculate the resulting temperature and pressure, if γ for the gas is 1.40.*

Solution. We have $\dfrac{V_2}{V_1} = 3$, $T_1 = 273$ K, $\gamma = 1.40$, $T_2 = ?$

From the relation,

$$T_1 V_1^{\gamma-1} = T_2 V_2^{\gamma-1}, \text{ we obtain}$$

$$T_2 = T_1 \left(\frac{V_1}{V_2}\right)^{\gamma-1} = 273\left(\frac{1}{3}\right)^{0.4}$$

On simplification, $T_2 = 176 \text{ K} = -97°\text{C}$

Again, from the relation

$$P_1 V_1^{\gamma} = P_2 V_2^{\gamma}, \text{ we get}$$

$$P_2 = P_1 \left(\frac{V_1}{V_2}\right)^{\gamma} = 1 \times \left(\frac{1}{3}\right)^{1.40}$$

Since $P_1 = 1$ atm.

On simplification, $P_2 = 0.2148$ atm.

Problem 4. *1 kg of water is boiled under a pressure of 2 atm. 120°C. If the volumes occupied by water and steam under given conditions are 10^{-3} m^3 and 0.824 m^3 respectively. Find (a) the work done and (b) the increase in internal energy. Given : $L = 2.20 \times 10^6$ J/kg, 1 atm $= 1.013 \times 10^5$ N m^{-2}.*

Solution. The work done against pressure is

$$dW = p\,dV$$
$$= (2 \times 1.013 \times 10^5) \times (0.824)$$
$$= 0.167 \times 10^6 \text{ J}$$

∴ The change in internal energy dV is, from the first law

$$dV = dQ - dW$$
$$= mL - dW$$
$$= 2.20 \times 10^6 - 0.167 \times 10^6$$
$$= 2.03 \times 10^6 \text{ J.}$$

Problem 5. *A Carnot's engine working between the temperature 100°C and 0°C absorbs 10^4 calories of heat at higher temperature. What will be the workdone when the cycle of operation is over?*

[Calcutta University]

Solution. We know that efficiency of Carnot engine's

$$\eta = 1 - \frac{T_2}{T_1} = 1 - \frac{273}{373} = \frac{100}{373}$$

$$\text{Again, } \eta = \frac{\text{Workdone by the engine (W)}}{\text{Head absorbed by the engine}} = \frac{W}{10^4 \times 4.2}$$

or,
$$W = \eta \times 10^4 \times 4.2 = \frac{100}{373} \times 10^4 \times 4.2 = 1.126 \times 10^4 \text{ Joule}.$$

Problem 6. *A reversible engine converts one sixth of heat input into work. When the temperature of the sink is reduced by 62 K its efficiency is doubled. Find the temperature of the source and sink.*

[Kanpur University, Pbi University, Bangalore University]

Solution. Here,efficiency $\eta = \frac{1}{6}$.

If T_1 and T_2 be the temperature of the source and sink respectively, then efficiency $\eta = \frac{1}{6} = 1 - \frac{T_2}{T_1}$

or,
$$\frac{T_2}{T_1} = 1 - \frac{1}{6} = \frac{5}{6}.$$

When the temperature of the sink is reduced by 62 K then efficiency

$$\eta' = 1 - \frac{T_2 - 62}{T_1}$$

If the efficiency is doubled,

then
$$\eta' = 2 \times \frac{1}{6} = \frac{1}{3}$$

$$\therefore \quad \frac{1}{3} = 1 - \frac{T_2}{T_1} + \frac{62}{T_1} = 1 - \frac{5}{6} + \frac{62}{T_1}$$

or $$T_1 = 6 \times 62 = 372$$

or $$T_2 = T_1 \times \frac{5}{6} = 372 \times \frac{5}{6} = 310 \text{ K}.$$

Problem 7. *A heat engine absorbs heat of 10^5 kilo-calories from a source which is at 127°C and rejects part of heat to sink at 27°C. Calculate the efficiency of the engine and work done by it.*

[Osamania University, 2004]

Solution. We know that the Efficiency of the engine

$$\eta = 1 - \frac{T_2}{T_1} = \frac{T_1 - T_2}{T_1} = \frac{(273+127)-(273+27)}{273+127}$$

$$= \frac{1}{4} = 25\%.$$

Again, we can write, $\eta = \dfrac{Q_1 - Q_2}{Q_1}$.

Therefore, workdone by the engine

$$W = Q_1 - Q_2 = \eta\, Q_1$$

or $$W = \frac{1}{4} \times 10^5 \times 1000 \times 4.2$$

$$= 1.05 \times 10^8 \text{ joules}.$$

Multiple Choice Questions

1. Increase in temperature results in –
   (*a*) isothermal expansion
   (*b*) adiabatic expansion
   (*c*) isothermal compression
   (*d*) adiabatic compression.
2. In an isochoric process –
   (*a*) P is constant
   (*b*) V is constant
   (*c*) T is constant.
3. Thermodynamic process which takes place at constant temperature is called –
   (*a*) isothermal process
   (*b*) isobaric process
   (*b*) adiabatic process
   (*d*) isochoric process.
4. During an adiabatic process –
   (*a*) T remains constant
   (*b*) S remains constant
   (*c*) U remains constant.
5. In a Carnot's engine when heat is taken by a perfect gas from the hot source, then temperature of the source –
   (*a*) decreases
   (*b*) increases
   (*c*) first decreases, then increases
   (*d*) remains constant.
6. The efficiency of a Carnot's engine working between temperatures 127°C and 27°C is
   (*a*) $\frac{100}{127}$ (*b*) $\frac{27}{127}$ (*c*) $\frac{100}{400}$ (*d*) $\frac{300}{400}$.

Answers

1. (*d*) 2. (*b*) 3. (*a*) 4. (*a*)
5. (*b*) 6. (*c*)

Review Questions

1. What is thermodynamics ? State the first law of thermodynamics and discuss its physical significances and limitations. [*Bangalore University, Nagpur University, Kerala University, Calcutta University, Lucknow University*]
2. What is concept of internal energy? On what state variables does it depend for a perfect gas.
3. Explain the terms adiabatic and isothermal changes. [*Bangalore University, Kanpur University, Calcutta University*]
4. Show that for an adiabatic change in a perfect gas PV^{γ} = constant, where γ is the ratio of specific heat at constant pressure and constant volume respectively. [*G.N.D. University, Calcutta University, Burdwan University*]
5. Discuss Carnot's reversible heat engine. Describe Carnot's cycle. Show how the work done in each operation is represented on a pressure volume diagram. Calculate the work done per cycle. Explain the term efficiency of a heat engine and deduce an expression for the efficiency of Carnot's reversible heat engine.

Unsolved Problems

1. A quantity of perfect gas at 10°C is a adiabatically compressed to one fourth of its volume. Calculate the final temperature (Y = 1.4, Li^4 = 1.74) [*Calcutta University*]
[**Ans.** 2145×10^7 ergs]
2. A certain volume of gas at 10^6 dynes/cm^2 pressure expands isothermally till its volume is four fold and adiabatically till its volume is again fourfold. Find the final pressure of the gas (V = 1.4). [*Kanpur University*]
[**Ans.** 3.6×10^4 dynes/cm^2]
3. The efficiency of a Carnot's engine is 0.25. On reducing the temperature of the sink by 60° the efficiency of the engine is 0.50. Find the temperature of the sink and source between which the engine is working. [**Ans.** T_1 = 240 K ; T_2 = 180 K]
4. A Carnot's engine whose sink is at 7°C has an efficiency of 50%. It is desired to increase the efficiency to 70%. By how many degrees the temperature of heat reservoir be increased. [*Bangalore University*]
[**Ans.** $\frac{2800}{3}$ °C]

C.V. Raman

3A

Chapter

Vibration and Waves

3A.1 INTRODUCTION

If the motion of a body or a particle is found to describe the same path repeatedly in some definite interval of time, its motion is said to be periodic, and the time taken by the body (or particle) to complete its path once is called its period. The number of times the path described by the body in a unit time is known as frequency. The periodic motion of a body is said to be oscillatory or vibratory when it is reversed in direction after a definite to interval of time. The to and fro motion of a pendulum is an example of motion of this type. A body moving uniformly in a circle executes a periodic motion, but the motion is not vibratory or oscillatory.

3A.1.1 Simple Harmonic Motion

It is the simplest type of oscillatory motion of a body with the following characteristics :

(*i*) The motion is oscillatory or vibratory and is repeated after equal intervals of time.
(*ii*) The restoring force acting on the body (or its acceleration) is always proportional to its displacement from some fixed point on its path called its mean position ;
(*iii*) Its acceleration is always directed towards its mean position.
(*iv*) The motion is linear *i.e.,* along a straight line.

A particle will vibrate if any displacement from its rest position calls into play a force (*i.e.*, restoring force or force of restitution) which tries to bring the particle back to its position of rest. This force may be expanded in powers of displacement x and can be written as

$$\text{Restoring force} = F(x) = a_0 + a_1 x + a_2 x^2 + a_3 x^3 + \ldots\ldots\ldots\ldots\ldots\ldots$$

When $x = 0$ *i.e.*, there is no force, we must have $a_0 = 0$. For small displacement, the next term $(a_1 x)$ varies directly as x. For restoring forces, a_1 is negative. The coefficients a_2, a_3, etc., are generally progressively smaller than a_1. In vibrations, when the restoring force is directly proportional to the displacement, it is the simplest in nature, and is called simple harmonic motion (SHM).

For small displacements, the motion of a simple pendulum, oscillation of a liquid in a U-tube, and the motion of a heavy body suspended by a elastic string are few examples of SHM. A particle in SHM is termed a Linear Harmonic Oscillator.

3A.1.2 Motion of a Linear Harmonic Oscillator

Suppose y be the displacement of a particle executing SHM at any instant, we may write $F = -Sy$ ...(3A.1)

where F is the force acting at the moment and S is the factor or constant of proportionality between force and displacement and will mean the restoring force per unit displacement. S is also referred to as the **stiffness factor or the spring constant.** The reciprocal of S is known as the **compliance.** The negative sign in equation (3A.1) implies that F and y are oppositely directed.

Using Newton's second law of motion

$$m\frac{d^2y}{dt^2} = -Sy \qquad [m = \text{mass of the moving body}]$$

or,

$$\frac{d^2y}{dt^2} + \omega^2 y = 0 \qquad \text{...(3A.2)}$$

putting

$$\omega^2 = \frac{S}{m}$$

Equation (3A.2) is known as the equation of motion of the Linear Harmonic Oscillator.

We take $y = e^{\alpha}$ as a trial solution of equation (3A.2)

Then $\omega^2 + \alpha^2 = 0$ or $\alpha = \pm j\omega$ where $j = \sqrt{-1}$

The solution is

$$y = C_1 e^{j\omega t} + C_2 e^{-j\omega t}$$

$$= (C_1 + C_2)\cos\omega t + (C_1 - C_2)\sin\omega t$$

$$= P\cos\omega t + Q\sin\omega t \qquad \text{...(3A.3)}$$

where $P = (C_1 + C_2)$ and $Q = (C_1 - C_2)$ are the two arbitrary constants.

By putting A cos δ for P and A sin δ for Q, equation (3A.3) may be written as

$$y = A\cos(\omega t - \delta) \qquad \text{...(3A.4)}$$

Also by putting A sin δ for P and A cos δ for Q, we may have

$$y = A\sin(\omega t + \delta) \qquad \text{...(3A.5)}$$

where A and δ are arbitrary constants. Here A denotes the maximum displacement of the body from the mean position. It is called the amplitude of oscillation.

$$A = \sqrt{P^2 + Q^2} \text{ and } \tan\delta = \frac{P}{Q}$$

Putting $\delta = \theta + \frac{\pi}{2}$ in equation (3A.5) we get

$$y = A\cos(\omega t + \theta) \qquad \text{...(3A.6)}$$

Any one of the above forms may be taken as the solution of equation (3A.2).

We may take any one of these forms to show that the motion is periodic, with a periodic time

$$T = \frac{2\pi}{\omega}$$

Let, y_t and y_{T+t} represent the values of y at times t and $(t + T)$ respectively. Then

$$y_t = A\cos(\omega t + \theta)$$

$$y_{T+t} = A\cos\{\omega(t + T) + \theta\}$$
$$= A\cos\{2\pi + (\omega t + \theta)\}$$
$$= A\cos(\omega t + \theta) = y_t$$

Thus the values of y repeat themselves at intervals of $T = \dfrac{2\pi}{\omega}$.

T may be also written as $T = \dfrac{2\pi}{\omega}$

or
$$= 2\pi\sqrt{\frac{m}{S}} = 2\pi\sqrt{\frac{\text{Mass of the moving particle}}{\text{Force per unit displacement}}}$$

$$= 2\pi\sqrt{\frac{1}{\text{Acceleration at unit distance}}}$$

The frequency of oscillation is $f = \dfrac{1}{T} = \dfrac{\omega}{2\pi}$

The angle $(\omega t + \delta)$ or $(\omega t + \theta)$ in equation (3A.5) or equation (3A.6) is known as the **phase angle**. The phase of a vibrating particle or a body at any instant determines the state of displacement and motion of the particle at that instant. Thus any quantity which can completely specify both the displacement of a vibrating point and its direction of motion at any time will serve as a measure of phase at that time. At $t = 0$ the phase angle is δ or θ.

The initial phase of a vibrating particle is called the **epoch.**

Dimensions of force constant : Consider a mass m attached to a spring executing SHM. If y is the displacement of the mass from its equilibrium position at any instant of time t, then the restoring forces F acting on the mass is given by

$$F = -Sy$$

where S is the force constant or stiffness of the spring

∴ Dimension of S is
$$= \frac{F}{y} = \frac{\text{Force}}{\text{Displacement}}$$

$$= \frac{MLT^{-2}}{L} = MT^{-2}$$

Stiffness is expressed in Newton/meter (Nm^{-1}).

Expression for the Velocity and Acceleration of a Simple Harmonic Oscillator

The displacement of a simple harmonic oscillator at any instant of time t, is given by

$$y = a\sin(\omega t + \theta)$$

Velocity is defined as the time rate of change of displacement.

∴ Velocity
$$V = \frac{dy}{dt} = a\omega\cos(\omega t + \theta)$$

$$= a\omega\sin\left(\omega t + \theta + \frac{\pi}{2}\right)$$

As
$$\sin(\omega t + \theta) = \frac{y}{a},\quad \cos(\omega t + \theta) = \sqrt{1 - \frac{y^2}{a^2}}$$

$\therefore \qquad V = a\omega \sqrt{1 - \frac{y^2}{a^2}}$ ...(3A.7)

The velocity of the oscillator is maximum when $\sin\left(\omega t + \theta + \frac{\pi}{2}\right) = 1$

$\therefore \qquad V_{max} = a\omega$

The value of $V = V_{max}$ when $y = 0$ *i.e.*, the particle executing SHM is in its mean position.

Now velocity $V = a\omega \cos(\omega t + \theta)$

$$\therefore \qquad \text{Acceleration} = \frac{dV}{dt} = -a\omega^2 \sin(\omega t + \theta)$$

$$= -a\omega^2 \sin(\omega t + \theta + \pi)$$

The acceleration of the oscillator is maximum when $\sin(\omega t + \theta + \pi) = 1$ and is given by

$$\left(\frac{d^2y}{dt^2}\right)_{max} = -a\omega^2$$

Expression for the Total Energy of a Harmonic Oscillator

The total energy of a simple harmonic oscillator at any time t is the sum of its kinetic and potential energies at that instant of time.

i.e., $\qquad E_{total} = \text{K.E.} + \text{P.E.}$

Kinetic Energy : The general solution of a simple harmonic oscillator is given by

$$y = a \sin(\omega t + \theta),$$

where y is the displacement of the oscillator at any time t, a is the amplitude, ω the angular frequency and θ the initial phase or epoch.

$$\therefore \quad \text{The velocity} \qquad V = \frac{dy}{dt} = a\omega \cos(\omega t + \theta)$$

$$\therefore \quad \text{Kinetic energy of the oscillator} = \frac{1}{2} m \left(\frac{dy}{dt}\right)^2$$

$$= \frac{1}{2} ma^2\omega^2 \cos^2(\omega t + \theta) \qquad \text{...(3A.8)}$$

Potential Energy : The potential energy is equal to the amount of work done in overcoming the restoring force from the mean position through a displacement y.

$$\text{Now acceleration} \qquad = \frac{d^2y}{dt^2} = -a\omega^2 \sin(\omega t + \theta) = -\omega^2 y$$

$\therefore$ Restoring force $= m\omega^2 y = Sy$ where S is the stiffness.

Hence, the work done by the force through a displacement y

$$= \int_{-y}^{y} Sy\, dy = \frac{1}{2} Sy^2 = \frac{1}{2} m\omega^2 y^2$$

$\therefore$ $$\text{P.E.} = \frac{1}{2} m\omega^2 y^2 = \frac{1}{2} m\omega^2 a^2 \sin^2(\omega t + \theta) \qquad \ldots(3A.9)$$

$\therefore$ Total energy, $$E = \text{K.E.} + \text{P.E.}$$

$$= \frac{1}{2} m\omega^2 a^2 \qquad \ldots(3A.10)$$

Graphical Representation of Energy

The variation of total energy, kinetic energy and potential energy of harmonic oscillator with displacement is shown in figure 3A.1.

(*i*) The K.E. is given by $\frac{1}{2}mv^2$. The velocity v and K.E. is zero at the extreme position $y = \pm a$ and maximum in the mean position $y = 0$. The K.E. decreases as the oscillator moves away from mean position and finally becomes zero at the extreme positions.

(*ii*) The P.E. is given by $\frac{1}{2}Sy^2$. For $y = 0$, P.E. $= 0$ and for $y = \pm a$, P.E. is maximum. The potential energy goes on increasing as oscillator move away from its mean position and becomes maximum at extreme position.

(*iii*) The total energy is given by $\frac{1}{2}Sa^2$, which is constant. The straight line marked total energy, parallel to the displacement axis, represents this.

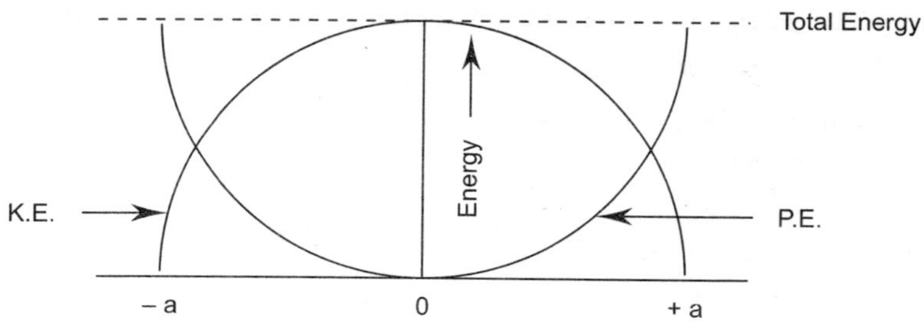

Figure 3A.1

3A.1.3 Superposition of Two Linear SHMs

Two simple harmonic vibrations of same frequency acting along the same line but differing in phase.

Let the two simple harmonic vibrations of angular frequency ω acting along the X-axis and having initial phases ϕ_1 and ϕ_2, amplitudes a_1 and a_2 be given by

$$x_1 = a_1 \sin(\omega t + \phi_1) \qquad \ldots(3A.11)$$

$$x_2 = a_2 \sin(\omega t + \phi_2) \qquad \ldots(3A.12)$$

The resultant displacement due to the two simple harmonic vibrations is given by

$$x = x_1 + x_2 = a_1 \sin(\omega t + \phi_1) + a_2 \sin(\omega t + \phi_2)$$

$$= \sin \omega t\,(a_1 \cos \phi_1 + a_2 \cos \phi_2) + \cos \omega t\,(a_1 \sin \phi_1 + a_2 \sin \phi_2)$$

Put $a_1 \cos \phi_1 + a_2 \cos \phi_2 = R \cos \theta$ and

$$a_1 \sin\phi_1 + a_2 \sin\phi_2 = \text{R} \sin\theta$$

$$\therefore \quad x = \text{R} \sin\omega t \cos\theta + \text{R} \cos\omega t \sin\theta$$

$$= \text{R} \sin(\omega t + \theta)$$

which gives the equation of the resultant simple harmonic vibration of amplitude R and initial phase θ where

$$\text{R}^2 = \text{R}^2 \cos^2\theta + \text{R}^2 \sin^2\theta = a_1^2 + a_2^2 + 2a_1a_2(\sin\phi_1 \sin\phi_2 + \cos\phi_1 \cos\phi_2)$$

$$= a_1^2 + a_2^2 + 2a_1a_2 \cos(\phi_2 - \phi_1)$$

$$= a_1^2 + a_2^2 + 2a_1a_2 \cos\phi \qquad \text{...(3A.13)}$$

where the phase difference $\phi = \phi_2 - \phi_1$

and $$\tan\theta = \frac{\text{R}\sin\theta}{\text{R}\cos\theta} = \frac{a_1 \sin\phi_1 + a_2 \sin\phi_2}{a_1 \cos\phi_1 + a_2 \cos\phi_2}$$

A particle is subjected simultaneously to two SHMs of the same period but of different amplitudes and phases in perpendicular direction.

The equation of an SHM is given by $x = a \sin(\omega t + \phi)$ where x is the displacement of the vibrating particle at any instant t, a the amplitude, ϕ the initial phase.

Let the displacement of two perpendicular SHMs of the same period (of frequency ω) taking place along X-axis and Y-axis respectively be represented by

$$x = a \sin(\omega t + \phi_1) \qquad \text{...(3A.14)}$$

$$y = b \sin(\omega t + \phi_2) \qquad \text{...(3A.15)}$$

where a is the amplitude of vibration along X-axis, b is the amplitude of vibration along Y-axis and ϕ_1, ϕ_2 are the initial phases along X and Y axes respectively. The phase difference between the two vibrations is

$$\phi_2 - \phi_1 = \phi$$

From (3A.14) we have $$\frac{x}{a} = \sin\omega t \cos\phi_1 + \cos\omega t \sin\phi_1 \qquad \text{...(3A.16)}$$

From (3A.15) we have $$\frac{y}{b} = \sin\omega t \cos\phi_2 + \cos\omega t \sin\phi_2 \qquad \text{...(3A.17)}$$

Multiplying (3A.16) by $\sin\phi_2$ and (3A.17) by $\sin\phi_1$ and subtracting, we have

$$\left(\frac{x}{a}\sin\phi_2 - \frac{y}{b}\sin\phi_1\right) = \sin\omega t\,(\cos\phi_1 \sin\phi_2 - \cos\phi_2 \sin\phi_1)$$

$$= \sin\omega t \sin(\phi_2 - \phi_1) \qquad \text{...(3A.18)}$$

Similarly, multiplying (3A.17) by $\cos\phi_1$ and (3A.16) by $\cos\phi_2$ and subtracting, we have

$$\left(\frac{y}{b}\cos\phi_1 - \frac{x}{a}\cos\phi_2\right) = \cos\omega t\,(\sin\phi_2 \cos\phi_1 - \sin\phi_1 \cos\phi_2)$$

$$= \cos\omega t \sin(\phi_2 - \phi_1) \qquad \text{...(3A.19)}$$

Squaring (3A.18) and (3A.19) and adding, we get

$$\left(\frac{x}{a}\sin\phi_2 - \frac{y}{b}\sin\phi_1\right)^2 + \left(\frac{y}{b}\cos\phi_1 - \frac{x}{a}\cos\phi_2\right)^2$$
$$= \sin^2(\phi_2 - \phi_1)\,[\sin^2\omega t + \cos^2\omega t]$$
$$= \sin^2(\phi_2 - \phi_1)$$

or,
$$\frac{x^2}{a^2} + \frac{y^2}{b^2} - \frac{2xy}{ab}\left(\sin\phi_2\sin\phi_1 + \cos\phi_2\cos\phi_1\right) = \sin^2(\phi_2 - \phi_1)$$

or,
$$\frac{x^2}{a^2} + \frac{y^2}{b^2} - \frac{2xy}{ab}\cos\left(\phi_2 - \phi_1\right) = \sin^2(\phi_2 - \phi_1)$$

Substituting $\phi_2 - \phi_1 = \phi$, we get,

$$\frac{x^2}{a^2} + \frac{y^2}{b^2} - \frac{2xy}{ab}\cos\phi = \sin^2\phi \qquad \text{...(3A.20)}$$

This is the equation giving the expression for resultant motion and represents the equation of an ellipse whose major and minor axes are inclined to the co-ordinate axes. This ellipse can be inscribed in a rectangle whose sides are $2a$ and $2b$.

Important Cases :

(*i*) **Straight line :** When $\phi = 0$, $\cos\phi = 1$ and $\sin\phi = 0$. Relation (3A.20) reduces to

$$\frac{x^2}{a^2} + \frac{y^2}{b^2} - \frac{2xy}{ab} = 0$$

or,
$$\left(\frac{x}{a} - \frac{y}{b}\right)^2 = 0$$

This represents, a pair of coincident straight lines lying in the first and third quadrant as shown in figure 3A.2 (*i*). The straight line is inclined to the X-axis at angle θ given by $\theta = \tan^{-1}\left(\frac{b}{a}\right)$.

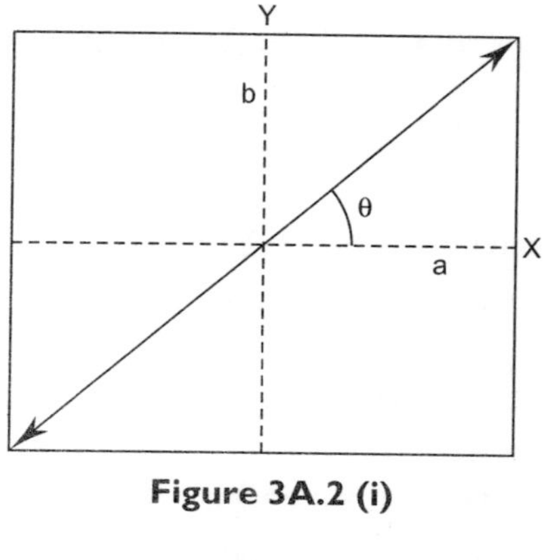

Figure 3A.2 (i)

(*ii*) **Oblique ellipse:** When $\phi = \frac{\pi}{4}$, $\cos\frac{\pi}{4} = \frac{1}{\sqrt{2}}$ and $\sin\frac{\pi}{4} = \frac{1}{\sqrt{2}}$

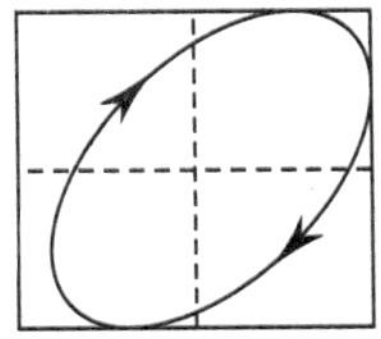

Figure 3A.2 (ii)

Relation (3A.20) is then reduced to

$$\frac{x^2}{a^2} + \frac{y^2}{b^2} - \frac{\sqrt{2}xy}{ab} = \frac{1}{2}$$

This represents an oblique ellipse lying in first and third quadrant as shown in figure 3A.2 (*ii*).

(*iii*) (*a*) *Symmetrical ellipse:* When $\phi = \frac{\pi}{2}$, $\cos\frac{\pi}{2} = 0$ and $\sin\frac{\pi}{2} = 1$

Relation (3A.20) is then reduced to

$$\frac{x^2}{a^2} + \frac{y^2}{b^2} = 1$$

This represents a symmetrical ellipse whose major and minor axes coincide with the co-ordinate axes.

(*b*) *Circle:* If $a = b$ the resultant vibration is circular and is represented by $x^2 + y^2 = a^2$ as shown in figure 3A.2 (*iii*).

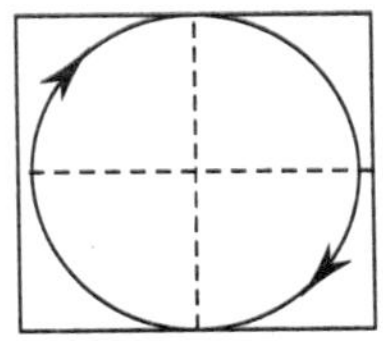

Figure 3A.2 (iii)

(*iv*) **Oblique ellipse:** When $\phi = \frac{3\pi}{4}$, $\cos\frac{3\pi}{4} = -\frac{1}{\sqrt{2}}$ and $\sin\frac{3\pi}{4} = +\frac{1}{\sqrt{2}}$

Relation (3A.20) reduced to

$$\frac{x^2}{a^2} + \frac{y^2}{b^2} + \frac{\sqrt{2}xy}{ab} = \frac{1}{2}$$

This again represents an oblique ellipse lying in second and fourth quadrant as shown in figure 3A.2 (*iv*).

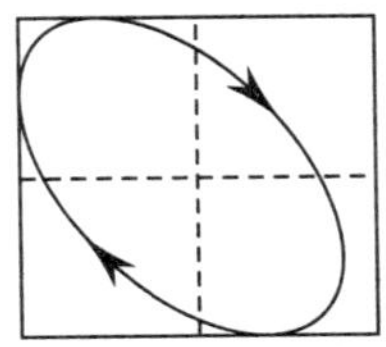

Figure 3A.2 (iv)

(*v*) When $\phi = \pi$, $\cos\phi = -1$, $\sin\phi = 0$, then equation (7) reduces to

$$\frac{x^2}{a^2} + \frac{y^2}{b^2} + \frac{2xy}{ab} = 0 \Rightarrow \left(\frac{x}{a} + \frac{y}{b}\right)^2 = 0.$$

This represents the pair of straight lines lying in the second and fourth quadrant and inclined at any angle $\theta = \tan^{-1}\left(-\frac{b}{a}\right)$ as shown in figure 3A.2 (*v*).

(*vi*) When $\phi = \pi$ and 2π the whole cycle is repeated in reverse order.

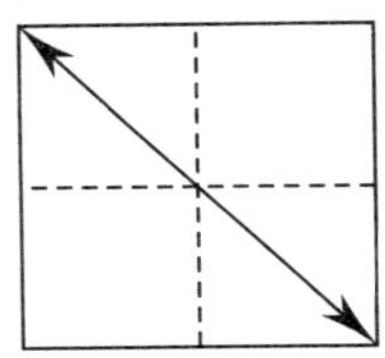

Figure 3A.2 (v)

3A.1.4 Lissajous Figures

The curves or figures which represent the resultant of two simple harmonic motions acting at right angles to each other are known as Lissajous figures. The constituent simple harmonic vibrations may have different frequencies and also an initial phase difference. Figures 3A.2. [(*i*), (*ii*), (*iii*), (*iv*), (*v*)] are examples of Lissajous figures.

The ratio of the frequencies of the component vibrations can be obtained from the Lissajous figures. Let us draw a horizontal, and vertical line to intersect the curve and count the number of cuts each line makes with the curve. Now the ratio of the number cuts made by horizontal line to that made by the vertical line is the ratio of a vertical to the horizontal frequency. The ratio of sides of the rectangle enclosing the figure represents the ratio of the amplitudes of the component oscillations.

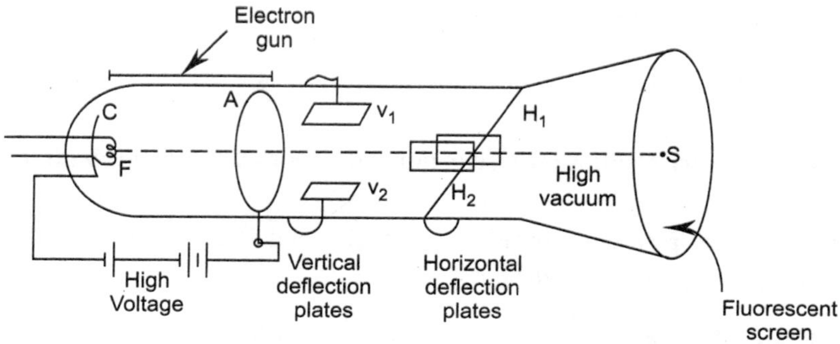

Figure 3A.3

Lissajous figures may be obtained experimentally by several ways like Blackburn's pendulum, Optical method, Cathode ray oscillograph etc. For demonstration of Lissajous figures to an audience the cathode-ray oscilloscope (CRO) is the best. It has three main components, namely, the electron gun, the horizontal deflection plates (H_1, H_2) and the vertical deflection plates (V_1, V_2). The electron gun creates a narrow beam of high speed electrons toward a fluoresent screen, a visible spot S is formed on the screen where the beam falls. On its way from the electron gun to the fluorescent screen, the beam passes through two sets of parallel deflection plates arranged at right angles to each other.

When an alternating potential difference is applied across the plates V_1, V_2 alone, spot of light produced by the electron beam on the screen will trace a vertical line. Again repeating the process across the plates H_1, H_2 alone, the spot will trace a horizontal line. If two sinusoidel alternating potential differences are simultaneously applied across the two pairs, the spot on the fluorescent screen trace out the various desired Lissajous figures by adjusting the intensities, frequencies and initial phases of the two field.

3A.1.5 Damped Vibration

We have noticed that the total energy of harmonic oscillator remains constant. Once started, the oscillations continue forever with a constant amplitude and a constant frequency. This is an ideal situation and is not observed unless energy is supplied to the vibratinmg body at a constant rate. Vibrations of a body when it is completely free from external forces is called **free vibration.** In all vibrations observed in nature, *e.g.*, those of a tuning fork, pendulum or of a string etc., the amplitude gradually diminishes and becomes imperceptible after some lapse of time. The presence of resistance to vibration implies that frictional or damping force acts on the system. The resistance offered by a damping force is known as **damping**. The simple harmonic vibrations where amplitude decreases with the passage of time are called **damped simple harmonic vibrations**. For small damping, the damping force is proportional to the velocity of the vibrating body.

Analytical Treatment of Damped Vibration

We shall investigate the effect of damping on the harmonic oscillations of the simplest system. One such system is shown in figure 3A.4.

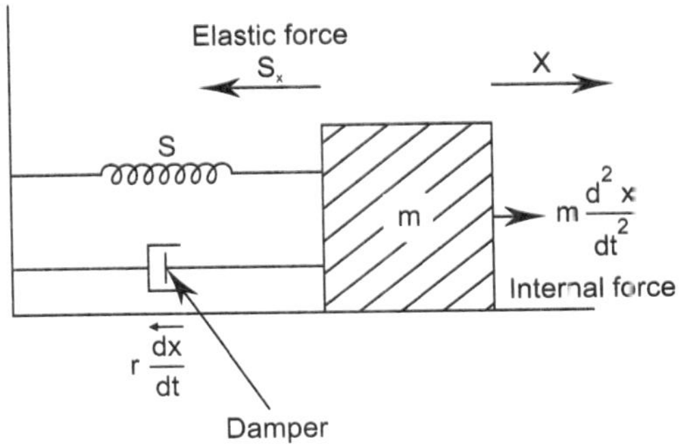

Figure 3A.4

When the system of mass m is stretched by an instantaneous displacement x from equilibrium state and released, it begins to move. In such a case there are two forces acting on the system.

(*i*) The restoring force proportional to the displacement x is given by $-Sx$, where S is a constant known as stiffness constant or spring constant. The – ve sign implies that the direction of the restoring force is opposite to that of displacement.

(*ii*) A retarding or damping force $-r\frac{dx}{dt}$, where r is the coefficient of damping force. The – ve sign again shows that the retarding force is also opposite to that of the direction of motion of the system.

From Newton's law for a rigid body in translation these forces must balance Newtonian force $m\frac{d^2x}{dt^2}$.

So the equation of motion of the particle may be written as

$$m\frac{d^2x}{dt^2} = -r\frac{dx}{dt} - Sx$$

Rearranging and dividing by m, we obtain

$$\frac{d^2x}{dt^2} + \frac{r}{m}\frac{dx}{dt} + \frac{S}{m}x = 0$$

or

$$\frac{d^2x}{dt^2} + 2k\frac{dx}{dt} + \omega_0^2 x = 0 \qquad \text{...(3A.21)}$$

where $2k = \dfrac{r}{m}$ = resisting force per unit mass per unit velocity.

$\omega_0{}^2 = \dfrac{S}{m}$ = force of restitution per unit mass per unit displacement.

Equation (3A.21) is known as the differential equation of damped simple harmonic motion,

Let the solution of the differential equation (3A.21) be

$$x = A\, e^{\alpha t}$$

Then
$$\frac{dx}{dt} = A\alpha\, e^{\alpha t} = \alpha x$$

and
$$\frac{d^2x}{dt^2} = A\alpha^2\, e^{\alpha t} = \alpha^2 x$$

Equation (3A.21) then becomes

$$A\alpha^2 e^{\alpha t} + 2k\, A\alpha\, e^{\alpha t} + \omega_0{}^2 A\, e^{\alpha t} = 0$$

or
$$\alpha^2 + 2k\alpha + \omega_0{}^2 = 0 \qquad [\text{Since } e^{\alpha t} \neq 0]$$

$$\alpha = \frac{-2k \pm \sqrt{4k^2 - 4\omega_0^2}}{2}$$

$$= -k \pm \sqrt{k^2 - \omega_0^2}$$

The general solution of the differential equation (3A.21) may be written as

$$x = A_1\, e^{\left(-k + \sqrt{k^2 - \omega_0^2}\right)t} + A_2\, e^{\left(-k - \sqrt{k^2 - \omega_0^2}\right)t}$$

$$= e^{-kt}\left[A_1\, e^{\sqrt{k^2 - \omega_0^2}\, t} + A_2\, e^{-\sqrt{k^2 - \omega_0^2}\, t}\right] \qquad \text{...(3A.22)}$$

where A_1 and A_2 are the arbitrary constants whose values are determined from the boundary conditions.

Putting $k = \dfrac{r}{2m}$ and $\omega_0{}^2 = \dfrac{S}{m}$, we get

$$x = e^{-\frac{rt}{2m}}\left[A_1\, e^{\sqrt{\frac{r^2}{4m^2} - \frac{S}{m}}\, t} + A_2\, e^{-\sqrt{\frac{r^2}{4m^2} - \frac{S}{m}}\, t}\right] \qquad \text{...(3A.23)}$$

This is the displacement equation of mass m executing damped simple harmonic motion. The nature of motion depends upon the relative values of k and ω_0 *i.e.*, whether $\sqrt{k^2 - \omega_0^2}$ or $\sqrt{\dfrac{r^2}{4m^2} - \dfrac{S}{m}}$ is + ve, 0 or, – ve

Case I : *When $k > \omega_0$, it is a case of heavy damping.*

In this case $$\sqrt{k^2 - \omega_0^2} = \sqrt{\frac{r^2}{4m^2} - \frac{S}{m}} = p \text{ (real)}$$

Putting in equation (3A.23), we have

$$x = e^{-kt}\left[A_1 e^{pt} + A_2 e^{-pt}\right]$$

Let $$A = A_1 + A_2 \text{ and } B = A_1 - A_2$$

then, $$A_1 = \frac{A+B}{2}\ ;\ A_2 = \frac{A-B}{2}$$

$\therefore$ $$x = e^{-kt}\left[\frac{A+B}{2} e^{pt} + \frac{A-B}{2} e^{-pt}\right]$$

$$= e^{-kt}\left[A \cdot \frac{\left(e^{pt} + e^{-pt}\right)}{2} + B\frac{\left(e^{pt} - e^{-pt}\right)}{2}\right]$$

$$= e^{-kt}\left[A \cosh pt + B \sinh pt\right]$$

This displacement consists of two terms both dying off exponentially to zero. The values of A and B can be find out putting the initial conditions.

If $x = 0$, at $t = 0$, then $A = 0$

$$x = e^{-kt} B \sinh pt$$

$$= B\ e^{-\frac{r}{2m}t} \sinh\left(\sqrt{\frac{r^2}{4m^2} - \frac{S}{m}}\right) t \qquad \ldots(3A.24)$$

This equation represents aperiodic or non oscillatory motion. Under very heavy damping, there is no vibration at all and the particle passes through its equilibrium position at the most once before going asymptotically to rest. It is represented in practice by such examples as a **dead beat** moving coil galvanometer or a pendulum vibrating in highly viscous liquid like thick oil. The variation of displacement (x) with time (t) for a highly damped simple harmonic oscillator is shown in figure 3A.5.

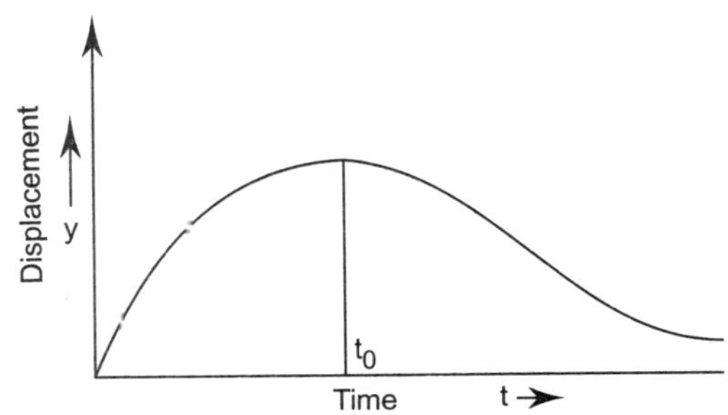

Figure 3A.5

The oscillator is disturbed by giving it a jerk at $t = 0$ and the system starts moving. At this stage the term e^{-kt} is very nearly equal to one ($e^{\circ} = 1$). The displacement increases with time up to $t = t_0$ when the system has maximum displacement. Thereafter the factor e^{-kt} dominates and the displacement decreases exponentially with time.

Case II : *Critically damped motion : $k = \omega_0$*

The oscillator is said to be critically damped if

$$k^2 - \omega^2 = 0, \; i.e., \; \frac{r^2}{4m^2} - \frac{S}{m} = 0$$

Putting this value in equation (3A.23), we get the displacement

$$x = e^{-kt}(A_1 + A_2) = C\,e^{-kt} \text{ where } C = A_1 + A_2 \qquad \ldots(3A.25)$$

In this case equation (3A.25) contains only one constant but a solution of second degree differential equation must contain two constants. Thus the solution given by equation (3A.25) is only a partial solution.

To find the complete solution, we take $\sqrt{k^2 - \omega_0^2} = \delta$ in equation (3A.23), where δ is a small quantity.

Equation (3A.23) becomes

$$x = e^{-kt}\left[A_1 e^{\delta t} + A_2 e^{-\delta t}\right] = e^{-kt}\,[A_1(1 + \delta t) + A_2(1 - \delta t)]$$

$$= e^{-kt}\,[(A_1 + A_2) + \delta t\,(A_1 - A_2)]$$

[Expanding and neglecting the terms higher than the term linear in δt]

$$\therefore \qquad x = e^{-kt}(A + Bt) \qquad \ldots(3A.26)$$

where $\quad A = A_1 + A_2$ and $B = \delta\,(A_1 - A_2)$

Here the constant A has the dimension of length and constant B that of velocity. Equation (3A.26) shows that here the motion is also aperiodic or nonoscillatory, but x approaches zero quicker than the overdamped case. This is illustrated in figure 3A.6. Therefore for $k = \omega_0$, the motion is said to be **critically damped**. The property of critical damping is made use of in measuring with instruments like Ballistic Galvanometers.

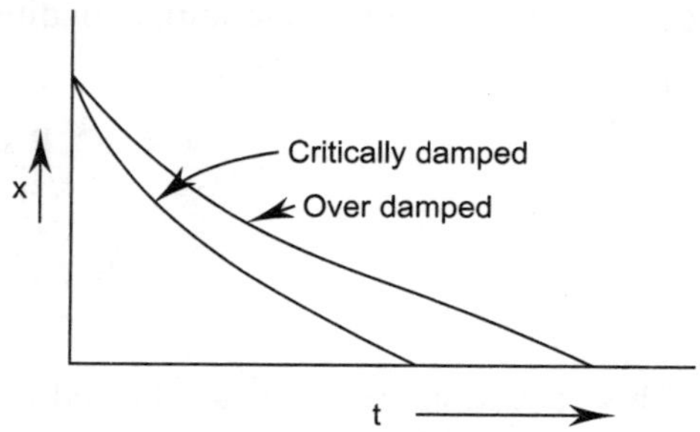

Figure 3A.6

Case III : *Light damping when $k < \omega_0$*

In this case we can write

$$\sqrt{k^2 - \omega_0^2} = j\sqrt{\omega_0^2 - k^2} \qquad [j = \sqrt{-1}]$$

$$= j\omega' \qquad [\text{where } \omega' = \sqrt{\omega_0^2 - k^2}\,]$$

Hence equation (3A.22) reduces to

$$x = e^{-kt}\left[A_1 e^{jw't} + A_2 e^{-jw't}\right]$$

Put $\qquad A_1 = \frac{A_0}{2j} e^{j\phi}$ and $A_2 = \frac{A_0}{2j} e^{-j\phi}$

where A_0 and ϕ are also constants to be determined from the initial conditions.

$$x = A_0\,e^{-kt}\,\frac{\left[e^{j(\omega' t + \phi)} - e^{-j(\omega' t + \phi)}\right]}{2j}$$

$$= A_0\,e^{-kt}\sin(\omega' t + \phi) \qquad \ldots(3A.27)$$

This is the equation of a damped oscillatory simple harmonic motion with amplitude $A_0 e^{-kt}$ which goes on decreasing with time and with a new angular frequency $\omega' = \sqrt{\omega_0^2 - k^2}$ which is less than $\omega_0 = \sqrt{\frac{S}{m}}$.

To find the value of A_0 and ϕ, let the displacement $x = x_0$ at time $t = 0$, so that the velocity $\frac{dx}{dt} = 0$

Substituting in equation (3A.27), we have

$$x_0 = A_0 \sin \phi$$

and differentiating equation (3A.27), we get

$$\frac{dx}{dt} = -A_0 k e^{-kt} \sin(\omega' t + \phi) + A_0 \omega' e^{-kt} \cos(\omega' t + \phi)$$

Putting $\frac{dx}{dt} = 0$ at $t = 0$,

$$0 = -A_0 k \sin\phi + A_0 \omega' \cos\phi$$

$\therefore$
$$\tan\phi = \frac{\omega'}{k} = \frac{\sqrt{\omega_0^2 - k^2}}{k}$$

and
$$\sin\phi = \frac{\omega'}{\sqrt{\omega'^2 + k^2}}$$

$\therefore$
$$x_0 = A_0 \frac{\omega'}{\sqrt{\omega'^2 + k^2}} = A_0 \frac{\omega'}{\omega_0} \qquad [\omega'^2 = \omega_0^2 - k^2]$$

or
$$A_0 = \frac{x_0 \omega_0}{\omega'}$$

Substituting the values in equation (3A.27), we have

$$x = \frac{x_0 \omega_0}{\omega'} e^{-kt} \sin(\omega' t + \phi)$$

$$= \frac{x_0 \omega_0}{\sqrt{\omega_0^2 - k^2}} e^{-kt} \sin\left\{\sqrt{\omega_0^2 - k^2}\, t + \phi\right\} \qquad \text{...(3A.28)}$$

$$= A_0 e^{-kt} \sin\left(\sqrt{\omega_0^2 - k^2}\, t + \phi\right)$$

$$= A \sin\left(\sqrt{\omega_0^2 - k^2}\, t + \phi\right) \qquad \text{...(3A.29)}$$

where
$$A_0 = \frac{x_0 \omega_0}{\sqrt{\omega_0^2 - k^2}} \text{ and } A = A_0 e^{-kt} \qquad \text{...(3A.30)}$$

Equation (3A.29) shows that the motion is oscillatory with an angular frequency $\omega' = \sqrt{\omega_0^2 - k^2}$ and an amplitude A (= $A_0 e^{-kt}$). The amplitude of the oscillations, therefore goes on decreasing with time. The motion is damped oscillatory.

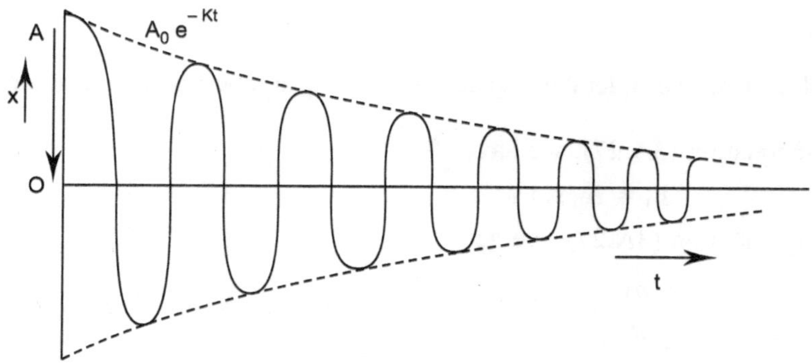

Figure 3A.7

The variation of x with time being displayed in figure 3A.7

$$\text{The time period } T' = \frac{2\pi}{\sqrt{\omega_0^2 - k^2}} = \frac{2\pi}{\sqrt{\frac{S}{m} - \frac{r^2}{4m^2}}}$$

and the frequency of damped oscillation

$$f' = \frac{\omega'}{2\pi} = \frac{1}{2\pi}\sqrt{\frac{S}{m} - \frac{r^2}{4m^2}}.$$

but undamped natural frequency of oscillation is

$$f_0 = \frac{1}{2\pi}\sqrt{\frac{S}{m}}.$$

Thus f' is less than f_0. The difference depends upon the damping factor r. Larger the value of r higher is the difference between f_0 and f'.

3A.1.6 Logarithmic Decrement

The amplitude of the damped oscillation at any time t is given by

$$A = A_0 e^{-kt}$$

Let us consider the oscillation to start from the mean position, then after a time $t = \frac{T}{4}$, where T is time period, the oscillating particle goes to extreme position. Let the first amplitude be denoted by A_1, then

$$A_1 = A_0 e^{-\frac{kT}{4}}$$

Again the particle comes back to the extreme position on the same side after a time T, *i.e.*, $\left(T + \frac{T}{4}\right)$

Let the second amplitude be A_2. Then

$$A_2 = A_0 \, e^{-k\left(T + \frac{T}{4}\right)}$$

Similarly the successive amplitudes of the 3rd, 4th, 5th etc., oscillations will be given by

$$A_3 = A_0 \, e^{-k\left(2T + \frac{T}{4}\right)}$$

$$A_4 = A_0 \, e^{-k\left(3T + \frac{T}{4}\right)}$$

....

....

$$A_n = A_0 \, e^{-k\left\{(n-1)T + \frac{T}{4}\right\}}$$

Hence, $$\frac{A_1}{A_2} = \frac{A_2}{A_3} = \frac{A_3}{A_4} = \ldots = \frac{A_{n-1}}{A_n} = e^{kT}$$

Taking natural logarithms, we have

$$\log_e \frac{A_1}{A_2} = \log_e \frac{A_2}{A_3} = \ldots = kT = \lambda \text{ (say)}.$$

It is defined as the natural logarithm of the ratio between the two successive amplitudes on the same side of the mean position of a damped oscillation separated by one full period.

∴ Logarithmic decrement for a mechanical oscillator

$$\lambda = kT = \frac{r}{2m} T$$

r being the damping constant and m the mass of the oscillator.

Here, $$T = \frac{2\pi}{\sqrt{\omega_0^2 - k^2}} = \frac{2\pi}{\omega'}$$

∴ $$\lambda = \frac{r}{2m} \cdot \frac{2\pi}{\omega'} = \frac{\pi r}{m\sqrt{\omega_0^2 - k^2}}$$

$$= \frac{\pi r}{m\sqrt{\frac{S}{m} - \frac{r^2}{4m^2}}}$$

So the logarithmic decrement gives a method of finding the damping co-efficient k and also r, the constant of proportionality for retarding force.

Logarithmic decrement has a very important physical significance. It gives the measure of the rate at which the amplitude of vibration decays.

3A.1.7 Quality Factor

The quality factor measures the rate of decay of energy of the damped harmonic oscillator. For an undamped simple harmonic oscillator the total energy at any time is given by

$$E = \frac{1}{2} m \omega_0^2 A^2$$

and the energy of the damped oscillator is given by

$$E' = \frac{1}{2} m \omega'^2 A_0^2 e^{-2kt} \qquad [\text{where } A = A_0 e^{-kt}]$$

where $$E_0 = \frac{1}{2} m \omega'^2 A_0^2$$

Putting $$t = \frac{1}{2k}, \qquad E' = E_0 e^{-2k \cdot \frac{1}{2k}}$$

$$= \frac{E_0}{e} \qquad \left[t = \frac{1}{2k} = \frac{1}{2}\frac{2m}{r} = \frac{m}{r}\right]$$

Thus the energy of the oscillator decays to $\frac{E_0}{e}$ from E_0 at the time $t = \frac{m}{r}$ where r is the damping constant. During this time the oscillator will have vibrated through an angle θ which is given by

$$\theta = \omega' t = \omega' \frac{m}{r} \text{ radians.}$$

The quantity $\frac{\omega' m}{r} = Q$ gives the quality factor or Q value. The quality factor is defined as the angle through which the damped oscillator oscillates as its energy decays from E_0 to $\frac{E_0}{e}$.

∴ Quality factor for the oscillator $Q = \frac{\omega' m}{r} = \frac{\omega'}{2k}$ *i.e.*, quality factor Q is large if the damping co-efficient k is small. The quality factor, therefore represents the efficiency of the oscillator. Thus lower the damping, higher is the quality factor. Further, when damping is small (*i.e.*, Q is large) then k^2 is negligible as compared to ω_0^2 and hence $\omega' = \omega_0$ $\left[i.e., \omega' = \sqrt{\omega_0^2 - k^2} = \omega_0\right]$.

∴ Quality factor $$Q = \frac{m\omega'}{r} = \frac{m\omega_0}{r}$$

$$= \frac{m}{r}\sqrt{\left(\frac{S}{m}\right)} = \frac{\sqrt{Sm}}{r}$$

i.e., for large quality factor damping has little or no effect on the frequency.

3A.1.8 Miscellaneous Relations in Damped Vibration

(*i*) **Relaxation Time :** It is defined as the time in which the amplitude of the damped oscillations falls to $\frac{1}{e}$ of its original value. It is usually denoted by the symbol τ. From the expression

$$A = A_0\, e^{-kt}$$

where A = the amplitude of the damped oscillator at any time t.

A_0 = the original amplitude of a damped oscillator.

if τ is the relaxation time, then $A = A_0\, e^{-k\tau} = A_0 \,.\, e^{-1}$

$$k\tau = 1, \quad \text{or} \quad \tau = \frac{1}{k} = \frac{2m}{r} \qquad \left[\because k = \frac{r}{2m}\right]$$

The heavier the mass or smaller the damping, the larger is the relaxation time and slower is the rate of fall of amplitude. Hence slower is the rate of dissipation of energy for higher relaxation time.

(*ii*) **Relation between Logarithmic Decrement and Relaxation Time :**

Relaxation time $\tau = \frac{1}{k}$

and logarithmic decrement $\lambda = k\text{T}$

$$\therefore \qquad \lambda = \frac{\text{T}}{\tau}$$

Thus the logarithmic decrement may also be defined as the ratio of the time period of vibration T to the modulus of decay or relaxation time.

(*iii*) **Fractional Loss of Energy Per Cycle :** Let x_1 and x_2 are the successive displacements on the same side of the swing after the time period T. At the end of the swing the velocity is momentarily zero and its energy is entirely potential. Therefore, corresponding potential energies are $\frac{1}{2}\text{S}x_1^2$ and $\frac{1}{2}\text{S}x_2^2$.

$\therefore$ The average fractional loss of energy per cycle is

$$\frac{\Delta\text{E}}{\text{E}} = \frac{\frac{1}{2}\text{S}x_1^2 - \frac{1}{2}\text{S}x_2^2}{\frac{1}{2}\text{S}x_1^2} = 1 - \left(\frac{x_2}{x_1}\right)^2$$

$$= 1 - e^{-2\lambda}$$

$$\approx 2\lambda \qquad \text{If } \lambda \text{ (= logarithmic decrement) is small}$$

3A.1.9 Forced Vibration and Resonance

A vibrating system, gradually loses its amplitude, energy due to damping forces. To maintain the system in vibration, energy has to be supplied from outside. If an external periodic force acts on a vibrating system, the system tends to vibrate with its own natural frequency, while the applied force tries to make the body vibrate with the frequency of the periodic force. In course of time, the natural vibration dies away due to damping forces that operate and finally, the vibration due to the impressed driving force are fully established. Such a vibration where the system oscillates with a frequency being the same as that of an externally impressed periodic force is called as the **forced vibration** of the system.

If we strike a tuning fork and hold its stem in the hand, the intensity of sound produced is low and prongs of the tuning fork set only a small volume of air near them into vibration. The energy is dissipated slowly. If the stem of the fork is pressed on the table, the table vibrates and we hear the intensified sound. The table is forced to vibrate with the frequency of the tuning fork. Intensity of sound now dies off quickly. This however, does not violate the principle of conservation of energy. The fork vibrates for a longer time if it is not put in contact with the table whereas when it is pressed on the table, its energy is drained very quickly to set the table and the larger volumes of air to vibrate. Thus, though the sound becomes louder for a short time, it keeps the total energy constant.

When the frequency of the impressed periodic force is the same as the natural frequency of the vibrating body, the vibrations build up quickly and a relatively large amplitude is set up within the body. This phenomenon of setting up vibrations of large amplitude in a system when the frequency of the applied force agrees with the natural frequency of the forcing system is known as **resonance** or **sympathetic vibration**.

There are two types of resonances:

(*i*) **amplitude resonance** and (*ii*) **velocity or energy resonance**.

In type **(*i*)**, the amplitude of the forced vibration acquires a maximum value. Since the potential energy is proportional to the square of the amplitude, the system will have maximum potential energy under this condition. In type **(*ii*)**, the velocity is maximum. This corresponds to maximum energy transfer between the forcing system and the forced system and gives the system maximum kinetic energy. The two resonances occur at slightly different frequencies when there is appreciable damping.

Analytic treatment of Forced Vibration

Let a particle of mass m be displaced from its equilibrium position and released, it then experiences a restoring force $-\,Sx$, where S is the spring constant and x is the instantaneous displacement. If there were no other force, the mass would oscillate harmonically with its natural frequency $\omega_0 = \sqrt{\frac{S}{m}}$.

Under the influence of restoring force, the mass begins to move. If the instantaneous velocity is $\frac{dx}{dt}$, the mass also experiences a frictional force $-\,r\,\frac{dx}{dt}$ where r is the coefficient of damping force.

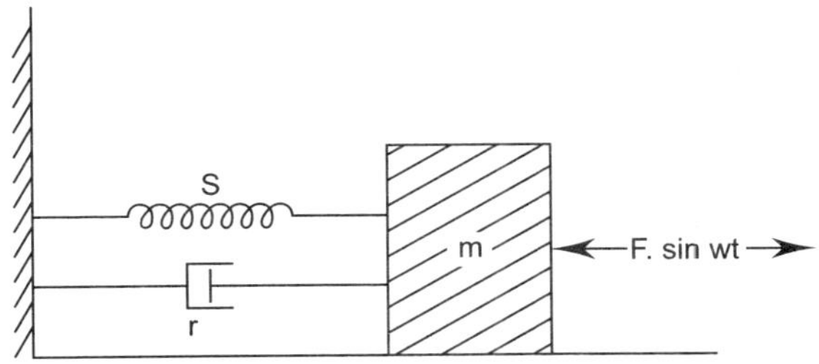

Figure 3A.8 *Mechanical forced oscillators with force F . sin ωt applied to damped oscillator.*

Finally the mass is subjected to a driving periodic force F sin ωt which opposes the restoring force as well as the frictional force and helps the motion.

Hence the equation of motion is

$$m \frac{d^2x}{dt^2} = -Sx - r\frac{dx}{dt} + F \sin \omega t$$

or, $$\frac{d^2x}{dt^2} + \frac{S}{m}x + \frac{r}{m}\frac{dx}{dt} = \frac{F}{m}\sin \omega t$$

or, $$\frac{d^2x}{dt^2} + 2k\frac{dx}{dt} + \omega_0^2 x = f \sin \omega t \qquad \text{...(3A.31)}$$

where $2k = \frac{r}{m}$, $\omega_0^2 = \frac{S}{m}$ and $f = \frac{F}{m}$. The quantity ω_0 is the undamped natural angular frequency of the particle and k is the decay constant.

Now let a value $x = x_1$ be found out which satisfies the equation

$$\frac{d^2x_1}{dt^2} + 2k\frac{dx_1}{dt} + \omega_0^2 x_1 = 0$$

and also let another value of $x = x_2$ be the particular solution of that equation, so that

$$\frac{d^2x_2}{dt^2} + 2k\frac{dx_2}{dt} + \omega_0^2 x_2 = f \sin \omega t \qquad \text{...(3A.32)}$$

$$\therefore \quad \frac{d^2}{dt^2}(x_1 + x_2) + 2k\frac{d}{dt}(x_1 + x_2) + \omega_0^2 (x_1 + x_2) = f \sin \omega t$$

The solution of x_1 is the same as that for the damped simple harmonic oscillation and the value of x_1 is

$$x_1 = A_1 e^{-kt} \sin\left(\sqrt{\omega_0^2 - k^2}\, t + \phi\right) \qquad \text{...(3A.33)}$$

where A_1 and ϕ are constants.

To find x_2, let us take a trial solution as $x_2 = A \sin(\omega t - \alpha)$

We can suppose this on the ground that the system will vibrate with the same frequency as that of the impressed sustained harmonic force.

Since $$x_2 = A \sin(\omega t - \alpha)$$

or $$\frac{dx_2}{dt} = A\omega \cos(\omega t - \alpha)$$

$\therefore$ $$\frac{d^2x_2}{dt^2} = -A\omega^2 \sin(\omega t - \alpha)$$

Putting this value in equation (3A.32), we get

$$-A\omega^2 \sin(\omega t - \alpha) + 2kA\omega \cos(\omega t - \alpha) + \omega_0^2 A \sin(\omega t - \alpha) = f \sin \omega t$$

or, $$A(\omega_0^2 - \omega^2) \sin(\omega t - \alpha) + 2kA\omega \cos(\omega t - \alpha)$$
$$= f \sin(\omega t - \alpha + \alpha)$$
$$= f \sin(\omega t - \alpha) \cos \alpha + f \sin \alpha \cos(\omega t - \alpha)$$

Since the above equation is true for all values of t. So we can equate the co-efficient of sin $(\omega t - \alpha)$ and cos $(\omega t - \alpha)$ from both sides of the equation.

$\therefore$ $$f \cos \alpha = A(\omega_0^2 - \omega^2)$$

and $$f \sin \alpha = 2kA\omega \qquad \text{...(3A.34)}$$

Squaring and adding, we have

$$f^2 = A^2 \{(\omega_0^2 - \omega^2)^2 + 4k^2\omega^2\}$$

$\therefore$ $$A = \frac{f}{\sqrt{(\omega_0^2 - \omega^2)^2 + 4k^2\omega^2}} \qquad \text{...(3A.35)}$$

and $$\tan \alpha = \frac{2k\omega}{\omega_0^2 - \omega^2} \qquad \text{...(3A.36)}$$

Thus the complete solution is

$$x = x_1 + x_2$$

$$= A_1 e^{-kt} \sin\left(\sqrt{\omega_0^2 - k^2}\, t + \phi\right) + \frac{f \sin(\omega t - \alpha)}{\sqrt{(\omega_0^2 - \omega^2)^2 + 4k^2\omega^2}} \qquad \text{...(3A.37)}$$

The first part of the solution for x, *i.e.*, x_1 represents a free oscillation with natural frequency set up in a damped system by the impressed force and the amplitude is decaying at a rate determined by the damping factor k. The second part represents the forced vibration *i.e.*, the response of the particle to the external impressed force or driving force which maintains the vibration. The inclusion of the first term agrees with the observation of an apparent initial irregularity of motion. At the initial stage, the effects of both the terms are present. After a lapse of time, the natural motion dies down due to the factor e^{-kt}, *i.e.*, the first term reduces to zero and we are left with the steady state motion represented by the second part. If $\sqrt{\omega_0^2 - k^2}$ and ω are nearly equal, the natural vibration will interfere with the forced vibration of the initial stage and produce beats. Such beats are called **transient beats.**

Transients appear both when the driving force is applied and when it is removed. Practically for musical instruments, they occur every time a note is started or stopped.

3A.1.10 Velocity Resonance

The motion of a particle in the steady state under the action of a periodic force is given by

$$x = A \sin(\omega t - \alpha)$$

where $$A = \frac{f}{\sqrt{\left(\omega_0^2 - \omega^2\right)^2 + 4k^2\omega^2}} \text{ and } \tan\alpha = \frac{2k\omega}{\omega_0^2 - \omega^2}$$

$\therefore$ $$\frac{dx}{dt} = A\omega \cos(\omega t - \alpha)$$

Hence the kinetic energy of the forced system at any instant is

$$\frac{1}{2} m \left(\frac{dx}{dt}\right)^2 = \frac{1}{2} mA^2 \omega^2 \cos^2(\omega t - \alpha).$$

Since the motion is steady harmonic motion, the total energy of the system at any instant is equal to the maximum kinetic energy (K.E.).

$\therefore$ The total energy of the system, E = Maximum K.E.

$$= \frac{1}{2} mA^2 \omega^2$$

$$= \frac{1}{2} \frac{mf^2 \omega^2}{\left(\omega_0^2 - \omega^2\right)^2 + 4k^2\omega^2}$$

$$= \frac{1}{2} \frac{mf^2}{\omega_0^2 \left(\frac{\omega_0}{\omega} - \frac{\omega}{\omega_0}\right)^2 + 4k^2}$$

$$= \frac{1}{2} \frac{mf^2}{\omega_0^2 \Delta^2 + 4k^2} \quad \text{...(3A.38)}$$

where, $$\Delta = \frac{\omega_0}{\omega} - \frac{\omega}{\omega_0}$$

If $\omega = \omega_0$, $\Delta = 0$, the energy of the system is maximum for any given value of k. Thus, when the frequency of the forcing system coincides with the natural frequency of the forced system (without damping), the energy of the forced system is maximum. This phenomenon is known as velocity resonance or energy resonance or simply resonance.

Energy at resonance,

$$E_{max} = \frac{1}{2}\frac{mf^2}{4k^2} \qquad \ldots(3A.39)$$

It is clear from equation (3A.38) that the decrease in energy due to mistuning *i.e.*, lack of coincidence between ω and ω_0 is same for a given ratio of frequencies whether the frequency of the forcing system is smaller or greater than that of the forced system. Since the changes of the ratio to its reciprocal simply changes the algebraic sign of Δ, leaving its numerical value unchanged therefore, its square is unchanged. If we plot E with Δ, we shall get response curves for different values of k as given in figure 3A.9. The curves will be symmetrical about $\Delta = 0$ with its maximum value at $\Delta = 0$, at $\omega = \omega_0$. From equation (3A.38) it indicates that for smaller value of k, greater will be the value of E for any particular value of Δ.

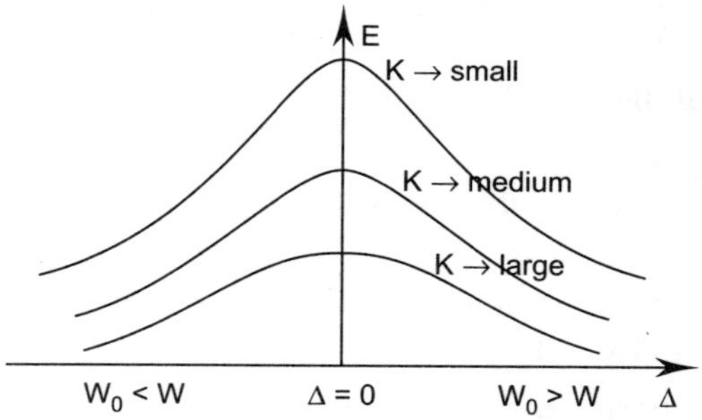

Figure 3A.9

3A.1.11 Amplitude Resonance

Let us find the condition when the amplitude of vibration of the driven system is greatest. The amplitude of forced vibration is

$$A = \frac{f}{\sqrt{\left(\omega_0^2 - \omega^2\right)^2 + 4k^2\omega^2}} \quad \text{[for a given system } \omega_0 \text{ and } k \text{ are constants]}$$

$$\therefore \quad \frac{dA}{d\omega} = -\frac{f}{2} \cdot \frac{2\left(\omega_0^2 - \omega^2\right).(-2\omega) + 8k^2\omega}{\left\{\left(\omega_0^2 - \omega^2\right)^2 + 4k^2\omega^2\right\}^{\frac{3}{2}}}$$

$$= f\omega \,.\, \frac{2\left(\omega_0^2 - \omega^2\right) - 4k^2}{\left[\left(\omega_0^2 - \omega^2\right)^2 + 4k^2\omega^2\right]^{\frac{3}{2}}}$$

for maximum amplitude $\frac{dA}{d\omega} = 0$

$$\therefore \quad \omega_0^2 - \omega^2 = 2k^2$$

or $\omega^2 = \omega_0^2 - 2k^2$, or $\omega = \omega_0\left(1 - \frac{2k^2}{\omega_0^2}\right)^{\frac{1}{2}} = \omega_0 - \frac{k^2}{\omega_0}$ [If k is small]

Putting this value of ω^2, we get the maximum displacement amplitude

$$A_{max} = \frac{f}{\sqrt{4k^4 + 4k^2\left(\omega_0^2 - 2k^2\right)}}$$

$$= \frac{f}{2k\sqrt{\omega_0^2 - k^2}}$$

Thus the angular frequency for amplitude resonance is slightly smaller than that at velocity or energy resonance.

Now for any value of damping the amplitude

$$A = \frac{f}{\omega_0^2} \text{ at } \omega = 0.$$

The relation between the angular frequency ω of the forcing system and the amplitude of oscillation of the driven at different damping constants is explained by the curve drawn in figure 3A.10. Now if k is very large, the increase in $2k\omega$ as ω increases towards ω_0 is larger than the decrease in the value of $\omega_0^2 - \omega^2$ and thus the denominator in the expression for A increases with increase of ω. Hence in these cases A will be maximum at $\omega = \omega_0$. But if k is comparatively small, the maximum amplitude occurs at $\omega^2 = \omega_0^2 - 2k^2$, *i.e.*, an angular frequency smaller than the natural angular frequency by an amount determined by the value of k. For a very small value of k, ω is almost equal to ω_0. If damping is absent, the amplitude becomes infinite at $\omega = \omega_0$, the resonance frequency.

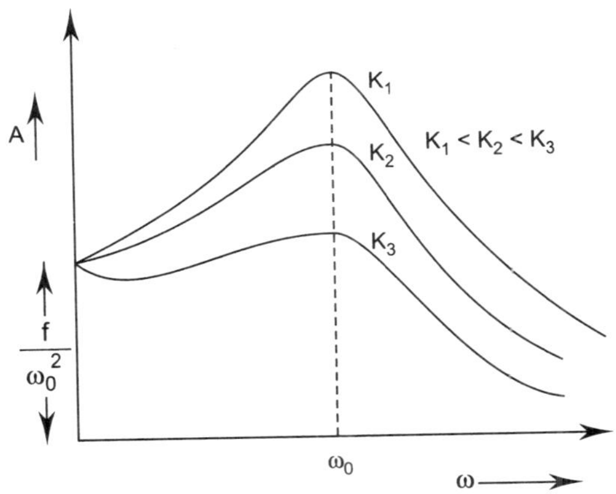

Figure 3A.10

3A.1.12 Sharpness of Resonance

We have noticed that the amplitude A of the forced vibration is controlled by the factor $\omega_0^2 - \omega^2$. If we draw a graph between $\omega_0^2 - \omega^2$ along x-axis and amplitude A along y-axis, we get different curves for different values of damping coefficient k (figure 3A.11).

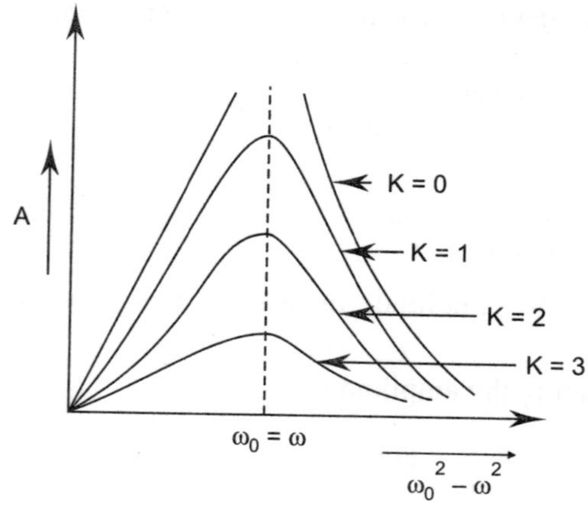

Figure 3A.11

It is clear that A is maximum when $\omega_0^2 - \omega^2 = 0$, *i.e.*, when $\omega_0 = \omega$, the natural frequency of the oscillation coincides with the frequency of the applied force and all the curves are symmetrical about this point.

(*a*) When $k = 0$, *i.e.*, there is no damping, *i.e.*, amplitude becomes infinite, which is not practically possible.

(*b*) The maximum value of A is different for different values of *k*, diminishing as the value of *k* gradually increases.

(*c*) For small values of *k*, the fall in the curves about $\omega_0^2 - \omega^2 = 0$ is steeper than for large values indicating that smaller the values of *k*, greater is the departure of the amplitude of the forced vibration from the maximum value and vice-versa. Thus, in the case of high damping (say $k = 3$) the amplitude remains more or less at its peak value over an appreciable range. The resonance in this case, is therefore said to be flat. So here it is difficult to find the exact position of resonance.

On the other hand if the damping is low (say $k = 1$) the steep fall of the curve on either side of the point ω_0 suggests that the oscillator responds only to the frequency exactly equal to the natural frequency ω_0 and to no other. This is sharp resonance.

3A.1.13 Mechanical and Electrical Analogues

Previously we have discussed about the resonant behaviour of a simple mechanical system consisting of a mass and a spring that is driven harmonically by the external periodic force. Now we shall discuss similarity between mechanical system and electrical system. One of the most popular and very important resonant system is an electrical system consisting of a capacitance (C), an inductance (L) and a resistance (R) which is shown in figure 3A.12.

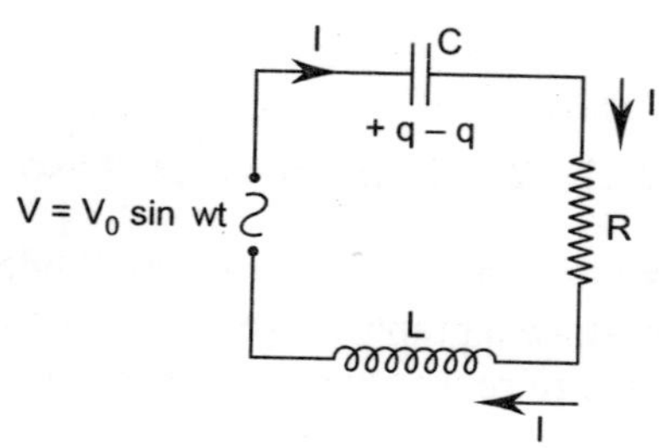

Figure 3A.12

Here $V = V_0 \sin \omega t$ is applied voltage. It is distributed among the three components as,

$$V_C = \frac{q}{C},\ V_R = IR \text{ and } V_L = L\frac{dI}{dt}$$

Here, $$V_L + V_R + V_C = V$$

i.e., $$L\frac{dI}{dt} + IR + \frac{q}{C} = V_0 \sin \omega t$$

Since $$I = \frac{dq}{dt}$$

i.e., $$L\frac{d^2q}{dt^2} + R\frac{dq}{dt} + \frac{q}{C} = V_0 \sin \omega t$$

or, $$\frac{d^2q}{dt^2} + \frac{R}{L}\frac{dq}{dt} + \frac{q}{CL} = \frac{V_0}{L}\sin \omega t \quad \text{...(3A.40)}$$

Equation (3A.40) is similar to equation (3A.31) of motion given in Art. 3A.1.9

$$\frac{d^2x}{dt^2} + 2k\frac{dx}{dt} + \omega_0^2 x = f \sin \omega t$$

where $$2k = \frac{r}{m},\ \omega_0^2 = \frac{S}{m} \text{ and } f = \frac{F}{m}$$

It seems to be seen that the mass m is analogous to self inductance L. r to the electrical resistance R, compliance $\frac{1}{S}$ to the electrical capacitance C, the force F to the voltage V_0, the displacement x to the charge q, and the velocity $\frac{dx}{dt}$ to the electrical current $I\left(=\frac{dq}{dt}\right)$.

Here the mechanical quantities and electrical analogues are given below :

Mechanical Quantities	*Electrical Analogues*
Displacement (x)	Charge (q)
Velocity $\left(\frac{dx}{dt}\right)$	Current $\left(I = \frac{dq}{dt}\right)$
mass (m)	Inductance (L)
Force (F)	Voltage (V_0)
Damping constant (r)	Resistance (R)
Stiffness constant (S)	Reciprocal of capacitance $\left(\frac{1}{C}\right)$
Quality factor, $Q = \frac{\omega' m}{r}$	Quality factor $Q = \frac{\omega' L}{R}$
For mechanical oscillator damping co-efficient $k = \frac{r}{2m}$	damping co-efficient, $k = \frac{R}{2L}$
Mechanical Impedance $Z = \sqrt{\left(\frac{S}{\omega} - m\omega\right)^2 + r^2}$	Impedance (Z) $= \sqrt{\left(\frac{1}{\omega C} - \omega L\right)^2 + R^2}$
Relaxation time $\left(\frac{1}{k}\right) = \frac{2m}{r}$	Relaxation time $\left(\frac{1}{k}\right) = \frac{2L}{R}$
Potential energy $= \frac{1}{2}Sx^2$	Electrostatic energy $= \frac{q^2}{2C}$
Kinetic energy of moving mass $= \frac{1}{2}m\dot{x}^2$	Electromagnetic energy of moving charge $= \frac{1}{2}L\dot{q}^2$

3A.1.14 Progressive Wave Equation and its Differential Form

A wave propagating from one point to another in a medium without being subjected to any boundary condition is termed as progressive wave. A progressive wave has the following characteristics :

(*i*) The waves are *produced by the periodic vibrations of the particles* in the medium.

(*ii*) Each particle of the medium *oscillates* about its mean position with *the same amplitude and period i.e., the vibrations are simple harmonic.*

(*iii*) The phase difference between two vibrating particles on the line of propagation is proportional to the distance between the particles.

(*iv*) The medium as a whole does not move along with the wave. Only the energy is carried in the direction of propagation of the wave.

(*v*) The wave velocity in a given medium is a constant determined by the density and elasticity of the medium.

However, the velocity of the vibrating particle goes on changing – maximum in the mean position and minimum at the two extreme points.

Equation of a Progressive Wave

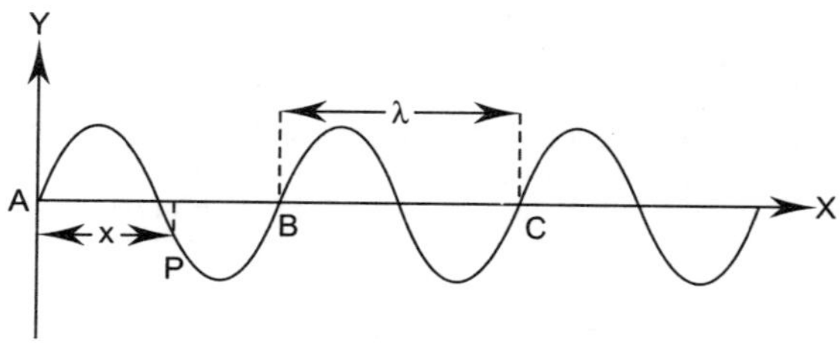

Figure 3A.13

Let a transverse wave in a string travel along the +ve x direction from left to right starting from the point A. A particle on the right of A will start vibrating a time later than one on the left.

As each particle executes SHM, the equation of motion of any particle, say A, is given by

$$y = a \sin \omega t$$

where, a is the amplitude, y the displacement at time t and ω the angular frequency.

$\therefore$ $$\omega = 2\pi n = \frac{2\pi}{T}, \qquad n = \text{frequency of vibration}$$
$$T = \text{Time period}$$

$\therefore$ $$y = a \sin 2\pi n t$$

when A is in its mean position of vibration, if the particles B, C etc., are also at their mean positions then A, B, C etc., are in the same state of vibration i.e., in same phase.

The distance BC between two consecutive particles in the same phase is known as the wavelength λ. The phase changes by 2π in going from A to B or from B to C etc.

So, in going from A to a point P at a distance x, the phase change would be

$$\varepsilon = \frac{2\pi}{\lambda} x \quad \text{[as in covering a distance } \lambda \text{, the phase change is } 2\pi]$$

So, the displacement of P is given by

$$y = a \sin(\omega t - \varepsilon)$$

$$= a \sin\left(2\pi n t - \frac{2\pi}{\lambda}x\right)$$

$$= a \sin\left(2\pi . \frac{V}{\lambda} . t - \frac{2\pi}{\lambda}x\right) \qquad [\because \text{ the wave velocity } V = n\lambda]$$

$$a = \sin\frac{2\pi}{\lambda}(Vt - x) \qquad \text{...(3A.41)}$$

This represents a general equation giving the displacement of any particle at a distant x from the fixed point A at any time t.

So this is the general equation of a progressive wave.

Equation (3A.41) may also be written as

$$y = a \sin\left(\frac{2\pi V}{\lambda}t - \frac{2\pi}{\lambda}x\right)$$

$$= a \sin(\omega t - kx) \qquad \text{...(3A.42)}$$

where, $\omega = \frac{2\pi}{T} = 2\pi n = \frac{2\pi V}{\lambda}$ and $k = \frac{2\pi}{\lambda}$ = propagation constant.

Differential Form

Now we have

$$y = a \sin\frac{2\pi}{\lambda}(Vt - x) \qquad \text{...(3A.43)}$$

here we have three variables y, x and t

Differentiating (3A.43) w.r.t. t we get

$$\frac{dy}{dt} = \frac{2\pi V a}{\lambda}\cos\frac{2\pi}{\lambda}(Vt - x)$$

$$\therefore \quad \frac{d^2y}{dt^2} = -\left(\frac{2\pi V}{\lambda}\right)^2 a \sin\frac{2\pi}{\lambda}(Vt - x)$$

$$= -\frac{4\pi^2V^2}{\lambda^2}y \qquad \text{...(3A.44)}$$

Now differentiating (3A.43) with respect to x we get

$$\frac{dy}{dx} = -a . \frac{2\pi}{\lambda}\cos\frac{2\pi}{\lambda}(Vt - x)$$

$$\therefore \quad \frac{d^2y}{dx^2} = -a . \left(\frac{2\pi}{\lambda}\right)^2 \sin\frac{2\pi}{\lambda}(Vt - x)$$

$$= -\frac{4\pi^2}{\lambda^2} . y \qquad \text{...(3A.45)}$$

Comparing (3A.44) and (3A.45), we get

$$\frac{d^2y}{dt^2} = V^2 \frac{d^2y}{dx^2}$$

This is the differential equation of a progressive wave, travelling with a velocity V.

3A.1.15 Difference between Elastic (Mechanical) Wave and Electromagnetic Wave

In the case of an elastic wave, there is some quantity which changes its value with time and space. The oscillations of the particle in the medium are similar but a phase difference exists between the oscillations of the different particles. In this case, waves propagate from one point to another in a material medium due to elasticity and are called elastic waves. Sound waves are elastic waves. Indeed, there are two types of wave motion: (*a*) **transverse** and (*b*) **longitudinal.** Transverse wave motion is that wave motion in which the particles of the medium vibrate about their mean positions in a direction at right angles to the direction of propagation of the wave.

The wave propagation along a stretched string is the example of transverse wave motion.

When the particles of the medium vibrate parallel to the direction of propagation of the wave, it is termed as longitudinal waves. Sound waves in gases, liquids and solids are examples of longitudinal waves.

The electromagnetic waves are transverse in nature. These waves need no material medium to travel. They can propagate in vacuum. It can travel very very faster than elastic waves. The electric field vector $\left(\vec{E}\right)$ and the magnetic field vector are mutually perpendicular to each other as well as perpendicular to the direction of propagation of the electromagnetic waves. If the wave propagates along Z direction and $\left(\vec{E}\right)$ is directed along X-direction, then $\vec{H}$ is directed along Y-direction. If $\vec{E}$ is directed along Y direction, $\vec{H}$ is directed along X direction. Light, X-rays, γ-rays are the examples of electromagnetic waves.

IN FOCUS

- **Simple Harmonic Motion**

  Motion of the Linear Harmonic Oscillator.

$$\frac{d^2y}{dt^2} + \omega^2 y = 0$$

where y is the displacement of the particle, ω is the angular frequency

Solving the above, the displacement equation is

$$y = A \sin(\omega t + \theta)$$

where $$\omega = \frac{2\pi}{T}$$ [T is the periodic time]

A : maximum displacement of the body from mean position or the amplitude of oscillation.

θ : initial phase

$$T = \frac{2\pi}{\omega} = 2\pi\sqrt{\frac{m}{S}} = 2\pi\sqrt{\frac{\text{Mass of the moving particle}}{\text{Force per unit displacement}}}$$

$$\text{velocity } V = a\,\omega \cos(\omega t + \theta)$$

$$\text{Acceleration} = -a\,\omega^2 \sin(\omega t + \theta + \pi)$$

- **Total energy of a Harmonic Oscillator**

Kinetic energy : $\frac{1}{2} ma^2 \omega^2 \cos^2(\omega t + \theta)$

Potential energy : $\frac{1}{2} m\omega^2 a^2 \sin^2(\omega t + \theta)$

Total energy E = K.E. + P.E. = $\frac{1}{2} m\omega^2 a^2$

- **Superposition of two linear SHM:** Let two simple harmonic vibrations of angular frequency ω, initial phases ϕ_1 and ϕ_2 and amplitudes a_1 and a_2 be given by

$$x_1 = a_1 \sin(\omega t + \phi_1)$$
$$x_2 = a_2 \sin(\omega t + \phi_2)$$

The resultant displacement

$$x = R \sin(\omega t + \theta)$$

θ : initial phase

R : amplitude of vibration

$$R^2 = a_1^2 + a_2^2 + 2a_1a_2 \cos\phi$$

$$\text{phase difference } \phi = \phi_2 - \phi_1 = \frac{a_1 \sin\phi_1 + a_2 \sin\phi_2}{a_1 \cos\phi_1 + a_2 \cos\phi_2}$$

Expression for resultant motion

$$\frac{x^2}{a^2} + \frac{y^2}{b^2} - \frac{2xy}{ab}\cos\phi = \sin^2\phi$$

where a is the amplitude of vibration along X-axis and b is the amplitude of vibration along Y-axis.

Different Cases :

(*i*) when $\phi = 0$, then

$$\left(\frac{x}{a} - \frac{y}{b}\right)^2 = 0$$

represents, a pair of coincident straight line

(*ii*) when $\phi = \frac{\pi}{4}$, then

$$\frac{x^2}{a^2} + \frac{y^2}{b^2} - \frac{\sqrt{2xy}}{ab} = \frac{1}{2}$$

represents oblique ellipse

(*iii*) when $\phi = \frac{\pi}{2}$, and $a \neq b$ then

$$\frac{x^2}{a^2} + \frac{y^2}{b^2} = 1$$

represents a symmetrical ellipse.

If $\phi = \frac{\pi}{2}$ and $a = b$ then

$$x^2 + y^2 = a^2$$

represents circular vibration.

(*iv*) when $\phi = \frac{3\pi}{4}$, then

$$\frac{x^2}{a^2} + \frac{y^2}{b^2} + \frac{\sqrt{2xy}}{ab} = \frac{1}{2}$$

again represents an oblique ellipse

(*v*) when $\phi = \pi$ and 2π the whole cycle is repeated in reverse order.

■ **Damped Vibration:** The equation of motion of particle executing damped vibration :

$$m\frac{d^2x}{dt^2} = -r\frac{dx}{dt} - Sx$$

where m : mass

r : co-efficient of damping force

S : stiffness constant

Solving the equation, we get

$$x = e^{-\frac{r}{2m}t}\left[A_1 e^{\sqrt{\frac{r^2}{4m^2} - \frac{S}{m}}\,t} + A_2 e^{-\sqrt{\frac{r^2}{4m^2} - \frac{S}{m}}\,t}\right]$$

This is the displacement equation of mass m executing damped harmonic motion.

Case I : when $k > \omega_0$

$$x = e^{-kt}\,[\cosh pt + B \sinh pt]$$

Now if $x = 0$, at $t = 0$, then A = 0.

$$\therefore \quad x = e^{-\frac{r}{2m}t} \sinh\left(\sqrt{\frac{r^2}{4m^2} - \frac{S}{m}}\right)t$$

This equation represents a periodic or non oscillating motion.

Case II : when $k = \omega_0$: Critical damped motion

$$x = e^{-kt}\,(A + Bt)$$

where $A = A_1 + A_2$ and $B = \delta(A_1 - A_2)$, $\delta = \sqrt{k^2 - \omega_0^2}$

Case III : when $k < \omega_0$: light damping

$$x = A \sin\left(\sqrt{\omega_0^2 - k^2}\; t + \phi\right)$$

where $A = \dfrac{x_0\,\omega_0}{\sqrt{\omega_0^2 - k^2}}\,e^{-kt}$

- **Logarithmic Decrement:** It is defined as the natural logarithm of the ratio between two successive amplitudes on the same side of the mean position of a damped oscillation separated by one full period.

$$\lambda = kT = \frac{r}{2m}T$$

where r is the damping constant, m is the mass of the oscillator, T is the time period.

- **Quality Factor:** It measures the rate of decay of energy of damped harmonic oscillator.

$$\therefore \quad \text{Quality factor } Q = \sqrt{\frac{Sm}{r}}$$

- **Forced Vibration:** The equation of motion is

$$m\frac{d^2x}{dt^2} = -Sx - r\frac{dx}{dt} + F\sin\omega t$$

where S : spring constant, r : co-efficient of damping force; F sin ωt : periodic force.

The complete solution

$$x = A_1 e^{-kt}\sin\left(\sqrt{\omega_0^2 - k^2}\,t + \phi\right) + \frac{f\sin(\omega t - \alpha)}{\sqrt{\left(\omega_0^2 - \omega^2\right)^2 + 4k^2\omega^2}}$$

The first part of the solution represents a free oscillation with natural frequency and the second part represents the forced vibration.

- **Velocity Resonance or Energy Resonance**

$$E_{max} = \frac{1}{2}\frac{mf^2}{4k^2}$$

- **Amplitude resonance**

$$A_{max} = \frac{f}{2k\sqrt{\omega_0^2 - k^2}}$$

- **Equation of Progressive Wave**

$$y = a\sin(\omega t - kx)$$

where $\omega = \dfrac{2\pi}{T} = 2\pi n = \dfrac{2\pi V}{\lambda}$ $\quad [V = n\lambda]$

$$k = \frac{2\pi}{\lambda}$$

In differential form $\dfrac{d^2y}{dt^2} = V^2\dfrac{d^2y}{dx^2}$

Solved Problems on Simple Harmonic Motion

Problem 1. *(a) An oscillatory motion of a body is represented by* $y = a\,e^{j\omega t}$ *where the symbols have their usual meaning. Show that the motion is simple harmonic.* [Panjab University. 1995]

(b) The displacement of a moving particle at any time is given by $y = a \cos \omega t + b \sin \omega t$. *Show that the motion is simple harmonic.*

Solution. (*a*) Given $y = a\,e^{j\omega t}$

Differentiating with respect to '*t*' ; we get $\dfrac{dy}{dt} = aj\omega e^{j\omega t}$

Again differentiating,
$$\frac{d^2y}{dt^2} = -a\omega^2 e^{j\omega t}$$
$$= -\omega^2 y \qquad [j^2 = -1]$$

i.e.,
$$\frac{d^2y}{dt^2} + \omega^2 y = 0$$

This is the differential equation of SHM,

So $y = ae^{j\omega t}$ represents SHM

(*b*) $y = a \cos \omega t + b \sin \omega t$

Differentiating with respect to '*t*', we get

$$\frac{dy}{dt} = -a\omega \sin \omega t + b\omega \cos \omega t$$

$$\frac{d^2y}{dt^2} = -a\omega^2 \cos \omega t - b\omega^2 \sin \omega t$$
$$= -\omega^2 (a \cos \omega t + b \sin \omega t)$$
$$= -\omega^2 y$$

or,
$$\frac{d^2y}{dt^2} + \omega^2 y = 0$$

So, $y = a \cos \omega t + b \sin \omega t$ is the equation of SHM

Problem 2. *A particle of mass 10 gm, is placed in a potential field given by* $V = (50x^2 + 100)$ *erg/gm. Calculate the frequency of oscillation.*

[Agra University, Bundelkhand University, Rajasthan University]

Solution. Given potential field

$$V = (50x^2 + 100) \text{ erg./gm.}$$

$$\text{Potential energy } U = 10\,(50x^2 + 100) \text{ erg.}$$

The force acting upon the particle is given by

$$F = -\frac{dU}{dx} = -1000x \text{ dyne}$$

By Newton's law
$$F = m\frac{d^2x}{dt^2} = -1000x$$

$$\frac{d^2x}{dt^2} = -100x \qquad [\because m = 10 \text{ gm}]$$

i.e., acceleration is proportional to its displacement.

We know for SHM $$\frac{d^2x}{dt^2} = -\omega^2 x$$

i.e.,. $$\omega^2 = 100, \omega = 10 \; [\omega = \text{angular frequency}, f = \text{frequency}]$$

or, $$\omega = \frac{2\pi}{T} = 2\pi f = 10$$

$$f = \frac{10}{2\pi} = 1.58/\text{sec.}$$

Problem 3. *A body executing SHM has velocities 80 cm/sec and 60 cm/sec when displacements are 3 cm and 4 cm respectively. Calculate the amplitude of vibration and the time taken to travel 2.5 cm from the +ve extremity of the oscillation.* [Pune University]

Solution. Velocity $u = \omega\sqrt{a^2 - x^2}$

Where ω = angular velocity, a = amplitude and x = displacement from the mean position.

$$\therefore \quad 80 \text{ cm/sec} = \omega\sqrt{a^2 - 3^2} \qquad \ldots(i)$$

$$60 \text{ cm/sec} = \omega\sqrt{a^2 - 4^2} \qquad \ldots(ii)$$

$$\therefore \quad \frac{80}{60} = \frac{\omega\sqrt{a^2 - 3^2}}{\omega\sqrt{a^2 - 4^2}} \text{ or, } \left(\frac{4}{3}\right)^2 = \frac{a^2 - 9}{a^2 - 16}$$

or, $$a = 5 \text{ cm}$$

Substituting $a = 5$ cm in (i) we get

$$80 = \omega\sqrt{5^2 - 3^2} \text{ or, } \omega = 20 \text{ rad/sec}$$

Now $$x = a \sin \omega t$$

$\therefore$ Time taken to reach the +ve extremity is given by

$$5 = 5 \sin \omega t$$

or, $$\omega t = \frac{\pi}{2}$$

or, $$t = \frac{\pi}{2\omega} = \frac{\pi}{40} \text{ sec}$$

Distance of the point 2.5 cm from +ve extremity, from the mean position = 5 – 2.5 = 2.5 cm

$$2.5 = 5 \sin \omega t,$$

or, $$\sin \omega t = \frac{1}{2} = \sin \frac{\pi}{6}$$

or, $$t = \frac{\pi}{6\omega} = \frac{\pi}{120} \text{ sec}$$

Time taken to travel from 2.5 cm from +ve extremity $= \dfrac{\pi}{40} - \dfrac{\pi}{120} = \dfrac{\pi}{60} = 0.052$ sec.

Problem 4. *A tunnel is dug across the earth passing through its centre. If a body of mass 1 kg is dropped into it, will it come out at the other end ? If not, then what type of motion does it execute ? If the radius of earth is 6400 km and g on the surface of the earth is 10 m/s², then calculate the possible time period of the body ?* [C.U. 1980, G.N.D.U.'1994, P.U.'1991, Vidya Sagar University.'92]

Solution. Let R be the radius of the earth. Let a particle of mass m is at a point P at any instant of time t in the cavity and at a distance r from the center of earth and x from the center c of the cavity.

Now the force acting on the particle at P along the center of earth is

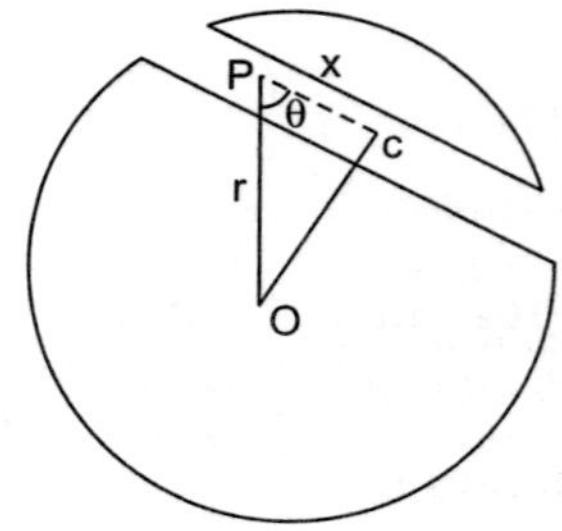

$$\frac{G \cdot 4\pi r^3 \rho m}{3r^2} \text{ along PO}$$

$$= \frac{4}{3}\pi G r m \rho$$

where ρ : average density of earth's material

Component of this force along the center (c) of the cavity is

$$\frac{4}{3} G\pi\, r m \rho \cdot \cos\theta$$

$$= \frac{4}{3} G\pi\, r m \rho \cdot \frac{x}{r}$$

$$= \frac{4}{3} G\pi\, m \rho x$$

∴ Equation of motion of the particle is given by

$$\frac{m d^2 x}{dt^2} = -\frac{4}{3} G\pi\, m\rho x$$

or,
$$\frac{d^2 x}{dt^2} + \frac{4}{3} G\pi\, \rho x = 0$$

This is the equation of a simple harmonic motion, the frequency of oscillation of which is

$$\omega^2 = \frac{4}{3} G\pi\rho$$

or,
$$\omega = \sqrt{\frac{4}{3} G\pi\rho}$$

∴ Time period of oscillation is

$$T = \frac{2\pi}{\sqrt{\frac{4}{3} G\pi\rho}} = \frac{2\pi}{\sqrt{\frac{4}{3} G \cdot \pi \cdot \frac{3M}{4\pi R^3}}} \qquad \text{(M : mass of the earth)}$$

$$= 2\pi\sqrt{\frac{R^3}{GM}} = 2\pi\sqrt{\frac{R^3}{gR^2}} \qquad \left(\because g = \frac{GM}{R^2}\right)$$

$$= 2\pi\sqrt{\frac{R}{g}}$$

here $\quad R = 6400 \text{ km} = 6400 \times 10^3 \text{ m}$

$g = 10 \text{ m/sec}^2$

Time period will be

$$= 2 \times 3.14 \times \sqrt{\frac{6400 \times 10^3}{10}}$$

$$= 2 \times 3.14 \times 8 \times 10^2$$

$$= 5024 \text{ sec.}$$

$$= 83.7 \text{ min.}$$

The body will execute simple harmonic motion the time period of which is 83.7 min.

Problem 5. *Three simple harmonic motions of the same frequency act on a particle simultaneously in the same direction. Their amplitudes are 1.0, 1.5 and 2.0 cm respectively. The phase angle of the second with respect to the first is 60°, and that of the third with respect to the second is 30°. Obtain the resultant amplitude and phase angle relative to the first.* [Burd. University]

Solution. The resultant motion is

$$X = a_1 \cos(\omega t + \phi_1) + a_2 \cos(\omega t + \phi_2) + a_3 \cos(\omega t + \phi_3)$$

Let $\quad \phi_1 = 0$, then $\phi_2 = 60°$, $\phi_3 = 90°$

Here $\quad a_1 = 1.0$ cm, $a_2 = 1.5$ cm, $a_3 = 2.0$ cm.

$$X = \cos \omega t + 1.5 \cos(\omega t + 60°) + 2 \cos(\omega t + 90°)$$

$$= \cos \omega t + 1.5 \{\cos \omega t \cos 60° - \sin \omega t \sin 60°\} + 2 \{\cos \omega t \cos 90° - \sin \omega t \sin 90°\}$$

$$= \left(1 + \frac{3}{4}\right) \cos \omega t - \left(\frac{3\sqrt{3}}{4} + 2\right) \sin \omega t.$$

Let $\quad A \cos \phi = \dfrac{7}{4}$ and $A \sin \phi = \dfrac{8 + 3\sqrt{3}}{4}$

$\therefore \quad X = A \cos \phi \cos \omega t - A \sin \phi \sin \omega t = A \cos(\omega t + \phi)$

and

$$A = \sqrt{\left(\frac{7}{4}\right)^2 + \left(\frac{8 + 3\sqrt{3}}{4}\right)^2} = \sqrt{13.95} = 3.73 \text{ cm.}$$

$$\tan \phi = \frac{8 + 3\sqrt{3}}{7} = 1.885$$

or, $\quad \phi = 62°$

So the resultant amplitude is 3.73 cm and phase angle relative to the first is 62°.

Problem 6. *Prove that if at any instant of motion the Kinetic Energy of an SHM is suddenly increased by a blow in the ratio $(1 + \beta) : 1$, the amplitude of the swing is increased in the ratio*

$$\sqrt{a + \frac{\beta v^2}{a^2 \omega^2}} : 1$$

Solution. Before blow, K.E. of the particle $= \dfrac{1}{2} mv^2$ [m = mass of the particle, v = Velocity before the blow] and $$\text{P.E.} = \frac{1}{2} m\omega^2 x^2$$

$$\text{Total energy before the blow} = \frac{1}{2} mv^2 + \frac{1}{2} m\omega^2 x^2 = \frac{1}{2} ma^2\omega^2 \qquad (\because \quad v = \omega a)$$

where a is the initial amplitude

$$\text{K.E. after the blow} = \frac{1}{2} mv^2 (1 + \beta)$$

If A is the amplitude after the blow then

$$\text{Total energy after the blow} = \frac{1}{2} m\text{A}^2 \omega^2$$

$$= \frac{1}{2} mv^2 + \frac{1}{2} m\omega^2 x^2 + \frac{1}{2} mv^2 \beta$$

$$= \frac{1}{2} ma^2 \omega^2 + \frac{1}{2} mv^2 \beta$$

or, $$\frac{\text{A}^2}{a^2} = 1 + \frac{\beta v^2}{a^2\omega^2}$$

$$\text{A} : a = \sqrt{1 + \frac{\beta v^2}{a^2 \omega^2}} : 1$$

Problem 7. *Two simple harmonic motions act in the same direction simultaneously on a particle. The SHMs have the same frequency 5Hz ; amplitudes 0.005 m and 0.002 m respectively ; and the second component leads the first by 45°. Determine the amplitude of the resultant displacement and its phase with respect to the first component. Give the expression for the resultant displacement as a function of time.* [Burd. University]

Solution. The equations of the SHMs may be written as

$$x_1 = 0.005 \sin (2\pi.5)\, t = 0.005 \sin 10\pi t$$

$$x_2 = 0.002 \sin (10\pi t + 45°)$$

The resultant displacement will be

$$x = x_1 + x_2 = 0.005 \sin 10\pi t + 0.002 \sin \left(10\pi t + \frac{\pi}{4}\right)$$

$$x = 0.005 \sin 10\pi t + 0.002 \left\{\sin 10\pi t \,.\, \frac{1}{\sqrt{2}} + \frac{1}{\sqrt{2}} \cos 10\pi t\right\}$$

$$= 0.0064 \sin 10\pi t + 0.0014 \cos 10\pi t$$

$$= \text{A} \sin (10\pi t + \theta)$$

where A cos θ = 0.0064 and A sin θ = 0.0014

$$A = \sqrt{(0.0064)^2 + (0.0014)^2} = 0.00656 \text{ m}$$

and $$\tan\theta = \frac{(0.0014)}{(0.0064)} = 0.21875$$

∴ Phase difference $\theta = 12.33°$

∴ The resultant displacement $x = 0.00656 \sin[10\pi t + 12.33°]$

Problem 8. *The displacement of a simple harmonic oscillator is given by*

$$x = a \sin(wt + \phi)$$

If the oscillations started at time $t = 0$ from a position x_0 with velocity $\dot{x} = v_0$. Show that tan ϕ $= \frac{\omega x_0}{v_0}$ and $a = \left[x_0^2 + \frac{v_0^2}{\omega^2}\right]^{\frac{1}{2}}$

Solution. Given $x = a\sin(wt + \phi)$; at $t = 0$, $x = x_0$

$$x_0 = a \sin\phi$$

$$\frac{dx}{dt} = \dot{x} = a\,\omega\cos(\omega t + \phi) \text{ ; at } t = 0,\ \dot{x} = v_0$$

$$v_0 = a\,w\cos\phi$$

Hence, $$\tan\phi = \frac{\sin\phi}{\cos\phi} = \frac{x_0}{a} \times \frac{a\,\omega}{v_0} = \frac{\omega x_0}{v_0}$$

and $$a^2\sin^2\phi + a^2\cos^2\phi = x_0^2 + \frac{v_0^2}{\omega^2}$$

or, $$a = \left[x_0^2 + \frac{v_0^2}{\omega^2}\right]^{\frac{1}{2}}.$$

Problem 9. *What is the ratio of kinetic energy at displacement one fourth to one third of the amplitude in case of simple harmonic motion ?*

Solution. Suppose m is the mass of the particular executing S.H.M., a be the amplitude and ω be the angular velocity, then

$$\text{Total energy} = \frac{1}{2} m\omega^2 y^2$$

If the displacement is y, Potential energy $= \frac{1}{2} m\omega^2 y^2$.

$$\therefore \quad \text{Kinetic energy} = \text{Total energy} - \text{Potential energy}$$

$$= \frac{1}{2} m\omega^2 (a^2 - y^2)$$

At displacement $y = \frac{a}{4}$, kinetic energy $E_1 = \frac{1}{2} m\omega^2 \left(a^2 - \frac{a^2}{16}\right)$

At displacement $y = \frac{a}{3}$, kinetic energy $= \frac{15}{32} m\,\omega^2 a^2$

$$E_2 = \frac{8}{18} m\,\omega^2 a^2$$

Therefore, $$\frac{E_1}{E_2} = \frac{15}{32} \times \frac{18}{8} = \frac{135}{128} = 1.055.$$

Review Questions On Simple Harmonic Motion

1. Define simple harmonic motion. Derive a general differential equation of motion of a simple harmonic oscillator and obtain its various solutions.
[C.U. 1992, Vidya Sagar U. 1992, Burdwan U. 1992, Kan. U. 1991, 95]
2. Simple harmonic motion is called sinusoidal or co-sinusoidal. Justify. *[P.U. 90]*
[**Hint** : SHM can be represented by the relations
$$x = a \sin(\omega t + \phi)$$
or, $x = \cos(\omega t + \phi)$ This is why a SHM is called sinusoidal]
3. What are the dimensions of force constant of a vibrating spring? *[P.U. 1991]*
4. (*a*) Velocity of simple harmonic oscillator at any time t leads the displacement by a phase angle $\frac{\pi}{2}$ radian. Explain why ?
(*b*) Show that for the body executing simple harmonic motion the acceleration leads the velocity by $\frac{\pi}{2}$ and the displacement by π.
5. Derive an expression for the total energy of a harmonic oscillator and show that it is a constant and proportional to the square of the amplitude. *[P.U. - 1990, Lucknow U. - 1993]*
6. Discuss the formation of Lissajous figures by the superposition of two simple harmonic motions when (*i*) the periods are the same but the amplitudes and phases are different. (*ii*) The periods are in the ratio 1 : 2 and the amplitudes and phases are different. *[C.U. 1983, 91, 99]*
7. A particle is subjected simultaneously to two SHMs of the same period but of different amplitudes and phase in perpendicular directions. Find the expression for the resultant motion. For what conditions the paths may be a straight line, ellipse or circle? Discuss the different important cases.
[Vidya Sagar U. 1991, C.U. - 1991, Burdwan U. - 1991, Lucknow University 94, 91]
8. A particle under two simple harmonic motions are at right angles to each other. The two motions are represented by $x = a \sin \omega t$ and $y = b \sin(\omega t + \delta)$
(*i*) Show that the resultant motion is an ellipse given by $\frac{x^2}{a^2} + \frac{y^2}{b^2} - \frac{2xy}{ab} \cos\delta = \sin 2\delta$
(*ii*) What is the locus of the particle when $d = \frac{\delta}{2}$ and $a = b$. *[C.U. 2002]*

Additional Miscellaneous Problems on SHM

1. A particle of mass 4 gm lies in a potential well given by $V = (200x^2 + 150)$ erg/gm. Deduce the frequency of oscillations. *[Agra University 76]* [**Ans.** 3.18 sec^{-1} [Ref. Problem 2]]
2. A particle in SHM has velocities U_1 and U_2 when its displacements from the mean position are x_1 and x_2 respectively. Calculate the period, amplitude and maximum speed of the particle.
[Kanpur University 83, Rohilkhand University 81]

$$\left[\textbf{Ans. } 2\pi\sqrt{\frac{x_2^2 - x_1^2}{u_1^2 - u_2^2}}, \sqrt{\frac{x_2^2 u_1^2 - x_1^2 u_2^2}{u_1^2 - u_2^2}}, \sqrt{\frac{x_2^2 u_1^2 - x_1^2 u_2^2}{x_2^2 - x_1^2}}\right]$$

(Ref. Problem 3)

3. Calculate the displacement to amplitude ratio for a S.H.M. when the KE is 90% of the total energy. [*P.U. 1995*] [**Ans.** $x : a = 0.316 : 1$] (Ref. Problem 6)
4. What is the ratio of the kinetic energies at displacements one fourth to one third of the amplitudes in case of simple harmonic motion? [*P.U. 1995*] [**Ans.** $E_1 : E_2 = 135 : 128$]
5. A simple harmonic oscillator is characterised by $y = a \cos \omega t$. Calculate the displacement at which its kinetic energy is equal to its potential energy. [*P.U. 1996*] [**Ans.** $x = \pm \frac{a}{\sqrt{2}}$]
6. Two vibrations, at right angles to each other, are described by the equations $x = 3 \cos 4\pi t$ and $y = 3 \cos\left(8\pi t + \frac{\pi}{3}\right)$, where x and y are expressed in centimetres and t in seconds. Construct the Lissajous figure of the combined motion. [*Burd. University 1994*]

$$\left[\textbf{Ans. } x^2\left(1 - \frac{x^2}{9}\right) = \left[y^2 - \left(x^2 - \frac{9}{2}\right)\right]^2\right]$$

Solved Problems on Damped and Forced Vibrations

Problem 1. *A massless spring whose upper end is fixed to a rigid support carries a horizontal disc of mass 100 gm at the lower end. It is observed that the system oscillates with a frequency of 10 Hz and the amplitude of the damped oscillation reduces to half its undamped value in one minute. Calculate (i) the resistive force constant ; (ii) the relaxation time of the system ; (iii) Its quality factor and (iv) the force constant of the spring.* [Burd. University 95]

Solution. The angular frequency of the oscillation

$$\omega' = \sqrt{\omega_0^2 - k^2} = 2\pi f = 20\pi \text{ Hz}.$$

Let the undamped amplitude be A_0.

So according to condition, $A = A_0 e^{-kt}$

$$\frac{A_0}{2} = A_0 e^{-60k}$$

or

$$k = \frac{1}{60} \ln 2 \text{ nepers/sec.}$$

(*i*) The resistive force constant $r = 2\,k.m.$ $\left[k = \frac{r}{2m}\right]$

$$= 2 \times \frac{1}{60} \ln 2 \times 100$$

$$= \frac{10}{3} \ln 2 \text{ dyne sec/cm}$$

$$= 2.31 \text{ dyne sec/cm}$$

(*ii*) The relaxation time $\tau = \frac{1}{k} = \frac{60}{\ln 2}$ sec.

(*iii*) Quality factor $\qquad Q = \dfrac{m\omega'}{r}$ where $\omega' = \sqrt{\omega_0^2 - k^2}$

$$= \frac{100 \times 20\pi \times 3}{10 \ln 2} = \frac{600\pi}{\ln 2}$$

(*iv*) The force constant of spring $s = \omega_0^2\, m$

$$= m\left(\omega'^2 + k^2\right)$$

$$= 100\,[400\pi^2 + (2.31)^2]$$

$$= 3.95 \times 10^5 \text{ dyne/cm}^2.$$

Problem 2. *In a one dimensional motion of mass 10 gm, it is acted on by a restoring force 10 dyne/cm and a resisting force 2 dyne-sec/cm. Find (i) whether the motion is a periodic or oscillatory (ii) the resisting force per unit velocity which will make the motion critically damped; (iii) the mass for which the given forces will make the motion critically damped.* [C.U. 1990, 92]

Solution. (*i*) Here, mass m = 10 gm, restoring force S = 10 dyne/cm and resisting force r = 2 dyne.sec/cm.

Now, $k = \dfrac{r}{2m}$ or $2k = \dfrac{r}{m} = \dfrac{2}{10} = 0.2$ nepers/sec and $\omega_0^2 = \dfrac{s}{m} = \dfrac{10}{10} = 1$ rad^2/sec^2.

$$\therefore \qquad \frac{k^2}{\omega_0^2} = 0.01 < 1$$

Now,

If $k > \omega_0$, the motion is aperiodic or non-oscillatory.

If $k = \omega_0$, the motion is critically damped.

If $k < \omega_0$, the motion is oscillatory or SHM

Here, $k < \omega_0$, *i.e.*, the motion is oscillatory,

(*ii*) for critical damping, $k = \omega_0$

i.e., $\qquad k^2 = \omega_0^2$

$$\frac{r^2}{4m^2} = \frac{s}{m}.$$

Hence $\qquad r = 2\sqrt{ms}$

$$= 2\sqrt{10 \times 10}$$

$$= 20 \text{ dyne-sec/cm}$$

(*iii*) Here $\qquad m = \dfrac{r^2}{4s} = \dfrac{4}{4 \times 10} = 0.1$ gm.

Problem 3. *Show that the fractional change in the natural frequency of a damped simple harmonic oscillator is* $\dfrac{1}{8Q^2}$ *where Q is the quality factor.* [G.N.D.U, 1996; Pune Univ, 1996]

Solution. Let ω_0 be in the angular frequency of the undamped oscillator and ω' that of the damped S.H.M. oscillator.

Then $\omega' = \sqrt{\omega_0^2 - k^2}$ where $k = \dfrac{r}{2m}$ = damping co-efficient and r is the damping constant.

Hence
$$\omega' = \left(\omega_0^2 - \frac{r^2}{4m^2}\right)^{\frac{1}{2}} = \omega_0\left(1 - \frac{r^2}{4m^2\,\omega_0^2}\right)^{\frac{1}{2}}$$

If r is small then $\dfrac{r^2}{4m^2\,\omega_0^2}$ is very very small.

Therefore,
$$\omega' = \omega_0\left\{1 - \frac{1}{2}\cdot\frac{r^2}{4m^2\,\omega_0^2} + \ldots\right\}$$
$$\frac{\omega'}{\omega_0} = 1 - \frac{1}{2}\cdot\frac{r^2}{4m^2\,\omega_0^2}$$

or,
$$\frac{\omega_0 - \omega'}{\omega_0} = \frac{1}{8}\cdot\frac{r^2}{m^2\,\omega_0^2}$$

But we know, Quality factor $Q = \dfrac{m\,\omega_0}{r}$

$\dfrac{\omega_0 - \omega'}{\omega_0}$ = The fractional change in the natural frequency of the oscillator.

$$\frac{\omega_0 - \omega'}{\omega_0} = \frac{1}{8Q^2} \text{ (proved)}$$

Problem 4. *A mass of 1 kg is suspended from a spring of stiffness constant 25 N/m. If the undamped frequency is $\frac{2}{\sqrt{3}}$ times the damped frequency, calculate the damping factor.* [H.P.University 1991]

Solution. Damped frequency $f' = \dfrac{1}{2\pi}\sqrt{\dfrac{S}{m} - \dfrac{r^2}{4m^2}}$ and undamped frequency $f_0 = \dfrac{1}{2\pi}\sqrt{\dfrac{S}{m}}$.

Now
$$\frac{f_0}{f'} = \frac{2}{\sqrt{3}}$$

or,
$$\frac{\sqrt{\dfrac{s}{m}}}{\sqrt{\dfrac{s}{m} - \dfrac{r^2}{4m^2}}} = \frac{2}{\sqrt{3}}$$

or, $\quad r^2 = s\,m$

Here $\quad s = 25$ N/m, and $m = 1$ kg

$\therefore \quad r = 5$ kg/sec.

Problem 5. *A mass of 10 gm is acted on by a restoring force 5 dyne/cm and resisting force 2 dyne-sec/cm. Find whether the motion is aperiodic or oscillatory. Find the value of resisting force for which the motion is critically damped.* [C.U. - 1994, 1999]

Solution. Here $m = 10$ gm, resisting force, $r = 2$ dyne/cm and restoring force $S = 5$ dyne/cm.

Now $$k = \frac{r}{2m} = \frac{2}{2 \times 10} = 0.1 \text{ neper/sec.}$$

and $$\omega_0^2 = \frac{S}{m} = \frac{5}{10} = 0.5 \text{ rad}^2/\text{sec}^2$$

$$\therefore \quad \frac{k^2}{\omega_0^2} = \frac{0.01}{0.5} = \frac{1}{100} \times \frac{10}{5} = 0.02 < 1 \quad i.e., \; k^2 < \omega_0^2$$

So the motion is oscillatory.

For critically damped, $$k = \omega_0$$

i.e., $$k^2 = \omega_0^2$$

or, $$\frac{r^2}{4m^2} = \frac{S}{m}, \quad \text{or} \quad r = 2\sqrt{mS}$$

$$\therefore \quad r = 2\sqrt{10 \times 5} = 10\sqrt{2} \text{ dyne-sec/cm.}$$

Problem 6. *A massless spring of spring constant 10 N/m is suspended from a rigid support and carries a mass of 0.1 kg at its lower end. The system is subjected to a resistive force – pk, where p is a constant and k is the velocity. It is observed that the system performs damped oscillatory motion and its energy decays to $\frac{1}{e}$ of its initial value in 50 sec.*

(a) What is the value of p ?

(b) What is the Q value of the oscillator ?

Solution. (*a*) $m = 0.1$ kg. The decay of the damped oscillator is given by $E' = E_0 e^{-kt}$, where E_0 is the initial energy and $k = \frac{p}{m}$, p being the constant of the resistive force and $t = 50$ sec.

or, $$\frac{E'}{E} = \frac{1}{e}, \text{ or, } e^{-50k} = e^{-1}$$

or, $$k = \frac{1}{50} \text{ / sec.}$$

Hence, $$p = mk = 0.1 \times \frac{1}{50} \text{ kg/sec} = 2 \times 10^{-3} \text{ N-s/meter.}$$

Since dimensionally, $p = \frac{\text{Force}}{\text{Velocity}}$, the unit of p is newton second per metre

(*b*) Since $S = 10 \text{ Nm}^{-1}$, the angular frequency ω_0 in the absence of damping is

$$\omega_0 = \sqrt{\frac{S}{m}} = \sqrt{\frac{10}{0.1}} = 10 \text{ rad/sec.}$$

The angular frequency of damped oscillator is

$$\omega' = \sqrt{\omega_0^2 - \frac{k^2}{4}} = \sqrt{(10)^2 - 10^{-4}} \approx 10 \text{ rad/sec.}$$

The quality factor $$Q = \frac{\omega'}{k} = \frac{10}{\frac{1}{50}} = 500.$$

Problem 7. *The amplitude of an oscillator of frequency 200 cycles/sec falls to $\frac{1}{10}$ of its initial value after 2000 cycles. Calculate (i) its relaxation time (ii) its quality factor (iii) time in which its energy falls to $\frac{1}{10}$ of its initial value (iv) damping constant,* [Rajasthan University 86, 83].

Solution. The oscillator having a frequency 200 cycles/sec, completes 2000 cycles in 10 sec. The instantaneous amplitude of the (damped) oscillator is $A_0 e^{-kt}$, where k is the damping constant

$$\frac{A_0}{10} = A_0 e^{-kt} = A_0 \cdot e^{-10k}$$

$$10k = \ln 10 = 2.3 \log_{10} 10 = 2.3$$

$$k = \frac{2.3}{10} = 0.23$$

(*i*) Now, the relaxation time $\tau = \frac{1}{2k} = \frac{1}{0.23} \times \frac{1}{2} \text{ sec.} = \frac{1}{0.46} \text{ sec.}$

(*ii*) The quality factor is $$Q = \frac{m\omega_0}{r}, \qquad \because \quad k = \frac{r}{2m}$$

$$= \frac{\omega_0}{2k}, \qquad 2k = \frac{r}{m}$$

$$= \omega_0 \tau = 2\pi \times 200 \times \frac{1}{0.23 \times 2}$$

$$= 435$$

(*iii*) The energy of the oscillator

$$E = E_0 e^{-2kt}$$

$$\frac{E_0}{10} = E_0 \cdot e^{-2kt} \quad \text{or} \quad t = \frac{1}{2k} \ln 10$$

or $$t = \frac{1}{2 \times 0.23} \times 2.3 \log_{10} 10 = \frac{10}{2} \text{ sec} = 5 \text{ sec.}$$

(*iv*) The damping constant $k = 0.23$

Problem 8. *If the relaxation time of a damped harmonic oscillator is 50 second, find the time in which (i) the amplitude falls to $\frac{1}{e}$ times the initial value, (ii) energy of the system falls to $\frac{1}{e}$ times the initial value, (iii) energy falls to $\frac{1}{e^4}$ of the initial value.*

Solution. The equation of a damped harmonic oscillator is

$$x = ae^{-rt} \sin(\omega' t + \phi),$$

where r is a damping constant. The relaxation time is

$$\tau = \frac{1}{2r} = 50 \text{ second}$$

or,
$$r = \frac{1}{100} \text{ sec}^{-1}.$$

(*i*) The damped amplitude at any instant is ae^{-rt}. Let a_0 be the amplitude at $t = 0$, and $\frac{a_0}{e}$ at t.

Then we get

$$a_0 = a$$

and
$$\frac{a_0}{e} = ae^{-rt} = a_0 e^{-rt}$$

From this, we have

$$e^{-1} = e^{-rt}$$

∴
$$t = \frac{1}{r} = 100 \text{ second}.$$

(*ii*) The energy (or velocity) falls to $\frac{1}{e}$ times the initial value in the relaxation time. Thus, in this case

$$t = \tau = 50 \text{ second}.$$

(*iii*) The energy of a damped harmonic oscillator is given by

$$E = E_0 e^{-2rt}$$

where E_0 is the energy at $t = 0$. Let t' be the time during which the energy falls to $\frac{E_0}{e^t}$.

Then, we have

$$\frac{E_0}{e^4} = E_0 e^{-2rt'}$$

or
$$e^{-4} = e^{-2rt'}$$

or,
$$4 = 2rt'$$

∴
$$t' = \frac{2}{r} = 2 \times 100 = 200 \text{ second}.$$

Problem 9. *The oscillating of a tuning fork of frequency 200 cps die away to* $\frac{1}{e}$ *times their amplitude in 1 second. Show that the reduction in frequency due to air damping is exceeding small.*

Solution. The damped amplitude at any time t is ae^{-rt}, where a is a constant. If a_0 be the amplitude at $t = 0$ and $\frac{a_0}{e}$ at time t, then we have

$$\frac{a_0}{e} = a_0 e^{-rt}$$

these gives, $t = \frac{1}{r}$.

Here $t = 1$ second

$\therefore$ $r = 1 \text{ sec}^{-1}$.

The angular frequency in presence of damping is

$$\sqrt{(\omega^2 - r^2)} = \sqrt{(\omega^2 - 1)}.$$

Now, $\omega = 2\pi n = 2 \times 3.14 \times 200 = 1256$, so that $\omega^2 \approx 1.6 \times 10^6$

$$\therefore \quad \text{damped frequency} = \sqrt{(1.6 \times 10^6 - 1)}$$

$$\simeq \sqrt{(1.6 \times 10^6)} \cong \sqrt{\omega^2} \approx \omega.$$

which is almost same as undamped frequency.

Thus, the effect of damping on frequency is exceedingly small.

Review Questions on Damped and Forced Vibrations

1. A particle executes a one-dimensional motion along X′OX under : (*i*) a restoring force proportional to its displacement from the equilibrium position 'O' and (*ii*) a damping force proportional to its velocity. Set up the equation of motion of the particle, and solve it. [*Cal. University 1997,*]
2. Define Damped Vibration.
   The differential equation for a one-dimensional damped harmonic oscillator is
   $$m\frac{d^2x}{dt^2} + k\frac{dx}{dt} + Sx = 0$$
   Explain the significance of each term in the equation. Solve the equation in the case of critical damping. [*Cal. University 2004, WBUT 2005*]
3. What is resonance ? Draw the plots of the displacement amplitude and the velocity amplitude against the frequency of the sinusoidal force driving a mechanical oscillator for different values of damping.
4. Discuss the phenomenon of resonance. Distinguish between amplitude resonance and velocity resonance. How is the sharpness of resonance defined. [*Cal. University 94*]
5. (*a*) Define damped vibration. (*b*) Write down the differential equation for damped vibratory motion explaining the physical significance of each term in the equation? (*c*) Obtain solution for the damped oscillator motion. [*WBUT*]

Problems on Damped and Forced Vibration

1. A massless spring whose upper end is fixed to a rigid support carries a horizontal disc of mass 100 g at the lower end. It is observed that the system oscillates with a frequency of 10 Hz and the amplitude of the damped oscillations reduce to half its undamped value in one minute. Calculate (*i*) the resistive force constant (*ii*) the relaxation time of the system (*iii*) its quality factor (*iv*) the force constant of the spring.
[**Hints:** See Solved Problem 1]
2. A vibrator of mass 1 gm is acted upon by a restoring force of 10^7 dyne/cm, a retarding force of 4×10^3 dyne-sec/cm and driving force of 10^{17} cos ωt dyne. Find the maximum possible amplitude.
[**Ans.** 1.02 cm] [*Cal. University 2001*]
3. A mass of 10 gms is acted on by a restoring force 5 dyne per cm and a resistive force 2 dyne sec/cm. Find whether the motion is periodic or oscillatory. Find the value of resisting force for which the motion is critically damped.
[See Solved Problem 5] [*Cal. University 94, 99*]
[**Ans.** $10\sqrt{2}$ dyne-sec/cm.]
4. A 2 gm particle is subjected to an elastic force of 30 dyne/cm and a frictional force of 5 dynes cm sec^{-1}. If it is displaced through 2 cm and then released, find whether the resulting motion is oscillatory or not. If so, find its period.
[**Hints:** See Solved Problem 2] [**Ans.** 1.71 sec]
5. If the relaxation time of damped harmonic oscillator is 50 second, find the time in which (*i*) the amplitude falls to $\frac{1}{e}$ times the initial value, (*ii*) energy of the system falls to $\frac{1}{e}$ times the initial value (*iii*) energy falls to $\frac{1}{e^4}$ of the initial value.
[**Hints :** See Solved Problem 7] [**Ans.** (*i*) $\frac{1}{100}$ sec^{-1} (*ii*) 50 sec (*iii*) 200 seconds]

Solved Problems on Progressive Waves

Problem 1. *A simple progressive wave is represented by equation y = 0.5 sin (314 t – 12.65x) where y and x are expressed in meters and t in sec. Find : (a) amplitude (b) wavelength (c) speed of the wave (d) and frequency.*

Solution. Given that $y = 0.5 \sin (314t - 12.65x)$

This equation can be rearranged in the following form

$$y = 0.5 \sin 12.56 \left(\frac{314}{12.56} t - x\right) \qquad \ldots(i)$$

Now we have the standard progressive wave equation

$$y = a \sin \frac{2\pi}{\lambda} (vt - x) \qquad \ldots(ii)$$

Comparing (*i*) and (*ii*), we get

(*a*) amplitude $a = 0.5$ meters

(*b*) $\frac{2\pi}{\lambda} = 12.56$, wavelength $\lambda = \frac{2\pi}{12.56} = 0.5$ meters (*c*) $v = \frac{314}{12.56}$, velocity = 25 meters/sec.

(*d*) Frequency, $\nu = \frac{v}{\lambda} = \frac{25}{0.5}$ = 50 per sec.

Problem 2. *A simple harmonic wave travelling along X axis is given by*

$$y = 5 \sin 2\pi (0.2t - 0.5x)$$

Calculate the amplitude, frequency, wavelength, wave velocity, [x in meters, t in seconds]

[Pune University]

Solution. Given that the $y = 5 \sin 2\pi (0.2t - 0.5x)$

We have the standard progressive wave equation

$$y = a \sin \frac{2\pi}{\lambda}(vt - x) \quad \ldots(i)$$

Comparing with the equation,
amplitude (a) = 5 meter,

or, $$2\pi \frac{v}{\lambda} = 2\pi \times 0.2$$

or, $$\frac{v}{\lambda} = 0.2$$

or, frequency $n = 0.2$ / sec.

or , $$\frac{2\pi}{\lambda} = 2\pi \times 0.5, \text{ wavelength } \lambda = \frac{1}{0.5} = 2 \text{ meter}$$

wave velocity $(v) = n\lambda = 0.2 \times 2 = 0.4$ meter/sec.

Problem 3. *Equation of a wave moving on a string is* $y = 8 \sin \pi (0.02x - 4.00t)$. *Here y and x are in cms and t in sec. Find the amplitude, frequency and velocity of the wave. Two particles at any instant are situated 20 cms apart, calculate the phase difference between the two particles.*

Solution. Given

$$y = 8 \sin \pi (0.02x - 4.00t) \quad \ldots(i)$$

We have the progressive wave equation,

$$y = a \sin \frac{2\pi}{\lambda}(vt - x) = -a \sin \frac{2\pi}{\lambda}(x - vt) \quad \ldots(ii)$$

Comparing (*i*) and (*ii*), we get

amplitude a = 8 cm,

and $$\pi \times 4 = 2\pi \times \frac{v}{\lambda} = 2\pi n$$

∴ frequency n = 2/sec.

Again, $$\frac{2\pi}{\lambda} = \pi \times 0.02, \quad \therefore \quad \lambda = 100 \text{ cm}$$

∴ velocity of the wave (v) $= n\lambda = 2 \times 100 = 200$ cm/sec.

phase difference between the two particles

$$= \frac{2\pi}{\lambda} \times \Delta x = \frac{2\pi}{100} \times 20 = \frac{2\pi}{5} \text{ radian} = \frac{2}{5} \times 180° = 72°,$$

[Δx = Distance between the two particles]

Review Questions on Progressive Waves

1. State the characteristics of progressive wave? Derive an equation for the progressive wave of wavelength λ and amplitude a moving with a velocity v in the +ve X direction. *[Lucknow University, Burd. University]*
2. Prove that the expression $y = a \sin \frac{2\pi}{\lambda}(vt - x)$ represents a train of waves of amplitude a and wavelength λ moving along the X axis with a velocity v.

Problems on Progressive Waves

1. If a wave of frequency 500 Hz is travelling with a velocity of 200 m/sec, then find the change in phase at a given point in space between a time interval of 10^{-3} sec. Also find the path difference between two points which differ in phase by $\frac{\pi}{2}$ radian. *[Pune University]*

   [Ans. π radian, 0.1 m**]**

2. The equation for a wave travelling in x direction on a string is

   $$y = (3.0 \text{ cm}) \sin [(3.14 \text{ cm}^{-1})\, x - (314\, s^{-1})\, t]$$

   (*a*) Find the maximum velocity of a particle of the string.
   (*b*) Find the acceleration of a particle at $x = 6.0$ cm at time $t = 0.11$ s.

   [Ans. (*a*) $v_{max} = 9.4$ m/sec (*b*) acceleration = 0**]**

Multiple Choice Questions

1. Sound wave is
   (*a*) An electromagnetic wave (*b*) An elastic wave (*c*) A radio wave.
2. The velocity of a particle executing S.H.M. is minimum at a point where the displacement is
   (*a*) Maximum (*b*) Zero
   (*c*) Midway between zero and maximum.
3. The S.I. unit of force constant or spring constant is
   (*a*) Nm (*b*) Nm^{-1} (*c*) Nm^{-2}.
4. For a particle executing S.H.M. the phase difference between displacement and velocity is
   (*a*) $\frac{\pi}{2}$ (*b*) π (*c*) 0.
5. The potential energy of a particle executing S.H.M is equal to its kinetic energy when the displacement of the particle is
   (*a*) $\pm a$ (*b*) $\pm \frac{a}{\sqrt{2}}$ (*c*) $\pm \frac{a}{2}$.

6. Which is not essential for S.H.M.
(a) Elasticity (b) Inertia and restoring force (c) Gravity.

7. Relaxation time is defined as the time during which the amplitude of the damped oscillator
(a) Grows to e times the initial value.
(b) Decays to $\frac{1}{e}$ times the initial value
(c) Decays to $\frac{1}{e^2}$ times the initial value.

8. The quality factor Q of an electrical oscillator is
(a) $\frac{\omega R}{L}$ (b) $\frac{\omega}{LR}$ (c) $\frac{\omega L}{R}$.

9. If ω_0 is the natural frequency of a body and k is the damping constant, then its quality factor is
(a) $\frac{\omega_0}{k}$ (b) $\frac{\omega_0}{2k}$ (c) $\frac{2\omega_0}{k}$.

10. The relaxation time τ for a mechanical oscillator is related to damping constant r as
(a) $\tau = \frac{r}{2m}$ (b) $\tau = \frac{r}{m}$ (c) $\tau = \frac{2m}{r}$.

11. For a particle executing S.H.M. the phase difference between acceleration and displacement is
(a) $\frac{\pi}{2}$ (b) 0 (c) π

12. Which of the following motion is periodic but is not oscillatory ?
(a) The motion of the earth around the sun.
(b) The motion of a simple pendulum.
(c) The oscillations of a liquid column in U tube.

13. A damped mechanical oscillator of frequency ω is acted upon by an external force $F = F_0 \sin \omega t$. If the natural frequency of free oscillation be ω_0, the frequency of oscillations in steady state will be
(a) ω_0 (b) ω (c) $\omega - \omega_0$.

14. The resonant frequency of an oscillator is given by
(a) $\nu = 2\pi \sqrt{LC}$ (b) $\nu = \frac{2\pi}{\sqrt{LC}}$ (c) $\nu = \frac{1}{2\pi \sqrt{LC}}$.

15. What is the value of phase angle ϕ when the driving force frequency is equal to the natural frequency of the undamped oscillator?
(a) $-\frac{\pi}{2}$ (b) 0 (c) $\frac{\pi}{2}$.

16. What is the phase difference between driving force and velocity of forced oscillator?
(a) ϕ (b) $\frac{\pi}{2} + \phi$ (c) $\frac{\pi}{2} - \phi$.

17. The displacement of a particle in simple harmonic motion in one time period is
(a) A (b) 2A (c) 4A (d) Zero.

18. The distance moved by a particle in simple harmonic motion in one time period is
(a) 2A (b) 4A (c) Zero.

19. The average energy in one time period in simple harmonic motion is
(a) $\frac{1}{2} m\omega^2 A^2$ (b) $\frac{1}{4} m\omega^2 A^2$ (c) $m\omega^2 A^2$.

20. A particle moves in a circular path with a uniform speed. Its motion is
(*a*) Periodic (*b*) Oscillatory (*c*) Simple harmonic.

21. A particle executes SHM with a frequency n. The frequency with which the kinetic energy oscillates is
(*a*) 0 (*b*) n (*c*) $2n$.

22. The average acceleration in one time period in a simple harmonic motion is
(*a*) Zero (*b*) $\frac{A\omega^2}{\sqrt{2}}$ (*c*) $A\omega^2$.

23. The time period of a particle in S.H.M is equal to the time between consecutive appearances of the particle at a particular point in its motion. This point is
(*a*) The mean position
(*b*) An extreme position
(*c*) Between the mean position and the positive extreme.

24. The total mechanical energy of a spring-mass system in SHM is $E = \frac{1}{2} m\omega^2 A^2$. Suppose the oscillating particle is replaced by another particle of double the mass while the amplitude A remains the same. The new mechanical energy will
(*a*) Become $\sqrt{2}$ E (*b*) Become $\frac{E}{2}$ (*c*) Remain E.

25. The motion of a particle is given by $x = A \sin \omega t + B \cos \omega t$. The motion of the particle is
(*a*) Simple harmonic with amplitude A + B
(*b*) Simple harmonic with amplitude $\frac{(A+B)}{2}$
(*c*) Simple harmonic with amplitude $\sqrt{A^2 + B^2}$

26. A particle moves on the X-axis according to the equation $x = x_0 \sin^2 \omega t$. The motion is simple harmonic
(*a*) With amplitude x_0 (*b*) With time period $\frac{2\pi}{\omega}$ (*c*) With time period $\frac{\pi}{\omega}$.

27. On superimposing two mutually perpendicular SHMs, we get circular Lissajous figures when their phase difference (ϕ) and amplitudes (P, Q) are as follows :
(*a*) $\phi = 0$, P = Q (*b*) $\phi = \frac{\pi}{2}$, P = Q (*c*) $\phi = 0$, P ≠ Q.

Answers

1. (*b*)	2. (*a*)	3. (*b*)	4. (*a*)	5. (*b*)	6. (*c*)
7. (*b*)	8. (*c*)	9. (*b*)	10. (*c*)	11. (*c*)	12. (*a*)
13. (*b*)	14. (*c*)	15. (*b*)	16. (*a*)	17. (*d*)	18. (*b*)
19. (*a*)	20. (*a*)	21. (*c*)	22. (*a*)	23. (*b*)	24. (*c*)
25. (*c*)	26. (*c*)	27. (*b*).			

J.C. Bose

3B Chapter

Polarisation of Light

3B.1 POLARISATION OF LIGHT

According to Huygens' light consists of waves set up by the vibrations of ether particles. These vibrations take place in all possible directions, and some of the optical phenomenon can be explained on this theory. But in explaining those phenomenon we need not have to consider the nature of vibrations, *i.e.*, whether the vibrations take place along the direction of propagation of light or at right angles to it. There are two kinds of waves which can be set up in an elastic medium, namely, longitudinal and transverse. It has been found that the nature of vibration of light is transverse. The transverse nature of light cannot be decided by the phenomenon of interference and diffraction, because longitudinal waves such as sound waves, also show these effects. But this question can be readily decided by the phenomenon of polarisation.

Polarisation is a restriction of the vibrations in a transverse wave. Light waves are transverse in nature. These are electromagnetic radiations. The time-varying electric field $\left(\vec{E}\right)$ and magnetic field $\left(\vec{B}\right)$ are perpendicular to each other and also to the direction of propagation $\left(\vec{k}\right)$ known as Poynting Vector – $\vec{E}$ and $\vec{B}$ being Electric Vector and Magnetic Vector respectively. The vectors $\vec{E}$ and $\vec{B}$ oscillate in magnitudes. If the directions of vibrations are restricted in any way, the radiation is said to be polarised. When, the oscillations of $\vec{E}$ is retricted in a single plane (say Y-Z, in this case) it is called a *plane-polarised* electromagnetic wave. It is to be remembered that the perception of light is effected by the oscillations of the vector $\vec{E}$ and not the oscillations of $\vec{B}$ – oscillations of $\vec{B}$ takes place due to that of $\vec{E}$ in a plane perpendicular to the plane of vibration of $\vec{E}$ as shown in figure 3B.1.

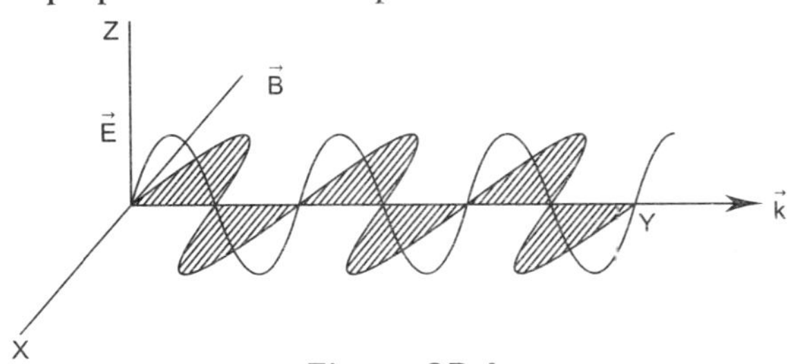

Figure 3B.1

3B.1.1(a) Unpolarised Light

Unpolarised or natural light means light which has vibrations in all possible and in fact, infinite directions at right angles to the direction of its propagation. The average effect of these random vibrations give rise to the impression that in natural or unpolarised light, the vibrations are symmetrically distributed in a plane perpendicular to the direction of propagation of light.

3B.1.1(b) Polarised Light

If somehow, the symmetrical distribution of the vibrations gets or is made lost, the light is partially polarised. If vibrations in all directions except one are removed, the light is said to be completely linearly polarised or plane polarised. Apart from linearly polarised light, there are circularly and elliptically polarised light.

An Experiment-showing Polarisation of Light

Let us take a tourmaline crystal (T_1) cut parallel to the crystallographic axis, [all crystals possess certain axes to which their shape and other properties may be referred in the simplest possible manner — these are called the crystallographic axis] and allow a beam of unpolarised light to pass normally through it. On rotating the crystal about the ray as the axis, the character of the light transmitted by it remains the same. Let us now bring another similar crystal (T_2) of tourmaline parallel to the first. It is observed that, light is almost completely transmitted by the second crystal so that the intensity is maximum in this position of the crystal-pair.

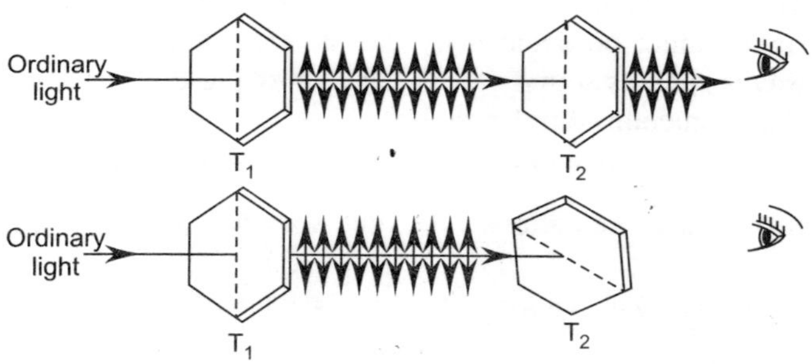

Figure 3B.2

If now T_2 is rotated about an axis parallel to the incident light (*i.e.*, the axes of the crystals are inclined to each other), the transmitted beam decreases in intensity and becomes zero, when the axes are at right angles to each other. This shows that no light is transmitted through the second crystal.

On further rotation of the second crystal, light again starts to be transmitted and the intensity becomes maximum when both T_1 and T_2 are parallel again.

This simple experiment clearly demonstrates that on its passage through the first crystal, light has acquired such a property as to prevent its passage through the second when it is at a definite orientation. This is because when light passes through the first crystal, it absorbs vibration in all direction except those parallel to the crystallographic axis. Hence the unabsorbed vibrations are only transmitted and the transmitted light has vibrations in one direction only *i.e.*, the transmitted light is polarised.

3B.1.2 Plane of Vibration and Plane of Polarisation

When ordinary light is passed through a tourmaline crystal, the light is polarised and the vibrations occur only in one direction perpendicular to the direction of propagation of light.

The plane in which the vibration of polarised light are confined is known as the plane of vibration. The plane ABCD as shown in figure 3B.3 which contains the direction of vibrations and also the direction of propagation is the plane of vibration.

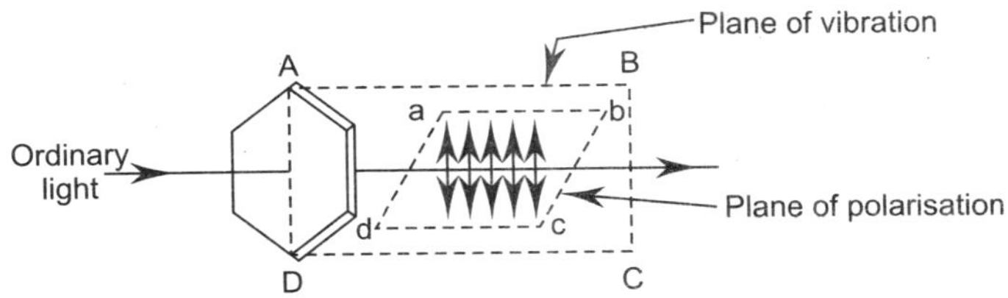

Figure 3B.3

The plane at right angles to the plane of vibration and that does not contain any vibration is known as the plane of polarisation. The plane *abcd* which passes through the direction of propagation and is perpendicular to the plane of vibration is the plane of polarisation.

The plane of polarised light can be produced in some of the ways grouped below. Polarisation by (*i*) **Reflection**, (*ii*) **Refraction**, (*iii*) **Double refraction**, (*iv*) **Selective absorption**, (*v*) **Scattering**. Here we will discuss on polarisation by reflection and on double refraction only.

Polarisation by Reflection

The simplest method of getting plane polarised light is by reflection. This method was discovered by Malus in 1808. He found that when a beam of ordinary light is reflected from the surface of a transparent medium like glass or water, the reflected light is partially polarised, and the degree of polarisation varies with the angle of incidence. The percentage of polarised light in the reflected beam from the surface of a given transparent medium, is greatest for a particular angle of incidence. This angle of incidence for which the reflected polarised light has the greatest intensity is known as the angle of polarisation. The angle of polarisation for ordinary glass is 57.5° and varies slightly with wavelength, since refractive index depends upon the wavelength. Thus complete polarisation is possible only with monochromatic light.

A simple instrument that can both produce and detect plane polarised light by reflection is Biot's polariscope. The principle of this is discussed below.

Light from a monochromatic source S is allowed to fall at the polarising angle of incidence upon a transparent glass plate A which is capable of rotation about a horizontal axis. From A it is reflected to a black glass plate B which can be rotated both about the horizontal and vertical axes. The ray after reflection from B is sent in the direction BC.

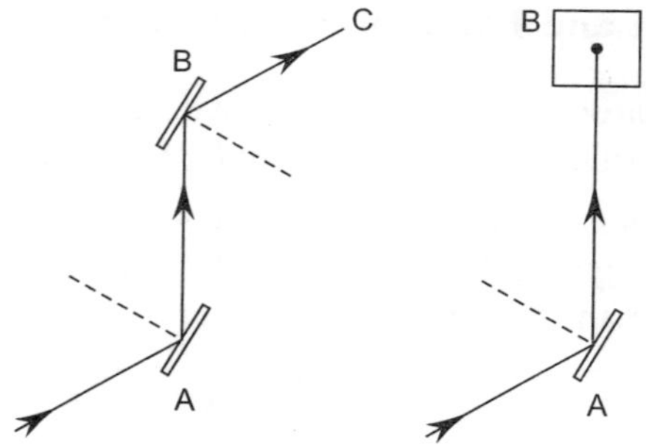

Figure 3B.4

Both the plates A and B are held equally inclined to the horizontal axis so that light always falls upon them at the same angle of incidence. This means that the planes of incidence of the two reflecting surfaces A and B are parallel to each other. In this position the intensity of the reflected beam along BC is maximum.

When the plate B is rotated about the vertical axis the intensity of the reflecting beam goes on decreasing until when the planes of incidence of A and B are at right angles to each other. At this position the intensity of the reflected beam is minimum and for this polarising angle of incidence, the intensity in this position of B is zero.

It is evident that in the first case the vibrations of the ray ABC are parallel to the plane of incidence of both A and B while in the second case the vibrations of the ray AB are perpendicular to the plane of incidence of A and parallel to the plane of incidence of B and so the light is not reflected from B. Thus the plate A polarises the incident light and the plate B analyses it.

The vibrations of the ordinary incident light which are supposed to be in all possible planes can be resolved into two mutually perpendicular vibrations, one in the plane of incidence and the other perpendicular to the plane of incidence. The vibrations in the plane of incidence make different angles with the reflecting surface with change in angle of incidence. When the angle of incidence equals the polarising angle for the medium, most of these vibrations are transmitted and not reflected. While the other set of vibrations always remains parallel to the reflecting surface and so the condition of reflection for them remains unchanged whatever be the angle of incidence. Thus the reflected light is completely plane polarised while the transmitted light is not as it also contains some of the vibrations perpendicular to the plane of incidence.

3B.1.3 Malus's Law

When a beam of light, polarised by reflection at one plane surface is allowed to fall on the second plane surface at the polarising angle, the intensity of the twice reflected beam varies with the angle between the planes of the two surfaces. From experiments it was found that the intensity of the twice reflected beam is maximum when the two planes are parallel and zero when the two planes are at right angles to each other. The same is also true for the twice transmitted beam from the polariser and analyser. The law of Malus states that the intensity of the polarised light transmitted through the analyser varies as the square of the cosine of the angle between the plane of transmission of the analyser and the plane of the polariser.

The proof of the law is based on the fact that any polarised vibration may be resolved into two rectangular components — (*i*) parallel to the plane of transmission of the analyser, (*ii*) at right angles to it.

Let, OP $= a$, is the amplitude of the vibrations transmitted or reflected by the polariser and θ be the angle between the planes of the polariser and the analyser.

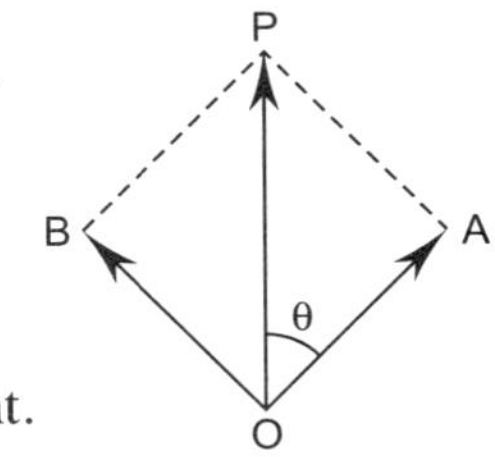

Figure 3B.5

The two components of OP are, $a \cos\theta$ along OA and $a \sin\theta$ along OB.

Only the $a \cos\theta$ component is transmitted through the analyser.

∴ Intensity of the transmitted light through the analyser is —

$$E_1 = (a \cos\theta)^2 = a^2 \cos^2\theta$$

But $\quad E = a^2$ = intensity of incident polarised light.

∴ $\quad E_1 = E \cos^2\theta$

i.e., $\quad E_1 \propto \cos^2\theta$

So when $\theta = 0$ *i.e.*, the two planes are parallel,

then $\quad E_1 = E \qquad [\because \cos 0 = 1]$

and when $\quad \theta = \frac{\pi}{2}$ *i.e.*, the two planes are perpendicular

then $\quad E_1 = E\left(\cos\frac{\pi}{2}\right)^2 = 0.$

3B.1.4 Qualitative Discussion : Circularly and Elliptically Polarised Light

Let a beam of monochromatic plane polarised light of wavelength λ and amplitude A be incident normally upon a uniaxial double refracting crystal, with its faces cut parallel to the optic axis. Suppose the incident light is vibrating along PQ making an angle θ with the optic axis of the crystal. On entering the crystal the incident light will be split up into two components — ordinary and extra ordinary. The ordinary component will have vibrations along PO and extra-ordinary along PE. The amplitudes of vibration of E-ray and O-ray are A cos θ and A sin θ respectively along the optic axis of the crystal and perpendicular to it.

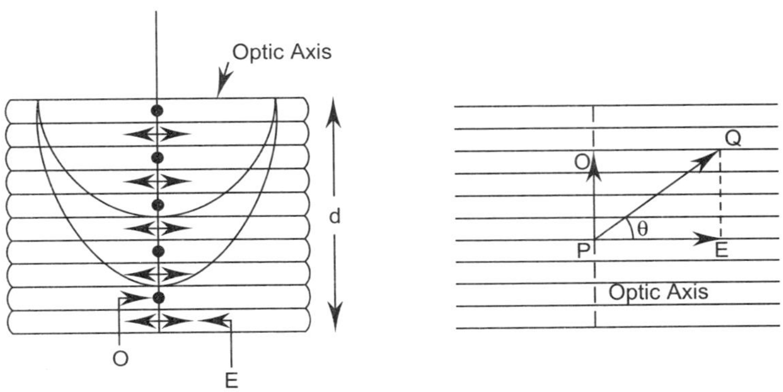

Figure 3B.6

Both the components travel in the same direction but with different speeds depending upon the nature of the crystal. In a negative crystal such as calcite, E-ray travels faster than O-ray ($\mu_o > \mu_e$), while in positive crystal such as quartz, E-ray travels slower than O-ray ($\mu_e > \mu_o$). Thus on emergence the two rays will have path difference depending upon the thickness, and nature of the crystal and also on the wavelength of the incident light. The incident plane polarised light on leaving the crystal will consist of two simple harmonic vibrations, in two mutually perpendicular planes, having same period but different amplitudes and phases. Thus the vibration of the emergent light will be linear or circular or elliptically polarised depending upon the phase difference introduced by the crystal and the angle which the vibration in the incident plane polarised light makes with the optic axis.

Let δ be the phase difference introduced between the components in passing through the crystal. Then their displacements can be represented by

for E-ray, $x = A \cos \theta \sin (\omega t + \delta)$

for O-ray, $y = A \sin \theta \sin (\omega t)$

which can be put as $x = a \sin (\omega t + \delta)$

$y = b \sin \omega t$

where, $a = A \cos \theta$ and $b = A \sin \theta$.

On compounding these, the path of the resultant vibration will be given by

$$\frac{x^2}{a^2} + \frac{y^2}{b^2} - \frac{2xy}{ab} \cos \delta = \sin^2 \delta$$

which represents the general equation of an ellipse with its major and minor axes inclined to the co-ordinate axes at certain angles depending upon their relative phase.

If the phase ahead of the x-component is such that $0 < \delta < \pi$, the resultant vibration is an elliptical one rotating in the counter-clockwise direction with respect to an observer towards whom the light moves. The light is then said to be left-handed elliptically polarised. But if $\pi < \delta < 2\pi$, the ellipse is described in clockwise direction. Light is then said to be right handed elliptically polarised.

Case I. When $\delta = 0$, the equation of the resultant motion becomes,

$$y = \frac{b}{a} x$$

which is the equation of a straight line passing through the origin and having a slope $\left(\frac{b}{a}\right)$. This indicates that the resultant light is plane polarised in the same plane as the incident light.

Case II. (*i*) When $\delta = \frac{\pi}{2}$ or $\frac{(2n+1)\pi}{2}$ and $a \neq b$ (*i.e.*, $\theta \neq 45^\circ$), the equation of resultant motion reduces to

$$\frac{x^2}{a^2} + \frac{y^2}{b^2} = 1$$

which is the equation of an ellipse.

The resultant amplitude of vibration is given by

$$R = \sqrt{x^2 + y^2} = \sqrt{a^2 \sin^2 \left(\omega t + \frac{\pi}{2}\right) + b^2 \sin^2 \omega t}$$

$$= \sqrt{a^2 \cos^2 \omega t + b^2 \sin^2 \omega t}$$

and is inclined to the direction of x (*i.e.*, with the optic axis) by an angle ϕ is given by

$$\tan \phi = \frac{b \sin \omega t}{a \sin (\omega t + \delta)}$$

$$\text{At } t = 0,\ R = a \text{ and } \tan \phi = 0 \ i.e.,\ \phi = 0$$

$$\text{At } t = -\frac{\pi/2}{\omega},\ R = b \text{ and } \tan \phi = -\infty \ i.e.,\ \phi = -\frac{\pi}{2}$$

Thus the resultant vibration rotates periodically and at the same time amplitude of vibration varies in magnitude. Thus the resultant light is elliptically polarised.

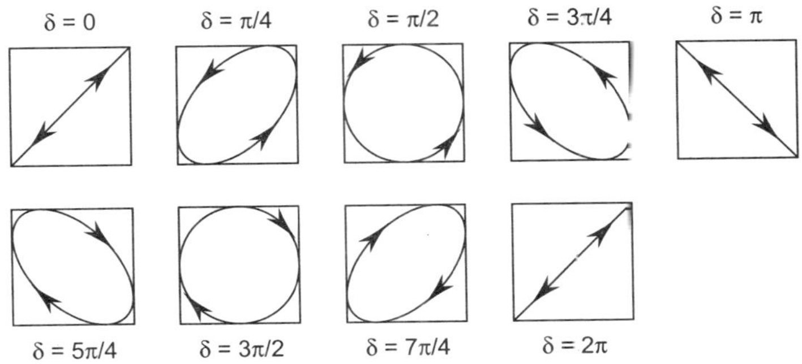

Figure 3B.7

(*ii*) A special case arises when $\theta = 45°$, we then have $a = b$ and the equation reduces to

$$\frac{x^2}{a^2} + \frac{y^2}{b^2} = 1$$

or,
$$x^2 + y^2 = a^2$$

which is the equation of a circle of radius a. The emergent vibrations will be circular, of constant amplitude but with constantly rotating plane of vibration, and the emergent beam, is then said to be circularly polarised. The resultant vibration will be left handed or right handed circular vibrations according as

$$\delta = \frac{\pi}{2} \quad \text{or} \quad \frac{3\pi}{2}$$

Case III. If $\delta = \pi$ or $(2n + 1)\pi$, the equation of the resultant motion reduces to,

$$y = -\frac{b}{a} x$$

which is again the equation of a straight line through the origin with the slope $\left(-\frac{b}{a}\right)$. In this case the direction of vibration of the transmitted light makes an angle equal to $2 \tan^{-1}\left(\frac{b}{a}\right)$ with that of the incident light.

3B.1.5 Polarisation through Reflection and Brewster's Law

When light is reflected from the surface of a refracting medium, the angle of polarisation is found to depend on the refractive index of the medium. According to Sir David Brewster, the tangent of the polarising angle is equal to the refractive index of the refracting medium. This is known as Brewster's law.

Let XY be the surface of separation of a transparent medium. PQ is the incident beam, QR the reflected beam and QS the refracted beam. Let μ be the refractive index of the transparent medium and *i* and *r* are respectively the angle of incidence and refraction. If *i* be equal to the polarising angle then by Brewster's law,

$$\mu = \tan i = \frac{\sin i}{\cos i}$$

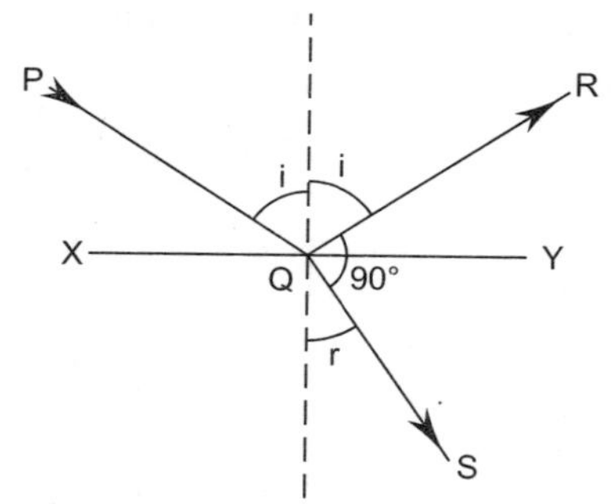

Figure 3B.8

Again from Snell's law we have

$$\mu = \frac{\sin i}{\sin r}$$

$$\therefore \quad \frac{\sin i}{\sin r} = \frac{\sin i}{\cos i}$$

or, $$\cos i = \sin r$$

or, $$\sin (90° - i) = \sin r$$

$$\therefore \quad 90° - i = r$$

$$\therefore \quad i + r = 90°$$

Thus when a beam of light falls at the polarising angle, the reflected and refracted beams are at right angles to each other. Brewster's law enables us to calculate the proper angle for complete polarisation for any substance if its refractive index is given. But Brewster's law does not hold good for polished metallic surface.

3B.1.6 Double Refraction or Birefringence—Ordinary and Extraordinary Rays

Refracting medium which have cubic arrangements of their atoms are isotropic. These substances have same physical properties in all directions and the secondary wavelets in such a medium are spherical surfaces. But other crystalline substances which show that their physical properties are different in different directions are called anisotropic and the behaviour of light through them is affected by the direction which the light takes through them. Examples are quartz, tourmaline, calcite etc.

When a beam of ordinary unpolarised light is allowed to pass through a calcite or quartz crystal, there are two refracted beams in place of usual one as in glass. This phenomenon is called **double refraction.** A narrow beam of light from a point source will produce two images on a screen after passing through a crystal of calcite.

The phenomenon of double refraction can also be illustrated in a simple manner by putting a black dot on a sheet of paper and viewing it through a calcite crystal. Two images will be seen. On rotating the crystal one image will remain stationary while the other rotates round the first. Moreover the first image appears nearer to the eye than the other image. The stationary image is known as the ordinary image and the refracted rays which produce this image are known as the ordinary rays, because they obey ordinary laws of refraction. The other image is called the extra-ordinary image and the refracted rays which give rise to this image are called the extra-ordinary rays because they do not obey the ordinary laws of refraction.

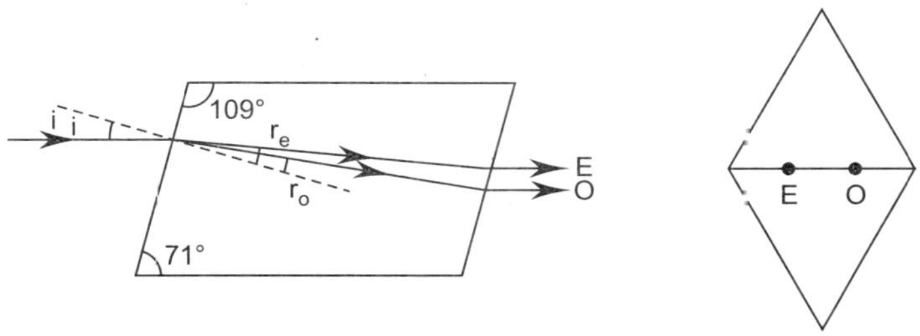

Figure 3B.9

For normal incidence the ordinary rays will pass straight through without deviation whereas the extra-ordinary rays will be refracted at some angle and will come out parallel to and displaced from the incident ray. On rotating the crystal about the incident beam as axis, the E-ray rotates about the fixed O-ray. The line joining the O– and E– images always remains parallel to the shorter diagonal of the emergent crystal face.

The ordinary and the extra-ordinary rays are plane polarised and their planes of polarisation are at right angles to one another. This can be verified by viewing the images through a tourmaline plate. On rotating the tourmaline plate the intensities of both the images undergo a change. If the intensity of ordinary image is decreasing then that of the extra-ordinary image is increasing and vice-versa. And so if one is at its brightest, the other is not visible. In short, the ordinary and the extra-ordinary images alternately are at their brightest twice in one complete rotation.

3B.1.7 Optic Axis

An important feature concerning the behaviour of calcite or quartz (*i.e.*, double refracting crystals) is that there is only one direction through the crystal in which there is no double refraction. This particular direction is called the optic axis. It may be clearly noted that the optic axis is a direction and not a particular line.

A ray passing along the optic axis does not break up into ordinary (O–) and extra-ordinary (E–) rays *i.e.*, they travel with the same velocity and in the same direction. Crystals like calcite, tourmaline, quartz have only one optic axis. These are called uniaxial crystals. There are other minerals like borax and mica which have two optic axes. These are called biaxial crystals.

Calcite is transparent to visible as well as to ultraviolet light. It is chemically calcium carbonate ($CaCO_3$) and occurs in nature in different forms. If a piece of calcite of any shape is struck with a hammer, it shows a remarkable tendency to cleave. The fragments formed by cleavage being rhombohedron. Each of the six faces of the rhombohedron is a parallelogram having angles of 78° and 102° nearly. At the two opposite corners A and G, the angles of three faces meeting these are all obtuse while at its remaining six corners one angle is obtuse and two are acute.

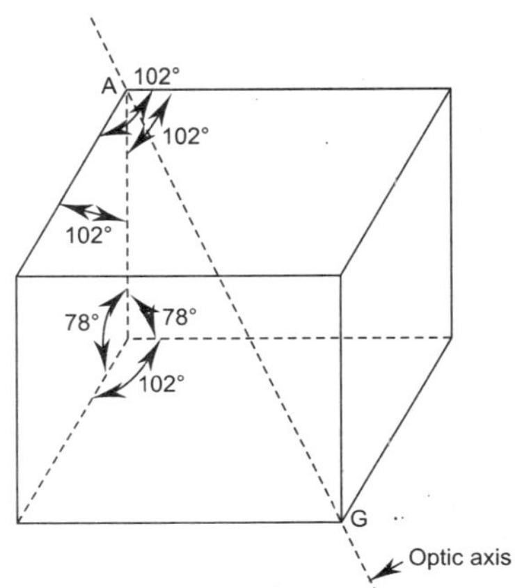

Figure 3B.10

The corners A and G are known as blunt corners. The line joining the two opposite blunt corners (AG) coincide with principal crystallographic axis of the crystal or any line parallel to it gives the direction of the optic axis.

3B.1.8 Principal Section and Principal Plane

A plane containing the optic axis of the crystal and perpendicular to the opposite refracting faces is called a principal plane or principal section of the crystal. The line joining the two images O– and E– is always parallel to the principal section of the crystal. For every point there are three principal sections, one for each pair of opposite crystal faces. A principal section always cuts the surfaces of a calcite crystal in a parallelogram having angles 71° and 109°.

The ordinary ray is plane polarised in the principal plane of the crystal while the extra-ordinary ray is plane polarised in a plane at right angles to it, *i.e.*, the ordinary ray vibrations are perpendicular to the principal plane of the crystal while the extra-ordinary ray vibrations are in its principal plane.

3B.1.9 Huygens' Theory of Double Refraction

Propagation of light through uni-axial crystals *viz.*, calcite or quartz gives rise to double refraction. Huygens explained the phenomenon of double refraction with the help of his principle of secondary wavelets. A point source of light in a doubly refracting medium is the origin of two wavefronts. For the ordinary ray which obey the ordinary laws of refraction and travel with the same velocity in all the directions, the wave-front is spherical. For extra-ordinary rays which do not obey the laws of refraction and travel with different velocities in different directions, the wave front is an ellipsoid of revolution. Since no double refraction occurs along the optic axis *i.e.*, the velocities of the ordinary and extra-ordinary rays are the same along the optic axis. The wave surfaces of the ordinary ray and extra-ordinary ray touch in two points. The line joining these two points gives the direction of the optic axis.

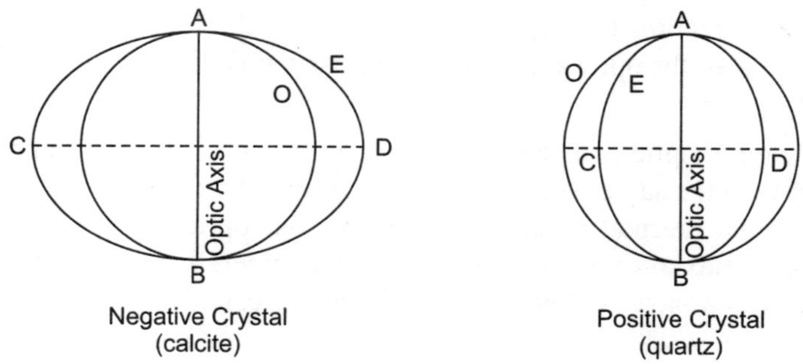

Figure 3B.11

In some crystals, like quartz, O-ray travels faster than E-ray. The spherical wave-front for the O-rays lie outside the ellipsoidal wave front for the E-rays. The diameter of the spherical surface is equal to the major axis of the ellipsoidal surface. These crystals are called positive crystals.

In some crystals like calcite, E-ray travels faster than O-ray, the spherical wavefront for the O-rays be inside the ellipsoidal wavefront for the E-rays. The diameter of the spherical surface is equal to the minor axis of the ellipsoidal surface. Such crystals are known as negative crystals.

3B.1.10 Polaroid

Very recently a method has been developed in which polarising materials are artificially produced and arranged in the form of a large sheet, which acquires the power of producing a strong beam of polarised light of any cross-section without much cost. These polarising sheets are known as Polaroids.

The material employed to construct these polarising sheets is a mineral consisting of crystals of per-iodide of quinine sulphate. These crystals exhibit strong dichroism, *i.e.*, they, like tourmaline, selectively absorb one of the two rectangular components of ordinary light and allow the other component to pass through with little loss of intensity. These crystals are ultra-microscopic and hence they should be arranged side by side, with their optic axes all parallel to one another, so that they may behave as a single crystal of large dimension. The process consists in first breaking the mineral into needle-shaped micro-crystals in nitro-cellulose. This paste is then forced through a horizontal slit and the ribbon issuing through this slit is mounted between two thin sheets of glass forming what is known as polaroid.

Very recently a better method has been found in which individual molecules, rather than micro-crystals, are employed to construct a polaroid known as H-polaroid.

Polaroids have wide applications in every-day life. They are employed in automobile headlights, screens, sun-glasses, windows of railway trains and aeroplanes and in viewing stereoscopic motion pictures.

3B.1.11 Nicol Prism

Nicol prism designed by William Nicol and known after his name is one of the most common and efficient methods of getting plane-polarised light. It is a useful optical device made from calcite crystal for producing and analysing plane-polarised light. In the former case it is called a polariser and in the latter an analyser. It is used in many optical instruments for producing and analysing plane polarised light.

A beam of ordinary unpolarised light in passing through a double refracting crystal, is split up into two plane polarised beams — ordinary and extra ordinary, polarised in two mutually perpendicular planes. If one of the beams can be got rid of, then we get a single plane-polarised light.

The nicol prism is so made that one of the rays *i.e.*, the O-ray is eliminated by total internal reflection so that we are left with only E-ray whose vibrations lie in the principal plane. In the end view these vibrations (*i.e.*, of the E-ray) are parallel to the shortest diagonals of the end faces.

(*a*) One of the most common forms of the nicol prism is made by taking a crystal about three times as long as it is wide. The two end faces are grounded away until they make an angle of 68° instead of 71° with the longitudinal faces. This piece is then cut into two parts along the plane passing through the shortest diagonal and perpendicular to both the principal section and the end faces. The two cut faces are grounded and optically polished. They are then cemented together with Canada balsam whose refractive index lie mid-way between the index of O-ray and E-ray. The two end faces of the prism are kept open while its sides are coated with lamp black and kept covered by a brass tube.

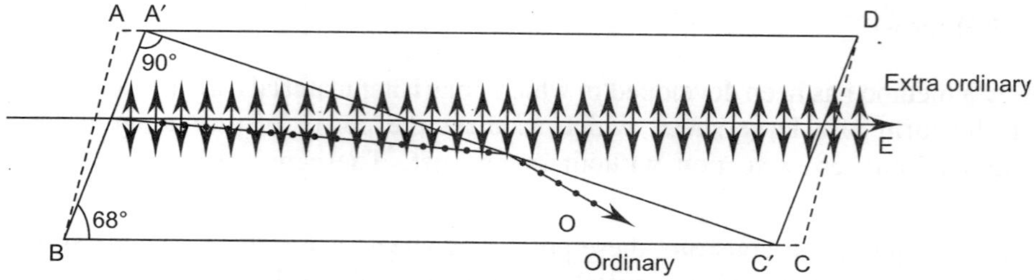

Figure 3B.12

(*b*) A ray of unpolarised light entering the crystal through one of the end faces and moving nearly parallel to its length is divided into O– and E– rays. Optically canada balsam is more denser than calcite for E-ray and less denser for O-ray. [For sodium light $\mu_e = 1.48641$, $\mu_0 = 1.65837$ and μ for Canada balsam = 1.55]. The ordinary ray whose vibration is perpendicular to the principal plane, on reaching the layer of the Canada balsam is totally reflected provided the angle of incidence at the balsam layer is greater than the critical angle of the two media. This totally reflected ray is absorbed by the lamp black. The extra-ordinary ray travels from an optically rarer to denser medium and so emerges out of the crystal. Hence the light coming out of the crystal is a plane polarised one.

To bring about total internal reflection of the ordinary beam the angle of incidence of the unpolarised light on one of the end faces of the crystal should be about 14°, canada balsam is optically denser than E-ray. The limits of the angle of incidence on the end faces of the crystal that the E-ray is not totally reflected, is also to about 14°. Thus to avoid the transmission of O-ray and total internal reflection of E-ray, the angle between the extreme incident ray is limited to about 28°. Hence Nicol prism cannot be used in highly convergent or divergent light.

3B.1.12 Retardation Plates

A plate cut from a doubly refracting crystal (like calcite) by sections parallel to the optic axis and employed to introduce a given phase difference between the O-rays and the E-rays in transmission normally through it, is called RETARDATION PLATE.

There are two most useful retardation plates – (*a*) Half-wave plate and (*b*) Quarter-wave plate.

Half-wave plates are of such thickness that in travelling through the plate a relative phase difference of π comes in between O and E waves and accordingly one wave drops behind the other by just one half a wave length.

The path difference between the rays can be obtained as

$$\Delta = (\mu_O - \mu_E)\, t$$

$\mu_O > \mu_E$ (for negative crystal)
t : thickness of the plate (calcite say)

The corresponding phase difference is

$$\delta = \frac{2\pi}{\lambda}\Delta = \frac{2\pi}{\lambda}(\mu_O - \mu_E)\, t \qquad (\lambda : \text{wavelength of the incident ray})$$

NOTE The incidence is normal to the optic axis.

For positive crystal [Quartz, say] where $\mu_E > \mu_O$

$$\Delta = (\mu_E - \mu_O)\, t, \quad \delta = \frac{2\pi}{\lambda}(\mu_E - \mu_O)\, t$$

For Half-Wave Plate

$$\delta = \pi \quad \text{or} \quad \pi = \frac{2\pi}{\lambda}(\mu_O - \mu_E)\,t$$

$\therefore$
$$t = \frac{\lambda}{2(\mu_O - \mu_E)} \text{ (negative crystal)}$$

and
$$t = \frac{\lambda}{2(\mu_E - \mu_O)} \text{ (for positive crystal)}$$

In particular, a plate whose thickness satisfies the equation, $t(\mu_O - \mu_E) = (2n+1)\frac{\lambda}{2}$ behaves like a half-wave plate, where $n = 1, 2, 3\ldots$

For Quarter-Wave Plate

$$\delta = \frac{\pi}{2} \qquad \text{[One wave lags behind the other by } \frac{\lambda}{4}\text{]}$$

and so,
$$t = \frac{\lambda}{4(\mu_O - \mu_E)} \qquad \text{(for negative crystal)}$$

and
$$t = \frac{\lambda}{4(\mu_E - \mu_O)} \qquad \text{(for positive crystal).}$$

$t(\mu_O - \mu_E) = (4n + 1)\frac{\lambda}{4}$ behaves like quarter wave plate, where $n = 1,2,3\ldots$

3B.1.13 Polarimetry (Analysis of Polarised Light)

If a plane polarised light is made to pass through certain substance, it is found that the plane of polarisation of the emergent light is not the same as that of the incident light and it is rotated through certain angle. This property of crystals or substances is called OPTICAL ACTIVITY and the phenomenon is OPTICAL ROTATION.

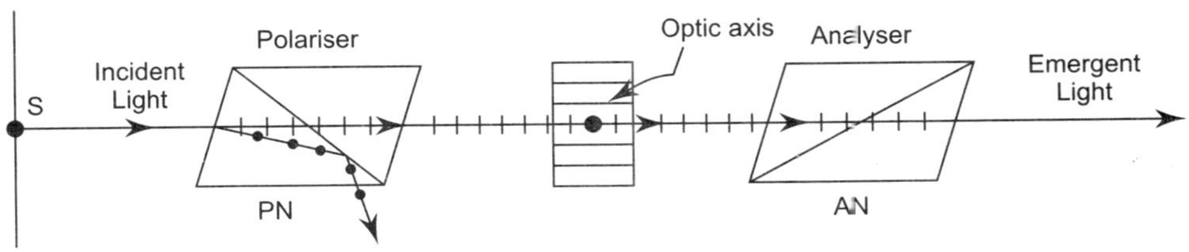

Figure 3B.13

There are two types of optically active substances :

(*a*) *Dextrorotatory (right-handed)* which produce clockwise rotation.

(*b*) *Laevorotatory (left-handed)* which produce anticlockwise rotation.

If a monochromatic light from a source S is made to pass through two Nicols PN and AN but in crossed positions, no emerged light can be received by the eye. In this case the polarising Nicol PN renders the beam polarised with its vibrations in the principal plane. In the analysing Nicol the position is set with its principal plane perpendicular to the direction of these vibrations – thus no light emerges from AN. If a quartz crystal cut with its refracting faces perpendicular to the optic axis is placed between PN and AN, the two crystals still being crossed, it is seen that the light is no longer cut off by analysing nicol. In order to cut off the light the analysing nicol ought to be rotated through some angle. So the emergent light is still plane polarised but its plane of polarisation has been rotated.

The amount of rotation is found to be dependent on

(*i*) The thickness and density of the crystal or substance placed.

(*ii*) Temperature and wavelength of the light used. $\left[\text{Rotation} \propto \dfrac{1}{\lambda^2}\right]$

(*iii*) Concentration in case of solutions.

Specific Rotation

Specific rotation at a given temperature T and for light of wavelength λ

$$S_{T\lambda} = \frac{\theta}{ld} = \frac{\text{Rotation in degrees}}{\text{length in decimeter} \times \text{concentration}\left(\dfrac{\text{gm}}{\text{cc}}\right)}$$

where, θ is produced by "*l*" decimeter length of solution of density (or concentration) "*d*" $\dfrac{\text{gm}}{\text{cc}}$

IN FOCUS

- **Malus's Law**

  Intensity of transmitted light through the analyser

  $$E_1 = E\cos^2\theta$$

  where $E = a^2$ = intensity of incident polarised light

  a = Amplitude of the vibration transmitted

  θ = angle between the plane of polariser and analyser.

- **Circularly and Elliptically polarised light**

  The resultant wave of superimpose of O-ray and E-ray is

  $$\frac{x^2}{a^2} + \frac{y^2}{b^2} - \frac{2xy}{ab}\cos\delta = \sin^2\delta$$

  where $x = a\sin(wt + d)$ [for E-ray]

  $y = b\sin(wt)$ [for O-ray]

  when $\delta = 0$,

  $$y = \frac{b}{a}x$$

  So the resultant light is plane polarised light

  When $\delta = \dfrac{\pi}{2}$ or $\dfrac{(2n+1)\pi}{2}$ and a ¹ b

  $$\frac{x^2}{a^2} + \frac{y^2}{b^2} = 1$$

represents an elliptically polarised light.

For this elliptically polarised light when $\theta = 45°$ (or $a = b$)

$$x^2 + y^2 = a^2$$

represents circularly polarised light

When $\delta = \pi$ or $(2n + 1)\pi$

$$y = -\frac{b}{a}x$$

which is again an equation of straight line through the origin with the slope $\left(-\frac{b}{a}\right)$.

Brewster's law

Refractive index m = tan i

where i is equal to the polarising angle.

Direct deduction from Brewster's Law

$$i + r = 90°$$ [i is the incident angle and r is the refracting angle]

Retardation Plates

For Half-wave plate $t = \dfrac{\lambda}{2(\mu_E - \mu_0)}$ [positive crystal]

For Quarter-wave plate $t = \dfrac{\lambda}{4(\mu_E - \mu_0)}$ [positive crystal]

where t = thickness of the plate.

μ_E = refractive index of E-ray

μ_0 = refractive index of O-ray.

Specific Rotation

Specific rotation at a given temperature T and wavelength λ

$$S_{T\lambda} = \frac{\theta}{l d} = \frac{\text{Rotation in degrees}}{\text{length in decimeter} \times \text{concentration}\left(\frac{\text{gm}}{\text{cc}}\right)}$$

Solved Problems

Problem 1. *Refractive index of glass is 1.5. Calculate the Brewster's angle for it. Also calculate the angle of refraction.* [H.P. University, B.H. University]

Solution. From Brewster's angle or polarising angle (i_p), we know $\tan i_p = \mu = 1.5$

or, $i_p = \tan^{-1}(1.5) = 56.3°$,

But we know, $i_p + r = 90°$

$$r = 90° - 56.3° = 33.7°.$$

Problem 2. *Find the polarising angle for light incident from (i) air to glass (ii) glass to air (iii) water to glass (iv) glass to water (v) air to water and (vi) water to air.*

Given μ of glass = 1.54 and μ of water = 1.33.

Solution. (*i*) When the light is incident from air to glass,

$$\mu = \frac{\text{refractive index of glass}}{\text{refractive index of air}} = \frac{1.54}{1} = 1.54$$

∴ Polarising angle from Brewster's law

$$\tan i_p = \mu$$

or, $$i_p = \tan^{-1}(\mu) = \tan^{-1}(1.54) = 57°.$$

(*ii*) When light is incident from glass to air

$$\mu = \frac{\text{R.I. of air}}{\text{R.I. of glass}} = \frac{1}{1.54} = 0.6494$$

∴ $$i_p = \tan^{-1}(\mu) = \tan^{-1}(0.6494) = 33°.$$

(*iii*) When light is incident from water to glass

$$\mu = \frac{\text{R.I. of glass}}{\text{R.I. of water}} = \frac{1.54}{1.33} = 1.158$$

∴ $$i_p = \tan^{-1}(\mu) = \tan^{-1}(1.158) = 49°\ 12'.$$

(*iv*) When light is incident from glass to water

$$\mu = \frac{\text{R.I. of water}}{\text{R.I. of glass}} = \frac{1.33}{1.54} = 0.8636$$

∴ $$i_p = \tan^{-1}(0.8636) = 40°\ 49'.$$

(*v*) When light is incident from air to water

$$\mu = \frac{\text{R.I. of water}}{\text{R.I. of air}} = \frac{1.33}{1} = 1.33$$

∴ $$i_p = \tan^{-1}(\mu) = \tan^{-1}(1.33) = 53°\ 4'.$$

(*vi*) Similarly when light is incident from water to air

$$\mu = \frac{1}{1.33} = 0.7519$$

∴ $$i_p = \tan^{-1}(0.7519) = 36°54'.$$

Problem 3. *A solution of camphor in alcohol in a tube 25 cm long is found to, affect the rotation of the plane of vibration of light passing through it by 36°. What must be the density of camphor in g/cc of the solution ? Specific rotation of camphor is 56°.* [Gurunanak University]

Solution. Specific rotation (S) in given temperature $S = \dfrac{\theta}{ld}$.

Here, rotation $\theta = 36°$, length of the solution = 25 cm and S = 56°,

∴ The density of the camphor $d = \dfrac{\theta}{Sl} = \dfrac{36}{2.5 \times 56} = 0.257$ gm/cc.

Problem 4. *Critical angle for refraction for glass to air is 40°. Calculate the polarising angle for glass.* [H.P. University]

Solution. If C is critical angle, then,

$$\mu = \frac{1}{\sin C} = \frac{1}{\sin 40°} = \frac{1}{0.6428} = 1.56$$

Now, $$\tan i_p = 1.56$$

∴ $$i_p = \tan^{-1}(1.56) = 57.3°$$

Problem 5. *When light is incident at an angle of 60° to the normal, the reflected light is plane polarised. What is the refractive index of the transparent reflecting medium ? What is the angle of refraction corresponding to the angle of incidence of 60°. What is the angle between the reflected and refracted components ?* [Pune University]

Solution. Given that angle of incidence $i = 60°$

If the reflected light is plane polarised, the angle of incidence is equal to the polarising angle according to Brewster's law.

$$\therefore \quad \text{Refractive Index } \mu = \tan i_p = \tan 60° = \sqrt{3} = 1.732.$$

Again according to Brewster's law, the reflected and transmitted beams are at right angles to each other when the light is incident at the polarising angle. If r is the reflection angle,

$$i + r = 90°$$

$$\therefore \quad \text{Angle of refraction} = 90° - 60° = 30°$$

$$\therefore \quad \mu = \frac{\sin i}{\cos i} = \frac{\sin i}{\sin r}$$

$$\therefore \quad \cos i = \sin r = \cos (90° - r) \therefore i + r = 90°.$$

$\therefore$ reflected and transmitted components are at right angles to each other.

Problem 6. *An analysing Nicol examines two adjacent plane polarised beams A and B whose planes of polarisation are mutually perpendicular. In one position of the analyser, the beam shows zero intensity. From this position, a rotation of 30° shows the two beams of equal intensity. Calculate the intensity ratio* $\frac{I_A}{I_B}$ *of the two beams.* [Agra University]

Solution. If intensity I of plane polarised light is incident on analyser, then intensity of emergent light from Malus law is given by

$$I_\theta = I \cos^2 \theta$$

where θ is the angle between the principal sections of the analyser and the polariser.

According to the problem, the beam B shows zero intensity. It implies that for beam B, $\theta = 90°$ and for beam A, $\theta = 0°$.

When the analyser is rotated through 30°, then

$$\theta = 30° \text{ for beam A}$$

$$\theta = 60° \text{ for beam B.}$$

The intensities of beams A and B transmitted by the analyser, which are equal, are given by

$$I' = I_A \cos^2 30° = I_B \cos^2 60°$$

$$\therefore \quad \frac{I_A}{I_B} = \frac{\cos^2 60°}{\cos^2 30°} = \frac{1 \times 4}{4 \times 3} = \frac{1}{3}.$$

Problem 7. *Calculate the thickness of a half-wave plate for sodium light, given* $n_O = 1.54$ *and the ratio of velocity of O-component and E-component is 1.007. Is the crystal positive or negative?*

[Poona University]

Solution. We know for half wave plate, the thickness

$$t = \frac{\lambda}{2(\mu_e - \mu_o)}.$$

We can consider for sodium light $\lambda = 5893 \times 10^{-10}$ meter and $\mu_0 = 1.54$.

According to the problem $\frac{v_o}{v_e} = 1.007$

or $$\frac{\mu_e}{\mu_o} = 1.007, \textit{ i.e., } \mu_e = 1.007\,\mu_o$$

or $$\mu_e = 1.007 \times 1.54 = 1.55$$

i.e., $\mu_e > \mu_o$. So crystal is positive

$$\therefore \quad t = \frac{5893 \times 10^{-10}}{2(1.55 - 1.54)} \simeq 0.268 \times 10^{-4} \text{ m.}$$

Problem 8. *Two Nicol prisms are so arranged that the amount of light transmitted through them is maximum. What will be the percentage reduction in the intensity of the incident light when the analyser is rotated through (i) 30° (ii) 45° (iii) 60° and (iv) 90°.* [Pune University]

Solution. From Malus's law

$$I_1 = I_0 \cos^2 \theta$$

Now percentage reduction of intensity

$$= \frac{(I_0 - I_1)}{I_0} \times 100 = \left(1 - \frac{I_1}{I_0}\right) \times 100$$

$$= (1 - \cos^2 \theta) \times 100$$

(*i*) When $\theta = 30°$

Percentage reduction of intensity $= (1 - \cos^2 30°) \times 100$

$= 25$

(*ii*) When $\theta = 45°$,

Percentage reduction of intensity $= \left(1 - \frac{1}{2}\right) \times 100 = 50$

(*iii*) When $\theta = 60°$,

Percentage reduction of intensity $= \left(1 - \frac{1}{4}\right) \times 100 = 75.$

(*iv*) When $\theta = 90°$,

Percentage reduction of intensity $= (1 - 0) \times 100 = 100.$

Problem 9. *Two polaroids are adjusted so as to obtain maximum intensity. Through what angle should one polaroid be rotated to reduce the intensity to (i) half (ii) one fourth.*

[Pune University, G.N.D. University, H.P. University]

Solution. From Malus's law,

$$I_1 = I_0 \cos^2 \theta$$

or,
$$\frac{I_1}{I_0} = \cos^2 \theta$$

(*i*)
$$\frac{I_1}{I_0} = \frac{1}{2} = \cos^2 \theta,$$

$$\cos \theta = \frac{1}{\sqrt{2}} = \cos 45°,$$

∴
$$\theta = 45°.$$

(*ii*)
$$\frac{I_1}{I_0} = \frac{1}{4} = \cos^2 \theta$$

or,
$$\cos \theta = \frac{1}{2} = \cos 60°,$$

∴
$$\theta = 60°.$$

Problem 10. *Calculate the thickness of* $\frac{\lambda}{4}$ *plate for light of* $\lambda = 5893$ *Å, given* $n_o = 1.544$ *and* $n_e = 1.553$. [Calcutta University, Himachal University]

Solution. Here $\frac{\lambda}{4}$ *i.e.*, for Quarter wave plate, thickness

$$t = \frac{\lambda}{4(n_e - n_o)} \text{ for positive crystal.}$$

Given $n_e = 1.553$ and $n_o = 1.544$, *i.e.*, $n_e > n_o$. So, this plate can be treated as a uniaxial positive crystal.

Thickness,
$$t = \frac{5893 \times 10^{-10} \ m}{4(1.553 - 1.544)} \simeq 0.164 \times 10^{-4} \text{ m.}$$

Problem 11. *Find the state of polarisation when the x and y components of the electric field are :*

(*i*) $E_x = E_o \sin(\omega t + kz)$; $E_y = E_o \cos(\omega t + kz)$

(*ii*) $E_x = E_o \cos(\omega t + kz)$; $E_y = \left(\frac{E_0}{\sqrt{2}}\right) \cos(\omega t + kz + \pi)$. [Calcutta University]

Solution.

(*i*) It can be represented as a

$$E_o^2 = E_x^2 + E_y^2 \text{ ; so it is the indication for circular vibration.}$$

At any position, say $z = 0$; the vector can be represented as :

$$\vec{E} = \hat{i} E_x + \hat{j} E_y = \hat{i} E_o \sin \omega t + \hat{j} E_o \cos \omega t$$

So, the rotation is clockwise and the light is circularly by polarized.

(*ii*) Given
$$E_y = \left(\frac{E_o}{\sqrt{2}}\right) \cos\{\omega t + kz + \pi\}$$

$$= -\left(\frac{E_o}{\sqrt{2}}\right)\cos(\omega t + kz) = -\frac{E_x}{\sqrt{2}}.$$

So, this illustrates a linearly polarised light making an angle $\tan^{-1}\left(-\frac{1}{\sqrt{2}}\right)$ with the x-axis.

Multiple Choice Questions

1. Which of the following phenomena proves the transverse characteristics of light ?
(*a*) Interference (*b*) Dispersion (*c*) Polarisation
[*W.B.U.T. - 2004*]

2. Nicol Prism is formed by
(*a*) Quartz crystal (b) Calcite crystal (c) Both of them (d) None of them.

3. The place of vibration and the plane of polarisation of a beam of plane polarised light.
(*a*) are identical to each other
(*b*) are orthogonal to each other
(*c*) make an angle which depends on the colour of the light.

4. E-rays and O-rays have
(*a*) Equal velocities along optical axis (*b*) Equal velocities perpendicular to the optical axis
(*c*) Equal refractive indices along optical axis (*d*) Refractive indices less than one. [*WBUT 2007*]

5. A light ray is passing through a calcite crystal. If the plane of vibration of the light ray is perpendicular to the optical axis then the ray
(*a*) is an O-ray (*b*) is an E-ray.
(*c*) has both non-zero O and E components. (*d*) cannot propagate through the crystal.
[*WBUT - 2008*]

6. Nicol prism is based on
(*a*) Double refraction (*b*) Scattering (*c*) Refraction (*d*) None of these.

7. If θ be the polarisation angle, the refractive index n of the material is given by
(*a*) sin θ (*b*) cos θ (*c*) tan θ (*d*) cot θ.

8. Which one of the following waves cannot be polarised
(*a*) X-ray (*b*) Sound wave (*c*) Radio-wave (*d*) None of these.

9. Calcite crystal is
(*a*) Dixto-rotatory (*b*) Laevo-rotatory (*c*) Both of these (*d*) None of these.

10. Inside the positive crystal, the velocity of E-ray (v_e) and O-ray (v_o) are observed.
(*a*) $v_e = v_o$ (*b*) $v_e > v_o$ (*c*) $v_e < v_o$ (*d*) $v_e = 2v_o$.

11. Inside the negative crystal, the velocities of E-ray (v_e) and O-ray (v_o) are observed.
(*a*) $v_e = v_o$ (*b*) $v_e < v_o$ (*c*) $v_e > v_o$ (*d*) $v_e = \sqrt{2}\,v_o$.

Answers

1. (*c*)	**2.** (*b*)	**3.** (*b*)	**4.** (*b*)	**5.** (*a*)	
6. (*a*)	**7.** (*c*)	**8.** (*b*)	**9.** (*b*)	**10.** (*c*)	**11.** (*c*)

Review Questions

1. What do you understand by 'Polarisation of Light' ? *[W.B.U.T.]*
2. Distinguish clearly between polarized and unpolarized light: *[Garhwal, Jabalpur, P.E.C., Kanpur]*
   [Hints : Refer to Art. 3B.1.1(*a*) & (*b*)]
3. Define plane of vibration and plane of polarisation. *[Garhwal, Kanpur]*
   [Hints : Refer to Art. 3B.1.2]
4. Explain the method of producing plane polarised light by reflection.
   [Hints : Refer to Art. 3B.1.2] *[Pune University, Kanpur, B.U., P.E.C.]*
5. State and explain law of Malus. **[Hints :** Refer to Art. 3B.1.3] *[Rohilkhand, Agra]*
6. State Brewster's law and hence prove that the angel between the reflected and refracted rays is 90°.
   [Hints : Refer to Art. 3B.1.5] *[WBUT, Rohilkhand, Garhwal, H.P.U., Pune Univ, Pbi. Univ., B.U.]*
7. What is birefringence? How does it explain double refraction? *[Pune Univ., G.N.D. Univ., Kanpur]*
   [Hints : Refer to Art. 3B.1.6]
8. Describe the phenomenon of double refraction in uni-axial crystal. Distinguish between the negative and positive crystals. Give Huygen's theory of double refraction ? *[Agra, Kanpur, Punjab, Garhwal]*
   [Hints : Refer to Art. 3B.1.9]
9. Write a short notes on
   (*i*) Nicol prism as polarizer and analyzer. **[Hints :** Refer to Art. 3B.1.9] *[W.B.U.T.]*
   (*ii*) Polaroid and its use. **[Hints :** Refer to Art. 3B.1.10] *[Agra]*
10. Describe the construction of Nicol prism. *[Rohilkhand, Rajasthan]*.
    [Hints : Refer to Art. 3B.1.11]

Unsolved Problems

1. The refractives index of a glass plate is 1.6. Calculate polarization angle and corresponding angle of refracted ray. *[WBUT]*
2. A ray of light incident on the surface of a glass plate of refractive index $\sqrt{3}$ at the polarising angle. Calculate the angle of refraction of the ray. *[Rohilkhand]* **[Ans.** 30°]
3. A ray of light is incident on the surface of a glass plate of refractive index 1.55 at the polarising angle. Calculate the angle of refraction. **[Ans.** 32°50′]
4. Light rays are incident on a glass plate (μ = 1.54) so that the refracted rays are completely polarized. Find the angle of incidence, angle of refraction and critical angle.
   [Bundelkhand] **[Ans.** 57°, 33°, 41.50°]
5. Two polarizing sheets have their polarizing directions pass so that the intensity of the transmitted light is maximum through what angle must either sheet be twin so that the intensity becomes one half the initial values. [$\theta = \pm 45°, \pm 135°$]

Chapter 3C

Interference of Light

3C.1 INTERFERENCE OF LIGHT

Propagation of light has two facets : Corpusculave nature and wave nature

The wave nature of light has been established primarily by polarisation and also by interference and diffraction. Interference deals with superposition of light waves from more than one source causing variation of intensity of light which are observed as alternate dark and bright spaces in the *penumbral* regions of the two light sources.

Thus the non-uniform distribution of the intensity of light due to superposition of two independent light waves of specific property is called interference of light.

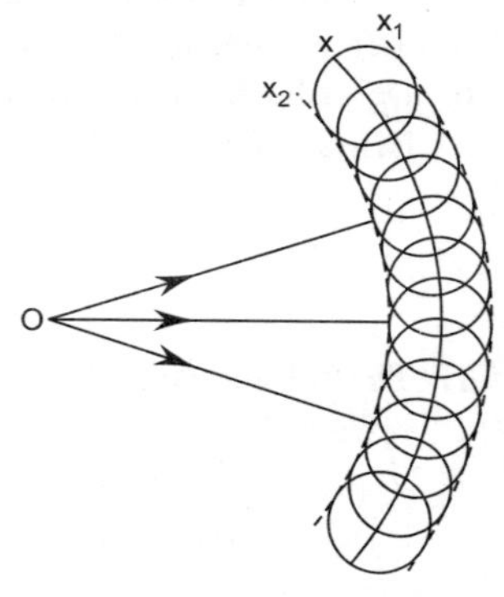

Figure 3C.1

3C.1.1 Huygens' Construction of Wave Surface and Wave Fronts

If a vibrating particle be placed at O in a homogeneous isotropic medium which extends in all directions then the motion of this particle will be communicated to the neighbouring particles in space. This disturbance will travel with equal velocity in every direction and will therefore reach simultaneously to all particles lying on the surface of a sphere drawn with the particle O as center. Such a sphere is called the wave front.

The wave front at any instant of time is thus the locus of all the neighbouring particles in the medium which are being just disturbed at that instant of time and are consequently in the same phase of vibration.

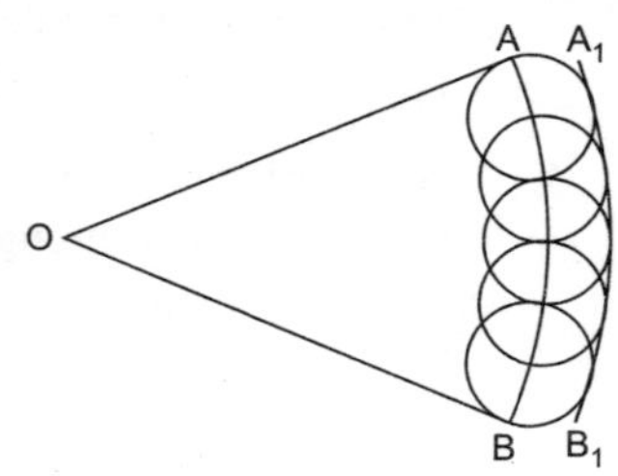

Figure 3C.2

As the wave advances, the wave front moves parallel to itself in a direction normal to the wave front. Such a wave is called *spherical wave*, the section of which by a plane passing through the center is a circle. When this spherical wave proceeds a long distance so that *the radius of sphere is large*, *it becomes a plane wave*.

Let, AB be the wave front at any instant of time t. In order to find the wave front after a time δ, Huygens' gave the following construction :

Particles on the wave front AB will be the sources of secondary disturbances and secondary waves or wavelets will begin to form with those particles as centres. *The size of each secondary wave after a time δ will be a sphere of radius $c\delta$, where c is the velocity of light*. The sections of these secondary waves by a plane will be circles as shown in figure 3C.2. The arc A_1B_1, which is the wave front envelope of these circles, will represent the section of the next wave front after a time δ. By this type of construction we can construct one wavefront from another.

Let, monochromatic waves of light from two sources A and B proceed in the same direction and superpose on a particle of the medium on a screen. The displacement of the particle and hence the intensity of light there, will be maximum or minimum according as the waves meet the particle in the same or in the opposite phases respectively. This phenomenon is known as interference of light.

The waves meet in the same phase, when the crest or trough of one wave superpose on the crest or trough of another wave. The waves meet in opposite phases when the crest of one wave superpose on the trough of another.

3C.1.2 Interference of Electromagnetic Waves

When two or more sets of waves pass through and cross one another in the same medium they behave as if other waves are not present *i.e.*, *the presence of one is totally independent of the presence of the other*. As a consequence of the superposition of two such waves of same wavelength moving in the same direction, *the resultant displacement at any point in the region depends upon the phase difference in the waves and is determined by the algebric sum of the instantaneous displacement of the two waves. If the two waves meet in the same phase then the resultant amplitude will be the sum of the amplitudes of the two component vibrations*. When the waves are in opposite phase then the resultant amplitude will be the difference of the two component amplitudes Now as the intensity of light depends on the square of the amplitude so the resultant intensity will be maximum or minimum depending on the relative phase difference. This phenomenon is known as *interference*.

Relation between Phase Difference and Path Difference

Let the displacement $$y = a \sin \frac{2\pi}{\lambda}(ct - r)$$

Phase part $$= \frac{2\pi r}{\lambda}$$

Let along a wave direction there are two waves at distances r_1 and r_2.

$\therefore$ The phases are $\frac{2\pi}{\lambda} r_1$ and $\frac{2\pi}{\lambda} r_2$.

So, the phase difference, $$\delta = \frac{2\pi}{\lambda}(r_2 - r_1)$$

$$(r_2 - r_1) = \text{Path difference}$$

$\therefore$ $$\text{Phase difference} = \frac{2\pi}{\lambda} \times \text{path difference}$$

Two particles are in same phase of vibration when the phase difference $= 2\pi n$, where $n = 0, 1, 2, 3, 4, 5, 6, \ldots,$

$$\therefore \quad \frac{2\pi}{\lambda}(r_2 - r_1) = 2\pi n$$

$$\therefore \quad r_2 - r_1 = 2n\frac{\lambda}{2}$$

So when path difference is even multiple of $\frac{\lambda}{2}$ then the waves are in same phase of vibration.

For opposite phase $\frac{2\pi}{\lambda}(r_2 - r_1) = (2n+1)\frac{\lambda}{2}$

$$\therefore \quad r_2 - r_1 = (2n + 1)\frac{\lambda}{2}$$

i.e., odd multiple of $\frac{\lambda}{2}$.

3C.1.3 Coherence

Coherence means that two or more electromagnetic waves are in a fixed and predictable phase relationship to each other. In general the phase between two electromagnetic waves can vary from point to point (in space) or change from instant to instant (in time). There are thus two independent concepts of coherence namely temporal coherence and spatial coherence.

(*i*) **Temporal Coherence.** This type of coherence refers to the correlation between the field at a point and the field at the same point at a later time *i.e.*, the relation between E (x, y, z, t_1) and E (x, y, z, t_2). *If the phase difference between the two fields is constant during the period normally covered by observations, the wave is said to have temporal coherence.* If the phase difference changes many times and in an irregular way during the shortest period of observation, the wave is said to be non-coherent.

(*ii*) **Spatial Coherence.** *The waves at different points in space are said to be space coherent if they preserve a constant phase difference over any time t.* This is possible, even when two beams are individually time coherent, as long as any phase change in one of the beams is accompanied by a simultaneous equal phase change in the other beam. With the ordinary light source, this is possible only if the two beams have been produced in the same part of the source.

Time coherence is a characteristic of a single beam of light whereas space coherence concerns the relationship between two separate beams of light.

3C.1.4 Analytical Treatment of Interference

When waves from two sources A and B proceed in the same direction and meet a particle at P, the resultant displacement and hence the intensity at P will be permanently maximum or minimum according as the waves meet the particle in the same or in the opposite phases for all times. This phase relationship of the waves meeting at P will be governed by two factors, *viz.*, (*i*) the path difference of the point P from the two sources A and B and (*ii*) phase difference of waves at the instant of their starting from the sources.

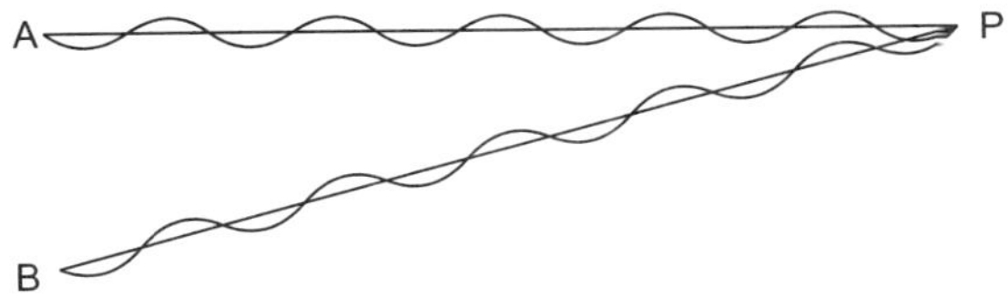

Figure 3C.3

Let monochromatic waves of light (of wavelength λ) from two sources A and B proceed in the same direction and superpose on a particle P of the medium.

If y_1 and y_2 be the displacements of the particle P at time t, when the waves from the sources A and B act on it independently then we may write,

$$y_1 = a \sin \frac{2\pi}{\lambda}(ct - r) = a \sin \theta \text{ (say)}$$

$$y_2 = b \sin \left\{\frac{2\pi}{\lambda}(ct - r) + \delta\right\} = b \sin (\theta + \delta) \text{ (say)}$$

where, $$\theta = \frac{2\pi}{\lambda}(ct - r).$$

The quantities a and b represent the amplitudes of vibration of the particle P when the waves from A and B fall on it, while the quantities C, λ and δ represent the velocity, wavelength and phase difference of the two waves respectively.

So, the resultant displacement of the particle at time t is given by

$$\begin{aligned} y &= a \sin \theta + b \sin (\theta + \delta) \\ &= a \sin \theta + b \sin \theta \cos \delta + b \cos \theta \sin \delta \\ &= (a + b \cos \delta) \sin \theta + (b \sin \delta) \cos \theta. \end{aligned}$$

or, $$y = A \sin (\theta + \phi) \quad \text{...(3C.1)}$$

where, $$A \sin \phi = b \sin \delta$$

$$A \cos \phi = a + b \cos \delta$$

Squaring and adding , we get

$$A^2 = a^2 + b^2 + 2ab \cos \delta \quad \text{...(3C.2)}$$

and $$\tan \phi = \frac{b \sin \delta}{a + b \cos \delta} \quad \text{...(3C.3)}$$

Equation (3C.1) gives the resultant displacement y of the particle at any time t while equation (3C.2) gives the square of the amplitude A of the resultant vibration of the particle.

Now we know that the intensity I is proportional to the square of the amplitude and hence

$$I = \text{constant.} (a^2 + b^2 + 2ab \cos \delta) \quad \text{...(3C.4)}$$

Condition I

The phase difference δ of the two waves which superpose on the particle P is dependent on two factors, *viz.*, (*i*) the phase difference of the waves just at the time when they leave the source, (*ii*) the path difference of the point P from the two sources.

If the two sources which are sending waves are derived from the same source, *i.e.*, if the two sources are coherent, then the phase relationship of the two waves at start from their sources will remain constant for all times and hence the path difference of the point from the two sources is the only deciding factor whether the waves will meet in the same or in the opposite phases. Thus a given point, whose path difference from the two sources is constant, will remain permanently bright or dark when the sources are coherent. Hence the essential condition for a point to remain permanently bright or dark, is that the two sources which are sending waves must be coherent.

Condition II

Let r_1 and r_2 are the distances of the particle P from the two sources. So the path difference of P from the two sources $= (r_2 - r_1)$.

So the corresponding phase difference is

$$\delta = \frac{2\pi}{\lambda}(r_2 - r_1)$$

Now, if $\delta = 2\pi n$ where $n = 0, 1, 2, 3, \ldots$

then, $$I = (a + b)^2 \text{ i.e., maximum}$$

and $$\phi = 0.$$

i.e., $$\frac{2\pi}{\lambda}(r_2 - r_1) = 2\pi n$$

or, $$(r_2 - r_1) = \frac{2\pi}{2\pi} n\lambda = n\lambda, \text{ where, } n = 0, 1, 2, 3, \ldots$$

or, $$(r_2 - r_1) = 2n \cdot \frac{\lambda}{2}$$

So we see that when the path difference of the point from the two sources is even multiple of $\frac{\lambda}{2}$, the intensity of light at that point becomes maximum. For $n = 0$, we get the central bright band for which the path difference is zero. When $n = 1, 2, 3, \ldots$, etc.we respectively get first order, second order, third order, ..., etc., bright bands for which the path difference is $\lambda, 2\lambda, 3\lambda \ldots$, etc.

Again if $\delta = (2n + 1)\pi$ then from equation (3C.4), we get

$$I = (a - b)^2 \text{ i.e., minimum}$$

and $$\phi = 0.$$

i.e., $$\frac{2\pi}{\lambda}(r_2 - r_1) = (2n + 1)\pi$$

or, $$(r_2 - r_1) = (2n + 1)\frac{\lambda}{2}$$

where, $$n = 0, 1, 2, 3, \ldots, \text{etc.}$$

Thus when the path difference of the point from the two sources is odd multiple of $\frac{\lambda}{2}$ the intensity of light at those points would be minimum.

It is now clear that on both sides of the central bright band, there are bands which are alternately of maximum and minimum intensity. This phenomenon is known as interference of light.

Special Case

When $a = b$, the amplitudes of vibration of the particle under the action of the two waves are equal. In that case, the minimum intensity is zero and we get alternately bright and dark bands.

This type of interference is called destructive interference.

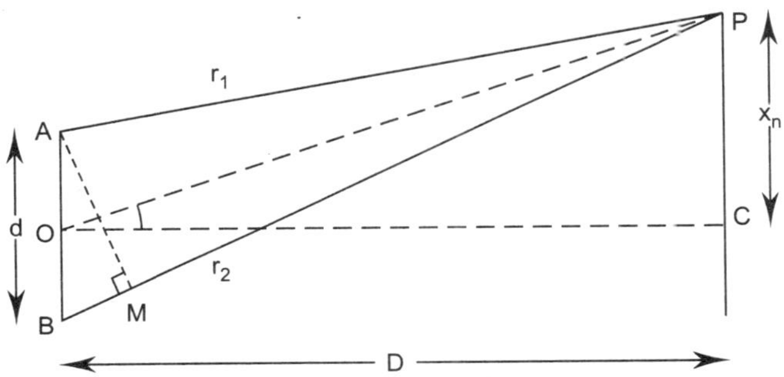

Figure 3C.4

Now, let the nth order bright band be formed at P whose distance from C is x_n (say).

Now since ΔPOC and ΔABM are similar, we get

$$\frac{BM}{AM} = \frac{PC}{OC}$$

or,
$$BM = \frac{PC}{OC} \cdot AM \simeq \frac{PC}{OC} \cdot AB$$

[$\because$ the two sources are close to each other we may take AM $\simeq$ AB = d]

$\therefore$
$$BM = \frac{x_n}{D} \cdot d$$

or,
$$x_n = \frac{D}{d} BM = \frac{D}{d}(r_2 - r_1)$$

or,
$$x_n = \frac{D}{d} \cdot n\lambda$$

which is the position of the nth order bright fringe.

And the distance of the $(n + 1)^{th}$ order bright fringe is

$$x_{n+1} = \frac{D}{d}(n+1)\lambda .$$

So the distance between two consecutive bright bands is given by

$$\beta = x_{n+1} - x_n = \frac{D}{d} \cdot \lambda$$

Again if P be the position of the nth order dark band at a distance x_n' from the central bright band then,

$$x_n' = \frac{D}{d}(r_2 - r_1)$$

$$= \frac{D}{d}(2n + 1)\frac{\lambda}{2}$$

So, the distance of the $(n + 1)^{th}$ order bright fringe is

$$x_n'{}_{+1} = \frac{D}{d}[2(n + 1) + 1]\frac{\lambda}{2}$$

So the distance between two consecutive dark bands is

$$\beta' = x_{n+1}' - x_n' = \frac{D}{d}\lambda = \beta.$$

Thus the distance between any two consecutive dark bands is equal to that between two bright bands. This distance is called the fringe width and is independent of the order of the fringe.

3C.1.5 Condition for Sustained or Continuous Interference of Light

For the production of sustained or continuous interference pattern the following conditions are to be satisfied

(*i*) The two sources should continuously emit waves of same wavelength or frequency since whatever we observe is the result of millions of waves and so the resultant displacement will be constant only if the time period remains the same.

(*ii*) For obtaining interference fringes, the amplitude of the two interference wave trains should be nearly equal, which will reduce general illumination. For complete darkness at the position of minimum intensity the amplitude of the two waves must be exactly equal.

(*iii*) The two interfering sources must be coherent *i.e.*, the two sources must have same frequency and the phase difference of the waves just at the time when they leave the two sources must be zero or have a constant phase difference all through. For in coherent sources there will be rapid alternation of brightness and darkness and general illumination results.

(*iv*) The two sources should be very narrow. A broad source is equivalent to a large number of narrow sources lying side by side. Each set of these sources will produce an interference pattern of its own which will overlap one another and the general illumination results.

(*v*) The two sources emitting a set of interfering beam should be very close to each other. Otherwise since the wavelength of light is extremely small, the path difference between the interfering wave trains will be large and the interference bands will be very close to each other and will not be sharp. They may overlap upon each other.

(*vi*) For complete annulment or for position of minima, the phase difference between the waves from the two sources must be odd multiple of π and for the position of maxima the phase difference must be an even multiple of π.

Explain why light from two different candles are not seen to interfere

When the light waves from two different candles meet at a point, the point will be bright or dark accordinng as the waves meet in the same or opposite phases. This phase difference of the waves meeting at the point will be dependent on (*i*) the path difference of the point from the two sources and (*ii*) the phase relationship of the waves when they are emitted from the two sources. The path difference of the point in question is the same for all times but the phase relation of the two waves when they are emitted by the two different candles cannot be the same for all times. In fact, this phase difference of the waves as they start from the two sources is changing very rapidly with time, and the result is that at every point there is a rapid alternation of brightness and darkness and hence we get general illumination.

Explain whether light energy is destroyed or not by interference

By interference of light we get alternate brightness and darkness. The light energy in the dark region is not destroyed but is simply redistributed in the neighbouring bright regions. Thus in interference, there is no destruction of light energy but it is a case of redistribution of energy. The energy in the bright region is greater when interference takes place, than when there was general illumination. Thus the phenomenon of interference is consistent with the principle of conservation of energy.

3C.1.6 Classification of Different Methods of the Production of Interference Fringes

All the devices for producing interference fringes may be classified as follows :

(*i*) **Division of Wavefront.** We know that a wave-front at any instant is the locus of all the particles of the medium which are in same phase of vibration at that instant. Hence if a wavefront is divided laterally into two parts and the directions of these two divided parts are changed either by reflection or by refraction or by diffraction then those two resulting wave-fronts will have a constant phase relationship between them. The two resulting wave-fronts re-unite, after travelling different paths and produce interference fringes. In these cases we require a very narrow source of light. *Example* Fresnel's Biprism, Lloyd's single mirror, Fresnel's double mirror, Rayleigh's interferometer etc.

(*ii*) **Division of Amplitude.** In this method, the wave front is divided into two parts by a combination of both reflection and refraction. The two resulting wave-fronts, which are derived from the same source will maintain the original width but their amplitudes will be reduced. In this case we need an extended source of light and as the interference effects corresponding to different points of the source are superposed, we get brighter fringes. *Example* : Thin film, Newton's rings, Michelson's interferometer, Fabry-Perot interferometer.

3C.1.7 Interference Fringes by Fresnel's Biprism

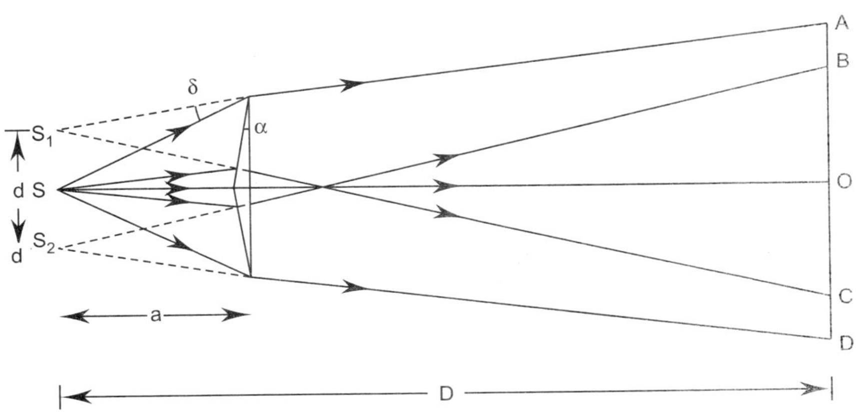

Figure 3C.5

A convenient laboratory arrangement of producing interference fringes is the Fresnel's biprism.

Let a slit S illuminated by a monochromatic source of light is placed close to and parallel to the edge of a biprism. Light from the source S is refracted by the halves of the biprism and illuminates the screen. The portion of the screen AC is illuminated by the light coming from the image S_1 of S due to refraction at the upper half of the biprism. And the portion BD of the screen is illuminated by light coming from the image S_2 of S due to refraction at the lower half of the biprism. Hence the portion BC of the screen is illuminated by both the virtual sources S_1 and S_2. Now since S_1 and S_2 are derived from the same source S then S_1 and S_2 act as coherent sources. Hence in the region BC interference pattern will be produced.

Let, O be the point where the line joining S and the edge of the prism meet the screen. So, O is equidistant from the virtual sources S_1 and S_2 and hence at O central bright fringe will be produced. On either side of O alternate dark and bright fringes will be formed. And the fringe width of the interference pattern is given by

$$\beta = \frac{D}{d}\lambda$$

where, λ = wavelength of light illuminating the source.

D = distance between the source and the screen.

d = separation between the virtual sources S_1 and S_2.

The wavelength λ of the source of light illuminating the slit can be calculated by knowing D and measuring β and d.

To measure d, a convex lens whose focal length is smaller than $\frac{D}{4}$ is interposed between the biprism and the screen. For two distinct positions of the lens sharp images of S_1 and S_2 are formed on the screen. If d_1 and d_2 are the distances between the images in the two positions of the lens and m_1 and m_2 are the magnifications in the two positions then,

$$m_1 = \frac{d_1}{d} \text{ and } m_2 = \frac{d_2}{d}$$

But,

$$m_1 \cdot m_2 = 1$$

$$\frac{d_1}{d} \cdot \frac{d_2}{d} = 1$$

∴

$$d^2 = d_1 \cdot d_2$$

or,

$$d = \sqrt{d_1 \cdot d_2}$$

The separation between two virtual sources may also be found out by making use of the fact that the deviation δ produced in a ray, by a thin prism of small refracting angle is given by

$$\delta = (\mu - 1)\,\alpha$$

where, μ = Refractive index of the material of the prism.

α = Refracting angle on either side of the prism producing the biprism.

Now since the deviation produced by a thin prism is very small, then from the figure we can write,

$$\delta = \frac{d/2}{a} = \frac{d}{2a}$$

where, a = Distance between the source and the biprism.

So, $$\frac{d}{2a} = (\mu - 1)\,\alpha$$

$\therefore$ $$d = 2a\,(\mu - 1)\,\alpha$$

3C.1.8 Determination of the thickness of a thin sheet of a transparent material by Fresnel's Biprism

The biprism experiment can be used to determine the thickness of very thin transparent sheets such as of mica, glass etc.

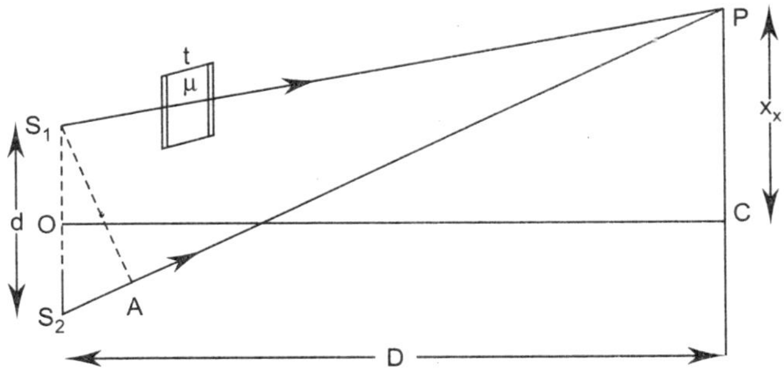

Figure 3C.6

Let S_1 and S_2 be two virtual coherent sources of light which are producing interference fringes on the screen so that C is the position of the central bright band, because the optical path S_1C and S_2C are equal. Now let a thin transparent sheet of thickness t and refractive index μ is introduced in the path of light from one of the sources. Then due to the difference in the velocity of light in air and in the transparent medium, there is a change in path difference and so the central bright band will be shifted.

If the sheet is placed in the path S_1P then the fringes will be shifted upwards and vice versa. Suppose C_a is the velocity of light in air and C_m its velocity in the medium.

$\therefore$ $$\frac{S_2P}{C_a} = \frac{S_1P - t}{C_a} + \frac{t}{C_m}$$

or, $$S_2P = S_1P - t + \frac{C_a}{C_m}\,t \qquad \left[\text{But } \frac{C_a}{C_m} = \mu\right]$$

or, $$S_2P - S_1P = -t + \mu t$$
$$= (\mu - 1)\,t$$

If P is the point originally occupied by the nth fringe then the path difference

$$S_2P - AP = n\lambda \quad \text{as } d \text{ is very small, } S_1P = AP$$

or $$(\mu - 1)\,t = n\lambda$$

Also the distance x through which the fringe is shifted

$$= \frac{n\lambda D}{d}$$

where $\frac{\lambda D}{d} = \beta$, the fringe width

or, $$x = \frac{n\lambda D}{d}$$

or, $$n\lambda = \frac{xd}{D}$$

or, $$(\mu - 1)\,t = \frac{xd}{D}$$

Hence, knowing x, the distance through which the central fringe is shifted, D, d and μ, the thickness of the transparent plate can be calculated.

3C.1.9 Interference from Thin Films (Wedge-Shaped) [DIVISION OF AMPLITUDE]

AB and CD : bounding surfaces of the film (in this case wedge shaped)

α : angle between the surfaces (wedge angle)

P : Extended source (*i.e.*, the width of the beam is large)

NOTE The film is thin and as such QO and SO are very close to each other and combine to form interference fringes.

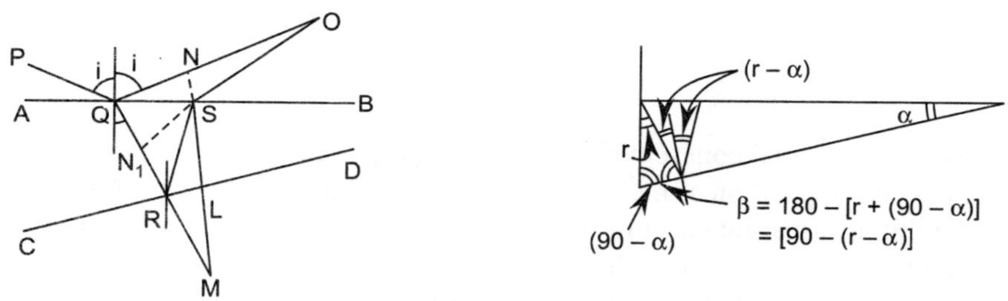

Figure 3C.7

[Here, the wavefront is divided into two parts by reflection and refraction (both) – the resulting wavefronts maintain the original width but their amplitudes get reduced causing reduction in intensities]. Hence this type of interference is due to division of amplitude. This requires extended source.

The path difference between QO and SO is given by

$$l = \mu\,QN_1 + \mu\,N_1R + \mu\,RS - QN$$

$$= \mu\,(N_1R + RS) \qquad (\because \mu\,QN_1 = QN)$$

$$= \mu\,N_1M \quad (\because RS = RM)$$

or $$l = \mu\,2t\cos(r - \alpha) \qquad [\because N_1M = 2t\cos(r - \alpha)]$$

t : the thickness of the film at S

(*i.e.*, SL) and SL = LM

$$\angle QSN_1 = \angle r,\ \angle QSN = \angle i$$

$$\mu = \frac{\sin i}{\sin r} = \frac{QS\sin i}{QS\sin r}$$

$$= \frac{QN}{QN_1}$$

For maxima

$$l = 2n \cdot \frac{\lambda}{2} + \frac{\lambda}{2}$$

the $\frac{\lambda}{2}$ has been added due to reflection at the surface CD. [Since, when a ray is reflected from a surface between a denser and a rarer medium, and the ray is incident *from the denser medium to the surface of separation* a phase change of π takes place for the ray which means a path difference of $\frac{\lambda}{2}$].

Hence, $\quad l = (2n+1)\frac{\lambda}{2} \quad$ or $\quad 2\mu t \cos(r - \alpha) = (2n+1)\frac{\lambda}{2}$

For parallel thin film $\alpha = 0$ and $\quad 2\mu t \cos r = (2n+1)\frac{\lambda}{2}$.

Fringes of equal thickness

The maxima and minima of interference in reflected light from a thin film are given by

$$2\mu + \cos r = (2n + 1)\frac{\lambda}{2} \quad \ldots \text{ maxima}$$

$$2\mu t \cos r = n\lambda \quad \ldots \text{ minima}$$

If μ, r, λ are kept constant for monochromatic parallel beam of light, then the fringes both dark and bright would have the same thickness. Such fringes are fringes of equal width or thickness.

Fringes of equal inclination

This is another type of fringe to be available in case of a thin air film. If μ, t and λ are also held constant, then different order numbers n satisfying maxima or minima will be determined by the values of r and i (angle of incidence).

Since i is remaining the same for a particular order number n, these fringes are referred to as fringes of equal inclination. They are also known as *Haidinger's fringes.*

3C.1.10 Theory and Application of Newton's Ring Experiment

The arrangement is shown in figure 3C.8

L_1OQL_2 : Plano-convex lens in contact with an optically plane glass G_1G_2.

Thus, a wedge-shaped thin film of air is created between the curved surface of the lens and the plane glass surface. The thickness of the air film all over the circumference of a particular circle of radius Q_1Q on the glass plate would be the same.

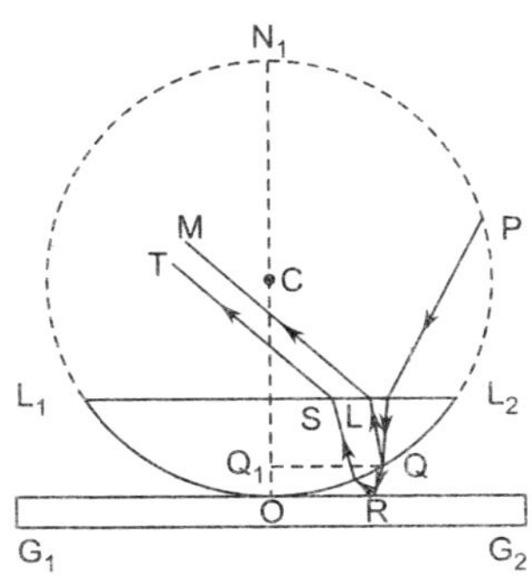

Figure 3C.8

The two rays QLM and RST reflected from the curved surface and the glass plate respectively are derived from the same incident ray PQ and hence are coherent. These two rays interfere and form fringes.

The path difference $l = 2\mu t \cos(r - \alpha)$

where, $\mu = 1$ (for air) in the interspace.

t : thickness of air at the point where the reflection at curved surface takes place. (In this case $t = Q_1O$)

α : wedge angle.

r : angle of refraction of the ray.

Phase Change by Reflection

The first reflection at Q being from a rarer medium there will be no phase change at Q but a phase change π occur at R the point of second reflection from a denser medium (*in this case*, glass) which is equivalent to path difference $\pm \frac{\lambda}{2}$

Hence, $\lambda = 2\mu t \cos r \pm \frac{\lambda}{2}$ (α : considered to be negligibly small)

Evaluation of λ

If the point Q fulfils the condition of brightness then all the points on the circumference of the circle of radius Q_1Q will be bright. Thus $\rho_n = Q_1Q$ (n : order of bright ring)

For very small t, (as compared to 2R)

$$(Q_1Q)^2 = \rho_n^{\,2} = 2Rt \qquad [\because Q_1Q^2 = Q_1O \,.\, Q_1N_1 = t \times (2R - t)\,]$$

$$t = \frac{\rho_n^2}{2R} \qquad [\, t \text{ (at Q)}]$$

and

$$l = \frac{\mu \,.\, \rho_n^2}{R} \cos r \pm \frac{\lambda}{2}$$

For maximum brightness at a point,

$$\frac{\mu \,.\, \rho_n^2}{R} \,.\cos r \pm \frac{\lambda}{2} = \text{even multiple of } \frac{\lambda}{2}$$

$$= 2n \,.\, \frac{\lambda}{2}$$

or

$$\rho_n^{\,2} = \frac{R\,.(2n+1)}{\mu \cos r} \,.\, \frac{\lambda}{2} \qquad [\rho_n : \text{the radius of } n^{\text{th}} \text{ bright ring}]$$

Thus the diameters of bright rings

$$D \propto \sqrt{\text{odd natural numbers}}$$

For

$$\rho_{(n+s)}^2 = \frac{R}{\mu \cos r} \,.\, \frac{[2\,(n+s)+1]}{2} \,.\lambda$$

or

$$\rho_{(n+s)}^2 - \rho_n^2 = \frac{Rs\,\lambda}{\mu \cos r} \qquad \text{(Difference between } (n+s)^{\text{th}} \text{ and } n^{\text{th}} \text{ bright ring)}$$

Hence $$\lambda = \frac{\left[\rho_{(n+s)}^2 - \rho_n^2\right]}{Rs}\cos r \qquad \text{[for air } \mu = 1]$$

For *minimum* brightness at a point (*i.e.*, darkness)

$$\frac{\rho_n^2 \,.\, \mu \cos r}{R} \pm \frac{\lambda}{2} = \text{odd multiple of } \frac{\lambda}{2}$$

or $$\frac{\rho_n^2\ \mu \cos r}{R} = 2n\,\frac{\lambda}{2} \text{ and } \rho_n^2 = \frac{Rn\lambda}{\mu \cos r}$$

and $$\lambda = \frac{\rho_{(n+s)}^2 - \rho_n^2}{Rs}\,.\cos r \qquad [\mu = 1 \text{ for air}]$$

In terms of ring diameters, the wavelength of light

$$\lambda = \frac{D_{(n+s)}^2 - D_n^2}{4Rs}\,.\cos r$$

For *normal incidence* of light from the source,

$$r = 0, \quad \text{or} \quad \cos r = 1$$

hence,

$$\lambda = \frac{D_{(n+s)}^2 - D_n^2}{4Rs}$$

For central fringe,

$$n = 0$$

and $$\rho_n = 0$$

which means that the central fringe will be dark (*i.e.*, at the contact point between the glass plate and the plane convex lens).

Experimental Arrangement

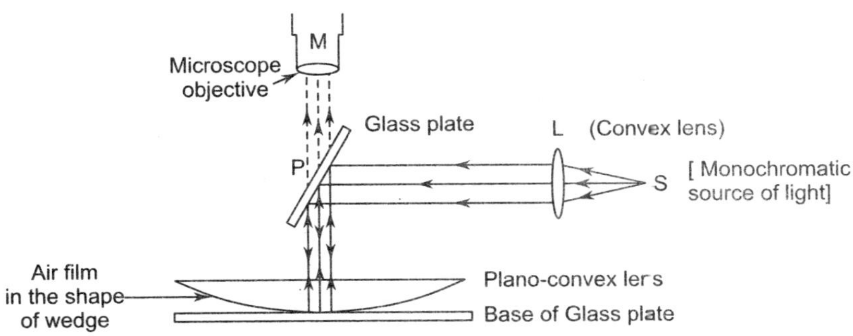

Figure 3C.9

Derivation of the expression for β (the fringe width)

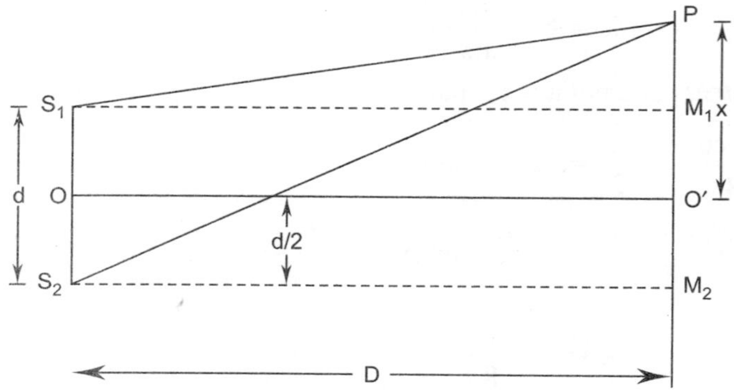

Figure 3C.10

Let S_1 and S_2 be the two virtual sources produced by the biprism and XY is the screen on which the interference fringes are obtained. The point O is equidistant from S_1 and S_2. Therefore the waves from S_1 and S_2 reach O′ in the same phase and reinforce each other at O′. Hence O′ is the center of a bright fringe.

The illumination at any other point P can be obtained by calculating the path distance $S_2P - S_1P$. Let us join S_2P and S_1P and draw perpendiculars S_1M_1 and S_2M_2 on the screen. Let $S_1S_2 = d$

$$S_1M_1 = S_2M_2 = D \text{ and } O'P = x$$

Then
$$(S_2P)^2 = (S_2M_2)^2 + (PM_2)^2$$

$$= D^2 + \left(x + \frac{d}{2}\right)^2$$

$$= D^2\left[1 + \frac{\left(x + \frac{d}{2}\right)^2}{D^2}\right]$$

$$S_2P = D\left[1 + \frac{\left(x + \frac{d}{2}\right)^2}{D^2}\right]^{\frac{1}{2}}$$

$$= D\left[1 + \frac{1}{2}\frac{\left(x + \frac{d}{2}\right)^2}{D^2}\right] \qquad [\because (x + d) << D]$$

$$= D + \frac{1}{2}\frac{\left(x + \frac{d}{2}\right)^2}{D}$$

Similarly $$S_1P = D + \frac{1}{2}\frac{\left(x - \frac{d}{2}\right)^2}{D}$$

$$\therefore \qquad S_2P - S_1P = \frac{1}{2}\frac{\left(x + \frac{d}{2}\right)^2}{D} - \frac{1}{2}\frac{\left(x - \frac{d}{2}\right)^2}{D} = \frac{xd}{D}$$

The resultant intensity at a point is max^m or min^n according as the path difference between the waves is an integral multiple of wavelength or an odd multiple of half wavelength.

Thus for P to be a bright fringe we must have

$$S_2P - S_1P = \frac{xd}{D} = n\lambda \qquad \text{where } n = 0, 1, 2 \ldots$$

or $$x = \frac{D}{d}(n\lambda)$$

for P to be the dark fringe

$$S_2P - S_1P = \frac{xd}{D} = (2n+1)\frac{\lambda}{2}$$

$$x = \frac{D}{d}(2n+1)\frac{\lambda}{2}$$

Let x_n and x_{n+1} denote the distances of the nth and $(n + 1)^{\text{th}}$ bright fringes. Then the distance between $(n + 1)$ and n^{th} bright fringes is given by

$$\beta = x_{n+1} - x_n = \frac{D}{d}(n+1)\lambda - \frac{D}{2d}n\lambda = \frac{D}{d}\lambda$$

The same result holds for dark fringes. The distance $\frac{D\lambda}{d}$ is called the 'fringe width'.

IN FOCUS

- **Intensity of Interference of Light**

$$I = \text{constant.}\ (a^2 + b^2 + 2ab\cos\delta)$$

where a = amplitude of first wave

b = amplitude of second wave

δ = phase difference of two rays

$$= \frac{2\pi}{\lambda}(r_2 - r_1)$$ [r_1 and r_2 are distance of any particle P from the two sources]

Condition for Maxima

If $\delta = 2pn$

where $n = 0, 1, 2, 3 \ldots$

$$I = (a + b)^2$$

So $$(r_2 - r_1) = 2n\frac{\lambda}{2} = n\lambda$$

Condition for Minima

If $\delta = (2n + 1)\pi$ where $n = 0, 1, 2 \ldots$

$$I = (a - b)^2$$

So $$(r_2 - r_1) = (2n + 1)\frac{\lambda}{2}.$$

Distance between two consecutive bright fringes

$$\beta = x_{n+1} - x_n = \frac{D}{d}.\lambda$$

where D = Distance between source and the screen

d = Distance between two sources

x_{n+1} = From centre distance of $(n + 1)^{th}$ bright fringe

x_n = From centre distance of nth bright fringe.

- **Fresnel's Biprism:** The deviation δ produced in a ray, by a thin prism of small refracting angle α is given by

$$\delta = (\mu - 1)\alpha$$

where μ = refractive index of the material of the biprism.

or $$d = 2a(m - 1)\alpha$$

where $$\delta = \frac{d}{2a}$$

d = distance between two sources

a = distance between source and biprism

Thickness of a thin sheet of a transparent material

$$t = \frac{xd}{D(\mu - 1)}$$

where x = the distance through which the central fringe is shifted.

Interference From Thin Film

Path difference $l = 2\mu t \cos(r - \alpha)$

where μ = refractive index

t = thickness of the material

r = refracting angle

α = angle between the surfaces (wedge angle)

Fringes of Equal Thickness

For maxima $2\mu t \cos r = (2n + 1)\frac{\lambda}{2}$

For minima $2\mu t \cos r = n\lambda$

Newton's Ring

Path difference $l = 2\mu\, t \cos (r - \alpha)$

Evaluation of t and λ

$$t = \frac{\rho_n^2}{2R}$$

where ρ_n = Radius of the circular ring.

R = Radius of curvature of the lens.

$$\lambda = \frac{D^2 (n + s) - D_n^2}{4RS}$$

λ : wavelength of monochromatic light

R : Radius of curvature of the lens.

D_n : Diameter of nth bright ring

D_{n+s} : Diameter of $(n + s)^{th}$ bright ring.

Solved Problems

Problem 1. *Two coherent sources of intensity ratio α interfere. Prove that in the interference pattern*

$\frac{I_{max} - I_{min}}{I_{max} + I_{min}} = \frac{2\sqrt{\alpha}}{1+\alpha}$ *where the symbols have the usual meaning.*

[Agra University, Meerut University, Rohilkhand University]

Solution. Let a and b be the amplitudes of the interfering waves then we know that the maximum and minimum intensities are

$$I_{max} = (a + b)^2 \quad [\because \text{ Intensity } I = a^2 + b^2 + 2ab \cos \delta]$$

and

$$I_{min} = (a - b)^2$$

Here,

$$\alpha = \frac{a^2}{b^2} = \frac{I_{max}}{I_{min}} \quad [\because \text{ Intensity (I)} \propto \text{(amplitude)}^2, I \propto a^2]$$

Now,

$$\sqrt{\alpha} = \frac{a}{b}$$

$$\therefore \quad \frac{I_{max}}{I_{min}} = \frac{(a+b)^2}{(a-b)^2}$$

or, $$\frac{I_{max} - I_{min}}{I_{max} + I_{min}} = \frac{(a+b)^2 - (a-b)^2}{(a+b)^2 + (a-b)^2} = \frac{4ab}{2(a^2+b^2)}$$

$$= \frac{2.\left(\frac{a}{b}\right)}{1+\left(\frac{a}{b}\right)^2} = \frac{2\sqrt{\alpha}}{1+\alpha} \quad \textbf{(Proved)}$$

Problem 2. *In a Llyod's mirror experiment, calculate the ratio of intensities at the interference maxima and minima if the mirror reflects only 75% of the light incident on it.* [Calcutta University]

Solution. Reflectivity, $R = 75\% = \frac{3}{4}$.

Here amplitude reflect co-efficient $r = \sqrt{R} = \sqrt{\frac{3}{4}} = \frac{\sqrt{3}}{2}$

We know that, $$\frac{I_{max}}{I_{min}} = \left(\frac{a_1 + a_2}{a_1 - a_2}\right)^2.$$

Consider a_1 be the amplitude of the direct beam then the amplitude of the reflected beam $a_2 = \frac{\sqrt{3}}{2} a_1$.

$$\therefore \quad \frac{I_{max}}{I_{min}} = \left(\frac{a_1 + \frac{\sqrt{3}}{2} a_1}{a_1 - \frac{\sqrt{3}}{2} a_1}\right)^2 = 194.$$

Problem 3. Two *coherent sources whose intensity ratio is 81 : 1 produce interference fringes. Deduce the ratio of maximum to minimum intensity in fringe system.*

[Agra University, Rohilkhand University, Garhwal University].

Solution. Here, given that,

$$\frac{I_1}{I_2} = \frac{81}{1} = \frac{a^2}{b^2}$$

$$\therefore \quad \frac{a}{b} = \frac{9}{1}$$

Now, $$\frac{I_{max}}{I_{min}} = \frac{(a+b)^2}{(a-b)^2} = \frac{b^2\left(1+\frac{a}{b}\right)^2}{b^2\left(\frac{a}{b}-1\right)^2} = \frac{(9+1)^2}{(9-1)^2} = \frac{100}{64} = \frac{25}{16}$$

$$\therefore \quad I_{max} : I_{min} = 25 : 16.$$

Problem 4. *Two straight and narrow parallel slits 1 mm apart are illuminated by monochromatic light. Fringes formed on the screen held at a distance of 100 cm from the slits are 0.50 mm apart. What is the wavelength of light?* [Delhi University]

Solution. We know that the separation between two slits (d) is the separation between two coherent sources.

i.e., $d = 1 \text{ mm} = 0.1 \text{ cm}$

Distance between the source (slit) and the screen D = 100 cm.

Here, fringe width (β) = 0.50 mm = 0.05 cm

We know, $$\beta = \frac{\lambda D}{d},$$

i.e., $$\lambda = \frac{\beta d}{D} = \frac{0.05 \times 0.1}{100} = 5 \times 10^{-5} \text{ cm}$$

$$= 5000 \text{ Å}. \qquad [\because 1 \text{ Å} = 10^{-8} \text{ cm}]$$

Wavelength of light is 5000 Å

Problem 5. *Two straight and narrow parallel slits 3 mm apart are illuminated by a monochromatic light of wavelength 5900 Å. Fringes are obtained on a 0.60 m distant screen from the slits. Find the value of the fringe width.* [Rohilkhand University].

Solution. Here $d = 3 \text{ mm} = 0.3 \text{ cm}$

$$\lambda = 5900 \text{ Å} = 5900 \times 10^{-8} = 59 \times 10^{-6} \text{ cm}$$

and distance between screen and slits are D = 0.60 m = 60 cm

Now fringe width $$\beta = \frac{59 \times 10^{-6} \times 60}{0.3} = 118 \times 10^{-4} \text{ cm}.$$

Problem 6. *A biprism is placed 5 cm from a slit illuminated by sodium light (λ = 5890 Å). The width of the fringe obtained on a screen 75 cm from the biprism is 9.424×10^{-2} cm. What is the distance between the two coherent sources?* [Nagpur University]

Solution. Here $\lambda = 5890 \text{ Å} = 5890 \times 10^{-8} \text{ cm}$

Width of the fringe (β) = 9.424×10^{-2} cm

$\therefore$ Distance between the slits and the screen is D = (75 + 5) cm

$$= 80 \text{ cm}$$

We know $$\beta = \frac{\lambda D}{d}$$

$$d = \frac{\lambda D}{\beta} = \frac{5890 \times 10^{-8} \times 80}{9.424 \times 10^{-2}} = 0.05 \text{ cm}.$$

$\therefore$ Distance between the two coherent sources is 0.05 cm.

Problem 7. *In Young's double slit experiment, the distance between the two slits is 0.5 mm. The wavelength of the light used is 5000 Å and the separation between the source and the screen is 50 cm. Calculate the fringe width in this case.* [Kalyani University Engineering Examination]

Solution. We know fringe width, $\beta = \dfrac{D\lambda}{d}$.

Here, d = distance between the slits = 0.5 mm = 0.05 cm.

D = Distance between the source and the screen = 50 cm

and λ = Wavelength of the light used = 5000 Å = 5000×10^{-8} cm

$$= 5000 \times 10^{-8} \text{ cm.}$$

$$\therefore \quad \beta = \frac{50 \times 5000 \times 10^{-8}}{0.05} = 0.05 \text{ cm.}$$

So fringe width is 50 cm.

Problem 8. *Interference fringes are produced by Fresnel's biprism in the focal plane of a reading microscope which is 1.0 m from the slit. A lens interposed between the biprism and the microscope gives two images of the slit in two positions. If the images of the slits are 4.05 mm in one position 2.90 mm in the other position and the wavelength of the sodium light is 5893 Å find the distance between the consecutive interference bands.*

[V.S. University, Pune University, Bhopal University, Meerut University]

Solution. Here $\lambda = 5893 \text{ Å} = 5893 \times 10^{-8} \text{ cm}, \text{D} = 1 \text{ m} = 100 \text{ cm}$

$$d_1 = 4.05 \text{ mm} = 4.05 \times 10^{-1} \text{ cm}, d_2 = 2.90 \text{ mm} = 2.90 \times 10^{-1} \text{ cm}$$

$$\therefore \quad d = \sqrt{d_1 \cdot d_2} = \sqrt{4.05 \times 2.90 \times 10^{-2}} = 3.427 \times 10^{-1} \text{ cm}$$

We know

$$\beta = \frac{\lambda \text{D}}{d} = \frac{5893 \times 10^{-8} \times 100}{3.427 \times 10^{-1}} \text{ cm} = 0.0172 \text{ cm}$$

∴ distance between the consecutive interference bands *i.e.*, fringe width is 0.0172 cm.

Problem 9. *In a biprism experiment the eye-piece is placed at a distance of 1.2 m from the source. The distance between the virtual sources was found to be 7.5×10^{-4} m. Find the wavelength of light if the eye piece is to be moved transversely through a distance of 1.888×10^{-2} m for 20 fringes.*

[Delhi University]

Solution. Here, given that $\text{D} = 1.2 \text{ m}, d = 7.5 \times 10^{-4} \text{ m}$

$$n = 20, \lambda = ?$$

∴ Fringe width

$$\beta = \frac{1.888 \times 10^{-2}}{20} = 0.944 \times 10^{-3} \text{ m}$$

$$\therefore \quad \lambda = \frac{\beta d}{\text{D}} = \frac{0.944 \times 10^{-3} \times 7.5 \times 10^{-4}}{1.2}$$

$$= 5900 \times 10^{-10} \text{ m} = 5900 \text{ Å}.$$

Problem 10. *A biprism is set with sodium light (λ = 5893 Å) and 40 fringes are seen in the field of view of the eye piece. If sodium lamp is replaced by mercury lamp with violet filter (λ = 4358 Å), then find the number of fringes obtained with this wavelength.*

[Garhwal University, Rohilkhand University]

Solution. We know, $\beta = \dfrac{\lambda \text{D}}{d}$

Here D is the distance of the field of view of the eye piece from source. Here the setting of arrangement remains the same. So we can say $\beta \propto \lambda$.

Let us consider fringewidth β_1 and β_2 for wavelength λ_1 and λ_2 respectively are given by

$$\beta_1 = \frac{\text{D}}{d} \cdot \lambda_1$$

and $$\beta_2 = \frac{D}{d}\lambda_2$$

∴ $$\frac{\beta_1}{\beta_2} = \frac{\lambda_1}{\lambda_2}$$

If 40 fringes are viewed with λ_1 and say x are viewed with λ_2 then since field of view remains the same.

$$x\,\beta_2 = 40\,\beta_1$$

or, $$x = 40\,\frac{\beta_1}{\beta_2} = \frac{40\times\lambda_1}{\lambda_2} = \frac{40\times 5893}{4358} \simeq 54.$$

∴ The number of fringes obtained with the wavelength = 54.

Problem 11. *The distance between the slit and the biprism and the screen are 50 cm each. The angle of the biprism is 179° and its refractive index is 1.5. If the distance between the successive fringes is 0.0135 cm. Calculate the wavelength of light.* [Meerut University, Agra University]

Solution. According to this problem $\beta = 0.0135$ cm, D = 100 cm, and $\alpha = \left(\frac{1}{2}\right)^{o} = \frac{\pi}{360}$ radian and $\mu = 1.5$

We know that $\beta = \frac{\lambda D}{d}$, where

$d = 2x(\mu - 1)\alpha$, Here x is the distance between the slit and biprism, α is the refracting angle of the prism and μ is the refractive index of biprism glass.

∴ $$x = 50\text{ cm},\ \alpha = \left(\frac{1}{2}\right)^{o} = \frac{\pi}{360}\text{ radian}$$

Therefore, $$\lambda = \frac{\beta d}{D}$$

$$= \frac{0.0135\times 2\times 50\times(1.5-1)\,\pi}{360\times 100}$$

$$= 5893\times 10^{-8}\text{ cm} = 5893\text{ Å}.$$

Problem 12. *Calculate the separation between the coherent sources formed by a prism, whose inclined faces makes angles of 2° with its base. The slit source being 10 cm away from the biprism ($\mu = 1.50$).* [Delhi University]

Solution. We know that the $d = 2a(\mu - 1)\alpha$

where d = distance between two coherent sources.

and μ = refractive index of the material of biprism

α = base angle = 2° = $\frac{2\times\pi}{180}$ radian = $\frac{\pi}{90}$ radian

a = distance between source and biprism = 10 cm

∴ $$d = \frac{2\times(1.5-1)\times\pi\times 10}{90} = 0.35\text{ cm}.$$

Problem 13. *The inclined faces of a biprism (μ = 1.5) make angles of 1° with the base of the prism. The slit is 10 cm from the biprism and is illuminated by light of wavelength 5900 Å. Find the fringe width observed at a distance of one meter from the biprism.* [K. University]

Solution. Given that $\mu = 1.5$, base angle $\alpha = 1° = \dfrac{\pi}{180}$ radian.

Distance between the source and screen = 100 cm + 10 cm = 110 cm

$$\lambda = 5900 \text{ Å} = 5900 \times 10^{-8} \text{ cm} = 59 \times 10^{-6} \text{ cm}$$

Fringe width, $\beta = \dfrac{\lambda D}{d}$, where $d = 2x(\mu - 1)\alpha$

$$\beta = \frac{59 \times 10^{-6} \times 110 \times 180}{2 \times \pi \times 0.5 \times 10} = 0.0372 \text{ cm.}$$

[Here x = Distance of the biprism from the source = 10 cm]

Problem 14. *In Newton's ring experiment the diameter of the nth and $(n + 14)^{th}$ rings are 4.2 mm and 7.0 mm respectively. Radius of curvature of the plano-convex lens is 1 m. Calculate the wavelength of light.* [Rohilkhand University]

Solution. If D_n and D_{m+n} are the diameters of nth and $(m + n)^{th}$

Newton's rings respectively formed by a plano-convex lens of radius of curvature R by light of wavelength λ, then

$$\lambda = \frac{D_{m+n}^2 - D_n^2}{4mR}$$

Here $D_n = 4.2 \text{ mm} = 0.42 \text{ cm}, D_{m+n} = 0.7 \text{ cm}$

∴ $m = 14$, and

radius of curvature $R = 1 \text{ m} = 100 \text{ cm}$

we get,

$$\lambda = \frac{(0.7)^2 - (0.42)^2}{4 \times 14 \times 100} = 5.6 \times 10^{-5} \text{ cm}$$
$$= 5600 \text{ Å.}$$

Problem 15. *In a Newton's ring experiment, the diameter of the 5th dark ring is 0.336 cm and the diameter of the 15th dark ring is 0.590 cm. Find the radius of the plano-convex lens if the wavelength of light used is 5890 Å.* [WBUT]

Solution. Here $D_5 = 0.336$ cm and $D_{10+5} = 0.590$ cm

$$\lambda = 5890 \text{ Å}, \quad R = ?$$

Now

$$R = \frac{D_{10+5}^2 - D_5^2}{4 \times 10 \times \lambda} = \frac{(0.590)^2 - (0.336)^2}{4 \times 10 \times 5890 \times 10^{-8}} \text{ cm}$$
$$= 99.83 \text{ cm.}$$

Problem 16. *In a Newton's ring experiment the diameter of the 23rd ring was found to be 0.501 cm and that of the 3rd ring was 0.181 cm. If the radius of curvature of the plano-convex lens is 0.5 m, calculate the wavelength of light used?* [Kolkata University]

Solution. Here $D_{23} = 0.501 \text{ cm}, D_3 = 0.181 \text{ cm}, m = 23 - 3 = 20$

and $R = 0.5 \text{ m} = 50 \text{ cm}, \quad \lambda = ?$

We know that $\lambda = \dfrac{D^2_{m+n} - D^2_n}{4mR} = \dfrac{(0.501)^2 - (0.181)^2}{4 \times 20 \times 50}$

$= 5456 \times 10^{-8}$ cm $= 5456$ Å.

Problem 17. *In a Newton's ring experiment, the diameter of the 4th and 12th rings are 0.4 cm and 0.7 cm respectively. Find the diameter of the 20th dark ring.* [P.E.C. University]

Solution. $D_{12} = 0.7$ cm, $D_4 = 0.4$ cm, $m = 12 - 4 = 8$,

We know that, $\lambda = \dfrac{D^2_{m+n} - D^2_n}{4mR}$

i.e., $\lambda R = \dfrac{(0.7)^2 - (0.4)^2}{4 \times 8} = 1.031 \times 10^{-4}$ cm^2

∴ The diameter of the 20th ring

$$\lambda = \frac{D^2_{20}}{4 \times 20 \times R} \quad \therefore D_{20} = \sqrt{80\,\lambda R}$$

$$= \sqrt{80 \times 1.031 \times 10^{-4}}$$

$$= 0.908 \text{ cm.}$$

Problem 18. *If the diameter of the 10th Newton dark ring is 0.50 cm and the wavelength of the light used is 6250 Å, calculate the radius of curvature of the lens and the thickness of the air film.*

[Kolkata University]

Solution. Given that $D_{10} = 0.50$ cm. $n = 10$, $\lambda = 6250 \times 10^{-8}$ cm

We know that, $D_n^2 = 4n\lambda R$, $\therefore R = \dfrac{D^2_n}{4n\lambda}$

or, $R = \dfrac{(0.50)^2}{4 \times 10 \times 6250 \times 10^{-8}}$ cm $= 100$ cm

Again, $2t = \dfrac{r^2_n}{R} = \dfrac{D^2_n}{4R}$ [t = thickness of the air film]

∴ $t = \dfrac{(0.50)^2}{4 \times 100 \times 2} = 0.0003125$ cm

Problem 19. *Newton's rings are observed in reflected light of wavelength 5.9×10^{-7} m. The diameter of the 10th dark ring is 0.5 cm. Find the radius of curvature of the lens and the thickness of the air film.* [Calcutta University, Delhi University, Punjab University]

Solution. Here, given that, $\lambda = 5.9 \times 10^{-7}$ m. $D_{10} = 0.5$ cm

$= 5 \times 10^{-3}$ meter

Now, $D^2_n = 4n\lambda R$,

$$R = \frac{D^2_n}{4n\lambda} = \frac{(5)^2 \times 10^{-6}}{4 \times 10 \times 5.9 \times 10^{-7}} = 1.059 \text{ meters}$$

Thickness of air film at 10th dark ring

$$t = \frac{r_n^2}{2R} = \frac{D_n^2}{8R} = \frac{5 \times 10^{-3} \times 5 \times 10^{-3}}{8 \times 1.059} = 2.95 \times 10^{-6} \text{ meter.}$$

Problem 20. *A thin parallel film of glass (μ = 1.5) is illuminated by sodium light (λ = 5893 × 10^{-10} m) in such a way that angle of refraction of light in the film is 60°. Calculate the smallest thickness of the film in which it appears dark in reflected light. How will it appear in transmitted light?*

[Meerut University]

Solution. We know that $2\mu t \cos r = n\lambda$ for destructive interference in reflected light of wavelength λ by a parallel film of thickness t, refractive index (μ) and angle of refraction (r).

For smallest thickness, the film appears dark when $n = 1$,

$$t = \frac{\lambda}{2\mu \cos r} = \frac{5893 \times 10^{-10} \, m}{2 \times 1.5 \times \cos 60°} = 3928.7 \times 10^{-10} \text{ m}$$

$$= 3928.7 \text{ Å}$$

Problem 21. *A soap film is illuminated by white light incident at an angle of 30°. The reflected light is examined by a spectroscope in which dark corresponding to λ = 6 × 10^{-7} m is found. Find the thickness of the film, given n of film = 1.33.* [Kurukshetra University]

Solution. Fringes of equal thickness for minima or dark band, $2\mu t \cos r = \lambda$

Given incident angle $i = 30°$, ∴ $\mu = n$

$$\therefore \quad \frac{\sin i}{\sin r} = 1.33 \Rightarrow \sin r = \frac{\sin 30°}{1.33} \approx 0.376$$

Given wavelength $\lambda = 6 \times 10^{-7}$ m,

$$\therefore \quad \text{The thickness of the film, } t = \frac{6 \times 10^{-7} \text{ m}}{2 \times 1.33 \times \sqrt{1 - (0.376)^2}} = 2.434 \times 10^{-7} \text{ m.}$$

Multiple Choice Questions

1. Light waves from two coherent sources of same intensity I interfere. The intensity of light of maxima is

   (*a*) Zero (*b*) $\frac{\lambda}{2}$ (*c*) λ (*b*) 2λ

   [*Himachal University*]

2. When interference of light takes place at some region the light energy is

   (*a*) created (*b*) destroyed (*c*) redistributed

   [*W.B.U.T.*]

3. On reflection from a denser medium, the path difference introduced is

   (*a*) zero (*b*) $\frac{\lambda}{2}$ (*c*) λ (*d*) 2λ

   [*Himachal University*]

4. Newton's rings are examples of interference fringes of

   (*a*) Equal inclination and coherence is obtained by division of amplitude
   (*b*) Equal thickness and coherence is obtained by division of amplitude
   (*c*) Equal thickness and coherence is obtained by division of wavefront. [*W.B.U.T.*]

5. In the phenomenon of interference, the law of conservation of energy be
   (*a*) Never violated (*b*) Always violated
   (*c*) Sometime violated (*d*) Violated depending on frequency.
6. The centre of the Newton's rings for the reflected system of a monochromatic source of light is
   (*a*) bright (*b*) dark (*c*) none [*W.B.U.T.* (*Special*)]
7. Radii of Newton's ring vary as
   (*a*) Square root of natural number
   (*b*) Square of the natural number
   (*c*) None. [*WBUT*]
8. Phase difference of π corresponds to path difference
   (*a*) 2λ (*b*) λ (*c*) $\frac{\lambda}{2}$ (*d*) $\frac{\lambda}{4}$.
9. The source of light are said to be coherent of the waves produced by them have the same
   (*a*) Wavelength
   (*b*) Amplitude
   (*c*) Wavelength and constant phase difference
   (*d*) Amplitude and the same wavelength.

Answers

1.	**2.** (*c*)	**3.** (*b*)	**4.** (*a*)	**5.** (*a*)
6. (*b*)	**7.** (*a*)	**8.** (*c*)	**9.** (*c*)	

Review Questions

1. What is interference of light ? Discuss the necessary conditions for interference of light?
   [**Hints :** Refer to Art. 3C.1, 3C.1.2 and 3C.1.5]
2. What are coherent sources ? [*WBUT, Pune University*]
   [**Hints :** Refer to Art. 3C.1.3]
3. Discuss the conditions necessary for producing interference of light? How did Fresnel obtain those conditions in his biprism experiment to determine the wavelength of monochromatic light?
   [**Hints :** Refer to Art. 3C.1.5, and 3C.1.7] [*Agra University*].
4. Find an expression for fringe width of bright and dark fringes?
   [**Hint :** Refer to Art. 3C.1.4] [*WBUT, V.S. University, G.N.D. University*]
5. Show that laws of conservation of energy is not violated in the interference.
   [**Hints :** Refer to Art. 3C.1.5] [*WBUT*]
6. Describe in detail how you would find the wavelength of a monochromatic source using a Fresnel's biprism.[**Hints :** Refer to Art. 3C.1.7] [*Mysore University, Punjab University, Agra University*]
7. Explain clearly the theory and the experimental arrangement of Newton's ring experiment.
   [**Hints :** Refer to Art. 3C.1.10] [*Vikram University*]
8. Describe the formation of Newton's rings obtained due to reflection of monochromatic light. Obtain expressions for the diameters of dark and bright rings. Why the central fringe is black?
   [**Hints :** Refer to Art. 3C.1.10]
9. Write a short note on Fresnel's biprism.
   [**Hints :** Refer to Art. 3C.1.7] [*WBUT*]
10. What are coherent sources ? How are they produced ? Explain. [*Cochin University*]
11. Explain analytically the interference of light on the basis of the wave theory. Explain briefly the phenomenon in relation to the law of conservation of energy. [*Himachal University*]

12. Show that the expression of the radius of curvature of the convex lens used in Newton's ring apparatus '6'

$$R = \frac{D_{n+m}^2 - D_n^2}{4m\lambda}$$ [*WBUT*]

13. (*a*) Derive an expression for the bright and dark fringe width formed by the biprism. [*WBUT, B. OPT*]

(*b*) Hence discuss how the wavelength of unknown light can be measured. [*Calcutta University*]

Unsolved Problems

1. Two coherent sources whose intensity ratio is 25 : 100, produce interference fringes. Calculate the intensity ratio of bright and dark fringes of the system. **[Ans.** 9 : 1]

2. In a Young's double slit experiment the distance between the two coherent sources is 1.15 mm. Calculate the fringe width that would be observed on a screen placed at a distance 85 cm from the source. The wavelength of light used is 5893 Å. [*WBUT*].

3. A biprism of angle 1° and $\mu = 1.5$ forms interference fringes on a screen 80 cm away from it. If the distance between the source and the biprism is 30 cm, then find the fringe width for light of wavelength 5000 Å. **[Ans.** $\beta = 1.05 \times 10^{-2}$ cm]

4. A fresnel biprism having angle of 1° and refractive index 1.5 forms interference fringes on a screen placed 80 cm from the biprism. If the distance between the source and the biprism is 20 cm, find the fringe separation when the wavelength of light used is [*Lucknow University*]

(*a*) 6900 Å (*b*) 4600 Å

[Ans. (*a*) 1.975×10^{-2} cm (*b*) 1.317×10^{-2} cm]

5. In a Newton's ring experiment the diameter of the 5th ring was 0.336 cm and the diameter of 15th ring was 0.590 cm. Find the radius of curvature of the plano convex lens if the wavelength of light used is 5890 Å. **[Ans.** 99.8 cm]

6. In an experiment of Newton's rings the diameters of nth and $(n + 8)^{th}$ order bright rings are 4.2 mm and 7.0 mm respectively. The radius of curvature of the convex lens is 2 meters. Find the wavelength of light. **[Ans.** 4900 Å]

[*Bundelkhand University*]

7. The radii of the nth and $(n + 5)^{th}$ Newton's rings are 4 mm and 6 mm respectively. The radius of the lens is 10 m. Find the wavelength of the light. [*Calcutta University*] **[Ans.** 4000 Å]

[Hints : Here $m = 5$, $d_{m+n} = 2 \times 0.6$ cm, $d_n = 2 \times 0.4$ cm

$R = 10\ m = R = 1000$ cm

$$\lambda = \frac{d_{mi+n}^2 - d_n^2}{4m\lambda} = 4000 \text{ Å}]$$

8. What should be the thickness of a non-reflecting layer to be deposited on glass surface corresponding wavelength 6000 Å ? Refractive index of the layer is 1.35.

[*Banaras Hindu University*] **[Ans.** 1110 Å]

[Hints : Using the equation $\lambda = 4\mu t$ where t is the thickness]

3D

Chapter

Diffraction of Light

3D.1 DIFFRACTION OF LIGHT

When light from a narrow linear slit is incident on the sharp edge of an obstacle, it will be found that there is illumination to some extent within the geometrical shadow of the obstacle. Thus light can bend around an obstacle and this bending of light is known as **Diffraction.** This bending is greater for light of longer wavelengths.

According to Huygens' principle each point on a wave front is a source of secondary wavelet. A wave front is the locus of all points which are in the same phase of vibration. Hence the secondary sources lying on a wave front are coherent sources and can interfere with each other. Such interference of secondary wavelets from a particular wave front gives rise to Diffraction phenomenon.

3D.1.1 Classes of Diffraction

Fresnel Diffraction

This is the type of diffraction in which either the source or the point of observation or both are at **finite** distances **from** the diffraction opening or obstacle. Here the **incident** wave front is divergent. **Fraunhofer Diffraction** is the one in which both the source and the point of observation are **effectively at infinite distance from the diffraction obstacle** or opening. Here the **incident** wave front is plane.

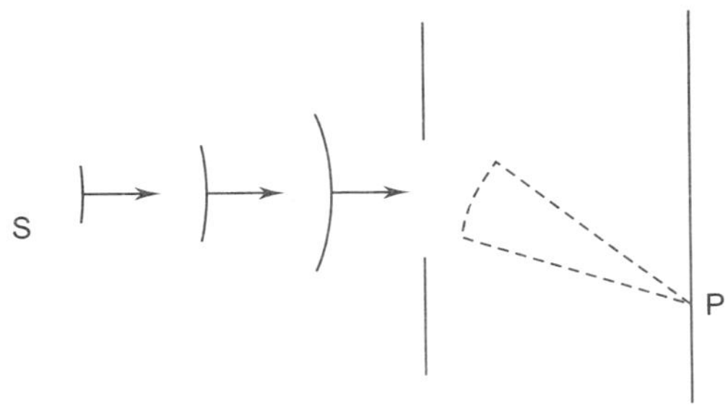

Figure 3D.1 *Fresnel diffraction.*

Table 3D.1 Difference between Interference and Diffraction

Interference	*Diffraction*
(*i*) Interference is the result of interaction of light coming from two different wave fronts.	(*i*) Diffraction pattern is the result of interaction of light coming from different parts of the same wave front.
(*ii*) Interference fringes may or may not be of same width.	(*ii*) Diffraction fringes are not of the same width.
(*iii*) Points of minimum intensity are perfectly dark.	(*iii*) Points of minimum intensity are not perfectly dark.
(*iv*) All bright bands are of uniform intensity.	(*iv*) All bright bands are not of same intensity.

3D.1.2 Fraunhofer Diffraction Pattern due to Single Slit

A parallel beam of monochromatic light (*i.e.*, a plane wave) of wave length λ is incident normally on a narrow slit of width "a" (= PQ). The parts of the **secondary waves** which proceed normal to the slit will get converged at 0 by the lens L to form the principal maximum. The parts of the secondary waves which proceed along the direction making an angle θ with the normal will meet at a point O_1 producing either minimum or maximum depending on their **phase relationship.** The **path difference** of the rays from the extreme points P and Q of the slit is PR = $a \sin\theta$ and hence their **phase** difference is

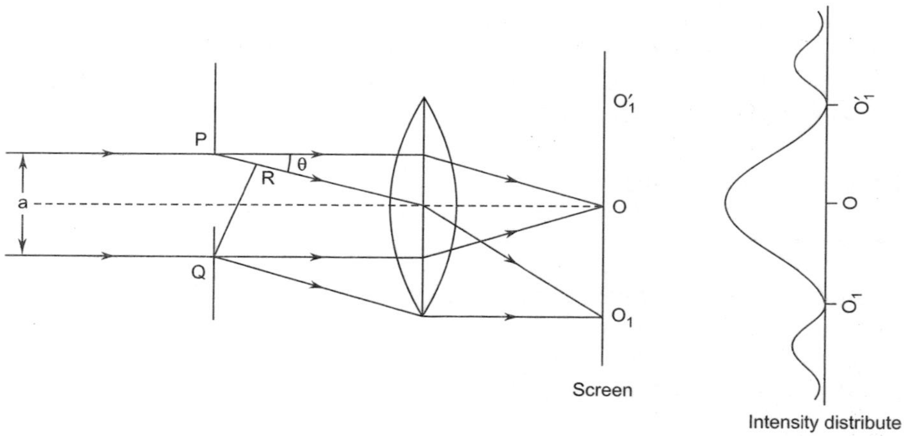

Figure 3D.2 *Fraunhofer diffraction pattern due to single slit.*

$$\frac{2\pi}{\lambda} \cdot a \sin\theta = 2\alpha \qquad \text{(say)}$$

or
$$\alpha = \frac{\pi}{\lambda} \cdot a \sin\theta$$

For oblique incidence making an angle i with the normal

$$\alpha = \frac{\pi a}{\lambda} (\sin i \pm \sin\theta)$$

When all the vibrations from PQ meet at 0 in the same phase, the resultant amplitude at O = ne, where n is the number of vibrations (very large) and e is the amplitude of each vibration. Denoting ne by A_0 it can be shown that the resultant intensity at any point (say, O_1) is given by,

$$I = I_0 \cdot \frac{\sin^2 \alpha}{\alpha^2} \quad \text{where } I_0 = A_0^2$$

$$\alpha = \frac{\pi a}{\lambda} \text{ for normal incidence}$$

and $$\alpha = \frac{\pi a}{\lambda} (\sin i \pm \sin \theta) \text{ for incidence at } \textbf{angle } \boldsymbol{i}.$$

For maximum and minimum intensity,

$$\frac{dI}{d\alpha} = 0$$

or $$\sin^2 \alpha \, (\alpha \cot \alpha - 1) = 0$$

Hence either $$\sin \alpha = 0 \quad \text{or, } \alpha = \frac{1}{\cot \alpha} = \tan \alpha$$

Putting $\sin \alpha = 0$ and $\alpha = \tan \alpha$ in the expression for $\frac{d^2I}{d\alpha^2}$ it is found that $\frac{d^2I}{d\alpha^2}$ is positive, when $\sin \alpha = 0$ and negative when $\alpha = \tan \alpha$

Hence, for minima, [when $\sin \alpha = 0$, $\alpha = s\pi$]

$$\frac{\pi a}{\lambda} \sin \theta = s\pi \quad \text{or, } a \sin \theta = s\lambda \qquad s : \text{any integer}$$

for maxima, the condition is $\alpha = \tan \alpha$.

The values can be obtained graphically by plotting $y = \alpha$ (45° straight line) and on the same graph $y = \tan \alpha$.

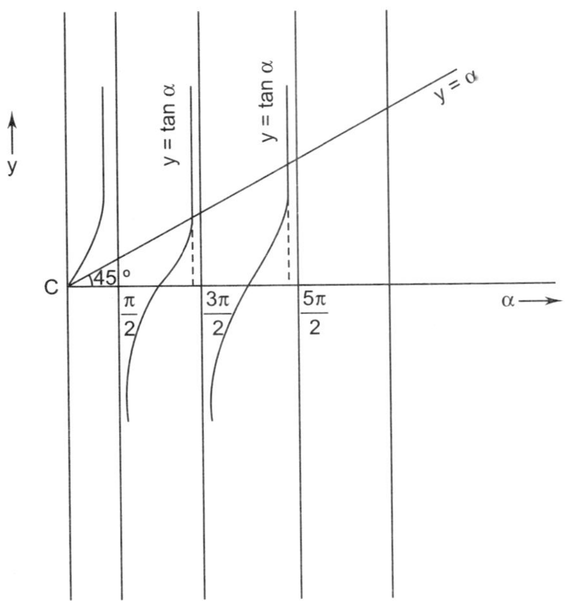

Figure 3D.3 *Graphical plot for tan α = α.*

From the graph one obtains,

The maximum intensity is for $\alpha = 0$, which is the principal maximum and other values of α which will give maxima are less than but gradually approaching towards $\pm \frac{3\pi}{2}, \pm \frac{5\pi}{2}$ etc. These maxima are known as **secondary maxima**.

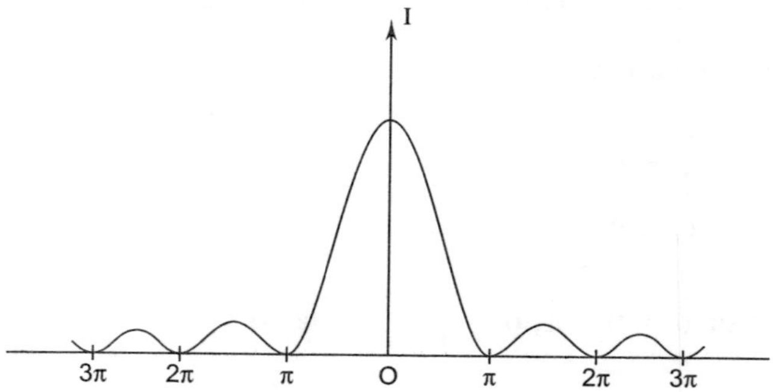

Figure 3D.4 *Intensity distribution.*

3D.1.3 Expression for the Resultant Intensity at Any Point due to All the Vibrations from PQ

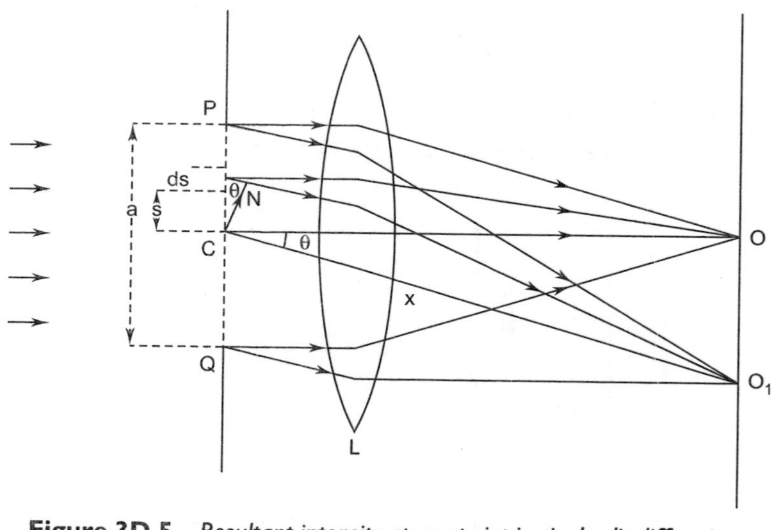

Figure 3D.5 *Resultant intensity at any point in single slit diffraction.*

Δ : path difference between the rays CO_1 and the oblique ray from *ds* through lens L to O_1

$$CO_1 = x, \angle OCO_1 = \angle PCN = \theta$$

CN : Perpendicular from C on the oblique ray from the element *ds*.

a : Width of the slit (PQ). The slit is perpendicular to the plane of the paper and is of infinite length compared to the width "*a*".

ds : Width of an element of the wave front in the plane of the slit

s : Distance of *ds* from the center (or origin) C. (C is midway between P and Q).

The explanation of the single slit pattern lies in the interference of the Huygen's secondary wavelets which can be thought of as sent out from every point on the wavefront at the instant it occupies the plane of the slit. The parts of each secondary wave which travel normal to the plane of the slit will be focussed at O while those travel at an angle θ will reach O_1. Considering the wavelet emitted by the element ds at s from the center the amplitude at O_1 is directly proportional to ds and inversely to the distance.

[In spherical waves, the intensity decreases as the inverse of the distance from the source. This law follows directly from the fact that provided there is no conversion of the energy into other forms, the same amount must pass through any sphere with the source as its center. Since the area of the sphere increases as the square of its radius, the intensity (*i.e.*, energy per unit area at a distance r from the source) will vary as $1/r^2$. The amplitude must then vary as $\frac{1}{r}$ $(\because A^2 = I)$.

Hence the equation of a spherical wave can be written as,

$$y = \frac{b}{r} \sin(\omega t - kr)$$

b : amplitude at unit distance from the center C

At O, therefore, the wavelet will produce an **infinitesimal** displacement

$$dy_0 = \frac{b\,ds}{x} \sin(\omega t - kx)$$

As the position of ds is varied, the displacement it produces at O_1 will vary in phase because of the different path length to O_1.

When it is above the origin at a distance s the contribution will be

$$dy_s = \frac{b\,ds}{x} \sin\left[\omega t - k(x + \Delta)\right] \quad \Delta : \text{path difference.}$$

$$= \frac{b\,ds}{x} \sin\left[\omega t - kx - ks\sin\theta\right]$$

The sum of the effects of all elements from one edge of the slit to the other will give the resultant amplitude of vibration at O_1. Now, $dy = dy_s + dy_{-s}$ is the contribution from a pair of elements symmetrically placed at s and $-s$,

Wherefrom,

$$dy_{o_1} = \frac{b\,ds}{x}\left[\sin(\omega t - kx - ks\sin\theta) + \sin(\omega t - kx + ks\sin\theta)\right]$$

$$= \frac{b\,ds}{x}\left[2\cos(ks\sin\theta)\sin(\omega t - kx)\right]$$

To get y_{o_1} the expression for dy_{o_1} has to be integrated from $s = 0$ to $\frac{a}{2}$. However, the distance x may be regarded as constant, in so far as it affects the amplitude.

Hence,

$$= \frac{2b}{x}\left[\sin(\omega t - kx)\right]\left[\frac{\sin(ks\sin\theta)}{k\sin\theta}\right]_0^{\frac{a}{2}}$$

$$= \frac{2b}{x}\left(\frac{a}{2}\right)\frac{\sin\left(\frac{ka\sin\theta}{2}\right)}{\frac{ka\sin\theta}{2}}\sin(\omega t - kx)$$

$$= \left[\frac{ab}{x}\cdot\frac{\sin\left(\frac{ka\sin\theta}{2}\right)}{\left(\frac{ka\sin\theta}{2}\right)}\right].\sin(\omega t - x)$$

The resultant vibration will therefore be simple harmonic, the amplitude of which varies with the position of O_1, since it is determined by θ.

The amplitude may be represented as,

$$A = A_0\frac{\sin\alpha}{\alpha}, \text{ where } \alpha = \frac{1}{2}ka\sin\theta$$

$$= \frac{\pi a\sin\theta}{\lambda} \qquad \left(\because k = \frac{2\pi}{\lambda}\right)$$

and $$A_0 = \frac{ab}{x}$$

The quantity α signifies one-half the phase difference between the contribution coming from opposite edges of the slit.

The intensity on the screen is then,

$$I \sim A^2 = A_0^2\frac{\sin^2\alpha}{\alpha^2}.$$

If the light, instead of being incident on the slit perpendicular to its plane, makes an angle 'i', then

$$\alpha = \frac{\pi a}{\lambda}(\sin i \pm \sin\theta)$$

3D.1.4 Fraunhofer Diffraction in Double Slits

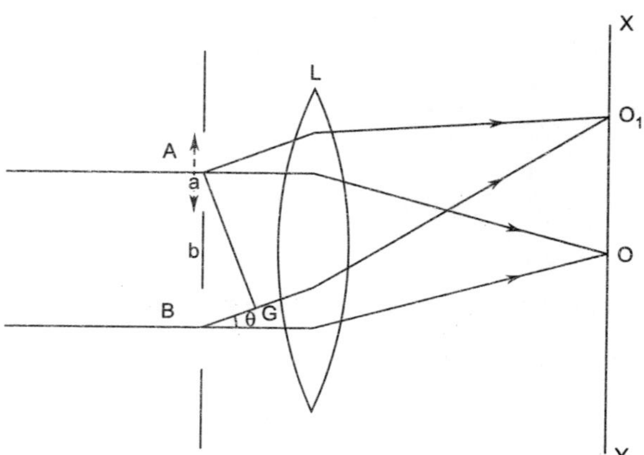

Figure 3D.6 *Double slits Fraunhofer diffraction.*

Let parallel rays of a monochromatic light fall on double slits.

The path difference BG of the rays which are diffracted at an angle θ from the corresponding points A and B of the two slits is $(a + b) \sin \theta$ and hence the phase difference between the two rays from A and B is,

$$2\beta = \frac{2\pi}{\lambda}(a+b)\sin\theta \qquad \text{where } a : \text{width of the slit}$$

$$b : \text{width of the obstacle}$$

or
$$\beta = \frac{\pi}{\lambda}(a+b)\sin\theta = \frac{\pi}{\lambda} d \sin\theta, \quad \text{where } d = (a+b)$$

The resultant displacement of the diffracted rays from the first slit at a given point O_1 can be written as,

$$y_1 = A \sin \omega t$$

where,
$$A = A_0 \frac{\sin\alpha}{\alpha} \quad \text{(as in single slit)}$$

A : Resultant amplitude of rays diffracted from the first slit at an angle θ.

A_0 : Resultant amplitude of direct rays from the slit (first).

2α : Phase difference between the two extreme rays from the first slit.

$$= \frac{2\pi}{\lambda} \times \text{path difference} = \frac{2\pi}{\lambda} a \sin\theta$$

The resultant displacement at a given point O_1 due to the diffracted rays from the second slit is, (extreme ends at B and A).

$$y_2 = A \sin(\omega t + 2\beta)$$

Hence, the resultant displacement at the given point P due to the diffracted rays from the two slits together at the focal plane of the lens L is,

$$\begin{aligned} y = y_1 + y_2 &= A \sin \omega t + A \sin(\omega t + 2\beta) \\ &= 2A \sin(\omega t + \beta) \cos\beta \\ &= [2A \cos\beta] \sin(\omega t + \beta) \end{aligned}$$

Thus the amplitude of the resultant displacement is given by $2A \cos \beta$

The resultant intensity, $I = (2A \cos\beta)^2 = 4A^2 \cos^2\beta$

∵
$$A^2 = A_o^2 \frac{\sin^2\alpha}{\alpha^2}$$

$$I = 4A_o^2 \left(\frac{\sin\alpha}{\alpha}\right)^2 \cos^2\beta$$

where $\beta = \frac{\pi}{\lambda} d \sin\theta$ and $d = (a + b)$

Discussion :

1. In this case two systems of fringes appear – one system is due to diffraction by each individual slit whose intensity is given by A^2 while another system is due to the interference of the beams coming from each slit and the intensity due to which is given by $\cos^2 \beta$. The two systems superpose.
2. Condition for minima :

$$I = 0 \text{ when any of the factors is Zero.}$$

(*a*) The factor $\cos^2 \beta$ will make $I = 0$

when $$\beta = \frac{\pi}{\lambda} d(\sin i \pm \sin \theta) = (2n + 1)\frac{\pi}{2}$$

(*i*) $$d(\sin i \pm \sin \theta) = (2n + 1)\frac{\lambda}{2} \text{ (for oblique incidence)}$$

and (*ii*) $$d \sin \theta = (2n + 1)\frac{\lambda}{2} \text{ (for normal incidence)}$$

(*b*) The factor $A_o^2 \dfrac{\sin^2 \alpha}{\alpha^2}$ will make $I = 0$ when $\alpha = n\pi$

(n = integer other than zero)

Now, $$\alpha = \frac{\pi}{\lambda} a(\sin i \pm \sin \theta)$$

Hence for (*i*) oblique incidence, $\dfrac{\pi}{\lambda} a(\sin i \pm \sin \theta) = n\pi$

or $$a(\sin i \pm \sin \theta) = n\lambda$$

and (*ii*) normal incidence, $a \sin \theta = n\lambda$

These minima are known as diffraction minima.

3. Condition for maxima :
When the width of the slit is very narrow and maxima near the center of the pattern is considered, variation of $\dfrac{\sin^2 \alpha}{\alpha^2}$ is negligible. Hence under this condition the maxima will be solely controlled by $\cos^2 \beta$ and thus for maxima

$$\frac{\pi}{\lambda} d \sin \theta = \beta = n\pi$$

or, $$d \sin \theta = n\lambda \text{ (for normal incidence)}$$

and $$d(\sin i \pm \sin \theta) = n\lambda \text{ (for oblique incidence at an angle } i)$$

3D.1.5 Intensity Distribution due to Double Slit Diffraction

The maximum intensity of the strong central band comes at the point O, where evidently all the secondary wavelets arrive in phase because the path difference $\Delta = 0$. For this point $\alpha = 0$ and although $\left(\frac{\sin\alpha}{\alpha}\right)$ becomes indeterminate for $\alpha = 0$, it will be remembered, that $\sin\alpha$ approaches α for small angles and $\sin\alpha = \alpha$ (when α vanishes). Hence for $\alpha = 0$ $\left(\frac{\sin\alpha}{\alpha}\right) = 1$ and $A = A_0$, the amplitude when all the wavelets arrive in phase at O. The value of the maximum intensity at O *i.e.*, at the center of the pattern is A_0^2. From this principal maximum the intensity falls to Zero at $\alpha = \pm\pi$, then passes through several secondary maxima with equally spaced points of zero intensity at $\alpha = \pm\pi, \pm 2\pi, \pm 3\pi$ etc., *i.e.*, $\alpha = m\pi$.

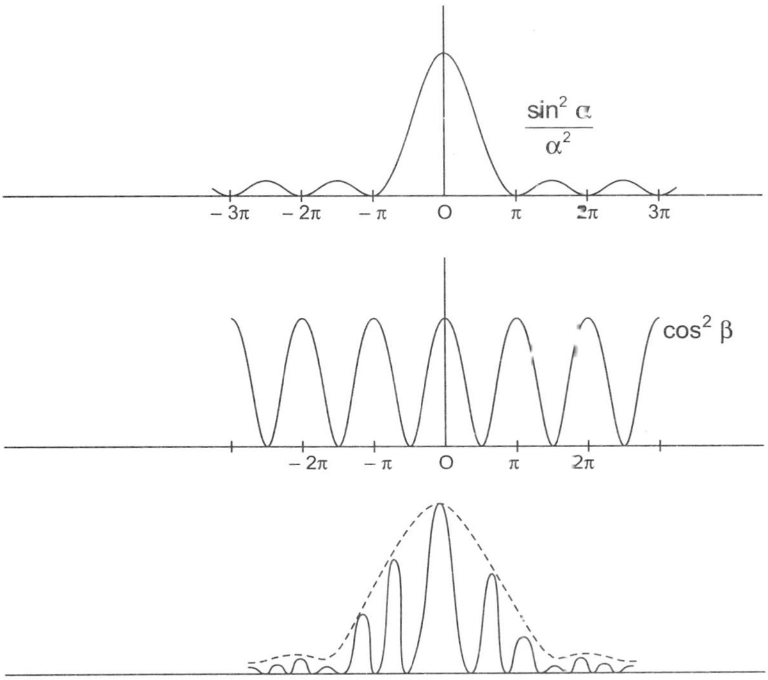

Figure 3D.7 *Intensity distribution pattern.*

The secondary maxima do not fall half-way between these points but are displaced towards the center by an amount which decreases with increasing m.

3D.1.6 Plane Diffraction Grating (Transmission)

This consists of a number of parallel and equidistant lines ruled on an optically plane and parallel glass plate. Each ruled line behaves as an obstacle and the transparent portion between the two consecutive lines behaves as a slit. If "a" be the width of the transparent space and "b" the width of the ruling then $(a + b)$ is the Grating Constant or the Grating Element. The two points on the consecutive clear spaces separated by a distance $(a + b)$ are the corresponding points.

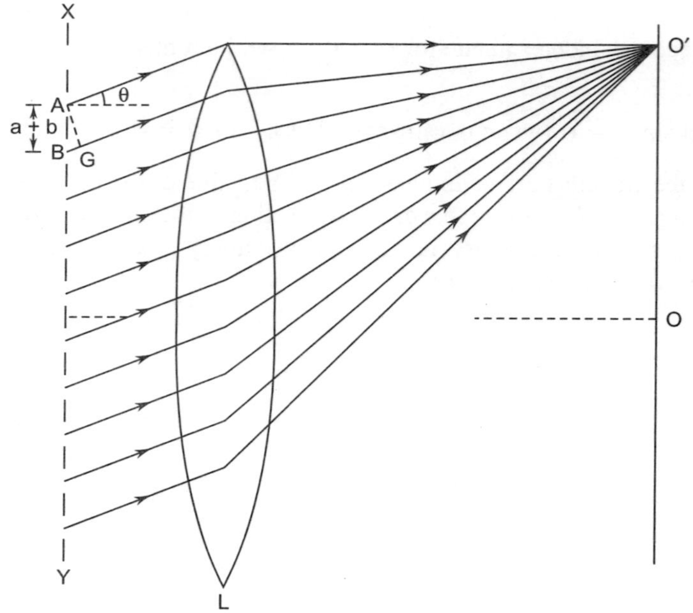

Figure 3D.8 *Diffraction due to grating.*

A parallel beam of monochromatic light incident normally on a grating surface gets its rays diffracted in various directions and these diffracted rays after passing through the converging lens L converges at its focal plane and form a number of bright bands separated from each other by dark regions.

Consider A and B to be the corresponding points separated by $(a + b)$, the grating constant and N be the number of clear spaces. A particular set of rays get diffracted from all the clear spaces of the grating in a direction making an angle θ with the grating normal.

The path difference of the rays so diffracted is $BG = (a + b) \sin\theta$. These rays will reinforce and produce brightness when BG is even multiple of $\frac{\lambda}{2}$. Thus for brightness the condition is,

$$BG = (a + b)\sin\theta = n\lambda \qquad (\lambda : \text{wavelength})$$

If there be "*m*" rulings per unit length of the grating then $(a + b) = 1/m$.

Hence, $\sin\theta = mn\lambda$, where n : any integer including zero.

If N : the total number of clear spaces be even, then the rays diffracted at an angle θ from the pairs of consecutive clear spaces will mutually reinforce and produce brightness satisfying the condition,

$$\sin\theta = mn\lambda$$

If N be odd then we may disregard the last clear space, since light from a single clear space can produce no appreciable effect.

The resultant intensity

$$I = A_0^2 \frac{\sin^2\alpha}{\alpha^2} \frac{\sin^2 N\beta}{\sin^2\beta}$$

where

$$\beta = \frac{\pi}{\lambda} d \sin\theta$$

$$\alpha = \frac{\pi}{\lambda} a \sin\theta$$

The point O in figure 3D.8 corresponds to $\theta = 0$ which is the central bright band. For values $\theta = \theta_1, \theta_2, \theta_3$ etc., by putting $n = 1, 2, 3$, etc., we get the 1st order, 2nd order, 3rd order etc., bright band. These values of $n = 1, 2, 3$ etc. causes maxima known as Principal Maxima. Between two principal maxima we get darkness due to interference of diffracted rays. For gratings having N number of rulings there are (N – 1) secondary minima and (N – 2) secondary maxima between any two principal maxima. The secondary maxima being practically invisible we get, between two principal maxima one dark region.

Conditions for Maxima and Minima :

For Maxima: $$(a + b) \sin\theta = n\lambda$$

For Minima: $$(a + b) \sin\theta = (2n + 1)\frac{\lambda}{2}$$

where $n = 0, 1, 2, 3$, etc.

The above condition holds for all the pairs of rays from the corresponding points of the total surface

Maxima:
- For $n = 0$, $\theta = 0$, $(a + b) \sin\theta = 0 \rightarrow$ Central maximum at O
- $n = 1$, $(a + b) \sin\theta = \lambda \rightarrow$ First principal maxima on either side of central maximum
- $n = 2$, $(a + b) \sin\theta = 2\lambda \rightarrow$ second order principal maxima on either side of O

Minima:
- For $n = 0$, $(a + b) \sin\theta = \frac{\lambda}{2} \rightarrow$ First order minima on either side of O.
- $n = 1$, $(a + b) \sin\theta = \frac{3\lambda}{2} \rightarrow$ Second order minima on either side of O.

So alternate bright and dark bands are obtained.

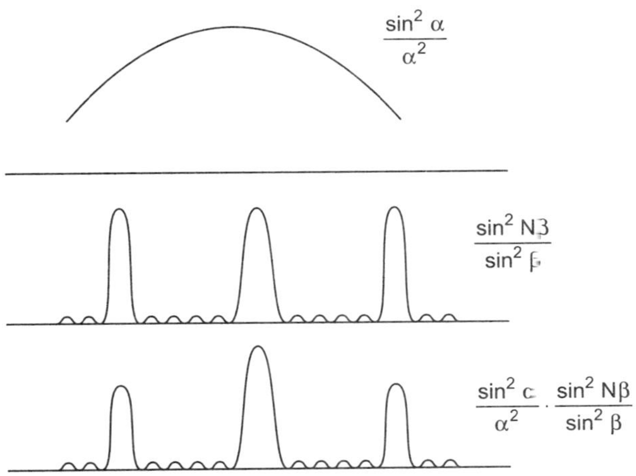

Figure 3D.9 *Intensity Distribution.*

For alternate Maxima and Minima :

If for a small variation of the angle of diffraction in any order maxima, the path difference between the extreme rays from a grating change by λ, then the change in path difference between the rays from any pair of the corresponding points from two adjoining spaces will be $\frac{\lambda}{N}$ (N : number of rulings in the grating). Considering the grating to be divided into two halves, the light from one half will differ in path by $\frac{\lambda}{2}$ coming from the corresponding spaces in the other half and so they will mutually cancel each other's effect. This will give the condition for the first secondary maxima. After the principal maxima the condition for the secondary minima is,

$$(a + b) \sin (\theta + d\theta) = n\lambda + \frac{\lambda}{N}.$$

Missing orders in the double-slit pattern :

For a given value of θ, when the conditions of interference maxima and diffraction minima are simultaneously satisfied, the interference maxima will be missing.

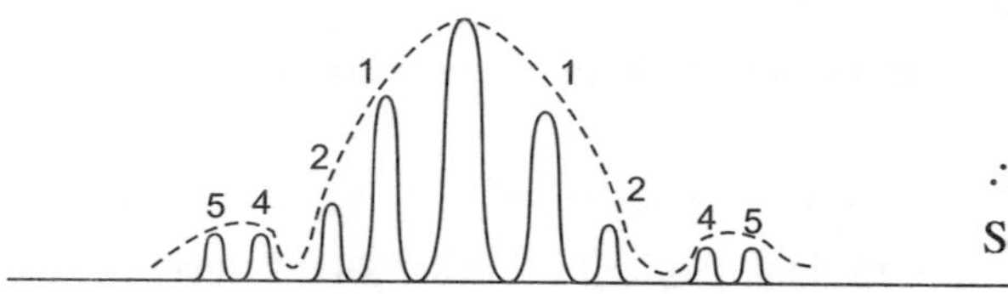

$b = 2a$

$\therefore \quad n = 3m$

So, 3,6,9... are missing

Figure 3D.10 *Missing order.*

3D.1.7 Resolving Power of a grating

The resolving power of a grating is its ability to show two neighbouring lines in a spectrum as separate.

It is also defined as :

The ratio of the wavelength of any spectral line to its difference of wavelengths between this line and the neighbouring line such that the two lines can be just seen as separate.

Two spectral lines of wavelength λ and $\lambda + d\lambda$ will be just resolved if the central maxima due to $\lambda + d\lambda$ falls on the first minima of λ.

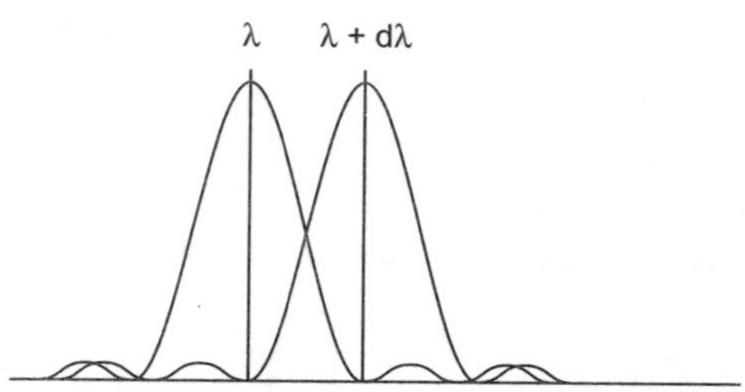

Figure 3D.11 *Just resolved.*

The resolving power of a grating $= \dfrac{\lambda}{d\lambda}$.

The condition for the first secondary minimum after the nth order maximum is given by,

$$(a + b) \sin (\theta + d\theta) = n\lambda + \frac{\lambda}{N}$$

n : order of the spectrum

N : total number of lines on the grating.

Now, let the nth order principal maximum corresponding to the wavelength $(\lambda + d\lambda)$ has the same direction as that of the first secondary minimum of the nth order principal maximum corresponding to the wavelength λ, then,

$$(a + b) \sin (\theta + d\theta) = n (\lambda + d\lambda)$$

Hence, $$n (\lambda + d\lambda) = n\lambda + \frac{\lambda}{N}$$

or, $$n\, d\lambda = \frac{\lambda}{N}$$

wherefrom, $\dfrac{\lambda}{d\lambda} = N \times n =$ (Total number of lines on the grating × order of the spectrum)

N.B. This shows that $\dfrac{\lambda}{d\lambda}$ is independent of $(a + b)$ [*i.e.*, the grating element].

IN FOCUS

Fraunhofer Diffraction Pattern due to Single Slit

Intensity $I = I_0 \dfrac{\sin^2 \alpha}{\alpha^2}$ where $I_0 = A_0^2$ [A = amplitude of the wave]

$$\alpha = \frac{\pi a}{\lambda}$$ for normal incidence

[a = width of the slit, λ = wavelength of light]

and $$\alpha = \frac{\pi a}{\lambda} (\sin i \pm \sin \theta)$$ for incidence at angle i

[θ = Angle with the normal]

For minimum intensity

$$a \sin \theta = s\, \lambda$$ where s = any integer

For maximum intensity

$$\alpha = \tan \alpha$$

Fraunhofer Diffraction in Double Slits

$$\text{Intensity } I = 4A_0^2 \left(\frac{\sin\alpha}{\alpha}\right)^2 \cos^2\beta$$

where A_0 = Resultant amplitude of direct rays from the slit.

2α = phase difference between the two extreme rays from the first slit

$$= \frac{2\pi}{\lambda} a \sin\theta$$

$$\therefore \quad \alpha = \frac{\pi}{\lambda} a \sin\theta$$

2β = phase difference between the two rays from the centers of the two slits.

$$= \frac{2\pi}{\lambda}(a+b)\sin\theta$$

$$\therefore \quad \beta = \frac{\pi}{\lambda} d \sin\theta \qquad [d = a + b]$$

a = width of the slit

b = width of the opaque space

Condition for Minima

(*i*) normal incidence $\quad a \sin\theta = n\lambda$

(*ii*) oblique incidence $\quad a(\sin i \pm \sin\theta) = n\lambda \quad$ [i = incidence angle]

Condition for Maxima

(*i*) normal incidence $\quad \delta \sin\theta = n\lambda$

(*ii*) oblique incidence $\quad d(\sin i \pm \sin\theta) + n\lambda$

Plane Diffraction Grating

If there are N number of ruling

then the resultant intensity will be

$$I = A_0 \frac{\sin^2\alpha}{\alpha^2} \frac{\sin^2 N\beta}{\sin^2\beta}$$

where $\alpha = \frac{\pi}{\lambda} a \sin\theta$

$$\beta = \frac{\pi}{\lambda} d \sin\theta$$

Condition for Maxima

$(a + b)\sin\theta = n\lambda \qquad n \rightarrow$ any integer

Condition for Minima

$$(a + b)\sin\theta = (2n + 1)\frac{\lambda}{2} \qquad n \rightarrow \text{any integer}$$

Resolving Power of Gratings

$$\frac{\lambda}{d\lambda} = N \times n$$

N → Total number of lines on the grating

n → order of the spectrum.

Solved Problems

Problem 1. *In Fraunhofer diffraction due to a narrow slit a screen is placed 2 m away from the lens to obtain the pattern. If the slit width is 0.2 mm and the first minima lie 5 mm on either side of the central maximum, find the wavelength of light.* [Delhi University]

Solution. In the case of Fraunhofer diffraction for a narrow rectangular aparture, $a \sin\theta = n\lambda$

Here $\sin\theta = \frac{x}{D}$,

where a = slit width = 0.2 mm = 0.02 cm

and D = Distance between the lens and screen = 2 m = 200 cm and x = 5 mm = 0.5 cm

$$n = 1$$

$\because$ $2x$ = Width of the central maxima.

$$\therefore \quad a\frac{x}{D} = n\lambda$$

$$\lambda = \frac{ax}{Dn} = \frac{0.02 \times 0.5}{200 \times 1} = 5 \times 10^{-5} \text{ cm} = 5000 \text{ Å}.$$

Problem 2. *A convex lens of focal length 20 cm is placed after a slit of width 0.6 mm. If a plane wave of wavelength 6000 Å falls on the slit normally, calculate the separation between the second minima on either side of central maxima.* [Pune University]

Solution. Given that, f = 20 cm, slit width a = 0.6 mm = 0.06 cm

$n = 2$, and λ = 6000 Å = 6×10^{-5} cm.

If the second minimum occurs in the direction making an angle θ_2 on either side of the central maxima, then

$$\sin\theta_2 = \frac{n\lambda}{a} = \frac{2 \times 6 \times 10^{-5}}{0.06} = 0.002$$

The diffraction pattern is formed on the screen placed in the focal plane of the lens placed close to the slit.

$$\sin\theta_2 = \tan\theta_2 = \frac{x}{f} = \frac{n\lambda}{a} = \frac{2\lambda}{a} \qquad [\because \theta_2 \text{ is very small}]$$

$$\therefore \qquad x = \frac{2\times 6\times 10^{-5}\times 20}{0.06} = 0.04 \text{ cm}$$

is the distance between the center of principal maximum and the second minimum.

Problem 3. *Find the distance between the central maximum and the first minimum on a screen, 100 cm away from a single slit, 0.1 mm wide and illuminated by a light of wavelength 5890 Å.*

[North Bengal University]

Solution. We know that the $\sin\theta_n = \frac{n\lambda}{a}$

where θ_n is the angle of diffraction for the nth minimum.

Further, if x_n is the distance of the nth minima from the central maxima, then $\theta_n = \frac{x_n}{D}$.

where D is the distance between the slit and the screen

$$\sin\theta_n = \frac{n\lambda}{a} = \frac{x_n}{D} \approx \theta_n \qquad (\text{where } \theta_n \text{ is very small})$$

or

$$x_n = \frac{n\lambda D}{a}$$

Now, for $n = 1$,

$$x_1 = \frac{1\times 5890\times 10^{-8}\text{ cm}\times 100}{0.01\text{ cm}} \simeq 0.589 \text{ cm}.$$

Problem 4. *A lens whose focal length is 40 cm forms a Fraunhofer diffraction pattern of a slit 0.3 mm wide. Calculate the distance of the first dark band from the axis. Wavelength of light used in 5890 Å.*

[Agra University]

Solution. The angle of diffraction corresponding to the 1st minimum is given by

$$a\sin\theta = n\lambda,$$

Here

$$n = 1$$

or,

$$\theta = \sin^{-1}\left(\frac{n\lambda}{a}\right) \approx \frac{\lambda}{a} = \frac{5890\times 10^{-8}\text{ cm}}{0.03\text{ cm}}$$

$$\theta = \frac{589}{3}\times 10^{-5} \text{ radian.}$$

If the linear distance of the first dark band from the axis is x cm, then

$$\frac{x}{f} = \theta$$

$$x = f\theta = 40\times\frac{589}{3}\times 10^{-5} = 0.0785 \text{ cm.}$$

Problem 5. *Light is incident normally on a grating 0.5 cm wide with 2500 lines. Find the angle of diffraction for the principal maxima of the sodium lines in the first order spectrum.*

$$\lambda_1 = 5890 \text{ Å and } \lambda_2 = 5896 \text{ Å}.$$ [Punjab University]

Are the two lines resolved?

Solution. We know that

$$(a + b) \sin \theta_n = n\lambda$$

Here it is given that width = 0.5 cm,

Total Number of lines = N = 2500 lines

$$\therefore \quad \text{Number of lines/cm} = N' = \frac{2500}{0.5} = 5000 \text{ lines/cm}$$

$$\text{Grating element } (a + b) = \frac{1}{N'} = \frac{1}{5000} \text{ cm.}$$

Case 1. $n = 1, \lambda = 5890 \text{ Å} = 5890 \times 10^{-8}$ cm

$$(a + b) \sin \theta_1 = 1 \times \lambda_1$$

or,
$$\sin \theta_1 = 5890 \times 10^{-8} \times 5000 = 0.2945$$

$$\therefore \quad \theta = 17.1°$$

Case 2. For $\lambda_2 = 5896 \times 10^{-8}$ cm, $n = 1$

$$(a + b) \sin\theta_1' = 1 \times \lambda_2$$

or,
$$\sin \theta_1' = 5896 \times 10^{-8} \times 5000 = 0.2948$$

or,
$$\theta_1' = 17.2°,$$

Now, the condition for just resolution is

$$\frac{\lambda}{d\lambda} = n\text{N}$$

or,
$$\frac{5890 \times 10^{-8}}{6 \times 10^{-8}} = 1 \,.\, \text{N}$$

or
$$\text{N} \simeq 982.$$

The two lines will appear well resolved because the total number of lines on the grating is 2500.

Problem 6. *A plane transmission grating gives an angle of diffraction of a line to be 30° in the first order, for a wavelength 6×10^{-5} cm. Find the number of lines per cm of the given grating.*

[Calcutta University]

Solution. For a plane transmission grating,

We have
$$\sin \theta_n = \frac{n\lambda}{a + b} = \frac{n\lambda}{\text{N}}$$

where N = number of lines/cm

Given that $\lambda = 6 \times 10^{-5}$ cm, $\theta_n = 30°$, $n = 1$.

$$\therefore \quad \text{N} = \frac{\sin 30°}{6 \times 10^{-5} \times 1} = 8333 \text{ lines/cm.}$$

Problem 7. *A parallel beam of sodium light is allowed to be incident normally on a plane grating having 4250 lines per centimeter and a second order spectral line is observed to be deviated through 30°. Calculate the wavelength of the spectral line?* [Vidya Sagar University]

Solution. We know that

$$(a + b) \sin \theta_n = n\lambda$$

Given that,

$$\text{Number of lines per cm } (a + b) = \frac{1}{4250}$$

and $$n = 2,$$

∴ $$\sin \theta_n = 2 \times \lambda \times 4250$$

or, $$\lambda = \frac{\sin 30°}{2 \times 4250} = \frac{1}{4 \times 4250}$$

$$= 5882 \text{ Å}.$$

Problem 8. *In a plane transmission grating the angle of diffraction for 2nd order maxima for wavelength 5×10^{-5} cm is 30°. Calculate the number of lines in one centimeter of the grating surface.* [Vidya Sagar University, Kolkata University]

Solution. We know that,

$$\text{grating element } (a + b) = \frac{1}{N}$$

where N is the number of lines per cm.

$$n = 2, \lambda = 5 \times 10^{-5} \text{ cm}.$$

$$N = \frac{1}{(a+b)} = \frac{\sin 30°}{n\lambda} \qquad [\because (a+b)\sin\theta_n = n\lambda]$$

$$= \frac{1}{2} \cdot \frac{1}{2 \times 5 \times 10^{-5}}$$

$$= 5000 \text{ per cm}.$$

Problem 9. *What is the highest order spectrum, which may be seen with monochromatic light of wavelength 6000 Å by means of a diffraction grating with 5000 lines/cm.* [Delhi University]

Solution. We know that $(a + b) \sin \theta_n = n\lambda$

Given that $$N = \frac{1}{a+b} = 5000 \text{ lines/cm}.$$

$\sin \theta_n = 1,$ $$\lambda = 6000 \times 10^{-8} \text{ cm} = 6 \times 10^{-5} \text{ cm}$$

$$n = \frac{(a+b)\sin\theta_n}{\lambda} = \frac{1 \times 10^5}{5000 \times 6} = \frac{20}{6} = 3.33$$

The highest order of the spectrum that can be visible is 3.33

Problem 10. *Show that in a diffraction grating with grating element 1.5×10^{-6} m and light of wavelength 550 nm, third and higher order principal maxima are not visible.* [G.N.D. University]

Solution. Given $\lambda = 550 \text{ nm} = 5500 \times 10^{-10}$ meter.

Grating element $(a + b) = 1.5 \times 10^{-6}$ m.

Let n be the maximum number of orders of the spectrum that can be observed with the grating. Since the maximum angle of diffraction is 90°.

$$\therefore \quad (a + b) \sin 90° = n\lambda$$

or,
$$n = \frac{1.5 \times 10^{-6}}{5500 \times 10^{-10}} = 2.72$$

i.e., the second order will be visible only. 3rd and higher orders are not possible.

Problem 11. *A plane diffraction grating at normal incidence gives a green line $\lambda = 5.4 \times 10^{-5}$ cm in a certain spectral order superposed on a violet line $\lambda = 4.05 \times 10^{-5}$ cm of the next higher order. If the angle of diffraction is 30°, obtain the number of lines per cm of the grating.*

[Calcutta University]

Solution. We have $\sin \theta_n = n\lambda \text{ N}$

Consider n^{th} order for green lines and $(n + 1)^{\text{th}}$ order for violet line.

For the green line, $\sin 30° = \text{N} \times n \times 5.4 \times 10^{-5}$ ...(*i*)

For the violet line, $\sin 30° = \text{N} \times (n + 1) \times 4.05 \times 10^{-5}$ ...(*ii*)

Comparing equations (*i*) and (*ii*) we get

$$\text{N} \times n \times 5.4 \times 10^{-5} = \text{N} \times (n + 1) \times 4.05 \times 10^{-5}$$

or $n = 3$,

Putting $n = 3$ in equation (*i*), we get

$$\text{N} = \frac{\sin 30°}{3 \times 5.40 \times 10^{-5}} \simeq 3087 \text{ per cm.}$$

Problem 12. *A diffraction grating used at normal incidence gives a line (5400 Å) in certain order superimposed on another line (4050 Å) of the next higher order. If the angle of diffraction is 30°, how many lines/cm are there on the grating?* [Pune University, Kolkata University]

Solution. Given,
$$\lambda_1 = 5400 \text{ Å} = 5400 \times 10^{-10} \text{ m}$$
$$\lambda_2 = 4050 \text{ Å} = 4050 \times 10^{-10} \text{ m},$$

According to the problem, n^{th} and $(n + 1)^{\text{th}}$ order superimpose.

$$(a + b) \sin \theta_1 = n\lambda_1 \qquad \text{...(i)}$$
$$(a + b) \sin \theta_2 = (n + 1) \lambda_2 \qquad \text{...(ii)}$$

Here
$$\theta_1 = \theta_2 = 30°$$

i.e.,
$$\frac{n\lambda_1}{(n+1)\lambda_2} = 1$$

or,
$$\frac{n+1}{n} = \frac{\lambda_1}{\lambda_2} = \frac{540}{405} = \frac{4}{3},$$

or, $$\frac{1}{n} = \frac{1}{3}, \quad \therefore n = 3$$

Now putting $n = 3$ in equation (*i*), we get

$$(a + b) = \frac{3 \times 5400 \times 10^{-10}}{\sin 30°} = 324 \times 10^{-8}$$

$$\therefore \quad \text{Number of lines/m } N = \frac{1}{324 \times 10^{-8}} = 308642\text{/meter}$$

$$\simeq 3086\text{/cm.}$$

Problem 13. *Calculate the minimum number of lines in a grating which will just resolve the sodium lines in the first order spectrum. The wavelengths are 5890 Å and 5896 Å.* [Delhi University]

Solution. We know that,

The resolving power, $$\frac{\lambda}{d\lambda} = n\text{N}$$

N = Number of lines in grating

Here $n = 1$, $$\lambda = 5890 \text{ Å} = 5890 \times 10^{-8} \text{cm}$$

and $$d\lambda = 5896 \text{ Å} - 5890 \text{ Å} = 6 \text{ Å} = 6 \times 10^{-8} \text{ cm}$$

$$\therefore \quad \text{N} = \frac{1}{n}.\frac{\lambda}{d\lambda} = \frac{1}{1}.\frac{5890 \text{ Å}}{6 \text{ Å}} = \frac{2945}{3} \simeq 982 \text{ (Approximately)}$$

Problem 14. *Obtain the design of a plane transmission diffraction grating capable of resolving a wavelength difference of 6 Å at a mean wavelength of 6000 Å in second order spectra.*

[Rohilkhand University]

Solution. Given that, $\lambda = 6000 \text{ Å} = 6000 \times 10^{-8} \text{ cm} = 6 \times 10^{-5} \text{ cm}$

$n = 2$, $d\lambda = 6 \text{ Å} = 6 \times 10^{-8}$ cm.

We know that, the resolving power of diffraction grating is given by

$$\frac{\lambda}{d\lambda} = n\text{N}$$

or, $$\text{N} = \frac{1}{n}.\frac{\lambda}{d\lambda} = \frac{1}{2}.\frac{6 \times 10^{-5}}{6 \times 10^{-8}} = 500$$

Hence, the minimum number of lines in the required diffraction grating is 500.

Problem 15. *Calculate the minimum number of lines per cm in a 2.5 cm wide grating which will just resolve the sodium lines (5890 Å and 5896 Å) in the second order spectrum?* [Delhi University]

Solution. We know $$\frac{\lambda}{d\lambda} = n\text{N}$$

Here, $\lambda = 5890$ Å and $d\lambda = 6$ Å ; $n = 2$,

$$N = ?$$

$$N = \frac{1}{n} \cdot \frac{\lambda}{d\lambda} = \frac{1}{2} \cdot \frac{5890 \text{ Å}}{6 \text{ Å}} \simeq 491$$

Width of the grating = 2.5 cm.

$$\text{Number of lines per cm} = \frac{491}{2.5} = 196.4.$$

Problem 16. *A plane transmission grating has 40,000 lines in all with grating element 12.5×10^{-5} cm. Calculate the maximum resolving power for which it can be used in the range of wavelength 5000 Å.* [Rohilkhand University]

Solution. We know that the resolving power of a grating is given by

$$\frac{\lambda}{d\lambda} = n\,N \qquad \ldots(i)$$

But for a grating we have,

$$(a + b) \sin \theta = n\,\lambda \qquad \ldots(ii)$$

From equations (*i*) and (*ii*), we get

$$\frac{\lambda}{d\lambda} = \frac{N\,.(a+b) \sin \theta}{\lambda}$$

For maximum, $\sin \theta = 1$,

$$\text{Maximum resolving power} = \frac{N\,(a+b)}{\lambda}$$

N = Total number of lines in grating = 40,000

Grating element = $(a + b) = 12.5 \times 10^{-5}$ cm.

λ = range of wavelength = 5000 Å = 5000×10^{-8} cm

$= 5 \times 10^{-5}$ cm.

$$\therefore \quad \text{Maximum Resolving power, } n_{max} = \frac{40{,}000 \times 12.5 \times 10^{-5}}{5 \times 10^{-5}} = 1{,}00{,}000.$$

But actually the maximum number of order available with the grating is 2. Therefore, the maximum resolving power of the grating = N n_{max} = $40{,}000 \times 2$ = 80,000.

Problem 17. *What is the maximum number of lines of a grating which will resolve the 3rd order spectrum of two lines having wavelengths 5890 Å and 5896 Å ?* [WBUT]

Solution. Here, $\frac{\lambda}{d\lambda} = n\,N$

$$n = 3, \; \lambda = 5890 \text{ Å}, \; d\lambda = 6 \text{ Å}$$

or
$$N = \frac{1}{n} \cdot \frac{\lambda}{d\lambda} = \frac{1}{3} \cdot \frac{5890 \text{ Å}}{6 \text{ Å}} = 327.22$$

Maximum Number of lines of a grating is 327.

Multiple Choice Questions

1. Light is
   (*a*) Particle phenomenon (*b*) Wave phenomenon
   (*c*) Both particle and wave phenomenon.
2. When interference of light takes place at some region the light is
   (*a*) Created (*b*) Destroyed (*c*) Redistributed.
3. The speed of light depends
   (*a*) On elasticity of the medium only.
   (*b*) On inertia of the medium only.
   (*c*) Neither on elasticity nor on inertia.
4. Which of the following phenomenon proves the transverse characteristic of light?
   (*a*) Interference (*b*) Dispersion (*c*) Polarisation.
5. The sources are called coherent if they produce waves
   (*a*) Of equal wavelength
   (*b*) Having same shape of wave front
   (*c*) Having a constant phase difference.
6. Which of the following sources gives best monochromatic light
   (*a*) A candle (*b*) A bulb (*c*) A laser.
7. Three light waves are represented by
   (*a*) $y = a_1 \sin \omega t$ (*b*) $y = a_2 \sin 2 (\omega t + \in)$ (*c*) $y = a_3 \sin 3 (\omega t + \in)$
   Interference fringes may be observed due to superposition of
   (*a*) (*i*) and (*ii*) (*b*) (*ii*) and (*iii*) (*c*) (*iii*) and (*i*).
8. Two coherent sources of different intensities send waves which interfere. The ratio of maximum intensity to the minimum intensity is 25. The intensities of the sources are in the ratio
   (*a*) 25 : 1 (*b*) 9 : 4 (*c*) 5 : 1 (*d*) 625 : 1
9. In Fraunhofer diffraction, the incident wave front is
   (*a*) Plane (*b*) Spherical (*c*) Cylindrical.
10. Which of the following belong to Physical optics ?
    (*a*) Dispersion (*b*) Refraction (*c*) Interference.
11. On reflection from a denser medium the path difference introduced is
    (*a*) Zero (*b*) $\frac{\lambda}{2}$ (*c*) λ.
12. Light waves from two coherent sources of same intensity I interfere, the intensity of light of maxima will be
    (*a*) I (*b*) 2I (*c*) 4I.
13. Huygens' principle of secondary wavelets may be used to
    (*a*) Find the velocity of light in vacuum.
    (*b*) Explain the particle behaviour of light.
    (*c*) Find the new position of a wavefront.

14. Which of the following properties of light conclusively support wave theory of light?
(*a*) Light obeys laws of reflection.
(*b*) Light shows interference.
(*c*) Light shows photoelectric effect.

15. When light is refracted, which of the following does not change?
(*a*) Wavelength (*b*) Frequency (*c*) Amplitude.

16. The wavefronts of a light wave travelling in vacuum are given by $x + y + z = c$. The angle made by the direction of propagation of light with the X-axis is
(*a*) 0° (*b*) 90° (*c*) $\cos^{-1}\left(\frac{1}{\sqrt{3}}\right)$

17. The wavefronts of light coming from a distant source of unknown shape are nearly
(*a*) Plane (*b*) Elliptical (*c*) Spherical.

18. If θ is the polarising angle, the refractive index μ of the material is given by :
(*a*) sin θ (*b*) tan θ (*c*) $\cos^{-1}\theta$.

19. The equation of a light wave is written as $y = A \sin(kx - \omega t)$. Here y represents,
(*a*) Displacement of other particles
(*b*) Density of the medium
(*c*) Electric field.

20. A thin transparent sheet is placed in front of Young's double slit. The fringe width will
(*a*) Remain same (*b*) Increase (*c*) Decrease.

21. If the source of light used in Young's double slit experiment is changed from red to violet,
(*a*) The fringe will become brighter
(*b*) Consecutive fringes will come closer
(*c*) The intensity of minima will increase.

22. Light waves travel in vacuum along the X-axis. Which of the following may represent the wavefronts.
(*a*) $x = c$ (*b*) $y = c$ (*c*) $z = c$.

23. The inverse square law of intensity (*i.e.*, the intensity $\mu \propto \frac{1}{r^2}$) is valid for a
(*a*) Point source (*b*) Line source (*c*) Plane source.

24. In Newton's rings experiment by reflected light, the center is
(*a*) Bright (*b*) Dark (*c*) Red.

25. In Newton's rings experiment by transmitted light, the center is
(*a*) Bright (*b*) Dark (*c*) Violet.

26. Polarisation of light establishes that light is essentially
(*a*) Corpuscular (*b*) Longitudinal (*c*) Transverse.

27. The angle of diffraction of the first order maxima in a diffraction grating —
(*a*) Decreases if the separation between the rulings is increased
(*b*) Increases if the separation between the rulings is increased
(*c*) Does not change if the separation between the rulings is increased. [*WBUT*]

Answers

1. (*c*)	**2.** (*c*)	**3.** (*c*)	**4.** (*c*)	**5.** (*c*)	**6.** (*c*)
7. (*a*)	**8.** (*b*)	**9.** (*a*)	**10.** (*c*)	**11.** (*b*)	**12.** (*c*)
13. (*c*)	**14.** (*b*)	**15.** (*b*)	**16.** (*c*)	**17.** (*a*)	**18.** (*b*)
19. (*c*)	**20.** (*a*)	**21.** (*b*)	**22.** (*a*)	**23.** (*a*)	**24.** (*b*)
25. (*a*)	**26.** (*c*)	**27.** (*a*).			

Review Questions

1. What is meant by diffraction of light ? Distinguish between Fresnel's and Fraunhofer classes of diffraction. *[Rohilkhand University, Agra University, WBUT]*
   [Hints : 3.4.1.]
2. Obtain an expression for the resultant intensity in a single slit Fraunhofer diffraction process and show the intensity pattern graphically. *[WBUT, Pune University, H.P. University, Cal. University]*
   [Hints : 3.4.2.]
3. Discuss Fraunhofer diffraction at a double slit. Find the positions of maxima and minima. *[Kanpur University]*
   [Hints : 3.4.4 and 3.4.5]
4. Give the theory of plane transmission grating and show how would you use it to determine the wavelength of light. *[Agra University, Pune University, Kolkata University]*
   [Hints : 3.4.6]
5. Find an expression for the resolving power of plane transmission grating. *[Lucknow University]*
   [Hints : 3.4.7]
6. What is the difference between interference and diffraction? *[WBUT]*
   [Hints : 3.4.1]
7. Discuss Fraunhofer diffraction at a narrow slit. Draw a curve indicating the distribution of light in the diffraction pattern. Is there any fundamental difference between interference and diffraction ? Give reasons. *[Delhi University]*
8. Discuss the phenomenon of Fraunhofer diffraction at a single slit and show that the intensities of successive maxima are nearly in the ratio.

$$\frac{4}{9\pi^2} : \frac{4}{25\pi^2} : \frac{4}{49\pi^2}.$$

[Kanpur University]

9. Explain the difference between Fresnel and Fraunhofer diffraction phenomena. *[WBUT, B. OPT]*
10. (a) Show that the intensity distribution for diffraction in a single slit is given by

$$I = I_0 \left(\frac{\sin^2 \beta}{\beta^2}\right) \qquad \left[\text{where } \beta = \frac{\pi d \sin \theta}{\lambda}\right]$$

where the symbol have their usual significance. Deduce also conditions for maxima and minima. *[Calcutta University, B.U. (Hons).]*

(*b*) Show that secondary maxima are given by the equations $\tan \beta = \beta$. *[WBUT]*

11. Obtain an expression for the resultant intensity in a single slit Fraunhofer diffraction process and show the intensity pattern graphically. *[WBUT]*

Unsolved Problems

1. A monochromatic beam of light on passing through a slit 2 mm wide falls on a screen held close to the slit. The screen is then gradually moved away and the middle of the path of light on it becomes dark when the screen is 80 cm from the slit. Calculate the wavelength of light. **[Ans.** 6250 Å]

2. A spectral line of λ = 5400 Å, falling on a plane grating produces first order images at an angle of 30°. How many lines are there in one cm of grating? **[Ans.** N = 10^4]
3. What should be the minimum number of lines in a grating which will just resolve in the first order the lines whose wavelengths are 5890 Å and 5896 Å? **[Ans.** 0.982]
4. Monochromatic light falls on a plane diffraction grating of 5000 lines/cm. The spectral line of second order is diffracted by 45°. Calculate the wavelength of light. **[Ans.** 7070 Å] [*Rohilkhand University*]
5. A grating is made with parallel and equidistant wires having 200 wires per cm. The diameter of each wire is 0.025 mm. Calculate the angle of diffraction of light of wavelength 6000 Å for the third order and also the width of the clear space between two wires. [*Pune University*]

   [Hints : $\sin\theta = 2\,mn\lambda$, Here $\sin\theta = 200 \times 3 \times 6000 \times 10^{-8}$,

   $$\therefore \qquad \theta = 2^\circ\, 4', \text{ Now } m = \frac{1}{a+b}\ ; \quad 200 = \frac{1}{a+0.025},\ \therefore a = 0.0025 \text{ cm}]$$
6. Light is incident normally on a grating of total ruled width 5×10^{-3} m with 2500 lines in all. Calculate the angular separation of the two sodium lines in the first order spectrum. Can they be seen distinctly. **[Ans.** 1′, yes] [*I.A.S. 83*]
7. Examine if two spectral lines of wavelength 5890 Å and 5896 Å can be clearly resolved in the (*i*) First order, and (*ii*) Second order by a diffraction grating of 2 cm width and 425 lines / cm. [*Delhi, B.S.C. (Hons.)*]

   [Hints : Total number of lines on the grating = 2 × 425 = 850.

   (*i*) For 1st order, $N = \dfrac{\lambda}{d\lambda} = \dfrac{5890 \times 10^{-8}}{6 \times 10^{-8}} = 982$ lines.

   Here 850 < 982. So the lines will not be resolved

   (*ii*) For 2nd order, $N = \dfrac{\lambda}{2d\lambda} = 491$ lines.

   Here 850 > 491. So the spectral lines will be resolved]

Albert Einstein

Chapter 4

Special Theory of Relativity

4.1 INTRODUCTION

It is not possible to describe the motion of a body without defining some well-defined co-ordinate system. This co-ordinate system is usually called frame of reference. A frame of reference must be chosen in such a way that the laws of nature may become fundamentally easier when expressed in terms of such frames of reference. Newton introduced the idea of absolute space but later it was discarded.

4.2 FRAMES OF REFERENCE—INERTIAL AND NON-INERTIAL

A frame of reference is a set of co-ordinates, which defines the position of a particle in two or three dimensional space. It can be of different types *e.g.*, Cartesian, Spherical, Cylindrical etc. In any case a frame of reference must be selected in such a way that the laws of nature may become fundamentally simpler when expressed in terms of such frame of reference.

There are two types of frames of reference :

(*i*) Inertial or unaccelerated frames

(*ii*) Non-inertial or accelerated frames.

4.2.1 Inertial Frames

A frame in which a body, not under the influence of external forces and initially at rest remains at rest or if initially be moving with some velocity, will move uniformly with the same velocity is called Inertial frame.

Alternatively, the frames in which Newton's laws are true are called Inertial frames.

It implies that all inertial frames in uniform translation relative to one another are equivalent in so far as the mechanical phenomena are concerned. Thus when mechanical laws are cited with reference to inertial frames they preserve the same form whatever may be the relative motion between the observers.

4.2.2 Non-inertial Frames

The frames in which a body is accelerated is called non-inertial frames. Newton's laws are not valid in these frames. To illustrate the motion of a body we should always select inertial frames, for if we select non-inertial frames, these involve different forces and hence it is much complicated to formulate the laws common to all observers.

4.3 GALILEAN TRANSFORMATIONS

These transformations are used to transform the co-ordinates of position and time from one frame to other.

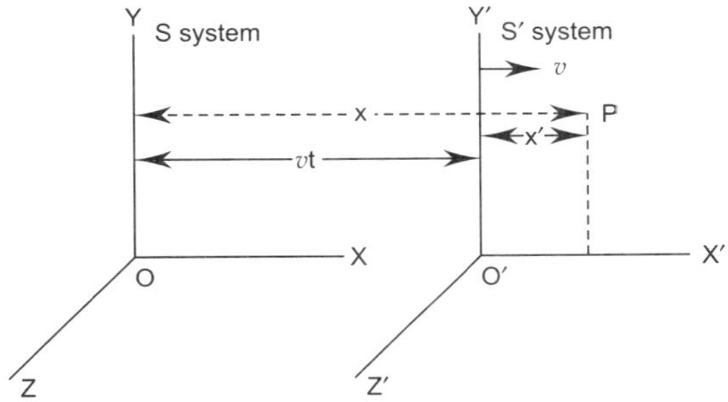

Figure 4.1

Consider we have two inertial frames of reference S and S′ are shown in figure 4.1. Let S be at rest and S′ frame be moving with uniform velocity v along the positive X-direction. Let the origins of S and S′ coincide initially ($t = 0$) and also consider two frames are parallel to each other. Consider some event occurs at p at any particular time. Let the co-ordinates of P with respect to O and O′ as origins be (x, y, z, t) and (x', y', z', t') and there is no relative motion between S and S′ along Y and Z directions. Thus we have $y' = y$ and $z' = z$. So the observations are taken by both the observers at any particular time t according to figure 4.1. Thus, we have

$$\left.\begin{aligned} x' &= x - vt \\ y' &= y \\ z' &= z \\ t' &= t \end{aligned}\right\} \quad \ldots(4.1)$$

These equations are known as the Galilean transformation equations which transform the co-ordinate system (x, y, z, t) of S frame into corresponding ones (x', y', z', t') of S′ frame.

If we want to transform from S′ system to S system, the conjugate set of transformation equations are

$$\left.\begin{aligned} x &= x' + vt \\ y &= y' \\ z &= z' \\ t &= t' \end{aligned}\right\} \quad \ldots(4.2)$$

This transformation is known as Inverse transformation.

When we consider the transformation equation for the velocities, we take the time derivatives. Now the transformation becomes (from equation (4.1))

$$\frac{dx'}{dt} = \frac{dx}{dt} - v \text{ i.e., } u' = u - v \quad \ldots(4.3)$$

$$\frac{dy'}{dt} = \frac{dy}{dt}, \frac{dz'}{dt} = \frac{dz}{dt}$$

Differentiating equation (4.3) with respect to t.

$$\frac{du'}{dt} = \frac{du}{dt} - \frac{dv}{dt} \quad [\because \; v = \text{constant}]$$

$$\therefore \quad \frac{du'}{dt} = \frac{du}{dt}, \text{ or, } a' = a$$

where a' and a are accelerations of particle in system S′ and S respectively. Thus the acceleration observed by the two observers in two different inertial frames are invariant. Therefore we can say that acceleration is invariant under Galilean transformation *i.e.*, Newton's laws are obeyed in every inertial frame.

4.4 THE ETHER HYPOTHESIS

Maxwell's equations predicts the existence of electromagnetic waves that propagate with a velocity not dependent on the frequency of wave motion. With the mechanical outlook of the nineteenth century, physicists felt that a material medium is necessity for the propagation of waves. The medium was given the name 'ether'. It was assumed that light propagates through 'ether' as the sound waves propagates through air. It was also assumed that it possessed some very strange properties, like massless, invisible, perfectly transparent, non-resistive and stays at absolute rest. The light passes through ether with fixed velocity 3×10^8 m/sec and earth passes through the stationary ether with velocity 3×10^4 m/sec.

4.5 MICHELSON-MORLEY EXPERIMENT

A number of experiments were performed to detect relative motion between earth and the ether. The most famous among them is the one performed by Michelson and Morley in 1887 using the Michelson interferometer to demonstrate the existance of the ether frame and determine the motion of the earth with respect to this frame.

The main motto of the experiment was used to measure the time difference between the two cases as seen from a frame fixed relative to the earth, in two perpendicular directions.

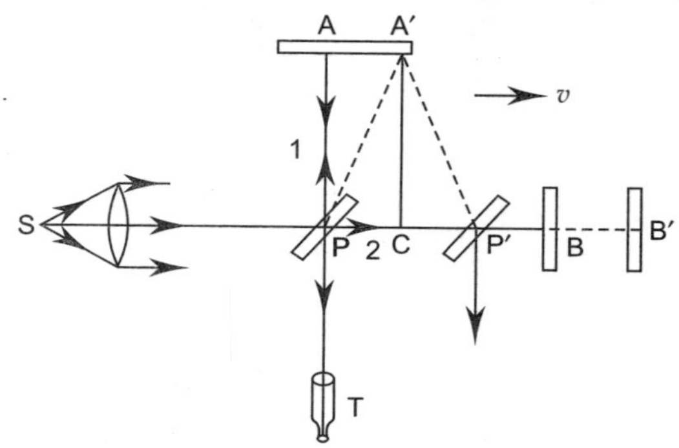

Figure 4.2

Michelson's interferometer consists of two plane mirrors (figure 4.2) A and B whose reflecting planes are mutually perpendicular. A beam of light from a monochromatic source S, falls upon the semisilvered glass plate P inclined 45° to the beam of light. The plate P, being semisilvered, divides the beam into two parts-a reflected beam (1) PA and refracted beam (2) PB. The beams 1 and 2 travelling at right angles to each other, fall normally on mirrors A and B which reflect them back to P. The two beams returned to P are directed towards a Telescope T in which interference fringes are observed. Let A and B be at the same distance 'l' from the plate 'P'. Thus if the apparatus is at rest in ether, the two waves, reflected and transmitted beams would take the same time to return to P. But actually the whole apparatus is moving with earth with a velocity v. Suppose the direction of motion of the earth is the direction of the initial beam of light. Due to the motion of the apparatus with earth, the optical path traversed by both the rays are not the same. Therefore, the reflections at mirrors A and B do not take place at A and B but at A′ and B′ respectively. The new position of the plate P is at P′. The time taken by the two beams on their journey are not equal. Let c and v be the velocities of light and the apparatus (earth) through ether respectively. The transmitted ray (beam 2) has velocity (c-v) relative to apparatus from P to B and ($c + v$) on the return trip from B′ to P′ as from P to B the beam 2 and the apparatus are moving in same direction while from B′ to P′ they are moving in opposite direction. If t_2 be the total time taken by this beam to go from P to B and back from B′ to P′, then

$$t_2 = \frac{l}{c-v} + \frac{l}{c+v} = \frac{2l}{c}\left(\frac{1}{1-\frac{v^2}{c^2}}\right)$$

The beam 1 (reflected ray from P) moving transversely with a velocity c strikes the mirror at A′ instead of A due to the motion of the earth. Consider it takes a time t to go from P to A′ travelling a distance ct. During the same time, the mirror A advances a distance vt.

From the right angled triangle PAA′, we have

$$PA'^2 = PA^2 + AA'^2$$

or

$$(ct)^2 = l^2 + v^2t^2$$

or

$$t = \frac{l}{\sqrt{c^2 - v^2}} = \frac{l}{c}\cdot\frac{1}{\left(1-\frac{v^2}{c^2}\right)^{\frac{1}{2}}}$$

If t_1 is the total time taken by the reflected ray to trace the total path PA′P′ (= PA′ + A′P′ = 2PA′) [Since PA′ = P′A′]

then

$$t_1 = 2t = \frac{2l}{c}\cdot\frac{1}{\left(1-\frac{v^2}{c^2}\right)^{\frac{1}{2}}}$$

Hence the difference between the times of travel of two beams (beam 1 and beam 2) is

$$\Delta t = t_2 - t_1 = \frac{2l}{c}\left[\frac{1}{\left(1-\frac{v^2}{c^2}\right)} - \frac{1}{\left(1-\frac{v^2}{c^2}\right)^{\frac{1}{2}}}\right]$$

$$= \frac{2l}{c}\left[\left(1+\frac{v^2}{c^2}\right)-\left(1+\frac{v^2}{2c^2}\right)\right]$$

$$= \frac{2l}{c}\cdot\frac{v^2}{2c^2} = \frac{lv^2}{c^3}$$

Using binomial expansion

$$\left[\left(1-\frac{v^2}{c^2}\right)^{-1} = 1+\frac{v^2}{c^2}\right]$$

and
$$\left[\left(1-\frac{v^2}{c^2}\right)^{-\frac{1}{2}} = 1+\frac{v^2}{2c^2}\right]$$

Thus the path difference *i.e.*, the distance travelled by light in $\Delta t = c\,\Delta t = c\,\frac{lv^2}{c^3} = \frac{lv^2}{c^2}$

If this corresponds to a shift of n fringes,

$$n\lambda = c\,\Delta t$$

or,
$$n = \frac{c\,\Delta t}{\lambda} = \frac{lv^2}{\lambda c^2} \quad [\lambda = \text{wavelength of light used}]$$

In the Michelson-Morley experiment, the whole apparatus, which was on a block of stone floated on mercury, was rotated through 90°, so that reflected and transmitted beams get interchanged. Thus, the path difference is in opposite direction $\left(-\frac{lv^2}{\lambda c^2}\right)$. So the resultant path difference becomes

$$\frac{lv^2}{\lambda c^2} - \left(-\frac{lv^2}{\lambda c^2}\right) = \frac{2lv^2}{\lambda c^2}$$

In this experiment, $l = 11$ meters $= 11 \times 10^2$ cm.

$v = 3 \times 10^6$ cm/sec, $c = 3 \times 10^{10}$ cm/sec, $\lambda = 5.5 \times 10^{-7}$ meter/sec.

$$\therefore \quad n = \frac{2 \times 11 \times \left(3 \times 10^4 \text{ m / sec}\right)^2}{5.5 \times 10^{-7} \times \left(3 \times 10^8 \text{ m / sec}\right)^2} = 0.4.$$

Taking all the necessary precautions, Michelson and Morley were extremely surprised to see that there was no shift in the fringes when the interferometer was rotated through 90°. The experiment was performed by several scientists at different points on the earth's surface and at different seasons of the year without detecting any measurable shift of the fringes. Thus this negative result compelled to suggest that it is impossible to measure the motion of the earth relative to the ether.

4.6 SIGNIFICANCE OF MICHELSON-MORLEY'S EXPERIMENTAL RESULTS

The null results of Michelson-Morley's experiment can be explained within the framework of classical physics by the following three explanations.

1. Lorentz-Fitzgerald contraction hypothesis
2. Light velocity hypothesis
3. 'Ether drag' hypothesis.

1. Lorentz-Fitzgerald Contraction Hypothesis: This hypothesis suggested that all material bodies contract, owing to their motion relative to ether by a factor $\sqrt{1-\frac{v^2}{c^2}}$ in the direction of motion. In the Michelson's experiment, the distance PA will remain unchanged but PB will get shortened to $l\sqrt{1-\frac{v^2}{c^2}}$. It is easily seen that such a contraction in the interferometer would equalize the times t_1 and t_2 and no fringe-shift would be expected. The hypothesis being totally adhoc in nature, could not be accepted.

2. Light Velocity Hypothesis: The velocity of light from a moving source is the vector sum of its natural velocity and velocity of the source. If these were valid, then the light from the source always remains the same constant velocity relative to the source and to the interferometer and the negative result could be explained. This explanation was ruled out because it was in conflict with the wave theory of light as well as with certain astronomical evidences regarding pairs of stars.

Einstein, who had given the proper explanation for the negative result of Michelson-Morley's experiment, concluded that the measured speed is the same for all directions in every inertial frames. This statement is called the principal constancy of the speed of light. Since the speed of light is c rather than $|c + v|$ in any frame.

3. Ether-drag Hypothesis: According to the ether drag hypothesis, the earth dragged along with it the ether in its immediate neighbourhood. Thus there was no relative motion between the two (earth and ether) and hence the question of shift does not arise. This is the explanation proposed by Michelson himself.

4.7 POSTULATES OF EINSTEIN'S SPECIAL THEORY OF RELATIVITY

In 1905, Einstein propounded his new principle of relativity — the special theory of relativity — based on the following two postulates.

Postulate 1. *The fundamental laws of physics, mechanical or electromagnetic, are the same (that is, covariant) in all inertial frames of reference having an uniform translational velocity with respect to one another.*

Postulate 2. *The velocity of light in free space is the same in all inertial frames and is independent of the relative motion of the source and the observer.*

The consequence of the postulate 1 is that all inertial frames are completely equivalent. This may however at first sight appear similar to the Newtonian Relativity but there is a fundamental difference. Newton assumed the existence of an absolute frame of reference in which his laws of motion were strictly valid and as a result they were also valid in all other frames in uniform translation relative to it. But Einstein considered only the frames in uniform translation with respect one another. Further, Newton's equivalence of the inertial frames was in respect of mechanical phenomena only. Einstein extended it to all the laws of physics, including electromagnetic phenomena. Because of this postulate, an observer cannot detect the motion of the system by performing any experiment — mechanical, optical, electrical etc., in the system.

This postulate required Einstein to modify either the Maxwell's equations or the Galilean transformation, for the two together imply contrary to the postulate. Einstein decided to retain Maxwell's equation unchanged and this was indicated by the second postulate. The second postulate follows directly from the null result of the Michelson-Morley's experiment. Due to this postulate, Einstein was forced to modify the Galilean transformation. This was a very bold move. First, the intuitive belief about the validity of Galilean transformation was too strong among the contemporary physicists. Secondly, any modification in Galilean transformation would at once require some compensatory change in Newton's equation in order that first postulate is satisfied for mechanics. The follow-up of this in fact forms an adventurous sojourn into one of the most exciting chapter of human achievement.

Insufficiency of Galilean Transformation

Consider two frames of references S and S′, S′ moving with respect to S with a uniform velocity v along x-direction. Let the origins of the systems O and O′, coincide at a particular instant $t = t' = 0$. Let at that instant a light wave originate at O and O′. In S-system, the wavefront of the light would be a sphere, given by

$$x^2 + y^2 + z^2 = c^2t^2 \qquad \ldots(i)$$

where c is the velocity of light. Equation (i) may be rewritten, using Galilean transformation equations, as

$$(x' + vt)^2 + y'^2 + z'^2 = c^2t'^2 \qquad \ldots(ii)$$

In S′ system clearly (ii) does not represent a sphere and is thus not in accord with the postulates of relativity. Further, it implies that by measuring the velocity of light in a number of directions it is possible, in principle, to determine the velocity of S′ system. The relativity principle demands that the wavefront in S′ system should also be a sphere, given by

$$x'^2 + y'^2 + z'^2 = c^2t'^2 \qquad \ldots(iii)$$

This is clearly an impossibility by the Galilean transformation equations. A new set of transformation formulae is required.

4.8 THE LORENTZ TRANSFORMATION EQUATIONS

We have to develop a set of transformation equations which are directly consistent with the new concept of the invariance of light velocity in free space. This transformation equations were propounded by Lorentz, and are known as "Lorentz transformations".

Derivation : Let two observers O and O′ exist in two inertial frames S and S′. Frame S′ is moving with uniform velocity v relative to the frame S along the +ve X-axis. For convenience, let us consider that the X axes of the two systems coincide permanently and the velocity of S′ is parallel to X axis. Also consider that the axes of the two co-ordinate systems coincide at $t = t' = 0$. Here, the event P is a light signal and is emitted when the origins of two frames coincide ($t = t' = 0$). The light pulse produced at $t = 0$ will spread out as a growing sphere and the radius of wavefront produced in this way will grow with speed c. After a time t the observer O in S frame will note that the light has reached a point P(x, y, z) as shown in figure 4.3.

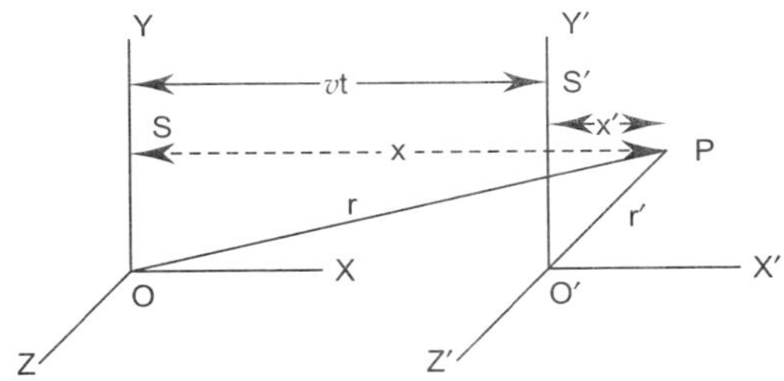

Figure 4.3

For him, the distance of the point P is given by $r = ct$. From figure 4.3, $r^2 = x^2 + y^2 + z^2$.

Therefore, $$x^2 + y^2 + z^2 = c^2t^2 \qquad \ldots(i)$$

Similarly, the observer O′ in frame S′ will note that the light has reached the same point P in a time t' with the same velocity c.

So, $$r' = ct'.$$

So the equation of spherical surface is

$$\therefore \quad x'^2 + y'^2 + z'^2 = c^2t'^2 \qquad \ldots(ii)$$

According to second postulate of special theory c is a constant for all inertial frames. Now equations (i) and (ii) must be equal since both the observers O and O′ are at the centers of the same expanding wavefront.

$$\therefore \quad x^2 + y^2 + z^2 - c^2t^2 = x'^2 + y'^2 + z'^2 - c^2t'^2 \qquad \ldots(iii)$$

Since there is no movement along Y and Z directions, $y' = y$ and $z' = z$. Therefore, equation (iii) gives

$$x^2 - c^2t^2 = x'^2 - c^2t'^2 \qquad \ldots(iv)$$

For the transformation equation relating to x and x', let us put

$$x' = k\,(x - vt) \quad \text{where } k \text{ is a constant.} \qquad \ldots(v)$$

The reason for taking the above relation is that, the transformation must reduce to Galilean for low speed ($v << c$ or $\frac{v}{c} \to 0$)

Similarly, let us consider that $t' = a\,(t - bx)$ where a and b are constants. ...(vi)

Putting these values for x' and t' in equation (iv) we have

$$x^2 - c^2t^2 = k^2\,(x - vt)^2 - c^2a^2\,(t - bx)^2$$

i.e., $$x^2 - c^2t^2 = (k^2 - c^2a^2b^2)\,x^2$$

$$-2\,(k^2v - c^2a^2b)\,xt - \left(a^2 - \frac{k^2v^2}{c^2}\right)c^2t^2 \qquad \ldots(vii)$$

Equating the co-efficients of corresponding terms in equation (*vii*)

$$k^2 - c^2a^2b^2 = 1, \quad k^2v - c^2a^2b = 0$$

and

$$a^2 - \frac{k^2v^2}{c^2} = 1.$$

Solving the above equations for k, a and b, we obtain

$$k = a = \frac{1}{\sqrt{1 - \frac{v^2}{c^2}}} \text{ and } b = \frac{v}{c^2}$$

we get

$$x' = \frac{x - vt}{\sqrt{1 - \frac{v^2}{c^2}}},\; y' = y,\; z' = z \text{ and } t' = \frac{t - \frac{vx}{c^2}}{\sqrt{1 - \frac{v^2}{c^2}}} \qquad \ldots(viii)$$

These are called Lorentz transformation equations.

The inverse Lorentz transformation equations are obtained by exchanging the co-ordinates and replacing v by $-v$ in equation (*viii*).

$$x = \frac{x' + vt'}{\sqrt{1 - \frac{v^2}{c^2}}},\; y = y',\; z = z' \text{ and } t = \frac{t' + \frac{vx'}{c}}{\sqrt{1 - \frac{v^2}{c^2}}} \qquad \ldots(ix)$$

i.e., the frame S moves relative to frame S′ with velocity v along the –ve direction of X-axis.

These equations convert measurements made in frame S′ into those in frame S.

The two important aspects of the Lorentz transformation equations are :

1. The measurements of position and time solely depend upon the frame of reference of the observer. Two events which are simultaneous when observed from one frame of reference are not simultaneous when observed from a second frame which is moving relative to the first.

2. If $v << c$ *i.e.*, $\frac{v}{c} \to 0$, the Lorentz transformation {equation (*viii*)} reduces to $x' = x - vt$, $y' = y$, $z' = z$, $t' = t$, which are Galilean transformations. Thus we can say that Lorentz transformations approach to Galilean for $v << c$. Therefore, the consequences of special relativity are not apparant unless enormous velocities are involved which are comparable to the velocity of light ($v < c$) and Galilean holds only for low velocities.

Example 1. *Show by means of Lorentz transformation equations that*

$$x'^2 - c^2\,t'^2 = x^2 - c^2\,t^2$$

[Kerala University, Pbi. University]

Solution. We have from Lorentz transformation equation

$$x' = \frac{x - vt}{\sqrt{1 - \frac{v^2}{c^2}}},\; y' = y,\; z' = z \text{ and } t' = \frac{t - \frac{vx}{c^2}}{\sqrt{1 - \frac{v^2}{c^2}}}.$$

Now from $$x'^2 - c^2 t'^2 = \frac{(x-vt)^2}{1-\frac{v^2}{c^2}} - c^2 \cdot \frac{\left(t - \frac{vx}{c^2}\right)^2}{\left(1-\frac{v^2}{c^2}\right)}$$

$$= \frac{\left(-x^2 + v^2t^2 - 2xvt - c^2\left(t^2 + \frac{v^2x^2}{c^4} - \frac{2tvx}{c^2}\right)\right)}{\left(1-\frac{v^2}{c^2}\right)}$$

$$= \frac{\left(x^2 - c^2t^2\right) - \frac{v^2}{c^2}\left(x^2 - c^2t^2\right)}{\left(1-\frac{v^2}{c^2}\right)}$$

$$= \frac{\left(x^2 - c^2t^2\right)\left(1-\frac{v^2}{c^2}\right)}{\left(1-\frac{v^2}{c^2}\right)}$$

$$= x^2 - c^2t^2.$$

Example 2. *In a frame S, the following two events occurs. Event 1 :* $x_1 = x_0$ *;* $t_1 = \frac{x_0}{c}$ *and* $y_1 = z_1 = 0$ *; Event 2 :* $x_2 = 2x_0$, $t_2 = \frac{x_0}{2c}$ *and* $y_2 = z_2 = 0$. *Find the velccity of S′frame (relative to S) in which the events occur at the same time and what is the value of corresponding time ?*

[Calcutta University]

Solution. According to the problem, the two events occur at same time in S′ rame.

$\therefore$ $$t_1' = t_2' \text{ or, } \frac{t_1 - \frac{vx_1}{c^2}}{\sqrt{1-\frac{v^2}{c^2}}} = \frac{t_2 - \frac{vx_2}{c^2}}{\sqrt{1-\frac{v^2}{c^2}}}$$

or, $$\frac{x_0}{c} - \frac{vx_0}{c^2} = \frac{x_0}{2c} - \frac{v \cdot 2x_0}{c^2} \text{ or } v = -\frac{c}{2}.$$

Now the value of the corresponding time $t_1' (= t_2') = \dfrac{t_1 - \frac{vx_1}{c^2}}{\sqrt{1-\frac{v^2}{c^2}}}$

$\Rightarrow$ $$t_1' = \frac{\frac{x_0}{c} - \left(-\frac{c}{2}\right)\cdot\frac{x_0}{c^2}}{\sqrt{1-\frac{c^2}{c_1}\cdot c^2}} = \frac{\sqrt{3}\,x_0}{c}. \quad \textbf{Ans.}$$

4.9 LENGTH CONTRACTION

Consider a rigid rod to be at rest in the S system parallel to X-axis. Let the co-ordinates of the end points of the rod in the system S measured at the time t_1 be x_1 and x_2.

Then the length of the rod as observed from S system is $L_0 = x_2 - x_1$.

Now suppose another observer measures the length of the rod from a frame of reference S′ which is moving with a velocity v relative to S frame. If the co-ordinates of the end points of the same rod, measured at the same instant, be x_1' and x_2' in the S′ system moving with velocity v along x' axis, parallel to x axis of S frame, then

$$x_2 = \frac{x_2' + vt_2'}{\sqrt{1 - \frac{v^2}{c^2}}} \text{ and } x_1 = \frac{x_1' + vt_1'}{\sqrt{1 - \frac{v^2}{c^2}}}$$

If L′ be the length of the rod in S′ system, then

$$L' = x_2' - x_1'$$

$$\therefore \quad L_0 = x_2 - x_1 = \frac{x_2' - x_1'}{\sqrt{1 - \frac{v^2}{c^2}}} \qquad (\because t_2' = t_1')$$

$$= \frac{L'}{\sqrt{1 - \frac{v^2}{c^2}}}$$

$$\therefore \quad L' = L_0 \sqrt{1 - \frac{v^2}{c^2}}$$

This shows that $L' < L_0$. Thus, the length of the rod seems to be contracted when measured by a moving observer along the direction of the length of the rod in the ratio $\left(1 : \sqrt{1 - \frac{v^2}{c^2}}\right)$. This contraction occurs only in the direction of relative motion. If v is parallel to x, then y and z dimensions of the rod are the same in both S and S′ systems.

4.10 TIME DILATION

Let us consider two systems S and S′, S′ moving with uniform velocity v relative to S along +ve direction of X-axis.

Consider t_1 and t_2 be the times of two events at the same place x in the S system. So the time interval between the events in S system is

$$\Delta t = t_2 - t_1$$

Let the same two events occur in the S′ system at times t_1' and t_2'. So the interval between the events in S′ system is $\Delta t' = t_2' - t_1'$

From Lorentz transformation, we have

$$t_1' = \frac{t_1 - \frac{vx}{c^2}}{\sqrt{1-\frac{v^2}{c^2}}} \; ; \; t_2' = \frac{t_2 - \frac{vx}{c^2}}{\sqrt{1-\frac{v^2}{c^2}}}$$

$$\therefore \quad \Delta t' = \frac{t_2 - t_1}{\sqrt{1-\frac{v^2}{c^2}}} = \frac{\Delta t}{\sqrt{1-\frac{v^2}{c^2}}}$$

$$\therefore \quad \Delta t' > \Delta t \quad \left(\because \frac{1}{\sqrt{1-\frac{v^2}{c^2}}} > 1 \right)$$

Therefore, the time interval Δt seems to the moving observer to be dilated or lengthened because the time interval in system S′ is greater than that in system S. *i.e.*, a moving clock appears to be slowed down to a stationary observer. This phenomenon is known as 'time dilation'.

Example 1. *A rod 1 meter long is moving along its length with a velocity 0.6 c. Calculate its length as it appears to an observer (a) on the earth ; (b) moving with the rod itself.*

Solution. According to the problem, 1 meter is the proper length (l_0) of the rod in its own moving frame of reference.

(*a*) Consider l be the length of the rod as it seems to an observer in the stationary reference frame of the earth.

Here $l_0 = 1\ m$; $v = 0.6\ c$;

$$\therefore \quad l = l_0 \sqrt{1-\frac{v^2}{c^2}} = 1 \} \sqrt{1-\left(\frac{0.6c}{c}\right)^2} = 0.8 \text{ m}$$

Thus, the observer on the earth will estimate the length of the rod be 0.8 meter.

(*b*) For an observer moving with the rod itself, the length of the rod is 1 meter.

The Twin Paradox

Consider two exactly identical twin brothers. Let one of them goes to a long space journey at a high speed in a rocket and the other stay behind on the earth. The clock in the moving rocket will seem to go slow than the clock on the surface of the earth (in accordance with $t = \frac{t_0}{\sqrt{1-\frac{v^2}{c^2}}}$). Therefore, when he comes back to the earth, he will find himself younger then the twin who stayed behind on the earth.

Example 2. *A clock in a spaceship emits signals at intervals of 1 second as observed by an astronaut in the spaceship. If the spaceship travels with a speed of 3×10^7 m/sec, what is the interval between successive signals as seen by an observer at the control center on the ground?*

Solution. Given that $t_0 = 1$ sec, $v = 3 \times 10^7$ m/sec and $c = 3 \times 10^8$ m/sec, $t = ?$

$$\therefore \quad t = \frac{t_0}{\sqrt{1-\frac{v^2}{c^2}}} = \frac{1}{\sqrt{1-\left(\frac{1}{10}\right)^2}} = \frac{1}{\sqrt{0.99}} = 1.005 \text{ sec.}$$

The interval between successive signals as seen by an observer at the control center on the ground is 1.005 sec.

4.11 ADDITION OF VELOCITIES

Let us consider there are two frames of reference S and S′, the frame S′ moving with a uniform velocity v relative to S along positive direction of X-axis. Suppose a particle is moving in the common direction of the X and X′ axes. Let its velocity, as measured by an observer in the system S, be u and as measured by an observer in S′, be u'.

Thus we have, $$u = \frac{dx}{dt} \quad \text{and} \quad u' = \frac{dx'}{dt}.$$

From inverse Lorentz transformations,

we have $$x = \frac{x' + vt'}{\sqrt{1 - \frac{v^2}{c^2}}} \quad \text{and } t = \frac{t' + \frac{vx'}{c^2}}{\sqrt{1 - \frac{v^2}{c^2}}}.$$

Differentiating, we get $$dx = \frac{dx' + vdt'}{\sqrt{1 - \frac{v^2}{c^2}}} \quad \text{and } dt = \frac{dt' + \frac{vdx'}{c^2}}{\sqrt{1 - \frac{v^2}{c^2}}}$$

But $\frac{dx}{dt} = u,$

Thus, $$u = \frac{dx}{dt} = \frac{dx' + v\,dt'}{dt' + v\frac{dx'}{c^2}} = \frac{\frac{dx'}{dt'} + v}{1 + \frac{v}{c^2}\frac{dx'}{dt'}} = \frac{u' + v}{1 + \frac{u'v}{c^2}}$$

This equation represents the relativistic law of addition of velocities. Whereas in classical mechanics it is simply $u = u' + v$. We can also express the velocity u' in term of u and v, *i.e.*, $u' = \dfrac{u - v}{1 - \frac{uv}{c^2}}$

There are some important consequences of the relativistic addition of velocities:

(*i*) If $u' = c$, *i.e.*, if a ray of light is emitted in the moving system S′ along the positive X-axis, an observer in system S will find the velocities as

$$u = \frac{u' + v}{1 + \frac{u'v}{c^2}} = \frac{c + v}{1 + \frac{cv}{c^2}} = c.$$

Thus both the observer find the same value for the velocity of light. So the velocity of light in free space is constant and is independent of the frame of reference.

(*ii*) When $u' = v = c$, then

$$u = \frac{c + c}{1 + \frac{c^2}{c^2}} = c.$$

This shows that the addition of velocity of light reproduces the velocity of light *i.e.*, the speed of light is same for all inertial frames whatever their relative velocity may be and no particle can attain a velocity greater than the velocity of light.

(*iii*) When u' and v are very very less than c *i.e.*, $u' << c, v << c$, then $\frac{u'v}{c^2} << 1$, and can be neglected, then

$$u = u' + v$$

which is the classical equation.

Example 1. *An experimenter observes a radioactive atom moving with a velocity of 0.25 c. The atom then emits a β particle which has a velocity of 0.9 c relative to the atom in the direction of its motion. What is the velocity of the β particle as observed by the experimenter?*

Solution. Here $v = 0.25\ c$ and $u' = 0.9\ c$; $u = ?$

$$u = \frac{u' + v}{1 + \frac{u'v}{c^2}} = \frac{(0.9 + 0.25)c}{1 + \frac{0.9\ c \times 0.25\ c}{c^2}} = 0.94\ c.$$

Example 2. *An electron is moving with a constant speed of 0.85 c in a direction opposite to that of a moving photon. Calculate the relative velocity of the photon with respect to the electron?*

Solution. Let the photon be moving along the +ve direction of X-axis and the electron in the –ve direction of X-axis. Then, the speed of the electron = – 0.85 c ; the speed of photon = c consider that the electron is at rest in system S. Then we assume that the system S′ (laboratory) is moving with velocity 0.85 c relative to system S (*i.e.*, electron) *i.e.*, $v = 0.85\ c$; $u' = c$; $u = ?$

$$u = \frac{u' + v}{1 + \frac{u'v}{c^2}} = \frac{c + 0.85\ c}{1 + \frac{c \times 0.85\ c}{c^2}} = c$$

4.12 VARIATION OF MASS WITH VELOCITY

Derivation : Consider two system S and S′, S′ is moving with a uniform velocity v relative to the system S in the + X direction (Figure 4.4). Suppose that in S′, two exactly similar elastic balls approach each other at equal speeds (*i.e.*, u and – u). Let the mass of each ball be m' in S′. At the moment of collision their velocity is zero in S′ system. But they still have a velocity v in S system. We shall describe collision in S and S′ system.

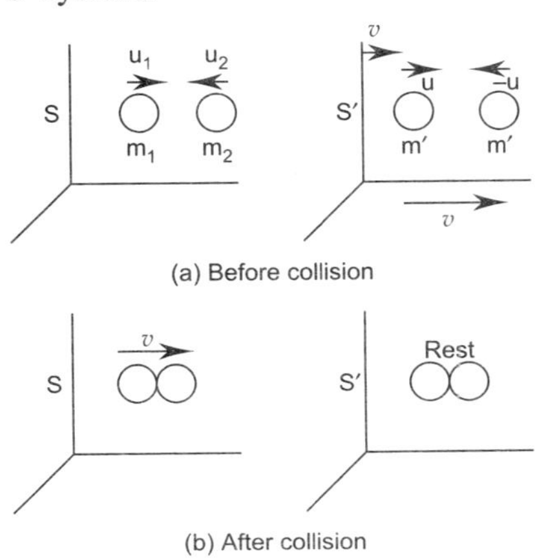

Figure 4.4

The conservation of momentum gives

$$m'u + m'(-u) = 0 \qquad \text{in S}' \text{ system} \qquad \ldots(i)$$

and

$$m_1u_1 + m_2u_2 = (m_1 + m_2)\,v \qquad \text{in S system} \qquad \ldots(ii)$$

The velocities u_1 and u_2 of the masses in S system are by velocity addition theorem,

$$u_1 = \frac{u+v}{1+\dfrac{uv}{c^2}}\;;\quad u_2 = \frac{-u+v}{1-\dfrac{uv}{c^2}} \qquad \ldots(iii)$$

Putting the values of u_1 and u_2 from (*iii*) in (*ii*)

or,

$$m_1\left(\frac{u+v}{1+\dfrac{uv}{c^2}} - v\right) = m_2\left(v - \frac{-u+v}{1-\dfrac{uv}{c^2}}\right)$$

or

$$m_1\frac{u\left(1-\dfrac{v^2}{c^2}\right)}{1+\dfrac{uv}{c^2}} = m_2\frac{u\left(1-\dfrac{v^2}{c^2}\right)}{1-\dfrac{uv}{c^2}}$$

or,

$$\frac{m_1}{m_2} = \frac{1+\dfrac{uv}{c^2}}{1-\dfrac{uv}{c^2}} \qquad \ldots(iv)$$

Again,

$$1-\frac{u_1^2}{c^2} = 1-\frac{\left\{\dfrac{u+v}{c}\right\}^2}{\left(1+\dfrac{uv}{c^2}\right)^2} = \frac{1+\dfrac{u^2v^2}{c^4}+\dfrac{2uv}{c^2}-\dfrac{u^2}{c^2}-\dfrac{v^2}{c^2}-\dfrac{2uv}{c^2}}{\left(1+\dfrac{uv}{c^2}\right)^2}$$

$$= \frac{\left(1-\dfrac{u^2}{c^2}\right)-\dfrac{v^2}{c^2}\left(1-\dfrac{u^2}{c^2}\right)}{\left(1+\dfrac{uv}{c^2}\right)^2} = \frac{\left(1-\dfrac{v^2}{c^2}\right)\left(1-\dfrac{u^2}{c^2}\right)}{\left(1+\dfrac{uv}{c^2}\right)^2} \qquad \ldots(v)$$

Similarly,

$$1-\frac{u_2^2}{c^2} = \frac{\left(1-\dfrac{u^2}{c^2}\right)\left(1-\dfrac{v^2}{c^2}\right)}{\left(1-\dfrac{uv}{c^2}\right)^2} \qquad \ldots(vi)$$

Dividing equation (*vi*) by equation (*v*)

$$\frac{1-\dfrac{u_2^2}{c^2}}{1-\dfrac{u_1^2}{c^2}} = \frac{\left(1+\dfrac{uv}{c^2}\right)^2}{\left(1-\dfrac{uv}{c^2}\right)^2}\;;\quad \text{or} \quad \frac{\sqrt{1-\dfrac{u_2^2}{c^2}}}{\sqrt{1-\dfrac{u_1^2}{c^2}}} = \frac{1+\dfrac{uv}{c^2}}{1-\dfrac{uv}{c^2}} \qquad \ldots(vii)$$

From equations (*vii*) and (*iv*), we obtain

$$\frac{m_1}{m_2} = \frac{\sqrt{1-\frac{u_2^2}{c^2}}}{\sqrt{1-\frac{u_1^2}{c^2}}}$$

or, $$m_1\sqrt{1-\frac{u_1^2}{c^2}} = m_2\sqrt{1-\frac{u_2^2}{c^2}} \qquad \ldots(viii)$$

The two balls have been assumed to be identical when they are at rest in the same frame.

Let m_0 be their mass under this condition,

then, $$m_1\sqrt{1-\frac{u_1^2}{c^2}} = m_2\sqrt{1-\frac{u_2^2}{c^2}} = m_0.$$

where m_0 is called the rest mass of the body and correspond to zero velocity.

Thus, $$m_1 = \frac{m_0}{\sqrt{1-\frac{v_1^2}{c^2}}}$$

If m denotes the mass of a body when it is moving with a velocity v,

then $$m = \frac{m_0}{\sqrt{1-\frac{v^2}{c^2}}} \qquad \ldots(ix)$$

Thus equation (*ix*) represents that a body moving with a speed v relative to an observer has a mass larger than when at rest relative to the observer by a factor $\dfrac{1}{\sqrt{1-\frac{v^2}{c^2}}}$.

Equation (*ix*) is the relativistic formula for the variation of mass with velocity.

Discussion: This formula leads to a very important conclusion that the mass of a body increases with increase of velocity.

Now, if $\frac{v}{c} \to 0$ then $m \to m_0$, the rest mass. If $\frac{v}{c} \to 1$ (*i.e.*, $v \to c$) $m \to \infty$. Thus, to make a material particle move at the speed of light ($v \to c$) its mass would require to be raised to infinity. This only indicates that c is the upper limit of velocity. Thus equation (*ix*) also leads to a conclusion that no material body can have the velocity equal to, or greater than the velocity of light.

The increase of mass with velocity has now been verified in "particle accelerators". Accelerators are basically used to accelerate various charged particles of matter to very high velocities. It has been experimentally verified that the electrons and protons accelerated in these machines to velocities close to velocity of light acquire increased masses exactly as predicted theoretically.

When $v << c$, $\frac{v^2}{c^2}$ can be neglected and we get $m \approx m_0$. It implies that under ordinary velocities the difference between m and m_0 is too small to be detectable.

4.13 MASS ENERGY EQUIVALENCE

According to Einstein's special theory of relativity, both mass and velocity are variable and it is given by

$$m = \frac{m_0}{\sqrt{1 - \frac{v^2}{c^2}}} \qquad \ldots(i)$$

where m_0 is the rest mass and c is the velocity of light. The variation of mass with velocity has subjected us to change our ideas about energy. Let us consider a particle of mass m acted upon by a force F in the same direction as the velocity.

Force is defined as rate of change of momentum,

i.e., $$\mathrm{F} = \frac{d}{dt}(mv) = m\frac{dv}{dt} + v\frac{dm}{dt} \qquad \ldots(ii)$$

Let the force F displace the body through a distance ds. Then the work done (F ds) is equal to the change in kinetic energy ($d\mathrm{E}_k$) of the body.

Hence $$d\mathrm{E}_k = \mathrm{F}\,ds = m\frac{dv}{dt}ds + v\frac{dm}{dt}ds$$

or, $$d\mathrm{E}_k = mv\,dv + v^2\,dm \left[\because \frac{ds}{dt} = v\right] \qquad \ldots(iii)$$

From equation (*i*), $$m = \frac{m_0}{\sqrt{1 - \frac{v^2}{c^2}}}$$

Squaring both sides, $$m^2 = \frac{m_0^2}{1 - \frac{v^2}{c^2}}$$

$$m^2c^2 = m_0^2c^2 + m^2v^2$$

Differentiating, the above expression,

$$c^2\,2m\,dm = m^2\,2v\,dv + v^2\,2m\,dm$$

or, $$c^2\,dm = mv\,dv + v^2\,dm \qquad \ldots(iv)$$

From equations (*iv*) and (*iii*) $dE_k = c^2\, dm$

Therefore, a change in kinetic energy dE_k is directly proportional to the change in mass dm. When a body is at rest, its velocity is zero, (K.E. = 0) and $m = m_0$. When its velocity is v, its mass becomes m. Therefore, integrating the equation we have,

$$E_k = \int_0^{E_k} dE_k = c^2 \int_{m_0}^{m} dm = c^2 (m - m_0)$$

$$\therefore \quad E_k = mc^2 - m_0 c^2$$

Thus the kinetic energy of a body is equal to the relativistic increase in mass of the body over the rest mass multiplied by the square of the velocity of light. When the body is at rest, the internal energy possessed by the body is $m_0 c^2$. m_0c^2 is called the rest mass energy. The total energy (E) of the body is therefore, the sum of kinetic energy (E_k) and rest mass energy (m_0c^2).

$$\therefore \quad E = E_k + m_0c^2 = (mc^2 - m_0c^2) + m_0c^2 = mc^2$$

$$\therefore \quad E = mc^2$$

This is the famous Einstein's mass energy relation.

This relation illustrates a universal equivalence between mass and energy. It signifies that mass may appear as energy and energy as mass.

The relation $E = mc^2$ is of fundamental importance in nuclear physics in understanding the mass defect (loss) and generation of energy is nuclear reaction. Energy evolved in nuclear fission, nuclear fusion etc., can be explained properly using the above relation. Basically this relation has established a new era in human civilization — the so called atomic or nuclear age

4.14 RELATIVISTIC RELATION BETWEEN THE TOTAL ENERGY, THE REST ENERGY AND THE MOMENTUM

The total relativistic energy (rest mass + kinetic energy) of a particle is

$$E = mc^2 = \frac{m_0c^2}{\left(1 - \frac{v^2}{c^2}\right)^{\frac{1}{2}}} \qquad \ldots(4.4)$$

The relativistic momentum of the particle is $p = mv$

$$\therefore \quad p = \frac{m_0 v}{\sqrt{1 - \frac{v^2}{c^2}}} \qquad \ldots(4.5)$$

$$\therefore \quad E = \frac{m_0c^2}{\left[1 - \frac{p^2}{m^2c^2}\right]^{\frac{1}{2}}} = \frac{m_0c^2}{\sqrt{\left[1 - \frac{p^2c^2}{m^2c^4}\right]}} = \frac{m_0c^2}{\sqrt{1 - \frac{p^2c^2}{E^2}}} \qquad [\because E = mc^2]$$

$$E^2 = \frac{m_0^2\, c^4}{\left(1 - \frac{p^2c^2}{E^2}\right)}$$

or, $$E^2 - p^2c^2 = m_0^2 c^4$$

∴ $$\boxed{E^2 = m_0^{\,2} c^4 + p^2c^2.}$$

This is the relativistic relation between total energy E and momentum p of the particle.

Example : *An electron of rest mass 9.1×10^{-31} kg is moving with a speed 0.99 c. What is its total energy? Find the ratio of Newtonian kinetic energy to the relativistic kinetic energy ? $c = 3 \times 10^8$ m/sec.*

Solution. Here $m_0 = 9.1 \times 10^{-31}$ kg, $v = 0.99\,c$ $m = ?$

$$m = \frac{m_0}{\sqrt{1 - \frac{v^2}{c^2}}} = 64.54 \times 10^{-31} \text{ kg}.$$

∴ Total energy of electron $E = mc^2 = 5.81 \times 10^{-13}$ J

Now, the newtonian kinetic energy is $\frac{1}{2} m_0v^2$ and the relativistic kinetic energy $= mc^2 - m_0c^2$

∴ The Required Ratio $= \dfrac{\frac{1}{2} m_0v^2}{mc^2 - m_0c^2} = 0.08.$

IN FOCUS

- **Galilean Transformation** (S′ frame → S frame)
- $x' = x - vt,\ y' = y,\ z' = z$ and $t' = t$

  Again Galilean Transformation (S frame → S′ frame)

  $$x = x' + vt',\ y = y',\ z = z' \text{ and } t = t'$$

- **Lorentz Transformation** (S′ frame → S frame)

  $$x' = \frac{x - vt}{\sqrt{1 - \frac{v^2}{c^2}}},\ y' = y,\ z' = z \text{ and } t' = \frac{t - \frac{vx}{c^2}}{\sqrt{1 - \frac{v^2}{c^2}}}$$

  For Inverse Lorentz Transformation (S frame → S′ frame)

  $$x = \frac{x' + vt'}{\sqrt{1 - \frac{v^2}{c^2}}},\ y' = y,\ z = z' \text{ and } t = \frac{t' - \frac{vx'}{c^2}}{\sqrt{1 - \frac{v^2}{c^2}}}$$

- **Length Contraction** $L' = L_0\sqrt{1 - \frac{v^2}{c^2}}$

  L′ : Length of the Rod in S′ system
  L : Length of the Rod in S system

- **Time Dilation** $\Delta t' = \dfrac{\Delta t}{\sqrt{1 - \frac{v^2}{c^2}}}$

Δt : Time interval between the events in S system.
$\Delta t'$: Time interval between the events in S′ system.

- **The relativistic law of Addition of velocities**

$$u = \frac{u' + v}{1 + \dfrac{u'v}{c^2}}$$

As measured by an observer in the system S, be u and as measured by an observer in S′ system be u'.
The velocity u' in terms of u and v, *i.e.*,

$$u' = \frac{u - v}{1 - \dfrac{uv}{c^2}}$$

- The relativistic formula for the variation of mass with velocity

$$m = \frac{m_0}{\sqrt{1 - \dfrac{v^2}{c^2}}}$$

- **Einstein's mass energy relation:** $E = mc^2$
- Relativistic relation between the total energy (E), the rest energy (m_0c^2) and momentum (p) :-
$$E^2 = m_0^2c^4 + p^2c^2$$

Solved Problems

Problem 1. *What will be the fringe-shift according to the ether theory in the Michelson- Morley's experiment, if the effective length of each path is 7 meters and light has 7000 Å wavelength? The velocity of earth is 3×10^4 m/sec.* [Rajasthan University, Agra University]

Solution. We know that the expected shift in Michelson-Morley's experiment is

$$n = \frac{2d}{\lambda} \cdot \frac{v^2}{c^2}$$

Here $d = 7$ meter, $v = 3 \times 10^4$ m/sec, $c = 3 \times 10^8$ m/sec.

and $\lambda = 7000$ Å $= 7000 \times 10^{-8}$ cm $= 7 \times 10^{-7}$ meter.

$$\therefore \quad n = \frac{2 \times 7 \times 9 \times 10^8}{7 \times 10^{-7} \times 9 \times 10^{16}} = 0.2.$$

Problem 2. *A rod of length 1 meter moves past an observer standing on the earth with a velocity 3×10^7 m/sec, along the direction of its length. What is its apparent length with respect to the observer?*

Solution. Here $L_0 = 1$ meter, $v = 3 \times 10^7$ m /sec.

Apparent length from Lorentz transformation is $L' = L_0 \sqrt{1 - \dfrac{v^2}{c^2}}$

$$= 1\sqrt{1-\left(\frac{3\times10^7}{3\times10^8}\right)^2} = \sqrt{1-0.01}$$

$$= 0.995 \text{ meter.}$$

Problem 3. *What will be the apparent length of a meter stick measured by an observer at rest when the stick is moving along its length with a velocity equal to c,* $\frac{c}{\sqrt{2}}, \frac{\sqrt{3}}{2}c, \frac{c}{2}$, *0.8 c.*

[Agra University, Kanpur University]

Solution. From Lorentz transformation, when a body moves with a velocity v relative to the observer, its length is contracted in the direction of motion by the factor $\sqrt{1-\frac{v^2}{c^2}}$. If L_0 be the true length (when at rest), the apparent length when in motion is

$$L = L_0\sqrt{1-\frac{v^2}{c^2}}$$

when $v = c$, the length of the stick would be

$$L = L_0\sqrt{1-1} = 0.$$

when $v = \frac{c}{\sqrt{2}}$, then

$$L = L_0\sqrt{1-\frac{1}{2}} = \frac{L_0}{\sqrt{2}} = 70.7 \text{ cm.}$$

when $v = \frac{\sqrt{3}}{2}c$, then

$$L = L_0\sqrt{1-\frac{3}{4}} = \frac{L_0}{2} = 50 \text{ cm.}$$

when $v = \frac{c}{2}$, then

$$L = L_0\sqrt{1-\frac{1}{4}} = \frac{\sqrt{3}}{2}L_0 = 86.5 \text{ cm}$$

when $v = 0.8\,c$, then

$$L = L_0\sqrt{1-0.64} = 0.6\,L_0 = 60 \text{ cm.}$$

Problem 4. (a). *A rigid bar of length $L_2 = 1.5$ m is at rest relative to system S′ (Figure 4.5). If the bar makes an angle $\theta_2 = 45°$ with respect to the x_2 axis what is the length L_1 and orientation of the bar θ_1 relative to S when $v = 0.98\,c$?*

Solution. If the length of the bar is resolved into the two components parallel to the x_2 and y_2 axes respectively, the corresponding lengths measured by S′ will be $L_{2x} = L_2 \cos\theta_2$

$$L_{2y} = L_2 \sin\theta_2$$

Now the vertical component is perpendicular to v and will not experience any contraction when viewed from S.

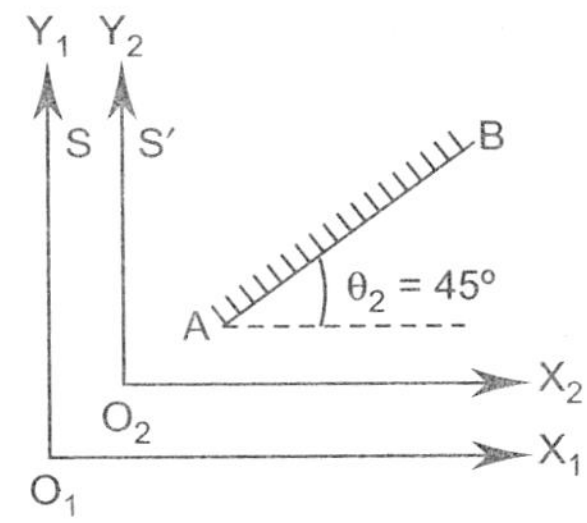

Figure 4.5

Thus,

$$L_{1y} = L_{2y} = L_2 \sin\theta_2$$

The horizontal component is parallel to v and will be contracted to give

$$L_{1x} = L_{2x}\sqrt{1-\frac{v^2}{c^2}}$$

$$= L_2\sqrt{1-\frac{v^2}{c^2}}\cos\theta_2$$

The length of the bar as measured by θ_1 will be

$$L_1 = \sqrt{L_{1x}^2 + L_{1y}^2} = \sqrt{L_2^2\left(1-\frac{v^2}{c^2}\right)\cos^2\theta_2 + L_2^2\sin^2\theta_2}$$

$$= L_2\sqrt{1-\frac{v^2}{c^2}\cos^2\theta_2} = 1.5\sqrt{1-\left\{\frac{0.98\,c}{c}\right\}^2\cos^2 45°}$$

$$= 1.08 \text{ m.}$$

The orientation relative to S will be given by

$$\tan\theta_1 = \frac{L_{1y}}{L_{1x}} = \frac{L_2\sin\theta_2}{L_2\sqrt{1-\frac{v^2}{c^2}}\cos\theta_2} = \frac{\tan\theta_2}{\sqrt{1-\frac{v^2}{c^2}}}$$

$$= \frac{\tan 45°}{\sqrt{1-\left(\frac{0.98\,c}{c}\right)^2}}$$

$$\therefore \quad \theta_1 = 78.7°.$$

Problem 4 (b). *Determine the length and position of a bar of 5 meter length in a reference frame which is moving with a velocity of 0.6 c at an angle 30° with the bar.*

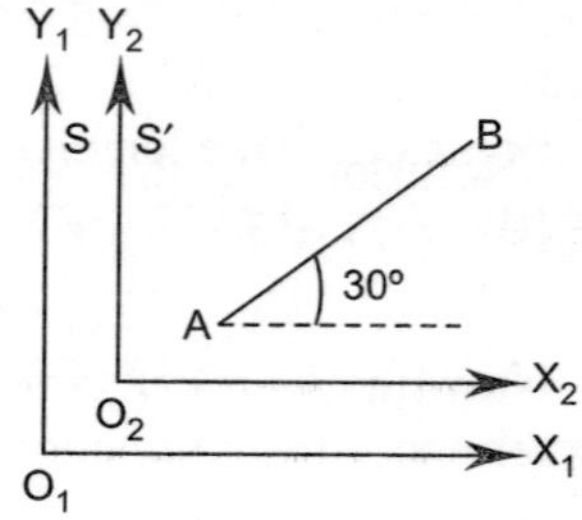

Figure 4.6

Solution. The length measured by S′ will be

$$L_{2x} = L_2 \cos 30°$$

$$L_{2y} = L_2 \sin 30°$$

Here, the vertical component is perpendicular to v and will not experience any contraction when viewed from S. Hence

$$L_{1y} = L_{2y} = L_2 \sin 30°.$$

The horizontal component is parallel to v and will get contracted

$$L_{1x} = L_{2x}\sqrt{1-\frac{v^2}{c^2}} = L_2\sqrt{1-\frac{v^2}{c^2}}\cos 30°$$

Length of the bar relative to S will be

$$L = \sqrt{L_{1x}^2 + L_{1y}^2} = \sqrt{L_2^2\left(1-\frac{v^2}{c^2}\right)\cos^2 30° + L_2^2 \sin^2 30°}$$

$$= L_2\sqrt{1-\frac{v^2}{c^2}\cos^2 30°} = 5\sqrt{1-(0.6)^2 \times \frac{3}{4}} = 4.2 \text{ meter}.$$

The orientation relative to S will be given by

$$\tan\theta = \frac{L_{1y}}{L_{1x}} = \frac{L_2 \sin 30°}{L_2\sqrt{1-\frac{v^2}{c^2}}\cos 30°} = \frac{\tan 30°}{\sqrt{1-0.36}} = \frac{5}{\sqrt{3}\times 4}$$

$$\theta = \tan^{-1}(0.72).$$

Problem 5. *What is the velocity of a π meson whose proper mean life is 2.5×10^{-8} sec and observed mean life is 2.5×10^{-7} sec?* [Rohilkhand University, 83]

Solution. According to Lorentz transformation, the observed mean life is $t = \dfrac{t_0}{\sqrt{1-\frac{v^2}{c^2}}}$, where t_0 is the proper mean life and v the speed of the particle. Thus,

$$\sqrt{1-\frac{v^2}{c^2}} = \frac{t_0}{t} = \frac{2.5\times 10^{-8}}{2.5\times 10^{-7}} = 0.1$$

or, $$\frac{v^2}{c^2} = 1 - 0.01, \text{ or } v = \sqrt{0.99}\, c = 0.995\, c.$$

Problem 6. *A man on the moon observes two spaceships coming towards him from opposite directions at speeds of 0.9 c and 0.8 c respectively. What is the relative speed of the ships as measured by an observer on either one?*

Solution. Let the moon is at rest in system S. Then we may assume that the spaceship (*i.e.*, the system S′) is moving with a velocity $u' = 0.9\,c$ with respect to the moon and $v = 0.8\,c$, $u = ?$

Now,
$$u = \frac{u' + v}{1 + \frac{u'v}{c^2}} = \frac{0.9\,c + 0.8\,c}{1 + \frac{0.8 \times 0.9\,c^2}{c^2}} = 0.988\,c.$$

Problem 7. *Two particles are moving in opposite directions each with a speed of 0.9 c in laboratory frame of reference. Find the velocity of one particle relative to other.*

Solution. The particles are moving with velocities $+ 0.9\,c$ and $- 0.9\,c$ in the laboratory frame, say S′ frame. Let S be a reference frame in which the particle with velocity $- 0.9\,c$ is at rest. Then the velocity of S′ (laboratory) relative to S is $v = 0.9\,c$. Therefore, the particle which is in S′ has velocity $u' = 0.9\,c$ has a velocity in S given by,

$$u = \frac{u' + v}{1 + \frac{u'v}{c^2}} = \frac{0.9\,c + 0.9\,c}{1 + \frac{0.9 \times 0.9\,c}{c^2}}$$
$$= 0.994\,c.$$

Problem 8. *A spaceship moving away from the earth with velocity 0.5 c fires a rocket whose velocity relative to the spaceship is 0.8 c. (a) away from the earth ; (b) towards the earth. What will be the velocity of the rocket as observed from the earth in the two cases.* [Kanpur University]

Solution. Consider earth as the S frame and the spaceship in the S′ frame and the rocket as the object whose velocity in the S frame we seek. Then,

(*a*) $u' = 0.8\,c$, $v = 0.5\,c$,

$$\therefore \qquad u = \frac{u' + v}{1 + \frac{u'v}{c^2}} = \frac{0.8\,c + 0.5\,c}{1 + \frac{0.8 \times 0.5\,c^2}{c^2}} = 0.93\,c$$

(*b*) Here $u' = - 0.8\,c$

$$u = \frac{-0.8\,c + 0.5\,c}{1 - 0.4} = -0.5\,c.$$

Problem 9. *In laboratory electron from two accelerators are projected with the same speed of 2×10^8 m/sec in opposite direction. What is their relative speed?* [WBUT]

Solution. Consider observer on the earth as a stationary frame S, and electrons are moving with 2×10^8 m/sec and $- 2 \times 10^8$ m/sec in the laboratory frame (S′) of reference.

Let S be a reference frame in which the electron with velocity $- 2 \times 10^8$ m/sec is at rest. Then the velocity of S′ relative to S is $v = 2 \times 10^8$ m/sec. Therefore, the particle which is in S′ has velocity $u' = 2 \times 10^8$ m/sec has a velocity in S is given by

$$u = \frac{u' + v}{1 + \frac{u'v}{c^2}} = \frac{2 \times 10^8 + 2 \times 10^8}{1 + \frac{4 \times 10^{16}}{\left(3 \times 10^8\right)^2}} = \frac{4}{13} \times 10^8 \times 9 = 2.77 \times 10^8 \text{ m/sec.}$$

Problem 10. *Two rods of proper length l_0 moves lengthwise towards each other parallel to the common axis with the same velocity v relative to the laboratory frame. Show that the length of each rod in the reference frame fixed to the rod is*

$$l = l_0 \frac{1 - \beta^2}{1 + \beta^2}, \beta = \frac{v}{c}$$ [Calcutta University, WBUT]

Solution. S, the relative velocity u in S frame

$$u = \frac{v + v}{1 - \frac{v \cdot (-v)}{c^2}} = \frac{2v}{1 + \frac{v^2}{c^2}}.$$

Now, the length of the rod in the reference frame fixed to the rod is

$$l = l_0 \sqrt{1 - \frac{v^2}{c^2}} = l_0 \sqrt{1 - \frac{4v^2}{c^2\left(1 + \frac{v^2}{c^2}\right)^2}}.$$

Putting $b = \frac{v}{c}$, we get $l = l_0 \sqrt{1 - \frac{4\beta^2}{\left(1 + \beta^2\right)^2}}$

$$= l_0 \sqrt{\frac{\left(1 - \beta^2\right)^2}{\left(1 + \beta^2\right)^2}} = l_0 \frac{1 - \beta^2}{1 + \beta^2}.$$

Problem 11. *If the total energy is exactly the thrice its rest energy, what is the velocity of the particle?*

Solution. We know that $E = mc^2$

Here, $E = 3\, m_0 c^2$

But, the variation of mass with velocity is expressed as,

$$m = \frac{m_0}{\sqrt{1 - \frac{v^2}{c^2}}}$$

$$\therefore \qquad \frac{m_0c^2}{\sqrt{1-\frac{v^2}{c^2}}} = 3m_0c^2 \text{ or } 1 - \frac{v^2}{c^2} = \frac{1}{9},$$

or,
$$v = \sqrt{\frac{8}{9}}\, c = 2.828 \times 10^{10} \text{ cm/sec.}$$

Problem 12. *Deduce the rest energy of an electron in joules and in electron-volts ($m_0 = 9.1 \times 10^{-31}$ kg, $c = 3 \times 10^8$ m/sec). Also deduce the speed at which the total relativistic energy becomes 1.25 times the rest energy.* [Meerut University]

Solution. Rest energy is $E_0 = m_0c^2 = (9.1 \times 10^{-31} \text{ kg}) \times (3 \times 10^8 \text{ m/sec})^2$

$$= 8.19 \times 10^{-14} \text{ joule}$$

Now 1 electron-volt $= 1.6 \times 10^{-19}$ joule.

$$\therefore \qquad E_0 = \frac{8.19 \times 10^{-14}}{1.6 \times 10^{-19}} = 0.51 \times 10^6 \text{ eV}$$

$$= 0.51 \text{ MeV}.$$

Now, according to the condition, the total relativistic energy E to the rest energy E_0 is 1.25.

$$\therefore \qquad \frac{E}{E_0} = \frac{mc^2}{m_0c^2} = \frac{m}{m_0} = \frac{m_0}{\sqrt{1-\frac{v^2}{c^2}}\, m_0} = 1.25$$

$$\frac{v^2}{c^2} = 1 - \frac{1}{1.5625}$$

or,
$$v \simeq 0.6\, c.$$

Problem 13. *An electron (rest mass 9.1×10^{-31} kg) is moving with speed 0.99 c. What is its total energy ? Find the ratio of Newtonian Kinetic energy to the relativistic energy. ($c = 3 \times 10^8$ meter/sec).* [Rajasthan University, Meerut University]

Solution. We know that, $m = \dfrac{m_0}{\sqrt{1-\frac{v^2}{c^2}}} = \dfrac{9.1 \times 10^{-31}}{\sqrt{1-(0.99)^2}} = 64.5 \times 10^{-31}$ kg.

$\therefore$ Total energy,
$$E = mc^2 = 64.5 \times 10^{-31} \text{ kg} \times (3 \times 10^8 \text{ m/sec})^2$$
$$= 5.8 \times 10^{-13} \text{ joule}$$

The relativistic kinetic energy

$$= \text{total energy} - \text{rest energy}$$
$$= mc^2 - m_0c^2$$

and Newtonian kinetic energy $= \dfrac{1}{2} m_0v^2$

$$\therefore \quad \frac{\text{Newtonian K.E.}}{\text{Relativistic K.E.}} = \frac{\frac{1}{2}m_0 v^2}{mc^2 - m_0 c^2} = \frac{m_0}{2(m - m_0)}\left(\frac{v}{c}\right)^2$$

$$\simeq 0.08.$$

Problem 14. *A particle of rest mass m_0 moves with speed $\frac{c}{\sqrt{2}}$. Calculate its mass, momentum, total energy and kinetic energy.* [Rajasthan University]

Solution. The relativistic mass of the particle is $m = \frac{m_0}{\sqrt{1 - \frac{v^2}{c^2}}} = \frac{m_0}{\sqrt{1 - \frac{1}{2}}}$

$$= \sqrt{2}\, m_0 = 1.41\, m_0$$

The momentum, $\quad p = mv = \sqrt{2}\, m_0 \times \frac{c}{\sqrt{2}} = m_0 c$

Total energy, $\quad E = mc^2 = \sqrt{2}\, m_0 \times c^2 = 1.41\, m_0 c^2$

Kinetic energy = Total energy – Rest energy

$$= mc^2 - m_0c^2 = 1.41\, m_0c^2 - m_0c^2$$

$$= 0.41\, m_0c^2.$$

Problem 15. *A proton has a total relativistic energy 900 MeV. If the rest mass of the proton is 1.6×10^{-27} kg. Find its speed and kinetic energy.* [Pbi. University, Pune University, G.N.D.U.]

Solution. Given that the total relativistic energy of the proton

$$E = mc^2 = 900 \text{ MeV} = 900 \times 10^6 \times 1.6 \times 10^{-19} \text{ J}.$$

Now relativistic mass of the proton,

$$m = \frac{E}{c^2} = \frac{900 \times 1.6 \times 10^{-13} \text{ J}}{(3 \times 10^8)^2} = 1.6 \times 10^{-27} \text{ kg}.$$

Here, rest mass of the proton = Relativistic mass.

∴ The Proton is at rest and its speed as well as kinetic energy is zero.

Problem 16. *If an electron and a positron annihilate, calculate the energy release, given that mass of an electron is 9.1×10^{-31} kg.* [WBUT]

Solution. Here, an electron and positron annihilate, so, total energy released $= 2\, m_e c^2$

[∵ mass of the electron (m_e) = mass of the positron ($m_{e'}$)]

$$= 2 \times 9.1 \times 10^{-31} \times (3 \times 10^8)^2 \text{ joule}$$

$$= \frac{54.6 \times 3 \times 10^{-15}}{1.6 \times 10^{-19}} \text{ eV}$$

$$= 1.024 \times 10^6 \text{ eV} = 1.024 \text{ MeV}.$$

Problem 17. *(a) Show that for small velocities the relativistic kinetic energy of a body reduces to the classical kinetic energy which is less than the rest energy.*

(b) If a particle could move with the velocity of light how much kinetic energy it would possess?

Solution. (*a*) We know relativistic kinetic energy (k) = Total energy – Rest energy

$$k = mc^2 - m_0c^2$$

$$= m_0c^2\left[\frac{1}{\sqrt{1-\frac{v^2}{c^2}}} - 1\right]$$

$$= m_0c^2\left[\left(1-\frac{v^2}{c^2}\right)^{-\frac{1}{2}} - 1\right]$$

$$= m_0c^2\left[\left(1+\frac{1}{2}\cdot\frac{v^2}{c^2}+\frac{3}{8}\cdot\frac{v^4}{c^4}+\ldots\right)-1\right]$$

$$= m_0c^2\left[1+\frac{1}{2}\frac{v^2}{c^2}-1\right] \qquad [\because v << c]$$

$$= \frac{1}{2}m_0v^2$$

i.e., the kinetic energy for classical physics.

∴ The classical kinetic energy is less than the rest energy because v is less than c.

(*b*)
$$k = m_0c^2\left[\left(1-\frac{v^2}{c^2}\right)^{-\frac{1}{2}} - 1\right]$$

Now, if $v \approx c$, then
$$k = m_0c^2\left[\frac{1}{0}-1\right]$$

$$= m_0c^2\left[\infty - 1\right] = \infty$$

i.e., kinetic energy tends to infinity.

Multiple Choice Questions

1. Michelson-Morley experiment was performed to
   (*a*) Prove the existence of ether
   (*b*) Measure the speed of light
   (*c*) Measure the speed of earth relative to ether.
2. The postulates of special theory of relativity are applicable to
   (*a*) Inertial frames (*b*) Accelerated frames (*c*) Stationary frames.

3. The speed of light in free space is
   (*a*) the same in all inertial frames of reference
   (*b*) different in all non-inertial frames
   (*c*) dependent on the nature of the frames.
4. The Lorentz equations reduce to the classical Galilean equations, when
   (*a*) $v << c$ (*b*) $v \approx c$ (*b*) $v > c$
5. Frame S′ moves along the positive X direction of S frame. A rod placed along Y′-axis if observed from S frame appears
   (*a*) elongated (*b*) unchanged (*c*) contracted.
6. Of the twin brothers, one goes on a relativistic tour and comes back. The brother on tour will
   (*a*) become older (*b*) become younger (*c*) be of the same age.
7. Decay of π meson supports
   (*a*) length contraction (*b*) time dilation (*c*) relativity of mass
8. Rest volume L_0^3 is connected to relativistic as
   (*a*) $L_0^3\left(1-\frac{v^2}{c^2}\right)^{\frac{3}{2}}$ (*b*) $L_0^3\left(1-\frac{v^2}{c^2}\right)$ (*c*) $L_0^3\left(1-\frac{v^2}{c^2}\right)^{\frac{1}{2}}$
9. A rocket has a velocity 0.6 *c*. Velocity of light with respect to the rocket is
   (*a*) 0.5 *c* (*b*) 1.6 *c* (*c*) *c*.
10. Two photons approach each other. Their relative velocity will be
   (*a*) 0 (*b*) less than *c* (*c*) *c*.
11. Two photons recede from each other. Their relative velocity will be
   (*a*) *c* (*b*) 2*c* (*c*) $\frac{c}{2}$
12. No material particle can have a velocity
   (*a*) equal to or greater than the velocity of light
   (*b*) equal to or less than the velocity of light
   (*c*) less than the velocity of light.
13. The rest mass of photon is
   (*a*) $\frac{E}{c^2}$ (*b*) Zero (*c*) $\frac{p}{c}$
14. A stone is raised to the top of Mount Everest. Its rest mass has
   (*a*) Increased (*b*) Decreased (*c*) Remains the same.
15. At what velocity the kinetic energy of a body is equal to its rest mass energy
   (*a*) $\sqrt{2}\,c$ (*b*) $\frac{c}{3}$ (*c*) $\frac{\sqrt{3}}{2}c$
16. A body of mass *m* falls through *h* meters. The decrease in its mass is equivalent to
   (*a*) M*gh* (*b*) $\frac{Mgh}{c^2}$ (*c*) $\frac{Mgh}{c}$
17. A body moves with velocity 0.2 *c*, the ratio of the moving mass to the rest mass is
   (*a*) 1.02 (*b*) 1.0 (*c*) 0.2
18. The relativistic relation between total energy E and momentum *p* of a particle is
   (*a*) $E = \sqrt{\left(m_0^2 c^4 + p^2c^2\right)}$ (*b*) $E = \sqrt{\left(m_0^2 c^2 + p^2c^4\right)}$ (*c*) $E = m_0^2 c^2 + p^2c^4$.

19. If two identical rods are at rest-one in S and the other in S′ then each of the observers find that
    (*a*) The other is longer than the rod of his own system.
    (*b*) The other is shorter than the rod of his own system.
    (*c*) The other is the same as the rod of his own system.

20. How fast a particle must travel so that its mass becomes twice its rest mass ? [*WBUT*]
    (*a*) 0.5c (*b*) 2c (*c*) $\frac{\sqrt{3}}{2}$.

21. Relative velocity of two particles moving with velocity of light in opposite direction is
    (*a*) 0 (*b*) 2c (*c*) *c*. [*WBUT*]

22. When one kilogram of mass is fully converted into energies, it produces
    (*a*) 9×10^6 J (*b*) 3×10^{10} J (*c*) 10^{20} J (*d*) 10^{16} J.
    [*Pune University*]

Answers

1. (*c*)	**2.** (*a*)	**3.** (*a*)	**4.** (*a*)	**5.** (*b*)	**6.** (*b*)
7. (*b*)	**8.** (*c*)	**9.** (*c*)	**10.** (*c*)	**11.** (*a*)	**12.** (*a*)
13. (*b*)	**14.** (*a*)	**15.** (*c*)	**16.** (*b*)	**17.** (*a*)	**18.** (*a*)
19. (*b*).	**20.** (*c*).	**21.** (*c*).	**22.** (*a*).		

Review Questions

1. What do you mean by inertial and non-inertial frames of reference? [*Agra University*]
   [Hints : 1.1]
2. Describe the Michelson-Morley experiment and discuss the various interpretations given to the negative result obtained thereform. **[Hints :** Refer to Art. 4.5 and 4.6]
3. State and explain the fundamental postulates of the special theory of relativity and deduce from them the Lorentz transformation equations. Show how Lorentz transformations are superior to Galilean transformations. [*Mysore University*]
4. What are the basic postulates of special theory of relativity? [*WBUT*]
5. Explain (*a*) length contraction and (*b*) time dilation in special theory of relativity. What are proper length and proper time interval? [*Agra University*]
6. Obtain the relativistic formula for the addition of velocities. Hence show that the velocity of light is an absolute constant independent of the frame of reference and that the addition of any velocity to the velocity of light merely reproduces the velocity of light.
7. Derive the formula for the variation of the mass of a particle with its velocity.
8. Establish Einstein's Mass-Energy relation? [*WBUT*]
9. Derive the formula $E^2 = m_0^2 c^4 + p^2c^2$,
   where the symbols have their usual meanings. [*Lucknow University*]
10. Write down the expression for the length contraction and time dilation as derived from the Lorentz transformation equations. [*C.U. 2000, 04*]
11. Using Lorentz transformation equations, show that the time measured by an observer moving with uniform velocity will appear dilated to a stationary observer. [*K.U.*]

12. Derive Einstein's velocity addition theorem

$$\beta_3 = \frac{\beta_1 + \beta_2}{1 + \beta_1\beta_2}$$

where symbols have their usual meaning. Show that the velocity β_3 remains always less than unity. *[C.U. (Hons)]*

13. On the basis of Lorentz transformations, discuss the following
(*i*) Length contraction and (*ii*) time dilation *[C.U.]*

14. Explain the terms "Frame of reference", event and observer in relativity. *[M.D. Univ, Pune Univ., H.P. Univ.]*

15. Describe briefly Michelson Moreley experiment and significance of the experimental results. *[WBUT., H.P. Univ., Pune Univ.]*

16. (*a*) State and explain the (basic) fundamental postulates of the special theory of relativity. *[H.P. Univ., WBUT]*

(*b*) A reference frame S′ moves with respect to another frame S with uniform velocity $\vec{v}$. Derive Lorentz space and time transformation equations giving x', y', z', t' in terms of x, y, z, t the moving frame coincides with stationary one at $t' = t = 0$. Prove that when $\vec{v}$ is much smaller than velocity of light Lorentz transformation reduces to Galelian transformation. *[Pune Univ, G.N.D. Univ, K. Univ, Pbi. Univ., Kerala Univ. WBUT]*

Unsolved Problems

1. The length of a rocketship is 100 meters on the ground. When it is in flight its length observed on the ground is 99 meters. Calculate its speed. **[Ans.** 4.2×10^7 m/sec.]

2. A rod 8.0 meter long is moving along its length with a velocity of 0.6 *c*. Calculate the length as it appears to a stationary observer. **[Ans.** 80 cm]

3. Muons have an average life-time of 2 micro-second in their own frame of reference. In a beam of cosmic ray muons has a speed of 0.98 *c* relative to earth, find the apparent average life and the distance travelled by the beam during the life interval. *[Kanpur University]*
[Ans. 10 micro-second, 2.94 km]

4. In the laboratory two particles are observed to travel in opposite directions each with speed 2.8×10^{10} cm/sec. Deduce the relative speed of the particles. *[Kanpur University]*
[Ans. 2.99×10^{10} cm/sec.]

5. The *two* particles A and B travel in opposite directions each with a velocity 0.9 *c*. What is their relative velocity (*i*) as observed by A and (*ii*) as observed by stationary observer.
[Kanpur University] **[Ans.** (*i*) and (*ii*) 0.99 *c*]

6. The rest mass of an electron is 9×10^{-31} kg. What will be its mass if it was moving with $\frac{4}{5}^{\text{th}}$ the speed of light? **[Ans.** 15×10^{-31} kg] *[Agra University]*

7. Calculate the energy equivalent to 1 gm of mass. *[WBUT]*
[Hints : $E = mc^2$

$$= 1 \text{ gm} \times (3 \times 10^{10} \text{ cm/sec})^2$$
$$= 9 \times 10^{20} \text{ erg.}$$
$$= 9 \times 10^{13} \text{ joule [1 Joule} = 10^7 \text{ erg]}$$

Again 1 electron volt $= 1.6 \times 10^{-19}$ joule.

$$E = \frac{9 \times 10^{13}}{1.6 \times 10^{-19}} \text{ eV} = 5.625 \times 10^{32} \text{ eV} = 5.625 \times 10^{26} \text{ MeV.]}$$

8. Mesons have a life time of 2.2×10^{-6} sec in its rest frame. If their velocities with respect to the laboratory frame is 0.998 c, then how far can they travel in the laboratory frame before decay?

[North Bengal University]

[Hints : The life time of mesons in laboratory frame.

$$\Delta t = \frac{2.2 \times 10^{-6}}{\sqrt{1 - \left(\frac{.998\,c}{c}\right)^2}} = \frac{2.2 \times 10^{-6}}{0.0632} \text{ sec}$$

and $u = 0.998\ c = 0.998 \times 10^8 \times 3$ meter.

The distance travelled by the particle $= u\Delta t = \dfrac{0.998 \times 3 \times 10^8 \times 2.2 \times 10^{-6}}{0.0632}$ sec.

$= 104.22 \times 10^{22}$ m]

9. Calculate the distance traversed by a particle during one mean life if it travels with a speed of 2.2×10^{10} cm/sec. Given proper mean life $= 2.5 \times 10^{-8}$ sec.

[Ans. 825.2 cm] *[Calcutta University]*

Erwin Schrödinger

Chapter 5

Quantum Mechanics

5.1 INTRODUCTION

The laws of Newtonian Mechanics were quite successful in the understanding of the dynamics of cellestical bodies and other earthly objects. Equations based on these laws are the simplest and they are suitable for the solution of simple dynamical problems such as motion of macroscopic bodies. Lagrange's equation, Hamilton's equation and Hamilton's principal are the fundamental principles of classical mechanics because they are consistant with each other and also with Newton's laws of motion.

However when the same laws were employed to the particles such as electrons, protons, neutrons etc., which are too small to be viewed through the highest powered microscopes, serious difficulties were encountered. It gives a signal that the classical concepts are not valid in the region of atomic dimension. The Newtonian law of microscopic world had to be modified to include the new concepts. This incorporation of new concepts gave rise to the birth of a new mechanics called Quantum Mechanics which was not successful in giving a satisfactory explanation to many observed facts but introduced revolutionary attitude in our way of thinking.

A short sketch of how a crisis in physical laws had developed at one stage in understanding the stability of an atom, discreteness in its energy and momenta and some other properties has been presented below.

(*i*) **Stability of the Atom :** Experiments of Rutherford (1910) had established that an atom consisted of a heavy positive charged nucleus of a very small dimension ($\approx 10^{-14}$ cm) surrounded by negatively charged electrons of very little mass. The system will be stable only if the negatively charge electrons are kept revolving around the positively charged heavy nucleus in a similar manner as the planets revolve round the sun. According to classical mechanics, the energy of the moving electron should decrease because the accelerated charged particles lose their energy in the form of electromagnetic wave and therefore their velocities should decrease continuously. Therefore, an oribiting electron comes closer and closer to the nucleus and ultimately coalesces with the nucleus. This shows the instability of the atom. This shows definite contradiction between the observed facts and the classical theory. Thus, the classical mechanics fails to explain the stability of the atom.

(*ii*) **Spectrum of the Hydrogen Atom :** Similar discrepancies like stability of the atom were observed in the spectral studies of the hydrogen spectrum. It was found that the hydrogen spectrum consisted of a discrete set of lines represented by

$$\bar{v} = R\left(\frac{1}{n_2^2} - \frac{1}{n_1^2}\right)$$

where $\bar{v}$ is the wave number, R is the Rydberg constant and n_1 and n_2 are integers such that $n_1 > n_2$.

According to classical mechanics, the excited atoms of hydrogen should emit electromagnetic radiations of all wavelengths continuously while it is observed that they emit radiation of certain wavelengths only.

This difficulty was later resolved by Bohr's theory of hydrogen spectra to certain extent.

So, it can be concluded from the above discussions that classical mechanics does not hold good in the region of atomic dimensions.

5.1.1 Inadequancy of Classical Physics

Black Body Radiation

A body which completely absorbs radiation of all wavelengths incident on it is called a perfectly black body and when such a body is heated, it emits radiations of all types of wavelengths. In practice, a perfectly black body is not available. There is no known surface which can be regarded as a perfectly black body. Lamp black is the nearest approach to a perfectly black body. It can absorb 96% of the radiant energy incident on it. Experimentally such a body is represented by a hollow cylinder with a very small hole in the wall as shown in figure 5.1. Its inner surface is coated with lampblack. Radiation entering the cavity through O is incident on its blackened inner surface and is partly absorbed and partly reflected and the process goes on repeatedly as there is very low probability of any part of the radiation to come out of the cavity through the orifice again, because of its smallness. At each reflection more than 98% of the incident beam is absorbed.

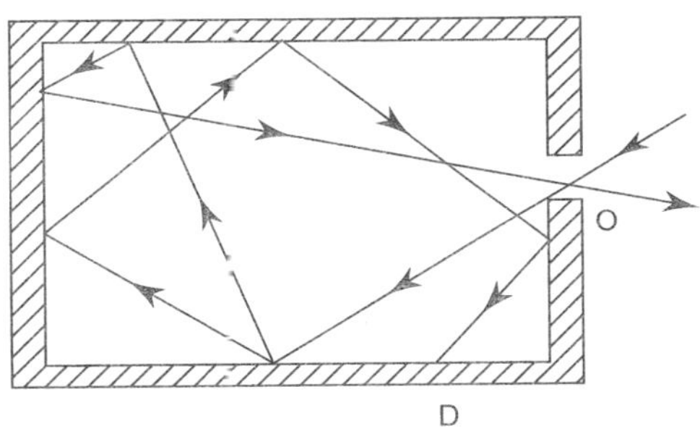

Figure 5.1

Now this enclosure behaves like a **black body.**

In 1859, Kirchhoff discovered the two laws about the black body which are of very vital importance in arriving at the properties possessed by black body radiation :

(*i*) A black body not only completely absorbs all the radiations falling on it but behaves as a perfect radiator when heated.

(*ii*) The radiation given out by a black body depends only on the temperature to which the black body is raised and is independent of the nature of the body.

In the study of black body radiation, a special interest was attached to the spectral distribution of energy *i.e.*, how the energy is distributed among the wavelengths and at what particular wavelength most of the energy is emitted.

Here the radiation was produced by heating this hollow enclosure. Its spectrum can be analysed by an infrared spectrometer. Thus the emissive power (E_λ) of the black body for different wavelengths can be determined (E_λ = quantity of radiation emitted per square cm per second at a given temperature T).

The variation of the intensity of the emitted radiation E_λ as a function of the wavelength is shown graphically in figure 5.2 for different temperatures of the black body.

From figure 5.2 the following conclusions can be arrived at :

(*i*) The energy is not uniformly distributed in the radiation spectrum of a hot body for a given temperautre.

(*ii*) The curve gets rapidly higher (increase in energy emission) as the temperature increases for all wavelengths.

(*iii*) The total energy emitted through radiation at any temperature is given by the area between the curve corresponding to that temperature and the horizontal axis.

(*iv*) The intensity of radiation increases with increase in wavelength for a given temperature and at a particular wavelength intensity of the radiation is maximum and with further increase in wavelength the intensity of heat radiation decreases.

(*v*) The wavelength corresponding to the maximum energy represented by the peak of the curve shifts towards shorter wavelengths as the temperature increases.

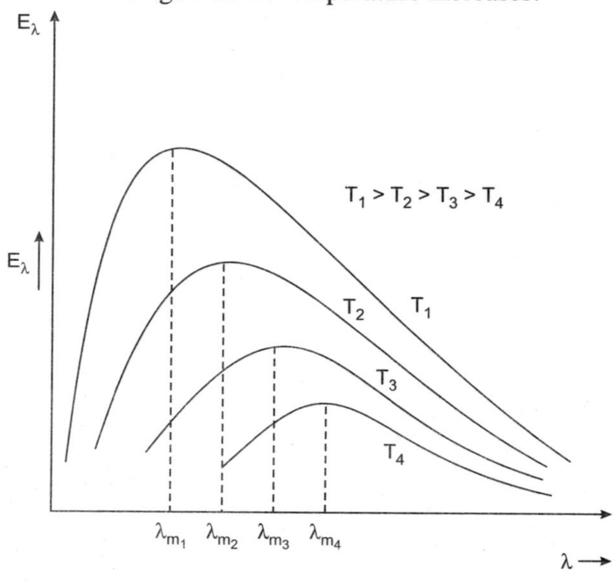

Figure 5.2

On the basis of classical theory to explain qualitatively and quantitatively, the shift in the peak of the intensity distribution curves as the temperature is changed is found to obey the relation

$$\lambda_m T = \text{constant}. \qquad \ldots(5.1)$$

This equation is known as **Wien's displacement law.** By measuring the wavelength λ_m spectroscopically the temperature of a black-body can be determined using equation (5.1). This method has been used extensively in determining temperatures of stars.

Wien from thermodynamical considerations proposed that the maximum value E_m of the emissive power for the wavelength λ_m and the absolute temperature T are connected by the relation

$$\frac{E_m}{T^5} = \text{constant.} \qquad \ldots(5.2)$$

The total radiation emitted per unit area per second by a blackbody at a given temperature T is called the total emissive power E of the body. For a blackbody it is found that

$$E = \int_0^{\infty} E_\lambda \, d\lambda = \sigma T^4 \qquad \ldots(5.3)$$

where $\sigma = 5.6697 \times 10^{-8}$ W/m^2 $^\circ$K^4 = Stefan's constant.

Equation (5.3) is called Stefan's law for the total radiation from a black body. According to this law, the intensity of total radiation E (the area covered under the curve) is proportional to the fourth power of the absolute temperature T.

Theoretical Laws of Black Body Radiation

Late in the nineteenth century various attempts were made to explain the spectral distribution of the intensity of radiation from a black body on the basis of classical mechanics and the electromagnetic theory.

1. Wien's Displacement law in terms of E_λ and T: Based on the classical electromagnetic theory and assuming that the oscillators from which the radiations are emitted are of molecular size, Wien showed that the energy $E_\lambda \, d\lambda$ radiated per unit area per second within the wavelength range λ and $\lambda + d\lambda$ of electromagnetic radiation by a blackbody at temperature T is given by

$$E_\lambda \, d\lambda = \frac{A}{\lambda^5} f(\lambda T) \, d\lambda \qquad \ldots(5.4)$$

and

$$E_\lambda \, d\lambda = A T^5 F(\lambda T) \, d\lambda \qquad \ldots(5.5)$$

where A is a constant and $f(\lambda T)$ and $F(\lambda T)$ are functions of the product λT.

From equations (5.4) and (5.5) it is clear that

$$T^5 F(\lambda T) = \frac{1}{\lambda^5} f(\lambda T)$$

or,

$$F(\lambda T) = (\lambda T)^{-5} f(\lambda T) \qquad \ldots(5.6)$$

Wien obtained equations (5.4) and (5.5) by considering the radiation field having thermodynamic properties.

Now equation (5.5) shows that for any black body,

$$\frac{E_\lambda}{T^5} = AF(\lambda T).$$

Thus if $E_{\lambda_1}, E_{\lambda_2}, \ldots,$ are the emissive powers for the respective wavelengths $\lambda_1, \lambda_2, \ldots$ and if T_1, $T_2, \ldots,$ are the corresponding absolute temperatures of the black body, then

$$\frac{E_{\lambda_1}}{T_1^5} = AF(\lambda_1 T_1)$$

$$\frac{E_{\lambda_2}}{T_2^5} = AF(\lambda_2 T_2) \ldots \text{and so on.}$$

If the wavelengths are inversely proportional to the corresponding temperatures, *i.e.*, if

$$\lambda_1 T_1 = \lambda_2 T_2 = \ldots, \text{ then we have,}$$

$$\frac{E_{\lambda_1}}{T_1^5} = \frac{E_{\lambda_2}}{T_2^5} = \frac{E_{\lambda_3}}{T_3^5} = \ldots = \text{constant.}$$

Thus the emissive power E_λ of a perfectly black body for wavelength λ is directly proportional to the 5th power of the absolute temperature, if the wavelength is inversely proportional to the absolute temperature. This is Wien's displacement law in terms of E_λ and λ.

It was found that equation (5.1) gave very good fitting of the experimental curves in the low wavelength region (figure 5.3) but did not reproduce the results at high wavelength region.

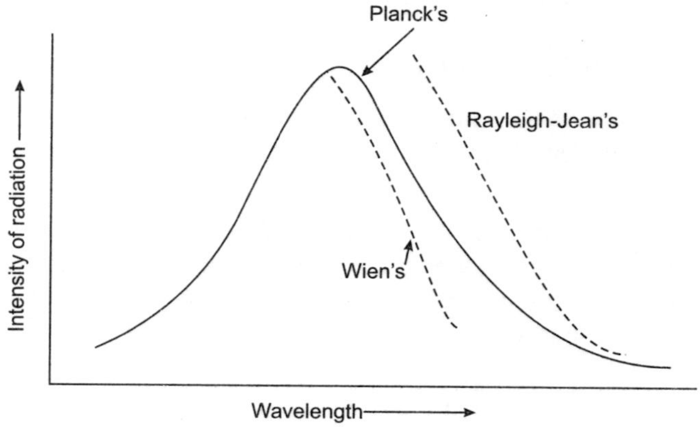

Figure 5.3

2. Rayleigh–Jeans Law : In 1900 Lord Rayleigh applied the principle of equipartition of energy to the electromagnetic vibrations. Then with a contribution from J.H. Jeans this attempt led to the deduction of a formula for energy per unit volume inside an enclosure with perfectly reflecting walls. This formula is called the Rayleigh-Jeans law.

According to this law the energy density $U_\nu\, d\nu$ *i.e.*, the amount of energy per unit volume of the enclosure in the frequency range from ν to $\nu + d\nu$ is given by

$$U_\nu\, d\nu = \frac{8\pi\nu^2 kT}{c^3}\, d\nu \qquad \ldots(5.7)$$

where U_ν is the energy per unit volume per unit frequency range at frequency ν, k is the Boltzmann's constant and c is the speed of light in free space.

The Rayleigh-Jeans formula can be altered in terms of the wavelength λ by using the relation

$$\nu = \frac{c}{\lambda}$$

and

$$d\nu = -\frac{c}{\lambda^2}\, d\nu.$$

The energy $U_\nu\, d\nu$ contained in a frequency interval between ν and $\nu + d\nu$ is equal to that contained in a corresponding wavelength interval between λ and $\lambda + d\lambda$, and an increase in frequency corresponds to a decrease in wavelength.

i.e.,

$$U_\lambda\, d\lambda = -U_\nu\, d\nu$$

$$= -\frac{8\pi\nu^2 kT}{c^3}\, d\nu$$

$$= -\frac{8\pi kT}{c^3}\left(\frac{c}{\lambda}\right)^2\left(-\frac{c}{\lambda^2}\right)d\lambda$$

$$= \frac{8\pi kT}{\lambda^4}d\lambda \qquad ...(5.8)$$

This is another form of the Rayleigh-Jeans law. This law explains the experimental facts for very long wavelenghts, but not for shorter wavelengths. According to this law as λ decreases, the energy density U_λ will continuously increase, and as λ tends to zero, U_λ approaches infinity. This is contrary to the experimental result.

This law leads to an absurd result. The total energy of radiation per unit volume of the enclosure for all wavelengths from zero to infinity is given by

$$U = \int_0^\infty U_\lambda \, d\lambda = \int_0^\infty \frac{8\pi kT}{\lambda^4}d\lambda$$

$$= 8\pi kT\left[-\frac{1}{3\lambda^3}\right]_0^\infty = \infty.$$

This result shows that for a given quantity of radiant energy, all the energy will finally be confined in vibrations of very small wavelengths. Thus if the classical treatment is correct, on opening a shutter in the black body cavity, we would be bombarded with radiations of extremely short wavelengths. But experimental results show that $U_\lambda \, d\lambda \to 0$ as $\lambda \to 0$. This discrepancy between the theoretical conclusion and the experimental result is called **ultraviolet catastrophe.** This indicates the limitations of classical mechanics.

From equation (5.8) it has been noticed that the experimental data fit at the high wavelength region but fails completely at low wavelength region and thereby leads to what is known ultraviolet catastrophe (figure 5.3).

3. Planck's Quantum Hypothesis : The failure of Rayleigh-Jeans formula to explain the observed energy distribution law of black body radiation showed that there was something wrong either with the equipartition law or with the classical electromagnetic theory or with both.

Keeping in view that the Wien's relation was applicable in the low wavelength region and that of Rayleigh-Jeans in the high wavelength region, Max Planck put forward the quantum theory (1900) of radiation. He assumed that the atoms in the walls of a blackbody behave like simple harmonic oscillators, and each has a characteristic frequency of oscillation. In his theory he made the following two assumptions about the atomic oscillators.

(*a*) A simple harmonic oscillator cannot have any arbitrary value of energy but has discrete energy equal to the integral multiple of some minimum energy.

i.e., $E = nh\nu$.

where $n = 0, 1, 2, 3, ...$ are permissible.

Here ν is the frequency of oscillation and h = Planck's constant = 6.625×10^{-34} Js. Here $h\nu$ is the basic unit of energy and it is called a quantum of energy.

(*b*) Planck further assumed that as long as the oscillator has energy equal to one of the allowed values given by the relation $E = nh\nu$, it cannot radiate or absorb energy. Therefore, the oscillator is said to be in a stationary state or a quantum state of energy. The emission or absorption of energy takes place only when the oscillator jumps from one energy state to another. If the oscillator jumps from a higher energy state of quantum number n_2 to a lower energy state of quantum number n_1, the energy emitted is given by

$$E_2 - E_1 = (n_2 - n_1)\, h\nu.$$

If, $n_2 - n_1$ = one unit then,

$$E_2 - E_1 = h\nu.$$

Similarly an oscillator jumps up to its next higher energy state when it absorbs a quantum of energy $h\nu$.

According to the postulates of Planck, the oscillators can exist in a set of discrete energy states 0, $h\nu$, $2h\nu$, $3h\nu$, etc., as shown in figure 5.4.

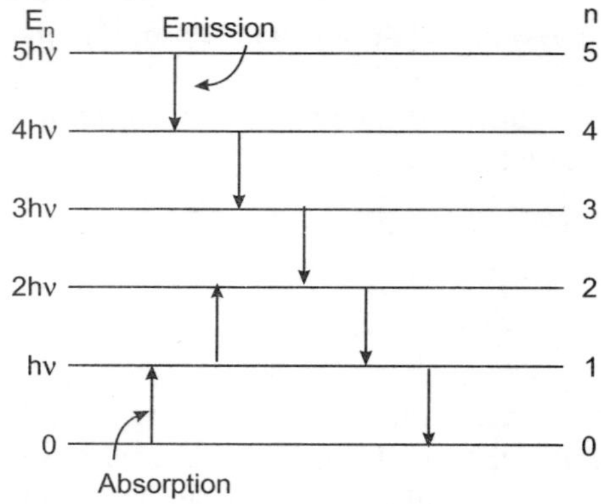

Figure 5.4

Now, the number of oscillators in an energy state $E_n = nh\nu$ is determined by the Maxwell-Boltzmann distribution function

$$N_n = N_0 \, e^{-\frac{E_n}{kT}} = N_0 \, e^{-\frac{nh\nu}{kT}}.$$

For $E_n = 0$, $N_n = N_0$, *i.e.*, where N_0 is the number of oscillators in the ground state. The number N_n decreases exponentially with increasing energy E_n.

We now calculate the mean energy $<E>$ of the oscillators according to the new concepts introduced by Planck. According to Planck, the energies of the oscillators can assume discrete values. So we have to replace integrals by sums over all possible oscillator energy states to determine $<E>$:

Thus we can write,

$$< E > = \frac{\sum_{n=0}^{\infty} N_n E_n}{\sum_{n=0}^{\infty} N_n}$$

$$= \frac{\sum_{n=0}^{\infty} N_0 E_n \, e^{-\frac{E_n}{kT}}}{\sum_{n=0}^{\infty} N_0 \, e^{-\frac{E_n}{kT}}}$$

$$= \frac{\sum_{n=0}^{\infty} nh\nu \, e^{-\frac{nh\nu}{kT}}}{\sum_{n=0}^{\infty} e^{-\frac{nh\nu}{kT}}} = \frac{h\nu x \left[1 + 2x + 3x^2 + 4x^3 + \ldots\ldots\right]}{1 + x + x^2 + x^3 + \ldots\ldots}$$

where, $x = e^{-\frac{h\nu}{kT}}$

or, $$< E > = h\nu x \cdot \frac{(1-x)^{-2}}{(1-x)^{-1}} = \frac{h\nu x}{1-x} = \frac{h\nu}{x^{-1}-1}$$

or, $$< E > = \frac{h\nu}{e^{\frac{h\nu}{kT}} - 1} \qquad \ldots(5.9)$$

Now it can be shown that the number of oscillations or degrees of freedom per unit volume in the frequency range ν and $\nu + d\nu$ is given by

$$N(\nu)\, d\nu = \frac{8\pi\nu^2}{c^3}\, d\nu \qquad \ldots(5.10)$$

Now, if $h\nu << kT$ then $< E > \simeq kT$ in the classical limit as $\frac{h\nu}{kT} << 1$.

$\therefore$ $$e^{\frac{h\nu}{kT}} \approx 1 + \frac{h\nu}{kT}$$

So, $$< E > = kT.$$

Hence, we have

$$u_\nu\, d\nu = \frac{8\pi\nu^2}{c^3}\, d\nu < E >$$

$$= \frac{8\pi}{c^3} \frac{h\nu^3\, d\nu}{\left[e^{\frac{h\nu}{kT}} - 1\right]} \qquad \ldots(5.11)$$

$$= \frac{8\pi hc}{\lambda^5} \cdot \frac{d\lambda}{\left[e^{\frac{hc}{\lambda kT}} - 1\right]} \qquad \ldots(5.12)$$

Equations (5.11) and (5.12) are the two forms of Planck's law of radiation. It can also be verified that it reduces to Rayleigh-Jeans law (equation 5.8) when $\lambda \to \infty$ and to Wien's law (equation 5.4) when $\lambda \to 0$.

5.1.2 Deductions from Planck's Radiation Law

1. Wein's Displacement Law : we have seen from the graph of variation of emissive power E_λ of a black body at a constant temperature that E_λ has a maximum value for a certain wavelength λ_m (say). So it means that for $\lambda = \lambda_m$, the energy density u_λ is maximum. Hence,

$$\left[\frac{d}{d\lambda}(u_\lambda)\right]_{\lambda = \lambda_m} = 0$$

Now, $$u_\lambda = \frac{8\pi hc}{\lambda^5} \frac{1}{e^{\frac{hc}{\lambda kT}} - 1}.$$

as, $$\ln u_\lambda = \ln(8\pi hc) - 5\ln\lambda - \ln\left(e^{\frac{hc}{\lambda kT}} - 1\right).$$

Differentiating the above equations w.r.t λ, we get

$$\frac{1}{u_\lambda}\frac{du_\lambda}{d\lambda} = -\frac{5}{\lambda} - \frac{1}{e^{\frac{hc}{\lambda kT}} - 1} \times e^{\frac{hc}{\lambda kT}} \times \left(-\frac{hc}{\lambda^2 kT}\right)$$

$$= -\frac{5}{\lambda} + \frac{hc}{\lambda^2 kT} \cdot \frac{e^{\frac{hc}{\lambda kT}}}{e^{\frac{hc}{\lambda kT}} - 1}$$

Now since at $\lambda = \lambda_m$, $\frac{du_\lambda}{d\lambda} = 0$. so,

$$-\frac{5}{\lambda_m} + \frac{hc}{\lambda_m^2 kT} \times \frac{e^{\frac{hc}{\lambda_m kT}}}{e^{\frac{hc}{\lambda_m kT}} - 1} = 0$$

or,
$$-5 + \frac{hc}{\lambda_m kT} \cdot \frac{e^{\frac{hc}{\lambda_m kT}}}{e^{\frac{hc}{\lambda_m kT}} - 1} = 0$$

Putting $\frac{hc}{\lambda_m kT} = x$, we get

$$-5 + x \cdot \frac{e^x}{e^x - 1} = 0$$

or,
$$x = \frac{5\left(e^x - 1\right)}{e^x} = 5\,(1 - e^{-x})$$

If we solve the above equation graphically then we can find that

$$x = 4.9651$$

or,
$$\frac{hc}{\lambda_m kT} = 4.9651$$

or,
$$\lambda_m\,T = \frac{hc}{4.9651 \times k} = \text{constant.}$$

or,
$$\lambda_m\,T = \frac{6.62 \times 10^{-34} \times 3 \times 10^8}{4.9651 \times 1.38 \times 10^{-23}}$$

or,
$$\lambda_m\,T = 0.0029 \text{ m° K} = \text{constant.}$$

This is Wien's displacement law.

2. Rayleigh-Jeans Law : We have Planck's law

$$u_\lambda\,d\lambda = \frac{8\pi hc}{\lambda^5} \cdot \frac{1}{e^{\frac{hc}{\lambda kT}} - 1}\,d\lambda.$$

In the long wavelength region,

$$\frac{hc}{\lambda kT} << 1$$

$\therefore \quad e^{\frac{hc}{\lambda kT}} \simeq 1 + \frac{hc}{\lambda kT}$ neglecting the higher power terms.

$$\therefore \qquad e^{\frac{hc}{\lambda kT}} - 1 = \frac{hc}{\lambda kT}$$

So,

$$u_\lambda \, d\lambda = \frac{8\pi hc}{\lambda^5} \times \frac{1}{\frac{hc}{\lambda kT}} \, . \, d\lambda$$

$$= \frac{8\pi hc}{\lambda^5} \times \frac{\lambda kT}{hc} \, . \, d\lambda$$

$$= \frac{8\pi kT}{\lambda^4} \, d\lambda \, .$$

which is Rayleigh-Jeans law.

Temperature of the Sun from Wien's Displacement Law :

According to Wien's displacement law

$$\lambda_m T = \frac{hc}{4.9651 \, K} = 0.0029 \text{ m}^\circ \text{ K} = \text{constant.}$$

Now it has been found that the wavelength corresponding to the maximum emissive power E_λ for radiation from the surface of the sun is

$$\lambda_m = 4900 \text{ Å}.$$

Hence the temperature of the sun's surface is given by

$$T = \frac{2.898 \times 10^{-3}}{4900 \times 10^{-10}} \, ^\circ K$$

$$= 5914 \, ^\circ K.$$

This is not the temperature of the sun but is the surface temperature of the sun from where the measured values of λ_m were emitted.

5.1.3 Photoelectric Effect

5.1.3.1 Introduction

Whenever electromagnetic radiations of sufficiently high frequency, such as ultraviolet light and X-rays, are incident on a metal surface, then electrons are emitted from it. This process of emission of electrons from a metal plate, when illuminated by light of suitable wavelength, is called the **photoelectric effect**. The electrons emitted are called **photoelectrons**.

According to the laws of classical mechanics, the energy of a linear harmonic oscillator can vary continuously, depending on the amplitude of its vibrations. But Planck's postulate regarding the discrete nature of the possible energy state of an oscillator marked a radical departure from the idea of classical physics. At the beginning Planck's ideas faced stiff opposition. But gradually Planck's ideas were found to be quite useful in explaining the phenomenon of photoelectric effect. It was observed that a plate of a metal such as zinc when exposed to ultraviolet radiation becomes positively charged which showed that it had lost negative charges from its surface. These negatively charged particles were later identified as electrons by Lenard. This phenomenon is known as **photoelectric effect**. This phenomenon cannot be explained in classical theory.

5.1.3.2 Lenard's Experiment

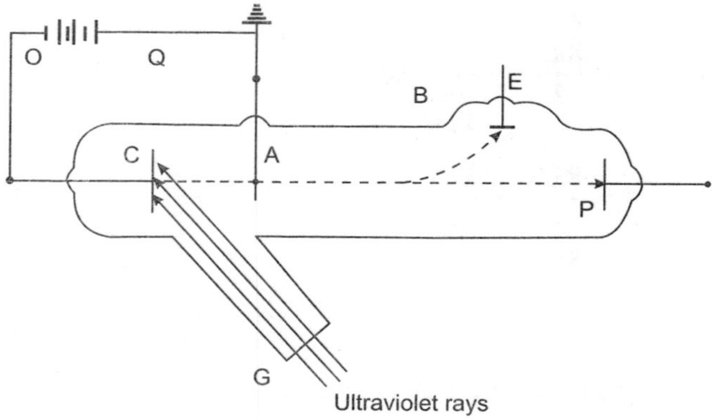

Figure 5.5

Lenard's apparatus for the measurement of *e*/*m* of the negatively charged particles emitted in photoelectric process is shown diagrammatically in figure 5.5. The whole apparatus is enclosed in the tube B in which a very low pressure is maintained with the help of a pump. C is a plane metal plate kept at a negative potential by means of a battery OQ. A is another earthed metal screen placed parallel to C inside the tube with a small central hole and is kept at zero potential. A is thus positive relative to C. At the center of A the beam of – ve ions emitted from C can pass. Ultraviolet radiation passes through Quartz Window G and is incident on the plate C, due to which – ve ions are emitted from C by photoelectric process. The – ve ions are attracted to A and a part of it passes through the hole in A and is finally collected at the collector P. The resulting feeble current can be detected with electrometer.

The enclosure B is placed between the poles of a magnet such that the magnetic field due to the magnets acts perpendicular to the plane of the diagram.

The field bends the path of the negative ions coming out from the hole into circular arcs so that the ions are now collected by the collector E, instead of P. The resulting electric current can be recorded by another electrometer.

From the arrangement, it is possible to determine the radius of curvature R of the – ve ions. From the values of R and the magnetic induction B, it is possible to determine the *e*/*m* for the – ve ions emitted from C. If V be the potential difference between C and A, then the kinetic energy gained by these negative ions will be

$$\frac{1}{2}mv^2 = eV \qquad \ldots(5.13)$$

where m = mass and v = velocity of electron.

Again equating the centripetal force to the magnetic force we get

$$Bev = \frac{mv^2}{R} \qquad \ldots(5.14)$$

Hence from (5.13) and (5.14) we get

$$\frac{2eV}{m} = \frac{B^2R^2e^2}{m^2}$$

or, $$\frac{e}{m} = \frac{2V}{B^2R^2} \quad \ldots(5.15)$$

Lenard's measurement of the e/m proved that the – ve ions emitted from C by the action of ultraviolet light were electrons.

In explaining the photoelectric effect, Einstein also postulated that electrons do not normally leave a metal surface of their own because they are bound in the metal. A certain minimum energy is required to release an electron from the metal surface.

Lenard also observed that if a +ve potential was applied to C instead of a –ve potential, the electric current recorded at C gets decreased. This is due to the fact that the negatively charged electrons emitted from C then experiences a repulsive force in going towards A which is now –ve with respect to C. As the +ve potential on C is increased, the current recorded at P ultimately becomes zero at a particular value of the potential V_s. V_s is called the stopping potential. Because electrons of all velocities upto the maximum v_m emitted from C are stopped from reaching A. Under this condition,

$$\frac{1}{2} m v_m^2 = eV_s.$$

Even if the experiment is repeated by increasing the intensity of the incident radiation, it will cause no electron to reach A, *i.e.*, the stopping potential remains same. However on making C – ve with respect to A, an increase in the intensity of the incident light causes the current to increase (figure 5.6).

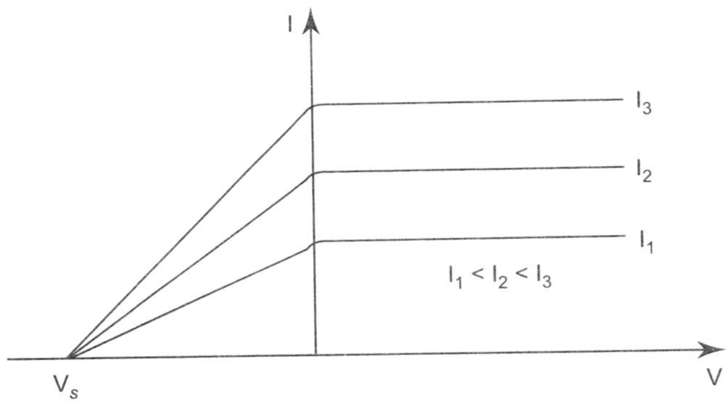

Figure 5.6 *Variation of photoelectric current with v for different intensities of the incident light.*

Keeping the intensity constant, if the experiment is repeated with λ of a monochromatic light of another frequency (wavelength), it is viewed that stopping potential will change. As the frequency of the incident light is increased, the stopping potential is increased (Figure 5.7).

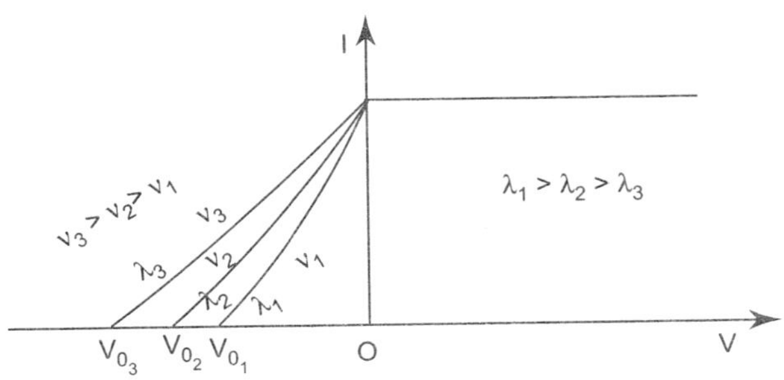

Figure 5.7 *Variation of photoelectric current for different wavelengths (frequency).*

5.1.3.3 Photoemission Characteristics

We can draw important conclusions from the results of the experiments on photoelectric effect which are given below :

1. The number of photoelectrons ejected per second, *i.e.*, photoelectric current is directly proportional to the intensity of the incident light but is independent of the frequency (wavelength) of light.
2. The velocities of the emitted photoelectrons may vary between zero to a definite maximum (V_m) which is independent of the intensity of the incident light but depends only on its frequency (wavelength). It is found that if ν is the frequency of light used, then maximum kinetic energy of the photoelectrons increases linearly with ν (figure 5.8).

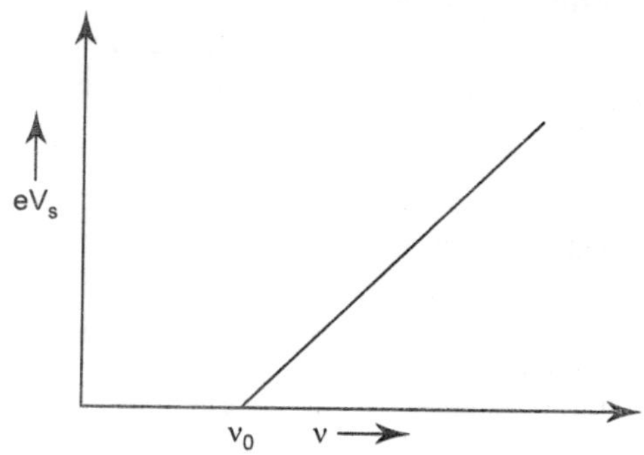

Figure 5.8 *Plot of stopping potential vs frequency.*

3. For each emitting surface, there is a certain minimum frequency (ν_0), called the **threshold frequency** of the incident radiation, below which photoelectrons emission does not take place, no matter how much is the intensity of light. The ν_0 value is different for different emitting surfaces.
4. Photoelectron emission is an instantaneous process. There is nearly no time gap ($< 10^{-8}$ sec) between the incidence of radiation and the emission of **photoelectrons.**

5.1.3.4 Failure of the (Classical) electromagnetic theory

The above experimental observations could not be explained on the basis of the electromagnetic theory of light.

According to this theory, light of higher intensity should impart greater velocity or kinetic energy to the ejected electrons. But the above conclusions are in disagreement with observed facts — the velocity of the ejected photoelectron depends on the frequency of the incident light.

Again, according to this theory, given sufficient time, all electrons will be able to acquire the necessary energy for ejection. But, experimentally we observe that electron ejection commences without delay.

The electromagnetic theory fails to explain why there is no time lag between the instant when light is incident and that of the emission of photoelectrons.

So the electromagnetic wave theory fails completely to explain the characteristics of the photoelectric emission.

5.1.3.5 Derivation of Einstein's Photoelectric Emission

In 1905, Einstein explained the laws of photoelectric emission on the basis of Planck's quantum theory of radiation. According to quantum theory, the radiation consists of packets of energy called **'photons'** or **'quanta'** where h is the Planck's constant and ν is the frequency of radiation.

Einstein proposed the following postulates :

1. Radiation is not only emitted and absorbed in discrete quanta, called photons, having energy $E = h\nu$, but also travels in quanta with the speed of light. Each photon is a bundle of localised energy in space.
2. The energy $h\nu$ of a photon is so concentrated that it can transfer the whole of its energy content to one electron (which thereby gains the quantum of energy).

When a photon of energy $h\nu$ is incident upon a metal surface, it can either be reflected according to the laws of optics or else absorbed. In later case, the energy is used up in two ways :

(*i*) A part of its energy is used up to eject the electron from the metal surface. The quantity of energy depends upon the nature of the metal and is called the work function (ϕ_0). The work function of a metal may be defined as the energy which is just sufficient to liberate electrons from the metal surface with zero velocity.

(*ii*) The rest part of the energy of the striking photon is used up in imparting kinetic energy $\frac{1}{2}mv^2$ to the ejected electron.

Thus,

$$h\nu = \phi_0 + \frac{1}{2}mv^2$$

where $\phi_0 = h\nu_0$, ν_0 being the threshold frequency. Therefore the equation becomes

$$\boxed{\frac{1}{2}mv^2 = h(\nu - \nu_0)}$$

This equation is the well known **Einstein's photoelectric equation.**

Example 1. *If a light of wavelength 6000 Å falls on a metal surface and emits photoelectrons with velocity 4×10^5 m/sec, what is its threshold wavelength ?* [WBUT]

According to Einstein's photoelectric emission

$$(\text{K.E.})_{max} = \frac{1}{2}m v_{max}^2 = h\nu - \phi = \frac{hc}{\lambda} - \frac{hc}{\lambda_0}$$

where ϕ = Work function

or,
$$\frac{1}{2} \times \frac{m v_{max}^2}{hc} = \left(\frac{1}{\lambda} - \frac{1}{\lambda_0}\right)$$

or,
$$\frac{1}{\lambda} - \frac{1}{\lambda_0} = \frac{1}{2} \times \frac{9.11 \times 10^{-31}\, kg \times (4 \times 10^5\, m)^2}{6.63 \times 10^{-34} \times 3 \times 10^8}$$

or, $$\frac{1}{\lambda_0} = \frac{1}{6000 \times 10^{-10}\ m} - 0.037 \times 10^7$$

or, $$\lambda_0 \simeq 7.692 \times 10^{-7} \text{ m}$$
$$= 7692 \text{ Å}.$$

∴ Threshold frequency is 7692 Å.

5.1.3.6 The Compton Effect

Compton discovered that when monochromatic X-rays were incident on a material of low atomic number such as carbon, they suffered a change in frequency on scattering. The scattered beam contains two wavelengths. In addition to the expected incident wavelength, there exists another line of longer wavelength or shorter frequency. The change of wavelength is due to loss of energy of the incident X-rays. This phenomenon is known as Compton Effect. In such a case an electron is also ejected with an energy depending upon its direction. This electron is known as Compton Recoil Electron.

Explanation

This effect was explained by Compton on the basis of corpuscular view (quantum theory of radiation), the process is regarded as a particle collision event between X-ray photon and a loosely bound electron of the scatterer. In this process both momentum and energy are conserved. In the photon-electron collision, a portion of the energy of photon is contributed to the electron. As a result, the X-rays proceed with less than the original energy (Actually separated photons are involved).

Calculations

Let the energy E of the photon before collision be $h\nu$. So its momentum p before collision is $\frac{h\nu}{c}$ where c is the velocity of light in vacuum. Let the energy E′ of the photon after collision be $h\nu'$. So its momentmum p' after collision is $\frac{h\nu'}{c}$. We consider the electron of rest before collision. So the initial momentum of the electron is zero and its initial energy is only the rest mass energy, $m_o c^2$.

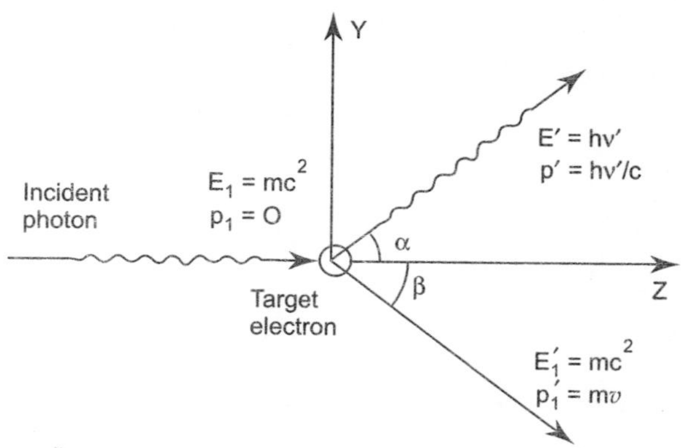

Figure 5.9

Let v be the velocity of the electron after collision so that its momentum p_1' is mv and the energy of the recoil electron is $E_1' = mc^2$.

Let α be the angle of scattering and β the angle of deviation of the electron.

Momentum, unlike energy is a vector quantity that incorporate direction as well as magnitude and in the collision momentum must be conserved along each of the two mutually perpendicular directions.

In the original photon direction,

$$\frac{h\nu}{c} + 0 = \frac{h\nu'}{c}\cos\alpha + mv\cos\beta,$$

and perpendicular to this direction.

$$0 = \frac{h\nu'}{c}\sin\alpha - mv\sin\beta \quad \text{for momentum conservation} \qquad \ldots(5.16)$$

and

$$h\nu + m_0c^2 = h\nu' + mc^2 \text{ \}for energy conservation} \qquad \ldots(5.17)$$

From equation (5.16) $h(\nu - \nu'\cos\alpha) = mvc\cos\beta$

and $h\nu'\sin\alpha = mvc\sin\beta$ ...(5.18)

Squaring the equations in (5.18) and adding

$$m^2v^2c^2(\cos^2\beta + \sin^2\beta) = m^2v^2c^2 = h^2\{\nu^2 + \nu'^2 - 2\,.\,\nu\nu'\cos\alpha\} \qquad \ldots(5.19)$$

From equation (5.17), we obtain,

$$mc^2 = h(\nu - \nu') + m_oc^2$$

$$\Rightarrow \quad m^2c^4 = h^2(\nu^2 + \nu'^2 - 2\nu\nu') + 2h\,m_0c^2(\nu - \nu') + m_0^2c^4 \qquad \ldots(5.20)$$

Subtracting (5.19) from (5.20), we get

$$m^2c^4\left(1 - \frac{v^2}{c^2}\right) = -2h^2\nu\nu'(1 - \cos\alpha) + 2\,m_0c^2h(\nu - \nu') + m_0^2c^4$$

or

$$m_0{}^2c^4 = -2h^2\nu\nu'(1 - \cos\alpha) + 2m_0c^2\,h(\nu - \nu') + m_0^2c^4 \quad \left[\because\ m = \frac{m_0}{\sqrt{1 - \frac{v^2}{c^2}}}\right]$$

$$\therefore \quad 2h^2\nu\nu'(1 - \cos\alpha) = 2m_0c^2h(\nu - \nu')$$

or,

$$1 - \cos\alpha = \frac{m_0c^2}{h}\frac{(\nu - \nu')}{\nu\nu'} = \frac{m_0c^2}{h}\left(\frac{1}{\nu'} - \frac{1}{\nu}\right)$$

But

$$d\lambda = \lambda' - \lambda = c\left(\frac{1}{\nu'} - \frac{1}{\nu}\right)$$

$$d\lambda = \frac{h}{m_0c}(1 - \cos\alpha) = \lambda_0(1 - \cos\alpha) \qquad \ldots(5.21)$$

where $\lambda_0 = \frac{h}{m_0c}$ is known as the universal constant as it involves the basic constants h, c and m_o. It has the dimension of length and is known as Compton wavelength of the electron having a value of 0.024 Å.

Discussion : Equation (5.21) shows that compton shift $d\lambda$ is independent of the wavelength λ of the incident radiation and also the nature of the scattering substance $d\lambda$ depends upon the angle of scattering (α) only.

If, $\alpha = 180°$, then $d\lambda = 2 \times \lambda_0 = 2 \times$ Compton wavelength and that is the maximum possible shift.

This treatment is valid for a free electron not bound to an atom. If the X-radiation collides with a bound electron, the whole atoms recoils as result of Compton shift and m_0 in the formula is to be altered by M_A, since $d\lambda \to 0$ for all values of α. Since $M_A >> m_0$, this represents the unmodified line observed by Compton.

5.2 WAVE PARTICLE DUALITY

To discuss about the wave and particle duality, it is necessary to understand what is a particle and what is a wave ? A particle can be located at some definite point and can be moved from one position to another position. Therefore a particle can be specified by its (*a*) mass (*b*) velocity (*c*) momentum and (*d*) energy. In the Einstein's explanation of photoelectric effect and the understanding of Compton scattering of X rays, it was established that electromagnetic (light, X-rays etc.) waves exhibit some properties that were particle like.

But wave is a form of disturbance which propagates in a medium and spreads out over a relatively large medium in space. It can be specified by its (*a*) frequency (*b*) wavelength (*c*) phase or wave velocity (*d*) amplitude and (*e*) intensity. The phenomena of interference and diffraction reveal the wave nature of light.

de Broglie first proposed that wave and particle duality was of a universal nature and extended the concept of material particles. According to him matter also has dual characteristics *i.e.*, particle like and wave like.

The material particles like electron, proton etc., should also act as a wave at some other times.

5.2.1 de Broglie Hypothesis

According to the Bohr's quantum theory, radiation of frequency ν consists of quanta or photons, each of energy

$$E = h\nu, \quad \text{where } h = \text{Planck's constant}$$

On the particle character alone energy E is given by

$$E^2 = p^2c^2 + m_0^2c^4 \text{ (Einstein's relativistic equation)}$$

As the rest mass of the photon is zero *i.e.*, $m_0 = 0$

$$E = pc$$

Hence photon momentum is given by

$$p = \frac{h\nu}{c}$$

where p = momentum, c = speed of photon in free space *i.e.*, speed of light.

$$\Rightarrow p = \frac{h}{\lambda} \qquad \ldots(5.22)$$

This represents wave (wavelength λ) particle (momentum p) relation for photon.

Louis de Broglie in 1924 assumed that equation (5.22) is generally valid and is applicable not only for photons but for the material particles as well.

He proposed —

A material particle will have a matter wave corresponding to it, just as a light quantum has a light wave.

Thus, a particle of mass m moving with a velocity v has an associated wavelength,

$$\lambda = \frac{h}{p} = \frac{h}{mv} \qquad \text{...(5.23)}$$

This is known as de Broglie wave equation and λ is called the **de Broglie wavelength**. The associated wave is termed — matter, guide, pilot or de Broglie wave. This relation connects the wave property of light, (*i.e.*, λ) to the particle property of photon.

de Broglie wavelength in terms of energy and temperature :

de Broglie wavelength according to relation (5.23) is given by $\lambda = \frac{h}{p} = \frac{h}{mv}$.

If E is the kinetic energy of the material particle, then $E = \frac{1}{2}mv^2 = \frac{1}{2}\frac{m^2v^2}{m} = \frac{p^2}{2m}$

or
$$p = \sqrt{2mE}$$

Hence, de Broglie wavelength $\lambda = \frac{h}{p} = \frac{h}{\sqrt{2mE}}$

According to kinetic theory of gases, the average kinetic energy of the material particle is given by

$$E = \frac{1}{2}mv^2 = \frac{3}{2}kT$$

where k is the Boltzmann constant $= 1.38 \times 10^{-23}$ J°K^{-1} and T = absolute temperature.

$$\therefore \qquad \lambda = \frac{h}{\sqrt{2mE}} = \frac{h}{\sqrt{3mkT}}$$

5.2.2 Experimental Study of Matter Waves : Davisson and Germer's Experiment

The first experimental evidence of de Broglie's hypothesis came from the experiments of two American Physicists, Davisson and Germer in 1927. This experiment conclusively demonstrated the wave character of electrons.

Experimental arrangement : Figure 5.10 represents the arrangement used by Davisson and Germer. Electrons are emitted by thermionic action of tungsten filament (F) by a low tension battery (L . T). By using thin aluminium diaphragms D_1 and D_2, the produced electrons are restricted to a fine parallel pencil. Due to the action of electric field across the gun, a fine stream of electrons emit out of this gun through an aluminium cylinder A to which required high tension (H.T.) can be applied. Now this pencil of electrons is allowed to reflect from the single crystal of Nickel at T. It is capable of rotating about its axis by a handle H and the rotation axis is parallel to the axis of the electron beam.

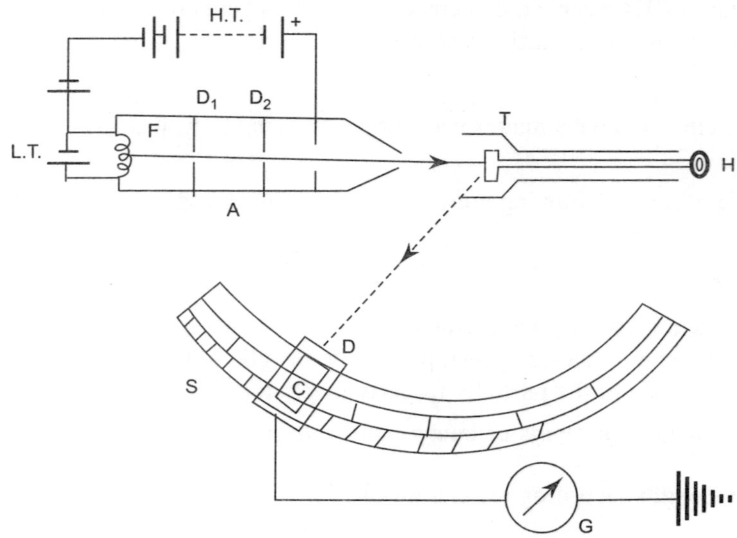

Figure 5.10

The electrons scattered by T in different directions were collected by a Faraday cup or cylinder C and the scattered intensity was measured by a sensitive galvanometer G. It can be moved along a circular scale S, so that it is able to locate the position of maxima and minima between the angle 20° and 90°. The inner (C) and outer (D) walls of the cylinder are insulated from each other and a retarding potential is applied between the two walls of the collector so that only the fastest moving electrons can enter the cylinder. The whole arrangement is kept in an evacuated chamber.

Two series of experiments were performed by Davisson and Germer.

(*a*) **For Normal Incidence :** Here the beam of electrons was made to fall normally on the surface of the crystal and then the collector (C) is moved to various positions on the scale S and the Galvanometer current at each position is noted. The current which was a measure of the intensity of the diffracted beam of electrons was plotted against the angle between the incident beam and the beam entering the collector, (colatitude). The observations are repeated for different accelerating voltages and several curves are drawn as shown in figure 5.11. For an accelerating potential of 40 volts, the curve is fairly smooth and at 44 volts, a small but distinct spur appears at an angle about 60°. As the accelerating voltage is increased, the length of the spur increases. The spur becomes most prominant at 54 volts accelerating potential and the corresponding angle is 50°. With further increases in the accelerating voltage, the length of the spur decreases until it disappears completely above about 68 volts.

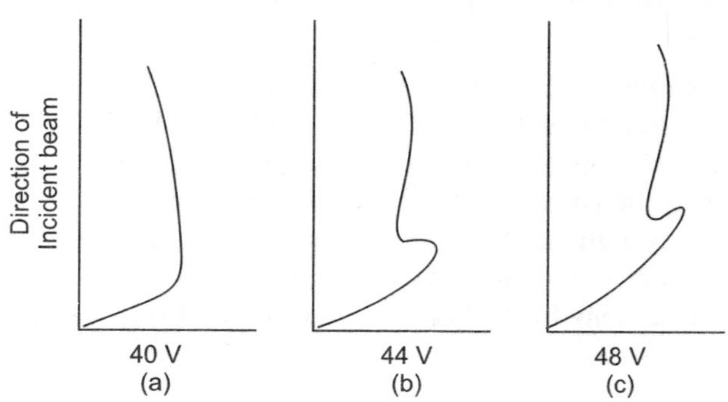

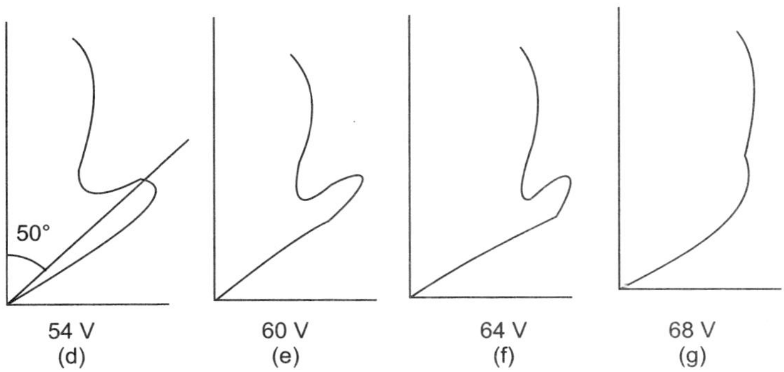

Figure 5.11

The spur or bump in its most important state of development offers a most convincing evidence for the existence of the electron waves.

The occurrence of pronounced spur at 50° with the electrons accelerated through 54 eV can be explained as due to the constructive interference of the electron waves, scattered in the direction, from regular spaced parallel planes in the Nickel crystal.

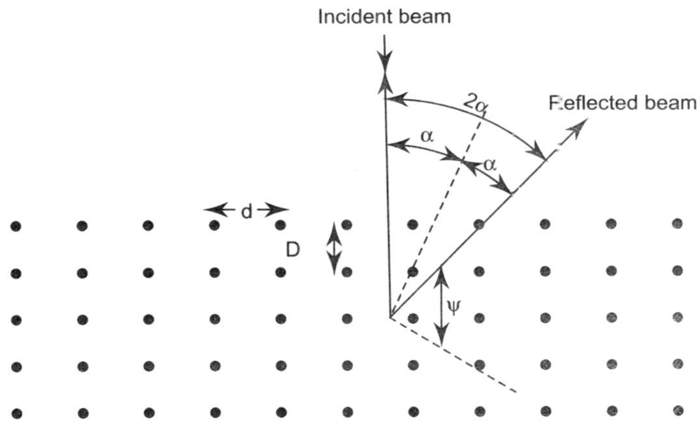

Figure 5.12

Such a regular array of atoms arranged in a lattice are shown in figure 5.12. Consider a beam of electrons to be incident normally upon the crystal face and it makes an angle α with the normal to the set of planes.

Now applying Bragg's law,

$$n\lambda = 2D \sin \psi = 2D \cos \alpha$$

where D = spacing between the atomic layers under consideration and $\psi = \frac{\pi}{2} - \alpha$. If d is the distance between the atoms in the surface layer then $D = d \sin \alpha$, so that we get

$$n\lambda = 2d \sin \alpha \cos \alpha = d \sin 2\alpha = d \sin \theta$$

where $\theta = 2\alpha$ is the angle between the incident and reflected beam. For the Nickel crystal, $d = 2.15$ Å.

In Davisson-Germer experiment $\theta = 50°$, $d = 2.15$ Å, when the accelerating potential is 50 volts.

For the 1st order maximum, $\lambda = 2.15 \times \sin 50° = 1.65$ Å,

From de Broglie's equation for electrons, we have

$$\lambda = \frac{h}{mv} = \frac{h}{p}$$

The momentum $$p = \sqrt{2E_k m} = \sqrt{2Vem}$$

$\therefore$ $$\lambda = \frac{h}{\sqrt{2eVm}} = 1.66 \text{ Å}$$

$\left[h = 6.6 \times 10^{-34} \text{ joule sec}, \; e = 1.6 \times 10^{-19} \text{ coulomb}, \; m = 9.11 \times 10^{-31} \text{ kg}, \; V = 54 \text{ volts} \right]$

Thus, the wavelength obtained from the diffraction experiment agrees very well with the value determined from the de Broglie hypothesis which implies the essential correctness of de Broglie's hypothesis.

(*b*) **Oblique Incidence at Crystal Face :** In this experiment the electron beam was incident at fixed oblique angle upon the crystal keeping the electron gun and collector only at one fixed position throughout the experiment. By changing the accelerating voltage V, the current is measured by galvanometer for each value of voltage. Measuring the intensity of reflected electron beam, the intensity is then plotted against the square root of the accelerating potential, which is shown in figure 5.13. As can be seen from figure 5.13, there are several sharp maxima in the intensity for the certain definite value of V. This is almost analogous to the behaviour of X-rays reflected from a crystal. This proves that electrons like X-rays exhibit wave nature. The distinct maxima at regular interval of $\sqrt{V}$ confirm de Broglie's theory. According to this theory $\lambda = \frac{h}{\sqrt{2meV}}$, or $\lambda \propto \frac{1}{\sqrt{V}}$.

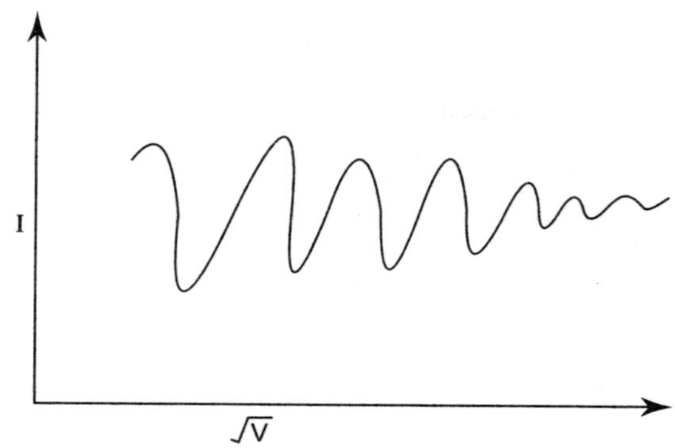

Figure 5.13

5.2.3 Notion of Wave Packets

Phase velocity

According to quantum mechanics, a particle is described by a wave packet. A group of waves or wave packets are developed by combining waves of slightly different wavelengths with phases and amplitudes selected such that they interfere constructively over only a small region of space, outside of which they create an amplitude that reduces to zero quickly as a result of destructive interference. Thus we can say that the wave packets are confined to a very small region of space and also may be considered as a point. As per de Broglie's concern, each particle of matter like proton, electron etc., may be described by the motion of wave packets. Each component wave travels with a definite velocity, called the phase velocity or curve velocity (v_P).

If λ is the wavelength of the wave of frequency ν,

Then, $$v_P = \nu\lambda$$

The quantum concept gives the energy of the wave,

$$E = h\nu \text{ and momentum } p = \frac{h}{\lambda}.$$

$$\therefore \quad v_P = \frac{E}{h}\cdot\frac{h}{p} = \frac{E}{p} \quad \left[\nu = \frac{E}{h} \text{ and } \lambda = \frac{h}{p}\right]$$

Again $E = \hbar\omega$ and $p = \hbar K$ where ω is the angular frequency, K is the wave vector and $\hbar$ is reduced Planck's constant.

Therefore, $$v_P = \frac{\hbar\omega}{\hbar K} = \frac{\omega}{K}.$$

This phase velocity (v_P) also holds true in the case of de Broglie waves associated with a moving particle.

Now we shall consider two special cases.

Case 1 : Non-relativistic Particle :

Here v is very very small compared to c ($v << c$)

So energy, $$E = \frac{p^2}{2m}$$

$$\therefore \quad v_P = \frac{E}{p} = \frac{p}{2m} = \frac{mv}{2m} = \frac{v}{2} \quad [p = mv]$$

So the phase velocity is half the particle velocity.

Case 2 : Relativistic Particle :

For the relativistic case, $$E = \sqrt{p^2c^2 + m_0^2 c^4}$$

or, $$\frac{E}{p} = \sqrt{c^2 + \frac{m_0^2 c^4}{p^2}} = c\sqrt{1 + \frac{m_0^2 c^2}{p^2}}$$

$$\therefore \qquad v_P = c\sqrt{1+\frac{m_0^2\, c^2\lambda^2}{h^2}} \qquad \left[\because \; p = \frac{h}{\lambda}\right]$$

This equation represents that the phase velocity (v_P) is greater than the velocity of light (c) and v_P increases with λ.

Again, we know that the velocity of the wave associated with particle photon is, $v_P = c$,

$$\therefore \qquad c = c\sqrt{1+\frac{m_0^2\, c^2\lambda^2}{h^2}}$$

or,
$$\frac{m_0^2\, c^2\lambda^2}{h^2} = 0,$$

$$\therefore \qquad m_o = 0$$

∴ The rest mass of photon is zero.

Group Velocity

When disturbances are created by a number of component waves, each travelling with slightly different velocities, then the resultant velocity can be treated as a periodicity. The velocity of advance of this periodicity is known as group velocity. So the group velocity is the velocity of the wave packet and is given by

$$v_g = \frac{d\omega}{dK} = \frac{dE}{dp}$$

where
$$E = \hbar\omega$$
$$p = \hbar K$$

v_g can be represented by two special cases —

(*i*) **Non-relativistic Particle :**

Here,
$$E = \frac{p^2}{2m} = \frac{1}{2}mv^2$$

or
$$\frac{dE}{dp} = \frac{2p}{2m} = \frac{p}{m} = v$$

So, we get $v_g = v$.

Thus the particle velocity is equal to the group velocity in case of a non-relativistic particle.

(*ii*) **Relativistic Particle :**

Here,
$$E = \sqrt{p^2c^2 + m_0^2\, c^4}$$

or
$$E^2 = p^2c^2 + m_o^2\, c^4$$

or
$$2E\frac{dE}{dp} = 2pc^2 \qquad [\because \; m_o \text{ and } c \text{ are constant}]$$

or
$$\frac{dE}{dp} = \frac{pc^2}{E} = \frac{mvc^2}{mc^2} = v \qquad [\because \; E = mc^2\,;\, p = mv]$$

Again, the group velocity is equal to the particle velocity.

5.2.4 Heisenberg's Uncertainty Principle

According to classical mechanics, a particle with definite value of the co-ordinates corresponds to an exact value of its momentum and it holds true for every instant of motion. But in quantum mechanics, fundamentally uncertainty exists in any simultaneous determination of the position in space and momentum of the particle. It arises due to the non-classical nature of microparticles. This inequality contains the statement of the Heisenberg's uncertainty principle which we shall now consider in detail.

Statement : *It is impossible to determine accurately and simultaneously the values of both the members of particular pairs of physical variables which describe the behaviour of an atomic system.*

Quantitatively, this principle suggests that this product of the uncertainties in the values of the two canonically conjugate variables must be atleast equal to the reduced Planck's constant $\hbar$. Suppose A and B are two canonically conjugate quantities,

then $$\Delta \mathbf{A} \, . \, \Delta \mathbf{B} \geq \hbar.$$

where ΔA and ΔB are the uncertainties of A and B. If ΔA is small, ΔB is large and vice-versa. It shows that if one quantity is measured accurately, the other becomes less accurate.

In the Hamiltonian sense, some examples of the canonically conjugate variables which are involved in the uncertainty principle, are given below.

(*i*) The uncertainty Δx in measuring the position co-ordinate x of the particle is related to the uncertainty Δp_x in measuring the x component p_x of its momentum by Heisenberg uncertainty relation

$$\Delta x \, \Delta p_x \geq \hbar \qquad \left[\hbar = \frac{h}{2\pi}\right]$$

Similarly $$\Delta y \, \Delta p_y \geq \hbar$$

and $$\Delta z \, \Delta p_z \geq \hbar$$

(*ii*) The uncertainty ΔE in measuring the energy E of a microparticle is related to the uncertainty Δt in the instant t of the measurement, by the relation

$$\Delta \text{E} \, \Delta t \geq \hbar.$$

(*iii*) The uncertainty $\Delta\phi$ in measuring the angular position ϕ of a microparticle in the perpendicular X-Y plane is related to the uncertainty ΔJ_x of the corresponding angular momentum component J_x by the relation

$$\Delta\phi \, \Delta \text{J}_x \geq \hbar$$

The uncertainty principle is beautifully explained with the help of gamma-ray microscope which we consider below :

5.2.5 Gamma-Ray Microscope

Consider we try to measure the position and linear momentum of an electron using an imaginary microscope with a very high resolving power (*i.e.*, the smallest distance between the two points that can be just resolved by the microscope) which are shown in figure 5.14.

Let due to very short wavelength of γ-rays, the γ-ray microscope has been used in this hypothetical method of measuring the position very accurately.

To see the particle in the microscope, it is necessary that at least one photon must strike the electron and scatter inside the microscope with minimum energy.

Let the particle at any instant be at O and the cone of rays scattered from it through the objective has an angle of 2α. (α = half angle subtended by the objective lens of the microscope).

If λ is the wavelength of light used for illuminating the object, then the uncertainty Δx in the measurement of the x-coordinate is (using the concept of the resolving power of a microscope),

$$\Delta x = \frac{\lambda}{2 \sin \alpha}.$$

Due to the impact of the photon on the particle (electron), the electron receives a recoil momentum of order of magnitude $\frac{h}{\lambda}$.

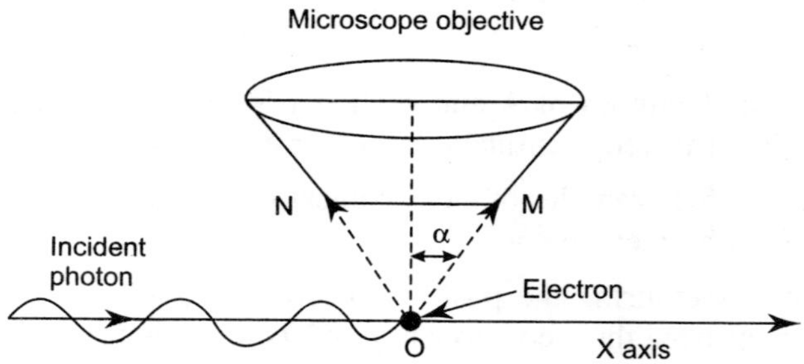

Figure 5.14

The component of this momentum along OM is $\frac{h}{\lambda} \sin \alpha$ and along ON is $-\frac{h}{\lambda} \sin \alpha$.

Hence, the uncertainty in the momentum along in +ve X direction is

$$\Delta p_x = \frac{h}{\lambda} \sin \alpha - \left(-\frac{h}{\lambda} \sin \alpha\right) = \frac{2h}{\lambda} \sin \alpha$$

$$\therefore \qquad \Delta x \, \Delta p_x = \frac{\lambda}{2 \sin \alpha} \cdot \frac{2h}{\lambda} \sin \alpha = h.$$

This shows that the product of the uncertainties in the position of the particle and its momentum (as observed by a gamma-ray microscope) is of the order of Planck's constant.

A more sophisticated approach will illustrate that

$$\Delta x \, \Delta p_x \geq \hbar .$$

5.2.6 Applications of the Uncertainty Principle

(*i*) **Non-existence of free electron in an atomic nucleus :** The radius of a typical atomic nucleus is less than 10^{-14} m. If an electron exists inside a nucleus, the uncertainty in the position should exceed 10^{-14} m. Now if electron exists in the nucleus, the uncertainty of its position

$$\Delta x \geq 10^{-14} \text{ m}.$$

According to Heisenberg's uncertainty principle, the uncertainty of the electron momentum is

$$\Delta p_x = \frac{\hbar}{\Delta x} = \frac{6.62 \times 10^{-34}}{2 \times 10^{-14} \times \pi} = 0.527 \times 10^{-20} \text{ kg m / sec.}$$

Thus, the minimum momentum of the electron, if it exists in the nucleus would be

$$\Delta p_{\min} = \Delta p = 0.527 \times 10^{-20} \text{ kg m/sec.}$$

According to the theory of relativity, the total relativistic energy E of the electron is given by

$$\text{E} = \sqrt{m_0^2 c^4 + p^2c^2}$$

$$= \sqrt{(9.1 \times 10^{-31})^2 (3 \times 10^8)^4 + (1.1 \times 10^{-20})^2 (3 \times 10^8)^2}$$

$$= 1.58 \times 10^{-12} \text{ Joule} = \frac{1.58 \times 10^{-12}}{1.6 \times 10^{-19}} \text{ eV} \approx 9.9 \text{ MeV.}$$

This shows that if a free electron exists in the nucleus, kinetic energy of the free electron must be more than 9.9 MeV. But the energy of the electron emitted from a radioactive nucleus is about 4 MeV which is less than the calculated value of minimum kinetic energy of the electron. Thus we conclude that free electrons cannot reside within nuclei.

(*ii*) **First Bohr radius of the hydrogen atom :** Applying the uncertainty principle, the minimum momentum of the electron is

$\Delta p = \frac{\hbar}{\Delta x} = \frac{\hbar}{r}$, where r is orbital radius of the electron revolving around the nucleus in the hydrogen atom.

The minimum possible value of the momentum would be

$$p_{\min} = \Delta p = \frac{\hbar}{r}.$$

Thus, the kinetic energy of the electron is

$$= \frac{p_{\min}^2}{2m} = \frac{\hbar^2}{2mr^2}$$

and the potential energy in S.I. unit is

$$\text{P.E.} = -\frac{e^2}{4\pi\varepsilon_0 r}$$

∴ The total energy of the electron is

$$\text{E} = \text{K.E.} + \text{P.E.} = \frac{\hbar^2}{2mr^2} - \frac{e^2}{4\pi\varepsilon_0 r}$$

E would attain a minimum value $\text{E}_{\min}$, when

$$\left(\frac{d\text{E}}{dr}\right)_{r = a_o} = 0$$

or, $$\left[-\frac{\hbar^2}{mr^3} + \frac{e^2}{4\pi\varepsilon_0 r^2}\right]_{r = a_o} = 0,$$

or $$\frac{e^2}{4\pi\varepsilon_0 a_o^2} = \frac{\hbar^2}{m a_0^3}$$

or $$a_0 = \frac{4\pi\varepsilon_0\hbar^2}{me^2} = \frac{\varepsilon_0 h^2}{\pi m e^2}$$

Putting $\varepsilon_0 = 8.85 \times 10^{-12}$ F/m, $h = 6.62 \times 10^{-34}$ J. sec and $m = 9.11 \times 10^{-31}$ kg

and $e = 1.6 \times 10^{-19}$ C, we get $a_0 = 5.3 \times 10^{-11}$ m.

Example 1. *Assume that an electron is inside a nucleus of radius 10^{-15} m. Calculate from the uncertainty principle the minimum kinetic energy of the electron.* [WBUT]

Solution. We know from Heisenberg's Uncertainty Principle

$$(\Delta\phi)_{\min}\,(\Delta x)_{\max} \simeq h \simeq \frac{h}{2\pi}$$

or, $$(\Delta p)_{\min} = \frac{6.63 \times 10^{-34}}{2\pi \times 2 \times 10^{-15}\ m} \times 2.63 \times 10^{-20} \text{ kg m/sec.}$$

Again, $$E_{\min}^2 = \phi_{\min}^2 c^2 + m_0^2 c^4$$

or, $$E_{\min}^2 = (2.63 \times 10^{-20})^2 \times (3 \times 10^8 \text{ m})^2 + (9.11 \times 10^{-31})^2 \times (3 \times 10^8)^2$$

or, $$E_{\min} = 7.9 \times 10^{-12} \text{ J} = \frac{7.9 \times 10^{-12}}{1.6 \times 10^{-19}} \text{ eV}$$

$$\simeq 4.9 \times 10^7 \text{ eV} = 49 \times 10^6 \text{ eV}.$$

$$= 49 \text{ MeV}.$$

5.3 QUANTUM MECHANICS

5.3.1 Formation of Quantum Mechanics and Basic Postulates

We have seen that the Bohr theory can not explain why certain spectral lines are more intense than others. It can not account for the observation that many spectral lines actually consist of several separate lines whose wavelength differ slightly. So the laws of classical mechanics can not explain the origin of quantized motion in the atomic or sub-atomic domain and also the old quantum theory is also inadequate in this respect. So, discarding the classical concept and old quantum theory we should try to develop an atomic mechanics on the basis of such concepts which are actually observable quantities. Hence the new atomic mechanics should be developed on the basis of quantities like energies of the atomic states, the intensities of the spectral lines and not to the positions or velocities of the particle (electron).

Such an approach was developed in 1925–1926 by Erwin Schrödinger, Heisenberg, Max Born, Dirac and others with the apt name of quantum mechanics. "The discovery of quantum mechanics was nearly a total surprise. It described the physical world in a way that was fundamentally new. It seems to many of us a miracle" stated Engine Wigner, one of the early workers in the field.

The concepts which de Broglie introduced to relate dynamical variables linked with corpuscular motion with characteristics of the waves were extended by Schrödinger in a more general way to a quantum system to develop a wave equation for describing the motion of such systems. The Schrödinger wave equation has the foundation of non-relativistic quantum mechanics or wave mechanics. The foundation of relativistic quantum mechanics were also laid about this time by Dirac (1928).

The Schrödinger equation cannot be deduced from theoretical considerations but one has to guess the equation. Its success will depend on some fundamental characteristics of the equation which are given below :

(*i*) The equation must be linear and the solution of the equation ensures the validity of the superposition principle.

(*ii*) The differential equation satisfied by the wave function must be a first order in time, so that if ψ is known at any point r at the initial instant t_0, then at any instant t, the wave function $\psi(r, t)$ at the same point will be uniquely determined.

(*iii*) The wave equation must be consistent with Bohr's correspondence principle and de Broglie's hypothesis.

5.3.2 Schrödinger's Equation

Time dependent form : Consistent with de Broglie's wave concept as an inherent property of each particle, Schrödinger proposed that the wave be represented by a function $\psi(x, t)$ depending on the space and time variables. It may be a complex function. Let us assume that ψ for a particle moving freely in the positive direction of x-axis is given by

$$\psi = Ae^{-i\omega\left(t - \frac{x}{v}\right)} \qquad \text{...(5.24)}$$

where $\omega = 2\pi\nu$ and $v = \lambda\nu$ [ν = frequency, λ = wavelength]

Now equation (5.24) becomes

$$\psi = A\,e^{-2\pi i\left(\nu t - \frac{x}{\lambda}\right)} \qquad \text{...(5.25)}$$

Now ψ can be described in terms of the total energy E and momentum (p) of the particle where $E = h\nu$ and $p = \dfrac{h}{\lambda}$. Substituting this in equation (5.25) we obtain

$$\psi = A\,.\,e^{-\left(\frac{2\pi i}{h}\right)(Et - px)} \qquad \text{...(5.26)}$$

Thus equation (5.26) illustrates the wave equivalent of an unrestricted particle of total energy E and momentum (p) moving in the positive X-direction.

Differentiating equation (5.26) twice with respect to x, we obtain

$$\frac{\partial^2\psi}{\partial x^2} = A\left(\frac{2\pi ip}{h}\right)e^{-\left(\frac{2\pi i}{h}\right)(Et - px)}$$

and

$$\frac{\partial^2\psi}{\partial x^2} = -\frac{4\pi^2 p^2}{h^2}\psi \qquad \text{...(5.27)}$$

Again, differentiating (5.26) once with respect to t

$$\frac{\partial\psi}{\partial t} = \left(-\frac{2\pi iE}{h}\right)A\,e^{-\left(\frac{2\pi i}{h}\right)(Et - px)}$$

$$= -\frac{2\pi i E}{h}\psi \qquad \text{...(5.28)}$$

At speeds small compared with that of light, the total energy can be expressed by the sum of its potential energy (V) and kinetic energy $\left(\frac{p^2}{2m}\right)$. V is in general a function of position x and time t.

$$\therefore \qquad E = \frac{p^2}{2m} + V \qquad \text{...(5.29)}$$

The function V describes the influence of the rest of the universe of the particle.

Multiplying both sides of equation (5.29) by ψ, we get

$$E\psi = \frac{p^2}{2m}\psi + V\psi \qquad \text{...(5.30)}$$

Equation (5.27) and (5.28) can be represented as

$$p^2\psi = -\frac{h^2}{4\pi^2}\frac{\partial^2\psi}{\partial x^2} = -\hbar^2\frac{\partial^2\psi}{\partial x^2}$$

and

$$E\psi = -\frac{\hbar}{i}\frac{\partial\psi}{\partial t} \qquad \left[\hbar = \frac{h}{2\pi}\right]$$

Substituting these values ($E\psi$ and $p^2\psi$) in equation (5.30) we obtain

$$-\frac{\hbar}{i}\frac{\partial\psi}{\partial t} = -\frac{\hbar^2}{2m}\frac{\partial^2\psi}{\partial x^2} + V\psi$$

or

$$\boxed{i\hbar\frac{\partial\psi}{\partial t} = -\frac{\hbar^2}{2m}\frac{\partial^2\psi}{\partial x^2} + V\psi} \qquad \text{...(5.31)}$$

Equation (5.31) represents the time dependent form of Schrödinger's equation in one dimension. In three dimensions the time independent form of Schrödinger's equation is

$$i\hbar\frac{\partial\psi}{\partial t} = -\frac{\hbar^2}{2m}\left(\frac{\partial^2\psi}{\partial x^2} + \frac{\partial^2\psi}{\partial y^2} + \frac{\partial^2\psi}{\partial z^2}\right) + V\psi \qquad \text{...(5.32)}$$

Steady state form : In too many situations the potential energy of a particle does not depend on time explicitly. The forces that act in it vary with the position of the particle only. If it is true, the Schrödinger's equation may be simplified by removing all references of t. The one dimensional wave function ψ of an unrestricted particle may be represented in the form in terms E and p

$$\psi = Ae^{\left(-\frac{2\pi i}{h}\right)(Et - px)}$$

$$= Ae^{-\frac{i}{\hbar}(Et - px)}$$

or, $$\psi = A\, e^{-\left(\frac{iE}{\hbar}\right)t} . e^{\left(\frac{ip}{\hbar}\right)x} \qquad \left[\hbar = \frac{h}{2\pi}\right]$$

or, $$\psi = \psi_0\, e^{-\left(\frac{iE}{\hbar}\right)t} \qquad \text{...(5.33)}$$

where $\psi_0 = A\, e^{\frac{ip}{\hbar}x}$.

Thus ψ is the product of time dependent function $e^{-\left(\frac{iE}{\hbar}\right)t}$ and a position dependent function ψ_0. Differentiating equation (5.33) once with respect to t, we obtain

$$\frac{\partial \psi}{\partial t} = \psi_0 \left(-\frac{iE}{\hbar}\right) e^{-\left(\frac{iE}{\hbar}\right)t} \qquad \text{...(5.34)}$$

Again, differentiating equation (5.33) twice with respect to x we get

$$\frac{\partial^2 \psi}{\partial x^2} = \frac{\partial^2 \psi_0}{\partial x^2} . e^{-\left(\frac{iE}{\hbar}\right)t} \qquad \text{...(5.35)}$$

Now, we can substitute these values in the time dependent Schrödinger's equation (5.31). We obtain

$$i\hbar . \psi_0 \left(-\frac{iE}{\hbar}\right) e^{-\left(\frac{iE}{\hbar}\right)t} = -\frac{\hbar^2}{2m} . \frac{\partial^2 \psi_0}{\partial x^2} e^{-\left(\frac{iE}{\hbar}\right)t} + V \psi_0\, e^{-\left(\frac{iE}{\hbar}\right)t}$$

Dividing through by the common exponential factor gives

$$E \psi_0 = -\frac{\hbar^2}{2m} \frac{\partial \psi_0}{\partial x^2} + V \psi_0$$

or, $$\frac{\partial^2 \psi_0}{\partial^2 x^2} + \frac{2m}{\hbar^2} (E - V) \psi_0 = 0 \qquad \text{...(5.36)}$$

Equation (5.36) represents the steady state form of Schrödinger's equation,

In three dimensions, it can be represented as

$$\frac{\partial^2 \psi_0}{\partial x^2} + \frac{\partial^2 \psi_0}{\partial y^2} + \frac{\partial^2 \psi_0}{\partial z^2} + \frac{2m}{\hbar^2} (E - V) \psi_0 = 0$$

or, $$\nabla^2 \psi_0 + \frac{2m}{\hbar^2} (E - V) \psi_0 = 0$$

Usually it may be written as

$$\nabla^2 \psi + \frac{2m}{\hbar^2} (E - V) \psi = 0 \qquad \text{...(5.37)}$$

5.3.3 Operator Correspondence

Any mathematical operation like addition, subtraction, integration, differentiation etc., can be described by certain symbols known as operators.

Suppose $\hat{A}$ is an operator, operates on a function $f(x)$,

i.e., $$\hat{A}\, f(x) = g(x).$$

Thus, $\hat{A}$ changes the function $f(x)$ to another function $g(x)$.

Differentiation with respect to x may also be considered as an operation.

i.e., $$\frac{d}{dx}(x^2 + 1) = 2x$$

Here, operator $\hat{A} = \frac{d}{dx}$ operates on the function $f(x) = x^2 + 1$, whereby it changes $f(x)$ to $g(x) = 2x$. In case of classical physics, certain definite functions of proper variables are linked with each observable quantity. Usually (x, y, z) are linked with position (r), mv is linked with momentum (p) and $\frac{1}{2} mv^2$ is associated with kinetic energy E_k and so on. Similarly in case of classical mechanics and quantum mechanics, certain operators are linked with observable quantities, such operators are given below :

Quantity	*Classical Definition*	*Quantum Operator*
(*i*) Position	$\vec{r}$	$\vec{r}$
(*ii*) Momentum	$\vec{p}$	$-i\hbar\vec{\nabla}$
(*iii*) Angular momentum	$\vec{r}\times\vec{p}$	$-i\hbar\,\vec{r}\times\vec{\nabla}$
(*iv*) Kinetic energy	$\frac{p^2}{2m}$	$-\frac{\hbar^2}{2m}\nabla^2$

When the operator operates on the suitable wave function it gives the eigen value of the operator. Suppose an eigen value equation is

$$i\hbar\frac{\partial\psi}{\partial t} = E\,\psi$$

Here, operator $i\hbar\frac{\partial}{\partial t}$ operates on the eigen function ψ and gives the corresponding eigen value E.

5.3.4 Physical Interpretation of Wave Function ψ

The wave function $\psi(x, y, z, t)$ is the solution of the Schrödinger's wave equation and gives a complete quantum-mechanical description of the behaviour of a moving particle of mass m. The function can be either real or complex. The function $\psi(\vec{r}, t)$ can not be measured directly by any physical experiment. However the position, momentum, potential energy and kinetic energy can be obtained using suitable mathematical operations on the wave function ψ. The function ψ is itself of no physical significance but the quantity $P = \psi^* \psi = |\psi|^2$ gives the probability density (ψ^* is the complex conjugate of ψ) for the existence of the particle at a particular point (x, y, z) of space at time t. This is possible only if the function ψ is well behaved.

To summarize the conditions for good behaviour of ψ

1. ψ must be continuous, finite and single valued over all the interval of space variables as well as at every instant of time.
2. $\dfrac{\partial \psi}{\partial x}, \dfrac{\partial \psi}{\partial y}, \dfrac{\partial \psi}{\partial z}$ must be continuous and single valued everywhere.
3. The function ψ and its derivatives shall asymptotically tend to zero as $x, y, z \to \pm \infty$.
4. If $\psi_1(x), \psi_2(x), \ldots, \psi_n(x)$ are solutions of Schrödinger equation then the linear combination
$$\psi(x) = a_1 \psi_1(x) + a_2 \psi_2(x) + \ldots + a_n \psi_n(x)$$
must be a solution.
Apart from the above four conditions, two more conditions are illustrated here.
5. **Normalization :** The probability that the particle will be found somewhere in the region is unity. This implies that the quantity $\int |\psi(x, y, z, t)|^2 \, dx\, dy\, dz$ integrated at some point in space and this measure, in probability focus should be unity.

   i.e., $\int |\psi(x, y, z, t)|^2 \, dx\, dy\, dz = 1$

   The above equation gives the condition of normalization and the wave function which satisfies this condition is said to be normalized. The nomalization accomplished by adjusting the numerical co-efficient of ψ so that the integral is unity. If the ψ function does not fulfill this condition, the procedure is
$$\int_{-\infty}^{\infty}\int_{-\infty}^{\infty}\int_{-\infty}^{\infty} \psi(x, y, z, t)\, dx\, dy\, dz = N^2$$
   where N is constant (real or complex). But N^2 is always real and positive. Thus one may consider for the normalized state function, the form,
$$\frac{1}{N} \psi(x, y, z, t)$$
   Since this will yield the value of unity for the integral in space. The quantity N is called the Normalization constant.
6. **Orthogonality :** This conditions states that if $\psi_m(x, y, z, t)$ and $\psi_n(x, y, z, t)$ be two linearly independent state functions which are solutions of the Schrödinger equations, then the integral
$$\int_{-\infty}^{\infty}\int_{-\infty}^{\infty}\int_{-\infty}^{\infty} \psi_m^*(x, y, z, t)\, \psi_n(x, y, z, t)\, dx\, dy\, dt = 0$$
   $m \neq n$.

   The state function ψ_m which satisfy both the unitary and the orthogonality conditions are said to form an "Orthonormal".

5.3.5 Expectation Values

The expectation value is the mathematical expectation for the result of a single measuremeant or it is average of a large number of measurements on independent systems.

In general dynamical quantities like position, momentum, kinetic energy and potential energy which form the objectives of measurement in experimental physics are called observables. According to quantum mechanics, each observable is illustrated by an operator which acts on the wave function ψ to generate a new function.

The expectation value or average value $\langle A \rangle$ of any observable A is defined by

$$\langle A \rangle = \int \psi^* \hat{A} \psi \, dx \, dy \, dz$$

where $\hat{A}$ is the operator associated with the observable A and ψ^* is the complex conjugate of ψ.

The expectation value of momentum and energy may be defined by using corresponding differential operator.

Thus the expectation value of momentum is

$$\langle \vec{p} \rangle = \int_{-\infty}^{\infty} \psi^* \left(-i\hbar \vec{\nabla}\right) \psi \, dx \, dy \, dz$$

and expectation value of energy is $\left[\hbar = \dfrac{h}{2\pi}\right]$

$$\langle E \rangle = \int_{-\infty}^{\infty} \psi^* \left(i\hbar \frac{\partial}{\partial t}\right) \psi \, dx \, dy \, dz$$

5.3.6 Application of Schrödinger's Equation

5.3.6.1 *Particle in an infinite square well potential (one dimensional)*

Consider a particle's motion is restricted to travelling along the X-axis between $x = 0$ and $x = L$ by infinite hard walls. Here the particle may collide with the walls but can not exchange energy as the potential is infinite. Hence the potential of a particle inside the walls is constant – Say zero (0) for convenience – on the inside figure 5.15.

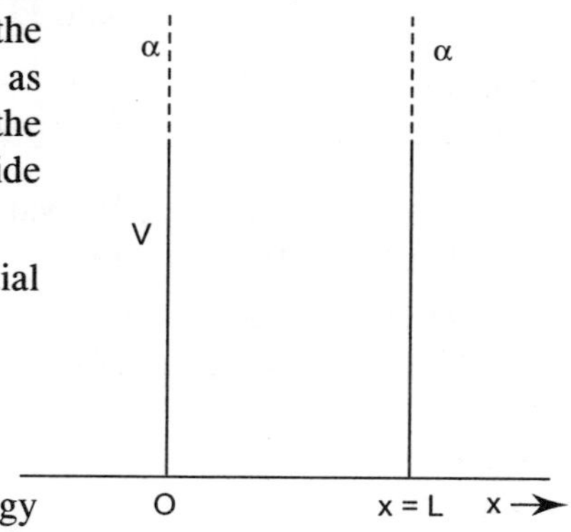

Figure 5.15

In terms of the boundary conditions imposed by the potential function is

$$V = 0 \text{ for } x < 0 \text{ and } x > L$$

$$V = \alpha \text{ for } x \le 0 \text{ and } x \ge L$$

Thus, the particle can not exist outside the walls where its energy would become infinite. Hence the state function $\psi(x, t)$ vanishes in the regions between $x \le 0$ and $x \le L$.

We have the Schrödinger's equation in one dimension

$$\frac{d^2\psi}{dx^2} + \frac{2m}{\hbar^2}(E - V)\,\psi = 0$$

With in the box Schrodinger equation becomes (V = 0)

$$\frac{d^2\psi}{dx^2} + \frac{2m}{\hbar^2}E\,\psi = 0$$

or,

$$\frac{d^2\psi}{dx^2} + K^2\,\psi = 0 \qquad \text{...(5.38)}$$

where,

$$K^2 = \frac{2mE}{\hbar^2}$$

So the general solution of equation (5.38) is

$$\psi = A \sin Kx + B \cos Kx \qquad \text{...(5.39)}$$

Using the boundary conditions, the constants A and B need to be evaluated.

$$\psi = 0, \text{ at } x = 0, \qquad B \cos Kx = 0$$

$$\therefore \qquad B = 0$$

$$\psi = 0, \text{ at } x = L, \therefore \quad A \sin Kx = 0$$

Since $A \neq 0$, $KL = n\pi$, or, $K = \dfrac{n\pi}{L}$

where n is an integer. ($n = 1, 2, 3, ...$)

$$\therefore \qquad \psi_n(x) = A \sin \frac{n\pi x}{L} \qquad \text{...(5.40)}$$

Again,

$$K^2 = \frac{2mE}{\hbar^2}$$

or,

$$E = \frac{K^2\hbar^2}{2m} = \frac{n^2\pi^2\hbar^2}{2mL^2}$$

Thus the energy of the particle can have only certain values which are called **eigen values** of the state function (eigen function) defined by

$$E_n = \frac{n^2\pi^2\hbar^2}{2mL^2}\ ;\ n = 1, 2, 3, ... \qquad \text{...(5.41)}$$

So, inside the box, the particle can only have discrete energy values specified by equation (5.41) and can not have zero energy.

Again, we use, A has to be determined by the normality of the function ψ_n. For normalization of the wave function

$$\int_{-\infty}^{\infty} |\psi_n|^2\, dx = \int_0^L |\psi_n|^2\, dx = 1$$

or, $\quad A^2 \int_0^L \sin^2\left(\frac{n\pi x}{L}\right) dx = 1$

or, $\quad \frac{A^2}{2}\left[\int_0^L dx - \int_0^L \cos\left(\frac{2\pi nx}{L}\right) dx\right] = 1$

or, $\quad \frac{A^2}{2} . L - 0 = 1,$

or, $\quad A = \sqrt{\frac{2}{L}}$

Therefore, the normalized wave functions of the particle are

$$\psi_n = \sqrt{\frac{2}{L}} \sin \frac{n\pi x}{L} \qquad \text{...(5.42)}$$

where $n = 1, 2, 3, \ldots$

The value of the first three functions, ψ_1, ψ_2 and ψ_3 are shown in figure 5.16 along with the probability density values $|\psi_1|^2$, $|\psi_2|^2$ and $|\psi_3|^2$. These wave functions resemble the vibrations of a string fixed at both ends because the wave function have the same form. However, a particle in the lowest energy state of $n = 1$ is most likely to be at the centre of the potential box while a particle in the next higher state of $n = 2$ is never there ! But classical physics suggests that same probability for the particle being anywhere in the box.

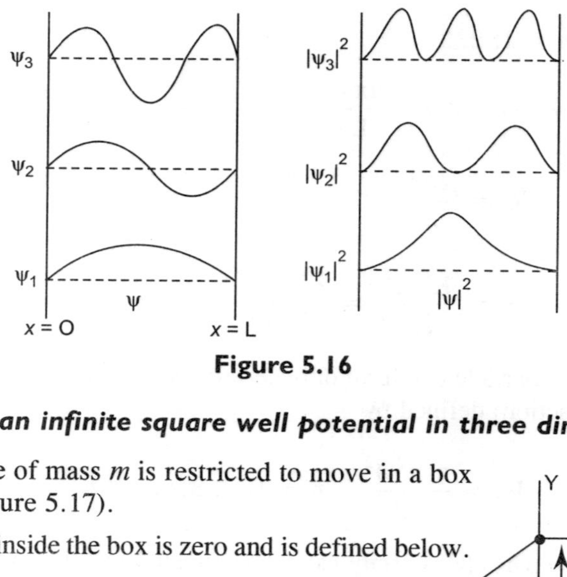

Figure 5.16

5.3.6.2 *A particle in an infinite square well potential in three dimensions*

We assume that a particle of mass m is restricted to move in a box with sides a, b and c (figure 5.17).

The potential function inside the box is zero and is defined below.

$$V \neq (x, y, z) = 0 \text{ for } \left.\begin{matrix} 0<x<a \\ 0<y<b \\ 0<z<c \end{matrix}\right\}$$

The potential outside the box is infinite.

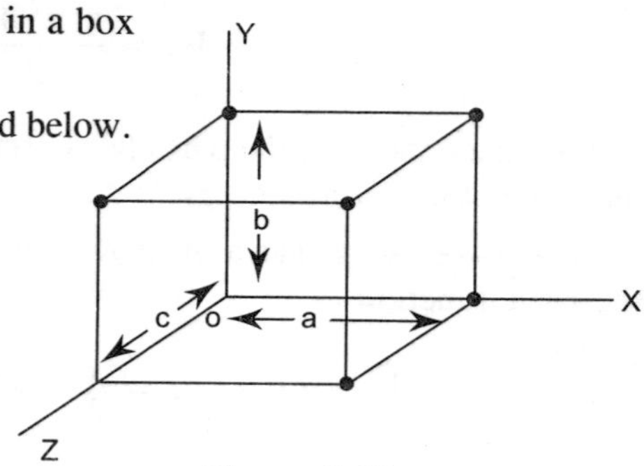

Figure 5.17

The Schrödinger equation becomes in three dimensions

$$\frac{\partial^2 \psi}{\partial x^2} + \frac{\partial^2 \psi}{\partial y^2} + \frac{\partial^2 \psi}{\partial z^2} + \frac{2mE}{\hbar^2}\,\psi = 0 \qquad \text{...(5.43)}$$

Here, to solve this equation, we have considered the method of separation of variables.

Putting $\psi(x, y, z) = X(x)\,Y(y)\,Z(z)$

and simplifying, $\dfrac{1}{X}\dfrac{d^2X}{dx^2} + \dfrac{1}{Y}\dfrac{d^2Y}{dy^2} + \dfrac{1}{Z}\dfrac{d^2Z}{dz^2} = -\dfrac{2mE}{\hbar^2}$

These equations can be written as

and

$$\left.\begin{aligned} \frac{1}{X}\frac{d^2X}{dx^2} &= -\alpha^2 \\ \frac{1}{Y}\frac{d^2Y}{dy^2} &= -\beta^2 \\ \frac{1}{Z}\frac{d^2Z}{dz^2} &= -\gamma^2 \end{aligned}\right\} \qquad \text{...(5.44)}$$

with the condition, $\alpha^2 + \beta^2 + \gamma^2 = \dfrac{2mE}{\hbar^2}$.

Solutions of equation (5.44) yield the simple functions

$$\left.\begin{aligned} X &= A_1 \cos \alpha x + B_1 \sin \alpha x \\ Y &= A_2 \cos \beta y + B_2 \sin \beta y \\ Z &= A_3 \cos \gamma z + B_3 \sin \gamma z \end{aligned}\right\} \qquad \text{...(5.45)}$$

where A_1, B_1, A_2, B_2, A_3 and B_3 are constants. The values of the constants can be found by applying boundary conditions. The boundary conditions require that the wave function vanishes at the box walls where the potential is infinite.

$$\psi(0, y, z) = \psi(x, 0, z) = \psi(x, y, 0) = 0 \qquad \text{...(5.46)}$$

and

$$\psi(a, y, z) = \psi(x, b, z) = \psi(x, y, c) = 0 \qquad \text{...(5.47)}$$

Applying boundary conditions (5.46), to the above equations

$$A_1 = A_2 = A_3 = 0$$

for nontrivial solutions therefore $B_1 \neq 0$, $B_2 \neq 0$, and $B_3 \neq 0$.

Applying boundary conditions (5.47) to the above equations

$$\left.\begin{aligned} \sin \alpha a = 0 \;&;\; \alpha a = n_x \pi \;;\; \alpha = \frac{n_x \pi}{a} \\ \sin \beta b = 0 \;&;\; \beta b = n_y \pi \;;\; \beta = \frac{n_y \pi}{b} \\ \sin \gamma c = 0 \;&;\; \gamma c = n_z \pi \;;\; \gamma = \frac{n_z \pi}{c} \end{aligned}\right\} \qquad \text{...(5.48)}$$

Here n_x, n_y and n_z are integers of which none is equal to zero.

$$X = B_1 \sin \frac{n_x \pi x}{a} \quad \ldots(5.49)$$

$$Y = B_2 \sin \frac{n_y \pi y}{b} \quad \ldots(5.50)$$

$$Z = B_3 \sin \frac{n_z \pi z}{c} \quad \ldots(5.51)$$

So the state function is,

$$\psi_n (x, y, z) = B_1 B_2 B_3 \sin \frac{n_x \pi x}{a} \sin \frac{n_y \pi y}{b} \sin \frac{n_z \pi z}{c}$$

$$= B \sin \frac{n_x \pi x}{a} \sin \frac{n_y \pi y}{b} \sin \frac{n_z \pi z}{c} \quad \ldots(5.52)$$

where B is the normalization constant ($B = B_1 B_2 B_3$),
and it can be found out using normalization condition

$$\int_0^a \int_0^b \int_0^c \psi \psi^* \, dx \, dy \, dz = 1,$$

or, $$A^2 \int_0^a \int_0^b \int_0^c \sin^2 \frac{n_x \pi x}{a} \sin^2 \frac{n_y \pi y}{b} \sin^2 \frac{n_z \pi z}{c} \, dx \, dy \, dz = 1.$$

or $$A^2 . \frac{a}{2} . \frac{b}{2} . \frac{c}{2} = 1$$

or, $$A = \frac{2\sqrt{2}}{\sqrt{abc}}$$

Thus the normalized wave function becomes

$$\psi_n (x, y, z) = \frac{2\sqrt{2}}{\sqrt{abc}} \sin \frac{n_x \pi x}{a} \sin \frac{n_y \pi y}{b} \sin \frac{n_z \pi z}{c} \quad \ldots(5.53)$$

we have $$\alpha^2 + \beta^2 + \gamma^2 = \frac{2mE}{\hbar}$$

and $$nx + ny + nz = n$$

or $$\frac{n_x^2 \pi^2}{a^2} + \frac{n_y^2 \pi^2}{b^2} + \frac{n_z^2 \pi^2}{c^2} = \frac{2mE}{\hbar^2}$$

or, $$E_n = \frac{\hbar^2 \pi^2}{2m} \left[\frac{n_x^2}{a^2} + \frac{n_y^2}{b^2} + \frac{n_z^2}{c^2} \right] \quad \ldots(5.54)$$

If $a \neq b \neq c$, there can be no such combination, and each value of E_n will have only one state function ψ_n. Such states are completely non-degenerate states.

If the particle is confined in a cubical region, then $a = b = c$. The energy level of equation (5.54) becomes

$$E_{cube} = \frac{\hbar^2\pi^2}{2ma^2}\left[n_x^2 + n_y^2 + n_z^2\right].$$

For the ground state, $n_x = n_y = n_z = 1$

or $$E_{111} = \frac{3\pi^2\hbar^2}{2ma^2}$$

There is only one set of quantum numbers that gives an unique energy state and this level is said to be non-degenerate.

5.3.7 Discussion of Degenerate States

If any two or all three sides are equal ($e.g.$, $a = b = c$), then there can be combinations of the type mentioned before and some energy states E_n may have more than one unique state function. Such states are called degenerate states. Example :

There are three possibilities for the first excited state

$$n_x = 2,\ n_y = n_z = 1$$
$$n_y = 2,\ n_x = n_z = 1$$
$$n_z = 2,\ n_x = n_y = 1.$$

We obtain $E_{211} = E_{121} = E_{112} = \dfrac{3\pi^2\hbar^2}{ma^2}$. Thus, several distinct quantum states possess the same energy. This property is called degeneracy. For the case $a = b = c$. the energy level diagrams are shown in figure 5.18.

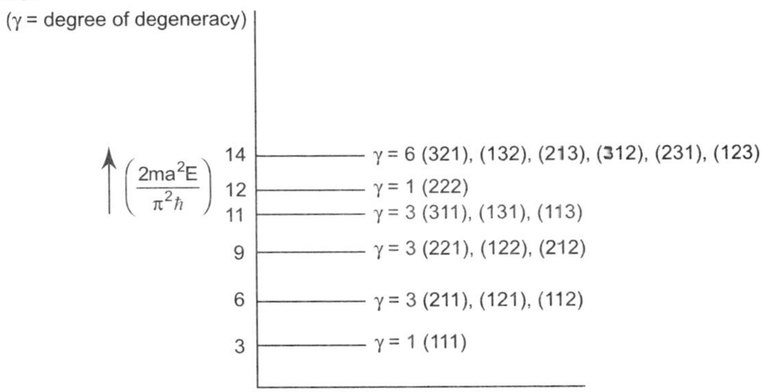

Figure 5.18

Suppose $\psi_{n_1}, \psi_{n_2}, \ldots, \psi_{n_r}$ are distinctly different linearly independent wave functions corresponding to the same eigen energy E_n, then the state is described r-fold degenerate and the quantity ν $(\le n)$ is called the degree of degeneracy of the state. In this case general state function for the state may be described as

$$\psi_n = \sum_r a_r\ \psi_{n_r} = a_1\ \psi_{n_1} + a_2\quad \psi_{n_2} + \ldots + a_n\psi_{n_r}$$

where $a_1, a_2, \ldots, a_n$ are complex numbers. Then $|\ a_r\ |^2$ will give the probability that the particle should be found in the state ψ_{n_r} of the eigen energy E_n when a measurement is carried out.

IN FOCUS

- **Wien's displacement law :** $\lambda_m T$ = constant

  Wien's displacement law in terms of E_λ and T : $\frac{E_\lambda}{T^5}$ = constant

- **Rayleigh-Jeans law :** $U_\nu d\nu = \frac{8\pi\nu^2 kT}{c^3} d\nu$

  U_ν : Energy density of the enclosure in the frequency range from ν to $\nu + d\nu$.

  Again in terms of wavelength $U_\lambda d\lambda = \frac{8\pi kT}{\lambda^4} d\lambda$

  This is another form of Rayleigh-Jeans law.

- **Planck's Quantum hypothesis :** $E_n = nh\nu$

  n : quantum number = 0, 1, 2, 3 ... are permissible.
  h : Planck's constant = 6.625×10^{-34} Js
  ν : frequency of oscillation

- **Planck's law for Radiation :**

$$U_\nu d\nu = \frac{8\pi h\nu^3}{c^3} \cdot \frac{d\nu}{\left[e^{\frac{h\nu}{kT}} - 1\right]}$$

and

$$U_\lambda d\lambda = \frac{8\pi hc}{\lambda^5} \cdot \frac{d\lambda}{\left[e^{\frac{hc}{\lambda kT}} - 1\right]}$$

- **Einstein's photoelectric equation :**

$$\frac{1}{2}mv^2 = h(\nu - \nu_0)$$

where $\phi = h\nu_0$ = work function

and ν_0 = threshold frequency.

- **Compton shift :** $d\lambda = \frac{h}{m_0 c}(1 - \cos\alpha) = \lambda_0 (1 - \cos\alpha)$

  λ_0 : compton wavelength $= \frac{h}{m_0 c} = 0.024$ Å

  α : Angle of scattering of photon.

- **de Broglie Wavelength :** $\lambda = \frac{h}{p}$

  p : momentum of the particle
  If E is the kinetic energy of the material particle, then

$$\lambda = \frac{h}{\sqrt{2mE}} \text{ where } p = \sqrt{2mE}$$

According to kinetic theory of gases, the average kinetic energy of the material particle,

$$E = \frac{1}{2}mv^2 = \frac{3}{2}kT$$

then $$\lambda = \frac{h}{\sqrt{3mkT}}$$ k : Boltzmann constant

- **Group velocity :** $v_g = \dfrac{d\omega}{dK} = \dfrac{dE}{dp}$ where $E = \hbar\omega$ and $p = \hbar K$

- **Phase velocity :** $v_p = \dfrac{\omega}{K}$

- **Heisenberg's Uncertainty Principle :**
  - The product of uncertainties in determining the position (Δx) and momentum (Δp) of a particle at the same instant is $\Delta x\, \Delta p \geq \hbar$
  - The uncertainty in determining energy (ΔE) and time (Δt) is given by $\Delta E \Delta t \geq \hbar$
  - The uncertainty in determining angular momentum (ΔJ) and angle ($\Delta\theta$) is given by $\Delta J \Delta\theta \geq \hbar$

Applications :

1. Non-existence of free electron in an atomic nucleus
2. Radius of the Bohr's first orbit.

- **Schrödinger Wave Equation :**

1. Time independent Schrödinger equation

$$\Delta^2\psi + \frac{2m}{\hbar^2}(E - V)\,\psi = 0$$

2. Time dependent Schrödinger equation

$$\left(-\frac{\hbar}{2m}\nabla^2 + V\right)\psi = i\hbar\frac{\partial\psi}{\partial t}$$

where $$\nabla^2 \equiv \frac{\partial^2}{\partial x^2} + \frac{\partial^2}{\partial y^2} + \frac{\partial^2}{\partial z^2}$$

- **Expectation values :**
  - Expectation value $\langle A \rangle$ of any observable A is defined by

$$\langle A \rangle = \int \psi^* \hat{A}\psi\, dx\, dy\, dz$$

  - Expectation value of momentum is $\langle p \rangle = \displaystyle\int_{-\alpha}^{\alpha} \psi^*\left(-i\hbar\vec{\nabla}\right)\psi\, dx\, dy\, dz$

and Expectation value of energy is $\langle E \rangle = \displaystyle\int_{-\alpha}^{+\alpha} \psi^*\left(+i\hbar\frac{\partial}{\partial t}\right)\psi\, dx\, dy\, dz$

Here $p \rightarrow \left(-i\hbar\,\vec{\nabla}\right)$

and $E \rightarrow +\, i\hbar\dfrac{\partial}{\partial t}$

Solution of the Schrödinger equation : one dimensional consideration

1. Orthogonality condition : $\int_{-\infty}^{+\infty} \psi_n{}^* \, \psi_n \, dx = 0$

2. Normalisation condition : $\int_{-\infty}^{+\infty} |\psi_n|^2 \, dx = 1$

3. Orthonormality condition : $\int_{-\infty}^{+\infty} \psi_k^* \, \psi_n \, dx = \delta_{kn}$

$$\delta_{kn} = 0, \; k \neq n = 1, \; k = n$$

Particle in an infinite square well potential (One dimensional) :

- The allowed energy levels are given by $E_n = \dfrac{n^2\pi^2\hbar^2}{2mL^2}$

  where $n = 1, 2, 3 \ldots$
- The corresponding eigen functions are

$$\psi_n = \sqrt{\frac{2}{L}} \sin \frac{n\pi x}{L}$$

Particle in an infinite square well potential in three dimensions

- Energy levels $= \dfrac{\hbar^2}{2m}\left[\dfrac{n_x^2}{a^2} + \dfrac{n_y^2}{b^2} + \dfrac{n_z^2}{c^2}\right]$

- Wave function $\psi_{n_x n_y n_z} = \dfrac{2\sqrt{2}}{\sqrt{abc}} \sin \dfrac{n_x \pi x}{a} \sin \dfrac{n_y \pi y}{b} \sin \dfrac{n_z \pi z}{c}$

Solved Problems

Problem 1. *Assuming Planck's law derive (i) Rayleigh-Jeans law ; (ii) Wien's displacement law.*

[Bangalore University, Banaras Hindu University, H.P. Univeristy, G.N.D. University]

Solution. We know from Planck's law for blackbody radiation in terms of wavelength

$$E_\lambda = \frac{8\pi hc \, \lambda^{-5}}{e^{\frac{hc}{\lambda kT}} - 1} \qquad \ldots(i)$$

(*i*) Considering for long wavelength $\dfrac{hc}{\lambda kT} < 1$

$$e^{\frac{hc}{\lambda kT}} - 1 \simeq 1 + \frac{hc}{\lambda kT} - 1 \simeq \frac{hc}{\lambda kT}$$

$$\therefore \qquad E_\lambda = 8\pi hc \, \lambda^{-5} \times \frac{\lambda kT}{hc} = \frac{8\pi kT}{\lambda^4}$$

$$\therefore \qquad \boxed{E_\lambda \, d\lambda = \frac{8\pi \, kT}{\lambda^4} \, d\lambda} \qquad \ldots(ii)$$

This is the Rayleigh-Jeans law derived from Planck's law only for long wavelength.

(*ii*) Differentiating w.r.t λ in equation (*i*), we get

$$\frac{dE_\lambda}{d\lambda} = \frac{8\pi hc\,(-5)\,\lambda^{-6}}{e^{\frac{hc}{\lambda kT}}-1} + \lambda^{-5}\,.\left[e^{\frac{hc}{\lambda kT}}-1\right]^{-2}.\,e^{\frac{hc}{\lambda kT}}\,\frac{hc}{kT}\,\lambda^{-2} = 0$$

or, $$e^{\frac{hc}{\lambda kT}} - 1 = \frac{e^{\frac{hc}{\lambda kT}}\,hc}{5\lambda\,kT}$$

or, $$e^{\frac{hc}{\lambda kT}}\left[1-\frac{hc}{5\lambda\,kT}\right] = 1 \qquad \left[\text{Putting } \frac{hc}{\lambda kT} = x\right]$$

or, $$e^{x}\left[1-\frac{x}{5}\right] = 1 \qquad \ldots(iii)$$

We get from this equation, $x = 0$ and 4.965

But $x \neq 0$ as $x = \frac{hc}{\lambda kT}$, which is a finite quantity

i.e., $$\frac{hc}{\lambda kT} = 4.965$$

or $$\lambda\,T = \frac{hc}{k \times 4.965} = \text{Constant} = 2.898 \times 10^{-3}\ \text{m}^\circ\ \text{K}$$

For this wavelength we get the maximum energy emission. It is denoted by λ_m.

$\therefore\quad \lambda_m\,T = \text{Constant} = b$ (say). This is Wien's displacement law.

Problem 2. *If the seen emits maximum energy at wavelength 4753 Å, calculate the temperature of its surface. Wien's constant b = 0.288 cm°C.* [Lucknow University]

Solution. We have from Wien's displacement law.

$$\lambda_m\,T = b \text{ (Wien's constant)}$$

Here given that $\lambda_m = 4753\ \text{Å} = 4753 \times 10^{-8}$ cm

$= 4753 \times 10^{-10}$ m

and $T = ?$ Wien's constant $= 0.288$ cm° C

$= 2.88 \times 10^{-3}$ m° K

$$\therefore\quad T = \frac{b}{\lambda_m} = \frac{2.88 \times 10^{-3}}{4753 \times 10^{-10}} = 6.06 \times 10^{3}$$

$= 6060°$ K

Problem 3. *What is the wavelength of the maximum intensity of radiation radiated from a source at temperature 3000°C ?*

Wien's constant = 2.898 × 10^{-3} m° K [Kanpur University]

Solution. According to Wien's displacement law,

$$\lambda_m\,T = \text{Constant} = b \text{ (say)}$$

Here,
$$T = 3000°C = 3273° K$$
$$b = \text{Wien's constant} = 2.898 \times 10^{-3} \text{ m° K}$$

$$\therefore \quad \lambda_m = \frac{b}{T} = \frac{2.898 \times 10^{-3} \text{ m°K}}{3273}$$
$$= 8.854 \times 10^{-7} \text{ m} = 8854 \times 10^{-10} \text{ m}$$
$$= 8854 \text{ Å}.$$

Wavelength for maximum intensity radiation (λ_m) is 8854 Å.

Problem 4. *The wavelength of the photoelectric threshold of tungsten is 2300 Å. Determine the kinetic energy of electron ejected from the surface by ultraviolet light of wavelength 1800 Å.*

[Delhi University]

Solution. The work function ϕ is given by $h\nu_0 = \frac{hc}{\lambda}$

$$= \frac{6.62 \times 10^{-34} \times 3 \times 10^8}{2300 \times 10^{-10}}$$
$$= 0.86 \times 10^{-18} \text{ J}$$

The energy of the incident radiation is

$$h\nu = \frac{hc}{\lambda} = \frac{6.62 \times 10^{-34} \times 3 \times 10^8}{1800 \times 10^{-10}} = 1.10 \times 10^{-18} \text{ J}$$

Kinetic energy of the ejected electron from the surface is given by

$$E = \frac{1}{2} m v_{max}^2 = h\nu - \phi$$
$$= 1.10 \times 10^{-18} \text{ J} - 0.86 \times 10^{-18} \text{ J}$$
$$= 0.24 \times 10^{-18} \text{ J} = \frac{0.24 \times 10^{-18}}{1.6 \times 10^{-9}} \text{ eV} = 1.5 \text{ eV}.$$

Problem 5. *Is it possible to liberate an electron from a metal surface having work function 4.8* eV *with an incident radiation of wavelength : (i) 2000 Å and (ii) 5000 Å*

Solution. The energy of the incident radiation is $= h\nu = \frac{hc}{\lambda}$

Case (*i*) Here, energy
$$= \frac{6.62 \times 10^{-34} \times 3 \times 10^8}{2000 \times 10^{-10}} \text{ Joules.}$$
$$= \frac{6.62 \times 10^{-16} \times 3}{2000 \times 1.6 \times 10^{-19}} \text{ eV}$$
$$= 6.21 \text{ eV}.$$

This energy is greater than the work function (4.8 eV). So the wavelength (λ = 2000 Å) will be able to liberate an electron.

Case (*ii*) Here the incident energy $E = \frac{hc}{\lambda}$

$$= \frac{6.62 \times 10^{-34} \times 3 \times 10^{8}}{5000 \times 10^{-10} \times 1.6 \times 10^{-19}} \text{ eV}$$

$$= 2.485 \text{ eV}.$$

This energy (2.485 eV) is less than the work function (4.8 eV) for the metal. So the wavelength (λ = 5000 Å) will not be able to liberate an electron.

Problem 6. *A ray of ultraviolet light of wavelength 3000 Å falling on the surface of a material whose work function is 2.28* eV, *ejects an electron. What will be the velocity of the emitted electron ?*

[Calcutta University (Hons)]

Solution. Given that, the work function $\phi = 2.28$ eV

$$= 2.28 \times 1.6 \times 10^{-19} \text{ Joule}$$

$$= 3.648 \times 10^{-19} \text{ J}$$

Wavelength $\lambda = 3000 \text{ Å} = 3000 \times 10^{-10}$ m

$\therefore$ Frequency $\nu = \frac{c}{\lambda} = \frac{3 \times 10^{8} \text{ m}}{3000 \times 10^{-10}} = 10^{15}$ Hz

$\therefore$ Kinetic energy of the electron is given by

$$\frac{1}{2} mv^2 = h\nu - \phi$$

$$= 6.62 \times 10^{-34} \times 10^{15} - 3.648 \times 10^{-19} \text{ J}$$

$$= 2.972 \times 10^{-19} \text{ J}$$

Velocity of the emitted electron $v = \sqrt{\frac{2 \times 2.972 \times 10^{-19}}{9.11 \times 10^{-31}}} = 8.13 \times 10^5 \text{ ms}^{-1}$.

Problem 7. *Light of wavelength 2000 Å falls on a photo-sensitive material having work function 4.2* eV. *What is the kinetic energy of the fastest and slowest photoelectron ? Also calculate the stopping potential. Given* $h = 6.62 \times 10^{-34}$ *Js,* $e = 1.6 \times 10^{-19}$ *C.*

Solution. Given that the work function, $\phi = 4.2$ eV, $\lambda = 2000 \times 10^{-10}$ m.

$\therefore$ Incident energy of photon, $E = \frac{hc}{\lambda} = \frac{6.62 \times 10^{-34} \times 3 \times 10^{8}}{2000 \times 10^{-10} \text{ m}}$

$$= 9.93 \times 10^{-19} \text{ J}$$

and $\phi = 4.2 \times 1.6 \times 10^{-19}$ J

$\therefore$ Kinetic energy of the fastest electron

$$\frac{1}{2} mv^2 = h\nu - \phi = (9.93 \times 10^{-19} - 6.72 \times 10^{-19}) \text{ J}$$

$$= 3.21 \times 10^{-19}\ \text{J} = \frac{3.21 \times 10^{-19}\ \text{J}}{1.6 \times 10^{-19}} = 2\ \text{eV}.$$

The kinetic energy of the slowest electron = 0

Let us consider V to be the stopping potential then

$$h\nu = eV + \phi$$

$$eV = h\nu - \phi = 3.2 \times 10^{-19}\ \text{J}$$

$$V = \frac{3.2 \times 10^{-19}\ \text{J}}{1.6 \times 10^{-19}} = 2\ \text{Volt}.$$

Problem 8. *The work function of a metal is 3.31×10^{-12} erg. Calculate the maximum wavelength of radiation which would start the emission of photoelectrons ($h = 6.62 \times 10^{-27}$ erg sec, $c = 3 \times 10^{10}$ cm/sec.)* [Meerut University]

Solution. Consider ν_0 be the threshold frequency of radiation. Then the threshold energy of the photoelectrons will be $h\,\nu_0$.

∴ Work function $\phi = h\nu_0$

$$= h\frac{c}{\lambda_0}$$

where λ_0 is the threshold wavelength,

$$\frac{1}{\lambda_0} = \frac{\phi}{hc} = \frac{3.31 \times 10^{-12}}{6.62 \times 10^{-27} \times 3 \times 10^{10}}\ /\text{cm}$$

$$= \frac{1}{6} \times \frac{1}{10^{-5}}\ /\text{cm}$$

$$\lambda_0 = 6 \times 10^{-5}\ \text{cm} = 6000\ \text{Å}.$$

The maximum wavelength is 6000 Å.

Problem 9. *Ultraviolet light of wavelength 350 nm and intensity 1 watt/m² is directed at a potassium surface. Find (i) The maximum kinetic energy of photoelectron. (ii) If 0.5% of incident photons produce photoelectrons, how many photoelectrons are emitted per second if the surface of potassium is 1 cm² ? Work function of potassium is 2.1 eV.* [Pune University]

Solution. Given that, the wavelength of light $\lambda = 350$ nm

$$= 350 \times 10^{-9}\ \text{m}.$$

Work function $\phi = 2.1$ eV

$$= 2.1 \times 1.6 \times 10^{-19}\ \text{J}$$

(*i*) The maximum K.E. of the electron

$$\frac{1}{2}mv^2_{max} = h\nu - \phi = \frac{hc}{\lambda} - \phi$$

$$= \frac{6.62 \times 10^{-34} \times 3 \times 10^{8}}{350 \times 10^{-9}} - 2.1 \times 1.6 \times 10^{-19} \text{ J}$$

$$= 2.313 \times 10^{-19} \text{ J}$$

$$= \frac{2.313 \times 10^{-19}}{1.6 \times 10^{-19}} \text{ eV} = 1.45 \text{ eV}.$$

(*ii*) Given that, the energy incident per second on 1 cm^2 surface of potassium = 10^{-4} Joule.

∴ Effective energy per second on 1 cm^2 surface = $\frac{0.5}{100} \times 10^{-4}$ J = 5×10^{-7} J.

∴ Number of photoelectrons emitted per second if the surface of potassium is 1 cm^2.

$$= \frac{5 \times 10^{-7} \text{ J}}{2.313 \times 10^{-19} \text{ J}} = 2.16 \times 10^{12}.$$

Problem 10. *The maximum energy of photoelectrons emitted by a metal surface is 3.62×10^{-19} J when the incident radiation is 3000 Å. When the incident radiant is 5000 Å, the maximum energy of ejected electrons is 0.972×10^{-19} J. Calculate the Planck's constant and the threshold wavelength of metal.*

Solution. According to the photoelectric emission,

$$(\text{K.E.})_{max} = \frac{1}{2} mV_{max}^2 = h\nu - \phi$$

or

$$3.62 \times 10^{-19} \text{ J} = \frac{h \cdot c}{\lambda} - \phi$$

$$= h \cdot \frac{3 \times 10^{8}}{3000 \times 10^{-10}} - \phi \quad \ldots(i)$$

Again,

$$0.972 \times 10^{-19} \text{ J} = h \cdot \frac{3 \times 10^{8}}{5000 \times 10^{-10}} - \phi \quad \ldots(ii)$$

∴ Solving equations (*i*) and (*ii*), we get

$$h = 6.62 \times 10^{-34} \text{ J sec.}$$

and

$$\phi = 3 \times 10^{-19} \text{ J.}$$

But we know that the work function $\phi = \frac{hc}{\lambda_0}$ where λ_0 is the threshold wavelength.

$$\lambda_0 = \frac{hc}{\phi} = \frac{6.62 \times 10^{-34} \times 3 \times 10^{8}}{3 \times 10^{-19}} = 6.62 \times 10^{-7} \text{ m}$$

$$= 6620 \text{ Å}.$$

Problem 11. *An X-ray photon is found to have doubled its wavelength on being scattered by 90°. Find the energy and wavelength of incident photon.* [Lucknow University]

Solution. Wavelength shift after collision is

$$d\lambda = \frac{h}{m_0 c}(1 - \cos\alpha) \text{ where } \alpha \text{ is the scattering angle.}$$

When $\alpha = 90°,\ d\lambda = \dfrac{h}{m_0 c} = \dfrac{6.62 \times 10^{-34}}{9.1 \times 10^{-31} \times 3 \times 10^{8}} = 0.0242 \times 10^{-10}$ m

If λ is the incident photon, then wavelength of the scattered photon is $\lambda + d\lambda$.

According to the conditions,

$$\lambda + d\lambda = 2\lambda$$

$$d\lambda = \lambda = 0.0242 \times 10^{-10} \text{ m.}$$

Energy of the incident photon $= h\nu$

$$= \frac{hc}{\lambda} = \frac{6.62 \times 10^{-34} \times 3 \times 10^{8}}{0.0242 \times 10^{-10}} \text{ J}$$

$$= \frac{19.9 \times 10^{-16}}{0.0242 \times 1.6 \times 10^{-19}} \text{ eV}$$

$$= 0.51 \times 10^{6} \text{ eV} = 0.51 \text{ MeV.}$$

Problem 12. *An x-ray photon of wavelength 0.1 Å is reflected at an angle of 90° with its original direction after collision with an electron at rest. Find the energy it loses on collision.*

Solution. Here, the incident wavelength $\lambda = 0.1$ Å $= 0.1 \times 10^{-8}$ cm

$$\alpha = 90,$$

∴ Wavelength shift after the collision is

$$d\lambda = \frac{h}{m_0 c}(1 - \cos 90°) = \frac{h}{m_0 c} = 0.024 \text{ Å}$$

∴ Wavelength of the scattered radiation

$$\lambda' = \lambda + d\lambda = 0.1 \text{ Å} + 0.024 \text{ Å} = 0.124 \text{ Å}$$

Energy loss in collision is

$$\Delta E = h\nu - h\nu' = \frac{hc}{\lambda} - \frac{hc}{\lambda'} = h\left(\frac{3 \times 10^{8}}{0.1 \times 10^{-10}} - \frac{3 \times 10^{8}}{0.124 \times 10^{-10}}\right)$$

$$= 6.6 \times 10^{-34}(3 - 2.149) \times 10^{19} \text{ J}$$

$$= \frac{6.6 \times 10^{-34} \times 0.851 \times 10^{19}}{1.6 \times 10^{-19}} \text{ eV} = 2.39 \times 10^{4} \text{ eV}$$

$$= 23.9 \text{ KeV.}$$

Problem 13. *(a) Find the change in wavelength of X-ray photon when it is scattered through an angle of (i) 90° ; (ii) 60° ; (iii) 45° and (iv) 180°.*

(b) X-rays of wavelength 0.2 Å are scattered from a target, calculate the wavelength of X-ray scattered through 45°. Also find the maximum kinetic energy of the recoil electron.

Solution. (*a*) Change in wavelength after collision is $d\lambda = \frac{h}{m_0 c}(1 - \cos \alpha)$ where α is the angle of scattering.

(*i*) When $\alpha = 90°$; $d\lambda = \frac{h}{m_0 c} = 0.024$ Å

(*ii*) When $\alpha = 60°$; $d\lambda = \frac{h}{m_0 c}(1 - \cos 60°) = 0.024 \text{ Å} \times \frac{1}{2} = 0.012$ Å.

(*iii*) When $\alpha = 45°$; $d\lambda = \frac{h}{m_0 c}(1 - \cos 45°) = 0.024 \text{ Å} \times \left(1 - \frac{1}{\sqrt{2}}\right) = 0.0071$ Å.

(*iv*) When $\alpha = 180°$; $d\lambda = \frac{h}{m_0 c}(1 - \cos 180°) = 0.024 \text{ Å} \times 2 = 0.048$ Å.

(*b*) Given, wavelength of X-rays $\lambda = 0.2$ Å.

We know that the change in wavelength for scattering angle $\alpha = 45°$, $d\lambda = 0.007$ Å,

∴ Wavelength of scattered X-rays

$$\lambda' = \lambda + d\lambda$$
$$= 0.2 \text{ Å} + 0.007 \text{ Å}$$
$$= 0.207 \text{ Å}.$$

Maximum kinetic energy of the recoil electron is

$$E = h\nu - h\nu' = \frac{hc}{\lambda} - \frac{hc}{\lambda'}$$

Therefore we can say, when λ' is maximum, then K.E. is maximum. λ' will be maximum when $\alpha = 180°$.

$$\lambda' = \lambda + d\lambda = 2 \text{ Å} + 0.048 \text{ Å} = 2.048 \text{ Å},$$

$$\therefore \quad \text{Maximum K.E.} = hc\left(\frac{1}{\lambda} - \frac{1}{\lambda'}\right)$$

$$= 6.62 \times 10^{-34} \left\{\frac{1}{0.2 \times 10^{-10}} - \frac{1}{0.248 \times 10^{-10}}\right\} \text{ J}$$

$$= 1.94 \times 10^{-15} \text{ J}.$$

Problem 14. *(a) An incident photon of frequency* ν *gets scattered through an angle* ϕ *from an electron at rest. Obtain an expression for the scattered photon energy.* [H.P. University].

(b) A photon of frequency ν *is scattered through an angle of 90° in compton scattering. Show that the scattering frequency is*

$$\nu' = \left[\frac{m_0 c^2}{h\nu + m_0 c^2}\right] \nu$$ [Pune University]

Solution. (*a*) From compton scattering, the change in wavelength $d\lambda = \dfrac{h}{m_0 c}(1 - \cos\phi)$

Let the energy of the incident photon $E = h\nu$
and energy of the scattered photon $E' = h\nu'$

$$\therefore \qquad d\lambda = \frac{c}{\nu'} - \frac{c}{\nu} = \frac{h}{m_0 c}(1 - \cos\phi)$$

or,
$$\nu - \nu' = \frac{h\nu\nu'}{m_0 c^2}(1 - \cos\phi)$$

or,
$$h(\nu - \nu') = \frac{h\nu \, . \, h\nu'}{m_0 c^2}(1 - \cos\phi)$$

or,
$$E - E' = \frac{EE'}{m_0 c^2}(1 - \cos\phi)$$

or,
$$E = E'\left\{1 + \frac{E(1 - \cos\phi)}{m_0 c^2}\right\}$$

or,
$$E' = \frac{E}{1 + \dfrac{E(1 - \cos\phi)}{m_0 c^2}}$$

or,
$$\boxed{E' = \frac{h\nu}{1 + \dfrac{2h\nu}{m_0 c^2} \, . \sin^2 \dfrac{\phi}{2}}}$$

This is the expression for the scattered photon energy.

(*b*) We have the energy of the scattered photon as

$$E' = \frac{E}{1 + \dfrac{E}{m_0 c^2}(1 - \cos\phi)}$$

Here $\qquad E' = h\nu', \; E = h\nu$ and $\phi = 90°$

$$h\nu' = \frac{h\nu}{1 + \dfrac{h\nu}{m_0 c^2}(1 - \cos 90°)}$$

or,
$$\nu' = \left[\frac{m_0 c^2}{m_0 c^2 + h\nu}\right]\nu$$

Problem 15. *Find the de Broglie wavelength associated with*

(i) A 46 gm golf ball with velocity 36 m / sec.
(ii) An electron with a velocity 10^7 m/sec.
Which of these two show wave character and why ? [Garhwal University]

Solution. According to de Broglie, the wavelength

$$\lambda = \frac{h}{p} = \frac{h}{mv}, \qquad h = \text{Planck's Constant}$$

$$mv = \text{momentum}$$

Case (*i*) Here, $v << c$ *i.e.*, we say $m = m_0$

$$\therefore \qquad \lambda = \frac{6.62 \times 10^{-34}\ \text{J.s}}{(0.046\ \text{kg}) \times (36\ \text{m/sec})} \simeq 4 \times 10^{-34}\ \text{meter}.$$

So, the wavelength associated with the golf ball is smaller as compared to its dimension. Therefore no wave aspect can be expected in its behaviour.

Case (*ii*) Again,

$$\lambda = \frac{h}{mv} = \frac{6.62 \times 10^{-34}\ \text{Js}}{\left(9.11 \times 10^{-31}\right) \times 10^{7}} = 7.3 \times 10^{-11}\ \text{m}.$$

This wavelength can be compared with the atomic dimensions. Therefore a moving electron may exhibit wave character.

Problem 16. *What voltage must be applied to an electron microscope to produce electrons of wavelength 0.40 Å ?* [Burdwan University]

Solution. According to de Broglie, the wavelength

$$\lambda = \frac{h}{p}$$

Kinetic energy
$$E_K = \frac{1}{2} mv^2 = 0 = \frac{m^2 v^2}{2m} = \frac{p^2}{2m}$$

or
$$p = \sqrt{2m\,E_K}$$

$$\therefore \qquad \lambda = \frac{h}{\sqrt{2m\,E_K}}$$

If a charged particle carrying a charge e is accelerated through a potential difference V volts, then kinetic energy

$$E_K = eV$$

$$\therefore \qquad \lambda = \frac{h}{\sqrt{2m\,eV}}$$

or, $$0.40 \times 10^{-10}\, m = \frac{6.62 \times 10^{-34}}{\sqrt{2 \times 9.11 \times 10^{-31} \times 1.6 \times 10^{-19}}} \times \frac{1}{\sqrt{V}}$$

or, $$\sqrt{V} = 30.6$$

or $$V = 936.3 \text{ Volt.}$$

Problem 17. *(a) Show that the de Broglie wavelength of a particle of rest mass m_0 and kinetic energy E_K is given by*

$$\lambda = \frac{h}{\sqrt{2m_0\, E_K}} \left(1 + \frac{E_K}{2m_0 c^2}\right)^{-\frac{1}{2}}$$ [North Bengal University]

(*b*) Hence find the de Broglie wavelength of an electron having an energy of 1 MeV.

Solution. Relativistically, the total energy E of the electron is given by $E = \sqrt{p^2c^2 + m_0^2\, c^4}$ where p is the momentum and m_0 is the rest mass.

The kinetic energy is $E_K = E - m_0\, c^2 = \sqrt{p^2c^2 + m_0^2 c^4} - m_0 c^2$

or, $$p^2c^2 + m_0^2\, c^4 = E_K^2 + 2E_K\, m_0\, c^2 + m_0^2\, c^4$$

or $$p = \frac{\sqrt{E_K\left(E_K + 2m_0 c^2\right)}}{c}$$

Therefore, the de Broglie wavelength is

$$\lambda = \frac{h}{p} = \frac{hc}{\sqrt{E_K\left(E_K + 2m_0 c^2\right)}}$$

$$= \frac{h}{\sqrt{2m_0\, E_K}} \left(1 + \frac{E_K}{2m_0 c^2}\right)^{-\frac{1}{2}}$$

Problem 18. *Find the de Broglie wavelength of a neutron of energy 12.8 MeV.*
Given : Mass of neutron = 1.675×10^{-27} kg, h = 6.62×10^{-34} Js.
c = 3×10^8 m/sec and 1 eV = 1.6×10^{-19} J [Meerut University]

Solution. The rest energy of the electron is given by

$$m_0\, c^2 = 1.675 \times 10^{-27}\, kg \times (3 \times 10^8)^2 \text{ J}$$

$$= \frac{1.675 \times 9 \times 10^{-11}}{1.6 \times 10^{-19} \times 10^6} \text{ MeV}$$ [1 MeV = 10^6 eV]

$$= 942 \text{ MeV.}$$

But the given energy is 12.8 MeV. So this energy is much smaller than the rest energy. Therefore, we can neglect relativistic effect.

$$\text{According to de Broglie } \lambda = \frac{h}{p} = \frac{h}{\sqrt{2m\,eV}}$$

$$= \frac{6.62 \times 10^{-34} \text{ Js}}{\sqrt{2 \times 1.675 \times 10^{-27} \times 12.8 \times 10^{-6}}} = 0.8 \times 10^{-14} \text{ m}$$

$$= 0.8 \times 10^{-4} \text{ Å}$$

Problem 19. *A proton is confined in a nucleus of radius 5×10^{-15} m. Calculate the minimum uncertainty in its momentum. Also calculate the minimum kinetic energy the proton should have. The proton mass is 1.67×10^{-27} kg.* [Rajasthan University].

Solution. The minimum uncertainty in its position $\Delta x = 2 \times 5 \times 10^{-15}$ m $= 10^{-14}$ m.

According to uncertainty principle, its momentum Δp is given by

$$\Delta p\, \Delta x = \hbar = \frac{h}{2\pi}$$

or

$$\Delta p = \frac{6.62 \times 10^{-34} \text{ Js}}{2 \times 3.14 \times 10^{-14} \text{ m}} = 1.05 \times 10^{-20} \text{ kg.m/sec.}$$

Minimum uncertainty of proton's momentum is 1.05×10^{-20} kg m/sec.

∴ Minimum Kinetic Energy

$$= \frac{p^2}{2m} = \frac{(\Delta p)^2}{2m} = \frac{(1.05 \times 10^{-20})^2}{2 \times 1.67 \times 10^{-27}}$$

$$= 3.3 \times 10^{-14} \text{ J} = \frac{3.3 \times 10^{-14}}{1.6 \times 10^{-9}} \text{ eV} = 0.206 \text{ MeV.}$$

Problem 20. *An electron is confined to a box of length 10^{-9} m. Calculate the minimum uncertainty in the measurement of its velocity.* [Calcutta University].

Solution. From Heisenberg's position-momentum uncertainty

$$\Delta x\, \Delta p_x \approx h$$

When Δp_x is minimum, Δx is maximum,

$$\therefore \quad (\Delta p_x)_{\min} \cdot (\Delta x)_{\max} = h$$

Here

$$(\Delta x)_{\max} = 10^{-9} \text{ m}$$

$$\therefore \quad (\Delta p_x)_{\min} = \frac{6.62 \times 10^{-34}}{10^{-9} \text{ m}} \text{ Js}$$

$$= 6.62 \times 10^{-25} \text{ kg m/sec.}$$

or

$$m\,(\Delta v_x)_{\min} = 6.62 \times 10^{-25}$$

$$\text{Minimum Velocity} = \Delta v_x = \frac{6.62 \times 10^{-25}}{9.11 \times 10^{-31}} = 7.27 \times 10^5 \text{ m/sec.}$$

Problem 21. *If the uncertainty in the location of a particle is equal to its de Broglie wavelength, what is the uncertainty in its velocity ?*

Solution. We have $\Delta p \, \Delta x = h$ or $\Delta (mv) \, . \, \Delta x - h$

or
$$\Delta v = \frac{h}{m \, \Delta x}$$

According to this Problem, $\Delta x = \lambda$

$$\therefore \quad \Delta v = \frac{h}{m\lambda} = \frac{p}{m} = \frac{mv}{m} = v.$$

So the uncertainty in the velocity is the velocity itself.

Problem 22. *A particle of mass 10^{-6} gm has a speed of 1 m/sec. If the speed is uncertain by 0.01%. What is the minimum uncertainty in the position of the particle.* [Pbi. University]

Solution. Here, given $m = 10^{-6}$ gm $= 10^{-9}$ kg.

$$v = 1 \text{ m/sec.} \quad \therefore \quad \Delta v = 1 \times \frac{0.01}{100} = 0.0001 = 10^{-4}$$

From the uncertainty principle,

$$\Delta p \, \Delta x = h$$

or
$$m \, \Delta v \, \Delta x = h$$

or
$$\Delta x = \frac{h}{m \, \Delta v} = \frac{6.62 \times 10^{-34} \text{ Js}}{10^{-9} \times 10^{-4}} = 6.62 \times 10^{-21} \text{ m.}$$

Problem 23. *An electron of mass 9.1×10^{-31} Kg has a speed of 1 km/sec. with an accuracy of 0.05%. Calculate the uncertainty with which the position of the electron can be located.*

[Pune University]

Solution. Given $m = 9.1 \times 10^{-31}$ kg,

$$v = 1 \text{ km/sec.} = 10^3 \text{ m/sec.}$$

$$\therefore \quad \Delta v = 10^3 \times \frac{0.05}{100} = 0.5 \text{ m/sec.}$$

$$\text{Uncertain of the position } \Delta x = \frac{h}{m \, \Delta v} = \frac{6.62 \times 10^{-34}}{9.1 \times 10^{-31} \times 0.5}$$

$$= 1.45 \times 10^{-3} \text{ m.}$$

Problem 24. *The life time of an excited state of an atom is about 10^{-8} sec. Calculate the minimum uncertainty in the determination of the energy of the excited state.*

Solution. From Heisenberg's time energy uncertainty principle,

$$\Delta E \,.\, \Delta t \geq \frac{h}{2\pi}$$

$$\therefore \qquad \Delta E \geq \frac{h}{2\pi\, \Delta t} \qquad [\Delta t = 10^{-8} \text{ sec}]$$

$$\geq \frac{6.62 \times 10^{-34}}{2 \times 3.14 \times 10^{-8}} \geq 1.054 \times 10^{-26} \text{ J}$$

$$\Delta E = 6.5 \times 10^{-18} \text{ eV}$$

Uncertainty in the energy of the excited state is 6.5×10^{-8} eV.

Problem 25. *Life time of a nucleus in the excited state is 10^{-12} sec. Calculate the probable uncertainty in energy and frequency of γ ray photon emited by it.* [Lucknow University]

Solution. Given Δt = Life time of a nucleus in the excited state

$= 10^{-12}$ sec.

The uncertainty in the energy, ΔE, of nucleus is given by

$$\Delta E \geq \frac{h}{2\pi\, \Delta t}$$

$$\geq \frac{6.62 \times 10^{-34} \text{ Js}}{2 \times 3.14 \times 10^{-12} \text{ sec}} = 1.054 \times 10^{-22} \text{ J}$$

Now $\qquad E = h\nu$

$$\therefore \text{ Uncertainty in frequency } \Delta \nu = \frac{\Delta E}{h} = \frac{1.054 \times 10^{-22} \text{ J}}{6.62 \times 10^{-34} \text{ Js}}$$

$$= 1.59 \times 10^{11} \text{ Hz}$$

$$= 1.59 \times 10^{5} \text{ MHz.}$$

Problem 26. *Normalize the one-dimensional wave function given by*

$$\psi(x) = A \sin\left(\frac{\pi x}{a}\right) \qquad 0 < x < a$$

$$\psi(x) = 0 \quad \text{outside.}$$

[H.P. University]

Solution. If $\int_{-\infty}^{\infty} \psi(x)\, \psi^*(x)\, dx = 1$ ...(*i*)

the wave function satisfying the above equation is said to be normalized.

Here $\psi(x)$ exists only in the range $0 < x < a$ and is zero outside. So we rewrite the equation like

$$\int_0^a \psi(x)\, \psi^*(x)\, dx = 1$$

$$\text{or,} \quad \int_0^a A \sin\left(\frac{\pi x}{a}\right) A^* \sin\left(\frac{\pi x}{a}\right) dx = 1$$

or $$AA^* \int_0^a \sin^2\left(\frac{\pi x}{a}\right) dx = 1$$

or, $$\frac{AA^*}{2}\left[\int_0^a dx - \int_0^a \cos\frac{2\pi x}{a}\, dx\right] = 1 \qquad [AA^* = A^2]$$

or $$\frac{A^2}{2}[a - 0] = 1$$

or $$A = \sqrt{\frac{2}{a}}.$$

So the normalized one dimensional wave function is

$$\psi(x) = \sqrt{\frac{2}{a}} \sin\left(\frac{\pi x}{a}\right)$$

Problem 27. *The ground state and the first excited state wave functions of an atom are ψ_0 and ψ_1, respectively, the corresponding energies being E_0 and E_1. If the system has a 40% probability of being found in the ground state and 60% probability of being found in the first excited state,*

(i) What is the wave function of the atom ? [Calcutta University]

(ii) What is the average energy of the atom ?

Solution. (*i*) Suppose, the wave function of the atom be

$\psi = C_0\psi_0 + C_1\psi_1$; where C_0 and C_1 are constants.

Since the wave functions are orthonormal. Then

$$\int \psi\psi^* \, dv = 1$$

or, $$\int (C_0\psi_0 + C_1\psi_1)(C_0^*\psi_0^* + C_1^*\psi_1^*)\, dv = 1$$

or, $$C_0 C_0^* \int \psi_0\psi_0^* \, dv + C_1 C_1^* \int \psi_1\psi_1^* \, dv = 1$$

or, $$C_0 C_0^* + C_1 C_1^* = 1$$

According to the given conditions, the probability for the ground state is 0.4

i.e., $C_0 C_0^* = 0.4$ and the probability for the first excited state is 0.6., *i.e.*, $C_1 C_1^* = 0.6$. So we can say, the wave function of the atom is

$$\psi = \sqrt{0.4}\,\psi_0 + \sqrt{0.6}\,\psi_1.$$

(*ii*) If E_0 and E_1 are the energy eigen values for the eigen functions ψ_0 and ψ_1 for the ground state and for the first excited state respectively, the average energy of the atom is

$$E_{av} = 0.4\, E_0 + 0.6\, E_1.$$

Problem 28. *(a) Determine the energy levels and the corresponding normalized eigen functions of a particle in one dimensional potential well of the form*

$$V(x) = \infty \text{ for } x < 0 \text{ and for } x > a$$
$$= 0 \text{ for } 0 < x < a$$

(b) (i) What are the boundary conditions for the problem ?
(ii) Is the wave function continuous everywhere ?
(iii) Is the derivative of the wave function continuous everywhere ?
(c) Show that :

$$(\Delta x)^2 = \langle x^2 \rangle - \langle x \rangle^2 = \frac{a^2}{12}\left(1 - \frac{6}{x^2\pi^2}\right)$$

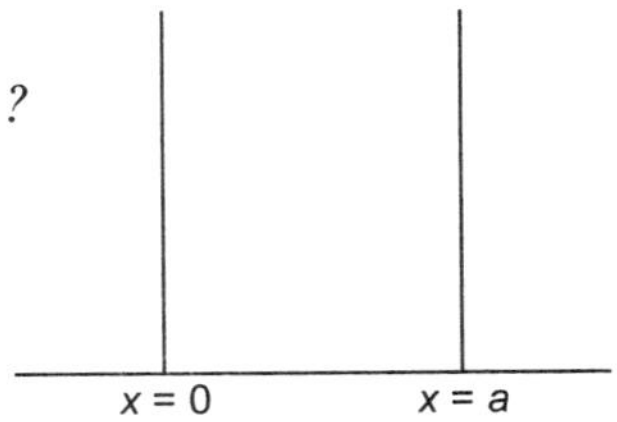

One dimensional potential well

Solution. (*a*) According to the Schrödinger wave equation,

$$\frac{d^2\psi}{dx^2} + \frac{2m}{\hbar^2} E\psi = 0$$

or, $$\frac{d^2\psi}{dx^2} + \alpha^2\psi = 0, \text{ where } \alpha^2 = \frac{2mE}{\hbar^2} \quad \ldots(i)$$

$$V(x) = \infty, \quad x < 0 \text{ and } x > a$$
$$= 0 \quad 0 < x < a.$$

The solution of the above equation is

$$\psi = A\cos\alpha x + B\sin\alpha x \quad \ldots(ii)$$

The boundary conditions are

$$\text{at } x = 0, \psi = 0$$
$$\text{at } x = a, \psi = 0$$

Applying these conditions in equation (*ii*)

$$A = 0, \quad \text{and} \quad B\sin\alpha x = 0, \quad [\because B \neq 0]$$

or $$\sin\alpha x = 0 = \sin n\pi$$

or $$\alpha = \frac{n\pi}{a}$$

or $$\alpha^2 = \frac{n^2\pi^2}{a^2} = \frac{2mE}{\hbar^2}$$

or, $$E = \frac{n^2\pi^2\hbar^2}{2ma^2}$$

$\therefore$ $$\psi = B\sin\alpha x.$$

Applying the normalizing condition, the value of B can be obtained.

i.e.,
$$\int \psi^* \psi \, dx = 1$$

or,
$$\int_0^a B^2 \sin^2 \alpha x \, dx = 1$$

or
$$\frac{B^2}{2} \int_0^a [1 - \cos 2\alpha x] \, dx = 1$$

or,
$$\frac{B^2}{2} [a - 0] = 1$$

or
$$B = \sqrt{\frac{2}{a}}$$

Therefore, the normalized wave function is

$$\psi = \sqrt{\frac{2}{a}} \sin \frac{2\pi x}{a}.$$

(*b*) (*i*) The boundary conditions are

$$\psi = 0 \text{ at } x = 0$$
$$\psi = 0 \text{ at } x = a.$$

Because ψ vanishes at the surface of infinite potential.

(*ii*) Wave functions are continuous except at $x = 0$ and $x = a$.

(*iii*) Derivatives of wave function are continuous except at $x = 0$ and $x = a$.

(*c*)
$$\langle x \rangle = \int \psi_n^* (x) \, x \, \psi_n (x) \, dx$$

$$= \int_0^a B^2 \, x \sin^2 \frac{n\pi x}{a} \, dx = \frac{2}{a} \int_0^a x \sin^2 \frac{n\pi x}{a} \, dx$$

$$= \frac{2}{a} \left[\left\{ \int \sin^2 \frac{n\pi x}{a} \, dx \right\}_0^a - \int_0^a \sin^2 \frac{n\pi x}{a} \, dx \right]$$

$$= \frac{2}{a} \left[\left\{ \frac{x}{2} \int_0^a \left(1 - \cos \frac{2n\pi x}{a} \right) dx \right\} - \int_0^a \frac{1}{2} \left(x - \frac{a}{2n\pi} \sin \frac{2n\pi x}{a} \right) dx \right]$$

$$= \frac{2}{a} \left[\frac{a^2}{2} - \frac{a^2}{4} \right] = \frac{2}{a} \cdot \frac{a^2}{4} = \frac{a}{2}.$$

and
$$\langle x^2 \rangle = \int \psi_n^* (x) \, x^2 \, \psi_n (x) \, dx$$

$$= B^2 \int_0^a x^2 \sin^2 \frac{n\pi x}{a} \, dx$$

$$= \frac{2}{a} \int_0^a x^2 \sin^2 \frac{n\pi x}{a} \, dx$$

Integrating by parts, we get

$$= \frac{2}{a}\left(\frac{a^3}{6} - \frac{a^3}{4n^2\pi^2}\right)$$

Now $$(\Delta x)^2 = \langle x^2 \rangle - \langle x \rangle^2$$

$$= \left(\frac{a^2}{3} - \frac{a^2}{2n^2\pi^2}\right) - \frac{a^2}{4} = \frac{a^2}{12} - \frac{a^2}{2n^2\pi^2}$$

$$= \frac{a^2}{12}\left(1 - \frac{6}{n^2\pi^2}\right).$$

Problem 29. *Calculate the value of the lowest energy of an electron in one dimensional force free region of length 4 Å.* [H.P. University]

Solution. For a particle in one dimensional force-free region of length l, the permitted electron energies are given by

$$E_n = \frac{n^2\pi^2\hbar^2}{2ml^2} = \frac{n^2\pi^2 h^2}{8ml^2\pi^2} \qquad \left[\because \hbar = \frac{h}{2\pi}\right]$$

The lowest energy state is for $n = 1$.

$$\therefore \qquad E_1 = \frac{1^2 h^2}{8ml^2} = \frac{(6.62 \times 10^{-34})^2}{8 \times 9.11 \times 10^{-31} \times (4 \times 10^{-10})^2}$$

$$= \frac{3.78 \times 10^{-19} \text{ J}}{1.6 \times 10^{-19}} = 2.36 \text{ eV}.$$

Problem 30. *The minimum energy possible for a particle entrapped in a one dimensional box is 3.2×10^{-18} J. What are the next three energies in eV, the particle can have ?* [Pune University]

Solution. For a particle in one dimensional force-free region of length l, the permitted energies are given by

$$E_n = \frac{n^2\pi^2\hbar^2}{2ml^2} = \frac{n^2h^2}{8ml^2} \qquad \left[\because \hbar = \frac{h}{2\pi}\right]$$

i.e., $$E_n \propto n^2,$$

For $n = 1$, the minimum energy $E_1 = 3.2 \times 10^{-18}$ J

$$= \frac{3.2 \times 10^{-8}}{1.6 \times 10^{-19}} = 20 \text{ eV}$$

Now, the next three energies are

$$E_2 = 2^2 \times 20 \text{ eV} = 80 \text{ eV}$$
$$E_3 = 3^2 \times 20 \text{ eV} = 180 \text{ eV}$$
$$E_4 = 4^2 \times 20 \text{ eV} = 320 \text{ eV}.$$

Problem 31. *A particle is moving in a one dimensional box of infinite height and width 10 Å. Calculate the probability of finding the particle within an interval of 4 Å at the centre of the box, when it in the state of least energy.* [Pune University].

Solution. The wave function of the particle in a one dimensional box of length l is given by

$$\psi_n = \sqrt{\frac{2}{l}} \sin \frac{n\pi x}{l}$$

or,

$$\psi_1 = \sqrt{\frac{2}{l}} \sin \frac{\pi x}{l}$$

The probability of finding the particle is unit interval at the centre of the box $\left(\frac{l}{2}\right)$ is

$$P = \psi_1^* \psi_1 = \psi_1^2 = \frac{2}{l} \sin^2 \frac{\pi x}{l} = \frac{2}{l} \sin^2 \left(\frac{\pi}{l} \cdot \frac{l}{2}\right)$$

$$= \frac{2}{l} \sin^2 \frac{\pi}{2} = \frac{2}{l}$$

The probability of finding the particle within an interval of Δx at the center of the box

$$W = P\,\Delta x = \psi_1^2\,\Delta x = \frac{2}{l} \times 4 \text{ Å}$$

$$= \frac{8 \text{ Å}}{10 \text{ Å}} = \frac{4}{5} = 0.8 \qquad [\because \Delta x = 4 \text{ Å} ; \text{ and } l = 10 \text{ Å}]$$

Problem 32. *Find the probability that a particle in a one dimensional box of length L can be found between 0.4 L to 0.6 L*

(i) for the ground state

(ii) for the first excited state

(iii) for the second excited state. [Pune University, G.N.D. University]

Solution. The wave function of the particle in one dimensional box of length L is given by

$$\psi_n = \sqrt{\frac{2}{L}} \sin \frac{n\pi x}{L}$$

We put for the ground state, first excited state and second excited states $n = 1$, $n = 2$ and $n = 3$ respectively.

Probability of finding the particle within an interval Δx about the mean position x is

$$W_n = P_n \Delta x = \psi_n^* \psi_n \Delta x$$

$$= \frac{2}{L} \sin^2 \frac{n\pi x}{L} \Delta x$$

Here mean position $$x = \frac{0.4\,L + 0.6\,L}{2} = 0.5\,L = \frac{1}{2} L$$

and $$\Delta x = 0.6\,L - 0.4\,L = 0.2\,L$$

Case (*i*) The probability for the ground state ($n = 1$)

$$= \frac{2}{L} \sin^2 \left(1 . \frac{\pi}{L} . \frac{L}{2}\right) . 0.2\,L \;= 0.4 \sin^2 \left(\frac{\pi}{2}\right)$$

$$= 0.4.$$

Case (*ii*) The probability for the 1st excited state ($n = 2$)

$$= \frac{2}{L} \sin^2 \left(2 . \frac{\pi}{2}\right) . 0.2\,L = 0$$

Case (*iii*) The probability for the 2nd excited state ($n = 3$)

$$= \frac{2}{L} \sin^2 \left(3 . \frac{\pi}{2}\right) . 0.2\,L = 0.4.$$

Problem 33. *(a) Find the expectation value $\langle x \rangle$ of the position of a particle trapped in a box L wide.*

(b) Find the expectation value $\langle p_x \rangle$ of the momentum of a particle trapped in a one dimensional box L wide.

Solution. Expectation value $$\langle x \rangle = \int_{-\infty}^{\infty} \psi_n^* \psi_n \, dx$$

$$= \int_{-\infty}^{\infty} x \,|\, \psi_n \,|^2 \, dx$$

Here $$\psi_n = \sqrt{\frac{2}{L}} \sin \frac{n\pi x}{L}$$

(*a*) $$\langle x \rangle = \frac{2}{L} \int_0^L x \sin^2 \frac{n\pi x}{L} \, dx$$

$$= \frac{2}{L} \left[\int_0^L \frac{x}{2} . 2 \sin^2 \frac{n\pi x}{L} \, dx \right]$$

$$= \frac{2}{L} \left[\int_0^L \frac{x}{2} \left(1 - \cos \frac{2n\pi x}{L} \, dx \right) \right]$$

or, $$\langle x \rangle = \frac{2}{L}\left[\frac{x^2}{4} - \frac{x \sin\left(\frac{2\pi n x}{L}\right)}{\frac{4n\pi}{L}} - \frac{\cos\left(\frac{2n\pi x}{L}\right)}{8\left(\frac{n\pi}{L}\right)^2}\right]_0^L$$

$$= \frac{2}{L}\cdot\frac{L^2}{4} = \frac{L}{2}$$

(*b*) Expectation value $\langle p_x \rangle$ of the momentum

$$\langle p_x \rangle = \int_{-\infty}^{\infty} \psi_n^* \left(-\frac{ih}{2\pi}\frac{d}{dx}\right)\psi_n \, dx$$

$$= -\frac{ih}{2\pi}\cdot\frac{2}{L}\cdot\frac{n\pi}{L}\int_0^L \sin\frac{n\pi x}{L} \,.\, \cos\frac{n\pi x}{L}\, dx$$

$$\left[\because \frac{d\psi_n}{dx} = \frac{d}{dx}\left\{\sqrt{\frac{2}{L}}\sin\frac{n\pi x}{L}\right\} = \frac{n\pi}{L}\cdot\sqrt{\frac{2}{L}}\cos\frac{n\pi x}{L}\right]$$

$$= -i\frac{nh}{L^2}\,.\,0 = 0.$$

Problem 34. *The wave function of a particle moving in a potential free region is given by $\psi(x) = A \cos kx$, where A and k are real constants. Is ψ an eigen state of the operators $\hat{H}$, $\hat{P}_x$ and $\hat{P}_x^2$? If so, find the corresponding eigen values.*

Solution. Given $\psi(x) = A \cos kx$

$$\hat{H}\psi = -\frac{\hbar^2}{2m}\frac{d^2}{dx^2}\psi(x) = -\frac{h^2}{8\pi^2 m}\cdot\frac{d}{dx}(-Ak\sin kx) = \frac{h^2}{8\pi^2 m}k^2\psi(x)$$

$$= \frac{\hbar^2 k^2}{2m}\psi(x). \qquad \left[\because \hbar = \frac{h}{2\pi}\right]$$

The eigen value of $\hat{H}$ is $\frac{\hbar^2 k^2}{2m}$.

$$\hat{P}_x\psi = -i\hbar\frac{d\psi(x)}{dx} = -i\hbar\frac{d}{dx}(A\cos kx) = i\hbar\, kA\sin kx$$

But it is not an eigen state of $\hat{P}_x$. So $\hat{P}_x\ \psi$ has no eigen value.

$$\hat{P}_x^2\,\psi(x) = -\hbar^2\frac{d^2\psi(x)}{dx} = -\hbar^2\frac{d}{dx}(-Ak\sin kx)$$

$$= \hbar^2 k^2 A\cos kx = \hbar^2 k^2\psi(x)$$

So eigen value of $\hat{P}_x^2$ is $\hbar^2 k^2$.

Problem 35. *The operator* $\left(x + \frac{d}{dx}\right)$ *has eigen value* α. *Derive the corresponding eigen function.*

Solution. Let, ψ be an eigen function, then we can write according to the condition

$$\left(x + \frac{d}{dx}\right)\psi = \alpha\psi$$

or
$$\frac{d\psi}{dx} = -(x - \alpha)\,\psi$$

or
$$\frac{d\psi}{\psi} = -(x - \alpha)\,dx$$

$$\ln \psi = -\left(\frac{x^2}{2} - \alpha x\right) + c$$

when $x = 0,\ \psi = \psi_0$

$\therefore$
$$\ln \psi_0 = c$$

Therefore,
$$\ln \frac{\psi}{\psi_0} = -\left(\frac{x^2}{2} - \alpha x\right)$$

$\therefore$
$$\psi = \psi_0 \exp\left[-\left(\frac{x^2}{2} - \alpha x\right)\right].$$

Problem 36. *If* $\psi_1\,(x, t)$ *and* $\psi_2\,(x, t)$ *are both solutions of Schrödinger's wave equation for a given potential* $V\,(x, t)$, *show that* $\psi = a_1\,\psi_1 + a_2\,\psi_2$ *in which* a_1 *and* a_2 *are arbitrary constants, is also a Solution.* [Pune University]

Solution. According to time dependent Schrödinger's wave equation

$$-\frac{\hbar^2}{2m}\frac{\partial^2\psi}{\partial x^2}(x,t) + V\,(x,t)\,\psi\,(x,t) = i\hbar\,\frac{\partial\psi}{\partial t}(x,t) \qquad \ldots(i)$$

$\psi_1\,(x, t)$ and $\psi_2\,(x, t)$ are both solutions of equation (*i*) according to given conditions. So we can write

$$-\frac{\hbar^2}{2m}\frac{\partial^2\psi_1}{\partial x^2}(x,t) + V\,(x,t)\,\psi_1\,(x,t) = i\hbar\,\frac{\partial\psi_1}{\partial t}(x,t) \qquad \ldots(ii)$$

and
$$-\frac{\hbar^2}{2m}\frac{\partial^2\psi_2}{\partial x^2}(x,t) + V\,(x,t)\,\psi_2\,(x,t) = i\,\hbar\,\frac{\partial\psi_2}{\partial t}(x,t) \qquad \ldots(iii)$$

Multiplying equation (*ii*) by a_1 and (*iii*) by a_2 and then adding, we get

$$-\frac{\hbar^2}{2m}\frac{\partial^2}{\partial x^2}\{a_1\psi_1 + a_2\psi_2\} + V(a_1\psi_1 + a_2\psi_2) = i\hbar\frac{\partial}{\partial t}(a_1\psi_1 + a_2\psi_2) \quad \ldots(iv)$$

Comparing equation (*iv*) with equation (*i*), we get

$$\psi(x, t) = a_1\psi_1 + a_2\psi_2$$

So, $(a_1\psi_1 + a_2\psi_2)$ is also a solution of Schrödinger's wave equation.

Multiple Choice Questions

1. The wavelength λ associated with a particle of mass moving with a velocity v is given by
   (*a*) $\lambda = \frac{h}{mv}$ (*b*) $\lambda = \frac{mv}{h}$ (*c*) $\lambda = \frac{hv}{m}$
2. The de Broglie hypothesis is associated with
   (*a*) Wave nature of radiation only
   (*b*) Wave nature of electron only
   (*c*) Wave nature of all material particles.
3. Which one of the following pairs of phenomenon illustrates the particle aspect of wave particle duality?
   (*a*) Photoelectric effect and Compton effect.
   (*b*) Compton effect and Bragg's Law.
   (*c*) Bragg's Law and photoelectric effect
4. Matter waves
   (*a*) Always travel with the speed of light
   (*b*) Show diffraction
   (*c*) Are longitudinal
5. The de Broglie wavelength of material particles which are in thermal equilibrium at temperature T is
   (*a*) $\frac{h}{\sqrt{2mkT}}$ (*b*) $\frac{h}{\sqrt{mkT}}$ (*c*) $\frac{h}{\sqrt{3mkT}}$
6. If a charged particle of mass *m* is accelerated through a potential difference of V volts, the de Broglie wavelength is proportional to
   (*a*) $V^{\frac{1}{2}}$ (*b*) $V^{\frac{-1}{2}}$ (*c*) V
7. The uncertainty relation holds for
   (*a*) Macroscopic particles only
   (*b*) Microscopic particles only
   (*c*) Both microscopic and macroscopic particles.
8. The uncertainty relation cannot hold for the following pairs
   (*a*) Linear momentum and angle
   (*b*) Energy and time
   (*c*) Position and momentum
9. Davisson and Germer's experiment relates to
   (*a*) Electron diffraction (*b*) Interference (*c*) Polarisation
10. Which of the following wave functions represents a free particle moving along +ve X-*axis*?
    (*a*) $Ae^{-i(kx - \omega t)}$ (*b*) $Ae^{i(kx - \omega t)}$ (*c*) $A\sin(kx - \omega t)$
11. According to wave mechanics, a free particle can possess
    (*a*) Only one single value of energy
    (*b*) Discrete energies
    (*c*) Continuous energies

12. A free particle has
(*a*) Definite energy and definite momentum.
(*b*) Definite energy but indefinite momentum.
(*c*) Definite momentum but indefinite energy.

13. The expression $|\psi|^2$ stands for
(*a*) Position probability density (*b*) Time probability density (*c*) Normalization

14. For stationary state the probability density is
(*a*) Independent of time
(*b*) Function of time
(*c*) Independent of space co-ordinate

15. Out of the following moving with the same velocity, the one which has largest wavelength is
(*a*) A photon (*b*) An electron (*c*) A neutron

16. The probability of finding a particle in a distance dx around a point x is
(*a*) $\psi\psi^*$ (*b*) ψ^* (*c*) $\psi\psi^* dx$

17. For a three dimensional case, which of the following is the normalization condition.
(*a*) $\psi_n^* \psi_m \, dv = 1$ for $n \neq m$
(*b*) $\psi_n^* \psi_m \, dv = 1$ for $n = m$
(*c*) $\psi_n^* \psi_m \, dv = 0$ for $n = m$

18. According to Schrödinger a particle is equivalent to a
(*a*) Light wave (*b*) Single plane wave (*c*) Wave packet

19. Schrödinger's wave equation for a moving particle contains
(*a*) 1st order time derivative
(*b*) 2nd order time derivative
(*c*) No time derivative at all

20. The energies of a particle in a box are given by
(*a*) $\dfrac{nh}{2\pi}$ (*b*) $\dfrac{n^2 \pi^2 h^2}{2m L^2}$ (*c*) $\dfrac{\pi^2 h^2}{2m L^2 n^2}$

21. Two wave functions $\psi_m(x)$ and $\psi_n(x)$ are orthogonal if
(*a*) $\int \psi_m^*(x)\, \psi_n(x)\, dx = 0$ (*b*) $\int \psi_m^*(x)\, \psi_n(x)\, dx = 1$ (*c*) $\int \psi_m(x)\, \psi_n(x)\, dx = 0$

22. For a particle trapped in a box of length ℓ, the average value of momentum $<p>$ is
(*a*) $\dfrac{h}{a\ell}$ (*b*) 0 (*c*) 1.

23. The emission power of a blackbody kept at an absolute temperature T is proportional to
(*a*) T^4 (*b*) T^3 (*c*) T^5 [*WBUT*]

24. The wavelength of emitted radiation at temperature T, at which the blackbody attains maximum value is proportional to
(*a*) $\dfrac{1}{T}$ (*b*) T^4 (*c*) T^3. [*WBUT*]

25. How fast particle must travel so that its mass becomes twice its rest mass ?
(*a*) $0.5\,c$ (*b*) $2\,c$ (*c*) $\dfrac{\sqrt{3}}{2}\,c$. [*WBUT*]

26. 1 milligram of matter after completely converted into energy will yields — [*WBUT*]
(*a*) 90 joules (*b*) 9×10^{10} joules (*c*) 900 joules

27. The energy equivalent of 1 gm of mass is —

(*a*) 9×10^{20} ergs (*b*) 9×10^{10} ergs (*c*) 90 ergs.

28. If visible light is used to study the compton scattering then the compton shift will be [*WBUT*]

(*a*) negative

(*b*) more positive than what is observed with X-ray

(*c*) zero.

29. Which of the following wave function is a solution of Schrödinger wave equation ? [*WBUT*]

(*a*) A cos *x* (*b*) A tan *x* (*c*) Ae^{-x^2}

30. What is the value of $\left[\hat{x}, \dfrac{\partial}{\partial x}\right]$? [W*BUT*]

(*a*) – 1 (*b*) 0 (*c*) – 2

Answers

1. (*a*)	**2.** (*c*)	**3.** (*a*)	**4.** (*b*)	**5.** (*a*)	**6.** (*b*)
7. (*c*)	**8.** (*a*)	**9.** (*a*)	**10.** (*b*)	**11.** (*c*)	**12.** (*a*)
13. (*a*)	**14.** (*a*)	**15.** (*b*)	**16.** (*c*)	**17.** (*b*)	**18.** (*c*)
19. (*a*)	**20.** (*b*)	**21.** (*b*)	**22.** (*b*)	**23.** (*a*)	**24.** (*a*)
25. (*c*)	**26.** (*b*)	**27.** (*a*)	**28.** (*c*)	**29.** (*c*)	**30.** (*a*).

Review Questions

1. Write short notes on

(*a*) Rayleigh-Jean's law

(*b*) Wien's law

(*c*) Planck's radiation law [*WBUT*]

2. (*a*) What is the photoelectric effect?

(*b*) State the laws of photoelectric emission.

(*c*) Derive Einstein's photoelectric equation. [*Lucknow University, Pune University*]

3. (*a*) Derive an expression for the compton shift in wavelength for a photon scattered from a free electron at an angle θ. [*WBUT*]

(*b*) At which angle the shift will be maximum. [*WBUT*]

(*c*) Calculate the compton wavelength (in Å) for an electron. [*WBUT*]

4. (*a*) What is the experimental evidence in favour of de Broglie's hypothesis of matter waves? [*WBUT*]

(*b*) Why in case of moving electrons quantum mechanics is used while for moving cars we use newtonian mechanics? Explain. [*WBUT*]

(*c*) Show that there cannot be any existence of free electrons inside the nucleus of an atom ? [*WBUT*]

5. What is Heisenberg's uncertainty principle? [*WBUT*]

6. (*a*) Give the physical interpretation of the wavefunction ψ (*x*).

(*b*) Obtain the expression for stationary energy levels for a particle of mass *m* which is free to move in a region of zero potential between two rigid walls of $x = 0$ and $x = L$. Are the energy levels degenerated? [*WBUT*]

(*c*) Find the corresponding wave function pertaining to (*b*) and normalize it.

7. Explain Heisenberg's uncertainty principle.
Devise an experiment to illustrate the principle. [*Burdwan University*]
8. What is de Broglie wave equation? A particle of mass m is at rest. What is the wavelength of the wave associated with it? [*Calcutta Univ.*]
9. Establish the time independent Schrödinger wave equation for a single particle. [*Calcutta Univ.*]
10. Establish the Schrödinger equation for a particle moving in one dimensional potential box with infinitely high sides. Determine the probable energy levels of the particle.
[*North Bengal University, Calcutta University*]
11. What is the significance of the wave function ψ?
12. When can a wave equation be called normalized?
13. State the laws of photo-electric emission.
[*H.P. Univ., Meerut Univ., Gharwal Univ., Lucknow Univ.*]
14. Derive Einstein's photoelectric equation. How does it explain the laws of photoelectric emission ? Why all the photo-electrons do not have the same energy ?
[*H.P. Univ., Meerut Univ., Lucknow Univ., Nagpur Univ.*]
15. (*a*) What is compton effect ? Derive an expression for compton shift and wavelength of scattered photon.
(*b*) What is compton wavelength ? Determine its value. Distinguish between compton shift and compton wavelength. What are the factors on which compton shift depends ?
[*H.P. Univ., M.D. Univ., Pbi Univ., Meerut Univ., Guwahati Univ.*]
16. Explain the concept of wave particle dualism. What lead de-Broglie to suggest that matter has wave characteristics ? [*G.N.D. Univ., M.D. Univ., Pune Univ., Madurai Univ.*]
17. State the de-Broglie hypothesis of matter waves. Derive an expression for de-Broglie wavelength of matter particle in terms of Kinetic energy and temperature.
[*Pune Univ, Nagpur Univ, Gharwal Univ, H.P. Univ, Pbi Univ.*]

Unsolved Problems

1. An electron is confined to a box of length 10^8 cm. Calculate the minimum uncertainty in the measurement of its velocity? [**Ans.** 1.02×10^8 cm/sec] [*Kolkata University*]
2. Show that ratio of the de Broglie wavelength to the compton wavelength of a particle is $\sqrt{\left(\frac{c}{v}\right)^2 - 1}$.

[**Hints :** de Broglie's wavelength $= \frac{h}{m_0 v}\sqrt{1-\frac{v^2}{c^2}}$, Compton wavelength $= \frac{h}{m_0 c}$].
[*Kolkata University*]
3. An electron has a de Broglie wavelength of 0.15 Å. Compute the phase and the group velocities of the de Broglie waves. Find the kinetic energy of the electron [Rest mass of the electron $= 0.511 \frac{\text{MeV}}{c^2}$]
[**Ans.** $v_g = 4.86 \times 10^7$ m/sec, $v_p = v_g/2$ (since $v_g << c$), 6700 eV]
[*Kolkata University*]
4. When ultraviolet rays of wavelength 800 Å and 760Å are made to incident on hydrogen atom at ground state, it emits electron of energies 1.8 eV and 4.0 eV respectively. Calculate Planck's constant. Using the value, determine the de Broglie wavelength associated with neutron having speed of 1.06×10^5 ms^{-1}. [**Hints :** $E = h\nu - \omega_0 = \frac{hc}{\lambda} - \omega_0$] [*Burd. Univ.*]
[**Ans.** $h = 6.57 \times 10^{-34}$ J.s, $\lambda = 0.0368$Å]

5. An electron has a speed of 200 m/sec accurate to 0.01%. With what fundamental accuracy can we locate the position of this electron? Mass of electron = 9.1×10^{-31} kg. **[Hints :** $\Delta x \Delta p_x \approx h$**]** **[Ans.** 0.036 m**]** [*Calcutta Univ.*]

6. A microscope using photon is employed to locate an electron in an atom to within a distance 1Å. What is the uncertainty in the momentum of the electron located in this way? $h = 6.62 \times 10^{-34}$ J.s. **[Hints :** $\Delta p_x = \dfrac{h}{\Delta x}$**]** **[Ans.** 6.62×10^{-23} kg m/sec**]**

[*North Beng. Univ.*]

7. If E be the kinetic energy of a particle with rest mass m_0, prove that the de Broglie wavelength is given by $\lambda = \dfrac{hc}{\sqrt{E\left(e + 2m_0c^2\right)}}$. **[Hints :** $mc^2 = m_0c^2 + E$**]**

8. An electron of mass m when accelerated through a potential differences V, has de Broglie wavelength λ. Find the de Broglie wavelength associated with a proton of mass m accelerated through the same potential difference.

[Ans. $\lambda\sqrt{\dfrac{m}{M}}$**]**

9. A beam of x-rays of wavelength 0.12 Å is scattered by a carbon target. Find the minimum wavelength of scattered photon. [*Pbi. Univ.*] **[Ans.** [0.12 Å]**]**

10. An electron has speed of 6.6×10^4 m/sec with an accuracy of 0.01%. Calculate uncertainty in position of electron. Given mass of electron = 9.1×10^{-31} Kg. [*Meerut Univ.*]
Planck's constant $h = 6.6 \times 10^{-34}$ Js **[Ans.** 1.59×10^{11} Hz**]**

11. The speed of a bullet of mass 50 gm is measured to be 300 m/sec with an uncertainty of 0.01%. With what accuracy can we locate the position of the bullet if it is measured simultaneously with its speed. [*H.P. Univ.*] **[Ans.** 7×10^{-32} m**]**

S.N. Bose

6

Chapter

Classical Statistics and Quantum Statistics

6.1 INTRODUCTION

Statistical Mechanics deals with the properties of macroscopic systems in thermal equilibrium—thereby links the mechanical properties and the thermal properties of systems. Statistical mechanics aims at studying the macroscopic parameters of a system in equilibrium from a knowledge of the microscopic properties of its constituent particles using the laws of mechanics.

Quantum statistics is the quantum form of statistical mechanics. Quantum statistics is based on quantum mechanical framework of energy levels and states in a single real system.

PROBABILITY is the basis of statistical physics.

Equilibrium Forms

1. **Mechanical Equilibrium** is established if there is no unbalanced force or torque in the interior of a system or between the system and its surroundings.
2. **Chemical Equilibrium** sets in, for a system initially in mechanical equilibrium, when no spontaneous change in it takes place by way of diffusion or chemical reaction or both.
3. **Thermal Equilibrium** for a system in both mechanical and chemical equilibrium sets in when there is no *spontaneous* change in the thermal parameters of the system which is kept separated by a heat insulated wall (*i.e.*, a diathermic wall)

When all the above three forms of equilibrium are totally fulfilled by a system, it is then in **thermodynamic equilibrium.**

Statistical Physics determines the *most probable* way in which a certain *total amount of energy E* is *distributed* among N number of particles *in the system in thermal equilibrium* at an absolute temperature T.

NOTE *Assumption* is that the system is composed of very very large number of particles existing in most probable state.

Most Probable State is the one in which maximum number of particles of the system exists. This is with reference to their energy states.

Classification of Statistical Mechanics are :

(*a*) **Classical Mechanics** where quantum principles do not apply.
(*b*) **Quantum Mechanics** which is based on quantum mechanical principles.

Predictions of statistical physics become *nearly exact* for a system that deals with *very very large* number of particles. Quantum statistics can be applied to quantum mechanical systems (notably photons in a cavity and free electrons in a metal) and it is one of the powerful tools of theoretical physics.

6.2 CONCEPT OF ENERGY LEVELS AND ENERGY STATES

The particles *forming atoms* obey the *laws of quantum mechanics*, according to which the quantities characterizing the motion of a particle in an atom can take on *only quite definite values*. Instead of the mechanical model of electrons moving in orbits one can speak of the *energy levels which it is at*. Consequently, an electron jump from one orbit to another is viewed as a *transition* of the atom *from one energy state to another*. The energy of the electron changes directly by a definite value (or quantum).

Bohr's Assumptions

In case of Hydrogen Atom (Single electron atom) –

1. Corresponding to the electron orbits the atom is said to be in *discrete* energy levels (or stationary states of energy)

   NOTE Atomic energy states (or quantum states) are characterized by its electronic energy states.

2. The *angular momentum* of the electron in motion in *allowed* circular orbits round the nucleus of the hydrogen atom (*i.e.*, mvr) $= \dfrac{nh}{2\pi}$ where $n = 1, 2, 3, \ldots$, etc., and "h" is the Planck's constant. Thus, $mvr = \dfrac{nh}{2\pi}\left(i.e.,\ 2\pi r = \dfrac{nh}{mv}\right)$

   [m : electronic mass and v : electron's orbital velocity]

3. When a higher energy state (*i.e.*, quantum state) E_{n_2} changes to another quantum state of *lower* energy E_{n_1}, the frequency ν of the radiation emitted by *the atom* is given by

$$\nu = \frac{E_{n_2} - E_{n_1}}{h}$$

An electron in orbit around a hydrogen nucleus will have a *de-Broglie wavelength* given by,

$$\lambda = \frac{h}{mv}$$

where h : Planck's constant
m : electron mass
v : orbital velocity of the electron

and 'v' is given *by*

$$v = \frac{e}{\sqrt{4\pi\varepsilon_0 mr}}$$

ε_0 : permittivity of the medium (*i.e.*, vacuum)
r : orbital radius

An electron can circle a nucleus *only if* its orbit contains an *integral* number de Broglie wavelengths so that the orbit becomes stable.

The condition for stability of the orbit (*i.e.*, stationary state) is,

$$n\lambda = 2\pi r_n$$

where r_n: radius of the orbit that contains "n" wavelengths

NOTE n is known as the quantum number

In case of hydrogen atom,

For a *dynamically stable orbit*,

$$F_c = F_e$$

F_c : Centrifugal force on orbiting electron

or

$$\frac{mv^2}{r} = \frac{1}{4\pi\varepsilon_0} \cdot \frac{e^2}{r^2}$$

F_e : Coulomb's electric force between positive nucleus and the negative electron $\left(i.e., \frac{1}{4\pi\varepsilon_0} \cdot \frac{q_1 \cdot q_2}{r^2}\right)$

ε_0 : permittivity of vacuum

Now,

$$n\lambda = 2\pi r_n$$

or

$$n\frac{h}{mv} = 2\pi r_n \quad \text{or} \quad nh \cdot \frac{\sqrt{4\pi\varepsilon_0 m r_n}}{me} = 2\pi r_n$$

The orbital radii of Bohr Atom is, therefore, obtained from

$$4\pi^2 r_n^2 = \frac{n^2h^2}{m^2e^2} \cdot (4\pi\varepsilon_0 m r_n)$$

or

$$r_n = \frac{n^2h^2\varepsilon_0}{\pi me^2}$$

The radius of the *innermost* orbit is called **Bohr Radius** of the hydrogen atom and is denoted by a_0.

Hence, $a_0 = r_1$ which is equal to 5.292×10^{-11} m

The other radii are $r_n = a_0 n^2$

where, $a_0 = \dfrac{h^2\varepsilon_0}{\pi me^2}$

$$h = 6.63 \times 10^{-34} \text{ Joule-sec}$$

$$e = 1.6 \times 10^{-19} \text{ Coulomb}$$

$$m = 9.1 \times 10^{-31} \text{ kg}$$

$$\varepsilon_0 = 8.85 \times 10^{-12} \text{ Farad/m}$$

The *Total Energy* E of the electron in a hydrogen atom is the sum of its K.E. and P.E. in the orbit which is,

$$E = \frac{1}{2} mv^2 + \left(- \frac{e^2}{4\pi\varepsilon_0 r} \right)$$

$$= \frac{1}{2} m \cdot \frac{e^2}{4\pi\varepsilon_0 mr} - \frac{1}{4\pi\varepsilon_0} \cdot \frac{e^2}{r}$$

$$= - \frac{1}{2} \cdot \left(\frac{e^2}{4\pi\varepsilon_0 r} \right) = - \frac{1}{8\pi\varepsilon_0} \cdot \frac{e^2}{r}.$$

Energy Levels

The various *permitted* electron orbits involve different *electron energies*. The electron energy E_n of the nth quantum number is

$$E_n = - \frac{e^2}{8\pi\varepsilon_0 r_n} \quad (r_n : \text{orbit radius})$$

The *Energy Levels* are, therefore,

$$E_n = - \frac{e^2}{8\pi\varepsilon_0} \cdot \frac{\pi me^2}{h^2 \varepsilon_0 n^2} = - \frac{me^4}{8\varepsilon_0^2 h^2} \left(\frac{1}{n^2} \right)$$

where $n = 1, 2, 3, \ldots$

or

$$E_n = E_1 \left(\frac{1}{n^2} \right) \qquad \left[\because E_1 = - \frac{me^4}{8\varepsilon_0^2 h^2} \right]$$

when evaluated,

$$E_1 = - 2.18 \times 10^{-18} \text{ Joule} = - 13.6 \text{ eV} \quad \left\{ \begin{aligned} 1 \text{ eV} &= 1.6 \times 10^{-19} \text{ J} \\ &= 1.6 \times 10^{-12} \text{ erg} \end{aligned} \right\}$$

N.B. E_1 is the lowest energy level of *hydrogen atom* and is known as the *GROUND STATE*. Other levels such as E_2, E_3 etc., are *EXCITED STATES*.

NOTE The drawbacks of Bohr's Theory are :

The quantization rules are arbitrary.
The rules cannot be applied to explain spectra of helium and other complex atoms. (*i.e.*, containing more than one electron)

6.3 PROBABILITY

Probability deals with *random events*, which cannot be predicted *beforehand with certainty*.

Mathematically, the probability P is

$$P = \lim_{N \to \infty} \frac{n}{N}$$

where, n : the number of occurrence of a *particular* event.

N : total number of *trials* for *the particular event* to occur.

Assumptions in the said definition :

1. Occurrence of the *particular event does not affect the occurrence* of other events (*i.e.*, all the occurrences are independent).
2. Occurrence of *any* of the *other alternatives* of events is equally likely (*i.e.*, have equal chance).

Mathematical Probability

If *an event* occurs in "i" ways but does not happen for "j" ways where "i" and "j" ways of happening are equally likely then the probability of the event "i" to occur is,

$$P = \frac{i}{i + j}$$

Thus, the probability of occurrence of a particular event is a *fraction*. When P = 1, the case of occurrence of the particular event is certain (or sure) and so P = 1 means *a certainty*.

(*i*) **Summation of Probabilities :**

Let P_1, P_2, P_3, .. etc., be the probabilities of several *mutually exclusive* events (*i.e.*, the occurrence of *one* of the alternatives of *different events does not affect* the occurrence of the others), then the probability that *one of them* will happen is the *sum* of the probabilities for happening of the *separate* events, which means,

$$P = P_1 + P_2 + P_3 + \ldots$$

This is SUMMATION OF PROBABILITIES.

Example : A box contains 20 identical balls - 5 white and 15 black. The probability of *extracting any* ball (black or white) is $\frac{1}{20}$ $\left(\because P = \frac{i}{i + j}\right)$.

However, the probability of extracting *a white* ball from the box is

$$P = \frac{1}{20} + \frac{1}{20} + \frac{1}{20} + \frac{1}{20} + \frac{1}{20} = \frac{5}{20}.$$

NOTE It is immaterial to know *which* of the *white balls* out of 5 is extracted (since they are identical and thereby indinguishable)

(*ii*) **Multiplication of Probabilities :**

The probability of the *simultaneous* happening of two or more *independent events* is equal to the product of the probabilities of *each* of them happening *separately*.

Example : Suppose there are 5 boys A, B, C, D and E who are interested in getting hold of a *red ball* (particular one). For A, the probability is $\frac{1}{5}$. If now A gets interested in *also* getting a particular *white* ball for which all others are also fighting, then for A the probability of getting hold of the white ball is again $\frac{1}{5}$. Thus, the probability of A to get the red and the white balls *simultaneously* is given by $P = \frac{1}{5} \times \frac{1}{5} = \frac{1}{25}$. (For there are now 25 *equally likely* ways in which the *composite event* takes place).

Thus the probability of two or more mutually independent events to occur *simultaneously* is $P = P_1 \times P_2$ and so on. This means that the probability of *composite event* to occur is the *product* of the probabilities of *mutually independent* events (*i.e.*, $P = P_1 \times P_2 \times P_3 \ldots$).

6.4 STATISTICAL MECHANICS : PHASE-SPACE, MICROSTATES, MACROSTATES

A complete knowledge of a *system* of particles *at any instant* of time is obtained if the position and momentum of *every* particle is *specified* in space. In terms of geometry, we construct a *six-dimensional* space whose *three* co-ordinates are the *three* position co-ordinates x, y, z of a particle and the *other three* co-ordinates are its *three momentum components*, which are p_x, p_y and p_z.

A particle located at any point P *in this space* has a particular set of values (x, y, z, p_x, p_y, p_z). This is then equivalent to the specification of the positions of *each* of the N particles in this space. This particular space is called the *phase space* or "μ-space". This phase space is purely a classical concept.

According to Quantum mechanics the position and momentum of a particle *cannot* be *exactly* determined simultaneously, the product of the *uncertainties* $\Delta x . \Delta p_x \geq h$ (according to the Heisenberg's Uncertainty Principle).

As there are three pairs of such *conjugate variables* it follows that the location of a particle in a region of *volume smaller* than h^3 in the phase space is *not* possible. In order to utilize the concept of a phase space, *even for* quantum mechanical systems, the phase space is divided into cells and this *cellular space* is considered in the development of the theory. The volume of each cell is h^3 and this is denoted by τ. The cells can be labelled as cell 1, cell 2 etc., and the N-particles of a system are placed within the cells in the phase space and defines a *state* :

1. The N particles are all *identical.*
2. According to classical theory, the particles are *distinguishable.*
3. The specification of the *state of a system* of N particles is to mention which particle is placed in which cell. The type (a, i) means that the particle designated by 'a' goes to the cell denoted by 'i'. Similarly the particle 'b' places itself in cell 'j' [*i.e.*, type (b, j)] and so on. This kind of specification is called microscopic description, which is the most detailed description of the system. This is *MICROSTATE* of a system.

If the concept of distinguishability of particles does not exist then the microscopic description is no longer meaningful. We still have to consider for any system, the *corresponding microstate.*

A *MACROSTATE* is a macroscopic description of a system which specifies the *number of particles* in each cell.

Example : Let there be *three particles a, b* and *c* and let there be *two cells* 1 and 2. Then we can have

a, b	c	;	a, c	b	and	b, c	a
cell 1	cell 2		cell 1	cell 2		cell 1	cell 2

as *three different microstates* but belong to the *same* macrostate which considers *only* the *number of molecules* (not the actual individual molecules) contained in cells 1 and 2.

As we change from one system to another, the microstates will change. Hence the macrostates are also likely to change. But in all cases a macrostate will contain several microstates. It is *assumed* that all the microstates are *equally likely* to occur. Hence, the **probability** of a macrostate will be proportional to the number of microstates it contains (*i.e.*, if the macrostate A has 100 microstates and macrostate B has 300 microstates, then, the occurrence of the macrostate B will be **three times** more probable than that of A).

(*i*) ***MACROSTATE*** is, therefore, the *distribution* of a system of particles by the *compartmentwise number* in *each* compartment and **not by the identity** of the individual particles.

Example : Distinguishable particles in *two identical compartments* can be distributed as below

Compartment	*Number of Particles*				
1	0	1	2	3	4
2	4	3	2	1	0

Hence, there are 5 different macrostates — (0,4), (1, 3), (2, 2), (3, 1) and (4, 0).

For a system having n particles the distribution among 2-compartments are $(0, n)$ $[1, (n - 1)]$, $[2, (n - 2)]$, ... $(n, 0)$.

Which gives the total number of macrostates $= (n + 1)$

(*ii*) **MICROSTATE**, however, describes *each* **distinct** *arrangement* of the particles as **specified** by their *identity*, corresponding to ***a*** **particular** macrostate (given).

Example : Let the 4 particles as considered in the previous example be denoted by a, b, c and d. Consider a *particular* macrostate [say, (2, 2)]. Then, **corresponding** to *the macrostate (2, 2),* the different arrangements of the particles bearing their identities may be as below :

Macrostate	*Possible Arrangement*		*Number of*
	Compartment I	*Compartment* II	*MICROSTATES*
(2, 2)	*ab*	*cd*	6
	ac	*bd*	
	ad	*bc*	
	bc	*ad*	
	bd	*ac*	
	cd	*ab*	

NOTE All microstates are *equally probable* (*i.e.*, over a **long** period of **time** a microstate will occur as often as the other).

6.5 THERMODYNAMIC PROBABILITY OR THERMODYNAMIC FREQUENCY OF THE MACROSTATE

It is the *number* of **possible** *microstates* corresponding to *any given macrostate*. This is denoted by W.

The number is *always* an *integer*

W is generally a *very large* number.

Example : Consider 2-compartments i and j with 5 (five) *distinguishable* particles a, b, c, d and e and consider further that n_i and n_j are the number of particles in the compartments i and j respectively.

The possible *macrostates* and the **corresponding** W are :

	①	②	③	④	⑤	⑥		
n_i	0	1	2	3	4	5	MACROSTATES	6 macrostates
n_j	5	4	3	2	1	0		
W	1	5	10	10	5	1	No. OF MICROSTATES	(*i.e.*, Thermodynamic probability)

Corresponding to the macrostate (0, 5), the number of microstates is 1 (one)

$$(n_i, n_j) \to (0, abcde)$$

For macrostate (1, 4) — Number of microstates is 5 (five)

$$(n_i, n_j) \to \left.\begin{matrix}(a, bcde)\\(b, cdea)\\(c, deab)\\(d, eabc)\\(e, abcd)\end{matrix}\right\} 5$$

For macrostate (2, 3) → Number of microstates is 10 (ten)

$$(n_i, n_j) \to \left.\begin{matrix}(ab, cde)\\(ac, deb)\\(ad, ebc)\\(ae, bcd)\\(bc, dea)\\(bd, eac)\\(be, acd)\\(cd, eab)\\(ce, dab)\\(de, abc)\end{matrix}\right\} 10$$

Thus, corresponding to the macrostate (5, 0) the number of microstates is 1 (one), for (4, 1) it is 5 (five) and for the macrostate (3, 2) it is 10 (ten).

There will *in all be 32 microstates possible* corresponding to 6 (six) *different* macrostates.

6.5.1 Probability of Occurrence of a Macrostate

It is the ratio of the number of microstates in **the macrostate** to the total number of *microstates*.

In the above example, the probability of occurrence of the macrostate (0, 5) is $\frac{1}{32}$, $\left[\frac{\text{Number of microstates}}{\text{Total number of microstates}}\right]$ that of (1, 4) is $\frac{5}{32}$ and so on.

Further, the *total* number of *ways* of arranging n *distinguishable particles* in k compartments $= k^n$ – which is the *total number* of *microstates possible.*

NOTE Out of n *distinguishable* objects the 1st one can be placed in *any* of the k compartments in k ways. Similarly, for the 2nd object can also be placed in k ways. So the first two objects can be **arranged** in k compartments in $k \times k = k^2$ ways. Similarly for n objects it is,

$$k \times k \times k \times \ldots \text{ upto } n \text{ factors} = k^n \text{ ways.}$$

6.5.2 Determination of Thermodynamic Probability

Suppose that our specified macrostate containing in total N particles consist of n_1 particles in cell 1, n_2 particles in cell 2 and so on. We can choose the 1st particle to go into cell 1 in N-ways, the next particle in (N-1) ways, the third in (N-2) ways etc.

After the 1st cell (*i.e.*, cell 1) is filled we choose from the remaining particles to fill cell 2, cell 3 and so on until we come to the last position in the last cell — which can be filled up only in *one* way. This gives us N (N – 1) (N – 2) ... 3, 2, 1 or N ! ways of filling the cells. In the above arrangement we counted separately distributions in which the same particle fell into the same cell, but in *different sequence*. There are n_1 ! permutations of the particles in cell 1, which were counted separately above, but which may correspond to *one microstate*. As such W becomes > N ! So we must divide N ! by n_1 ! n_2 ! n_3 ! etc.

Thus we have,
$$W = \frac{N!}{n_1!\,n_2!\,n_3!\ldots}$$

where W : number of microstates for a *particular* macrostate.

Fundamental Hypothesis

1. A system of particles may be in *any microstate* – the probability of each microstate is the same.
2. The probability of *the system* being in a *certain macrostate varies* from one to another which is proportional to the *thermodynamic* probability W of *that macrostate*. For a system consisting of large number of particles the *maximum W* is most important.

Example : Let there be 10 particles in 5 cells. The *most probable distribution* (a priori) in this case is 2 particles in a cell.

$$W_{max} = \frac{10!}{2!2!2!2!2!} = 113{,}400$$

The probability of a *neighbouring* distribution is,

$$W = \frac{10!}{2!2!3!1!2!} = 75{,}600$$

Thus even for 10 particles the difference between a maximum and a neighbouring probability is quite large. This difference *will be* larger for larger number of particles.

6.6 EQUILIBRIUM MACROSTATE

At equilibrium, the system will be at that *macrostate* which has the maximum thermodynamic probability (*i.e.*, W_{max}).

This is subject to two restrictions :

(*i*) $n_1 E_1 + n_2 E_2 + n_3 E_3 + \ldots = U$ (Total Energy) = Constant

where, E_1, E_2, E_3 etc., denote the energy of a particle in the 1st, 2nd, 3rd, etc., respectively.

NOTE This necessitates from the conservation of energy principle.

(*ii*) $n_1 + n_2 + n_3 + \ldots = N$ = constant (*i.e.*, the total number of particles is constant)

NOTE This restriction (*ii*) is not always valid. As for example, in the case of photons - they can be absorbed or emitted.

SUMMARY

1. All microstates corresponding to *possible macrostates* are *equally probable*.
2. The *equilibrium state* of a system of particle *corresponds* to the *macrostate of maximum probability*.
   Provided that :
   (*i*) the *total number of particles* is constant.
   (*ii*) the *total energy of the system* is constant.

6.7 DISTRIBUTION FUNCTION

To determine the *average properties* of *a system* of particles it is *necessary* to know the *number* of particles occupying quantum states in the energy range E and E + *d*E. This number is *n* (E) *d*E expressed as,

$$n(E)\,dE = f(E)\,g(E)\,dE$$

where, $f(E)$: distribution function of the system

$g(E)\,dE$: number of quantum states in the energy range E and E + *d*E.

NOTE $f(E)$ is the *average* number of particles of the system which occupy *a single* quantum state *of energy* E.

STIRLING'S THEOREM (will be frequently used in this study)

This is

$$\boxed{\ln x! = x \ln x - x}$$

Proof :

$$\ln x! = \ln 1 + \ln 2 + \ln 3 + \ldots + \ln x$$

$$= \sum_{n_i} \ln n_i \text{ where } n_i = 1, 2, \ldots, x.$$

$$= \int_1^x \ln n \, dn \text{ (for large } x)$$

$$= [n \ln n]_1^x - [n]_1^x$$

$$= (x \ln x - x + 1)$$

$$= x \ln x - x \quad \text{when } x \text{ is very large.}$$

6.7.1 Boltzmann Distribution Law

We want to find the *equilibrium distribution* of a system of particles which are regarded as distinguishable. Here,

$$W = \frac{N!}{n_1! . n_2! . n_3! \cdots} = \frac{N!}{\Pi_i n_i!}$$

The *equilibrium distribution* is that distribution in which W is maximum under the conditions

$$\sum n_i = N = \text{constant and } \sum n_i E_i = U = \text{constant}$$

This means that at *equilibrium*,

$$\delta W = 0$$

$$\sum \delta n_i = 0$$

$$\sum E_i \delta n_i = 0$$

In above, δW is the small change in W as the occupation number (n) of each cell is changed by a small amount δn.

To find δW, we use Stirling's theorem,

$$\ln W = \ln N! - \sum_i \ln n_i!$$

$$= (N \ln N - N) - \sum_i (n_i \ln n_i - n_i)$$

$$= N \ln N - \sum_i (n_i \ln n_i) \qquad \left(\because N = \sum_i n_i\right)$$

$$\therefore \quad \delta(\ln W) = \frac{1}{W} . \delta W = \delta N . \ln N + \delta N - \sum_i \delta n_i \ln n_i - \sum_i \delta n_i$$

or

$$\frac{\delta W}{W} = - \sum_i \delta n_i \ln n_i \qquad \left(\because N = \text{constant and } \sum_i \delta n_i = 0\right)$$

As $\delta W = 0$, it follows that

$$\sum_i \ln n_i \,.\, \delta n_i = 0.$$

In the expression above the δn_i's are *not independent* but are subject to the conditions

$$\sum \delta n_i = 0 \text{ and } \sum E_i \,\delta n_i = 0$$

We can combine the above three equations and from one equation as :

$$\sum_i (\ln n_i + \alpha + \beta E_i)\,\delta n_i = 0$$

where α and β are the Lagrange's multipliers

In this expression δn_i's are effectively *all independent*.

$\therefore$ $\ln n_i + \alpha + \beta E_i = 0$

wherefrom we get, $n_i = e^{-\alpha}\, e^{-\beta E_i}$

or $n_i = A\, e^{-\beta E_i}$, where α and β are the same for all cells and $A = e^{-\alpha}$

This is *Boltzmann's Distribution Law.*

The law as given above allows us to determine the number of particles *per cell solely by its energy in that cell*, the number decreasing exponentially with energy. It, however, *does not* say that the *number* of particles having energy E_i is given by this formula.

To find the number of particles N_i having energy E_i we have to find the *number of cells* g_i having the energy E_i. Then,

$$N_i = A\, e^{-\beta E_i} \,.\, g_i \;\; i.e., \;\; N_i = A \,.\, g_i\, e^{-\beta E_i}$$

where g_i can be found out in specific cases.

A and β can also be found out in a similar way.

In fact, we can divide the *phase space* into a number of zones, each zone being characterized by a particular value of *a physical variable*, but the *cells of each zone* having almost the same energy E_i. Then if N_i is the number of particles in the ith *zone* and g_i, the number of cells in the ith *zone* then,

$$N_i = A\, g_i \,.\, e^{-\beta E_i} \quad \text{as obtained above.}$$

If each zone is characterized by a particular parameter (or variable) t (say) lying between t and $t + dt$. Then we can write,

$$N_t\, dt = A\, g_t\, e^{-\beta E_t}\, dt\,.$$

6.8 MAXWELL-BOLTZMANN STATISTICS [CLASSICAL STATISTICS]

Consider an assembly of N similar molecules which are sufficiently widely separated to be distinguished. The molecules may have energies $E_1, E_2, \ldots, E_i \ldots, E_N$. We are to find the distribution of molecules among the various cells of the phase space to find the energy distribution of molecules.

In this case, *more than one cell* of the phase space may correspond to a given energy thus giving rise to degeneracy. Furthermore in M.B. statistics there is no restriction on the number of molecules that may be accommodated in one cell. Let the degeneracy parameters be denoted by g's, that is, g_1 cells are with E_1 energy, g_2 cells are with energy E_2 and so on. Consider n_1, n_2, n_3 etc., are the number of molecules possessing energies E_1, E_2, E_3, etc., respectively.

If n_i are the number of molecules having energy E_i ; these n_i molecules are to be accommodated in g_i cells. The number of ways of accommodating n_i molecules in g_i cells is $(g_i)^{n_i}$. Hence, the number of ways in which N molecules can be distributed among the various cells in the product of the factors of the form $(g_i)^{n_i}$ which is

$$(g_1)^{n1}\ (g_2)^{n2}\ \ldots = \Pi_i\ (g_i)^{n_i} \qquad \ldots(6.1)$$

subject to the condition $\sum n_i = N$

[the notation Π_i is called the running product]

Equation (6.1), however does not give the *total* distribution as we must take into account the possible permutations of the molecules among the different energy levels which is

$$\frac{\underline{|N}}{\underline{|n_1}\ \underline{|n_2}\ \underline{|n_3}\ldots} = \frac{\underline{|N}}{\Pi_i\ \underline{|n_i}} \qquad \ldots(6.2)$$

Thus the number of ways of distribution is given by the product of (6.1) and (6.2), which is

$$W = \frac{\underline{|N}}{\Pi_i\ \underline{|n_i}}\ .\ \Pi_i\ (g_i)^{n_i} \qquad \ldots(6.3)$$

wherefrom,

$$\ln W = \ln \underline{|N} - \sum_i \ln \underline{|n_i} + \sum_i n_i \ln g_i$$

Using Stirling's formula we get,

$$\ln W = N \ln N - N - \sum_i (n_i \ln n_i - n_i) + \sum_i n_i \ln g_i$$

or

$$\ln W = N \ln N - N - \sum_i n_i \ln n_i + \sum_i n_i + \sum_i n_i \ln g_i$$

$$= N \ln N - \sum_i n_i \ln n_i + \sum_i n_i \ln g_i \qquad \ldots(6.4)$$

The condition for the distribution to be the most probable one is that small changes δn_i in any one of the n_i's does not affect the value of W.

Thus, $\delta \ln W = 0$

Using the above in equation (6.4) by setting $\delta \ln W = 0$,

$$\delta \ln W = -\sum_i n_i \frac{\delta n_i}{n_i} - \sum_i \ln n_i \ \delta n_i + \sum_i \ln g_i \ \delta n_i$$

$$= -\sum_i \ln n_i \ \delta n_i + \sum_i \ln g_i \ \delta n_i \ = 0 \left[\text{Since} \sum_i \delta n_i = \delta N = 0 \right]$$

or

$$\delta \ln W = \sum_i (-\ln n_i + \ln g_i)\, \delta n_i \ = 0 \qquad \ldots(6.5)$$

Since,

$$\sum_i \delta n_i \ = 0 \qquad \ldots(6.6)$$

and

$$\sum_i E_i \ \delta n_i \ = 0 \qquad \ldots(6.7)$$

We use Lagrange's method of undetermined multipliers for solving equations (6.5), (6.6) and (6.7). Multiplying equation (6.6) by – α and equation (6.7) by – β and adding these to equation (6.5) we get,

$$\sum_i (-\ln n_i + \ln g_i - \alpha - \beta E_i)\, \delta n_i \ = 0$$

In order that this equation holds, δn_i being arbitrary,

$$-\ln n_i + \ln g_i - \alpha - \beta E_i = 0$$

Hence,

$$\ln \frac{g_i}{n_i} = \alpha + \beta E_i$$

or

$$\frac{g_i}{n_i} = e^{(\alpha + \beta E_i)}$$

or

$$n_i = \frac{g_i}{e^{(\alpha + \beta E_i)}}$$

Thus the MB distribution function :

$$f_{MB}(E_i) = \frac{n_i}{g_i} = \frac{1}{e^{(\alpha + \beta E_i)}} = \frac{1}{e^{\alpha} \ . \ e^{\beta E_i}}$$

Now, $\beta = \frac{1}{kT}$ (to be proved later)

where k : Boltzmann Constant

T : Temperature of the system in Absolute Scale

Thus, the M.B. distribution function is

$$f_{MB}(E_i) = \frac{1}{e^{\alpha} \ . \ e^{\frac{1}{kT} . E_i}}$$

6.8.1 Limitations of the Maxwell-Boltzmann Method

Some of the important limitations of M.B. Statistics are :

1. It is applicable only to an isolated gas of identical molecules in equilibrium which satisfy the following conditions :
   (*a*) The mean potential energy due to mutual interaction between the molecules is very small compared their mean kinetic energy.
   (*b*) The number of molecules per unit volume is very small which means that the density of the gas is very small (*i.e.*, dilute gas).
2. It cannot be applied to a system of indistinguishable particles.
3. The statistics is not applicable for a system at low temperature.
4. The Maxwell-Boltzmann distribution number does not give correct expression for entropy of an ideal gas. The expression does not satisfy the third law of thermodynamics.
5. In M.B. distribution any number of particles are assumed to occupy the same quantum state while many particles such as electrons, obey Pauli's Exclusion principle which does not permit one quantum state to accept more than one particle.

6.9 QUANTUM STATISTICS

6.9.1 Bose-Einstein Statistics

Bose-Einstein statistics deals with the distribution of identical indistinguishable particles. These particles do not obey Pauli's Exclusion principle (*i.e.*, there is no restriction on the number of particles occupying a particular quantum state). Such particles possess integral spins or zero spin.

Consider N particles possessing various energies E_1, E_2, E_3, ..., E_i ..... etc. Let there be g_i cells in the phase space having the same energy E_i (the degeneracy parameter is represented by g's).

If the number of indistinguishable particles having the energy E_i is n_i, then the first step is to determine the number ways in which the n_i particles can be distributed in g_i cells. The problem can be easily solved by arranging the n_i particles in a row and by putting $(g_i - 1)$ partitions in between the particles, so that ●● ¦¦ ●● ¦ ●●● ¦ ● ¦ ●●●● ¦ ●●● ¦ ● the particles are now separated into g_i intervals. Thus we are considering $(n_i + g_i - 1)$ *objects* and there are $|\underline{(g_i + n_i - 1)}$ possible permutations of these objects. But neither of n_i particles and $(g_i - 1)$ partitions are distinguishable. Hence, the number of distinct arrangements are

$$\frac{|\underline{n_i + g_i - 1}}{|\underline{n_i}\;|\underline{(g_i - 1)}}$$

Thus the number of ways in which n_1 particles may have energy E_1, n_2 particles may have energy E_2, etc., is given by the product given by,

$$W = \Pi_i \frac{|\underline{n_i + g_i - 1}}{|\underline{n_i}\;|\underline{(g_i - 1)}} \qquad ...(6.8)$$

wherefrom,

$$\ln W = \sum_i \left[\ln |\underline{n_i + g_i - 1} - \ln |\underline{n_i} - \ln |\underline{g_i - 1}\right]$$

Using Stirling formula ($\ln \lfloor x = x \ln x - x$)

We obtain,

$$\ln W = \sum_i [(n_i + g_i - 1)\ln(n_i + g_i - 1) - (n_i + g_i - 1) - n_i \ln n_i + n_i - (g_i - 1)\ln(g_i - 1) + (g_i - 1)]$$

$$= \sum_i [(n_i + g_i - 1)\ln(n_i + g_i - 1) - n_i \ln n_i - (g_i - 1)\ln(g_i - 1)]$$

For the most probable distribution small changes δn_i in any of the individual n_i will not affect the value of W *i.e.*,

$$\delta(\ln W) = 0$$

So,

$$\delta \ln W = \sum \left[\frac{(n_i + g_i - 1)}{(n_i + g_i - 1)} . \delta n_i + \ln(n_i + g_i - 1) . \delta n_i - \frac{n_i}{n_i}\delta n_i - \ln n_i \, \delta n_i\right]$$

$$= \sum_i [\ln(n_i + g_i - 1) - \ln n_i]\,\delta n_i = 0 \quad \text{...(6.9)}$$

Since $n_i + g_i >> 1$

$$\delta(\ln W) = \sum_i [\ln(n_i + g_i) - \ln n_i]\,\delta n_i = 0 \quad \text{...(6.10)}$$

Further,

$$\delta N = \sum_i \delta n_i = 0 \quad (\because N = \text{constant}) \quad \text{...(6.11)}$$

and

$$\delta E = \sum_i E_i\,\delta n_i = 0 \quad (\because E = \text{constant}) \quad \text{...(6.12)}$$

Multiplying (6.11) by $-\alpha$ and (6.12) by $-\beta$ and adding the two to equation (6.10), we get

$$\sum_i [\ln(n_i + g_i) - \ln n_i - \alpha - \beta E_i]\,\delta n_i = 0 \quad \text{...(6.13)}$$

Since δn_i is arbitrarily chosen this is not zero.

Therefore,

$$\sum_i [\ln(n_i + g_i) - \ln n_i - \alpha - \beta E_i] = 0$$

or

$$\ln \frac{(n_i + g_i)}{n_i} = \alpha + \beta E_i$$

or

$$1 + \frac{g_i}{n_i} = e^{\alpha + \beta E_i}$$

or
$$\frac{g_i}{n_i} = e^{\alpha + \beta E_i} - 1$$

$$= e^{\alpha} . e^{\beta E_i} - 1 \qquad \ldots(6.14)$$

Hence, the B.E. distribution function is,

$$f_{BE}(E_i) = \frac{n_i}{g_i} = \frac{1}{e^{\alpha} . e^{\beta E_i} - 1} = \frac{1}{e^{\alpha} . e^{\frac{E_i}{kT}} - 1}$$

(Derivation of $\beta = \frac{1}{kT}$ is given later)

Example : *Two indistinguishable particles are distributed over three cells. Calculate the total number of microstates of the system.*

Solution. Since the particles are indistinguishable, therefore, their total number of microstates according to B.E. statistics is

$$\frac{\lfloor n_i + g_i - 1}{\lfloor n_i \lfloor (g_i - 1)}$$

Here $n_i = 2$ and $g_i = 3$

So,
$$\frac{\lfloor n_i + g_i - 1}{\lfloor n_i \lfloor (g_i - 1)} = \frac{\lfloor 2 + 3 - 1}{\lfloor 2 \lfloor (3 - 1)} = \frac{\lfloor 4}{\lfloor 2 \lfloor 2}$$

$$= 6$$

Here the particles are A and A

1	A	A	
2		A	A
3	A		A
4	AA		
5		AA	
6			AA

6.9.2 Fermi-Dirac Statistics

Fermi-Dirac statistics applies to particles which are indistinguishable and are governed by Pauli Exclusion principle. According to Pauli exclusion principle "no two particles can be in the same quantum state". Thus in F.D. statistics a phase cell can contain either a single particle or no particle at all. Such particles only possess half integral spins.

Let there be N particles, the particles have various energies E_1, E_2, E_3, ...

Let the number of cells in the phase space having the same energy E_i be g_i. If the number of indistinguishable particles having energy E_i is n_i then the first step to find the number of ways in which n_i particles can be distributed in g_i such that a cell does not contain more than one particle is as below.

The first particle can be arranged in g_i-ways. The second particle can be arranged in the remaining $(g_i - 1)$ cells in $(g_i - 1)$ ways, the third particle in $(g_i - 2)$ ways and so on. Thus the number of ways of arranging the n_i particles is,

$$g_i (g_i - 1)(g_i - 2) \ldots (g_i - n_i + 1) = \frac{g_i!}{(g_i - n_i)!}$$

Out of these arrangements, the permutation $n_i!$ of the n_i particles among themselves is irrelevant. Thus, the number of the particles, subject to exclusion principle is,

$$\frac{g_i!}{n_i!\,(g_i - n_i)!}$$

Now, the number of possible ways in which n_1 particles may have the energy E_1, n_2 particles may have energy E_2 etc., is given by the product of the terms like above. Thus the total number of ways of distribution is,

$$W = \Pi_i \frac{g_i!}{n_i!\,(g_i - n_i)!} \qquad \ldots(6.15)$$

wherefrom,

$$\ln W = \sum_i \left[\ln g_i! - \ln n_i! - \ln (g_i - n_i)!\right]$$

Using Stirling's formula, $\ln x! = x \ln x - x$ we get,

$$\ln W = \sum_i \left[g_i \ln g_i - g_i - n_i \ln n_i + n_i - (g_i - n_i)\ln(g_i - n_i) + (g_i - n_i)\right]$$

$$= \sum_i \left[g_i \ln g_i - n_i \ln n_i - (g_i - n_i)\ln(g_i - n_i)\right] \qquad \ldots(6.16)$$

For the most probable distribution, we must have,

$\delta \ln W = 0$ [δn_i's will not affect W]

Thus, from (6.16)

$$\delta \ln W = \sum_i \left[-\frac{n_i}{n_i}\delta n_i - \ln n_i\, \delta n_i - \frac{(g_i - n_i)}{(g_i - n_i)}(-\delta n_i) + \ln(g_i - n_i)\,\delta n_i\right]$$

$$= \sum_i \left[\ln(g_i - n_i) - \ln n_i\right]\delta n_i = 0 \qquad \ldots(6.17)$$

Further,

$$\delta N = \sum_i \delta n_i = 0 \qquad \textbf{(Conservation of number of particles)} \ldots(6.18)$$

and $$\delta E = \sum_i E_i \, \delta n_i = 0 \qquad \textbf{(Conservation of energy)} \quad ...(6.19)$$

To use the method of Lagrange's undetermined multipliers we multiply (6.18) by $-\alpha$ and (6.19) by $-\beta$ and add them to equation (6.17) to obtain,

$$\sum_i \left[-\ln n_i + \ln (g_i - n_i) - \alpha - \beta E_i\right] \delta n_i = 0 \qquad ...(6.20)$$

Since δn_i is arbitrarily chosen, the term in the bracket must be zero for each value of i and so,

$$-\ln n_i + \ln (g_i - n_i) - \alpha - \beta E_i = 0$$

or $$\ln \frac{(g_i - n_i)}{n_i} - \alpha - \beta E_i = 0$$

or $$\ln \frac{(g_i - n_i)}{n_i} = \alpha + \beta E_i \qquad \text{or} \qquad \frac{(g_i - n_i)}{n_i} = e^{\alpha + \beta E_i}$$

wherefrom,

$$\frac{g_i}{n_i} - 1 = e^{\alpha + \beta E_i} \qquad \text{or} \qquad \frac{g_i}{n_i} = e^{(\alpha + \beta E_i)} + 1$$

or $$\frac{g_i}{n_i} = e^{\alpha} \, . \, e^{\beta E_i} + 1$$

$\therefore$ $$n_i = \frac{g_i}{e^{\alpha} \, . \, e^{\beta E_i} + 1}$$

and the Fermi-Dirac distribution function

$$f_{FD}(E_i) = \frac{n_i}{g_i} = \frac{1}{e^{\alpha} \, . \, e^{\beta E_i} + 1}$$

$\frac{n_i}{g_i}$: the average number of particles in each of the quantum states of that energy

N.B. The value of $\beta = \frac{1}{kT}$ will be derived later where, k = Boltzmann constant.

Thus,

$$f_{FD}(E_i) = \frac{1}{e^{\alpha} \, . \, e^{\frac{E_i}{kT}} + 1}$$

N.B. $f(E_i)$ is also known as the occupation index of a state of energy E_i.

Example : *Two indistinguishable particles obeying Pauli's exclusion principle are distributed among three cells. Calculate the number of microstates.*

Solution. According to FD statistics the number of microstates is given by

$$\frac{\lfloor g_i}{\lfloor n_i \cdot \lfloor g_i - n_i}$$

Here, $n_i = 2 \text{ and } g_i = 3$

∴ The number of microstates are

$$\frac{\lfloor\underline{3}}{\lfloor\underline{2}\,\lfloor\underline{3-2}} = 3$$

The distribution diagram is shown here.

In this case the two indistinguishable particles are A and A, which obey Pauli's Exclusion Principle.

1	A	A	
2		A	A
3	A		A

6.9.3 Evaluation of the Constants α and β

In translational motion of molecules, the energy is continuous rather than discrete and quantized. If $n(\varepsilon)\, d\varepsilon$ is the number of molecules lying in the energy range ε and $\varepsilon + d\varepsilon$, then

$$n(\varepsilon)\, d\varepsilon = g(\varepsilon)\, e^{-\alpha} \,.\, e^{-\beta\varepsilon}\, d\varepsilon.$$

Since, $\varepsilon = \dfrac{p^2}{2m}$ | p : momentum; m : molecular mass

We have, in terms of momentum,

$$n(p)\, dp = g(p)\, e^{-\alpha} \,.\, e^{-\frac{\beta p^2}{2m}}\, dp \qquad \ldots(6.21)$$

$g(p)\, dp$ is the number of cells in phase space in which a molecule has momentum between p and $p + dp$.

Since the volume of a phase cell is h^3

$$g(p)\, dp = \frac{\iiint \iiint dx\, dy\, dz\, dp_x\, dp_y\, dp_z}{h^3}$$

Now, $\iiint dx\, dy\, dz = V$ = volume occupied by the molecules in ordinary position space and $\iiint dp_x\, dp_y\, dp_z = 4\pi p^2\, dp$ = volume of a spherical shell between the radii p and $p + dp$ in the momentum space.

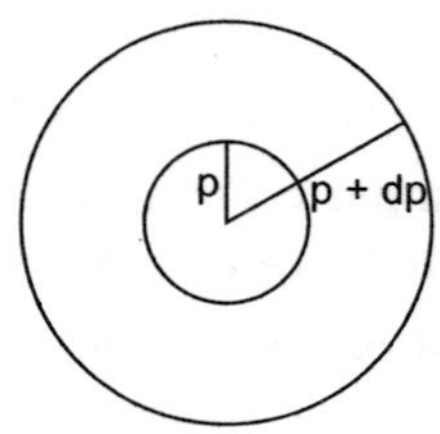

$$\therefore \qquad g(p)\, dp = \frac{4\pi p^2\, dp \,.\, V}{h^3}$$

Therefore, from equation (6.21)

$$n(p)\, dp = \frac{4\pi V}{h^3} \,.\, p^2\, e^{-\alpha} \,.\, e^{-\frac{\beta p^2}{2m}}\, dp \qquad \ldots(6.22)$$

The value of $e^{-\alpha}$ can then be calculated as below :

We know, $\int_0^\infty n(p)\,dp = \text{N}$

$$\therefore \quad \text{N} = \int_0^\infty n(p)\,dp = \frac{4\pi e^{-\alpha}\text{V}}{h^3}\int_0^\infty p^2 e^{-\frac{\beta p^2}{2m}}\,dp$$

$$= \frac{4\pi\text{V}}{h^3}\,.\,e^{-\alpha}\left[\frac{1}{4}\left(\frac{2m}{\beta}\right)^{\frac{3}{2}}.\,\pi^{\frac{1}{2}}\right]$$

Standard integral,

$$\int_0^\infty x^2 e^{-kx^2}\,dx = \frac{1}{4k}\sqrt{\frac{\pi}{k}}$$

here, $x = p\,;\, k = \dfrac{\beta}{2m}$

$$= \frac{e^{-\alpha}\,.\,\text{V}}{h^3}\left(\frac{2\pi m}{\beta}\right)^{\frac{3}{2}}$$

Hence, $e^{-\alpha} = \dfrac{\text{N}h^3}{\text{V}}\left(\dfrac{\beta}{2\pi m}\right)^{\frac{3}{2}}$

Substituting this result in equation (6.22), we get

$$n(p)\,dp = \frac{4\pi\text{V}}{h^3}\,.\,\frac{\text{N}h^3}{\text{V}}\left(\frac{\beta}{2\pi m}\right)^{\frac{3}{2}}.\,p^2\,e^{-\left(\frac{\beta}{2m}\right)p^2}\,.\,dp$$

$$= 4\pi\text{N}\left(\frac{\beta}{2\pi m}\right)^{\frac{3}{2}}.\,p^2\,e^{-\frac{\beta p^2}{2m}}\,dp \qquad \text{...(6.23)}$$

To obtain β, we calculate the *total energy* E of the system.

Now, $p^2 = 2m\varepsilon$ and $dp = \dfrac{m\,d\varepsilon}{p} = \dfrac{m\,d\varepsilon}{\sqrt{2m\varepsilon}}$ [ε is the energy corresponding to p.]

Equation (6.23) can then be written as

$$n(\varepsilon)\,d\varepsilon = 4\pi\text{N}\left(\frac{\beta}{2\pi m}\right)^{\frac{3}{2}}.\,2m\varepsilon\,.\,e^{-\beta\varepsilon}\,.\,\frac{m\,d\varepsilon}{\sqrt{2m\varepsilon}}$$

$$= 4\pi\text{N}\,.\,\frac{\beta}{2\pi m}\,.\,\frac{\sqrt{\beta}}{\sqrt{2\pi m}}\,.\,2m\varepsilon\,.\,\frac{m}{\sqrt{2m\varepsilon}}\,.\,e^{-\beta\varepsilon}\,d\varepsilon$$

$$= \frac{2\text{N}\,(\beta)^{\frac{3}{2}}}{\sqrt{\pi}}\sqrt{\varepsilon}\,e^{-\beta\varepsilon}\,d\varepsilon$$

The total energy is

$$E = \int_0^{\infty} \varepsilon\, n(\varepsilon)\, d\varepsilon$$

$$= \frac{2N\beta^{\frac{3}{2}}}{\sqrt{\pi}} \int_0^{\infty} \varepsilon^{\frac{3}{2}} e^{-\beta\varepsilon}\, d\varepsilon$$

$$= \frac{2N\beta^{\frac{3}{2}}}{\sqrt{\pi}} \cdot \frac{3}{4\beta^2}\sqrt{\frac{\pi}{\beta}} \qquad \left[\int_0^{\infty} x^{\frac{3}{2}} e^{-\beta x}\, dx = \frac{3}{4\beta^2}\sqrt{\frac{\pi}{\beta}}\right]$$

$$= \frac{3}{2} \cdot \frac{N}{\beta} \qquad \ldots(6.24)$$

Now, according to kinetic theory of gases the total energy of N molecules of an ideal gas at absolute temperature T is

$$E = \frac{3}{2} NkT \qquad \ldots(6.25)$$

Comparing equations (6.24) and (6.25)

$$\beta = \frac{1}{kT}$$

6.9.4 Fermi Distribution at Zero and Non-zero Temperatures

Fermi Energy (or Fermi Level)

We often express $\alpha = -\frac{E_F}{kT}$ where E_F is known as Fermi Energy

$$\therefore \qquad f(E_i) = \frac{1}{e^{\frac{E_i - E_F}{kT}} + 1}$$

At $T = 0$

$$f(E_i) = 1 \quad \text{when } E_i < E_F$$
$$= 0 \quad \text{when } E_i > E_F$$

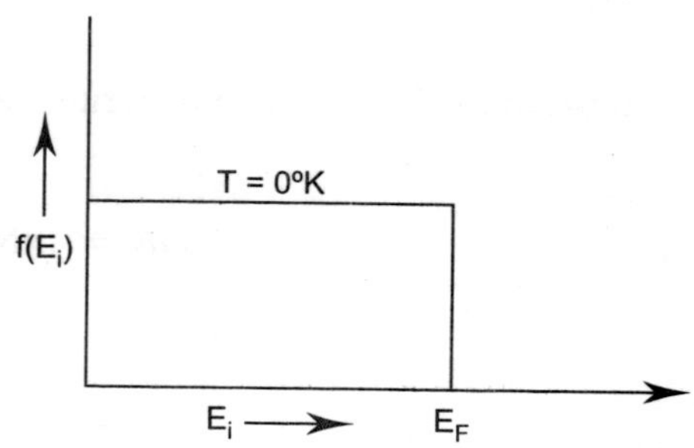

Thus at *absolute zero* of temperature, $f(E_i)$ is a step function.

In this case the probability of occupation of all states with energy *less than* $E_F = 1$ is *unity* and that of states with energy *higher than* $E_F = 1$ *is zero*.

Thus, the Fermi energy (or Fermi level) represents the *highest occupied* energy level at *absolute zero*. This is the energy value *upto which* all the energy states *are full at absolute zero*, and *above which* all the energy states are empty.

As the temperature increases, $f(E_i)$ changes from 1 to 0 more and more *gradually*. At *all non-zero temperatures* (*i.e.*, $T > 0$) the occupation index $f(E_i)$, *when* $E_i = E_F$ is given by,

$$f(E_i) = \frac{1}{e^0 + 1} = \frac{1}{2}$$

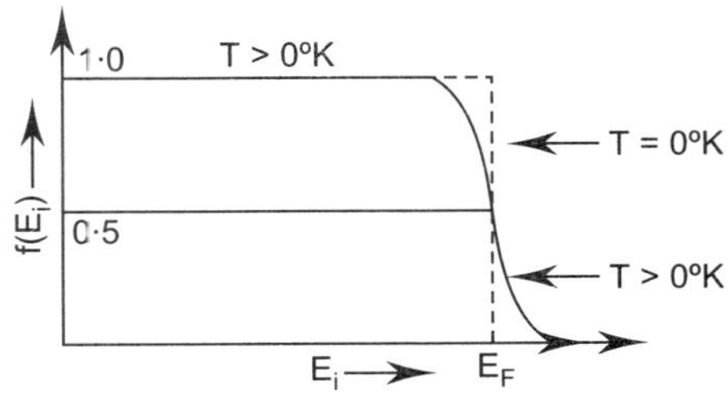

Hence, Fermi level is that energy level for which the probability of occupation (*i.e.*, occupation index) *at* $T > 0$ is $\frac{1}{2}$, that is 50% of the quantum states are occupied and 50% are empty. This gives the physical interpretation of Fermi energy *at a finite temperature*. If the temperature is not very large $f(E_i)$ varies *rapidly* from *about* unity to *about* zero over an energy range of a few times kT around E_F.

At *low* temperatures when $f(E_i)$ is *nearly* a step function the *distribution* is said to be *strongly degenerate*. When, however, the *steplike character is lost* (*i.e.*, at high temperatures) it is said to be nearly non-degenerate.

Energy Distribution of Free Electrons in Metal

The high electrical conductivity in metals is caused due to presence of free electrons in it. These free electrons move about freely and also collide with the fixed atoms in the metal and thus behave like *electron gas*. Electrons possess *half integral* spin angular momentum *i.e.*, $\frac{1}{2}\hbar \left(\text{or } \frac{h}{4\pi}\right)$. They are : Fermi particles and follow FD statistics.

Considering a continuous distribution of energy among the electrons we can write,

$$n(u)\,du = \frac{g(u)\,du}{e^{\alpha}\,.\,e^{\frac{E_i}{kT}} + 1} \quad \text{(from FD-statistics)}$$

where $n(u)$: number of electrons

$g(u)$: number of cells lying within the unit energy interval between $\left(u + \frac{1}{2}\right)$ and $\left(u - \frac{1}{2}\right)$

The number of cells in the momentum interval between p and $p + dp$ is,

$$g(p)\,dp = \frac{4\pi V p^2\,dp}{h^3}$$ in the case of *ordinary gas molecules* obeying MB statistics.

In case of *electrons*, obeying *FD statistics*, the value is given by,

$$g(p)\,dp = \frac{8\pi V}{h^3}\,.\,p^2\,dp$$

($\because$ the electrons have both left handed and right handed polarisation)

Under *non-relativistic case,*

$$u = \frac{1}{2}mv^2 \quad \text{and} \quad p = mv$$

$$\therefore \quad p^2 = m^2v^2 = 2m\,.\,\frac{1}{2}mv^2 = 2\,mu \quad \text{and so} \quad p^2 = 2\,mu$$

Hence, $$2p\,dp = 2m\,du \quad \text{or} \quad p\,dp = m\,du$$

Now, $$g(p)\,dp = \frac{8\pi V}{h^3} \cdot p \cdot p\,dp$$

and so, $$g(u)\,du = \frac{8\pi V}{h^3} \cdot (2mu)^{\frac{1}{2}} \cdot m\,du$$

$$= \frac{8\sqrt{2}\,\pi V}{h^3} \cdot m^{\frac{3}{2}}\, u^{\frac{1}{2}}\, du$$

From which, $$n(u)\,du = \frac{\dfrac{8\sqrt{2}\,\pi V}{h^3} \cdot m^{\frac{3}{2}}\, u^{\frac{1}{2}}\, du}{e^{\alpha} \cdot e^{\frac{E_i}{kT}} + 1}$$

$$= \frac{8\sqrt{2}\,\pi V}{h} \cdot \frac{m^{\frac{3}{2}}\, u^{\frac{1}{2}}\, du}{e^{\frac{(E_i - E_F)}{kT}} + 1}$$

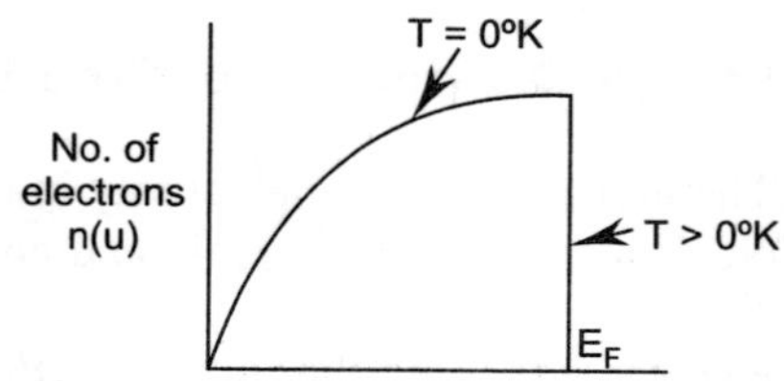

In the model of an electron gas in metals as T → 0°K, the electrons uniformly fill all the cells of the phase space within a sphere of volume $\frac{4}{3}\pi p_F^3 V$ where p_F is the *maximum momentum* of an electron at T = 0. The *number of states* in this sphere (taking into account the Exclusion principle and the two possible orientations of electron spin) is equal to *the number of electrons*.

Thus, $\frac{2}{h^3} \cdot \frac{4}{3}\pi p_F^3 V = N$, wherefrom $p_F = h\left(\frac{3}{8\pi} \cdot \frac{N}{V}\right)^{\frac{1}{3}}$

The maximum energy of the *electron* at T = 0°K is given by,

$$E_F = \frac{p_F^2}{2m} = \frac{h^2}{2m}\left(\frac{3}{8\pi} \cdot \frac{N}{V}\right)^{\frac{2}{3}} \qquad \left[\frac{N}{V} : \text{electron density}\right]$$

$$\therefore \qquad E_F^{3/2} = \frac{h^3}{(2m)^{3/2}} \cdot \frac{3}{8\Pi} \cdot \frac{N}{V} = \frac{3Nh^3}{16\sqrt{2}\Pi V m^{3/2}}$$

Average K.E. at 0°K

We have $$n(u)\,du = \frac{8\sqrt{2}\,\pi V}{h^3} \cdot \frac{m^{\frac{3}{2}}\, u^{\frac{1}{2}}\, du}{e^{\alpha} \cdot e^{\frac{E_i}{kT}} + 1} \text{(F.D. distribution of free electrons)}$$

or $$n(u)\,du = \frac{16\sqrt{2}\,\pi V}{3nh^3} \frac{3N}{2} \cdot m^{\frac{3}{2}}\, u^{\frac{1}{2}} \cdot \frac{du}{e^{\alpha} \cdot e^{\frac{E_i}{kT}} + 1}$$

Substituting the value of $\alpha \left(= \frac{-E_F}{kT}\right)$ and the value of $E_F^{3/2}$

we get, $$n(u)\,du = \frac{3}{2} N \cdot \frac{1}{(E_F)^{\frac{3}{2}}} \cdot u^{\frac{1}{2}} \frac{du}{e^{\frac{(E_i - E_F)}{kT}} + 1}$$

$$= \frac{3}{2} n (E_F)^{-3/2} u^{\frac{1}{2}} \frac{du}{e^{\frac{(E_i - E_F)}{kT}} + 1}$$

at T = 0°K, all electrons have *energies* $\leq E_F$

hence $(E_i - E_F)$ is *negative* and so $e^{\frac{(E_i - E_F)}{kT}} + 1 = 0$

and thus,

$$n(u)\, du = \frac{3}{2} N E_F^{-3/2} u^{\frac{1}{2}} du$$

If $\bar{u}$ be the average K.E. per electron at T = 0°K then,

$$\bar{u} = \frac{\int_0^{E_F} u \,.\, n(u)\, du}{n} = \frac{\frac{3}{2} N E_F^{-3/2} \int_0^{E_F} u^{\frac{3}{2}} du}{n}$$

$$= \frac{3}{2} \frac{N}{N} E_F^{-3/2} \int_0^{E_F} u^{\frac{3}{2}} du = \frac{3}{2} E_F^{-3/2} \left[\frac{2}{5} u^{\frac{5}{2}}\right]_0^{E_F}$$

$$= \frac{3}{2} E_F^{-3/2} \,.\, \frac{2}{5} E_F^{5/2}$$

or $$\bar{u} = \frac{3}{5} E_F \text{ (This is the average energy per electron at 0°K)}$$

Hence the TOTAL ENERGY at T = 0°K is

$$U = N \bar{u} = \frac{3}{5} N E_F$$

|Total number of electrons in a metal at T = 0°K|

The number of electrons that can have energy u is equal to the number of states $g(u)$ that have this energy (since, each state is limited to one electron)

$$\therefore \quad \int_0^{E_F} g(u)\, du = N \qquad \text{(N : number of free electrons contained in the metal)}$$

or $$\int_0^{E_F} g(u)\, du = \frac{8\sqrt{2}\,\pi V}{h^3} \,.\, m^{\frac{3}{2}} \int_0^{E_F} u^{\frac{1}{2}} du = N$$

wherefrom, $$N = \frac{16\sqrt{2}\,\pi V\, m^{\frac{3}{2}}}{3h^3} \,.\, E_F^{3/2}$$

and $$E_F(0) = \frac{h^2}{2m} \left(\frac{3N}{8\pi V}\right)^{\frac{2}{3}} \quad (E_F(0) \text{ : Fermi Energy at T = 0°K})$$

The Fermi energy is independent of the size and volume of the conductor as it only depends on $\frac{N}{V}$ (the number of electrons per unit volume *i.e.*, electron concentration).

Hence, $$E_F(0) = \frac{h^2}{8m}\left(\frac{3N_0}{\Pi}\right)^{2/3} \text{ where } N_0 = \frac{N}{V} \text{ (particle density at } T = 0° \text{ K)}$$

6.10 NECESSITY OF BOSE-EINSTEIN STATISTICS

Atomic particles possess spin angular momentum as one of their characteristic properties.

FD-statistics applies to those particles which have $\frac{1}{2}$ *integral spins*. In these cases the spin orientations are in opposite directions. These particles comply with the rules of quantum mechanics as: (1) Heisenberg's Uncertainty principle and (2) Pauli's Exclusion principle. Thus no two identical, indistinguishable particles stay in the same quantum state *i.e.*, only one particle can occupy a single quantum states. The types of particles are electrons $\left(s = \frac{1}{2\hbar}\right)$, positrons $\left(s = \frac{1}{2\hbar}\right)$, neutrons and protons with spin $\frac{1}{2}\hbar$. He^3, μ-meson (or muons) with spin $\frac{1}{2}\hbar$ etc. These are known as *Fermions*.

There are other types of particles whose *spin angular* momenta are: 0, $1\hbar$, $2\hbar$, $3\hbar$, ... etc., $\left(\text{where } \hbar = \frac{h}{2\pi}\right)$. *These, therefore possess integral spins*. They follow the uncertainty principle but *do not* comply with the exclusion principle. The distribution function for these particles are different from FD-distribution function. The particles of this above type are photons and phonons with spin = 1, deuterons (s = 1), π-mesons (or pions) with spin = 0, α-particles (*i.e.*, He^2-nucleus) with spin = 0 and so on. These are known as *Bosons*.

To obtain the distribution function for these particles one has to take recourse to *BE-statistics*, which is another form of statistics that *follows* the Heisenberg's uncertainty principle but *not* the Pauli's Exclusion principle. *In BE-statistics*, therefore, *more than one* particle can crowd into *one-single energy state*.

6.11 EQUIPARTITION OF ENERGY

If a system is in thermal equilibrium, each term of the energy that depends quadratically on a momentum or a spatial co-ordinate, contributes a mean energy of $\frac{1}{2}$ kT. This is known as the principle of equipartition of energy.

Let us consider a system which is described in terms of f position co-ordinates $(q_1, q_2, q_3, ..., q_f)$ and corresponding f momenta co-ordinates $(p_1, p_2, p_3, ..., p_f)$. The energy of the system can be written as

$$E = E(q_1, q_2, ..., q_f; p_1, p_2, ..., p_f).$$

Let the energy of a molecule of a system is quadratic in momentum p_i *i.e.*, $E_i(p_i) = a\,p_i^2$. Then the total energy can be broken into sum of two parts,

$$E = E_i(p_i) + E'(q_1, \ldots, p_f)$$

where $E'(q_1, \ldots, p_f)$ is a function of all position co-ordinates and momenta *except* p_i.

If the system is in thermal equilibrium at a particular temperature T, it will be characterised by a canonical distribution. The mean value of E_i can be expressed as

$$\overline{E_i} = \frac{\int_{-\infty}^{\infty} E_i\, e^{-\beta E(q_1 \ldots p_f)}\, dq_1 \ldots dp_f}{\int_{-\infty}^{\infty} e^{-\beta E(q_1 \ldots p_f)}\, dq_1 \ldots dp_f}$$

Since $$E = E_i + E'$$

$$\overline{E_i} = \frac{\int_{-\infty}^{\infty} E_i\, e^{-\beta E_i}\, dp_i \int_{-\infty}^{\infty} e^{-\beta E'}\, dq \ldots dp_f}{\int_{-\infty}^{\infty} e^{-\beta E_i}\, dp_i \int_{-\infty}^{\infty} e^{-\beta E'}\, dq \ldots dp_f}$$

The last integrals in the numerator and denominator extend over all q's and p's except p_i and are equal.

$\therefore$ $$\overline{E_i} = \frac{\int_{-\infty}^{\infty} E_i\, e^{-\beta E_i}\, dp_i}{\int_{-\infty}^{\infty} e^{-\beta E_i}\, dp_i}$$

Writing $$E_i = a\,p_i^2$$

$$\overline{E_i} = \frac{\int_{-\infty}^{\infty} e^{-\beta a p_i^2}\, a\,p_i^2\, dp_i}{\int_{-\infty}^{\infty} e^{-\beta a p_i^2}\, .dp_i}$$

Let us put $$y^2 = \beta a p_i^2$$

$\therefore$ $$y = \sqrt{\beta a}\; p_i \quad i.e.,\ dp_i = \frac{dy}{\sqrt{\beta a}}$$

Then, $$\overline{E_i} = \frac{\int_{-\infty}^{\infty} y^2\, e^{-y^2}\, dy}{\beta \int_{-\infty}^{\infty} e^{-y^2}\, dy}$$

Using standard formula of integration, we get

$$\overline{E_i} = \frac{1}{2\beta} = \frac{1}{2}\,kT, \text{ since } \beta = \frac{1}{kT}$$

This is the classical equipartition theorem.

6.12 DERIVATION OF PLANCK'S LAW APPLYING BOSE–EINSTEIN STATISTICS

Let us consider a black body radiation chamber of volume V. The chamber contains black body radiation in equilibrium with the walls of the chamber at temperature T. Thus inside the chamber we have a large number of indistinguishable photons of various frequencies. Photons do not obey Pauli exclusion principle, they are Bosons and follow Bose-Einstein distribution law. Hence we can apply Bose-Einstein statistics to photons.

Let $g(p)\,dp$ be the number of cells in phase space in which photons has momentum between p to $p + dp$. Since the volume of a cell in phase space is h^3,

$$g(p)\,dp = \frac{\iint \dots \int dx\,dy\,dz\,dp_x\,dp_y\,dp_z}{h^3} \quad \dots(6.26)$$

Now $\iiint dx\,dy\,dz = \text{V}$ = the volume occupied by photons in ordinary position space.

Also, $\iiint dp_x\,dp_y\,dp_z = 4\pi p^2\,dp$, the volume lying between spherical shells of radii p and $p + dp$ in the momentum space.

$$\therefore \quad g(p)\,dp = \frac{4\pi \text{V} p^2\,dp}{h^3} \quad \dots(6.27)$$

Now photon of a particular frequency has two states of polarisation. Hence the number of states in which photon may have a momentum between p to $p + dp$ is twice this number.

Hence,
$$g(p)\,dp = \frac{8\pi \text{V} p^2\,dp}{h^3} \quad \dots(6.28)$$

Since momentum of a photon is $p = \dfrac{h\nu}{c}$

$$\therefore \quad dp = \frac{h}{c}\,d\nu$$

$$\therefore \quad p^2\,dp = \frac{h^2\nu^2}{c^2}\frac{h}{c}\,d\nu = \frac{h^3\nu^2}{c^3}\,d\nu$$

$$\therefore \quad g(\nu)\,d\nu = \frac{8\pi \text{V}\nu^2\,d\nu}{c^3} \quad \dots(6.29)$$

We now introduce Bose-Einstein distribution law,

$$n(\nu)\,d\nu = \frac{g(\nu)\,d\nu}{e^{\alpha} e^{\frac{E_i}{kT}} - 1} \quad \dots(6.30)$$

where $n(\nu)\,d\nu$ is the number of photons lying between frequency ν to $(\nu + d\nu)$

Therefore from equation (6.29) and (6.30), we get

$$n(\nu)\, d\nu = \frac{8\pi V}{c^3} \cdot \frac{\nu^2\, d\nu}{e^{\alpha}\, e^{\frac{E}{kT}} - 1} \qquad \ldots(6.31)$$

Now the number of photons in the cavity is not conserved. Photons are created and destroyed, but the total radiant energy in the cavity is constant. As an example, two photons of energy $h\nu$ may be emitted by a simultaneous absorption of a single photon of energy $2h\nu$.

Thus $$\delta N = \sum_i \delta n_i \neq 0$$

This fact can be taken into account in Bose-Einstein distribution law by setting $\alpha = 0$

Hence $$n(\nu)\, d\nu = \frac{8\pi V}{c^3} \frac{\nu^2\, d\nu}{e^{\frac{h\nu}{kT}} - 1}, \text{ where } E = h\nu, \text{ and } e^{\alpha} = e^0 = 1$$

Hence the corresponding spectral energy density in the frequency range ν to $\nu + d\nu$ is

$$E(\nu)\, d\nu = \frac{h\nu}{V}\, n(\nu)\, d\nu$$

$$= \frac{8\pi h}{c^3} \frac{\nu^3\, d\nu}{\left(e^{\frac{h\nu}{kT}} - 1\right)} \qquad \ldots(6.32)$$

6.13 CLASSICAL LIMIT OF BOSE-EINSTEIN AND FERMI-DIRAC DISTRIBUTION FUNCTION

The Bose-Einstein and Fermi-Dirac distributions are given by

$$f_{BE}(E_i) = \frac{1}{e^{\alpha}\, e^{\frac{E_i}{kT}} - 1}$$

and $$f_{FD}(E_i) = \frac{1}{e^{\alpha}\, e^{\frac{E_i}{kT}} + 1}$$

Let us now suppose that $e^{\alpha} >> 1$. Then $e^{\alpha}\, e^{\frac{E_i}{kT}}$ becomes very very larger than unity so that both Bose-Einstein and Fermi-Dirac functions are reduced to the form

$$f(E_i) = \frac{1}{e^{\alpha}\, e^{\frac{E_i}{kT}}}$$

which is exactly the Maxwell-Boltzmann distribution function.

Thus quantum statistics gives the classical form in the limiting case $e^{\alpha} >> 1$.

Now
$$e^{-\alpha} = \frac{Nh^3}{V} \cdot \left(\frac{\beta}{2\pi m}\right)^{\frac{3}{2}}$$

or
$$e^{\alpha} = \frac{V}{N} \cdot \left(\frac{2\pi m\, kT}{h^2}\right)^{\frac{3}{2}}$$

Thus $e^{\alpha} >> 1$, if T is very large and $\frac{N}{V}$ is very small. Thus quantum distribution laws approach classical limit if temperature T is very high and the particle density $\frac{N}{V}$ is very low. e^{α} is known as Degeneracy Parameter. Let $e^{\alpha} = A$. If $A << 1$, the quantum distribution will not be according to Maxwell-Boltzmann statistics but according to either Bose-Einstein or Fermi-Dirac statistics. In this case the collection of particle is said to be in the state of degeneracy.

If $A >> 1$, *i.e.*, in the case of weak degeneracy, the quantum statistics will reduce to Maxwell-Boltzmann statistics.

6.14 DETERMINATION OF FERMI ENERGY AT 0°K

The number of phase cell in which electron may have momentum lying between p to $(p + dp)$ is

$$g(p)\, dp = \frac{\int \ldots \int dx\, dy\, dz\, dp_x\, dp_y\, dp_z}{h^3}$$

$$= \frac{4\pi V p^2\, dp}{h^3}$$

because $\iiint dx\, dy\, dz = V$ and $\iiint dp_x\, dp_y\, dp_z = 4\pi p^2\, dp$ = the volume of a spherical shell between the spheres of radii p and $p + dp$ in momentum space.

Now
$$p^2 = 2mE \Rightarrow p = \sqrt{2mE}$$

∴
$$dp = \frac{m}{p}\, dE = \frac{m dE}{\sqrt{2mE}}$$

∴
$$g(E)\, dE = \frac{4\pi V}{h^3} \cdot 2mE \frac{m dE}{\sqrt{2mE}}$$

$$= \frac{4\pi V}{h^3} \cdot \sqrt{2m} \cdot m \sqrt{E}\, dE$$

$$= \frac{4\sqrt{2}\,\pi V}{h^3} m^{\frac{3}{2}} \sqrt{E}\, dE$$

since the number of electrons having energy lying between E to E + dE is

$$N(E)\, dE = g(E) f(E)\, dE$$

$$= \frac{8\sqrt{2}\,\pi V}{h^3} m^{\frac{3}{2}} \sqrt{E}\, dE \cdot \frac{1}{e^{\frac{(E - E_F)}{kT}} + 1} \qquad \ldots(6.33)$$

The multiplication by 2 comes as the two electrons with opposite spin can be accomodated in a phase cell. So the total number of electrons N (integrating the above expression) is

$$N = \frac{8\sqrt{2}\,\pi V}{h^3} m^{\frac{3}{2}} \int_0^\infty \frac{\sqrt{E}\, dE}{e^{\frac{(E - E_F)}{kT}} + 1}$$

The above integral cannot be evaluated analytically at any arbitrary temperature T. It can be evaluated by numerical methods. However, we can evaluate the integral at T = 0 using

$$f(E) = 1 \text{ for } E < E_F(0)$$

$$f(E) = 0 \text{ for } E > E_F(0)$$

Hence the total number of electrons

$$N = \int_0^{E_F(0)} g(E) f(E)\, dE + \int_{E_F(0)}^{\infty} g(E) f(E)\, dE$$

$$= \int_0^{E_F(0)} g(E)\, dE \quad \text{[using the condition given above]}$$

$$= \int_0^{E_F(0)} \frac{8\sqrt{2}\,\pi V}{h^3} m^{\frac{3}{2}} E^{\frac{1}{2}}\, dE$$

$$= \frac{8\sqrt{2}\,\pi V}{h^3} m^{\frac{3}{2}} \cdot \frac{2}{3} [E_F(0)]^{\frac{3}{2}}$$

$$\therefore \quad [E_F(0)]^{\frac{3}{2}} = \frac{3Nh^3}{16\sqrt{2}\,\pi V\, m^{\frac{3}{2}}}$$

$$= \frac{3N}{\pi V} \left(\frac{h}{2\sqrt{2m}} \right)^3$$

$$\therefore \quad E_F(0) = \left(\frac{h}{2\sqrt{2m}} \right)^2 \left(\frac{3N}{\pi V} \right)^{\frac{2}{3}}$$

$$= \frac{h^2}{8m}\left(\frac{3N_0}{\pi}\right)^{\frac{2}{3}} \quad \text{where, } E_F(0) : \text{Fermi Energy at } 0° \text{ K}$$

$$N_o = \frac{N}{V}$$

Calculation of Fermi Level in Metals

To calculate the Fermi energy for Cu at T= 0, the molecular weight of Cu is 63.57 and density of Cu is 8.94 gm/cc.

Now of have
$$E_F(0) = \frac{h^2}{8m}\left(\frac{3N}{\pi V}\right)^{\frac{2}{3}}$$

Now
$$h = 6.6 \times 10^{-34} \text{ J.s.}$$
$$m = 9.1 \times 10^{-31} \text{ kg.}$$

Now
$$\frac{N}{V} = \frac{6.023 \times 10^{23} \times 10^6 \times 8.94}{63.57} \text{ m}^{-3}$$
$$= 8.47 \times 10^{28} \text{ m}^{-3} \qquad \text{(Assuming valancy of Cu = 1)}$$

$\therefore$
$$E_F(0) = \frac{(6.6 \times 10^{-34})^2}{8 \times 9.1 \times 10^{-31}} \cdot \left(\frac{3 \times 8.47 \times 10^{28}}{3.14}\right)^{\frac{2}{3}} \text{ J}$$
$$= \frac{43.56}{72.8} \times 10^{-37} \times 4.031 \times 10^{18} \sqrt[3]{100} \text{ J}$$
$$= 2.412 \times 4.642 \times 10^{-19} \text{ J}$$
$$= \frac{11.197}{1.6} \text{ eV}$$
$$= 6.998 \text{ eV.}$$

Fermi Temperature :

Fermi temperature, T_F is defined as

$$kT_F = E_F(0), \qquad k = \text{Boltzmann constant.}$$

T_F is of the order of 10^4 °K

6.15 AVERAGE ENERGY OF AN ELECTRON AT 0°K

The number of phases in which electron may have momentum lying between p to $(p + dp)$ is

$$g(p)\,dp = \frac{\int \cdots \int dx\, dy\, dz\, dp_x\, dp_y\, dp_z}{h^3}$$

Now $\iiint dx\, dy\, dz = V$

and $\iiint dp_x\, dp_y\, dp_z = 4\pi p^2\, dp$ = Volume of a spherical shell between the spheres of radii p and $p + dp$ in momentum space.

$$\therefore \quad g(p)\, dp = \frac{4\pi V p^2}{h^3}\, dp \qquad \ldots(6.34)$$

Now $$p^2 = 2mE \Rightarrow p = \sqrt{2mE}$$

$$\therefore \quad dp = \frac{m}{p}\, dE = \frac{m dE}{\sqrt{2mE}}$$

Therefore from equation (6.34)

$$g(E)\, dE = \frac{4\pi v}{h^3}\, 2mE \,.\, \frac{m dE}{\sqrt{2mE}}$$

$$= \frac{4\sqrt{2}\,\pi V\, m^{\frac{3}{2}}}{h^3} \,.\, \sqrt{E}\, dE$$

Now the number of electrons having energy lying between E to E + dE is

$$N(E)\, dE = \frac{8\sqrt{2}\,\pi V\, m^{\frac{3}{2}}\sqrt{E}\, dE}{h^3} \,.\, \frac{1}{e^{\frac{(E - E_F)}{kT}} + 1} \qquad \ldots(6.35)$$

The multiplication by 2 comes as two electrons with opposite spin can be accomodated in a phase cell.

Now the average energy of an electron is

$$\bar{U} = \frac{1}{N}\int_0^{\infty} E\, N(E)\, dE$$

$$= \frac{1}{N}\int_0^{\infty} \frac{8\sqrt{2}\,\pi V\, m^{\frac{3}{2}}}{h^3}\, E\sqrt{E}\, dE\; \frac{1}{e^{\frac{(E - E_F)}{kT}} + 1} \qquad \ldots(6.36)$$

The above integral can not be evaluated analytically at any arbitrary temperature T. It can be evaluated by numerical method. However we can evaluate this integral at T = 0.

Now, $$\left.\begin{aligned} f(E) &= 1 \quad \text{for} \quad E < E_F(0) \\ f(E) &= 0 \quad \text{for} \quad E > E_F(0) \end{aligned}\right\} \qquad \ldots(6.37)$$

Hence the average energy,

$$\bar{U} = \frac{1}{N}\int_0^{E_F(0)} E\, f(E)\, g(E)\, dE + \frac{1}{N}\int_{E_F(0)}^{\infty} E\, f(E)\, g(E)\, dE$$

$$= \frac{1}{N}\int_0^{E_F(0)} E\, g(E)\, dE \text{ (using condition given in equation (6.37))}$$

$$= \frac{1}{N}\int_0^{E_F(0)} \frac{8\sqrt{2}\,\pi V\, m^{\frac{3}{2}}}{h^3} E^{\frac{3}{2}}\, dE$$

$$= \frac{8\pi\sqrt{2}V\, m^{\frac{3}{2}}}{Nh^3}\ \frac{2}{5}\cdot[E_F(0)]^{\frac{5}{2}}$$

$$= \frac{16\sqrt{2}\,\pi V\, m^{\frac{3}{2}}}{3Nh^3}\cdot\frac{3}{5}\cdot E_F(0)\,.\,[E_F(0)]^{\frac{3}{2}}$$

$$= \frac{3}{5}\,.\,E_F(0)\,.\,\frac{[E_F(0)]^{\frac{3}{2}}}{[E_F(0)]^{\frac{3}{2}}} = \frac{3}{5}E_F(0),$$

substituting from equation (6.36) for calculation of $E_F(0)$.

Example. *Show that for an electron, the fraction of occupied energy states ΔE above the Fermi level is equal to the fraction of unoccupied energy states ΔE below the Fermi level.*

Solution. The distribution function for energy ΔE above Fermi level is

$$f(E_F + \Delta E) = -\frac{1}{e^{\frac{(E_F + \Delta E - E_F)}{kT}} + 1} = -\frac{1}{e^{\frac{\Delta E}{kT}} + 1} \quad \ldots(i)$$

The distribution function for the energy level ΔE below the Fermi level

$$f(E_F - \Delta E) = \frac{1}{e^{\frac{(E_F - \Delta E - E_F)}{kT}} + 1} = \frac{1}{e^{-\frac{\Delta E}{kT}} + 1}$$

Probability of the non-occupation of the energy state $(E_F - \Delta E)$ is

$$1 - f(E_F - \Delta E) = 1 - \frac{1}{e^{-\frac{\Delta E}{kT}} + 1} = \frac{e^{-\frac{\Delta E}{kT}}}{e^{-\frac{\Delta E}{kT}} + 1}$$

$$= \frac{1}{e^{\frac{\Delta E}{kT}} + 1} \quad \ldots(ii)$$

Now expressions (*i*) and (*ii*) are identical. Hence the fraction of occupied energy state ΔE above the Fermi level is equal to the fraction of unoccupied energy states below E_F the Fermi level.

6.16 COMPARISON OF M.B., B.E. AND F.D. STATISTICS

	MAXWELL-BOLTZMANN	BOSE-EINSTEIN	FERMI-DIRAC
1. Applies to systems of	Identical, *distinguishable* particles	Identical, indistinguishable particles that do not obey exclusion principle.	Identical, indistinguishable particles that obey exclusion principle.
2. Category of particles	Classical	Bosons	Fermions
3. Examples	Molecules of a gas	Photons in a cavity. Phonons in a solid, licuid helium at low temperatures.	Free electrons in a metal. Electrons in a star whose atoms have collapsed (white dwarf star)
4. Properties of particles	*Any spin*; particles far apart from each other (*i.e.*, dilute gas)	Spin 0, 1, 2, ... particles are very close wave functions are symmetric under interchange of two bosons.	Spin $\frac{1}{2}, \frac{3}{2}, \frac{5}{2}, \ldots$ Antisymmetric under interchange of two fermions.
5. Properties of distribution	No limit to number of particles per state	No limit to number of particles per state. At *low* energies the number of particles per state is more than M.B. approaches the value for M.B. at high energies	Never more than *one* particle per state. At low energies fewer particles per state than MB, approaches the value for MB at high energies.
6. Distribution function $f(E_i)$	$\frac{N}{V}\left(\frac{h^2}{2\pi m kT}\right)^{\frac{3}{2}} . e^{-\frac{E_i}{kT}}$	$\frac{1}{e^{\alpha} . e^{\frac{E_i}{kT}} - 1}$	$\frac{1}{e^{\frac{(E_i - E_f)}{kT}} + 1}$.

The three different laws can be summerised by a single equation

$$\frac{n_i}{g_i} = \frac{1}{e^{\alpha} e^{E_i/kT} + \nu}$$

where $\nu = 0$ for MB statistics

$\nu = -1$ for B.E statistics

and $\nu = +1$ for F.D statistics.

The ratio $\frac{n_i}{g_i}$ $\left[i.e., \frac{\text{(number of particles)}}{\text{(number of states)}} \text{ for the } i^{\text{th}} \textit{ energy level} \right]$

is known as the occupation-index of the ith state

when $\frac{n_i}{g_i} << 1$ all the three statistics give the *same* distribution and $\frac{n_i}{g_i} = \frac{1}{e^{\alpha} e^{E_i/kT}}$

IN FOCUS

- Number of microstates for a particular macrostate

  Thermodynamic Probability $W = \dfrac{N!}{n_1!\,n_2!\,n_3!\ldots}$

- **Maxwell-Boltzmann (MB) distribution function** is

$$f_{MB}(E_i) = e^{-\alpha}\, e^{-\frac{E_i}{kT}} = \frac{N}{V}\left(\frac{h^2}{2\pi mk\,T}\right) e^{-\frac{E_i}{kT}}$$

  Applicable for identical and distinguishable particles. The particle can have any spin and any number of particles can occupy a phase cell.

- **Bose-Einstein (BE) distribution function** is given by

$$f_{BE}(E_i) = \frac{1}{e^{\alpha}\, e^{\frac{E_i}{kT}} - 1} \quad \text{where } e^{\alpha} = \frac{V}{N}\left(\frac{2\pi mkT}{h^2}\right)^{\frac{3}{2}}$$

  Applicable for identical indistinguishable particles and the particles obey uncertainty principle but do not obey Pauli exclusion principle. The particles have zero or integral spins (0, 1, 2, ...) and any number of particles can occupy a phase cell. The particles are known as Bosons. At high temperature and/or at low density B.E. distribution approaches M.B. distribution.

- **Fermi-Dirac distribution function** is given by

$$f_{FD}(E_i) = \frac{1}{e^{\alpha}\, e^{\frac{E_i}{kT}} + 1}$$

  The particles are identical, indistinguishable and obey both uncertainty principle and exclusion principle. The particles have half integral spin $\left(\frac{1}{2}, \frac{3}{2}, \frac{5}{2} \ldots\right)$ and can accommodate only one particle per state. At high temperature and/or at low density the distribution approaches M.B. distribution.

- Fermi Energy at 0°K is given by $E_F(0) = \dfrac{h^2}{8m}\left(\dfrac{3N}{\pi V}\right)^{\frac{2}{3}}$

Solved Problems

Problem 1. *In a system of 14 distinguishable particles distributed in two equally probable halves of a box, determine the probability of distribution (10, 4), (14, 0), (7, 7).*

Solution. Given,

Total number of particles $n = 14$

Total number of compartments $C = 2$

$\therefore$ Total number of microstates $= C^n = 2^{14} = 16384$

(*i*) Number of microstates in the macrostate (10, 4)

$$W_{(10,\,4)} = {}^nC_{n_1} = {}^nC_{n_2}$$

$$= {}^{14}C_{10} = {}^{14}C_4 = \frac{14\,!}{10\,!\,4\,!} \qquad [\because n_1 + n_2 = 14 = n\,;\, n_2 = 14 - n_1]$$

$$= 1001$$

$\therefore$ Probability of the macrostate (10, 4)

$$P_{(10,\,4)} = \frac{W_{(10,\,4)}}{C^n} = \frac{1001}{2^{14}} = \frac{1001}{16384}.$$

(*ii*) Number of microstates in the macrostate (7, 7)

$$W_{(7,\,7)} = {}^{14}C_7 = \frac{14\,!}{7\,!\,7\,!} = 3432$$

$\therefore$ Probability of the macrostate (7, 7)

$$W_{(7,\,7)} = \frac{W_{(7,7)}}{C^n} = \frac{3432}{2^{14}} = \frac{3432}{16384}$$

(*iii*) Number of microstates in the macrostate (14, 0)

$$W_{(14,\,0)} = {}^{14}C_0 = \frac{14\,!}{14\,!\,.\,0\,!} = 1$$

$\therefore$ Probability of the macrostate (14, 0)

$$P_{(14,\,0)} = \frac{W_{(14,\,0)}}{C^n} = \frac{1}{2^{14}} = \frac{1}{16384}.$$

Problem 2. *Write down the various microstates and macrostates for a system of three particles in 2 compartments when*

(*i*) *The particles are distinguishable*
(*ii*) *The particles are indistinguishable*

Solution. Here Given,the number of particles $n = 3$

Number of compartments $C = 2$

Total number of macrostates $= n + 1 = 4$

These are [(0, 3), (1, 2), (2, 1), (3, 0)]

(*i*) When the particles are distinguishable. Let a, b, c

Number of microstates in macrostates (0, 3) $= {}^nC_{n_1} = {}^nC_{n_2}$

where $n_1 = 0,\ n_2 = 3 = {}^3C_0 = {}^3C_3$

$= 1$ [0, *abc*]

Number of microstates in the macrostate (1, 2)

$$= {}^3C_1 = {}^3C_2 = \frac{3!}{1!.2!} = 3$$

= 3 [(*a*, *bc*), (*b*, *ac*), (*c*, *ab*)]

Number of microstates in the macrostate (2, 1)

$$= {}^3C_2 = {}^3C_1 = \frac{3!}{2!.1!} = 3$$

= 3 [(*ab*, *c*), (*bc*, *a*), (*ca*, *b*)]

Number of microstates in the macrostate (3, 0) $= {}^nC_3 = {}^nC_0$

$= {}^3C_3 = 1 =$ [*abc*, 0]

Total number of microstates $= C^n = 2^3 = 8$

In the above case it is $= 1 + 3 + 3 + 1 = 8$

(*ii*) When the particles are indistinguishable say *a*, *a*, *a*. In such a case each macrostate will give rise to only one microstate as under :

[0, *aaa*] [*a*, *aa*] [*aa*, *a*] [*aaa*, 0]

Problem 3. *Calculate (a) the number of macrostates for four particles having a total energy of 6 $\in$, the energy levels are equaly spaced.*

(b) What will be the total number of microstates ?

(c) What will be the thermodynamic probability when the four particles are distributed as one in each cell ?

Solution. The system has four distinguishable particles and the total energy $6\in$. The energy levels are equally spaced. There are seven energy levels in all 0, $\in$, $2\in$, $3\in$, $4\in$, $5\in$, $6\in$ and we are to accommodate the given four particles in these energy levels so that the total energy becomes $6\in$.

So subject to these conditions, nine possible distributions are sketched below :

Row Number	*0*	$\in$	$2\in$	$3\in$	$4\in$	$5\in$	$6\in$	*Total energy*
I	3						1	$6\in$
II	2	1				1		$6\in$
III	2		1		1			$6\in$
IV	2			2				$6\in$
V	1	2			1			$6\in$
VI	1	1	1	1				$6\in$
VII	1		3					$6\in$
VIII		3		1				$6\in$
IX		2	2					$6\in$

(*a*) From the above sketch, the total number of macrostates = 9

(*b*) For Row I, the accommodation of four particles can be made in

$$^4C_1 \text{ ways} = 4$$

For Row II, in

$$^4C_2 \text{ ways} \times 2 = 6 \times 2 = 12 \text{ ways}$$

For Row III, in

$$^4C_2 \times 2 = 6 \times 2 = 12 \text{ ways}$$

For Row IV, in $^4C_2$ ways = 6 ways

For Row V, in $^4C_2 \times 2$ ways = 6×2 = 12 ways

For Row VI, in $\lfloor 4$ ways = 24 ways

For Row VII, in $^4C_3$ ways = 4 ways

For Row VIII, in $^4C_1$ ways = 4 ways

For Row IX, in $^4C_2$ ways = 6 ways

Hence, the total number of microstates

$$= 4 + 12 + 12 + 6 + 12 + 24 + 4 + 4 + 6 = 84.$$

(*c*) Here, one particle in each level represents the row number (VI). For this macrostate, microstates is 24. Hence the required thermodynamic probability is 24.

Problem 4. *Three distinguishable particles have a total energy of 9 units. But the particles are restricted to energy levels from 0 to 4. Calculate the number of macrostates and microstates.*

Solution. Let us consider the total energy be represented by 9ε and energy levels are equally spaced.

Let the particles be designated as A, B and C. There will be five different energy levels which are sketched here.

Thus,the number of macrostates = 3

In A, the number of microstates = $^3C_1 = 3$

In B, the number of microstates = $^3C_3 = 1$

In C, the number of microstates = $\lfloor 3 = 6$

Hence, the total number of microstates = 3 + 1 + 6 = 10

Energy Level	*A*	*B*	*C*
0			
ε	1		
2ε			1
3ε		3	1
4ε	2		1

Problem 5. *Six distinguishable particles are distributed over three non-degenerate levels of energies 0, ε, and 2ε. Calculate the total number of microstates of the system. Find the total energy of the distribution for which the probability is a maximum.* [Kolkata University]

Solution. Given that the levels are non-degenerate, so there is only one state associated with each energy. Since each of the 6 particles can be in any of the 3 energy levels, the total number of ways of arranging the particles, *i.e.*, the total number of microstates are $3^6 = 729$.

Let the number of particles in the three energy states be n_1, n_2 and n_3 respectively, where $n_1 + n_2 + n_3 = 6$, the total number of particles.

As the particles are distinguishable, the number of ways of choosing n_1, n_2 and n_3 particles from 6 particles is

$$W = \frac{6!}{n_1!\,n_2!\,n_3!},$$

where W gives the thermodynamic probability. It will be maximum when $n_1!\,n_2!\,n_3!$ is a minimum.

By inspection, we find that $n_1!\,n_2!\,n_3!$ is minimum when $n_1 = n_2 = n_3 = 2$.

∴ The corresponding total energy of the distribution is

$$= 0 \times n_1 + \varepsilon \times n_2 + 2\varepsilon \times n_3$$

$$= 2\varepsilon + 2\varepsilon \times 2 = 6\varepsilon.$$ **Ans.**

Problem 6. *Three distinguishable particles each of which can be in one of the ε, 2ε, 3ε, 4ε energy states have total energy 6ε. Find all possible number of distributions of all particles in the energy states. Find the number of microstates in each case.* [WBUT]

Solution. Here, the system has three distinguishable particles and have the total energy 6ε. There are four energy levels in all ε, 2ε, 3ε, 4ε and we have to accommodate the given three particles in these energy levels so that total energy becomes 6ε, so according to the conditions, three possible distribution are sketched below :

Row Number	*ε*	*2ε*	*3ε*	*4ε*	*Total Energy*
I	2	0	0	1	6ε
II	0	3	0	0	6ε
III	1	1	1	0	6ε

Here, the total number of macrostates = 3.

They are (2, 0, 0, 1), (0, 3, 0, 0) and (1, 1, 1, 0). Consider three particles are A, B and C.

Number of microstates in the macrostates are given below :

Row Number	*ε*	*2ε*	*3ε*	*4ε*	*Number of Microstates*
I	*ab* *bc* *ca*	0 0 0	0 0 0	*c* *a* *b*	3
II	0	*abc*	0	0	1
III	*a*	*b*	*c*	0	6
	b	*c*	*a*	0	
	c	*a*	*b*	0	
	b	*a*	*c*	0	
	c	*b*	*a*	0	
	a	*c*	*b*	0	

Total number of microstates = 10

Problem 7. *Five identifiable particles are distributed in three non-degenerate levels with energies 0, ε and 2ε. Determine the most probable distribution for a total energy 3ε.*

Solution. According to the conditions, let us consider the number of particles occupying the three energy states be n_1, n_2 and n_3 respectively, where $n_1 + n_2 + n_3 = 5$, the total number of particles. As the particles are identifiable, the number of ways of choosing the particle is $W = \dfrac{5!}{n_1!\,n_2!\,n_3!}$

The energy of the system

$$0 \times n_1 + \varepsilon \times n_2 + 2\varepsilon \times n_3 = 3\varepsilon$$

or,
$$n_2 + 2n_3 = 3 \qquad \ldots(i)$$

Thus, when W is a maximum subject to the conditions that,

If
$$n_2 = 1,\ n_3 = \frac{3-1}{2} = 1,\ n_1 = 5 - (1 + 1) = 3$$

or, if
$$n_2 = 3,\ n_3 = 0,\ n_1 = 5 - 3 = 2$$

No other distributions is possible.

For
$$n_1 = 3,\ n_2 = 1,\ n_3 = 1,\ W = \frac{5!}{3!1!1!} = 20$$

For
$$n_1 = 2,\ n_2 = 3,\ n_3 = 0,\ W = \frac{5!}{2!3!0!} = 10$$

So we can say, the most probable distribution is

$$n_1 = 3,\ n_2 = 1 \text{ and } n_3 = 1.$$

Problem 8. *Consider that there are a large number of particles each of mass 0.1 gm, all lying in a box at an equilibrium temperature 300° K. Find the expression for the probability that any one of them will spontaneously fly to a height of 1.0 Å. Assume that the particles obey MB statistics.*

Solution. We know from Maxwell Boltzmann (MB) statistics that the required probability is

$$P = e^{-\frac{E}{kT}}$$

where
$$E = mgh \text{ in which}$$

$$m = 0.1 \text{ gm},\ g = 980 \text{ cm/sec}^2,\ h = 1 \text{ Å} = 1 \times 10^{-8} \text{ cm}$$

$$k = 1.38 \times 10^{-16} \text{ erg/°K and } T = 300° \text{ K}$$

∴
$$P = \exp\left(-\frac{mgh}{kT}\right) = \exp\left\{-\frac{0.1 \times 980 \times 10^{-8}}{1.38 \times 10^{-16} \times 300}\right\}$$

$$= \exp\left[-\frac{9.8}{4.14} \times 10^6\right]. \textbf{ Ans.}$$

Problem 9. *Consider a two particles system each of which can exist in states ε_1, ε_2 and ε_3. What are the possible states if the particles are (i) bosons ; (ii) fermions.* [WBUT]

Solution.

(*i*) **Bosons :** In this case, the bosons particles are indistinguishable but any number can be accommodated in one cell. So, it has 6 possible distributions which are given below :

No. of Combinations	ε_1	ε_2	ε_3
I	*aa*	X	X
II	X	*aa*	X
III	X	X	*aa*
IV	*a*	*a*	X
V	*a*	X	*a*
VI	X	*a*	*a*

(*ii*) **Fermions :** Fermions are also indistinguishable but has some restrictions but has some restrictions, *i.e.*, one and only one can be accommodated in one cell. So it has three possible arrangement which are given below :

No. of Combinations	ε_1	ε_2	ε_3
I	*a*	*a*	X
II	*a*	X	*a*
III	X	*a*	*a*

Therefore, the total number of arrangements are 3.

Problem 10. *A system can take only three energy states, $E_1 = 0$, $E_2 = 1.38 \times 10^{-14}$ erg, $E_3 = 2.76 \times 10^{-14}$ erg. These three states occur in 2, 5 and 4 different ways, respectively. Calculate the probability that at temperature 100° K, the system may be*

(i) In one of the microstates of energy E_3, and (ii) In ground state E_1
Given, Boltzmann constant $k = 1.38 \times 10^{-16}$ erg/°K. Assume MB statistics.

Solution. Using Maxwell Boltzmann statistics, the probability of occurrence for each of the three states is given by

(*a*)
$$g_1\, e^{-\frac{E_1}{kT}} = 2 \times e^{-\frac{0}{kT}} = 2 \times 1 = 2$$

(*b*)
$$g_2\, e^{-\frac{E_2}{kT}} = 5 \times e^{-\frac{1.38 \times 10^{-14}}{1.38 \times 10^{-16} \times 100}} = 5e^{-1} = \frac{5}{e}.$$

$$= \frac{5}{2.716} = 1.83$$

(*c*)
$$g_3\, e^{-\frac{E_3}{kT}} = 0.4 \times \exp\left[-\frac{2.76 \times 10^{-14}}{1.38 \times 10^{-16} \times 100}\right]$$

$$= 4e^{-2} = \frac{4}{e^2} = 0.54$$

Therefore,
$$\sum_{i=1}^{3} g_i\, e^{-\frac{E_i}{kT}} = g_1\, e^{-\frac{E_1}{kT}} + g_2\, e^{-\frac{E_2}{kT}} + g_3\, e^{-\frac{E_3}{kT}}$$

$$= 2 + 1.83 + 0.54 = 4.37$$

(*i*) Thus the desired probability for the system to be in any one microstates of E_3 is

$$P(E_3) = \frac{g_3 e^{-\frac{E_3}{kT}}}{\sum_{i=1}^{3} g_i e^{-\frac{E_i}{kT}}} = \frac{0.541}{4.37} = 0.1236$$

(*ii*) The probability for the system to be in the ground state is

$$P(0) = \frac{g_1 e^{-\frac{E_1}{kT}}}{\sum_{i=1}^{3} g_i e^{-\frac{E_i}{kT}}} = \frac{2}{4.37} = 0.457$$

Problem 11. *A system consists of 5 particles arranged in two compartments. The first compartment is divided into 6 cells and the second into 8 cells. The cells are of equal size. Calculate the number of microstates in the macrostate (2, 3) if the particle obeys Fermi-dirac statistics.*

[G.N.D. University]

Solution. According to the Fermi-Dirac statistics, the number of microstates in the macrostate $(n_1, n_2, \dots n_k)$ or thermodynamic probability of the macrostate $W(n_1, n_2, \dots, n_k)$ is given by

$$W_{(n_1, n_2 \dots, n_k)} = \prod_{i=1}^{k} \frac{g_i!}{n_i!(g_i - n_i)!}$$

Here $k = 2$,

$$\therefore \quad W_{(n_1, n_2)} = \frac{g_1!}{n_1!(g_1 - n_1)!} \times \frac{g_2!}{n_2!(g_2 - n_2)!}$$

Here given $g_1 = 6$, $g_2 = 8$, $n_1 = 2$, $n_2 = 3$

$$\therefore \quad W_{(2, 3)} = \frac{6!}{2!.4!} \times \frac{8!}{3!.5!}$$

$$= 15 \times 56 = 840.$$

Problem 12. *A system has 7 particles arranged in two compartments. The first compartment has 8 cells and the second has 10 cells (all cells of equal size). Calculate the number of microstates in the macrostate (3, 4) when particles obey F.D. Statistics.* [Pune University]

Solution. According to Fermi-Dirac statistics, the number of microstates in the macrostate $(n_1, n_2, \dots, n_k)$ or thermodynamic probability of the macrostate $W_{(n_1, n_2, \dots, n_k)}$ is given by

$$W_{(n_1, n_2, \dots, n_k)} = \prod_{i=1}^{k} \frac{g_i}{n_i!(g_i - n_i)!}$$

$$\therefore \quad \text{Here } k = 2, \quad W_{(n_1, n_2)} = \frac{g_1!}{n_1!(g_1 - n_1)!} \times \frac{g_2!}{n_2!(g_2 - n_2)!}$$

$$W_{(3, 4)} = \frac{8!}{3!5!} \times \frac{10!}{4!6!} = \frac{8 \times 7 \times 6}{6} \times \frac{10 \times 9 \times 8 \times 7}{4 \times 3 \times 2}$$

$$= 56 \times 210 = 11760.$$

Problem 13. *Calculate the fermi energy at 0° K of metallic silver containing one free electron per atom. The density and atomic weight of silver is 10.5 g/cm³ and 108, respectively. Also determine the internal energy of the electron gas per unit volume at 0° K.* [Kolkata University]

Solution. According to Fermi, We have

$$E_F(0) = \frac{h^2}{8m}\left(\frac{3N}{\pi V}\right)^{\frac{2}{3}}$$

The number density of free electrons is given by

$$\frac{N}{V} = \frac{6.023 \times 10^{23} \text{ atom/mole}}{108 \text{ g/mole}} \times 10.5 \text{ g/cm}^3$$

$$= 5.9 \times 10^{22} \text{ cm}^{-3} = 5.9 \times 10^{28} \text{ m}^{-3}$$

$$\therefore \quad E_F(0) = \frac{(6.6 \times 10^{-34} \text{ j.s})^2}{8 \times 9.1 \times 10^{-31} \text{ kg}} \times \left(\frac{3 \times 5.9 \times 10^{28}}{3.14}\right)^{\frac{2}{3}}$$

$$= 8.8 \times 10^{-19} \text{ J} = \frac{8.8 \times 10^{-19}}{1.6 \times 10^{-19}} \text{ eV} = 5.5 \text{ eV}$$

The required internal energy is

$$U = \frac{3}{5}\left(\frac{N}{V}\right) E_F(0) = \frac{3}{5} \times 5.9 \times 10^{28} \text{ m}^{-3} \times 8.8 \times 10^{-19} \text{ J}$$

$$= 3.11 \times 10^{10} \text{ J/m}^3.$$

Problem 14. *Fermi energy of conduction electrons in silver is 5.48 eV. Calculate the number of such electrons per cm³ given that h = 6.62 × 10⁻²⁷ erg sec. and 1 eV = 1.62 × 10⁻¹² erg.*

Solution. We know that the Fermi energy

$$E_F(0) = \frac{h^2}{8m}\left(\frac{3N}{\pi V}\right)^{\frac{2}{3}} = \frac{h^2}{8m}\left(\frac{3}{\pi}\right)^{\frac{2}{3}} \cdot \left(\frac{N}{V}\right)^{\frac{2}{3}}$$

or,

$$\left(\frac{N}{V}\right)^{\frac{2}{3}} = \frac{8m}{h^2}\left(\frac{\pi}{3}\right)^{\frac{2}{3}} E_F(0)$$

or,

$$\frac{N}{V} = \left(\frac{8 \times m}{h^2}\right)^{\frac{3}{2}} \frac{\pi}{3} \cdot \{E_F(0)\}^{3/2}$$

$$= \left(\frac{8 \times 9.1 \times 10^{-28}}{6.62 \times 10^{-27} \times 6.62 \times 10^{-27}}\right)^{\frac{3}{2}} \cdot \frac{\pi}{3} \cdot (5.48 \times 1.62 \times 10^{-12})^{\frac{3}{2}}$$

$$= 5.9 \times 10^{22} \text{ cm}^{-3}.$$

Problem 15. *The number of conduction electrons per cc is 24.2 × 10²² in beryllium and 0.91 × 10²² in cesium. The fermi energy of conduction electrons in Be is 14.44 eV. Calculate that in cesium.*

[H.P. University]

Solution. Here given,

$$\text{Electrons per cc in Be} = \frac{N}{V} = 24.2 \times 10^{22}$$

and $$\text{Electrons per cc in Cs} = \frac{N}{V} = 0.91 \times 10^{22}$$

We have from fermi energy

$$E_F(0) = \frac{h^2}{8m}\left(\frac{3N}{\pi V}\right)^{\frac{2}{3}}$$

$$= \frac{h^2}{8m}\cdot\left(\frac{3}{\pi}\right)^{\frac{2}{3}}\cdot\left(\frac{N}{V}\right)^{\frac{2}{3}} = K\,.\left(\frac{N}{V}\right)^{\frac{2}{3}} \qquad \left[\text{Here } K = \frac{h^2}{8m}\cdot\left(\frac{3}{\pi}\right)^{\frac{2}{3}}\right]$$

$$\therefore \quad \frac{\text{Fermi energy for Be}}{\text{Fermi energy for Cs}} = \frac{K\,.(24.2\times 10^{22})^{\frac{2}{3}}}{K\,.(0.91\times 10^{22})^{\frac{2}{3}}} = \left(\frac{24.2}{0.91}\right)^{\frac{2}{3}}$$

or, $$\text{Fermi energy for Cs} = 14.44 \times \left(\frac{0.91}{24.2}\right)^{\frac{2}{3}}$$

$$= 1.587 \text{ eV}.$$

Problem 16. *Find out the number of possible arrangements of three particles in two cells assuming that the particles obey (i) M.B. statistics ; (ii) B.E. statistics ; and (iii) Fermi-Dirac statistics.*

[G.N.D. University]

Solution. Case (*i*) Maxwell-Boltzmann Statistics : In this case, the particles are distinguishable and any number of particles can be accommodated in one cell. So there are 2^3 or 8 possible arrangements as given below :

Cell 1	*abc*	—	*ab*	*bc*	*ac*	*a*	*b*	*c*
Cell 2	—	*abc*	*c*	*a*	*b*	*bc*	*ca*	*ab*

Case (*ii*) Bose-Einstein Statistics : In this case, the particles are indistinguishable but any number can be accommodated in one cell. So it has 4 possible arrangement as given below :

Cell 1	*aaa*	—	*aa*	*a*
Cell 2	—	*aaa*	*a*	*aa*

Case (*iii*) Fermi-Dirac Statistics : In this case, the particles are also indistinguishable but has some restriction *i.e.*, one and only one can be accommodated in one cell but here, the number of cells is less than the number of particles, so no arrangement is possible.

Problem 17. *Draw a simple diagram (Symbolic) to show all possible meaningful arrangements of three particles in three cells assuming that the particles obey —*

(i) M.B. Statistics *(ii) B.E. Statistics* *(iii) F.D. Statistics.*

[H.P. University]

Solution. Three distributions are sketched below :

Maxwell Boltzmann Statistics		
Cell 1	Cell 2	Cell 3
ABC	0	0
0	ABC	0
0	0	ABC
A	BC	0
B	CA	0
C	AB	0
AB	C	0
BC	A	0
CA	B	0
0	A	BC
0	B	CA
0	C	AB
AB	0	C
BC	0	A
CA	0	B
A	0	BC
B	0	CA
C	0	AB
0	AB	C
0	BC	A
0	CA	B
A	B	C
A	C	B
B	C	A
B	A	C
C	A	B
C	B	A

i.e., 27 cases

Bose-Einstein Statistics		
Cell 1	Cell 2	Cell 3
AAA	0	0
0	AAA	0
0	0	AAA
AA	A	0
A	AA	0
0	AA	A
0	A	AA
AA	0	A
A	0	AA
A	A	A

10 cases

Fermi-Dirac Statistics		
Cell 1	Cell 2	Cell 3
A	A	A

only one case

So we can say that there are 27 arrangements under M.B., 10 arrangements under B.E. and only one arrangement under F. D. statistics.

Multiple Choice Questions

1. Statistical methods give greater accuracy when the number of observations is
   (*a*) Very small (*b*) Very large
   (*c*) Neither very large nor very small.
2. The value of probability of an event cannot be
   (*a*) Zero (*b*) 1 (*c*) Negative.
3. The thermodynamic probability of a system in equilibrium is
   (*a*) Maximum (*b*) Minimum but not 1 (*c*) Zero.
4. Constraints imposed on a system
   (*a*) Increase number of inaccessible microstates.
   (*b*) Decrease the number of inaccessible microstates.
   (*c*) Have no effect.

5. In a system of 8 distinguishable particles distributed in two equal size compartments the probability of the macrostate (2, 6) is
   (a) $\frac{5}{32}$ (b) $\frac{7}{64}$ (c) $\frac{1}{8}$.
6. The dimension of phase space volume are
   (a) $(\text{Length} \times \text{momentum})^3$ (b) $(\text{Angular momentum})^3$ (c) $(\text{Length})^3$.
7. Pauli's exclusion principle is applied to
   (a) M.B. Statistics (b) F.D. Statistics (c) B.E. Statistics.
8. According to B.E. Statistics the particles have
   (a) Zero or integral spin (b) Half integral spin (c) Any spin.
9. A phase cell can accommodate only one particle according to
   (a) F.D. Statistics (b) B.E. Statistics (c) M.B. Statistics.
10. At high temperature or at low density B.E. distribution approaches to
   (a) M.B. distributions (b) F.D. distributions (c) None.
11. Planck's formula for black body radiation can be derived from the statistics as proposed by
   (a) F.D. (b) B.E. (c) M.B.
12. The average energy of an electron in fermi gas at 0° K is
   (a) $0.2\ E_F$ (b) $0.4\ E_F$ (c) $0.6\ E_F$.
13. The number of meaningful ways in which 4 (Fermions) can be arranged in 5 compartments is
   (a) 4 (b) 5 (c) 9.
14. In which probability the occupation index can tend to
   (a) Maxwell-Boltzmann (b) Fermi-Dirac (c) None of these.
15. The number of different arrangements of 6 indistinguishable particles among 4 cells of equal a priori probability, there being no restriction on the number of particles entering into a cell is
   (a) 84 (b) 210 (c) 30.
16. Average energy ε of an electron in a metal at T = 0 K
   (a) $\frac{5}{2}\varepsilon_F$ (b) $\frac{\varepsilon_F}{2}$ (c) $\frac{3}{5}\varepsilon_F$ [*WBUT*]
17. Bose-Einstein statistics is applicable for
   (a) electron (b) photon (c) ideal gas.

Answers

1. (b)	**2.** (c)	**3.** (a)	**4.** (a)	**5.** (b)	**6.** (c)
7. (b)	**8.** (a)	**9.** (a)	**10.** (a)	**11.** (b)	**12.** (c)
13. (b)	**14.** (c)	**15.** (a)	**16.** (c)	**17.** (b)	

Review Questions

1. Write a short notes on
   (a) Macro states and thermodynamic probability. [*WBUT*]
   (b) Microstates and Macrostates.
   (c) Phase space.
2. What is Bose-Einstein (B.E.) statistics? What difficulties were not explained with M.B. statistics and how they were explained? Derive an expression for the most probable distribution of particles of a system obeying B.E. statistics. [*G.N.D. University, Pune University, H.P. University*]

3. Write B.E. and FD distribution functions (no proof required). What are the basic assumptions used in the deviation of these distribution functions? Sketch the FD distribution function for T = 0° K and T > 0° K. *[Kolkata University]*
4. Starting from Bose-Einstein distribution formula establish Planck's law of blackbody radiation. *[Kolkata University]*
5. (*a*) Compare M.B., F.D. and B.E. statistics mentioning at least three characterstics. *[WBUT]*
   (*b*) Draw the Fermi distribution curve of (*i*) T = 0° K and (*ii*) T > 0° K. Explain their significance. *[WBUT]*
   (*c*) Indicate which statistics will be applicable for
   (*i*) $_1H^1$ (*ii*) e^- (*iii*) π° (*iv*) Photon *[WBUT]*
   [Hints : If the nucleus contains an odd number of FD particles, *i.e.*, if the mass number is odd it will be applicable for FD statistics. If the number is even, BE statistics is obeyed].
6. Write down the Fermi-Dirac distribution function and define Fermi level at absolute zero and at a finite temperature. *[Burdwan University]*
7. What is the Fermi temperature? Estimate it for a metal if the Fermi energy $E_f(\sigma) = 4$ eV. *[Burdwan University]*
8. Name the statistics obeyed by each of the following particles :
   Neutrino, alpha, muon. Photon, neutron, π-meson. *[Calcutta University]*
9. Write down the basic assumptions in deriving the (*i*) M.B. (*ii*) B.E. and (*iii*) F.D. statistics. How does the distribution functions differ in the three cases ? *[Calcutta University]*
10. Show that the probability of occupation of a state with energy ΔE above the Fermi level is equal to the probability of a state ΔE below the Fermi level. *[Burdwan University]*
11. Write down Bose-Einstein distribution function and obtain an expression for the total number of particles. *[Burdwan University]*
12. What are the main points of difference between classical and quantum statistics. *[Poorvanchal Univ., Lucknow Univ., Pbi. Univ., H.P. Univ., G.N.D. Univ.]*
13. Name the three kinds of statistics. Explain the distinguishing features of M.B., B.E. and F.D. Statistics. *[Pbi. Univ., Poona Univ., G.N.D. Univ.]*
14. Starting from Fermi Dirac distribution law derive the expression for energy distribution of free electrons in a metal. Bring out the concept of Fermi energy and Fermi energy level. Discuss the results. *[Poona Univ., Pbi. Univ., M.D. Univ., H.P. Univ., Calicut Univ.]*
15. What is Fermi energy of conduction electrons ? Calculate the value of Fermi energy. *[Puna Univ., Guwahati Univ., H.P. Univ., M.D. Univ., Pbi. Univ.]*
16. Distinguish between (or compare and contrast) Bose-Einstein and Fermi-Dirac statistics. *[Poorvanchal Univ., Madurai Univ., Guwahati Univ., Pbi. Univ., G.N.D. Univ.]*

Unsolved Problems

1. A system has *f* particles arranged in two compartments. The first compartment has 8 cells and second has 10 cells (all cells of equal size). Calculate the number of microstates in the macrostate (3,4) when particles obey F.D. statistics. **[Ans.** 11760] *[Pune University]*
2. Consider a system of 2 identical particles each of which can be in any one of 3 single particle states. How many states of the system are possible if the particles obey
   (*i*) Maxwell-Boltzmann statistics
   (*ii*) Fermi-Dirac statistics
   (*iii*) Bose-Einstein statistics. **[Ans** (*i*) 9 (*ii*) 3 (*iii*) 6]*[Kolkata University]*

3. The particles of a gas have mass 1.6×10^{-24} gm, the density of the gas is 10^{14} gm/cc and its temperature 10^9 K. Discuss the energy distribution according to Fermi statistics and illustrate it with diagram. [Calculate E_f (0) and show E_f (0) $>> kT$]
[*Kolkata University*]
4. The Fermi energy E_f for lithium is 4.72 eV at T = 0. Calculate the number of conduction electrons per unit volume in lithium. [*Calcutta University*]

   [**Hints :** $E_f(0) = \frac{h^2}{8m}\left(\frac{3n}{H}\right)^{\frac{2}{3}}$] [**Ans.** $4.67 \times 10^{28}\ m^{-3}$]
5. Six distinguishable particles are distributed over three non-degenocate level of energies 0, ε and 2ε.
   (*i*) Calculate the total number of microstates of the system.
   (*ii*) Find the total energy of the distribution for which the probability is maximum.
   [**Ans.** 729, 6ε] [*Calcutta University*]
6. Calculate the Fermi energy at 0° K of sodium containing one free electron per atom. The density and the atomic weight of sodium is 9.7×10^2 kg/m^3 and 23 respectively. Also find the fermi temperature.
   [**Ans.** 3.2 eV, 37101 K]
7. Prove that for a system at T > 0° K obeying F.D. statistics, the probability that a level lying ΔE below the Fermi level is unoccupied and is the same as the probability of occupation of a level lying ΔE above the Fermi level. [*Calcutta University*]
   [**Hints :** Consider the probability of occupancy of a level E is f(E), then probability of non-occupancy is 1 – f(E)]
8. Calculate the number of different arrangements of 10 indistinguishable particles in 15 cells of a equal priori probability considering that one cell contains only one particle. [*K. Univ.*]

   [**Ans. Hints :** $\frac{g_i\,!}{n_i\,!\,(g_i - n_i)\,!}$, where $g_i = 15$; $n_i = 10$]
9. Fermi energy of conduction electrons in silver is 5.48 eV. Calculate the number of such electrons per cm^3 given that $h = 6.62 \times 10^{-27}$ erg sec and 1 eV = 1.62×10^{-12} erg.
   [*G.N.D. Univ.*] [**Ans.** 5.9×10^{22} cm^{-3}]
10. Calculate the number of different arrangements of 6 Bosom among 4 cells of equal a priori probability.
    [**Ans.** 84]
11. Find out the number of possible arrangements of three particles in two cells for
    (*i*) M-B Statistics (*ii*) B-E Statistics (*iii*) F-D Statistics.
    [**Ans.** (*i*) 8 ; (*ii*) 4 ; (*iii*) no possible arrangements]

Wilhelm Konrad Roentgen

7

Chapter

Crystal Structure

7.1 ELEMENTARY IDEAS OF CRYSTAL STRUCTURE

Matter exists in nature in the solid, liquid and gaseous states. A solid of definite shape with its atom or molecules arranged in some regular three dimensional pattern is termed as a crystal. Briefly, crystal is a periodic array of atoms. The solids may be broadly classified as crystalline and non-crystalline or amorphous depending upon the arrangement of atoms and molecules.

The crystalline state of solids is characterized by regular or periodic arrangement of atoms or molecules. Most of the solids are crystalline structure. This is due to the reason that the energy release during the formation of a regular structure is more than that released during the formation of irregular structure. Thus crystalline state is a low energy state and is, therefore, preferred by most of the solids. Example of crystalline solids are diamond, quartz etc.

Ideal Crystal

It is constructed by the infinite repetition of identical structure in space. In other words we can say that there exists some smallest grouping that repeats itself exactly in all direction in the crystal, so that the environment at one location is identical in all respects to the environment at a corresponding location somewhere else.

The non crystalline or the amorphous solids are characterized by the completely random arrangement of atoms or molecules. Glass is an example of amorphous material.

Figure 7.1

Crystal Lattice

The regular periodic arrangement of atoms in a crystal is called the crystal lattice. It is constructed by the infinite repetition in space of identical structural units (atoms, molecules, ions). In two dimensions, any atom has got the same environment as any other atom. Such an array of atoms is called a **plane lattice**.

In three dimensional picture any atom will similarly have same environment as any other atom and this array is called **space lattice.**

Basis

It is defined as an assembly of atoms, ions or molecules identical in composition arrangement and orientation and which is attached to each lattice point in space to form the structure that of the crystal.

The smallest logical relation between lattice, basis and crystal structure can be expressed as

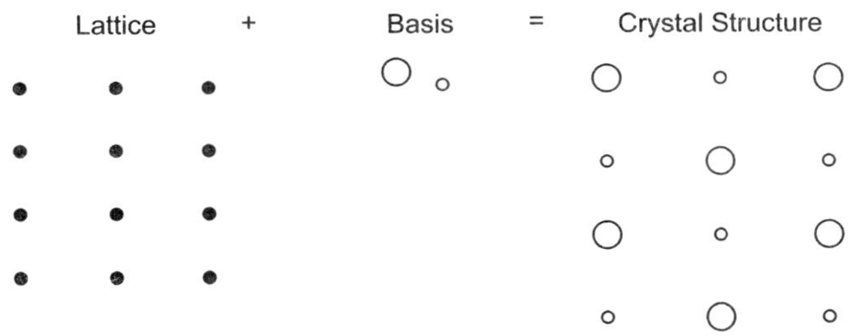

Figure 7.2

Unit Cell

A region of crystal defined by the basis vector [in two dimensions a and b of area $a \times b$ or the a, b and c in three dimensions of volume $(a\ .\ (b \times c))$] which, translated by any integral multiple of those vectors reproduces a similar region of the crystal, is called the unit cell of the crystal. The unit cell is the building block of the crystal. A crystal may not possess a unique unit cell. Many unit cells can exist for the same structure. For certain structures unit cell of completely different geometry are possible. Most of the structures have smallest unit cell.

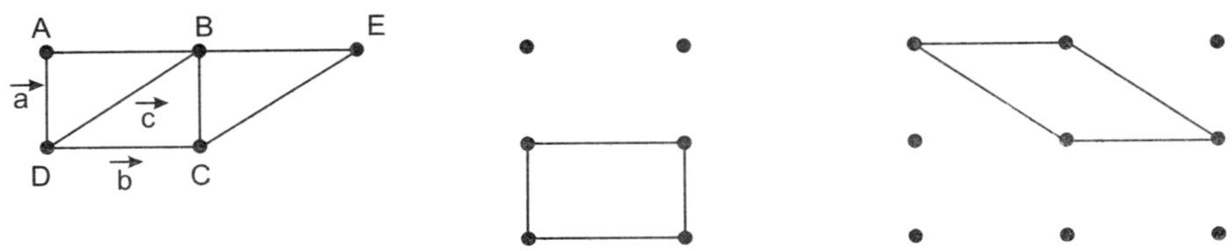

Figure 7.3

The choice of unit cell is arbitrary. From figure 7.3 we see that there are different choices of unit cell such as ABCD, BECD etc.

Primitive Unit Cell

In a primitive unit cell the lattice points or the atoms are at the corners of the unit cell. The effective number of atom in unit cell is one. The effective number of atoms in the unit cell is calculated by the share of atoms in a unit cell.

ABCD is a primitive unit cell. Any corner atom is shared by four unit cells and the total number of such atoms is 4. So the effective number of atoms in the unit cell is $\frac{1}{4} \times 4 = 1$

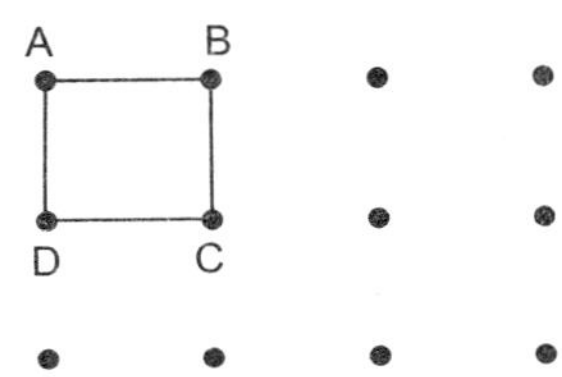

Figure 7.4

Non-primitive Unit Cell

The effective number of atoms in a unit cell is more than one.

MNOP (figure 7.5) is a non-primitive unit cell because the effective number of atoms in the cell is 2 $\left(\frac{1}{4} \times 4 + 1 = 2\right)$.

The conventional unit cell has the highest possible symmetry and the lowest possible volume. It may be primitive or non-primitive.

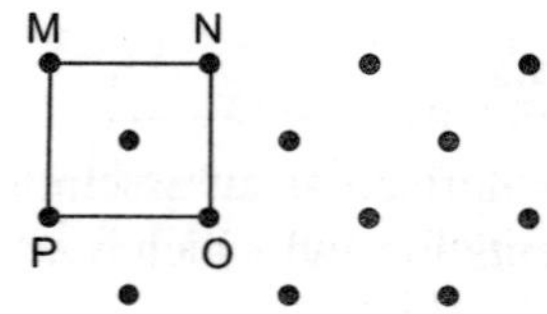

Figure 7.5

7.2 SYMMETRY OPERATION OF CRYSTAL

A symmetry operation is that which transforms the crystal to itself *i.e.*, a crystal remains invariant under a symmetry operation. In a crystal there are four types of symmetry operation :

1. Translation 2. Rotation 3. Reflection 4. Inversion

We shall consider a group of atoms or molecules called motif, which can be repeated periodically in a number of permissible ways to form the crystal. For simplicity we shall represent the motif in a crystal by a figure of 7.

1. *Translation.* The motif represented by a figure of 7 can be repeated by a translation operation as shown in figure 7.6(*a*).

   A lattice point r under translation operation T gives another point r' which is exactly identical to r *i.e.*, $r' = r + \text{T}$.

   Translation repeats the motif an infinite number of times.

2. *Rotation.* In this operation the motif can be rotated through an angle Φ about an axis of rotation as shown in figure 7.6(*b*). This operation repeats the motif a finite number of times.

   A lattice is said to possess rotational symmetry if its rotation by an angle Φ about an axis transforms the lattice to itself.

   As the lattice remains invariant under a rotation of 2π, the angle 2π must be integral multiple of Φ *i.e.*, $n\Phi = 2\pi$

3. *Reflection.* A lattice is said to possess reflection symmetry if there exists a plane (or line in two dimensions) in the lattice which divides it into two identical halves which are mirror images of each other.

   Reflection across a line or plane changes the character of the lattice from left handed to right handed and vice-versa [figure 7.6(*c*)].

4. *Inversion.* Inversion is a point operation which is applicable to three dimensional lattice only. The symmetry implies that each point located at r relative

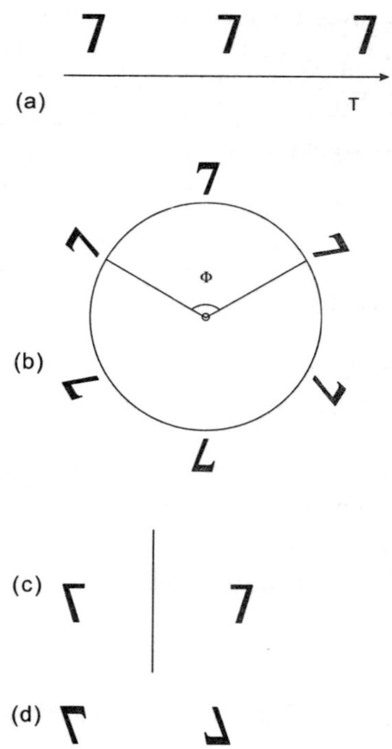

Figure 7.6

to a lattice point has an identical point located at $-r$ relative to the same lattice point. Inversion through a point is shown in figure [7.6(d)]. It also changes the characters of the motif from right handed to left handed.

We say a cube to be a symmetrical body. It means that if we consider an axis passing through the mid point of ABCD and perpendicular to it and if we rotate the cube about this axis (say XY) through 360°, then how many times will it occupy congruent position? The answer is four times. We say the axis XY is a four fold rotational symmetry axis.

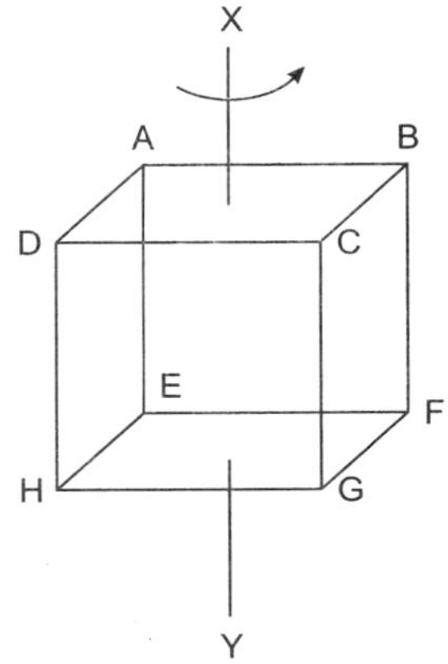

Figure 7.7

If we rotate a body about an axis by 360° and if it occupies congruent position 'n' times then we call this axis an 'n' fold symmetry axis. The angular rotation required to achieve one congruent position is called the *throw* of the axis.

We know $$\Phi = \frac{2\pi}{n}$$

Now if $n = 1$, we call that an identity axis,

if $n = 2$, we call that a diad axis,

if $n = 3$, we call that a triad axis,

if $n = 4$, we call that a tetrade axis,

if $n = 6$, we call that a hexad axis,

In a crystal only these values of symmetry axis exists.

A crystal cannot have a five fold symmetry.

To prove it mathematically let us consider one dimensional lattice, with lattice parameter t [Figure 7.8]. We associate 'n' fold symmetry axis to each point A, B, C, D.

As rotation by 360° bring the body into its congruent position, hence clockwise or anticlockwise rotations are equivalent. We rotate the lattice vector say BA clockwise through $\Phi = \frac{2\pi}{n}$.

Also rotate lattice vector CD, anticlockwise by the same angle $\Phi = \frac{2\pi}{n}$

If the structure has to be a crystal, the distance between A′ and D′ must be some integral of lattice parameter,

So, $$A'D' = mt$$

Draw A′E perpendicular to AB and D′F perpendicular CD,

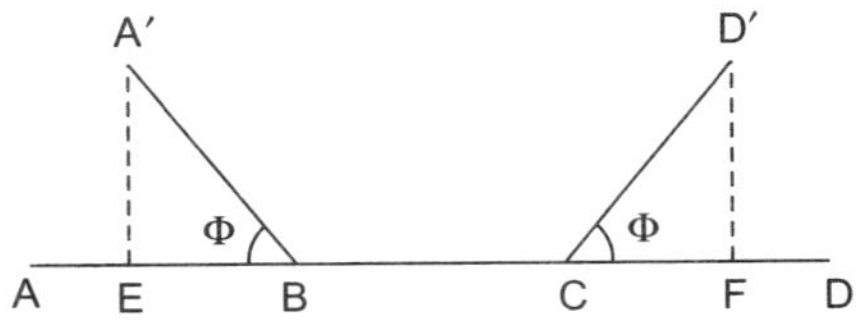

Figure 7.8

then $$EB + BC + CF = mt$$

or $$t\cos\Phi + t + t\cos\Phi = mt$$

or $$2t\cos\Phi = t(m-1)$$

$$\therefore \quad \cos\Phi = t\frac{(m-1)}{2t} = \frac{(m-1)}{2} = \frac{N}{2}$$

[Where $(m-1) = N$]

We know $-1 \le \cos\Phi \le 1$ so,

for $$N = -2, -1, 0, 1, 2$$

$$\Phi = \pi, \frac{2\pi}{3}, \frac{\pi}{2}, \frac{\pi}{3}, 0 \text{ respectively}$$

$$n = \frac{2\pi}{\Phi} = 2, 3, 4, 6, 1 \text{ respectively}$$

N	*cos* Φ	Φ	$n = \frac{2\pi}{\Phi}$
− 2	− 1	180°	2
− 1	$-\frac{1}{2}$	120°	3
0	0	90°	4
1	$\frac{1}{2}$	60°	6
2	1	0° or 360°	1

7.3 FUNDAMENTAL TYPES OF LATTICE

Bravais Lattices

There are 10 and 32 numbers of point groups in two and three dimensions respectively. These point groups form the basis for construction of different type of lattices. Only those lattices are permissible which are consistent with the point group operations. Such lattices are called the Bravais lattices. It can be stated that 10 and 32 point group in two and three dimensions produce only 5 and 14 different Bravais lattices respectively. These Bravais lattices further become parts of 4 and 7 distinct crystal systems respectively and are described below :

Crystal System in two Dimensions

Sl No.	*Crystal System*	*Unit Cell Characteristics*	*Picture*
1.	Oblique	$a \neq b, \quad \gamma \neq 90°$	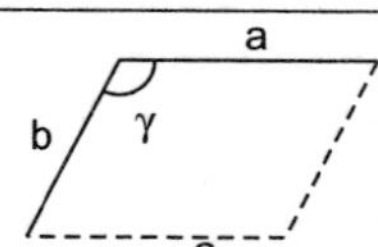
2.	Rectangular	$a \neq b, \quad \gamma = 90°$	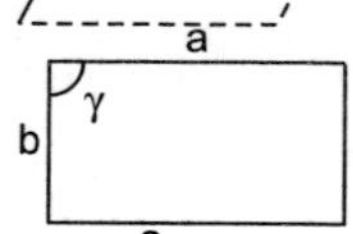
3.	Square	$a = b, \quad \gamma = 90°$	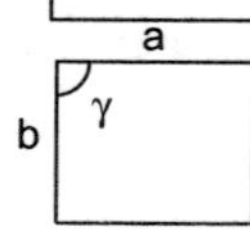
4.	Hexagonal	$a = b, \quad \gamma = 120°$	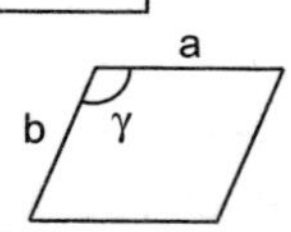

Crystal System in three Dimensions

Sl No.	*Crystal System*	*Unit Cell Characteristics*	*Picture*
1.	Triclinic	$a \neq b \neq c$, $\alpha \neq \beta \neq \gamma \neq 90°$.	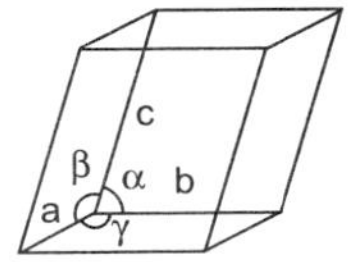
2.	Monoclinic	$a \neq b \neq c$, $\alpha = \gamma = 90° \neq \beta$	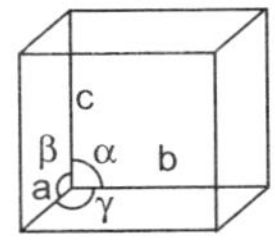
3.	Orthorhombic	$a \neq b \neq c$, $\alpha = \beta = \gamma = 90°$	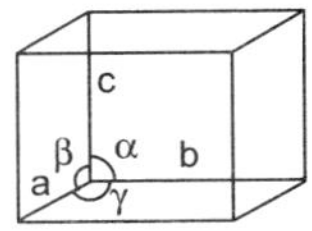
4.	Cubic	$a = b = c$, $\alpha = \beta = \gamma = 90°$	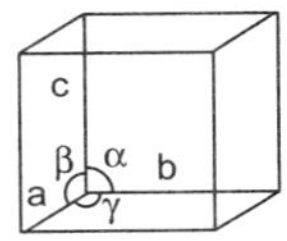
5.	Trigonal	$a = b = c$, $\alpha = \beta = \gamma \neq 90°$	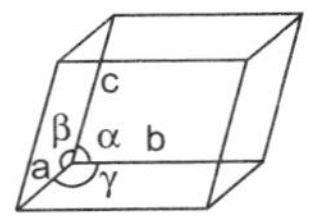
6.	Tetragonal	$a = b \neq c$, $\alpha = \beta = \gamma = 90°$	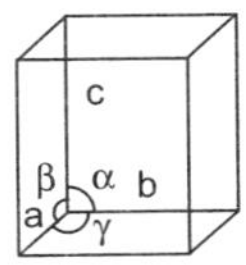
7.	Hexagonal	$a = b \neq c$, $\alpha = \beta = 90°, \gamma = 120°$	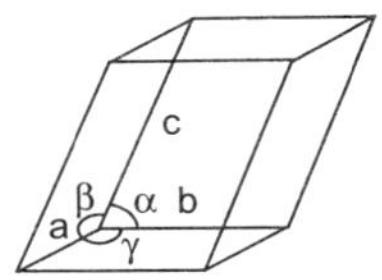

7.4 IDENTIFICATION OF CRYSTAL PLANE: MILLER INDICES

Any crystal is bounded by a number of planes. How can we identify the planes ?

Let one plane cut 2 units from axis a, 3 units from axis b and 4 units from axis c. We can designate the plane as ($2a$, $3b$, $4c$). But the general equation of the plane is,

$$\frac{x}{a} + \frac{y}{b} + \frac{z}{c} = 1$$

Here the intercepts occur in the denominator. Hence ($2a$, $3b$, $4c$) notation does not conform to mathematical formulation of planes.

To overcome this difficulty, Miller introduced the following procedure to represent a plane.

1. Write the co-efficient of each intercepts
2. Take their reciprocal
3. Clear off the fractions
4. Write the number in first brackets ( ) without comas

Now we can find the Miller indices of the intercepts 2, 3, 4 units from x, y, z axis respectively of our previous example :

1. Write the co-efficient of each intercepts — 2, 3, 4
2. Take their Reciprocal — $\frac{1}{2}, \frac{1}{3}, \frac{1}{4}$
3. Clear off the fractions — 6, 4, 3
4. Write the number in first brackets ( ) without comas — (643)

If a plane has an intercept from negative of an axis put a bar '—' on it. In general Miller indices are represented by (hkl). It should be noted that parallel plane have same Miller indices.

Inter-planar Spacing

We want to find out the spacing between two parallel planes. We shall take a simple unit cell in which co-ordinate axes are orthogonal.

Consider a set of parallel planes with indices (hkl). The origin is taken on a lattice point. In figure 7.9, OX, OY and OZ are orthogonal axes. Suppose the reference plane passes through the origin and next plane cut the intercepts $\frac{a}{h}, \frac{b}{k}$ and $\frac{c}{l}$ on X, Y, and Z axes respectively.

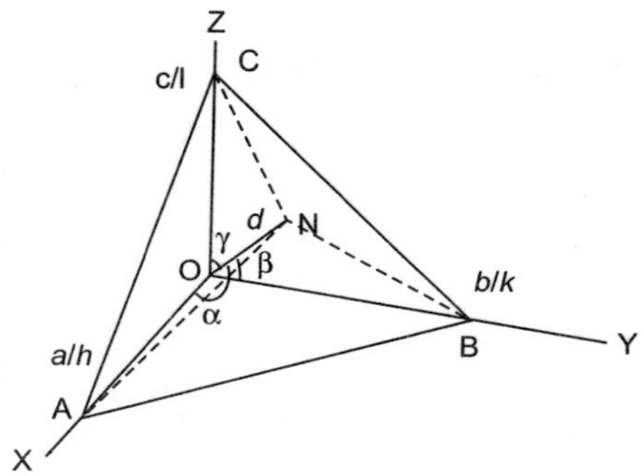

Figure 7.9

A normal ON is drawn to the plane ABC from the origin. The length d on this normal from origin to the plane will be the distance between adjacent planes (interplanar distance). We have to find an expression for d in terms of a, b, c and h, k, l.

Let α, β and γ are the angle between the normal and X, Y and Z be the axes. Because d is normal to the plane ABC, we write

$$\cos\alpha = \frac{ON}{OA} = \frac{d}{\frac{a}{h}}, \quad \cos\beta = \frac{ON}{OB} = \frac{d}{\frac{b}{k}} \text{ and } \cos\gamma = \frac{ON}{OC} = \frac{d}{\frac{c}{l}}$$

The law of directional cosines is

$$\cos^2\alpha + \cos^2\beta + \cos^2\gamma = 1$$

Substituting the values, we get

$$\frac{d^2}{\left(\frac{a}{h}\right)^2} + \frac{d^2}{\left(\frac{b}{k}\right)^2} + \frac{d^2}{\left(\frac{c}{l}\right)^2} = 1$$

or

$$d^2\left[\left(\frac{h}{a}\right)^2 + \left(\frac{k}{b}\right)^2 + \left(\frac{l}{c}\right)^2\right] = 1$$

or
$$d = \left[\frac{h^2}{a^2} + \frac{k^2}{b^2} + \frac{l^2}{c^2}\right]^{-\frac{1}{2}}$$

This relation is applicable to the primitive lattice in cubic orthorhombic and tetragonal systems. For cubic system, $a = b = c,$ so that

$$d = \frac{a}{\sqrt{\left(h^2 + k^2 + l^2\right)}}$$

so that for (1 0 0) and (1 1 2) planes in a cubic cell,

$$d_{100} = a \text{ and } d_{112} = \frac{a}{\sqrt{6}}$$

7.5 THE CUBE CRYSTAL

Following are three types of cube crystal :

1. **Simple Cube (sc).** The unit cell is a cube having one atom or molecule at each corner. So there are eight atoms at eight corners of the cube. Since each corner is shared by eight surrounding cubes, share of each cube comes to one-eight of an atom.

   Total Number of atoms $= \left(\frac{1}{8}\right) \times 8 = 1$ atom

2. **The body centered cubic structure (bcc).** Body centered cube has eight atoms at eight corners and one center atom. The number of atoms belonging to this cube are

   (*a*) One center atom and

   (*b*) $\left(\frac{1}{8}\right) \times 8 = 1$ corner atom

   Total Number of atoms $= 1 + 1 = 2$ atoms

3. **The face centered cubic structure (fcc).** The face centered cube has 6 face centered and eight corner atoms. The six face centered atoms at six faces of the cube are shared by their adjacent cubes.

   Hence, a total of $\frac{6}{2} = 3$ such atoms belong to the cube. As each corner atom is shared by eight surrounding cubes hence share of each cube comes to one eight of an atom.

   Total Number of atoms $= 3 + 1 = 4$ atoms

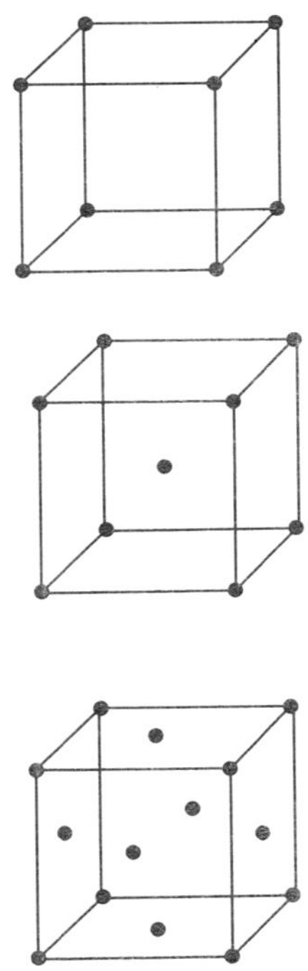

Figure 7.10

7.6 CO-ORDINATION NUMBER

The co-ordination number is defined as the number of equidistant neighbour that an atom has in the given structure. Greater is the co-ordination number, the more closely packed up will be the structure. The co-ordination number of three cubic lattice can be determined as follows :

1. **Simple cubic lattice.** In simple cubic lattice, there is one atom at each of the eight corners of the unit cell. If we take the atom at one corner as the center, it is observed that this atom is surrounded by six equidistant neighbours and hence the co-ordination number of simple cube is six (6).
2. **Body centered cubic lattice (bcc).** In case of bcc lattice there are eight atoms at each of the eight corners of the unit cell and one atom at the body center. The central atom is surrounded by eight equidistant neighbours and hence the co-ordination number is eight (8).
3. **Face centered cubic lattice (fcc).** In case of fcc lattice there are eight atoms at each of the eight corners of the unit cell and six atoms at the center of six faces. Considering the atom at the face center as origin, it can be observed that this face is common to two cubes and there are twelve points surrounding it situated at a distance equal to half the face diagonal of the unit cell. In this way we can say that the co-ordination number in fcc lattice is twelve (12).

7.7 ATOMIC PACKING FRACTION

Packing fraction is defined by the ratio of volume occupied by the atom present in the unit cell to the total volume of the unit cell. To find out the packing fraction of sc, bcc and fcc we need to consider the atomic radius of atom in the unit cell

$$\text{Packing fraction} = \frac{\text{Volume of the atoms per unit cell}}{\text{Volume of the unit cell}}$$

1. **Simple cube.** The atoms in the simple cube are arranged as follows :

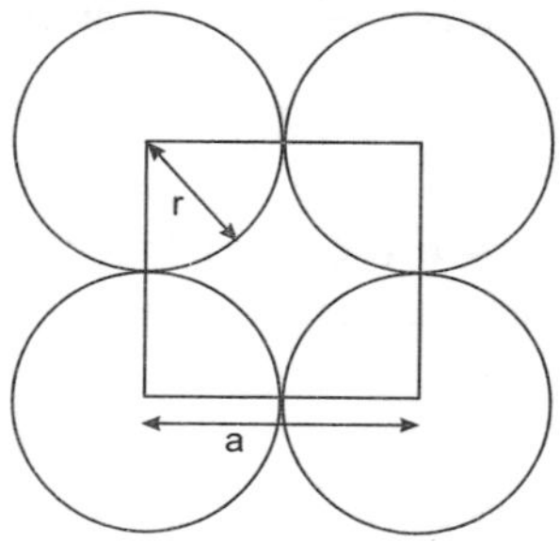

Figure 7.11 (a)

So the atomic radius related to the lattice parameter is $a = 2r$
The number of atoms in the unit cell is one.

So, $$\text{Packing fraction} = \frac{1 \times \frac{4}{3}\pi r^3}{a^3} = \frac{1 \times \frac{4}{3}\pi r^3}{(2r)^3} = 0.52$$

2. **Face centered cube (fcc).** The atoms in the fcc are arranged as follows :

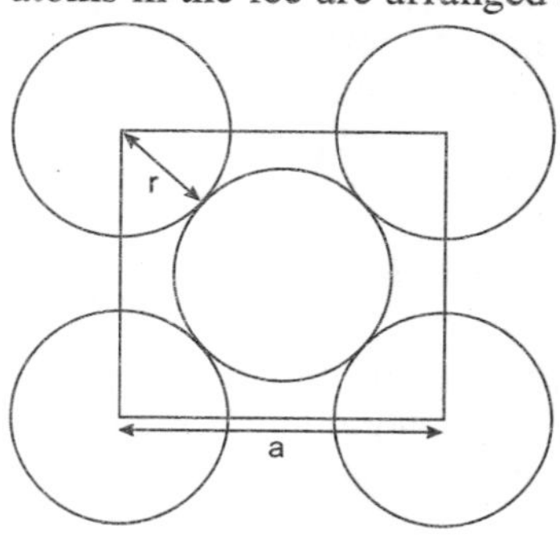

Figure 7.11 (b)

So the atomic radius related to the lattice parameter is $2a = 4r$
The number of atoms in the unit cell is 4.
So,

$$\text{Packing fraction} = \frac{4 \times \frac{4}{3}\pi r^3}{a^3} = \frac{4 \times \frac{4}{3}\pi r^3}{\left(\frac{4r}{2}\right)^3} = 0.74$$

3. **Body centered cube (bcc).** The atoms in the fcc are arranged as follows :

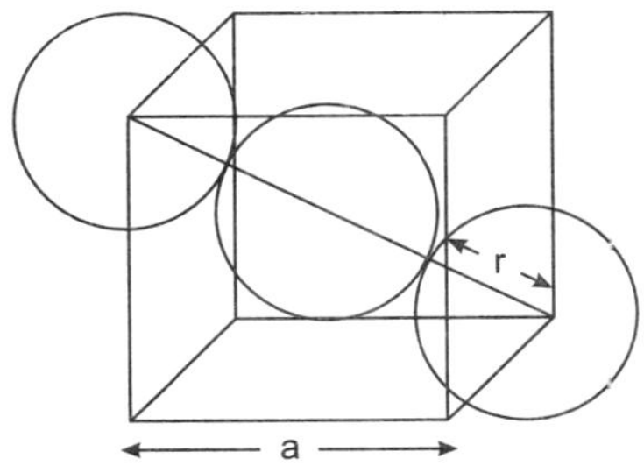

Figure 7.11 (c)

So the atomic radius related to the lattice parameter is $3a = 4r$

The number of atoms in the unit cell is 2.

So, $$\text{Packing fraction} = \frac{2 \times \frac{4}{3}\pi r^3}{a^3} = \frac{2 \times \frac{4}{3}\pi r^3}{\left(\frac{4r}{3}\right)^3} = 0.68$$

7.8 RELATION BETWEEN THE LATTICE CONSTANT *a* OF A CUBIC CRYSTAL AND THE DENSITY OF THE CUBE MATERIAL

If n be the number of molecules per unit cell and M the molecular weight of the crystal then we can say, in M gm of material there are N (Avogadro number $= 6.023 \times 10^{23}$) molecules.

So, weight of n molecules is $(n \times \text{M})/\text{N}$ gm

$$\text{We know that, density} = \frac{\text{Mass}}{\text{Volume}}$$

$$\text{Density } \rho = \frac{m}{a \times a \times a}.$$

$$\rho = \frac{n\text{M}}{\text{N}a^3}$$

$$a = \text{lattice constant} = \left[\frac{n\text{M}}{\text{N}\rho}\right]^{\frac{1}{3}}$$

7.9 THE ORIGIN OF X-RAYS

When electrons, accelerated under high potential difference (around 50 to 100 kV) bombard a heavy-metal target (known as anticathode) in an X-ray tube electro-magnetic waves of very short wavelength known as X-rays are produced.

The X-rays so generated by the electron impact with the metal target consist of (a) continuous (or white) X-rays and (*b*) characteristic X-rays.

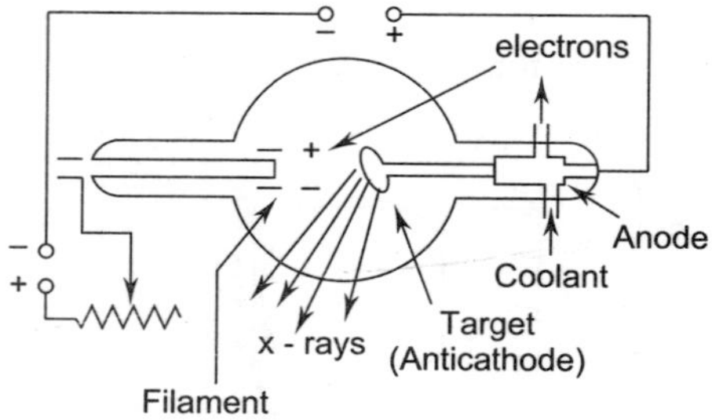

Figure 7.12 *X-ray tube (Coolidge type)*

Continuous X-rays

Continuous X-rays (or Bremsstrahlung) are generated when the high-energy electrons penetreating the surface of the metallic target are rapidly decelerated. The electrons lose their energy by making collisions with the target atoms and at each collision some or all of its lost energy appears as radiation. This is an "inverse photoelectric effect" as in this case an electron produces a photon. For each accelerating potential there exists a sharp short wavelength cut-off (λ_{min}) to the continuous radiation below which no radiation is produced. Considering each X-ray photon to be produced due to a particular collision of one electron, the photons of maximum energy are produced when an electron loses whole of its energy in a single collision.

Thus, $h\nu_{max} = \frac{1}{2}mv^2 = e\text{V}$ where h is the planck's constant, ν_{max} is the maximum frequency of the photon, V is the accelerating potential on the electron, m and v are the mass and the velocity of the colliding electron. Hence the minimum wavelength λ_{min} (corresponding to ν_{max}) is

$$\lambda_{min} = \frac{c}{\nu_{max}} = \frac{hc}{e\text{V}} = \frac{12400}{\text{V}} \text{ Angstrom.}$$

This is Duana and Hunt laws. The cut-off wavelength (λ_{min}) is only dependent on V, the electron accelerating potential, and diminishes as V increases. Change in target material though will cause change in the X-ray spectrum but will not change λ_{min} provided V remains the same. As the accelerated electron penetrates into the atoms it is attracted by the nuclei and thereby get deviated from the original path. This deviation is equivalent to its collision with the nucleus and due to this some energy loss of the electron takes place which appears as an X-ray photon (Figure 7.13).

If ν is the frequency of an X-ray photon then $h\nu = E_i - E_f$ where E_i and E_f are the initial and the final K.E.'s of the colliding electron. The electron may suffer several such collisions with various nuclei before coming to rest emitting number of X-ray photons of different frequencies which produce the ***continuous X-rays*** (Figure 7.14).

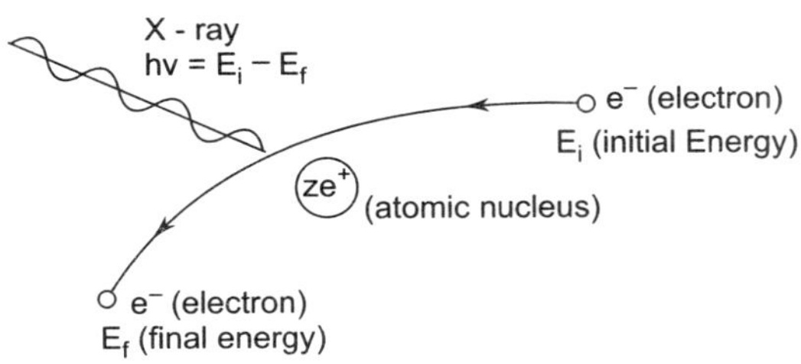

Figure 7.13

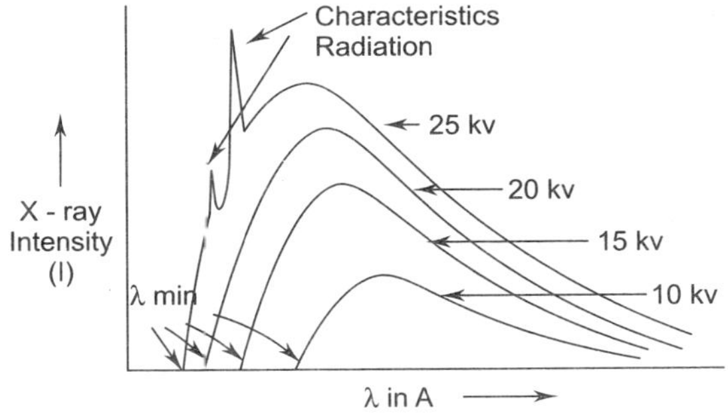

Figure 7.14

NOTE The Fig. 7.14 shows the λ_{min} to move towards higher values as the electron accelerating potential diminishes.

Characteristic X-rays

The continuous X-rays arise out of rapidly slowing down of high energy electrons while penetrating through a target atom. Another kind of collision also takes place when a high-energy bombarding electron knocks out an electron from one of the inner shells (*i.e.*, low lying) of the target atom. This causes the target atom to be left with a vacancy (or hole) in that particular tightly bound (*i.e.*, low lying) level. An electron from higher level falls to the lower level to fill in the hole thus created. This leads to the emission of a characteristic X-radiation photon frequency, $\nu = \dfrac{E_L - E_M}{h}$

Where E_L is the energy required to dislodge an electron from the Lth shell of the atom which is closer to the nucleus as compared to the electron in the Mth shell being at a further distance from the nucleus thereby a bit (Loosely) bound. E_M is the energy needed to dislodge the electron from the Mth shell. Thus, $E_L > E_M$. This phenomenon is known as Characteristic X-rays. In this case the characteristic X-rays constitute L-series.

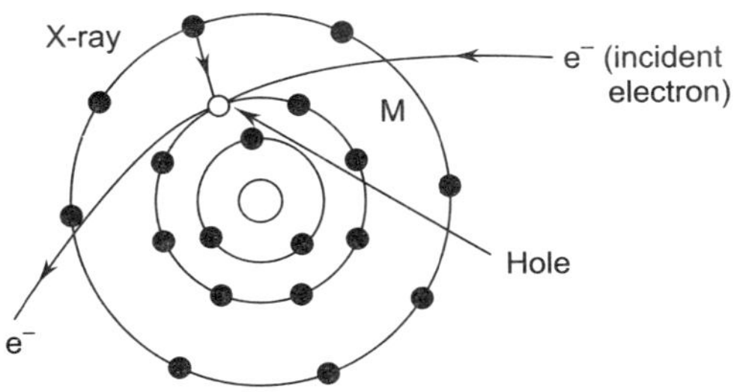

Figure 7.15

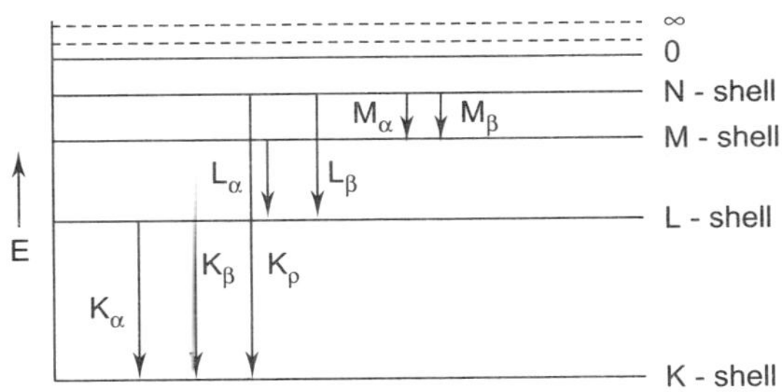

Figure 7.16

Moseley's Law

H.G.I. Moseley made a systematic study on the characteristic X-radiations emitted by different target elements which revealed the following facts regarding line spectra.

(*a*) The characteristic X-ray spectra (*i.e.*, line spectra) of different elements are similar to each other in the sense that each consists of K, L and M-series. However, the lines produced by the different target elements possess different wavelengths.

(*b*) For a particular line, the square root of the frequency is proportional to the atomic number (z) of the target element.

$$\sqrt{\nu} = a(z - b)$$

where a and b are constants, ν is the frequency corresponding to the line spectrum. The constant 'b' is known as the ***nuclear screening constant***. For K_α-line (*i.e.*, when electron jumps from $n = 2$ shell to the $n = 1$ shell)

$$\nu = \frac{3cR}{4}(z-1)^2,$$

where $b = 1$, R is the Rydberg constant

$$\text{and} \quad a^2 = cR\left[\frac{1}{n_1^2} - \frac{1}{n_2^2}\right] = \frac{m_e e^4}{8\varepsilon_0^2 h^3}\left[\frac{1}{n_1^2} - \frac{1}{n_2^2}\right]$$

$$R = \frac{m_e e^4}{8\varepsilon_0^2 h^3 c} = 1.097 \times 10^7 \text{ m}^{-1}.$$

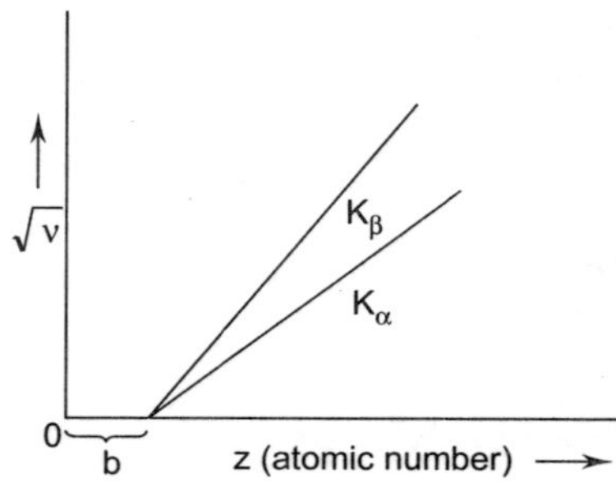

Figure 7.17

7.10 X-RAY DIFFRACTION

When an electron comes in the path of X-ray wave, it is set into forced vibration by the periodically changing electric field of X-ray passing by it. Due to such oscillations, the electron is accelerated. As an accelerated electrically charged particle behaves as a source of electromagnetic disturbance the electron acts as a source of electromagnetic wave of the same frequency and wavelength as the original X-ray wave. By this interaction the electron is said to scatter the original X-ray wave.

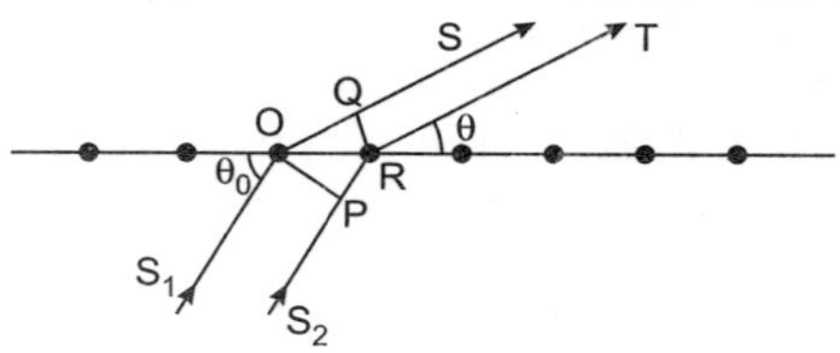

Figure 7.18

Let us now consider an one-dimensional regular array of points, spaced a apart as shown in figure 7.19. The atom scatters the X-radiation and produce about itself a new set of spherical wave envelops. Any line up of envelops constitute a combined wave moving in the direction of the common tangent. The cooperative combination of scattered wavelets is known as diffraction. When the tangent is parallel to the original wave front, the diffraction is known as zero order diffraction. When the tangent starts from the inner most spherical envelop of one atom, continuing through the second nearest envelop of next atom, through the third nearest envelop of next atom and so on, the wave built up along this front is known as first order diffraction. Similarly, when the envelop difference is two, we have second diffraction and so on.

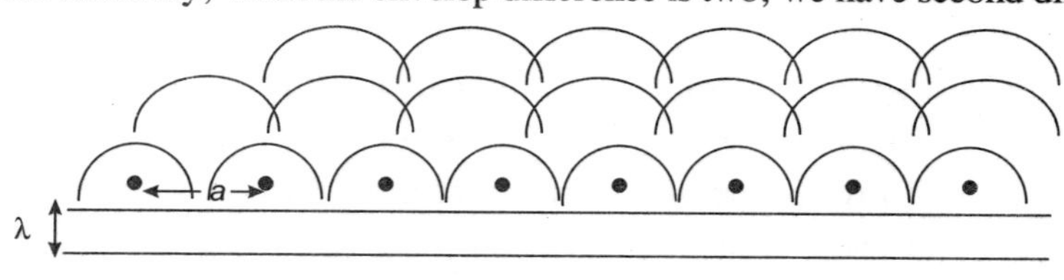

Figure 7.19

7.11 BRAGG'S LAW

X-rays are electromagnetic waves with wavelengths thousand times smaller then the visible light. In order to measure the wavelengths, a grating of corresponding dimension is required and hence simple gratings can not be used. Moreover it is rather impossible to prepare artificially a grating of such dimension. X-rays are diffracted from a crystal when certain geometrical condition is satisfied. Laue showed that atoms in a crystal are arranged in regular manner and the spacing between them is comparable to the wavelength of X-rays and hence the crystal can act as suitable natural grating for diffracting X-rays.

W.L. Bragg presented a simple explanation of the observed angles of the diffracting beams from a crystal. Consider an atomic array of the crystal in which the atoms are arranged in a given plane of the crystal. Suppose parallel beam of X-rays is incident in a direction making an angle θ with the surfaces of the planes. Each atom acts as a centre of disturbance and sends spherical wave-fronts by Huygens construction. As X-rays are much more penetrating then ordinary light, it is necessary to consider the rays reflected not from a single layer but from several layers together. There will be waves reflected from different planes in phase. Now they will reinforce themselves and an intense reflected beam will result. The condition for the reflected beam to be in the same phase is that the path–difference between the reflected wave from one layer and that from the next layer must be an exact wavelength or an integral multiple of it.

Let us consider two parallel rays LMN and PQR, which are reflected by two atoms M and Q in adjacent layers as shown in figure 7.20.

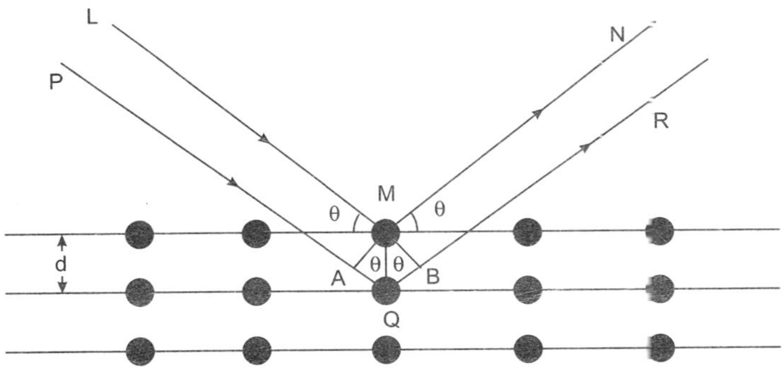

Figure 7.20

The length of the path PQR is greater than the length of the path LMN. The path difference is (AQ + QB) and according to the condition of reflection, we have

$$(AQ + QB) = n\lambda$$

But from the figure $AQ = BQ = d \sin\theta$

So $$2d \sin\theta = n\lambda \qquad \dots (7.1)$$

where n is an integer. This is known as Bragg's law and gives the condition for the reflection of X-rays from series of atomic layers in a given plane. Here, n represents the order of reflection. For $n = 0$, we get the zeroth order reflection which occurs for θ equal to zero. For the given values of d and λ, higher order of reflections appear for larger values of θ(glancing). The diffraction lines appearing for $n = 1$, 2 and 3 are called first, second and third order diffraction lines respectively and so on.

Bragg's reflection can occur only for wavelength $\lambda \leq 2d$. Due to this fact, visible light wavelength cannot be used in X-ray diffraction

7.12 THE VON LAUE TREATMENT : VALIDITY OF BRAGG'S TREATMENT

Von Laue treated the diffraction in a more general way by considering the scattering of X-rays from individual atoms in the crystal followed by their recombination to obtain the diffraction maxima. The diffraction maxima appear in some specific directions which obey certain conditions known as the Laue's equations. It also proves the validity of Bragg's treatment and the Bragg's law can be derived from the Laue's equation.

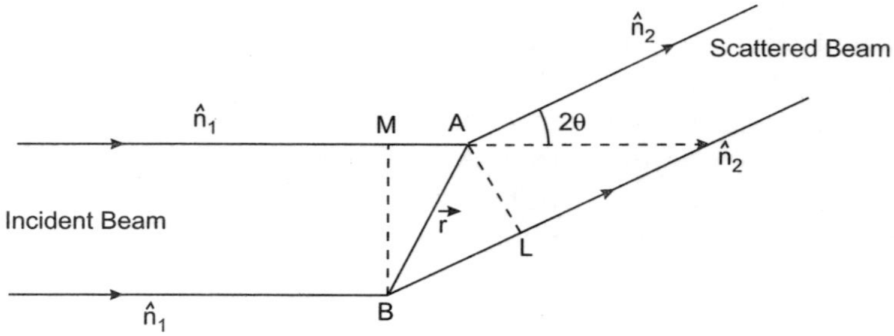

Figure 7.21

Consider the scattering of an incident beam from two identical scattering centers A and B placed at a distance $\vec{r}$ from each other in a crystal as shown in figure 7.21. Let n_1 and n_2 be the unit vectors in the direction of the incident and scattered beam respectively and let the angle between n_1 and n_2 be 2θ. Draw BM and AL perpendiculars to the direction of incident and scattering beam respectively. Then the path difference between the rays scattered from A and B is given by

$$\text{Path difference} = AM - BL = \vec{r} \cdot \hat{n}_1 - \vec{r} \cdot \hat{n}_2 = \vec{r} \cdot (\hat{n}_1 - \hat{n}_2) = \vec{r} \cdot \vec{N} \quad \text{...(7.2)}$$

where $\vec{N} = \hat{n}_1 - \hat{n}_2$. The vector $\vec{N}$ happens to be a normal to the reflecting plane. It is a plane which may be assumed to be reflecting the incident ray into the direction of the ray following the ordinary laws of reflection. This is one of the planes which form the basis of Bragg's treatment. From figure 7.22 we find $|N| = 2\sin\theta$

The phase difference between the ray scattered from A and B is

$$\Phi = \frac{2\pi}{\lambda}\left(\vec{r} \cdot \vec{N}\right) \quad \text{...(7.3)}$$

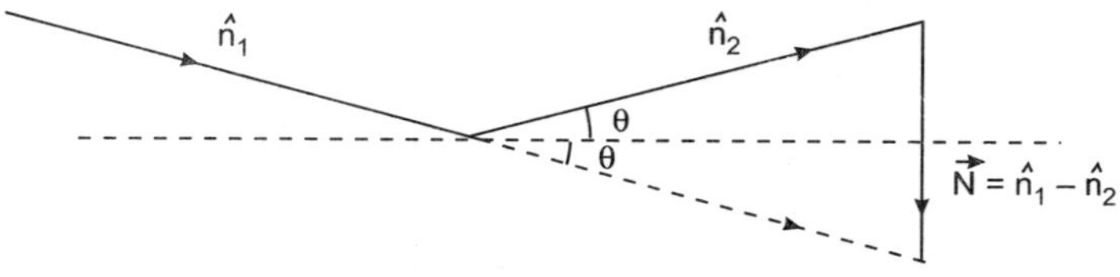

Figure 7.22

The radiations scattered by atoms A and B will interfere constructively only if the phase difference becomes an integral multiple of 2π. Due to the periodicity of the crystal, the other atoms placed in the same direction would also scatter the radiations exactly in phase with those scattered from A and B. In three-dimensional crystal, $\vec{r}$ may be coincide with any of the three crystallographic axes a, b and c. Thus for the occurrence of a diffraction maximum, the following three conditions must be satisfied simultaneously :

$$\frac{2\pi}{\lambda}\left(\vec{a}\,.\,\vec{\mathrm{N}}\right) = 2\pi h' = 2\pi nh$$

$$\frac{2\pi}{\lambda}\left(\vec{b}\,.\,\vec{\mathrm{N}}\right) = 2\pi k' = 2\pi nk$$

$$\frac{2\pi}{\lambda}\left(\vec{c}\,.\,\vec{\mathrm{N}}\right) = 2\pi l' = 2\pi nl \qquad \text{...(7.4)}$$

where h', k', l' represent any three integers. While obtaining the equations, it is assumed that atoms A and B are the nearest neighbours and so, the magnitudes a, b, and c represent the interatomic distances along their crystallographic directions. The integers h', k' and l' and h, k, l differ only by a common factor other than unity. Thus integers h, k and l can not have common factor other than unity and resemble the Miller indices of a plane which happens to be the reflection plane. Let α, β and γ be the angles between the scattering normal N and crystallographic axes a, b and c respectively.

Then,

$$\vec{a}.\vec{\mathrm{N}} = a\mathrm{N}\cos\alpha = 2a\sin\theta\cos\alpha, \text{ and so on.}$$

Therefore, the equations becomes

$$\left.\begin{aligned} \vec{a}\,.\,\vec{\mathrm{N}} &= 2a\sin\theta\cos\alpha = h'\lambda = nh\lambda \\ \vec{b}\,.\,\vec{\mathrm{N}} &= 2a\sin\theta\cos\beta = k'\lambda = nk\lambda \\ \vec{c}\,.\,\vec{\mathrm{N}} &= 2a\sin\theta\cos\gamma = l'\lambda = nl\lambda \end{aligned}\right\} \qquad \text{...(7.5)}$$

Equations (7.5) are known as Laue's equations and represent the conditions for diffraction to occur. In an orthogonal system α, β and γ also satisfy the condition

$$\cos^2\alpha + \cos^2\beta + \cos^2\gamma = 1 \qquad \text{...(7.6)}$$

where $\cos\alpha$, $\cos\beta$ and $\cos\gamma$ represent the direction cosines of the scattering normal. The equations yield the values of α, β, γ and θ for which diffraction takes place provided h, k, l and n are known. Thus for a given reflecting plane, equation (7.5) serve to determine unique values of θ and N which define a scattering direction.

7.13 ELEMENTARY IDEA ABOUT CRYSTAL BAND

In an atom for transition of electron from the inner orbit to the outer orbit, some energy is required. So, different orbits in an atom correspond to different states (levels) of potential energy. The electron try to occupy the inner orbit but according to Pauli's exclusion principal no more than two electrons having opposite spins in an atom can exist in same energy level.

If two similar atoms are brought together, there is an interaction between the orbits of their electron which causes splitting of each energy level into two slightly different levels. In a solid large number of atoms come close together. It cause perturbing effect *i.e.,* electrons of different atoms interact with each other, resulting in the formation of a large number of energy levels. The energy levels are so close together that they form a band, which is almost continuous. If an energy band consists of as much electrons as permitted by Pauli's exclusion principal, then it is said to be completely filled band. In such a band there will be no free electrons for conduction while for a pertly filled band conduction is possible.

Similar to the discrete energy levels of single atom, in solid, bands are separated by forbidden energy gap (E_g). E_g is the energy that should be imparted to the electron in valence band to jump to the conduction band. Electrons belonging to a certain band are mobile in the crystal are not confined to an individual atom. They belong to the crystal as a whole. The outermost band is called the conduction band and the band just below the conduction band is called the valence bond.

On the basis of the band theory, solids are classified as metal, insulator and semiconductor. In metal, valence and conduction bands overlap. The free electrons are available for electric conduction. Thus metals are good conductors. In an insulator and pure semiconductor (at 0° K) lower band is completely filled while the upper band is completely empty. In insulator the E_g is of the order of 5 to 10 eV and it is practically impossible to impart this amount of energy to the electron in the valence band to jump to conduction band. So conduction is not possible in insulator. On the other hand in semiconductor the E_g is of the order of 1 or 0.75 eV. With the rise of temperature more and more electrons from valence band jump to the conduction band and the material behaves like a intrinsic semiconductor. The examples are silicon, germanium. Thus it is evident that all the intrinsic semiconductors are insulators at 0° K and all the insulators may behave as semiconductors at temperature sufficiently higher than 0° K.

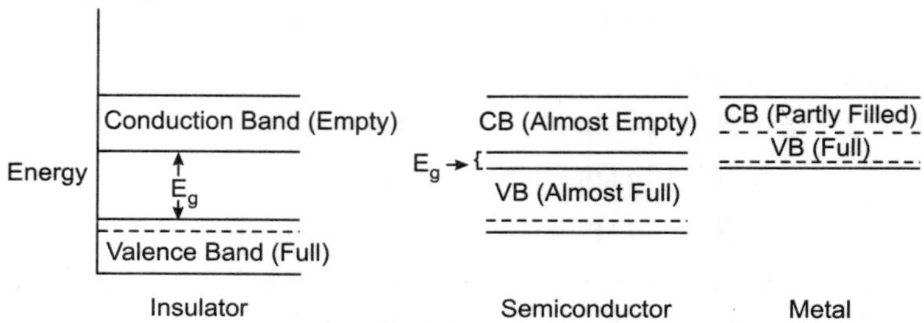

Figure 7.23

IN FOCUS

- Number of Atoms per unit cell are as under :

Types of Crystal Crystal structure	*Number of Atoms per unit cell*	*Atomic radius*	*Co-ordination Number*	*Atomic Packing Factor*
Simple Cube structure	1	$\frac{a}{2}$	6	0.52
Body centered cube structure	2	$\frac{\sqrt{3}a}{4}$	8	0.68
Face centered cube structure	4	$\frac{\sqrt{2}a}{4}$	12	0.74

- To determine the Miller Indices
  - Take the Reciprocals of the intersection of the axes of the point.

- Take the LCM of the denominators of the reciprocals.
- Multiply each of the reciprocals in (*i*) by the LCM as obtained in
  (*ii*) Inter-planar spacing

$$d = \left[\frac{h^2}{a^2} + \frac{k^2}{b^2} + \frac{l^2}{c^2}\right]^{-\frac{1}{2}}$$

where $(h\ k\ l)$ are miller indices and a, b, c intercepts from axis x, y, z

- Write the numbers in first bracket

■ **Bragg's Law:** $2d \sin \theta = n\lambda$

d = spacing between the consecutive Bragg planes of the crystal

θ = The glancing angle

n = Order of reflection

Solved Problems

Problem 1. *Calculate "Atoms per unit cell, 'co-ordination number' and 'atomic radius' and atomic packing fraction for the case of simple cubic, face centered cubic and body centered cubic cells.*

Solution. Atoms Per Unit Cell

(*a*) In the case of 'simple cubic' (*sc*), each of the eight corner atoms in cubic cell contributes only $\frac{1}{8}$ th of an atom to unit cell. So, there are in all 8 corner-atoms, their total contribution is equal to

$\frac{1}{8} \times 8\text{th} = 1$

So, the number of atoms per unit cell for simple cubic cell is one 1.

(*b*) For the case of face centered cubic (*fcc*), there are six faces of a cube and each face-centered atom is shared by two unit cells. There are in addition eight corner-atoms.

$\therefore$ The number of atoms per unit cell for the face centered cubic cell is equal to $\frac{6}{2} + \frac{8}{8} = 4$.

(*c*) In the case of body centered cubic cell (*bcc*), the contribution is full is one and this is in addition to eight corner-atoms.

$\therefore$ The number of atoms per unit cell for the body centered cubic cell is equal to $1 + \frac{8}{8} = 2$.

Co-ordination Number. The co-ordination number for simple cubic cell is 6, that of a *bcc* cell is 8 and that of a *fcc* cell is 12.

Atomic Radius. It is the distance between the centres of two neighbouring atoms.

In the case of simple cubic cell, if r is the atomic radius and a is the lattice parameter, then

$$r = \frac{a}{2}.$$

For the *bcc*, the atoms touch each other along the diagonal of the cube as shown in the figure then,

we have

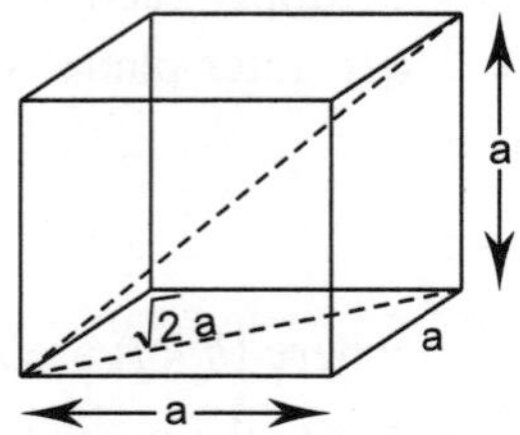

$$(4r)^2 = \left(\sqrt{2}a\right)^2 + a^2$$

or,
$$= 3a^2$$

$$4r = \sqrt{3}\,a$$

or,
$$r = \frac{\sqrt{3}}{4}\,a$$

In the case of *fcc*, the atoms are in contact along the diagonal of the faces.

$$(4r)^2 = 2a^2$$

or,
$$4r = \sqrt{2}a$$

$$r = \frac{\sqrt{2}}{2\sqrt{2}\,.\sqrt{2}}\,a = \frac{a}{2\sqrt{2}}$$

Atomic Packing Factor (or fraction). The ratio of the volume of the spheres in the unit cell to the volume of the unit cell is called the packing fraction of the structure.

In the case of simple cubic fraction, the packing fraction is

$$f = \frac{\frac{4}{3}\pi r^3}{a^3} = \frac{\frac{4}{3}\pi\left(\frac{a}{2}\right)^3}{a^3} = \frac{\pi}{6} = 52\% \qquad [\because \text{Here } r = \tfrac{a}{2}]$$

In the case of body-centered cubic lattice (*bcc*) [for *bcc* there are two atoms per unit cell]

$$f = \frac{2 \times \frac{4}{3}\pi\left(\frac{\sqrt{3}}{4}a\right)^3}{a^3} = \frac{2\sqrt{3}\pi}{16} = \frac{\sqrt{3}}{8}\pi = 68\%.$$

In the case of face-centered cubic lattice (*fcc*),

$$f = \frac{4 \times \frac{4}{3}\pi\left(\frac{a}{2\sqrt{2}}\right)^3}{a^3} = \frac{\pi}{3\sqrt{2}} = 74\%$$

[For *fcc*, there are four atoms per unit cell]

Problem 2. *For a cubic (fcc) crystal, lattice constant*

$$a = \left[\frac{4M}{\rho N}\right]^{\frac{1}{3}}$$

where M is the gm molecular weight of molecules at lattice points, ρ *is the density of crystal and N is Avogadro's number.* [Agra University]

Solution. Let *n* be the number of molecules in a unit cell.

Since mass per molecule $= \dfrac{M}{N}$

$\therefore$ mass of a unit cell $= n\dfrac{M}{N}$

Therefore, volume of a unit cell $v = \dfrac{nM}{\rho N}$

There are 4 lattice points per unit cell in the case of *fcc* crystal.

So, we can put $n = 4,$

If a is the lattice constant, $v = a^3$

or $$\frac{4M}{\rho N} = a^3$$

or $$a = \left(\frac{4M}{\rho N}\right)^{\frac{1}{3}}.$$

Problem 3. *Sodium chloride (NaCl) crystal has a cubic structure. If the molecular weight of NaCl is 58.46 and its density is 2.17 gm/cm³. Find the distance between two adjacent atoms in the NaCl crystal.* [Burd. University]

Solution. Mass of each NaCl molecule $= \dfrac{58.46}{6.023 \times 10^{23}} = \dfrac{M}{N}$

$$= 9.7 \times 10^{-23} \text{ gm}$$

Hence number of NaCl molecules per unit volume

$$= \frac{\text{Density}}{M} = \frac{2.17}{9.7 \times 10^{-23}}$$

$$= 2.237 \times 10^{22} \text{ molecules/cm}^3.$$

Since NaCl is diatomic, the number of atoms per unit volume will be $2 \times 2.237 \times 10^{22} = 4.47 \times 10^{22}$ atoms / cm³.

If a is the distance between two adjacent atoms, the volume a^3 will contain one atom due to cubic structure. Therefore $\dfrac{1}{a^3}$ is the number of atoms per unit volume.

$\therefore$ $$\frac{1}{a^3} = 4.47 \times 10^{22}, \text{ or, } a^3 = 2.237 \times 10^{-23} \text{ cm}^3.$$

$\therefore$ $$a = 2.82 \text{ Å}.$$

Problem 4. *Cs metal (atomic weight 130) has a cubic unit cell of side 6 Å. If the density of Cs is 2 g/cm³, determine whether the unit cell is simple, face-centered or body centered.*

[North Bengal University]

Solution. Given Atomic weight M = 130, N = 6.023×10^{23} = Avogadro's number

$$\rho = 2 \text{ gm/cm}^3.$$

Cs metal has a cubic unit cell of side $= a = 6$ Å $= 6 \times 10^{-8}$ cm

$\therefore$ Mass of each Cs atom $= \dfrac{M}{N}$.

Let x be the number of Cs atoms in cubic unit cell.

$\therefore$ Mass of a unit cell $= x\dfrac{M}{N}$

Therefore, volume of a unit cell $= \dfrac{xM}{\rho N}$

So, we can say $a^3 = \dfrac{xM}{\rho N}$

or,

$$x = \frac{\rho a^3 . N}{M}$$

$$= \frac{2 \times (6 \times 10^{-8})^3 \times 6.023 \times 10^{23}}{130}$$

$$= 2.$$

So the unit cell is the body centered.

Problem 5. *Calculate interplanar spacing 'd' of planes (1 1 1) in a simple cubic lattice of side 'a'.* [*WBUT*]

Solution. We know that the interplanar spacing 'd'

$$d_{hkl} = \frac{1}{\sqrt{\frac{h^2}{a^2} + \frac{k^2}{b^2} + \frac{l^2}{c^2}}}.$$

Here given $h = 1$, $k = 1$, $l = 1$ and for a simple cube $a = b = c$

$$\therefore \quad d_{111} = \frac{1}{\sqrt{\frac{1}{a^2} + \frac{1}{a^2} + \frac{1}{a^2}}} = \frac{a}{\sqrt{3}}. \quad \textit{Ans.}$$

Problem 6. *In a crystal, a lattice plane cuts intercepts of 2a, 3b and 6c along the axes, where a, b and c are primitive vectors of the unit cell. Determine the Miller indices of the given plane.*

Solution. Let the given plane cuts intercepts pa, qb, rc along the three axes, we have

$$pa : qb : rc = 2a : 3b : 6c$$

where a, b and c are primitive vectors of the unit cell.

$$\therefore \quad p : q : r = 2 : 3 : 6$$

or,

$$\frac{1}{p} : \frac{1}{q} : \frac{1}{r} = \frac{1}{2} : \frac{1}{3} : \frac{1}{6} = 3 : 2 : 1.$$

Therefore, Miller Indices of the plane are (3 2 1)

Problem 7. *Explain the meaning of (3, 2, 2) and (1, 1, 1) planes.* [Agra University]

Solution. The plane (3, 2, 2) has Miller indices 3, 2, 2. Its intercepts on $\vec{a}, \vec{b}$ and $\vec{c}$ axes are $\frac{1}{3}, \frac{1}{2}$ and $\frac{1}{2}$. Therefore, we can say, the intercepts are 2, 3, 3 *i.e.*, $2a$, $3b$ and $3c$.

The plane (1, 1, 1) has Miller indices 1, 1, 1. Its intercepts on $\vec{c}, \vec{b}$ and $\vec{c}$ axes are 1, 1 and 1. *i.e.*, a, b and c.

Problem 8. *Show that spacing d of a plane (h k l) in a simple cubic lattice of side a is* $d = \frac{a}{(h^2+k^2+l^2)^{\frac{1}{2}}}$ [Pune University, H.P. University, Agra University]

Solution. Given that d is the normal distance of plane from the origin.

$$\because \qquad hx + ky + lz = a$$

The direction cosines of the normal to this plane are $\frac{d}{a/h}, \frac{d}{a/k}, \frac{d}{a/l}$ *i.e.*, $\frac{dh}{a}, \frac{dk}{a}, \frac{dl}{a}$.

Since the sum of the squares of the direction cosines of the normal to plane is 1, we get

$$\frac{h^2d^2}{a^2} + \frac{k^2d^2}{a^2} + \frac{d^2l^2}{a^2} = 1$$

or,
$$d\sqrt{k^2+h^2+l^2} = a$$

or,
$$d = \frac{a}{\sqrt{k^2+h^2+l^2}}$$

Problem 9. *Lattice constant of a cubic lattice is a. Calculate the spacing between (2, 1, 1), (0, 0, 1), (1, 1, 0), (1, 0, 0), (1, 1, 1) and (1, 0, 1) planes.* [Pune University]

Solution. We know that the spacing d of plane ($h\ k\ l$) in a simple cubic lattice of side a is

$$d_{hkl} = \frac{a}{(h^2+k^2+l^2)^{\frac{1}{2}}}$$

For (2 1 1) plane

$$d_{2\,1\,1} = \frac{a}{\sqrt{2^2+1^2+1^2}} = \frac{a}{\sqrt{6}}$$

For (0 0 1) plane

$$d_{0\,0\,1} = \frac{a}{\sqrt{0^2+0^2+1^2}} = a$$

For (1 1 0) plane

$$d_{1\,1\,0} = \frac{a}{\sqrt{1^2+1^2+0^2}} = \frac{a}{\sqrt{2}}$$

For (1 1 1) plane

$$d_{111} = \frac{a}{\sqrt{1^2+1^2+1^2}} = \frac{a}{\sqrt{3}}$$

For (1 0 1) plane

$$d_{101} = \frac{a}{\sqrt{1^2+0^2+1^2}} = \frac{a}{\sqrt{2}}.$$

Problem 10. *The primitives of a crystal are 1.2 Å, 1.8 Å and 2 Å along the three axes where the Miller indices (2, 3, 1) cut intercept 1.2 Å along x-axis. What will be the lengths of intercept along y and z axis ?*

Solution. According to given conditions that,

$$h : k : l = 2 : 3 : 1$$

we have, $$\frac{1}{p} : \frac{1}{q} : \frac{1}{r} = h : k : l = 2 : 3 : 1$$

or, $$p : q : r = \frac{1}{2} : \frac{1}{3} : \frac{1}{1}$$
$$= 3 : 2 : 6$$

Hence, $$pa : qb : rc = (3 \times 1.2) : (2 \times 1.8) : (6 \times 2)$$
$$= 3.6 : 3.6 : 12$$

if lx, ly and lz be the lengths of intercepts along x, y and z axes respectively, then

$$lx : ly : lz = 3.6 : 3.6 : 12$$

If $lx = 1.2$ Å, $$ly = 1.2 \text{ Å} \times \frac{3.6}{3.6} = 1.2 \text{ Å}$$

and $$lz = \frac{12}{3.6} \times 1.2 \text{ Å} = 4 \text{ Å}.$$

Therefore, the length of intercepts along y and z axes are 1.2 Å and 4 Å respectively.

Problem 11. *Calculate the interplanar spacing for a (3 2 1) plane in a simple cubic lattice whose lattice constant is 4.2×10^{-10} m.* [Rohilkhand University]

Solution. Given that $a = 4.2 \times 10^{-10}$ m

We know that the spacing d of plane ($h\ k\ l$) in a simple cubic lattice a is

$$d_{hkl} = \frac{a}{\sqrt{h^2+k^2+l^2}}$$

For the plane (3 2 1), $h = 3$, $k = 2$, $l = 1$.

$$d_{321} = \frac{4.2 \times 10^{-10}}{\sqrt{3^2+2^2+1^2}} = 1.1 \times 10^{-10} \text{ m}.$$

Problem 12. *An X-ray tube operating at (i) 30 kV, (ii) 44 kV (iii) 50 kV and (iv) 200 kV emits a continuous spectrum with shortest wavelength (i) 0.414 Å, (ii) 0.284 Å, (iii) 0.284×10^{-10}m and (iv) 6.2×10^{-12} m. Calculate Planck's constant. Given : $e = 1.6 \times 10^{-19}$ c and $c = 3 \times 10^8$ m/sec.*
[G.N.D. University, Pbi. University, Lucknow University, Pune University]

Solution. Minimum wavelength is given by

$$eV = hv = h \frac{c}{\lambda_{min}} \qquad [h = \text{Planck's constant}]$$

or,
$$\lambda_{min} = \frac{ch}{eV}, \text{ or } h = \frac{eV}{c} \cdot \lambda_{min}.$$

Given : $e = 1.6 \times 10^{-19}$ C $= 3 \times 10^8$ m/sec.

(*i*)
$$h = \frac{1.6\times10^{-19}\ \text{C}\times30\times10^3\ \text{volt}\times0.414\times10^{-10}\ \text{m}}{3\times10^8\ \text{m/sec}} = 6.62 \times 10^{-34}\ \text{Js}$$

(*ii*)
$$h = \frac{1.6\times10^{-19}\ \text{C}\times44\times10^3\ \text{volt}\times0.284\times10^{-10}\ \text{m}}{3\times10^8\ \text{m/sec}} = 6.66 \times 10^{-34}\ \text{Js}$$

(*iii*)
$$h = \frac{1.6\times10^{-19}\ \text{C}\times50\times10^3\ \text{volt}\times0.284\times10^{-10}\ \text{m}}{3\times10^8} = 6.61 \times 10^{-34}\ \text{Js}$$

(*iv*)
$$h = \frac{1.6\times10^{-19}\ \text{C}\times200\times10^3\ \text{volt}\times6.2\times10^{-12}\ \text{m}}{3\times10^8} = 6.61 \times 10^{-34}\ \text{Js}.$$

Problem 13. *An X-ray tube operates at the voltage given below. Find the maximum speed of electrons striking at the anticathode and the shortest wavelength of X-ray produced.*

(i) 20 kV [Pune University, Lucknow University]

(ii) 100 kV [Pune University, G.N.D. University]

Solution. (*i*) We know that the maximum kinetic energy

$$\frac{1}{2} mv^2 = eV$$

or,
$$v = \sqrt{\frac{2eV}{m}}$$

where V is the voltage applied to the X-ray tube, e is the electronic charge and v is the maximum velocity acquired by the electron.

∴
$$v = \sqrt{\frac{2\times1.6\times10^{-19}\times20\times10^3}{9.11\times10^{-31}}} = 84.27 \times 10^6\ \text{m/sec.}$$

Again the shortest wavelength

$$\lambda_m = \frac{ch}{eV}$$

$$= \frac{3\times10^8\times6.62\times10^{-34}\ \text{Js}}{1.6\times10^{-19}\ \text{C}\times20\times10^3\ \text{volt}} = 0.62\ \text{Å}$$

(*ii*) For $V = 100 \text{ kV} = 100 \times 10^3$ volt

$$\therefore \quad v = \sqrt{\frac{2 \times 1.6 \times 10^{-19} \times 100 \times 10^3}{9.11 \times 10^{-31}}} = 188.45 \times 10^6 \text{ m/sec}$$

and

$$\lambda_{min} = \frac{3 \times 10^8 \times 6.62 \times 10^{-3} \text{ Js}}{1.6 \times 10^{-9} \text{ C} \times 100 \times 10^3} = 0.124 \text{ Å}$$

Problem 14. *The Bragg's angle corresponding to the first order reflection from (1 1 1) plane in a crystal is 30°, when X-rays of wavelength 1.75 Å are used. Calculate the interatomic spacing.*

Solution. We know that the spacing d of plane $(h\,k\,l)$ in a crystal of side a is

$$d_{h\,k\,l} = \frac{a}{\sqrt{h^2 + k^2 + l^2}}$$

$$d_{1\,1\,1} = \frac{a}{\sqrt{1^2 + 1^2 + 1^2}} = \frac{a}{\sqrt{3}}$$

From Bragg's law we know that

$$2d \sin \theta = n\lambda$$

Here $n = 1$, $\lambda = 1.75 \text{ Å} = 1.75 \times 10^{-10}$ m.

or,

$$2 \times \frac{a}{\sqrt{3}} \sin 30° = 1 \times 1.75 \times 10^{-10} \text{ m}$$

or,

$$a = \sqrt{3} \times 1.75 \times 10^{-10} \text{ m}$$

$$= 3.031 \text{ Å. } \textbf{Ans.}$$

Problem 15. *A beam of X-rays in an X-ray spectrometer is incident on a crystal of interplanar spacing 2.82 Å at an angle of 14°. The operating voltage is gradually increased. An intense reflection appears when the operating voltage is 9.1 kV. Find the value of Planck's constant.*

[Calcutta University]

Solution. Given that $d = 2.82 \text{ Å} = 2.82 \times 10^{-10}$ m, $\theta = 14°$

and according to the Problem, the intense reflection corresponds to $n = 1$.

Applying Bragg's law,

$$2d \sin \theta = n\lambda$$

or,

$$\lambda = \frac{2 \times 2.82 \times 10^{-10} \sin 14°}{1} \text{ m}$$

$$\therefore \quad \text{Frequency of X-ray, } \nu = \frac{c}{\lambda} = \frac{3 \times 10^8 \text{ m}}{2 \times 2.82 \times 10^{-10} \times 0.242 \text{ m}}$$

$$= 2.1989 \times 10^{18} \text{ Hz.}$$

Again, if V be the operating voltage and e the charge of electron (1.6×10^{-19} C), we have

$$eV = hv$$

$$h = \frac{eV}{v} = \frac{1.6 \times 10^{-19}\ \text{C} \times 9.1 \times 10^{3}\ \text{volt}}{2.1989 \times 10^{18}\ \text{Hz}}$$

$$= 6.62 \times 10^{-34}\ \text{Js}.$$

Problem 16. *If X-rays of wavelength 0.5 Å are diffracted at an angle of 5° in the first order, what is the spacing between the adjacent planes of the crystal? At what angle will the second maximum occur?* [North Bengal University]

Solution. Given that, $\lambda = 0.5\ \text{Å} = 0.5 \times 10^{-10}$ m, $\theta = 5°$, $n = 1$.

From the Bragg's law,

$$2d \sin\theta = n\lambda$$

The spacing between the adjacent planes of the crystal,

$$d = \frac{n\lambda}{2\sin\theta} = \frac{1 \times 0.5 \times 10^{-10}\ \text{m}}{2 \times \sin 5°} = 2.87\ \text{Å}$$

Let θ_1 be the angle when the second maximum will occur.

$$2d \sin\theta = n\lambda$$

or,

$$d = \frac{n\lambda}{2\sin\theta} = \frac{2 \times 0.5 \times 10^{-10}\ \text{m}}{2 \times \sin\theta_1} = 2.87 \times 10^{-10}\ \text{m}$$

or,

$$\sin\theta_1 = 0.1742$$

$$\theta_1 = 10.03°.$$

Problem 17. *The molecular weight of rock salt (a simple cubic) crystal is 58.5 kg per mole and its density is 2.16×10^{13} kg/m^3. Taking Avogadro's number = 6.023×10^{23}, calculate the grating space d_{100} of rock salt. Using this value of lattice constant calculate the wavelength of X-rays in second order if the angle of diffraction $\theta = 26°$.*

[Calcutta University, Bangalore University, Agra University]

Solution. Mass of each NaCl molecule $= \dfrac{58.5}{6.023 \times 10^{23}} = \dfrac{M}{N}$

$\therefore$ Volume of each NaCl molecule $= \dfrac{9.72 \times 10^{-26}}{2.16 \times 10^{3}} = \dfrac{M}{N\rho}$

$$= 4.5 \times 10^{-29}\ \text{m}^3.$$

There are two atoms for NaCl molecule.

Hence, the volume per atom $= \dfrac{4.5 \times 10^{-29}}{2} = 2.25 \times 10^{-29}\ \text{m}^3.$

Let d be the atomic spacing in the NaCl crystal, then the volume per atom is d^3.

$$d^3 = 2.25 \times 10^{-29}\ \text{m}^3$$

$\therefore$

$$d = 2.82\ \text{Å} = d_{100}$$

From the Bragg's law

$$2d \sin\theta = n\lambda$$

or,
$$\lambda = \frac{2d \sin\theta}{n} = \frac{2 \times 2.82 \times 10^{-10} \times \sin 26°}{2} \text{ m}$$

$$= 1.24 \times 10^{-10} \text{ m} = 1.24 \text{ Å}.$$

Problem 18. *The distance d_{100} between (1, 0, 0) plane in BCC structure is 0.232 nm. What is the size of the unit cell? What is the radius of the atom?* [WBUT]

Solution. For a BCC structure, atomic radius $r = \frac{\sqrt{3}}{4} a$

where a = lattice constant.

Again, for BCC crystal, $d_{100} = \frac{a}{2} = 0.232$ nm

or, $a = 0.464$ nm.

$$\therefore \quad r = \frac{\sqrt{3}}{4} a = \frac{\sqrt{3}}{4} \times 0.464 = 0.20 \text{ nm}.$$

Problem 19. *Copper has FCC structure and the atomic radius is 0.1278 nm. Calculate its density and inter-planar spacing for (321) planes. Take the atomic weight appear as 63.5.* [WBUT]

Solution. Here given that, M = atomic weight = 63.5

n = number of atom per unit cell for FCC = 4

For FCC crystal atomic radius $r = \frac{\sqrt{2}a}{4} = 0.1278$ nm.

or
$$a = \frac{4 \times 0.1278}{\sqrt{2}} = 2\sqrt{2} \times 0.1278 \text{ nm}.$$

We know for a cubic crystal, the lattice constant

$$a = \left(\frac{n\text{M}}{\rho\text{N}}\right)^{\frac{1}{3}} \text{ or, } \rho = \frac{n\text{M}}{\text{N}a^3}.$$

or,
$$\rho = \frac{4 \times 63.5}{6.023 \times 10^{23} \times \left(2\sqrt{2} \times 0.1278 \times 10^{-7} \text{ cm}\right)^3} = 8.926 \text{ gm/cm}^3.$$

The interplanar spacing for (3 2 1)

$$d_{hkl} = \frac{a}{\sqrt{n^2 + k^2 + l^2}} = \frac{2\sqrt{2} \times 0.1278 \times 10}{\sqrt{(3)^2 + (2)^2 + (1)^2}}$$

$$\simeq 0.097 \text{ nm}.$$

Problem 20. *The wavelength of the K_2 line of Ag is 0.563 Å. The radiation from a Ag target is analysed with a Bragg spectrometer using a calcite crystal (a simple cube of a lattice constant 3.02945 Å). Determine the angle of reflection for the first order. What is the highest order for which this line may be observed ?* [Calcutta University]

Solution. Given that, $\lambda_\alpha = 0.563 \text{ Å} = 0.563 \times 10^{-10}$ m,

$d = d_{100} = 3.02945 \text{ Å} = 3.02945 \times 10^{-10}$ m.

and $n = 1$.

According to Bragg's law,

$$2d \sin\theta = n\lambda$$

or,
$$\sin\theta = \frac{n\lambda}{2d} = \frac{1 \times 0.563 \times 10^{-10} \text{ m}}{2 \times 3.02945 \times 10^{-10} \text{ m}}$$

$$= 0.0929$$

$\therefore$ The angle of reflection for the 1st order is $\theta = 5.33°$.

If n^{th} is the highest order, then

$$2d \sin\theta = n\lambda$$

$$n = \frac{2d \sin\theta}{\lambda} \quad \text{[For the highest value } \theta = 90°]$$

$$= \frac{2 \times 3.02945 \times 10^{-10} \text{ m} \times \sin 90°}{0.563 \times 10^{-10} \text{ m}} = 10.8,$$

i.e., $n = 10$

The highest order is 10.

Problem 21. *The Bragg's angle for first order reflection from (1 1 1) plane in a crystal is 60°. Calculate the interatomic spacing if X-rays of wavelength* 1.8×10^{-10} *m are used.*

[K. University]

Solution. Given that $n = 1$, $\theta = 60°$, $\lambda = 1.8 \times 10^{-10}$ m.

According to Bragg's law,

$$2d \sin\theta = n\lambda$$

or
$$d = \frac{1 \times 1.8 \times 10^{-10} \text{ m}}{2 \sin 60°} = \frac{1.8 \times 10^{-10} \text{ m}}{\sqrt{3}}$$

$$= 1.039 \times 10^{-10} \text{ m}$$

Here $d = d_{1\,1\,1} = 1.039 \times 10^{-10}$ m.

We know that for a simple cubic lattice

$$d_{1\,0\,0} = a, \text{ then } d_{1\,1\,1} = \frac{a}{\sqrt{3}}$$

$$\therefore \quad \frac{a}{\sqrt{3}} = 1.039 \times 10^{-10} \text{ m}$$

$$a = 1.8 \times 10^{-10} \text{ m} = 1.8 \text{ Å}.$$

The interatomic spacing is 1.8 Å.

Problem 22. *X-rays from a tube operated at 50 kV are analysed with a Bragg's spectrometer using a calcite crystal cut along its cleavage plane. If the grating space of calcite be given as 3.02945 Å calculate the smallest angle between the crystal plane and the X-ray beam at which the shortest wavelength produced by the tube can be detected.* [Calcutta University]

Solution. Given that $V = 50 \text{ kV} = 50 \times 10^3$, $d = 3.02945$ Å

We know that,
$$\lambda_{\min} = \frac{hc}{eV} = \frac{6.62 \times 10^{-34} \times 3 \times 10^8}{1.62 \times 10^{-19} \times 50 \times 10^3}$$
$$= 0.248 \times 10^{-10} \text{ m} = 0.248 \text{ Å}.$$

If θ be the required smallest angle, then we have
$$2d \sin \theta = n\lambda$$
$$\sin \theta = \frac{n\lambda}{2d} = \frac{1 \times 0.248 \times 10^{-10} \text{ m}}{2 \times 3.02945 \times 10^{-10} \text{ m}}$$
$$= 0.0409$$
$$\theta = 2°20'$$

Problem 23. *In a X-ray diffraction experiment the second order glancing angle was 30°. Calculate the glancing angles for the 1st and third order.* [H.P. Univ.]

Solution. From the Bragg's law, we have
$$2d \sin \theta = n\lambda$$
Here,
$$n = 2,\ \theta = 30°,$$
$$2d \sin 30° = 2 \times \lambda$$
$$\text{as } \frac{\lambda}{d} = \frac{1}{2} \quad \ldots(i)$$

For the 1st order spectrum, $2d \sin \theta_1 = \lambda$

or
$$\sin \theta_1 = \frac{1}{2} \times \frac{\lambda}{d} = \frac{1}{2} \times \frac{1}{2} = 0.25$$
or
$$\theta_1 = 14°\ 29'.$$

For 3rd order spectrum, $2d \sin \theta_3 = 3\lambda$

or,
$$\sin \theta_3 = \frac{3}{2} \times \frac{\lambda}{d} = \frac{3}{4} = 0.75$$
or,
$$\theta_3 = 48°\ 35'. \quad \textbf{Ans.}$$

Problem 24. *The spacing of planes in a crystal is 1.2 Å and the angle for the 1st order Bragg's reflection is 30°. Determine the energy of the X-ray beam in eV.* [WBUT]

Solution. From the Bragg's equation, $2d \sin \theta_n = n\lambda$
$$\lambda = \frac{2d \sin \theta_n}{n} = \frac{2 \times 1.2 \text{ Å} \times \sin 30°}{1} = 1.2 \text{ Å}.$$

Now the energy of the incident beam $E = h\nu = \dfrac{hc}{\lambda}$
$$= \frac{6.63 \times 10^{-34} \times 3 \times 10^{48}}{1.6 \times 10^{-19} \times 1.2 \times 10^{-10} \text{ m}} eV$$
or,
$$E = 10.35 \times 10^3 \text{ eV} = 10.35 \text{ KeV}.$$

Multiple Choice Questions

1. The atomic radius for fcc lattice is

(*a*) $\frac{\sqrt{3}}{4}a$ (*b*) $\frac{\sqrt{2}}{4}a$ (*c*) $\frac{a}{2}$

2. The atomic radius for bcc lattice is

(*a*) $\frac{a}{2}$ (*b*) $\frac{\sqrt{3}}{4}a$ (*c*) $\frac{\sqrt{2}}{4}a$

3. The atomic radius for simple cube lattice is

(*a*) $\frac{a}{2}$ (*b*) $\frac{\sqrt{3}}{4}a$ (*c*) $\frac{\sqrt{2}}{4}a$

4. For the case of simple cubic lattice, the co-ordinate number is

(*a*) 6 (*b*) 8 (*c*) 12.

5. For the case of fcc, the co-ordinate number is

(*a*) 6 (*b*) 8 (*c*) 12.

6. For the case of bcc, the co-ordinate number is

(*a*) 6 (*b*) 8 (*c*) 12.

7. In a body centered cubic lattice (bcc), the packing fraction is

(*a*) 52% (*b*) 68% (*c*) 74%.

8. In a face centered cubic lattice (fcc), the packing fraction is

(*a*) 52% (*b*) 68% (*c*) 74%.

9. Which of the following has the highest frequency ?

(*a*) X-rays (*b*) Infra-red rays (*c*) Visible light.

10. The intensity of X-rays is determined by

(*a*) Filament current

(*b*) Size of the cathode

(*c*) Filament voltage.

11. In coolidge tube the electrons are focussed on the target by using

(*a*) Magnetic lens (*b*) A concave cathode (*c*) An electron gun.

12. Continuous X-rays are produced as a result of

(*a*) Transition of electrons of target from higher to lower orbits.

(*b*) Transition of electrons of target from lower to higher orbits.

(*c*) Loss of kinetic energy of incident electrons.

13. The separation between lattice planes (1 0 0) in simple cubic lattices is

(*a*) a (*b*) $\frac{a}{\sqrt{2}}$ (*c*) $\frac{a}{\sqrt{3}}$

14. The separation between lattice planes (1 0 0) in bcc is

(*a*) $\frac{a}{\sqrt{2}}$ (*b*) $\frac{a}{2}$ (*c*) $\frac{a}{\sqrt{3}}$

15. Which of the followings has the Bragg's angle having wavelength 1.54 Å in the 1st order, if the interplanar distance is 2.67 Å.

(*a*) 59.9° (*b*) 35.22° (*c*) 16.76°.

16. The continuous X-ray spectrum is

(*a*) Directly proportional to the atomic weight of the target

(*b*) Directly proportional to the atomic number of the target

(*c*) Independent of the nature of the target.

17. The Miller indices of a plane having intercepts 2, 3, 4 units along X, Y and Z axis respectively are

(*a*) (6, 4, 3) (*b*) (4, 3, 6) (*c*) (2, 3, 1) (*WBUT*)

18. For BCC crystal, the atomic radius is –

(*a*) $\frac{\sqrt{3}}{4}a$ (*b*) $\frac{\sqrt{2}}{4}a$ (*c*) a (*WBUT*)

19. Origin of X-rays is due to the process of –

(*a*) Ionisation (*b*) Inner orbital transition (*c*) Bremsstrahlung. [*WBUT*]

20. If λ_L and λ_K are wavelenghts of L and K series X-ray lines respectively then –

(*a*) $\lambda_L > \lambda_K$ (*b*) $\lambda_L < \lambda_K$ (*c*) $\lambda_L = \lambda_K$ [*WBUT*]

21. The minimum frequency limit of X-ray spectrum depends upon

(*a*) nature of target material (*b*) kinetic energy of incident electron

(*c*) degree of vacuum in coolidge tube.

Answers

1. (*b*)	**2.** (*b*)	**3.** (*a*)	**4.** (*a*)	**5.** (*c*)	**6.** (*b*)
7. (*b*)	**8.** (*c*)	**9.** (*a*)	**10.** (*a*)	**11.** (*c*)	**12.** (*c*)
13. (*a*)	**14.** (*b*)	**15.** (*c*)	**16.** (*c*)	**17.** (*a*)	**18.** (*a*)
19. (*c*)	**20.** (*a*)	**21.** (*b*)			

Review Questions

1. Define crystal lattice, basis, unit cells and primitive cell. Draw the two dimensional representation whenever necessary. [*Pune Univ., Kolkata Univ., Kalinga Univ.*]

2. Explain the terms "atoms per unit cell", "co-ordinate number", "atomic radius" and "atomic packing fraction" as applied to crystal structure, calculate their values for a simple cubic, face centered cubic and body centered cubic cell. [*G.N.D. Univ., H.P. Univ., Pb. Univ.*]

3. What are the Miller Indices ? How the orientation of a plane is specified by Miller indices ? Explain their importance. [*Kolkata Univ., H.P. Univ., G.N.D. Univ.*]

4. Show that in a cubic crystal of side '*a*' the spacing between consecutive parallel planes of the Miller indices (*h k l*) is

$$d_{hkl} = \frac{a}{\sqrt{h^2 + k^2 + l^2}}$$ [*WBUT*]

5. (*a*) Distinguish between continuous and characteristic X-ray spectra. (*b*) Explain and derive Bragg's law of X-ray diffractions from a crystal. [*WBUT*]

6. What is Bravais lattice ? [*WBUT*]

7. Write a short note on :

Energy band diagram of a crystal vis-a-vis metal, insulator and semiconductor.

8. Derive Laue's equations for diffraction of X-rays and obtain Bragg's diffraction condition from them. [*H.P. University, Kolkata University, G.N.D. University*]

9. Distinguish between a metal, a semiconductor and an insulator on the basis of their energy band structure. [*Burdwan Univ.*]

10. What are Miller indices and what is their utility? [*Calcutta Univ.*]

11. Explain the terms : Lattice, basis and unit cell.
12. Write down Bragg's equation for reflection of X-rays by a crystal and interpret the symbols. [*Calcutta Univ.*]
13. (*a*) What is meant by Miller Indices ? Explain with examples. What is the utility of Miller indices ? [*WBUT*]
14. Deduce the interplanar spacing of a simple cubic lattice of side a. [*WBUT*]
15. What is the important properties of X-ray ? [*WBUT*]
16. Explain and derive Bragg's law of X-ray diffraction from a crystal. [*WBUT*]

Unsolved Problems

1. The lattice constant of a b.c.c. lattice is 5.96 Å. Find the distance between a corner atom and the body centered atom in the cubic cell. [**Ans.** 5.16 Å]
2. The density of b.c.c. iron is 7.9 gm/cm^3 and its atomic weight is 56. Calculate the length of the side of the cubic unit cell and the nearest neighbour distance. [*Calcutta University*]
   [**Hints :** The b.c.c. lattice contains two atoms in the cubic unit cell of volume a^3, where a is the edge length. The nearest neighbour distance is that between the body-center atom and a corner atom *i.e.*, $\frac{\sqrt{3}a}{2}$] [**Ans.** 2.87 Å, 2.48 Å]
3. Consider X-rays of wavelength 1.54 Å incident on a single cubic crystal of lattice constant 4Å. Calculate the angle for the first order (n = 1) reflection. [*Calcutta University*]
   [**Ans.** 11°]
4. The Bragg angle for first order reflection from (1 1 1) plane in a crystal is 60°. Calculate the interatomic spacing if X-rays of wavelength 1.8×10^{-10} m are used. [*Kalinga University*]
   [**Ans.** 1.8×10^{-10} m]
5. A beam of X-rays of wavelength 0.842 Å is incident on a crystal at a glancing angle of 8.35′ when first order Bragg's reflection occurs. Calculate the glancing angle for the third order reflection. [*WBUT*]

   [**Hints :** From the Bragg's law $2d \sin \theta = n\lambda$, putting $n = 1$, $\lambda = 0.842$ Å, $\theta = 8°35'$, $d = 3.125$ Å, Let θ be the glancing angle for the third order (n = 3) reflection.

   $\sin \theta = \frac{3\lambda}{2d}\theta$, or, $\theta = 26°29'$.]
6. The spacing between successive (100) planes in NaCl is 2.82Å. X-rays incident on the surface of the crystal is found to give rise to first order Bragg reflection at glancing angle 8.8°. Calculate the wavelength of X-rays. [*Burdwan Univ.*]

   [**Hints :** Solved Problems 14 and 9]
7. Bragg found that for a KCl crystal, strong reflection from the sets of planes (1, 0, 0), (1, 1, 0) and (1, 1, 1) are obtained at the same order for angles 5°23′, 7°25′ and 9°25′ respectively. Show that the KCl crystal has a simple cubic crystal structure. [*C.U. (Hons.), 96*]
8. The grating space of calcite is 3.036 Å. Calculate the wavelength of X-rays that undergo first order reflection at a glancing angle of 12°. At what angle will the third order reflection take place for the same wavelength? [*Osmania Univ.*]

   [**Ans.** 1.263Å, 33°36′]
9. If a narrow beam of K_α radiation (λ = 1.541Å) from Cu is reflected from (1, 0, 0) planes of such a crystal, calculate the glancing angle corresponding to the first order spectrum. [*Calcutta Univ.*]

10. Find the wavelength of short-wave limit of X-ray continuous spectrum, if it shifts by 0.5Å when the voltage applied to X-ray tube is doubled. [*Calcutta Univ.*]

[Ans. 1Å]

11. An X-ray tube is operated at an anode potential to 10 kV and anode current of 15 mA, calculate (*i*) number of electrons hitting the anode per second and (*ii*) minimum wavelength produced by the X-ray tube. [(*i*) 9.4×10^{16} electrons / sec ; (*ii*) 1.24 Å] [*H.P. Univ., Lucknow Univ.*]

12. The spacing of X-ray of wavelength 0.842 Å is incident on a crystal at a glancing angle of 8° 35′ when the first order Bragg's reflection occurs. Calculate the glancing angle for the third order reflection. [*WBUT*]

[**Hints :** Using $2d \sin \theta_n = n\lambda$, put $n = 1$, $\theta_n = 8° 35'$, $d = 2.8$ Å, $\lambda = 0.842 \times 10^{-8}$ cm
For order number n = 3, corresponding glancing angle

$$\sin \theta_3 = \frac{n\lambda}{2d}$$

or $$\theta_3 = 26° 48'.$$

Albert Einstein-
Rabindra Nath Tagore

8

Chapter

Laser and Fibre Optics

8.1 INTRODUCTION

LASER is the acronym for '***Light Amplification by Stimulated Emission of Radiation***'.

Laser is the most exciting discovery of twentieth century. Laser is a beam of radiation with some unusual properties. Generally, the beam is coherent, monochromatic, parallel and intense. The laser first came on the scene in the beginning of 1960s.

It is different from the ordinary light in many ways. As we know, the ordinary light diffuses as it travels. The light emitted from a conventional source is said to be incoherent because the radiations emitted from different atoms do not, in general, bear any definite phase relationship. It is electromagnetic wave oscillating at random frequency from radio wave to X-ray. However, the laser light is coherent, oscillating at only one frequency.

The basic principle involved in lasing action is the phenomenon of stimulated emission which are predicted by Einstein. According to Einstein when an atom is in the excited state, it can make a transition to a lower energy state through emission of electromagnetic radiation. The emission can occur in two different ways – *Spontaneous emission* and *Stimulated emission*.

8.2 CHARACTERISTICS OF LASER

The most important properties of laser are :

1. **Monochromaticity.** Laser light is extremely monochromatic. It means it is a light of single frequency, namely of a single colour.
2. **Coherence.** The Laser light is spatially and temporally coherent.
3. **Directionality.** The energy carried out by the laser beam can easily be collimated and focussed to a small region.
4. **Brightness.** The brightness of a source is defined as the power emitted per unit surface per unit solid angle. Laser represents source of high brightness and the brightness is higher than any ordinary source of light or it is comparable with the sun.

8.3 PRINCIPLES OF LASER ACTION

The electrons in an atomic system move in special, discrete orbits or energy states. The electron can move from one state to the other through the photons. An electron remains in the ground state (or normal state) with minimum energy E_1 in the absence of external influence. On being subject to some action (irradiation of photon of right frequency) in transits to a higher state absorbing the energy of the radiation. The system can change from a higher state to a lower state by emission of photon. If an electron in the excited state emits radiation even in the absence of any incident radiation then it is called the **spontaneous radiation.** The frequency of absorbed or emitted photon is given by $v = (E_2 - E_1)/h$, where h is the planck's constant and E_1 and E_2 are the energies of the states occupied by the electron before and after the occurrence of radioactive transition.

When a photon of energy hv ($= E_2 - E_1$) irradiates an atom already in the excited state E_2 it de-excites the atom thereby releasing two photons simultaneously each of energy hv. This type of transition is called the **stimulated or induced emission of radiation.**

Schematically,

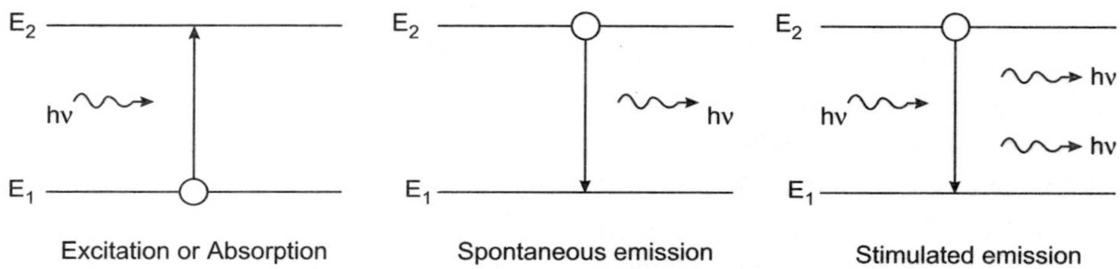

Figure 8.1

The directions of the incident and stimulated photons are in phase *i.e.*, coherent and add completely to the incident beam. If these two photons are now incident on two other photons and so on, it goes in chains. Hence if a large number of atoms is involved then the stimulated emission generates an intense beam of high coherent directivity. This device is known as laser.

Einstein showed that both spontaneous and stimulated emissions are necessary to obtain Planck's radiation Law.

Atomic Excitation

An atom can be excited to a state higher than its ground state by the following methods :

(*i*) By collision with another atom during which part of the kinetic energy of the photon is absorbed by the atom and an atom excited in this way returns to its ground state in time 10^{-8} sec by emitting one or more photons.

(*ii*) By establishing an electric field that accelerates electron and atomic ions till kinetic energies are sufficient to excite atom they collide with. Since the energy transfer is maximum when the colliding particles are of the same mass, so electrons in luminous discharge in a rarefied gas are more efficient than the ions in providing energy to atomic electrons.

Excitation by Absorption

In this kind of excitation an atom absorbs a photon of light whose energy is just the right energy to raise the atom to higher energy level. For example, a photon of $\lambda = 121.7$ mm is emitted when a hydrogen atom in $n = 2$ state comes to $n = 1$ state. The absorption of a photon of $\lambda = 121.7$ mm by hydrogen at ground state $n = 1$ will bring it to $n = 2$ state.

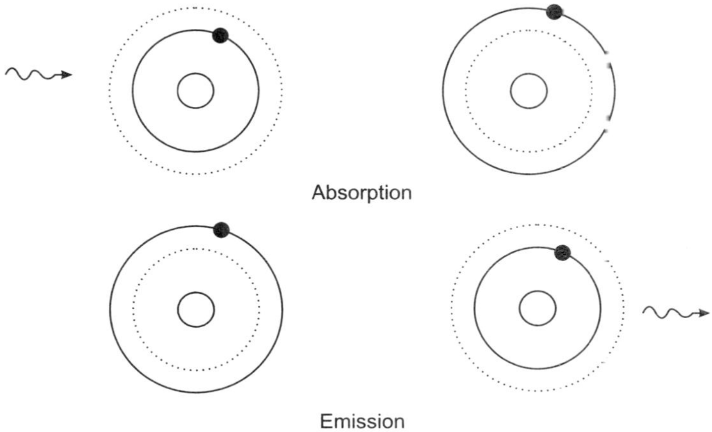

Figure 8.2

8.4 EINSTEIN'S THEORY — RELATION BETWEEN SPONTANEOUS AND STIMULATED EMISSION

Transition between various energy states is essentially a statistical process and one cannot predict which particular atom at a given instant will make transition from one energy state to the other. But if a very large number of atoms are involve, as in the case of atomic assembly, we may apply the theory of probability and calculate the rate of relative transitions between two energy states fairly and accurately.

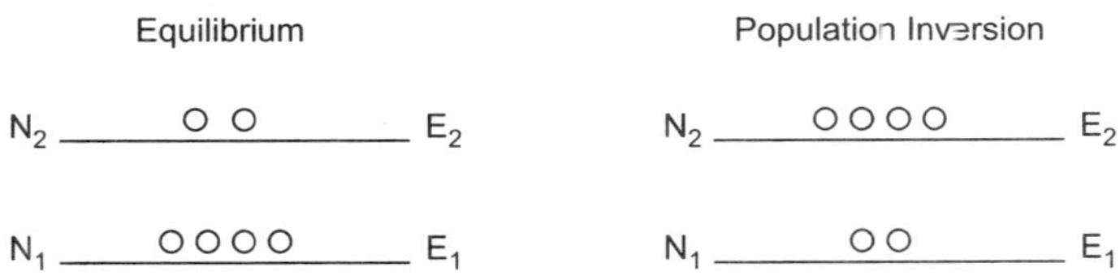

Figure 8.3

A two level energy system is depicted in figure 8.3. Here E_1 and E_2 are the energies of the respective levels, and N_1 and N_2 are the number of atoms per unit volume in each level. The ratio of N_2 to N_1 at thermal equilibrium is given by the Boltzmann equation.

$$N_2 = N_1 \exp(-(E_2 - E_1)/kT)$$

Here k is the Boltzmann constant and T is absolute temperature. It can be seen from the above relation that N_2 cannot exceed N_1 for an equilibrium situation which means $N_2 < N_1$ if $E_2 > E_1$.

The atoms in the higher excited state E_2 make spontaneous transitions to the lower energy state E_1 at a rate determined by the number of atoms per unit volume N_2 in the upper state and the probability A_{21} of spontaneous transition from E_2 to E_1. So N_2A_{21} gives the number of spontaneous transition per unit volume per unit time (rate per unit volume). This type of transition will give rise to emission of electromagnetic radiation of energy $h\nu = E_2 - E_1$.

Assume the atoms to be in equilibrium with the above radiation energy density U_ν. By absorption of energy from radiation, some of the atoms in the lower energy state E_1 may again make transition to the higher state E_2. The rate of such induced transition would be determined by N_1, energy density U_ν of radiation and the probability of induced transition B_{12}. Thus the product $N_1U_\nu B_{12}$ gives the rate of absorption of photon per unit volume.

Further there may, in addition to the above two, be a downward induced transition of some atoms from E_2 to E_1 due to the radiation field. This will result in stimulated emission of radiation of frequency $\nu = (E_2 - E_1)/h$

The number of such stimulated transitions per second per unit volume is $N_2U_\nu B_{21}$.

At equilibrium, assuming that only optical transitions occur, these rates are related as:

$$N_1U_\nu B_{12} = N_2A_{21} + N_2U_\nu B_{21}$$

$$U_\nu = \frac{N_2A_{21}}{N_1B_{12} - N_2B_{21}}$$

$$= \frac{A_{21}}{B_{21}} \cdot \frac{1}{\left[\frac{B_{12}}{B_{21}}\frac{N_1}{N_2} - 1\right]}$$

Now $$\frac{N_1}{N_2} = \exp\,[-\,(E_2 - E_1)/\,kT = \exp\,(h\nu/kT)$$

$\therefore$ $$U_\nu = \frac{A_{21}}{B_{21}} \cdot \frac{1}{\left[\frac{B_{12}}{B_{21}}\exp\,(h\nu/kT) - 1\right]}.$$

The above equation must be in conformity with Planck's equation for radiation, which is

$$U_\nu = \frac{8\pi h\nu^3}{c^3} \cdot \frac{1}{\exp\left[(h\nu/kT) - 1\right]}.$$

The above equation must be in conformity with Planck's equation for radiation, which is

$$U_\nu = \frac{8\pi h\nu^3}{c^3} \cdot \frac{1}{\exp\,(h\nu/kT) - 1}.$$

Thus by comparing, $B_{12} = B_{21}$ ...(8.1)

and $$\frac{A_{21}}{B_{21}} = \frac{8\pi h\nu^3}{c^3} \qquad \ldots(8.2)$$

Therefore the probability of absorption and stimulated emission are equal.

Again equation (8.2) shows that the probability of spontaneous emission increase as ν^3, which means that the probability of spontaneous emission increase rapidly with the energy difference between the involved states.

The A's and B's are respectively known as Einstein's A and B coefficient.

8.5 OTHER PRINCIPLES OF LASER

Population Inversion

The necessary condition for amplification of light intensity by the foregoing method is that the number of atoms in the upper energy state E_2 must outnumber the atoms in the lower state E_1. If N_1 be the number of atom per unit volume in the lower energy state E_1 and N_2 be the number of atom per unit volume in the upper energy state E_2 then the equilibrium condition is

$$N_2 = N_1 \exp(-(E_2 - E_1)/kT)$$

where k is the Boltzmann constant, T is the temperature. A population inversion refers to a non-equilibrium situation in which $N_2 > N_1$. It occurs as a result of the excitation process and has been referred to as a negative-temperature situation, since $N_2 > N_1$ if T is negative.

A population inversion is necessary for amplification of light to occur. Without a population inversion, a beam of light directed through the medium with photon energy $h\nu = (E_2 - E_1)$ will be absorbed.

Multilevel System

A two level of lasing is not workable due to the fact that the life time of an atom in an excited state is very small ~ 10^{-8} sec. This means that before an atom can be stimulated to emit a photon it may make a spontaneous emission.

We have $N_2 = N_1 \exp(-(E_2 - E_1)/kT)$.

For atomic radiation $(E_2 - E_1)$ must be greater than kT and so in the thermal equilibrium the population of higher state is smaller than that of ground state, with the result that there will be few stimulated emission of radiations.

A three level architecture is a solution of the problem. The atom that are originally in the ground state G are raised to a higher short-lived state by some external source of energy. However, the atom cannot remain in excited state H for long time and decay into an intermediate state M called **metastable state**. The lifetime of an atom in the metastable state is 10^{-3} sec, which is much higher than the lifetime of 10^{-8} sec for atoms in the excited state. A passing photon produces the lasing transition from M-state to G-state whereby population is made more efficient.

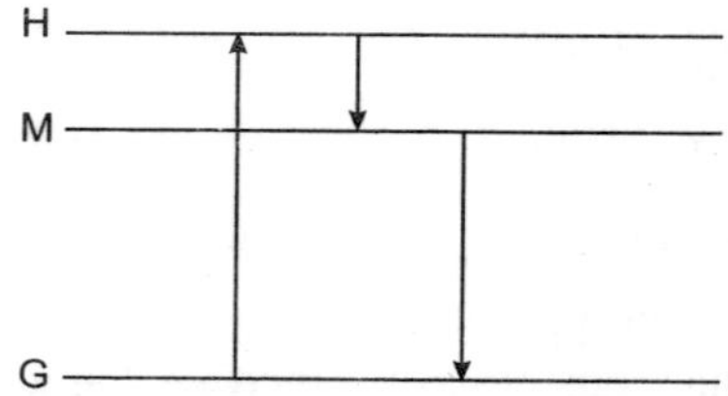

Figure 8.4

Laser Pumping

The process by which atoms are raised from lower level to the higher level is called the pumping process. There are broadly four different method of doing this and these are :

(*a*) Optical pumping or excitation by photon.
(*b*) Excitation with the help of electron.
(*c*) Inelastic collision between atoms.
(*d*) Excitation by chemical methods.

Active Medium

Active medium is one in which light gets amplified. This medium may be solid, liquid or gas. It is important that only a small fraction of a particular medium is responsible to stimulated emission and consequent amplification. These are called the **active centers** and the remaining part of the medium supports the active centers.

Optical Resonator

Light can be amplified by an active medium which is in a state of population inversion. A photon emitted spontaneously by an atom may cause stimulated emission. In order to sustain stimulated emission in a cumulative way, a positive feedback of the light is required. Also spontaneous emission of photon takes place in various directions. Again, an intense beam of light will be available only when optical gain of the system exceeds the optical losses.

The above requirements are met with the help of an optical cavity resonator which convert the active medium into light generator.

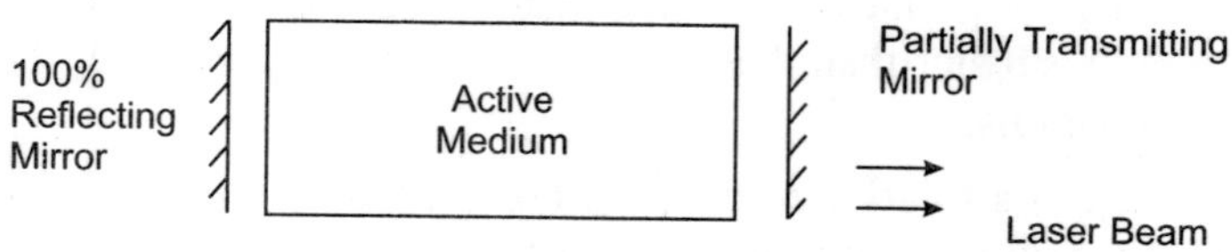

Figure 8.5

An optical resonator consists of two opposing plane parallel mirrors with the active material placed between them. One of the mirrors is 100% reflecting mirror and other is partially reflecting mirror. The mirrors are set perpendicular to the optic axis. The laser beam is coming out from the partially reflecting mirror.

Helium Neon Laser

The helium-neon laser consists of a long, narrow discharge tube (80 cm long and diameter 1 cm) filled with about 1 torr of helium and about 0.1 torr of neon. The gas mixture of helium-neon forms the lasing medium and the mixture is enclosed between a set of mirrors forming a resonant cavity.

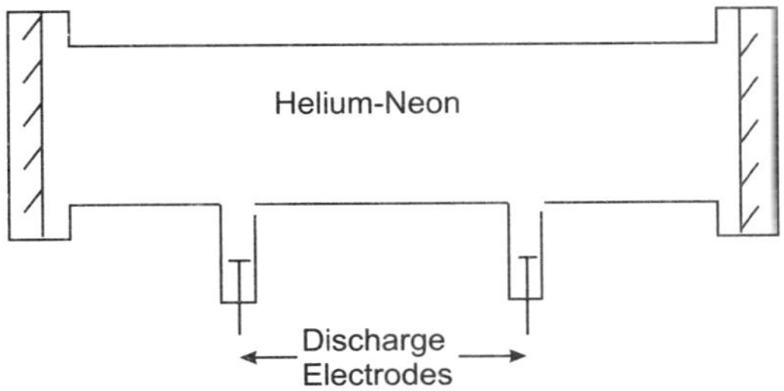

Figure 8.6

When a discharge is passed through the gaseous mixture, electrons are accelerated down the tube. These electrons collide with helium atoms and excite them to the higher energy state. The helium atoms excited from F_1 level to F_2 and F_3. These levels happen to be metastable and the helium atom spends a sufficiently long time before it radiates some of the neon excited levels approximately to the same energy as the excited levels F_2 and F_3 of helium. Thus, when helium atoms collide with neon atoms in ground state E_1, energy exchange take place. These results in excitation of neon atoms to the levels E_4 and E_6 and de-excite the helium atom to the ground level.

Because of the long lifetime of the levels F_2 and F_3 of helium, this process of energy transfer has a high probability. Thus discharge through the gas mixture continuously populates the neon excited energy level E_4 and E_6. This helps to create a state of population inversion in E_4 and E_6. The excited neon atoms drop down from level E_3 to level E_2 by spontaneous emission. At last neon atoms in E_2 de-excite to level E_1.

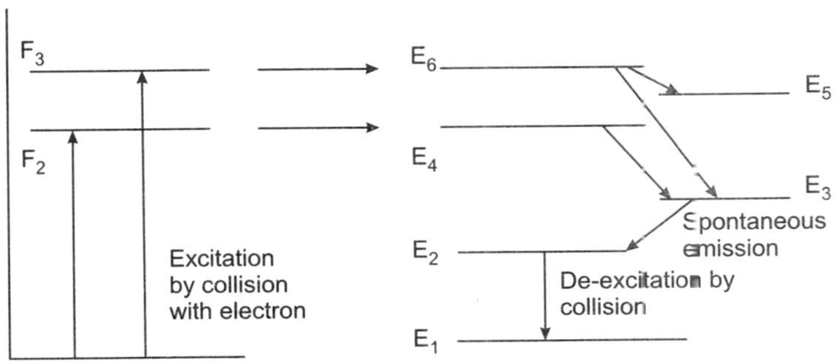

Figure 8.7 *Energy Level Diagram.*

8.6 CORE, CLADDING AND TOTAL INTERNAL REFLECTION IN OPTICAL FIBRE

It is a technology related to transportation of optical energy (light energy) in guiding media specially glass fibre. Optical fibres, a wave guide, consist of an inner cylindrical glass called **core**, surrounded by a cylindrical shell of glass of relatively lower refractive index called **cladding**. On the outside of the cylinder a protective skin is also used.

Optical fibres operate under the principle of 'total internal reflection'. As light passes through the fibre, it is constantly reflected through the inner core of the fibre.

When light ray is incident on the interface between the two dielectrics of differing refractive indices, both reflection and refraction occur. If the refractive index of incident medium (n_1) is higher than the refractive index (n_2) of the refracting medium *i.e.*, $n_1 > n_2$, the refracted light goes away from the normal. If we increase the incident angle then at the critical angle the refracted ray makes an angle 90°. If θ_c be the critical angle then from Snell's law the value of θ_c is given by

$$\sin \theta_c = \frac{n_1}{n_2}$$

At an angle of incidence greater than the critical angle light is reflected back into the incident medium and there will be no refraction. This phenomenon is called the total internal reflection.

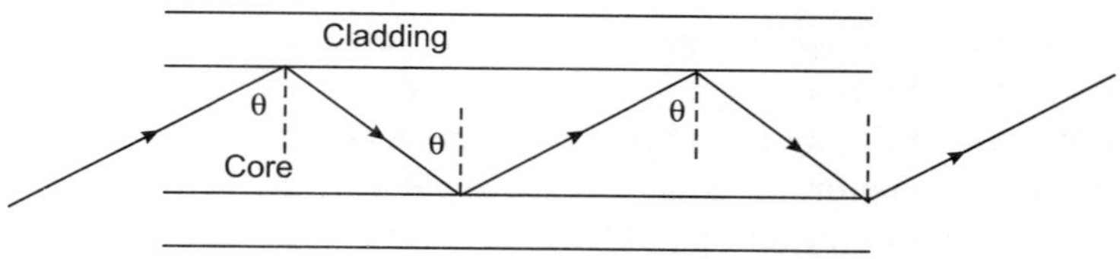

Figure 8.8

8.7 ACCEPTANCE ANGLE AND NUMERICAL APERTURE

Consider the light wave propagation in an optical fibre through total internal reflection; a light ray makes a critical angle within the core-cladding interface of the fibre. This light ray enters the fibre edge at an angle θ_a to the fibre and is refracted at air-core interface with an angle θ_r called the critical propagation angle. Hence rays which are incident into the fibre edge at angle greater than θ_a meet the core-cladding interface at an angle less than θ_c, and will not be totally internally reflected. Only the rays entering the fibre edge within the angle $2\theta_a$ will be accepted for total internal reflection. This an angle $2\theta_a$ is called the acceptance angle. In 3D it is acceptance cone.

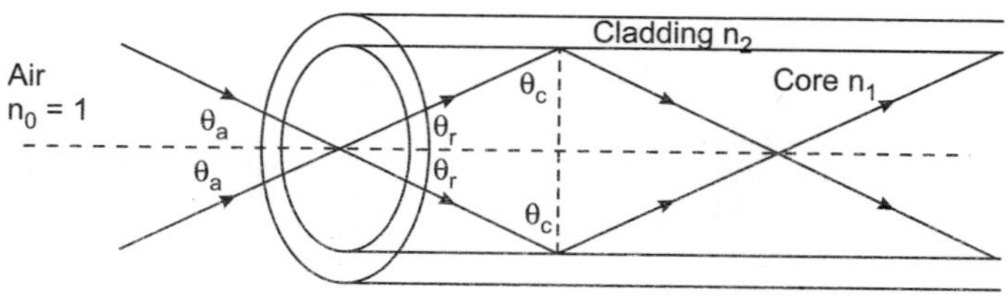

Figure 8.9

We can write from Snell's law at the edge of fibre

$$n_0 \sin \theta_a = n_1 \sin \theta_r \qquad (n_0 \text{ refractive index of air} = 1)$$

$$\sin \theta_a = n_1 \sin \theta_r$$

This term $\sin \theta_a$ is called Numerical Aperture (N.A)

$\therefore$ Numerical Aperture $= n_1 \sin \theta_r$

Numerical Aperture can be expressed in terms of the refractive index of core (n_1) and the refractive index of cladding (n_2).

From the above picture $\sin \theta_r = \cos \theta_c$

$$\text{N.A.} = n_1 \sin \theta_r = n_1 \cos \theta_c = n_1 \sqrt{(1 - \sin^2 \theta_c)}$$

Again,

$$\sin \theta_c = n_2 / n_1$$

$$\therefore \quad \text{N.A.} = \left(n_1^2 - n_2^2\right)^{\frac{1}{2}}.$$

IN FOCUS

- **Laser :** Light Amplification by Stimulated Emission of Radiation.

  Characteristics of Laser is

  1. Extremely monochromatic **2.** Spatially and temporarily coherent. **3.** Extremely directional and **4.** Highly intense and has a high power density when focussed.

- Relationship between the co-efficient of spontaneous emission (A_{21}) and the co-efficient of stimulated emission (B_{21}) is

$$\frac{A_{21}}{B_{21}} = \frac{8\pi h v^3}{c^3}$$

- **Population Inversion :** The situation in which the number of atoms in the higher state exceeds that in the lower state is known as Population Inversion.
- Optical fibre is a wave guide consisting of an inner cylindrical glass called core surrounded by a cylindrical shell of glass of relatively lower refractive index called cladding.
- **Numerical Aperture** $= \sqrt{n_1^2 - n_2^2}$

  where n_1 and n_2 are the refractive indices of the core and cladding respectively.

Solved Problems

Problem 1. *A laser beam of wavelength λ = 6000 Å, power 10 mW and angular speed 1.5×10^{-4} rad is focussed by a lens of focal length 10 cm. Find (i) Radius and (ii) Power density of Image (iii) The coherence width.* [Calcutta University]

Solution. (*i*) The radius of the image, $r = f\theta$, here f = 10 cm and $\theta = 1.5 \times 10^{-4}$ rad.

$$\therefore \quad r = f\theta = 10 \times 1.5 \times 10^{-4} = 1.5 \times 10^{-3} \text{ cm}.$$

(*ii*) Here, the power density of the image is $\dfrac{\text{Power}}{\text{Area}\left(\pi r^2\right)} = \dfrac{10 \times 10^{-3}\text{ W}}{\pi \times \left(1.5 \times 10^{-4}\right)^2} \simeq 1.4 \text{ kW/cm}^2$

(*iii*) The Coherence width $L_w = \frac{\lambda}{\theta} = \frac{6000 \times 10^{-8}}{1.5 \times 10^{-4} \text{ rad}}$ cm = 4 mm.

Problem 2. *In a He-Ne laser transition from 3S → 2P Level gives a laser emission of λ = 632.8 nm. If the 2P level has energy 15.2 × 10^-19 J, calculate the pumping energy required, assuming no loss.* [Calcutta University]

Solution. Here Pumping energy = energy of 2P level + Photon energy

$$\therefore \quad (E)_{Pump} = 15.2 \times 10^{-19} \text{ J} + h\left(\frac{c}{\lambda}\right)$$

$$= 15.2 \times 10^{-19} \text{ J} + \frac{6.62 \times 10^{-34} \times 3 \times 10^{8} \text{ J}}{632.8 \times 10^{-9}}$$

$$= 11.46 \text{ eV} \qquad [\because \; 1 \text{ eV} = 1.6 \times 10^{-19} \text{ J}]$$

Problem 3. *A relative population (Boltzmann ratio) of 1/e is representative of the ratio of populations in two energy states at room temperature (= 27° c). Determine the wavelengths of the radiation emitted at that temperature.*

Solution. We know that $\frac{N_2}{N_1} = e^{-\frac{(E_2 - E_1)}{kT}}$

given that $\frac{N_2}{N_1} = e^{-1}$

or, $\frac{E_2 - E_1}{kT} = 1$ or, $E_2 - E_1 = h\nu = kT = \frac{hc}{\lambda}$

or, $\lambda = \frac{ch}{kT} = \frac{3 \times 10^{8} \times 6.62 \times 10^{-34}}{1.38 \times 10^{-23} \times 300} = 4.8 \times 10^{-5}$ m.

Problem 4. *The Numerical Aperture of an optical fibre is 0.5 and the core refractive index is 1.54. Find the refractive index of the cladding.*

Solution. We know that Numerical Aperture (NA) $= \sqrt{n_1^2 - n_2^2}$

or $n_2^2 = n_1^2 - (\text{N.A.})^2 = (1.54)^2 - (0.5)^2$

or $n_2 = 1.46$

Problem 5. *Find the numerical aperture (N.A.) of a step-index fibre when the refractive index of the core is 1.51 and that of the material used for cladding is 1.47. Also calculate the angle of acceptance of the fibre when it is placed in air.*

Solution. Given that $n_1 = 1.51$ and $n_2 = 1.47$

$$\therefore \text{ Numerical aperture (N.A.)} = \sqrt{n_1^2 - n_2^2}$$

$$= \sqrt{(1.51)^2 - (1.47)^2}$$

$$= 0.3452$$

Suppose θ_a be the required angle of acceptance. Then

$$n_o \sin \theta_a = \text{N.A. Here } n_o = 1 \text{ for air.}$$

$$\theta_a = \sin^{-1} (\text{N.A.}) = \sin^{-1} (0.3452) = 20°11'$$

Multiple Choice Questions

1. For laser action to occur, the medium used must have at least

(*a*) 4 energy levels (*b*) 2 energy levels

(*c*) 3 energy levels.

2. In population inversion

(*a*) Large number of atoms are made in appropriate excited energy state

(*b*) Small number of atoms are made in appropriate excited energy state

(*c*) Zero number of atoms are made in appropriate excited energy state.

3. Light amplification by stimulated emission of radiation is known as

(*a*) Light amplifier (*b*) Laser

(*c*) Maser (*d*) none.

4. The ratio of Einstein's A, B co-efficient *i.e.*, $\frac{A_{21}}{B_{21}}$ say, is proportional to

(*a*) v (*b*) v^2

(*c*) v^3 (*d*) v^4

5. Metastable state has a mean life of more than

(*a*) 10^{-2} sec (*b*) 10^{-3} sec

(*c*) 10^{-4} sec (*d*) 10^{-5} sec

6. Fibre optics is based on

(*a*) Refraction (*b*) Polarisation

(*c*) Total internal reflection (*d*) Dispersion.

7. The numerical aperture (NA) of a step-index fibre is given by

(*a*) $\sqrt{n_1^2 - n_2^2}$ (*b*) $\sqrt{\frac{n_1^2 - n_2^2}{n}}$

(*c*) $n_1^2 - n_2^2$ (*d*) none of these

8. If n_1 and n_2 be the indices of the core and the cladding respectively, then

(*a*) $n_1 \geq n_2$ (*b*) $n_2 \geq n_1$

(*c*) $n_2 > n_1$ (*d*) $n_1 > n_2$

9. A measure of light gathering power of an optical fibre is the

(*a*) Numerical aperture (*b*) Acceptance angle

(*c*) Fraction index change (*d*) none of these.

Answers

1. (*b*)	**2.** (*a*).	**3.** (*b*)	**4.** (*c*)	**5.** (*b*)	**6.** (*c*)
7. (*a*)	**8.** (*d*)	**9.** (*a*)			

Review Questions

1. Explain the terms 'absorption', 'spontaneous emission' and 'stimulated emission' of radiation.
2. What are the important features of stimulated emission ? Discuss the essential requirements for producing laser action. [*Allahabad University*]
3. State the characteristics of the laser beam. [*WBUT*]
4. (*a*) What is the population inversion ? Explain.
   (*b*) How higher probability of stimulated emission compared to that of spontaneous emission is achieved in a laser ? [*WBUT*]
   (*c*) Calculate the numerical aperture of a fibre with a refractive index of the core and cladding 1.54 and 1.50 respectively. [*WBUT*]
5. Describe the principle, construction and working of He-Ne laser with necessary diagrams.
6. What is an optical fibre ? Describe a typical optical fibre with diagram.
7. What do you mean by 'Core' and 'cladding'? Whose refractive index is more and why? Give an idea of the diameter of core and cladding.
8. Define numerical aperture of a fibre and find out an expression for the same in a step index fibre. [*Burdwan University*]
9. Write short notes on :
   (*a*) Einstein's A and B co-efficients [*Calcutta University*]
   (*b*) Population Inversion [*Calcutta University*]
   (*c*) He-Ne Laser [*Calcutta University*]
   (*d*) Fibre Optics [*Calcutta University*]

Unsolved Problems

1. A laser beam has a power density of 100 mW. It has an aperture 5×10^{-3} m and units a light of $\lambda = 6943$ A. The beam is focussed with a lens of focal length 0.1 m. Calculate the area and intensity of the image. [**Ans.** 2.89×10^{-10} m^2, 3.46×10^8 W/m^2]
2. Calculate the ratio of stimulated to spontaneous emission rate at a temperature of 250°C for the sodium D lines?
   [**Hints :** $T = 273 + 250 = 523$ K, $\lambda_{Na-D} = 590$ nm

$$R = \frac{\text{Stimulated emission rate}}{\text{Spontaneous emission rate}} = \frac{1}{e^{\frac{h\nu}{kT}-1}} = \frac{1}{e^{\frac{he}{\lambda kT}-1}}$$

$$\text{or, } \frac{hc}{\lambda kT} = \frac{6.63 \times 10^{-34}\,(\text{Js}) \times 3 \times 10^8\,(\text{m/s})}{590 \times 10^{-9}\,(\text{m}) \times 1.38 \times 10^{-23}\,(\text{J/K}) \times 523} \simeq 46.7$$]

3. A relative population of two energy states at room temperature, T = 300 K is 1/e. Determine the wavelength of radiation emitted due to transition between the states. [**Ans.** 48 μm]
4. A step index optical fibre has refractive indices of the core and cladding as 1.44 and 1.40 respectively and is surrouded by air. Calculate the numerical aperture. [*GATE*] [**Ans.** 0.337]
5. A step-index fibre has a core of refractive index 1.5 and a cladding of index 1.4. The fibre is used in an environment of index 1.33. Calculate the numerical aperture and the acceptance angle. [**Ans.** 0.538; 23.9°]

Marie Curie

Chapter 9

Nuclear Physics

9.1 INTRODUCTION

In the beginning of the nineteenth century, the atomic view of matter was first propounded by the John Dalton and he assumed that atoms were the smallest part of the element identical in mass, while different elements had atoms of different masses. It took about hundred years before the internal structure of the atom could be understood. In the α-ray scattering experiments of Rutherford in 1911 established the existance of the atomic nucleus. He proposed on the basis of the studies on the scattering of α particles that atoms which had a radial dimension of the order of 10^{-10} m had a positively charged central nucleus of smaller dimensions around 10^{-14} m. The nucleus of an atom, like the atom itself is a bound quantum system and can exist in different quantum states based on their energies, angular momenta etc. The lowest energy state is referred as the ground state in which a nucleus normally exists. In surveying the general properties like its mass, size, shape, angular momentum, magnetic dipolemoment etc., we shall assume that the nucleus is in the ground states and properties in such state is termed as the static characteristics of the nuclei. The so-called dynamic characteristics are exhibited in the process of nuclear reactions, nuclear excitation and nuclear decay. In this module we shall illustrate a general review on both the characteristics.

9.2 CONSTITUENTS OF NUCLEI

The atomic nuclei are made up of two different types of elementary particles—*protons* and *neutrons*.

The neutrons and protons are collectively known as *nucleons*. Any particular type of nucleus, a species of nucleus, is known as *nuclide*. The proton can be considered as the nucleus of the hydrogen atom, *i.e.*, a hydrogen atom in which the single orbital electron has been removed. A proton has a positive charge of the same magnitude as that of an electron and has a mass about 1836 times the electronic mass (m_e). A neutron is electrically neutral and mass is slightly greater than the mass of the proton (m_p). Thus the electronic mass is negligibly small compared to the mass of the protons and the neutrons.

The number of protons in the nucleus determine the *atomic number* of the nuclide. It is also called Z-value or the *proton number*. The sum of the numbers of protons (Z) and neutrons (N) inside the nucleus is known as its *mass number* (A).

i.e., $$A = Z + N$$

A should be integer as Z and N are.

As an example, the oxygen nucleus ${}_8O^{17}$ has Z = 8 protons, A = 17 nucleous and N = 17 – 8 = 9 neutrons.

9.3 CLASSIFICATION OF NUCLEI, ISOTOPES, ISOBARS, ISOTONES AND MIRROR NUCLEI

Nuclei with the same atomic number Z, but different the mass number A are called *isotopes. e.g.,* ${}^6Li$ and ${}^7Li$.

Nuclei with the same atomic mass number A, but different atomic number Z are called *isobar. e.g.,* ${}_8O^{16}$ and ${}_7N^{16}$

Nuclei with the same number of neutrons N are known as *isotones. e.g.,* ${}_{11}Na^{23}$ and ${}_{12}Mg^{24}$.

The pairs of isobaric nuclei (such as ${}_6C^{11}$ and ${}_5B^{11}$), where in the proton number Z and neutron number N are interchanged and differ by *one unit* are known as *mirror nuclei*.

9.4 PROPERTIES OF NUCLEUS

Nuclear Sizes and Shapes

A nucleus can be considered as a sphere of constant density and its volume is proportional to the total number of nucleons in it or its mass number A.

$$\therefore \qquad \frac{4}{3}\pi r^3 \propto A \qquad \text{where } r \text{ is the radius of the nucleus.}$$

$$r \propto A^{\frac{1}{3}}$$

or,

$$r = r_0\, A^{\frac{1}{3}}$$

where $r_0 = 1.3 \times 10^{-15}$ m = 1.3 fermi [1 fermi (1 fm) = 10^{-15} m].

However, Rutherford's work on the scattering of α-particles estimated that the mean radius of an atomic nucleus is of the order of 10^{-14} to 10^{-15} m while that of the atom is about 10^{-10} m.

As an example, the radius of the ${}_{47}Ag^{107}$ nucleus is

$$r = (1.3)\,(107)^{\frac{1}{3}} = 6.2 \text{ fm}.$$

Assumption that nuclei are spherical is not generally true and many nuclei are ellipsoidal in shape.

Nuclear Charge

The charge of the nucleus is positive due to protons. Each proton has a positive charge of 1.6×10^{-19} C. The nuclear charge is Ze where Z is the atomic number of the nucleus. The value of Z is determined from X-ray scattering experiment, nuclear scattering of α particles and X-ray spectrum.

9.5 NUCLEAR MASS, MASS DEFECT AND BINDING ENERGY

We know that the nucleus consists of neutrons and protons. The assumed nuclear mass = $Zm_p + Nm_n$ where m_p and m_n are the proton and neutron masses respectively and N is the number of neutron. Again, the nuclear mass M_{nuc}. is obtained from the atomic mass M (A, Z) by subtracting the masses of Z orbital electrons.

$\therefore$ $$M_{nuc} = M(A, Z) - Zm_e.$$

Nuclear masses are experimentally determined accurately by mass spectrometer. These measurements expressed that real nuclear mass $< Zm_p + Nm_n$.

The departure or difference between the 2 measured atomic mass M (A, Z) from the mass number (A) is known as *mass defect* ΔM.

i.e., $$\Delta M = M(A, Z) - A.$$

As an example, the mass defect of $^4He_2 = 4.002603 - 4 = 0.002603$ and that of $^{16}O_8 = 15.994915 - 16 = 0.005085$ u. Thus, the mass defect can be both positive and negative.

It is observed that for very light atoms with $A < 20$ and for very heavy atoms with $A > 180$, the atomic masses are slightly greater than the corresponding mass number, *i.e.*, the mass defect is + ve and it is negative in the region between two.

The energy equivalent of the missing mass of a nucleus is called the binding energy of the nucleus. To break up a nucleus of Z protons and N neutrons completely so that they are all separate from one another, a certain *minimum amount* of *energy* is to be required. This energy is known as *binding energy* of the nucleus.

In forming a nucleus out of the constituent particles, a fraction of the total mass of the constituents disappear and the evolution of energy equal to binding energy (E_B) takes place. If ΔM be the mass disappeared, then the

Binding energy, $$E_B = \Delta Mc^2$$

M_H, M_n be the masses of hydrogen atom and the neutron respectively, then

$\Delta M = ZM_H + NM_n - M(A, Z)$, where M (A, Z) is the mass of atom of mass number A and atomic number Z.

[*Unit of Atomic Mass* is defined as $\frac{1}{12}^{th}$ of the mass of the atom of ^{12}C isotope and is known as unified atomic mass unit "represented by '*u*'

$$1\,u = \frac{1}{12} \times \frac{12 \times 10^{-3}}{N_A} = \frac{10^{-3}}{6.02205 \times 10^{23}}\ \text{kg}$$

or $$1\,u = 1.660566 \times 10^{-27}\ \text{kg}$$ [N_A : Avogadro Number]

[The energy equivalent of $1\,u = 931.501$ MeV] [1 u : 1 a.m.u.]

Binding Energy,
$$E_B = [ZM_H + NM_n - M(A, Z)]\,C^2$$
$$= [ZM_p + NM_n + Zm_e - M_{nuc}. - Zm_e]\,C^2$$
$$= [ZM_p + NM_n - M_{nuc}]C^2$$

The magnitude of the binding energy (E_B) of a nucleus determines its stability against disintegration.

1. If $E_B > 0$, the nucleus is stable and energy must be supplied to break up the nucleus.
2. If $E_B < 0$, the nucleus is unstable and it will disintegrate by itself.
3. If $E_B = 0$, it corresponds to dissociation of the nucleus into its constituent nucleus *i.e.*, neutrons and protons.

Binding Energy and Stability of the Nucleus

If the binding energy (E_B) of a nucleus is divided by the mass number (A), we get the binding energy per nucleon in the nucleus, which is known as the *binding fraction* (f_B) and is given by

$$f_B = \frac{E_B}{A} = \frac{ZM_H + NM_n - M(A, Z)}{A}$$

Here we assumed that the masses are expressed in energy unit. So C^2 term in right hand side has been omitted.

Binding Fraction vs Mass Number Curve

Here figure 9.1 represents the binding energy per nucleon or binding fraction (f_B) plotted against the various mass number (A). The curve rises steeply at first and then more gradually until it reaches a maximum of 8.79 MeV at 56, corresponding to the iron nucleus $_{26}Fe^{56}$. Thus, this is the most stable nucleus of them, since the greatest amount of energy must be needed to pull a nucleon away from it.

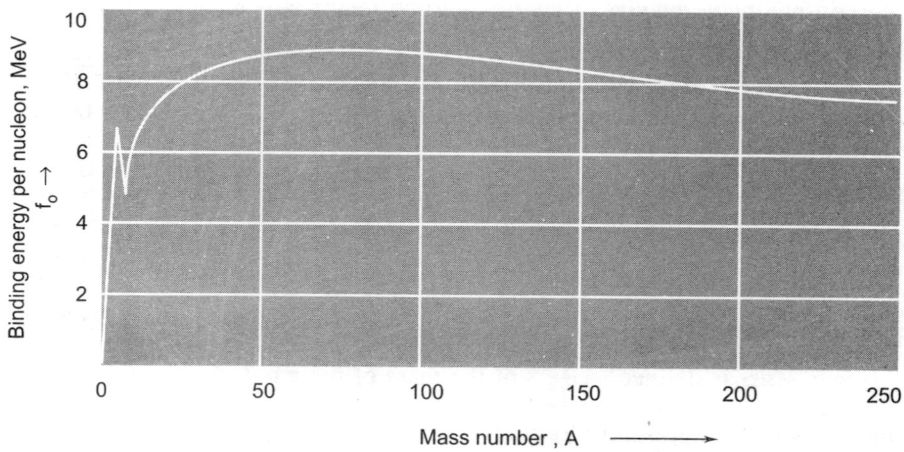

Figure 9.1

This fact suggests the following points :

(*i*) The change of F_B is very small for very light nuclei and goes on increasing rapidly with increasing A. There after the rise of the curve is much slower, reaching maximum value of 8.7 MeV per nucleon for A = 56. For further increase in A, the curve decreases slowly.

(*ii*) The variation of f_B is very small in the range $20 < A < 180$. In this region f_B is nearly 8.5 MeV/neu.

(*iii*) For $A > 180$, (for heavy nuclei), f_B decreases monotonically with increasing A and it is ~ 7.5 MeV/nucleon.

(*iv*) Peaks are observed in the region between $4 < A < 16$ for $A = 4n$ (where $n = 1, 2, 3$, etc.). These are even Z-even N nucleies. Peaks are also observed at Z or N equal to 20, 50, 82, 126. These are called Magic numbers. The nuclei corresponding to the peaks are more stable compared to those in the neighbourhood.

Some remarkable conclusions can be drawn from this curve which will bring forth the following points with immense practical importance.

(*a*) If we can somehow split a heavy nucleus into two medium sized ones, each of the new nuclei will have more amount of f_B than the original nucleus did and the extra energy will be evolve as predicted. This process is known as nuclear fission. For instance if the Uranium nucleus $\left({}_{92}U^{235}\right)$ is broken into lighter ones, the f_B is about 0.8 MeV and total energy evolve 188 MeV.

(*b*) If lighter nuclei can somehow be joined to form heavier ones, it gives more amount of f_B in the new nucleus. This process is known as nuclear fusion. For instance, if two deuterium nucleus combine to form a ${}_2He^4$ nucleus, over 23 MeV is evolved.

9.6 PACKING FRACTION (*f*) AND PACKING FRACTION CURVE

The packing fraction (*f*) is defined as the *mass defect per nucleon* in the nucleus, that is the mass defect of an atom divided by its mass number. So

$$f = \frac{\Delta M}{A} = \frac{M(A, Z) - A}{A}$$

$$= \frac{M(A, Z)}{A} - 1$$

or $$M(A, Z) = A(f + 1).$$

The *f* has the same sign as ΔM and it is thus positive for very light and very heavy atoms and it is negative for atoms in the intermediate range.

It is found that the packing fraction (*f*) varies with the mass number A in a systematic fashion as represented graphically in figure 9.2, is known as *packing fraction curve*. This graph shows following results :

1. *f* is positive for very light nuclide having mass number below 20.
2. The *f* is negative for elements having mass number between 20 and 180. It attains the minimum negative value at A ~ 60.
3. *f* becomes positive beyond 180.

The negative *f* implies that some mass has been converted into energy during the formation of that particular nucleus. Thus the nucleus is exceptionally stable. On the other hand, positive *f* implies that the nucleus is unstable.

N.B. The packing fraction curve and binding fraction curve are complementary of each other.

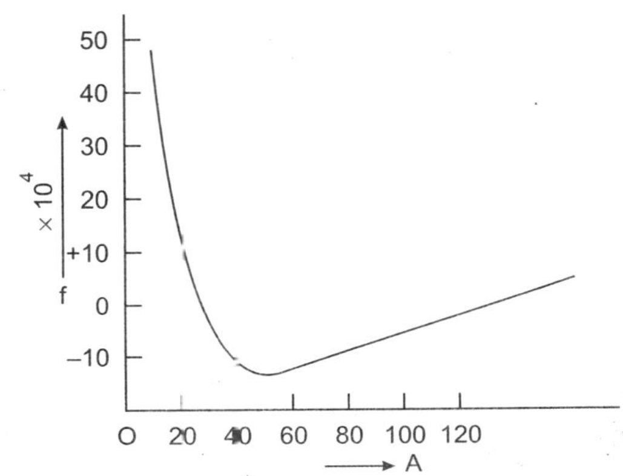

Figure 9.2

9.7 NATURE AND CHARACTERISTICS OF NUCLEAR FORCE

The positively charged protons being *closely spaced* within the nucleus should repel each other by *Coulomb-Force*. As the nucleus is a *stable structure* one should assume that there *must be* very *strong attractive short-range nuclear forces* to counteract the coulomb repulsion and keep the nucleons strongly bound to each other. The forces inside the nucleus binding N to N, P to P and N to P are known as *strong interactions forces* and are represented by *n-n*, *p-p* and *n-p* forces respectively.

According to Yukawa, the following are the *characteristics of the nuclear forces :*

1. They are *short-range* forces *i.e.*, effective only in short ranges.
2. They are *independent* of *nucleonic charge*.
3. They are *strongest known* forces in nature.
4. They get *readily saturated* by the surrounding nucleons.
5. They are *spin-dependent*.

The *strong forces* are $\sim 10^{40}$ times stronger the Gravitational Force. *Saturation* is the property which means that the nucleons located outside *the nucleon surrounded by other nucleons* do not feel the interaction of *the nucleon* as surrounded by others.

Short range forces operate over a distance ~ 1 Fermi or less (1 Fermi = 10^{-15} m). There is some evidence that at extremely short distances (~ 0.5 F) the attractive force between nucleons *turns into repulsion*, so that the nucleons do not get *too close* to each other and the nucleus remains stable.

9.8 CONSERVATION PRINCIPLE IN NUCLEAR REACTION, THE Q-VALUE AND THRESHOLD ENERGY

A *nuclear reaction* is a process that takes place when a nuclear particle (*viz.*, proton, neutron, deuteron, α-particle, a nucleons etc.) comes in close contact (within 10^{-15} m) *with another* and energy and momentum exchange occurs between them. The *final products* of the reaction are some *nuclear particle or particles* that leave the point of contact in *different directions*.

A general equation representing a nuclear reaction is of the form

$$X + x \longrightarrow Y + y$$

where X is the target nucleus, x, the bombarding particle, Y the residual product nucleus and y the ejected particle.

Types of nuclear reactions to mention are :

1. Elastic scattering,
2. Inelastic scattering,
3. Radiative capture,
4. Reaction of transformation,
5. Photodisintegration,
6. Stripping reactions and
7. Fission.

There are many other types of reactions not listed above.

Conservation Laws in Nuclear Reactions

The principal properties of nuclei are mass, charge, spin, linear and angular momenta. The principal conservation laws are :

(*a*) **Conservation of mass number :** The total number of neutrons and protons in the nuclei in a nuclear reaction *remains unaltered after the reaction*.
Thus in the reaction X (*x, y*)/Y, the sum of the mass numbers of X and *x must equal* the sum of the mass numbers of Y and *y*.

(*b*) **Conservation of atomic number:** The total number of protons in a nuclear reaction *remains unaltered after* the reaction (*i.e.*, Z-number remains constant in nuclear reactions).
Obviously, because of (*a*) and (*b*) the neutron number N also remains conserved in nuclear reactions.

(*c*) **Conservation of energy (including mass-energy):** The *sum* of the rest-mass energies and the kinetic energies of *all the nuclei* taking part in the reaction must be equal to *the sum* of the rest-mass energies and kinetic energies of the products.

Further conservation laws are obtained when quantum mechanical properties are considered alongwith the conservation of angular momentum. These are :

(*i*) conservation of isotopic spin and (*ii*) conservation of parity.

Q-Values

Consider a nuclear reaction represented by,

$$X + x \longrightarrow Y + y$$

where, X : the target nucleus of mass M_X

Y : the recoil (or product) nucleus of mass M_Y

x : the incident (or bombarding) particle of mass m_x

y : the ejected (or product) particle of mass m_y

Consider also the Kinetic energy (K.E.) of the particles X, Y, *x* and *y* as K_X, K_Y, K_x and K_y respectively.

For conservation of energy in a nuclear reaction,

$$(K_X + M_X c^2) + (K_x + m_x c^2) = (K_Y + M_Y c^2) + (K_y + m_y c^2)$$

Assuming X is initially at rest so that its $K_E \rightarrow K_X = 0$.

Hence the above equation becomes,

$$M_X c^2 + (K_x + m_x c^2) = (K_Y + M_Y c^2) + (K_y + m_y c^2)$$

or

$$(K_Y + K_y - K_x) = (M_X + m_x - M_Y - m_y) c^2$$

(c : Velocity of light in free space)

The L.H.S. of the above equation is represented by Q. Thus, Q (or Q-value) $= K_Y + K_y - K_x$

$$= (M_X + m_x - M_Y - m_y) c^2$$

In energy units, $Q = (M_X + m_x - M_Y - m_y)c^2$

NOTE M_X, M_Y, m_x, m_y are the rest masses.

Q is the *energy balance* of the nuclear reaction and referred to as *Q-Value of the reaction*.

There are *three* possibilities that can occur for Q—

(*i*) If $(M_X + m_x) > (M_Y + m_y)$, then $(K_Y + K_y) > K_x$ and $Q > 0$. Such a nuclear reaction is *Exoergic Reaction* (or exothermic), because it *releases energy* (Q is + ve).
In exoergic reaction there is thus a *net decrease* in mass *after* the nuclear reaction. This lost mass gets converted to energy [according to $E = mc^2$] which is *imparted* as K.E. to the *product particles*.

(*ii*) If $(M_X + m_x) < (M_Y + m_y)$ then $(K_Y + K_y) < K_x$ and $Q < 0$. Such a nuclear reaction is *ENDOERGIC REACTION* (or Endothermic) because some energy *is to be supplied from outside* for the reaction *to occur.*
THRESHOLD ENERGY — In endoergic reaction, therefore, a *minimum* amount of energy *equal to the Q-value (– ve)* of the reaction *must be* given *to the incident particle* so that the reaction occurs. Such an energy is called the *threshold* energy of the endoergic reaction.

(*iii*) If $(M_X + m_x) = (M_Y + m_y)$ then $Q = 0$. Such a nuclear reaction gives rise to an *elastic collision* where there is *no loss* of energy but *only a change in direction* takes place of the *reacting particles*.

9.9 NUCLEAR FISSION

The phenomenon of the *division of a heavy nucleus* into two *nearly equal* parts is termed *nuclear fission* or simply FISSION.

The schematic equation for the *fission process* is represented by,

$$^{235}_{92}\text{U} + {}^{1}_{0}n \longrightarrow {}^{236}_{92}\text{U} \longrightarrow \text{X} + \text{Y} + \text{neutrons}$$

where ${}^{1}_{0}n$ is a *slow neutron.*

${}^{236}_{92}\text{U}$ is a *highly unstable* isotope of Uranium.

Typical *Fission Reactions* are :

1. $^{235}_{92}\text{U} + {}^{1}_{0}n \longrightarrow \left({}^{236}_{92}\text{U}\right) \longrightarrow {}^{144}_{56}\text{Ba} + {}^{89}_{36}\text{Kr} + 3\,{}^{1}_{0}n + \text{Q}$

2. $^{235}_{92}\text{U} + {}^{1}_{0}n \longrightarrow \left({}^{236}_{92}\text{U}\right) \longrightarrow {}^{140}_{54}\text{Xe} + {}^{94}_{38}\text{Sr} + 2\,{}^{1}_{0}n + \text{Q}$

where Q : energy released in the reaction.

The energy released in the fission process is so high that the two main fission products fly apart in opposite directions with great speeds.

For a *rough calculation*, consider the type of equation (2) of the fission reaction, — (the end products are ${}^{140}_{58}\text{Ce}$, ${}^{94}_{40}\text{Zr}$, 6β, $2\,{}^{1}_{0}n$)

The isotopic mass *before fission* is,

$$\begin{aligned} {}^{235}_{92}\text{U} &= 235.0439 \text{ amu} \\ {}^{1}_{0}n &= 1.0087 \text{ amu} \\ \hline & 236.0526 \text{ amu} \end{aligned}$$

The isotopic mass *after fission* is,

$$\begin{aligned} {}^{140}_{58}\text{Ce} &= 139.9054 \text{ amu} \\ {}^{94}_{40}\text{Zr} &= 93.9036 \text{ amu} \\ 2\,{}^{1}_{0}n &= 2.0173 \text{ amu} \\ 6\beta^- &= 0.0033 \text{ amu} \\ \hline & 235.8296 \text{ amu} \end{aligned}$$

$\therefore$ Mass difference $= (236.0526 - 235.8296) = 0.223$ amu

This mass difference gives a Q-value, $Q = 0.223 \times 931$ MeV (1 amu = 931 MeV)

$= 208$ MeV

In *Joules*, ($1 \text{ MeV} = 1.6 \times 10^{-13}$ J)

$$208 \text{ MeV} = 208 \times 1.6 \times 10^{-13} \text{ J} = 3.32 \times 10^{-11} \text{ J}$$

The above energy is released due to the fission of *one* U-nucleus : 1g of U-235 would thus generate $\dfrac{6.02 \times 10^{23}}{235} \times 3.32 \times 10^{-11}$ J $= 8.50 \times 10^{10}$ J which is equivalent to about 20 tons of TNT (Tri-nitro toluene). If it is released *all at once* we will have a *nuclear bomb* which is popularly known as **ATOM BOMB**. If the *release is controlled*, we have a *nuclear reactor*.

9.10 CHAIN REACTION

As because fission leads to other neutrons being given off, a self-sustaining *sequence of fissions* should be possible.

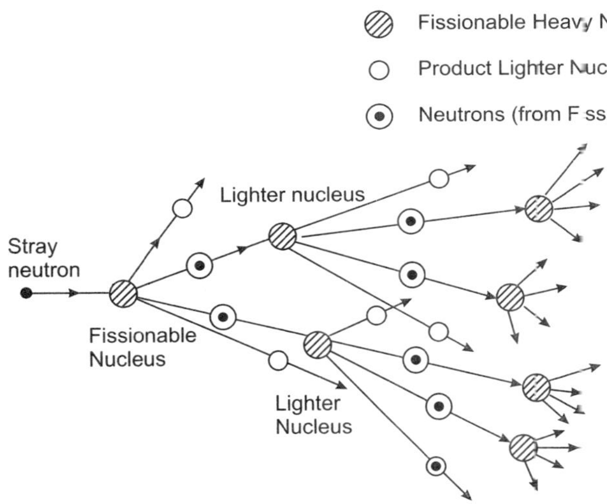

Figure 9.3

The condition for such a *chain reaction* to occur in an assembly of *fissionable material* is — *at least one* neutron produced during each fission *must*, on the average, cause another fission. If too few neutrons cause fissions, the reaction will slow down and stop. If, precisely, *one neutron per fission* causes *another* fission, energy will be released at a constant rate (as is in case of *nuclear reactor*) and if the frequency of fissions increases the energy will be released so rapidly that an explosion will occur (as is in the case of *atom bomb*). These situations are respectively called *sub critical*, *critical* and *super critical*.

If *two* neutrons from *each fission* in an *Atomic Bomb* induce further fissions in 10^{-8} sec, a *chain reaction starting with a single fission* will give off 2×10^3 *J* of energy is *less than* 10^{-6} sec.

9.11 FUSION: (THERMONUCLEAR REACTION)

If two *light nuclei* combine or *fuse together* to produce a *relatively* heavier nucleus, there would be a greater binding energy and *consequent decrease* in nuclear mass. This would, therefore, result in a *positive Q-value* and a *release* of energy.

This type of nuclear reactions is known as NUCLEAR FUSION, a process *opposite* to that of nuclear fission.

Examples :

$${}_{1}^{1}\mathrm{H} + {}_{1}^{1}\mathrm{H} \rightarrow {}_{1}^{2}\mathrm{H} + {}_{1}^{0}e + \nu\ ;\ \mathrm{Q} = 0.62\ \mathrm{MeV}$$

$${}_{1}^{2}\mathrm{H} + {}_{1}^{2}\mathrm{H} \rightarrow {}_{2}^{3}\mathrm{He} + {}_{0}^{1}n\ ;\ \mathrm{Q} = 3.3\ \mathrm{MeV}$$

(ν : neutrino), (Q $\rightarrow$ energy release)

Fusion depends for its action on the *collisions* of two *very* energetic nuclei, a subsequent rearrangement of nucleons and the release of energy as *kinetic energy* of the product particles and their *excitation* energy. For *fusion*, the K.E. of the colliding particles must be *high enough* to overcome the electrostatic coulomb repulsion of the nuclei. A *large* K.E. implies a high temperature such that fusion energy should be sufficient to provide enough energy to secondary particle to make the process *self-sustaining*. For this reason fusion reactions are also known as THERMO-NUCLEAR reactions.

Nuclei taking part in fusion are light and so the energy released *per unit mass* of the reaction material is *comparatively much greater*.

IN FOCUS

- The number of Protons (Z) and neutrons (N) together gives atomic numbers A = Z + N
- The nuclear radius $r = r_0 A^{1/3}$

  where A = atomic mass number and $r_0 = 1.3 \times 10^{-15}$ fermi
- **Mass defect** $\Delta M = M(A, Z) - A$

  where M (A, Z) = Measured Atomic Mass

  A = Mass number
- **Packing fraction** $= \dfrac{\Delta M}{A} = \dfrac{M(A, Z) - A}{A}$
- **Binding energy** $E_B = \Delta Mc^2$
- Binding Energy per nucleon is

$$f_B = \frac{E_B}{A} = \frac{ZM_H + NM_n - M(A, Z)}{A}$$

- The overall energy liberated or absorbed in a nuclear disintegration is called nuclear reaction energy Q.
- Those nuclear reactions in which energy is released or liberated are called exothermic or exoergic (Q = +ve).
- Those reactions in which energy is absorbed are called endothermic or endoergic (Q = –ve).
- The reaction in which, energy is not absorbed or released are called elastic collision.
- The division of a heavy nuclide into two rearly equal parts when hit by neutron is called nuclear fission.
- Nuclear fusion means fusing of two lighter nuclei into one stable and heavier nuclide. The energy released is called fusion energy.

Solved Problems

Problem 1. *Calculate the average binding energy per nucleon for* $_{28}Ni^{64}$ *having 63.9280 a.m.u. given that Z = 28, A = 64,* m_p *= 1.007825 a.m.u. and* m_n *= 1.008665 a.m.u.* [H.P. University]

Solution. $_{28}Ni^{64}$ nucleus contains 28 protons and (64 – 28) or 36 neutrons.

Thus, the total mass of the $_{28}Ni^{64}$ nucleus

$$= 28 \times 1.007825 + 36 \times 1.008665$$

$$\therefore \quad \text{Mass difference} = 28 \times 1.007825 \times 36 \times 1.008665 - 63.9280$$

$$= 0.6030 \text{ a.m.u.}$$

$$\therefore \quad \text{Total binding energy} = 0.6030 \times 931.501 \qquad [1 \text{ a.m.u.} = 931.501 \text{ MeV}]$$

$$= 561.8 \text{ MeV}$$

∴ Binding energy per nucleon

$$= \text{Average binding energy}$$

$$= \frac{561.8}{64} = 8.78 \text{ MeV/Nucleon.}$$

Problem 2. *Calculate the binding energy per nucleon for the deutron. Given* $m_n = 1.675 \times 10^{-27}$ *kg,* $m_p = 1.672 \times 10^{-27}$ *Kg and* $C = 3 \times 10^8$ *m/sec and* $M_D = 3.343 \times 10^{27}$. [Agra University]

Solution. A deutron nucleus contains one proton and one neutron. Mass of deutron nucleus

$$= 1.675 \times 10^{-27} + 1.672 \times 10^{-27}$$

$$= 3.347 \times 10^{-27} \text{ kg}$$

$$\text{Mass defect or difference } \Delta m = (3.347 - 3.343) \times 10^{-27} \text{ kg}$$

$$= 0.004 \times 10^{-27} \text{ kg}$$

$$\therefore \quad \text{Binding energy } E_B = \Delta mc^2$$

$$= 0.004 \times 10^{-27} \times 9 \times 10^{16} \text{ Joules}$$

$$= \frac{36 \times 10^{-14}}{1.6 \times 10^{-13}} \text{ MeV} = 2.25 \text{ MeV}$$

$$\text{Binding energy per nucleon} = \frac{2.25}{2} = \text{Average binding energy}$$

$$= 1.125.$$

Problem 3. *The mass of the hydrogen atom and of neutron are 1.008142 and 1.008982 a.m.u. respectively. Calculate the packing fraction and the binding energy per nucleon of* ^{16}O *nucleus.* [Calcutta University]

Solution. ^{16}O nucleus consists of 8 protons and 8 neutrons.

$$\therefore \quad \text{Mass of } ^{16}O \text{ nucleus} = (8 \times 1.008142 + 8 \times 1.008982) \text{ a.m.u.}$$

$$= 16.136992 \text{ a.m.u.}$$

$$\text{Atomic mass of } ^{16}O = 15.994915 \text{ a.m.u.}$$

$$\text{Mass Difference} = (16.136992 - 15.994915) \text{ a.m.u.}$$

$$\text{or} \quad \text{defect} = 0.142077 \text{ a.m.u.}$$

$\therefore$ Binding energy $= 0.142077 \times 931.501$ MeV

and

$$\text{Binding energy per nucleon} = \frac{132 \cdot 3}{16}\ \text{MeV} = 8.27\ \ \text{MeV}$$

$$\text{Mass defect } \Delta M = M(A,Z) - A$$

$$= 15.994915 \sim 16 = 0.005085 \text{ a.m.u.}$$

$$\text{Packing fraction} = \frac{\Delta M}{A} = \frac{0.005085}{16} = 3.178 \times 10^{-4}$$

Problem 4. *The binding energy of* $_{10}Ne^{20}$ *is 160.64 MeV. Find the atomic mass. Given mass of proton = 1.007825 a.m.u. and of neutron = 1.008665 a.m.u.* [G.N.D. University, H.P. University.]

Solution. The atomic mass of an atom consists of Z protons, (A – Z) neutrons and Z electrons is given by

$$M = Zm_p + (A - Z)m_n + Zm_e - E_B$$

where E_B is the binding energy in a.m.u.

Here Z = 10, A = 20, A – Z = 10, m_p = 1.007825 a.m.u., and m_n = 1.008665 a.m.u.

Hence,
$$M = 10 \times 1.007825 + 10 \times 1.008665 + \frac{10 \times 0.51}{931.501} - \frac{160.64}{931.501}$$

$$= 19.9979 \text{ a.m.u.}$$

$\therefore$ Atomic mass $= 19.9979$ a.m.u.

Multiple Choice Questions

1. What is the nature of force exerted by a neutron on proton?
   (*a*) Nuclear (*b*) Electromagnetic
   (*c*) Electromotive (*d*) None of these.
2. Energy release in nuclear fission of heavy element is possible because
   (*a*) Binding energy per nucleon decreases as the mass number increases when it is greater than 65.
   (*b*) Some nucleon are destroyed and their mass is transformed into energy.
   (*c*) More neutrons are releasing in fission.
   (*d*) The total kinetic energy of the incident neutrons and the atoms undergoing fission is given out as nuclear energy.
3. The nuclear fission fragments undergo decay because they have an excess of neutrons. After each decay
   (*a*) The mass number of fragments increased by one.
   (*b*) The atomic number of the fragments increases by one.
   (*c*) The atomic number of the fragments decreases by one.
   (*d*) The mass number of the fragments decreases by one.
4. In a nuclear fission, 1 microgram of U^{235} disappears. The velocity of light is 3×10^{10} cm/sec. The energy of this process is
   (*a*) 3×10^{10} ergs (*b*) 3×10^{12} ergs
   (*c*) 9×10^{20} ergs (*d*) 9×10^{24} ergs.
5. A nuclear reaction is called exoergic if Q value is
   (*a*) Negative (*b*) Positive
   (*c*) Zero (*c*) None of these.
6. If the binding energy per nucleon of Li^7 and He^4 nuclei are 5.60 MeV and 7.06 MeV respectively, then the energy of the reaction $Li^7 + P \longrightarrow {}_2He^4$ is
   (*a*) 17.3 MeV (*b*) 1.96 MeV (*c*) 2.4 MeV (*d*) 4.6 MeV

7. The process in which a heavy nucleus splits into two nucleus is called
   (*a*) Alpha decay (*b*) Beta decay (*c*) Fission (*d*) Fusion.
8. Source of steller energy is
   (*a*) Chemical reaction (*b*) Fusion reaction
   (*c*) Fission reaction. (*d*) None of these.

Answers

1. (*a*)	**2.** (*a*)	**3.** (*b*)	**4.** (*c*)	**5.** (*b*)	**6.** (*a*)
7. (*c*)	**8.** (*b*)				

Review Questions

1. Explain the terms mass defect, missing energy and binding energy of a nucleus and binding energy per nucleon. [*G.N.D. University, Lucknow University, Calcutta University*]
2. Discuss graphically the variation of average binding energy per nucleon with mass number and stability of the nucleus. Using the binding energy curve, explain the release of energy in fusion of light nuclei and fission of heavy nuclei.
   [*H.P. University, G.N.D. University, Bangalore University, A. University, Lucknow University, Calcutta University*]
3. (*a*) Explain the nuclear forces responsible for holding the nucleons together. Discuss the nature of these forces giving a qualitative account.
   [*G.N.D. University, Pune University, Bangalore University*]
   (*b*) What do you mean by charge impedance of nuclear forces. [*H.P. University*]
4. Explain with suitable diagram the variation of per nucleon binding energy with mass number. [*WBUT*]
5. (*a*) What is a nuclear reaction? [*Pune University, G.N.D. University*]
   (*b*) Discuss the various conservation laws in nuclear reaction.[*Pbi. University, Kanpur University*]
6. What do you mean by Q value of a nuclear reaction? Discuss the energetics of exoergic reaction. Define threshold energy and derive an expression for the threshold energy of an endoergic reaction. What is the significance of negative sign of Q. [*Pune University, G.N.D. University*]
7. Describe the phenomenon of nuclear fission.
   [*Banaras University, Magadh University, G.N.D. University*]
8. Hence explain why fusion takes place in the light mass region while fission occurs for heavy nuclei. [*WBUT*]
9. Why high temperature is needed to initiate fusion? [*WBUT*]
10. Write short note on chain reaction. [*WBUT*]

Unsolved Problems

1. Find the energy released if two ${}_1H^2$ nuclei can fuse together to form the ${}_2He^4$ nucleus. The binding energy per nucleon of H^2 and ${}_2He^4$ are 1.1 MeV and 7.0 MeV respectively. Given m_n = 1.007825 a.m.u., m_p = 1.008665 a.m.u. and ${}_2He^4$ = 4.002603. [**Ans.** 23.6 MeV] [*Lucknow University*]
2. Calculate the binding energy in MeV of ${}_2He^4$ from the following data : Mass of ${}_2H^4_e$ = 4.003875 a.m.u. ; mass of ${}^1H$ = 1.008145 a.m.u. and mass of neutron = 1.008986 a.m.u.
   [**Ans.** 28.29 MeV] [*Burdwan University*]
3. Calculate the binding energy per nucleon for ${}_{17}C^{35}$.
   Given m_p = 1.007825 a.m.u.
   m_n = 1.008665 a.m.u.
   M_N = 34.9800 a.m.u. [**Ans.** 8.22 MeV/nucleon] [*Pbi. University*]

Illustrative Examples

Example 1. *Find the average number of photons/cubic metre in a monochromatic beam of radiation of frequency 10 kHz and of intensity 10^4 W/m^2*

Solution. Let n : no. of photons per unit volume ; $h\nu$: photon energy ; A : area of cross section, l : beam length. $nAl\,h\nu$: total photon energy in vol. al t = (time taken by this beam to move through the distance l) $= \dfrac{l}{c}$

$$\text{I : intensity of the beam} = \frac{nAl\,h\nu}{At} = nh\nu c \qquad (\because \; t = l/c)$$

$$\therefore \qquad n = \frac{I}{h\nu c} = \frac{10^4}{h \times 10^4 \times c} = \frac{1}{6.63 \times 10^{-34} \times 3 \times 10^8}$$

$$= 5.03 \times 10^{24} \text{ photons}/m^2$$

Example 2. *Prove that in the photoelectric effect from a metal surface, the maximum velocity of the photo-electrons are related to the stopping potential by,*

$$v_{max} = 5.927 \times 10^5 \times \sqrt{V_o} \qquad \left| \begin{array}{l} v_{max} \text{ in m/sec.} \\ V_o \text{ in volts.} \end{array} \right.$$

Given : m_e (mass of electron) $= 9.11 \times 10^{-31}$ kg

e (charge on electron) $= 1.6 \times 10^{-19}$ coul.

Solution. The maximum K.E. related to stopping potential is,

$$\frac{1}{2} m_e v_{max}^2 = e\,V_o \quad \text{or} \quad v_{max} = \sqrt{\frac{2e\,V_o}{m_e}} \quad \text{or,} \; v_{max} = \sqrt{\frac{2e}{m}} \cdot \sqrt{V_o}$$

$$= 5.927 \times 10^5 \sqrt{V_o}$$

Example 3. *The photo-electric threshold wavelength of silver is 2762Å. Calculate (a) The maximum K.E. of the ejected electrons. (b) The maximum velocity of the electrons and (c) The stopping potential (in volts) for the electrons, when the silver surface is illuminated with ultraviolet ray of wavelength 2000Å.*

Solution. Here

$$\lambda = 2762 \times 10^{-10} \text{ m}$$
$$\lambda_o = 2000 \times 10^{-10} \text{ m}$$

(*a*)
$$E_{max} = h(\nu - \nu_o) = hc\left(\frac{1}{\lambda} - \frac{1}{\lambda_o}\right) = 2.745 \times 10^{-19} \text{J}$$

(*b*)
$$\frac{1}{2} m_e v^2_{max} = E_{max} \quad \therefore \quad v_{max} = \sqrt{\frac{2\,E_{max}}{m_e}} = 7.76 \times 10^5 \text{ m/sec}$$

(*c*)
$$e\,V_o = E_{max} \quad \therefore \quad V_o = \frac{E_{max}}{e} = 1.715 \text{ Volts.}$$

Example 4(a). *A metallic surface, when illuminated with light of wavelength λ_1, emits electrons with energies upto a maximum of E_1, and when illuminated with light of wavelength λ_2, where $\lambda_2 < \lambda_1$, emits electrons with energies upto maximum of E_2. Prove that Planck's constant "h" and the work function W_o of the metal are given by,*

$$h = \frac{(E_2 - E_1)\,\lambda_1\,\lambda_2}{c\,(\lambda_1 - \lambda_2)} \quad and \quad W_o = \frac{E_2\,\lambda_2 - E_1\lambda_1}{\lambda_1 - \lambda_2}$$

Solution. The maximum kinetic energy is $E_1 = h\nu_1 - W_o = \frac{hc}{\lambda_1} - W_o$

Similarly, $E_2 = \frac{hc}{\lambda_2} - W_o$

$$\therefore \quad (E_2 - E_1) = hc\left[\frac{1}{\lambda_2} - \frac{1}{\lambda_1}\right] = hc\left[\frac{\lambda_1 - \lambda_2}{\lambda_1\lambda_2}\right] \quad \text{or } h = \frac{(E_2 - E_1)\,\lambda_2\,\lambda_1}{c\,(\lambda_1 - \lambda_2)} \quad \ldots(A)$$

Again, $E_1\,\lambda_1 = hc - W_o\,\lambda_1$ and $E_2\,\lambda_2 = hc - W_o\,\lambda_2$

$$\therefore \quad E_2\lambda_2 - E_1\lambda_1 = W_o\,(\lambda_1 - \lambda_2) \quad \text{or} \quad W_o = \frac{E_2\lambda_2 - E_1\lambda_1}{(\lambda_1 - \lambda_2)} \quad \ldots(B)$$

Example 4(b). *A metallic surface, when illuminated with light of wavelength 3333 Å emits electrons with energies upto 0.6 V and when illuminated by light of wavelength 2400 Å emits electrons with energies upto 2.04 eV. Calculate planck's constant and the work function of the metal.*

Solution. Here, $\lambda_1 = 3333 \times 10^{-10}$ m, $E_1 = 0.6 \times 1.6 \times 10^{-19}$ J

(1 eV = 1.6×10^{-19} J)

$\lambda_2 = 2400 \times 10^{-10}$ m, $E_2 = 2.04 \times 1.6 \times 10^{-19}$ J

Substituting in above equations (A) and (B), we obtain – $h = 6.5 \times 10^{-34}$ J.s.

$W_o = 3.1$ eV

Example 5. *An X-ray photon of wavelength 0.3 Å is scattered through an angle 45° by a loosely bound electron. Find the wavelength of the scattered photon.*

Solution. Wavelength of scattered photon is $\lambda' = \lambda + \frac{h}{m_e c}(1 - \cos\phi)$

Here, $\lambda = 0.3$ Å, $h = 6.63 \times 10^{-3}$ Js, $\phi = 45°$

$m_e = 9.11 \times 10^{-31}$ kg, $c = 3 \times 10^{8}$ m/sec

$\therefore$ $\lambda' = 0.3071$ Å Ans.

Example 6(a). *If a narrow beam of electrons is accelerated by a potential difference V volts the kinetic energy K acquired by each electron is K = eV, e : electronic charge. Show that*

The de Broglie wavelength is

$$\lambda = \frac{h}{\sqrt{2m_e\,K\left(1 + K/2m_e c^2\right)}}$$

Solution. The total relativistic energy is given by,

$E = K + m_o c^2$ $\qquad m_o$: rest mass of the electron ($= m_e^2$ (here)

or, $\sqrt{p^2c^2 + m_e^2\,c^4} = K + m_e c^2$

or, $p^2c^2 + m_e^2\,c^4 = K^2 + 2Km_e c^2 + m_e^2\,c^4$

or, $$p^2 = \frac{K^2 + 2Km_ec^2}{c^2} \quad \text{or} \quad p = \frac{1}{c}\sqrt{K^2 + 2Km_ec^2}$$

or, $$p = \frac{\sqrt{K\,(K + 2m_ec^2)}}{c}$$

∴ $$\lambda = \frac{h}{p} = \frac{hc}{\sqrt{K\left(K + 2m_ec^2\right)}}$$

$$= \frac{hc}{c\sqrt{2m_eK\left(1 + K/2m_ec^2\right)}} = \frac{h}{\sqrt{2m_eK\left(1 + K/2m_ec^2\right)}}$$

Example 6(b). *The most rapidly moving valence electron in metallic sodium, at O°C temperature, has a K.E. equal to 3 eV. Show that its de Broglie's wavelength is 7 Å.*

Solution. Here $K = 3\ eV = 3 \times 1.6 \times 10^{-19}$ J

For electron the *rest mass energy* m_oc^2 *(i.e.,* m_ec^2) $= 9.11 \times 10^{-31} \times (3 \times 10^8)^2$

$= 5.12 \times 10^5$ eV

compared to this value K.E. of the electron is negligible (3 $eV <<< m_ec^2$) (Thus $K/2m_ec^2 \approx 0$)

∴ $$\lambda = \frac{h}{\sqrt{2m_eK}} = \frac{6.63 \times 10^{-34}}{\sqrt{2 \times \left(9.11 \times 10^{-31}\right) \times \left(3 \times 1.6 \times 10^{-19}\right)}} = 7.1 \text{ Å}.$$

Example 7. *Calculate the de Broglie wavelength of an electron moving with velocity* $\frac{3}{5}c$

Solution. Here, $v = \frac{3}{5}c$ or, $\frac{v}{c} = \frac{3}{5}$

v in this case is comparable to c

hence $$m = \frac{m_o}{\sqrt{1 - v^2/c^2}} = \frac{m_e}{\sqrt{1 - \frac{9}{25}}}$$

∴ $$v = \frac{3}{5}c = \left(\frac{3}{5} \times 3 \times 10^8\right) \text{m/sec}$$

Now, $$\lambda = \frac{h}{mv} = \frac{h}{mv}\left[\sqrt{1 - v^2/c^2}\right]$$

$$= \frac{6.63 \times 10^{-34}}{9.11 \times 10^{-31} \times \left(\frac{3}{5} \times 3 \times 10^8\right)}\sqrt{1 - \frac{9}{25}}$$

$= 0.323 \times 10^{-11}$ m $= 0.0323$ Å.

Example 8. *The average period of time that elapses between the excitation of an atom and the time it emits radiation is 10^{-8} sec. Find the uncertainty in the energy emitted and the uncertainty in the frequency of light emitted.* *(Given, $\hbar = 1.054 \times 10^{-34}$ J.s. and $h = 6.63 \times 10^{-34}$ J.s)*

Solution. The maximum uncertainty in the interval of time, $\Delta t = 10^{-8}$ sec.

Now, $\Delta E \,.\, \Delta t \geq \hbar$.

Hence the minimum uncertainty in energy $\Delta E = \frac{\hbar}{\Delta t} = \frac{1.054 \times 10^{-34}}{10^{-8}} = 1.054 \times 10^{-26}$ J.

We know, $E = h\nu$

or $\Delta E = h\Delta\nu \quad \therefore \quad \Delta\nu = \frac{\Delta E}{h} = \frac{1.054 \times 10^{-26}}{6.63 \times 10^{-34}} = 1.59 \times 10^{7}$ Hertz.

Example 9. *Find the shortest wavelength present in the radiation from an X-ray machine whose accelerating potential is 50,000 volt.*

Solution.
$$Ve = h\nu_{max} = \frac{hc}{\lambda_{min}}$$

$$\therefore \quad \lambda_{min} = \frac{hc}{Ve} = \frac{1.24 \times 10^{-6}}{V}.\ \text{Volt-m} = \frac{1.24 \times 10^{-6}}{5 \times 10^{4}}\ \text{m} = 2.48 \times 10^{-11}\ \text{m}$$

$$= .0248 \text{ nm (nano meter)}$$

Example 10. *A particle is in motion along a line between $x = 0$ and $x = a$ with zero Potential Energy and at points $x < 0$ and $x > a$ the potential energy is infinite. The wave function of the particle in the nth state is given by*

$$\psi_n = A \sin \frac{n\pi x}{a}$$

Find the expression for the normalized wave function.

Solution. we are to evaluate A within the range.

Now, for the nth state, the probability that the particle will be between x & $x + dx$ is given by

$$P_n\, dx = |\psi_n(x)|^2\, dx = A^2 \sin^2 \frac{n\pi x}{a}\, dx$$

There is only one particle in the region from $x = 0$ to $x = a$ and at all times it is in the same region.

$$\therefore \quad \int_o^a P_n dx = 1 \quad i.e., \int_o^a |\psi_n(x)|^2\, dx = 1$$

$$\text{or} \quad \int_o^a A^2 \sin^2 \frac{n\pi x}{a}\, dx = 1 \quad \text{or} \quad A^2 . \frac{1}{2}\int_o^a 2\sin^2 \frac{n\pi x}{a}\, dx = 1$$

$$\text{or} \quad \frac{A^2}{2}\int_o^a \left(1 - \cos\frac{2\pi nx}{a}\right) dx = 1 \quad \text{or} \quad \frac{A^2}{2}\left[x - \frac{a}{2\pi nx}.\sin\frac{2\pi nx}{a}\right]_o^a = 1$$

wherefrom,
$$\frac{A^2}{2}.a = 1 \quad \therefore \quad A^2 = \frac{2}{a} \quad \text{or } A = \sqrt{\frac{2}{a}}$$

$\therefore$ The normalized wavefunction $\psi_n = \sqrt{\frac{2}{a}} \sin \frac{n\pi x}{a}$

Example 11. *Calculate the Q-value of reaction* ${}^{14}_{7}N(\alpha, p)\,{}^{17}_{8}O$ which *occurred in Rutherford's α-range in Nitrogen experiment.*

Solution.

$$\begin{aligned} Q &= [(m_x + M_X) - (m_y + M_Y)]\,c^2 \\ &= [(4.002603 + 14.0031) \\ &\quad - (1.007825 + 16.9994] \times 931.5 \text{ MeV} \\ &= [-1.522 \times 10^{-3}] \times 931.5 \text{ MeV} \\ &= -1.418 \text{ MeV}. \end{aligned}$$

known, $1u = 931.5$ MeV
Given, $m_\alpha = 4.002603\,u$
$M_{N_2} = 14.0031\,u$
$m_p = 1.007825\,u$
$M_{o_2} = 16.9994\,u$

Example 12. *Obtain the threshold energy for the reactions*

(a) ${}^{209}_{83}Bi\,(p, d)\,{}^{208}_{83}Bi \begin{bmatrix} d \text{ is deuteron} \\ ({}^{2}_{1}d) \end{bmatrix}$ *(b)* ${}^{27}_{13}Al\,(\alpha, p)\,{}^{31}_{15}P$

Solution. By definition, the smallest value of projectile energy (*i.e.*, bombarding energy) at which an endoergic reaction can take place is the threshold energy for that reaction.

It can be obtained that

$$(E_x)_{\text{thresh}} = -Q\left(\frac{m_x + M_X}{M_X}\right)$$

For (*a*), the Q-value is,

$$Q = [(208.980394 + 1.007285) - (207.79731 + 2.014102)] \times 931.5 = -5.23 \text{ MeV}$$

Given,
${}^{209}_{83}Bi = 208.980394\,u$
$p\,(i.e., H) = 1.007825\,u$
${}^{208}_{83}Bi = 207.979731\,u$
${}^{2}_{1}d = 2.014102\,u$

$$\text{Now, } (E_x)_{\text{Threshold}} = -Q\left[\frac{m_x + M_X}{M_X}\right] = 5.23\left[\frac{1 + 209}{209}\right] = 5.26 \text{ MeV}.$$

$\begin{cases} m_x : \text{mass no. of } {}^{1}p \\ M_X : \text{mass no. of } {}^{209}Bi \end{cases}$

For (*b*)

$$\begin{aligned} Q &= [(25.986892 + 4.002603) \\ &\quad - (30.973765 + 1.007825)] \times 931.5 \\ &= -2.9 \text{ MeV} \end{aligned}$$

Given
${}^{27}_{13}Al = 25.986892\,u$
${}^{1}_{1}p = 1.007825\,u$
$\alpha\,(i.e., {}^{4}_{2}He) = 4.002603\,u$
${}^{31}_{15}p = 30.986892\,u$

$$(E_x)_{\text{Threshold}} = 2.9\left[\frac{27 + 4}{27}\right] \text{MeV} = 2.9 \times \frac{31}{27} = 3.33 \text{ MeV}.$$

$\begin{cases} m_x : \text{mass no. of He} \\ M_X : \text{mass no. of Al} \end{cases}$

Example 13. *Find out the unknown particles in the nuclear reactions given below :*

(i) ${}^{6}_{3}Li\,(d, \alpha)\,X$ *and* *(ii)* ${}^{122}_{52}Te\,(X, d)\,{}^{124}_{53}I$

Solution. In a nuclear reaction, the total charge and the mass number are conserved.

In (*i*), we have, ${}^{6}_{3}\text{Li} + {}^{2}_{1}d \rightarrow \text{X} + {}^{4}_{2}\text{He}$ $\left(\because \ \alpha \text{ is } {}^{4}_{2}\text{He}\right)$

$\therefore$ X must be ${}^{4}_{2}\text{X}$ *i.e.*, it is an α-particle.

In (*ii*), we have, ${}^{122}_{52}\text{Te} + \text{X} \rightarrow {}^{124}_{53}\text{I} + {}^{2}_{1}\text{d}$

$\therefore$ X must be ${}^{4}_{2}\text{X}$ *i.e.*, it is an α-particle.

Exercise : Complete the following nuclear reactions.

(*i*) ${}^{6}_{3}\text{Li} + {}^{2}_{1}\text{H} \rightarrow {}^{7}_{4}\text{Be} + \text{X}$

(*ii*) ${}^{23}_{11}\text{Na} + {}^{1}_{1}\text{H} \rightarrow {}^{20}_{10}\text{Ne} + \text{X}$

(*iii*) ${}^{35}_{17}\text{Cl} + \text{X} \rightarrow {}^{32}_{16}\text{S} + {}^{4}_{2}\text{He}$

Example 14. *X-rays of wavelength 10.0 × 10^{-12} m (pm i.e., picometer = 10^{-12} m) (i.e., 10.0 pm) are scattered from a target.*

(a) Find the wavelength of the X-rays scattered through 45°.
(b) Find the maximum wavelength present in the scattered X-rays.
(c) Find the maximum kinetic energy of the recoil electrons.

Solution. For (*a*) $\lambda' - \lambda = \lambda_c (1 - \cos\phi)$ where λ_c : compton wavelength **for electron**

$\lambda_c = 2.426 \times 10^{-12}$ m = 2.426 pm

or,
$$\begin{aligned}\lambda' &= \lambda + \lambda_c (1 - \cos\phi)\\ &= 10 \text{ pm} + (1 - \cos 45°) \,.\, \lambda_c\\ &= 10 \text{ pm} + (1 - 0.7071).\ 2.426 \text{ pm}\\ &= (10 + 0.293 \times 2.426) \text{ pm}\\ &= 10.71 \text{ pm} = 10.71 \times 10^{-12} \text{ m}\end{aligned}$$

For (*b*) $(\lambda' - \lambda)$ is maximum when $(1 - \cos\phi) = 2$ (*i.e.*, $\phi = 180°$ for maximum scattering)

$$\therefore \quad \lambda' = \lambda + 2\lambda_c = 10 \text{ pm} + 2 \times 2.426 \text{ pm}$$
$$= (10 + 4.852) \text{ pm} = 14.852 \text{ pm}.$$

For (*c*), KE_{max} is the difference between incident and scattered photons

$$\therefore \quad \text{KE}_{\text{max}} = h(\nu - \nu') = hc\left(\frac{1}{\lambda} - \frac{1}{\lambda'}\right)$$

$$\text{KE}_{\text{max}} = \frac{(6.63 \times 10^{-34} \text{ J.S})(3 \times 10^{8} \text{ m/s})}{10^{-12}} \cdot \left[\frac{1}{10} - \frac{1}{14.852}\right]$$

$$= \frac{96.506 \times 10^{-26}}{10^{-12}} \cdot \frac{1}{10 \times 14.852} = 0.6498 \times 10^{-14} \text{ J}$$

$$\therefore \quad \text{KE}_{\text{max}} = 6.498 \times 10^{-15} \text{ J}.$$

Example 15. *There are two particles in three quantum states $g_i = 1, 2, 3$. Distribute the particles in accordance with M.B., B.E., and F.D. statistics.*

Solution. Since there are three-states and two-particles, the number of microstates possible are –

M.B. $n = 3^2 = 9$

B.E. $n = \dfrac{(3+2-1)!}{3!\,2!} = 2$

F.D. $n = \dfrac{3!}{2!\,1!} = 3$

M.B. : Particles distinguishable *i.e.,* A, B

State	*Arrangements*								
1	AB	–	–	A	B	A	B	–	–
2	–	AB	–	B	A	–	–	A	B
3	–	–	AB	–	–	B	A	B	A

B.E. : Particles indistinguishable *i.e.,* A, A

State	*Arrangements*					
1	AA	–	–	A	A	–
2	–	AA	–	A	–	A
3	–	–	AA	–	A	A

F.D. : Particles indistinguishable *i.e.,* A, A & not more than ONE particle can stay in one state.

State	*Arrangements*				
	A	–	A		
	A	A	–		
	–	A	A		

Example 16. *Find the Miller indices for planes with each of the following sets of intercepts: (i) 3a, 3b, 2c (ii) a, 2b, ∞; where a, b, c are lattice parameter*

Solution. (*i*) Intercepts 3,3,2

Reciprocal $\frac{1}{3}, \frac{1}{3}, \frac{1}{2}$

Simplification 2, 2, 3

Miller indices **(2 2 3)**

(*ii*) Intercepts 1, 2, ∞

Reciprocal 1, $\frac{1}{2}$, 0

Simplification 2, 1, 0

Miller indices **(2 1 0)**

Example 17. *Draw a* $(1\,\bar{1}\,1)$ *plane.*

Solution.

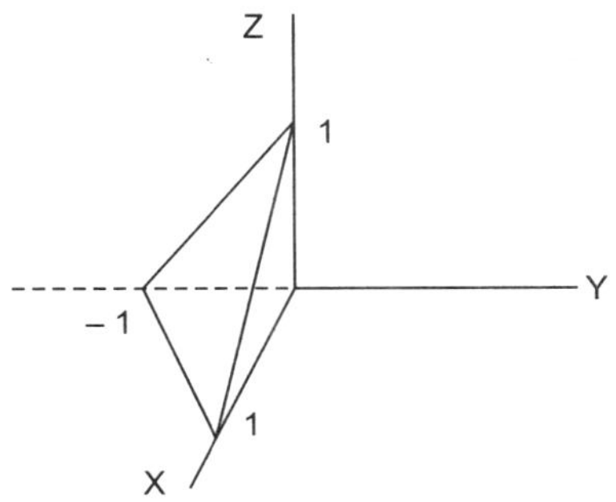

Example 18. *Calculate the distance between the adjacent parallel planes of the type (100), (110) and (111) in an fcc lattice of lattice constant a.*

Solution. Inter planer distance $d = \dfrac{a}{\left(h^2 + k^2 + l^2\right)^{1/2}}$

For (100) planed $= a / (1 + 0 + 0)^{1/2}$

$$d = a$$

For (110) planed $= a / (1 + 1 + 0)^{1/2}$

$$d = a / \sqrt{2}$$

For (111) planed $= a / (1 + 1 + 1)^{1/2}$

$$d = a / \sqrt{3}$$

Example 19. *A crystal reflects monochromatic X-ray strongly when the Bragg's reflecting angle for a first order reflection is 15°. What are the glancing angles for the second and third order reflection of the same order?*

Solution. Bragg's law $2d \sin \theta = n\lambda$

$$\sin \theta = n\lambda / 2d$$

1st order reflection at 15°

$$\sin 15° = 1 \times \lambda / 2d$$

$$\lambda / 2d = 0.2588$$

2nd order at $\sin \theta = 2 \times \lambda / 2d = 2 \times 0.2588$

$$\theta = 31.2°$$

3rd order at $\sin \theta = 3 \times \lambda / 2d = 3 \times 0.2588$

$$\theta = 50.9°$$

Miscellaneous Questions

1. (*a*) What do you mean by (*i*) Virtual displacement and (*ii*) holonomic constraint in classical dynamics?
   (*b*) State D'Alembert's principle in words.
   (*c*) Write down the Lagrange's equation of motion for a conservative holonomic system.

(*d*) What do you mean by cyclic co-ordinates ? Show that generalized momentum corresponding to a cyclic co-ordinate is a constant of motion.

(*e*) Define the Hamiltonian. When is it equal to the total energy of the system?

2. (*a*) For a general system of N-particles obtain the relation-

$$m_i a_i = \text{P}_i = \text{F}_i^{(e)} + \sum_{i \neq j} \text{F}_i^{\text{J}} \qquad \text{where} \begin{cases} i = 1, 2, \ldots \text{N} \\ \text{J} = 1, 2, \ldots \text{N} \end{cases}$$

a_i is the acceleration of the ith particle and P_i is its momentum.

(*b*) Discuss the principle of virtual work and obtain the expression of D'Alemberts principle.

(*c*) Apply D'Alembert's principle, obtain :

(*i*) The acceleration of the masses under gravity in the Atwoods machine.

(*ii*) The equation of motion of two unequal masses constrained to move on a smooth double inclined plane.

3. (*a*) What are generalized co-ordinates? What is the advantage of using them? Obtain the expression for generalized force.

(*b*) Write down the Lagrangian of a simple pendulum explaining each term. Hence obtain the equation of motion.

4. (*a*) Define field. Write down the classification of field.

(*b*) Show that $\vec{\text{A}} \cdot \dfrac{d\vec{\text{A}}}{dt} = \text{A}\dfrac{d\text{A}}{dt}$

(*c*) Show that the vector field $\vec{\text{A}} = \dfrac{2z^2y}{x^3}\hat{i} + \dfrac{z^2}{x^3}\hat{j} + \dfrac{2yz}{x^2}\hat{k}$ is irrotational.

(*d*) if $\phi = x^2 - y^2 + 2z$, find divgrad ϕ.

(*e*) Is the field $\vec{\text{A}} = x^2\hat{i} + y^2\hat{j} + z^2\hat{k}$ is solenoidal?

5. (*a*) What is the value of $\vec{\nabla} \times \vec{r}$?

(*b*) State Stoke's theorem. Prove that

$$\vec{\text{A}} \times (\vec{\text{B}} \times \vec{\text{C}}) = \vec{\text{B}}(\vec{\text{A}} \cdot \vec{\text{C}}) - \vec{\text{C}}(\vec{\text{A}} \cdot \vec{\text{B}})$$

(*c*) If $\vec{\text{A}} = x^2y\hat{i} - 2xz\hat{j} + 2yz\hat{k}$, find curl $\vec{\text{A}}$.

6. (*a*) What do you mean by gradient of a scalar field?

If $\phi = 3x^2y - y^3z^2$, find $\vec{\nabla}\phi$ at the point (1, – 2, – 1).

(*b*) Define vector field. Give an example.

(*c*) Prove $\vec{\text{A}} \times (\vec{\text{B}} \times \vec{\text{C}}) + \vec{\text{B}} \times (\vec{\text{C}} \times \vec{\text{A}}) + \vec{\text{C}} \times (\vec{\text{A}} \times \vec{\text{B}}) = 0$. Symbols have their usual meaning.

7. (*a*) State Gauss's law in electrostatics and deduce its differential form.

(*b*) In a region of space, the electric field is given by $\vec{\text{E}} = 8i + 4\hat{j} + 3\hat{k}$. Calculate the electric flux through a surface of area 100 unit in $x - y$ plane.

8. (*a*) State Ampere's Circuital Law.

(*b*) Find the magnetic field with direction at a distance r from a straight infinite wire carrying current.

(*c*) Write down its differential form.

9. (*a*) Write down Maxwell's equations in differential form and explain the physical significance of each equation.

(*b*) Show that the wave equation in free space for electric field $\vec{E}$ is given by

$$\nabla^2\vec{E} = \mu_0\varepsilon_0 \frac{\partial^2\vec{E}}{\partial t^2}$$

10. (*a*) State and prove Gauss's theorem in electrostatics.

(*b*) Apply Gauss theorem to calculate the electric field due to an infinitely long charged straight wire and a charged hollow cyllinder.

11. (*a*) From differential form of Gauss's law derive the Poisson's and Laplace's equation.

(*b*) Derive the equation of continuity in electrostatics.

12. (*a*) Write down the Faraday's law of electromagnetic induction.

(*b*) Derive an expression for the quantity of electricity induced in a coil through which the magnetic flux is varying.

13. (*a*) Write down Maxwell's equation in time varying field and explain the physical significance of each.

(*b*) What do you mean by magnetic vector potential? Why is it called so?

(*c*) A steel rod of radius 1 cm is bend to form a circular loop and wound with a coil having 200 turns. It mean radius of the loop is 10 cm, calculate the amount of current that will produce a flux of 0.5 mwb in the core. Given that for steel $\mu = 1000$.

14. (*a*) Deduce an expression for the energy of a simple harmonic oscillator of mass '*m*', amplitude 'A' and frequency '*v*; At what displacement is the energy, half - kinetic and half potential?

(*b*) What are stationary waves? State their characteristics.

15. (*a*) Derive the differential equation of a plane progressive wave and hence find an expression for energy density of the wave as it passes through a medium.

(*b*) Find the position of nodes and antinodes in a stretched string of length 1.0 meter for the fundamental mode of vibration. Find the same for an open pipe of same length.

16. (*a*) Write down the differential equation of motion of the damped vibration. Explain each term.

(*b*) Solve the differential equation of damped vibration.

(*c*) When the vibration of damped vibration should be oscillatory?

17. (*a*) Show that $x = (\mathrm{A} + \mathrm{B})\, e^{-rt/2\mathrm{m}}$ is the solution of differential equation

$$\frac{d^2x}{dt^2} + \left(\frac{r}{m}\right)\frac{dx}{dt} + \left(\frac{s}{m}\right)x = 0 \text{ for the critically damped oscillations.}$$

(*b*) Discuss the two cases :

(*i*) $k^2 > \omega^2$

(*ii*) $k^2 < \omega^2$.

Symbols have their usual meaning.

18. (*a*) Define logarithmic decrement and relaxation time of a damped oscillator.

(*b*) Establish the differential equation of simple harmonic motion and solve the equation.

(*c*) What is meant by critical damping?

19. (*a*) A particle is subjected simultaneously to two S.H.M of the same period but of different amplitudes and phases in perpendicular direction. Find the expression for the resultant motion.

(*b*) Find the resultant of '*n*' numbers of S.H. vibration of equal amplitude by successive phase angle δ.

(*c*) What do you meant by Lissajous figure? On which the Lissajous figure depends? Write down the condition for getting the elliptical Lissajous figure.

20. (*a*) Write down the differential equation of *e.m.f* of a L-C-R circuit and solve it.

(*b*) Is the circuit having L = 0.1 μH, C = 0.10 mF and R = 10 kΩ oscillatory?

21. (*a*) Write down the differential equation of a Forced vibration and explain its each term.

(*b*) Starting from the solution of the forced vibration explain the velocity resonance and amplitude resonance.

22. (*a*) Write and solve the differential equation of motion of a Forced vibration.

(*b*) What is meant by sharpness of resonance? What is the effect of damping constant on the sharpness of resonance show it schematically?

(*c*) Define Q-factor. Write its expression for mechanical and electrical oscillator.

23. (*a*) Monochromatic light from a narrow slit falls on two parallel slits and the interference fringes are obtained on a screen. Calculate the fringe width and show that is directly proportional to the wavelength of light used and distance between the parallel slit screen and inversely proportional to the distance between two parallel slits.

(*b*) A biprism is placed 5 cm away from a slit illuminated sodium light (λ = 5890 Å). The width of the fringes obtained on the screen 75 cm away from biprism is 9.424×10^{-2} cm.What is the distance between the coherent sources?

24. (*a*) What do you mean by interference of light? State the condition required for obtaining sustained interference pattern.

(*b*) Obtain an expression for the intensity as any 3 point P in a Young's double slit experiment and hence calculate the fringe width.

25. (*a*) What is coherence of light sources? Explain spatial and temporal coherence.

(*b*) Obtain the expression for the path difference of two interfering beams of light in the case of interference from wedge shaped thin films.

26. (*a*) Deduce the expression for wavelength of a monochromatic wave which forms Newton's Rings by normal incidence in an experimental arrangement.

(*b*) Show that interference of light obey the law of conservation of energy.

27. What is the effect of the following operation on the interference fringes in Young's double slit experiment?

(*a*) The screen is moved away from the plane of the slits S_1 and S_2.

(*b*) The monochromatic source at S is replaced by another monochromatic sources of lesser wavelength.

(*c*) The separation between S_1 and S_2 is decreased.

(*d*) The width of the source slit S is increased.

(*e*) The distance between S and the plane of two slits S_1 and S_2 is increased.

28. (*a*) Distinguish between Fraunhofer diffraction due to single slit and due to double slit.

(*b*) Show that the intensity of the first secondary maximum formed by a single slit Fraunhofer diffraction process is nearly 4.5% of that of the principal maximum.

29. (*a*) What are coherent and non-coherent sources of light. Outline the theory for the determination of wavelength of light by Fresnel's biprism.

(*b*) Distinguish between Fresnel and Fraunhofer diffraction.

(*c*) Distinguish between interference of light and diffraction of light.

30. (*a*) What do you mean by missing order spectrum?

(*b*) Discuss the nature of diffraction grating spectrum.

(*c*) Deduce an expression for resolving power of a grating.

31. (*a*) What do you mean by polarisation of light, plane of vibration and plane of polarisation?

(*b*) State and explain Brewster's law of polarisation indicating clearly the nature of polarisation of the reflected and refracted rays.

(*c*) Show that the reflected and refracted components are mutually perpendicular to each other.

32. (*a*) Discuss the phenomenon of double refraction. What do you mean by positive and negative crystal?

(*b*) A right circularly polarized beam is incident on a calcite halfwave plate. Show that the beam will be left circularly polarized.

(*c*) Define circularly and elliptical polarized light.

33. (*a*) State and discuss the Molus law of polarization of light.

(*b*) What is analyzer?

(*c*) How can we used Nichol prism to produce the polarization of light and also analyse the polarisation of light?

34. (*a*) Distinguish between M.B., F.D. and B.E. statistics.

(*b*) Define Fermi energy.

(*c*) Describe with a neat diagram the structure of an optical fiber.

(*d*) When does the Lorentz transformation transfer to Galilean transformation.

35. Obtain the number of macrostates and the number of microstates of a system of three particles arranged in three compartments when the particles are distinguishable.

OR

Find the Fermi energy of copper assuming that each coper atom contributes one free electron to the electron gas. The density of copper is 8940 kg/m^3 and its atomic mass is 63.5 *a.m.u.* (1 *a.m.u* = 1.66×10^{-27} kg, $h = 6.63 \times 10^{-34}$ J-sec, $m_e = 9.11 \times 10^{-31}$ kg).

36. Show that for a system of particles under most probable distribution, the distribution modulus $\beta = 1/KT$.

37. (*a*) Distinguish between spontaneous emission and stimulated emission of radiation.

(*b*) What is population inversion? How is it achieved?

38. Derive the expression for relativistic velocity addition or subtraction of a body when S′ frame moving with velocity v along X-axis with respect to S frame.

OR

Show $m = m_0/(1 - v^2/c^2)^{1/2}$, where the symbols have their usual significance.

39. (*i*) Define binding energy of a nucleus. Draw a diagram to show the variation of binding energy per nucleon with atomic mass. What pieces of information can you get from the study of such a curve?

(*ii*) Given the following isotopic masses :

$$^{7}_{3}\text{Li} = 7.016004 \text{ amu}$$

$$^{6}_{3}\text{Li} = 6.015125 \text{ amu}$$

$$^{1}_{0}n = 1.008665 \text{ amu}$$

calculate the binding energy of a neutron in the $^{7}_{3}\text{Li}$ nucleus. Express the result in MeV.

40. (*i*) What do you mean by

(*a*) virtual displacement and

(*b*) holonomic constraint in classical dynamics?

(*ii*) State D' Alembert's principle in words.

(*iii*) Write down the Lagrange's equation of motion for a conservative holonomic system.

(*iv*) What do you understand by cyclic co-ordinates? Show that the generalized momentum corresponding to a cyclic co-ordinate is a constant of motion.

(*v*) Define the Hamiltonian. When is it equal to the total energy of the system?

41. (*a*) With a ray diagram explain the mode of action of the plane transmission grating.

(*b*) Derive the expression for intensity distribution due to diffraction by a plane transmission grating and hence deduce grating formula.

(*c*) Parallel beam of sodium light having wavelengths 5890 Å and 5896 Å is made to be incident normally on a diffraction grating having 5000 lines/cm (per centimeter). Find the angular separation between the maxima for given wavelengths in the second order.

42. (*a*) Monochromatic light from a narrow slit falls on two parallel slits and the interference fringes are obtained on a screen. Calculate the fringe width and show that it is directly proportional to wavelength of light used and distance between parallel slits & screen and inversely proportional to the distance between two parallel slits.

(*b*) A biprism is placed 5 cm away from a slit illuminated by sodium light (λ = 5890 Å). The width of the fringes obtained on a screen 75 cm away from the biprism is 9.424×10^{-2} cm. What is the distance between the coherent sources ?

43. (*a*) Deduce an expression for the energy of a simple harmonic oscillator of mass '*m*', amplitude 'A' and frequency '*v*'. At what displacemet is the energy, half-kinetic and half-potential?

(*b*) A fork of unknown frequency gives 5 beats per second when sounded with another fork of frequency 250 Hz.

Now the known fork is loaded with paraffin and again 5 beats per second are heard. Calculate the frequency of the unknown fork.

44. (*a*) State and explain de Broglie's hypothesis and obtain appropriate expressions for the wave velocity and group velocity indicating clearly their physical significance.

(*b*) Calculate de Broglie wavelength for a beam of electrons accelerated through 50V. Derive the formula used.

45. Describe an arrangement for the production of X-rays in the laboratory.

46. (*a*) Discuss the origin of characteristic X-rays.

(*b*) If μ is the linear absorption coefficient of X-rays, prove that $\mu = \dfrac{0.693}{X_{1/2}}$

where $X_{1/2}$ is the half value thickness of the substance.

(*c*) An X-ray tube operates at 60 kV. What will be the minimum wavelength of X-ray emitted in it?

47. (*a*) Obtain Schrödinger time independent wave equation. What is the physical significance of ψ used in this equation?

(*b*) Consider a particle is trapped in an infinitely deep one dimensional potential well and is free to move within the potential walls. Starting from the Schrödinger wave equation show that the energy of the particle is quantised.

48. Derive the differential equation of a plane progressive wave anad hence find an expression for the energy density of the wave as it passes through a medium.

49. State and explain Gauss's theorem.

50. What do you mean by interference of light? State the conditions required for obtaining sustained interference pattern.

Obtain an expression for the intensity at any 3 point P in a Young's double slit experiment and hence calculate the fringe width.

51. (*a*) Distinguish between Fraunhofer's and Fresnel's diffraction. Discuss with necessary theory the diffraction due to a single slit (Fraunhofer type) and show the resultant intensity pattern.

(*b*) Light of wavelength 5890 Å falls normally on a diffraction grating having 5000 lines per centimeter. Calculate the angles of the first and the second order diffraction maxima. Also calculate the maximum number of maxima permissable.

52. What do you mean by polarisation of light, plane of vibration and plane of polarisation? State and explain 'Brewster's law' of polarisation indicating clearly the nature of polarisation of the reflected and refracted rays. Show that the reflected and refracted components are mutually perpendicular to each other.

53. (*a*) Describe an experiment to demonstrate that the electrons in motion behave as waves.

(*b*) Calculate the de Broglie wavelength of an electron having kinetic energy equal to 15 eV. [1 eV = 1.6×10^{-19} J]

(*c*) State and explain Heisenberg's Uncertainty Principle.

54. (*a*) What are the basic postulates of Quantum Mechanics?

(*b*) Deduce the time-dependent Schrödinger wave equation.

(*c*) Obtain solution for the one-dimensional rectangular potential well such that

$$V(x) = \infty \quad \text{for } x = 0 \quad \text{and } x = a$$

$$V(x) = 0 \quad \text{for } 0 < x < a.$$

55. A series LCR circuits with L = 5.0 H, C = 80 μF and R = 60 Ω is connected to a variable frequency source of 230 V.

(*a*) Determine the resonance frequency of the circuit.

(*b*) Obtain the impedance of the circuit and the amplitude of current at resonance.

56. Light of wavelength 6×10^{-7} m is diffracted by an aperture of width 3×10^{-3} m. For what distance travelled by the diffracted beam does the spreading due to diffraction becomes greater than the width of the aperture?

57. Apply Gauss's theorem to derive an expression for this electric field due to a uniformly charged spherical shell at a point (*i*) inside (*ii*) outside the shell.

58. (*a*) Distinguish between continuous and characteristic X-rays and explain their origin.

(*b*) Derive Bragg's law and describe briefly an experiment to verify it experimentally.

(*c*) In a Bragg's diffraction experiment, a crystal with interplanar spacing 0.855 Å gives first order diffraction at a glancing angle of 65°. Find the wavelength of the X-ray used for the experiment.

59. (*a*) State and explain de Broglie's hypothesis and obtain appropriate expression for the wave velocity and group velocity indicating clearly their physical significance.

(*b*) The energy of the slow neutrons in a collimated beam obeys a Maxwellian distribution law at a temperature 300 K. Calculate de Broglie wave of the neutron. Given : $k = 1.38 \times 10^{-23}$ J/mole K ; $m_n = 1.67 \times 10^{-27}$ kg.

60. (*a*) Why light nuclei is taken for nuclear fusion?

(*b*) Why nuclear fission reaction must be done first before nuclear fusion?

61. (*a*) State and explain Heisenbergs's uncertainty principle.

(*b*) Give an account of the γ-ray microscope experiment and discuss the results.

(*c*) Write down the Schrödinger equation for a particle of mass *m* in a rectangular box with infinitely hard walls whose edges are *a*, *b* and *c*. What are the boundary conditions? What conditions are responsible for the quantization of energy of the particle?

The wave functions of the particle are given by

$$U = A \sin\left(\frac{n_x \pi x}{a}\right) \sin\left(\frac{n_y \pi y}{b}\right) \sin\left(\frac{n_z \pi z}{c}\right)$$

where $n_x = n_y = n_z = 1, 2, 3, ...$

Find the value of the normalization constant A.

62. (*a*) State Gauss's divergence theorem.

(*b*) Evaluate $\iint_S \vec{r}.\hat{n}ds$, where *s* is a closed surface.

(*c*) Show that $F = (2xy + z^3)\,\hat{i} + x^2\hat{j} + 3xz^2\hat{k}$ is a conservative force field. Find also the scalar potential.

63. (*a*) What are Miller indices?

(*b*) Find the packing factor for sc and bcc lattice.

(*c*) On the basis of band theory distinguish between conductors, semiconductors and insulators.

64. Write short notes on

(*a*) Bose-Einstein Statistics.

(*b*) X- ray diffraction and Bragg's Law.

65. Answer the following questions:

(*a*) In what respect does a wire carrying a current differ from a wire, which carries no current?

(*b*) What is the value of $\vec{\nabla}.\vec{r}$?

(*c*) What are bosons? Give an example.

(*d*) Why unmodified line appear in Compton scattering?

(*e*) What is the physical interpretation of probability density?

(*f*) Evaluate expectation value $<X>$ for a onedimensional potential box of length L in the ground state.

(*g*) State Stefan-Boltzmann's Law.

(*h*) Calculate interplaner spacing '*d*' of planes (111) in a simple cubic lattice of side '*a*'.

66. What is photoelectric effect? Calculate the threshold frequency of photon for photoelectric emission from a metal of work function 0.1 eV.

67. Calculate the de Broglie wavelength of an electron beam accelerated through a potential difference of 60 V.

68. Consider interferance between two sources of intensities I and 4I. Obtain intensity at a point where the phase defference is $\frac{\pi}{2}$.

69. (*a*) What is coherence of light sources? Explain spatial coherence.

(*b*) Obtain the expression for the path difference of two interfering beams of light in the case of interference from wedge shaped thin films.

(*c*) Interference fringes are produced by Fresnel's biprism in the focal plane of the microscope which is 100 cm from the slit. A lens placed between the biprism and the microscope gives two images of the slit at two positions, the distance between the virtual slit so obtained are 4.05 mm and 2.90 mm at the positions mentioned. Calculate the width of the interference formed with sodium lamp of wavelength 5893Å.

70. Discuss the phenomenon of diffraction at a single slit. Discuss the results graphically.

71. Write down the Lorentz transformation equations. Explain time dilation and show that a moving clock appears to run slowly to a stationary observer.

72. (*a*) Can two vectors of different magnitudes be combined to give a zero resultant? Can three Vectors?

(*b*) Two straight lines drawn on the same displacement time curve make angles 30° and 60° with the time-axis respectively. Which line represents greater velocity? Find the ratio of the two velocities.

73. Explain 'forced vibration' and 'Resonance' with examples.

74. (*a*) Establish the equation of a plane progressive wave in one dimension.
(*b*) State the properties of a Progressive wave.

75. (*a*) What is the essential difference between X-rays and visible light?
(*b*) An X-ray tube is operated at 45 kV. Calculate the maximum frequency of the emitted X-Ray.
(*c*) What is atomic mass unit? What is its equivalent energy?

76. Answer the following questions:
(*a*) What is the value of $\vec{\nabla} \times \vec{r}$?
(*b*) What are the experimental evidences to indicate that the nuclear forces are charge-independent?
(*c*) Find the Miller indices of a plane which intercepts at a, $b/2$, $3c$ in a simple cubic unit cell.
(*d*) What is the value of $\left[\hat{x}, \frac{\partial}{\partial x}\right]$?
(*e*) What is meant by conservative force?
(*f*) What is Fermi energy?
(*g*) Write down Einstein's photoelectric equation.
(*h*) Two lines of force do not intersect- why?
(*i*) Can X-rays be diffracted by a plane grating?
(*j*) Why neutrons (and not protons) are mostly used for nuclear fission?
(*k*) Why is the speciality of energy bands in conductors?

77. (*a*) State Stoke's theorem.
(*b*) Prove that $\vec{A} \times (\vec{B} \times \vec{C}) = \vec{B}(\vec{A}.\vec{C}) - \vec{C}(\vec{A}.\vec{B})$
(*c*) If $\vec{A} = x^2y\hat{i} - 2xz\hat{j} + 2yz\hat{k}$ find curl $\vec{A}$.

78. (*a*) What are Fermions and Bosons? Give two examples of each.
(*b*) State the basic postulates of Fermi-Dirac Statistics and derive an expression for the probability distribution of particles obeying Fermi-Dirac Statistics.

79. (*a*) Write notes on mass defect and binding energy of a nucleus.
(*b*) Explain with the help of binding energy curve why light nuclei are taken for nuclear fusion.
(*c*) Find the binding energy of a deuterium, given that
mass of proton = 1.0081 amu, mass of neutron = 1.0087 amu and mass of deuterium = 2.0147 amu.

80. (*a*) Distinguish between Inertial and Non-Inertial frames of reference.
(*b*) Write down the relation between logarithmic decrement and relaxation time.
(*c*) Show that the unit of damping term K is S^{-1}.
(*d*) What are constraints? Write down the equation of constraint for a simple pendulum.
(*e*) Write down the two differences between progressive wave and electromagnetic wave.
(*f*) Draw the Lissajous figure. When $\phi = \pi/4$ and $a = b$.

81. Alternating current of peak value 3 A and frequency 50 Hz is flowing in a coil. Calculate the e.m.f. induced in a second circuit which has mutual inductance 1.0 H with first circuit.

82. Light of wavelength 5×10^{-7} m is diffracted by an aperture of width 2×10^{-3} m. For what distance travelled by the diffracted beam does the spreading due to diffraction become greater than the width of the aperture?

83. Red light of wavelength 6500 Å from a distant source falls on a slit 0.50 mm wide. What is the distance between the two dark bands on each side of the central bright band of the diffraction pattern observed on a screen placed 1.8 m from the slits?

84. Define work function and threshold frequency.

85. Light of wavelength 3500 Å is incident on two metals A and B. Which metal will yield photoelectrons if their work functions are 4.2 eV and 1.9 eV respectively? ($h = 6.6 \times 10^{-34}$ J s).

86. Distinguish between metals, insulators and semiconductors on the basis of energy band diagrams. Indicate against each of the following if it is a conductor, a semiconductor or an insulator : (*i*) silicon, (*ii*) iron, (*iii*) mica.

87. Write down the properties of wave function ψ.

(*a*) Calculate the de Broglie wavelength associated with an electron having a kinetic energy of 10 eV. [h = 6.6 × 10^{-34} Joule-sec, m_e = 9.1 × 10^{-31} kg, 1 eV = 1.6 × 10^{-19} Joule, 1 Joule = Newton-meter].

(*b*) Assume that an electron is inside a nucleus of radius 10^{-15} m. From the uncertainty principle ($\Delta X \, \Delta p \geq h/2\pi$), estimate the kinetic energy of electron.

(*c*) The particle is confined in a one-dimensional box of length L. The wave function of a particle is given by $\psi(X) = C \sin(\pi X/L)$ in the region $0 < X < L$ and $\psi(X) = 0$ outside this region. Calculate the probability of finding the particle in the region $0 < X < L$ and the expectation value of X.

88. (A) Select the correct answer from the given alternatives in each case :

(*i*) A system seleronomous when constraint equations
(*a*) do not contain velocity.
(*b*) do not contain time.
(*c*) do not contain velocity and time.

(*ii*) If $\hat{n}$ is the unit vector in the direction of $\vec{A}$, then

(*a*) $\hat{n} = \dfrac{\vec{A}}{|\vec{A}|}$.

(*b*) $\hat{n} = \vec{A}|\vec{A}|$.

(*c*) $\hat{n} = \dfrac{|\vec{A}|}{\vec{A}}$.

(*iii*) The value of $\vec{\nabla}.\vec{r}$ is
(*a*) 0. (*b*) 1. (*c*) 3.

(*iv*) Two vectors $\vec{A}$ and $\vec{B}$ are parallel when
(*a*) $\vec{A} \times \vec{B} = 0$. (*b*) $\vec{A}.\vec{B} = 0$. (*c*) $\vec{A}.\vec{B} = 1$.

(*v*) The displacement current arises due to
(*a*) positive charges only.
(*b*) negative charges only.
(*c*) time varying electric field.

(*vi*) Sound wave is
(*a*) an electromagnetic wave.
(*b*) an elastic wave.
(*c*) a radio wave.

(*vii*) Which of the following phenomenon proves the transverse characteristic of light?
(*a*) interference. (*b*) dispersion. (*c*) polarisation.

(*viii*) When Interference of light takes place at some region the light energy is
(*a*) created. (*b*) destroyed (*c*) redistributed.

(*ix*) In Fraunhofer diffraction, the incident wave front is
(*a*) plane. (*b*) spherical (*c*) cylindrical.

(*x*) If ω_0 is the natural frequency of a body and k is the damping constant, then its quality factor is
(*a*) ω_0/k (*b*) $\omega_0/(2k)$ (*c*) $2\omega_0/k$.

(B) Answer the following questions:

(*i*) What is a scalar field? Give example.
(*ii*) What is meant by "ignorable" or 'cyclic' co-ordinate in classical mechanics?
(*iii*) State Brewster's law in polarization.
(*iv*) Write down the Faraday's law of electromagnetic induction.
(*v*) Define simple harmonic motion.
(*vi*) Show that the potential function

$$V = V_0 (x^2 - 2y^2 + z^2) \text{ satisfy Laplace's equation}$$

where V_0 is a constant.

(*vii*) Why the center of Newton's rings is dark?
(*viii*) What is the difference between interference and diffraction of light?

89. (*a*) What are generalized co-ordinates? What is the advantage of using them? Obtain the expression for generalized force.

(*b*) Write down the Lagrangian of a simple pendulum explaining each term. Hence obtain the equation of motion.

90. (*a*) What do you mean by gradient of a scalar field?

$$\text{If } \phi(x, y, z) = 3x^2 y - y^3z^2, \text{ find } \vec{\nabla}\phi \text{ at the point } (1, -2, -1).$$

(*b*) Define vector field. Give an example.

(*c*) Prove: $\vec{A} \times (\vec{B} \times \vec{C}) + \vec{B} \times (\vec{C} \times \vec{A}) + \vec{C} \times (\vec{A} \times \vec{B}) = 0$ where symbols have their usual meaning.

(*d*) State Gauss's law in electrostatics and deduce its differential form.

(*e*) In a region of space, the electric field is given by $\vec{E} = 8\hat{i} + 4\hat{j} + 3\hat{k}$. Calculate the electric flux through a surface of area 100 unit in X – Y plane.

91. (*a*) State Ampere's Circuital law.

(*b*) Find the magnetic field with direction at a distance *r* from a straight infinite wire carrying current.

(*c*) If $\vec{F} = (2xy + z^3)\hat{i} + x^2\hat{j} + 3xz^2\hat{k}$, show that $\vec{F}$ is an irrotational vector.

92. (*a*) Write down Maxwell's equations in differential form and explain the physical significance of each equation.

(*b*) Show that the wave equation in free space for electric field $\vec{E}$ is given by $\nabla^2\vec{E} = \mu_0\varepsilon_0 \dfrac{\partial^2\vec{E}}{\partial t^2}$.

93. (*a*) Establish the differential equation of simple harmonic motion and solve the equation.

(*b*) Establish the differential equation of damped harmonic motion and explain the different terms in the expression.

(*c*) What is meant by critical damping?

94. (*a*) What are the conditions to get the sustained interference pattern?

(*b*) In an interference experiment '*d*' is the distance between the two coherent sources of light with wavelength λ, and D is the source-screen distance. Show that the separation between the two consecutive dark bands is given by $\beta = D\lambda/d$.

(*c*) In a Newton's ring experiment, the diameter of 5th dark ring is 0.336 cm and the diameter of the 15th dark ring is 0.590 cm. Find the radius of the plano-convex lens if the wavelength of light used is 5890 Å.

95. (*a*) Show that the intensity of the first secondary maximum formed by a single slit Fraunhofer diffraction process is nearly 4.5% of that of the principal maximum.

(*b*) What do you mean by the resolving power of an optical instrument?

(*c*) What is the maximum number of lines of a grating which will resolve the 3rd order spectrum of two lines having wavelengths 5890 Å and 5896 Å?

96. (*a*) What is polarization of light?
(*b*) What do you mean by positive crystal and negative crystal?
(*c*) How can you distinguish between an unpolarised light, a circularly polarized light and a plane polarized light?
(*d*) A ray of light is incident on the surface of a glass plate of refractive index 1.732 at the polarization angle. Calculate the angle of refraction of the ray.

97. Write short notes on any two of the following:
(*a*) Sharpness of resonance.
(*b*) Nicol prism and its use as polariser and analyzer.
(*c*) Hamiltonian of a dynamical system.
(*d*) Continuity equation in electromagnetism.

98. What is a non-inertial frame of reference? What are the basic postulates of Theory of Relativity? Using Lorentz transformation derive expressions for (*i*) length contraction and (*ii*) time dilation.

99. (*a*) What do you mean by polarization of light and plane of polarization?
(*b*) State and explain 'Brewster's Law' of polarization indicating clearly the nature of polarization of the reflected and refracted rays. Show that the refracted and reflected components are mutually perpendicular to each other.

100. Consider a diffraction grating of width 5 cm with slit width 0.0001 cm separated by a distance of 0.0002 cm. What is the corresponding grating element? How many orders would be observable if $\lambda = 5.5 \times 10^{-5}$ cm.?

101. (*a*) Discuss the phenomenon of double refraction. What do you mean by positive crystal and negative crystal?
(*b*) A right circularly polarized beam is incident on a calcite half wave plate. Show that the beam will be left circularly polarized.

102. (*a*) What are the characteristics of simple harmonic motion? Define phase. Prove the principle of conservation of energy in this case.
(*b*) A cubical block of side L cm and density d is floating in water of density p ($p > d$). The block is slightly depressed and released.
Show that it will execute simple harmonic motion. Determine the frequency of oscillations.

103. (*a*) Derive an expression for the velocity of transverse waves along a string under tension.
(*b*) Two identical wires are subjected to the same tension. Their lengths are 42.5 cm and 43 cm respectively. If the difference of frequencies is 3, determine their frequencies.

104. A vector is not changed if-
(*a*) it is rotated through an arbitrary angle
(*b*) it is multiplied by an arbitrary scalar
(*c*) it is cross multiplied by any arbitrary scalar
(*d*) it is slid parallel to itself

105. The resultant of **A** and **B** makes an angle α with A and β with B
(*a*) $\alpha < \beta$
(*b*) $\alpha < \beta$ if $\mathbf{A} < \mathbf{B}$
(*c*) $\alpha < \beta$ if $\mathbf{A} > \mathbf{B}$
(*d*) $\alpha < \beta$ if $\mathbf{A} = \mathbf{B}$

106. The component of a vector is
(*a*) always less than its magnitude
(*b*) always greater than its magnitude
(*c*) always equal to its magnitude
(*d*) none of these

107. A vector **A** points vertically upward and **B** points towards north. The vector product **A** × **B** is

(*a*) along west.

(*b*) along east

(*c*) zero

(*d*) vertically downward.

108. Let **C** = **A** + **B**

(*a*) |C| is always greater than |**A**|

(*b*) it is possible to have |**C**| < |**A**| and |**C**| < |**B**|

(*c*) C is always equal to **A** + **B**

(*d*) C is never equal to **A** – **B**

109. Match (C) and (D)

(C)	(D)
(*a*) X-component of |**A** × **B**|	(*i*) $A_X + B_X$
(*b*) X-component of |**A** . **B**|	(*ii*) $A_X B_y - A_y B_y$
(*c*) X-component of |**A** + **B**|	(*iii*) $A_y B_X$
(*d*) X-component of |**A** – **B**|	(*iv*) $A_y B_z - A_z B_y$
	(*v*) $A_z B_X - A_X B_z$
	(*vi*) $A_X B$
	(*vii*) $A_X B_X$
	(*viii*) $A B_X$

110. If **F** = 4*i** – 2*j** and *r* = 3*i** + 4*k** (here *i**, *j**, & *k** are unit vectors)
What is the maximum work done? Under this condition find also the angle between *r* & **F**.

111. (*a*) Calculate effective lattice points in a unit cell of fcc structure

(*b*) What do you mean by atomic packing fraction? Calculate it for fcc structure.

112. Find the scalar product of any two diagonal of an unit cube. What is the angle between them

113. ∇*r* is-

(*a*) 0 (*b*) 1

(*c*) *r* (unit vector) (*d*) *r*

114. ∇ . *r* is

(*a*) 0 (*b*) 1

(*c*) *r* (*d*) 3

115. ∇(1/*r*) is

(*a*) **0** (*b*) *r*

(*c*) $-\mathbf{r}/r^2$ (*d*) $-\mathbf{r}/r^3$

116. ∇ × *r* is –

(*a*) **0** (*b*) *r*

(*c*) $\mathbf{r}/r^2$ (*d*) $-\mathbf{r}/r^3$

117. A vector **A** is solenoidal when :

(*a*) div **A** = 0 (*b*) curl **A** = **0**

(*c*) curl curl **A** = **0** (*d*) grad (div **A**) = 0

118. A vector **A** is 'irrotational' at a point if :

(*a*) div **A** = 0 (*b*) curl **A** = **0**

(*c*) curl curl **A** = 0 (*d*) grad (div **A**) = 0

119. Gauss's divergence theorem permits change from :

(*a*) volume integral to surface integral and vice-versa

(*b*) volume to line integral and vice-versa

(*c*) surface integral to line integral and vice-versa
(*d*) none of these

120. Stoke's theorem permits change over from :
(*a*) volume integral to surface integral and vice-versa
(*b*) volume integral to line integral and vice-versa
(*c*) surface integral to line integral and vice-versa
(*d*) none of these

121. Discuss comparatively the basic postulates of Maxwell-Boltzmann(M.B.), Bose-Einstein(B.E.), Fermi-Dirac(F.D.) statistics. Name the statistics (B.E. or F.D.) obeyed by each of the following particles :
neutron, π-meson, muon, photon.

122. (*a*) Starting from Fermi statistics derive an expression for electron distribution in metal.
(*b*) Show that at T = 0, average energy ε of an electron in a metal is given by $\varepsilon = 3\varepsilon_F/5$, where ε_F is the Fermi energy.

123. Find the Fermi energy in copper on the assumption that each copper atom contributes one free electron to the electron gas. The density of copper is 8.94×10^3 kg/m^3 and its atomic mass is 63.5 *u*.

124. (*a*) What is Compton deffect? Derive a formula for Compton wevelength.
(*b*) (*i*) Find the change in wavelength of pm X-rays that are scattered 120° by a target
(*ii*) Find the angle between the direction of the recoil electron and incident photon
(*iii*) Find the energy of the recoil electron.

125. (*a*) Explain Hisenberg's uncertainty Principle.
(*b*) A Hydrogen atom is 5.3×10^{-11} m in radius. Use the uncertainty principle to estimate the minimum energy an electron can have in this atom.

126. (*a*) Which of the wave function in the following figures cannot have physical significance? Why not?

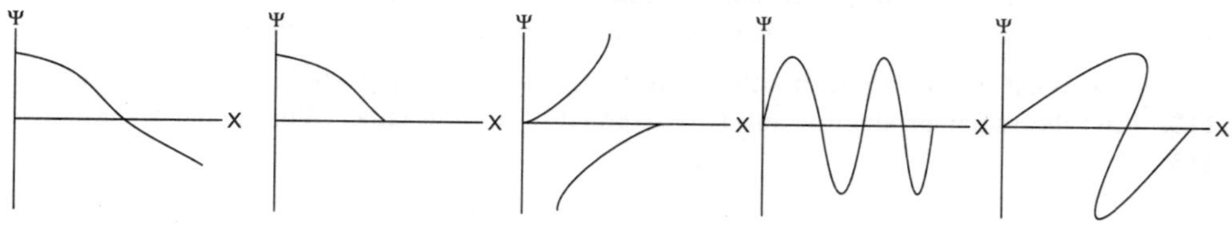

(*b*) A particle is in a cubic box with infinitely hard walls whose edges are L long. The wave function of the particles are given by

$$\psi = A \sin\frac{n_x \pi_x}{L} \sin\frac{n_y \pi_y}{L} \sin\frac{n_z \pi_z}{L}$$

where $n_x = 1, 2, 3,\ n_y = 1, 2, 3,\ n_z = 1, 2, 3$

Find the value of the normalization constant A.
(*c*) Find the expectation value $\langle x \rangle$ of the proton of a particle trapped in a box L inside.

127. Drive an expression for the magnetic field intensity at a point on the axis of a circular coil of wire carrying electric current.

128. (*a*) Write down four Maxwell's form equation of electromagnetic wave and show that they represent the formulation of electrostatics, magnetostatics and electromagnetism.
(*b*) Starting from the equation of continuing and assuming Ohm's law, show that the charge density in a conductor obeys the following relation.

129. Consider a co-axial cable which consists of an inner wire of radius "*a*" surrounded by an outer shell of inner and outer radii "*b*" and "*c*" respectively. The inner wire carries an electric current I and the outer shell carries an equal current in opposite direction. Find the magnetic field at a distance "*x*" from the axis where (*i*) $x < a$ (*ii*) $a < x < b$ (*iii*) $b < x < c$ and (*iv*) $x > c$. Assume that the current density is uniform in the inner wire and also uniform in the other shell.

130. Vector product of two true vectors is a pseudovector.

(*i*) True (*ii*) False

131. If $\boldsymbol{a} \times \boldsymbol{b} = \boldsymbol{c}$, then $\boldsymbol{c}$ is

(*a*) normal to $\boldsymbol{a}$ only; (*b*) parallel to $\boldsymbol{a}$ only ;

(*c*) parallel to $\boldsymbol{b}$ only; (*d*) normal to both $\boldsymbol{a}$ & $\boldsymbol{b}$;

132. Gravitational force field is :

(*a*) central force field (*b*) non-central force field

(*c*) non-conservative force field (*d*) repulsive force field

133. (*i*) Angular momentum is :

(*a*) Scalar (*b*) A polar vector

(*c*) An axial vector (*d*) none of these.

(*ii*) Which of the following is a scalar quantity-

(*a*) electric current (*b*) force

(*c*) temperature (*d*) pressure

134. Fill in the gaps with suitable words :

(*i*) The resultant of two vectors is maximum when the angle between them is ....

(*ii*) The resultant of two vectors is minimum when the angle between them is...........................

135. A vector is not changed if :

(*a*) it is rotated through an arbitrary angle

(*b*) it is multiplied by an arbitrary scalar

(*c*) it is cross multiplied by any arbitrary scalar

(*d*) it is slid parallel to itself

136. The resultant of **A** and **B** makes an angle α with **A** and β with **B**

(*a*) $\alpha < \beta$ (*b*) $\alpha < \beta$ if $\mathbf{A} < \mathbf{B}$

(*c*) $\alpha < \beta$ if $\mathbf{A} > \mathbf{B}$ (*d*) $\alpha < \beta$ if $\mathbf{A} = \mathbf{B}$

137. Distinguish between the total energy of a vibrating particle and the energy of vibration of the system. Obtain expressions for both.
A particle of mass 10 gms. lies in potential given by $V_x = 32x^2 - 0.2$, where V_x is in Joules and x is in meters. Write down the equation of motion and solve it. What is the frequency of oscillation?
[**Hints:** P.E. of oscillation $= 1/2\, k x^2$ + const.]

138. Explain free undamped,free damped and forced vibrations. Derive the expression for displacement in case of free damped vibration.

139. (*a*) Define field. Write down the classification of field.

(*b*) Write down the condition for sustained interference.

(*c*) Write down the difference between Interference and Diffraction.

(*d*) Show that $\vec{A}.d\vec{A}/dt = \mathbf{A}\, d\mathbf{A}/dt$.

140. (*a*) Show that the vector field $A = -2z^2y/x^3\hat{i} + z^2/x^2\hat{j} + 2yz/x^2\hat{k}$ is irrotational.

(*b*) If $\phi = x^2 - y^2 + 2z$ find divgrad ϕ.

(*c*) Is the vector field $A = x^2\hat{i} + y^2\hat{j} + z^2\hat{k}$ is solenoidal?

141. (*a*) Write down two drawbacks of classical mechanics.

(*b*) What do you mean by wave-particle duality?

(*c*) Define perfect black body.

(*d*) What is lattice?

(*e*) Draw a plane having Miller indices (101).

142. (*a*) Find the packing fraction and co-ordination number of body centered cube.
(*b*) What is the maximum number of possible Bravais lattice?
(*c*) Calculate the distance between the adjacent parallel plane of the type **(100), (110)** in an fcc lattice constant '***a***'.

143. (*a*) Obtain the Bragg's law from the Laues equation.
(*b*) A crystal reflects monochromatic X-rays strongly when the Bragg's glancing angle for a first order reflection is 15 degree. What is the glancing angle for the third order reflection of the same type?

144. (*a*) Derive the Planck's law of radiation for black body.
(*b*) Derive the Einstein's law of photoelectric effect.

145. Find the energy of copper assuming that each copper atom contributes one free electron to the electron gas. The density of copper is 89400 kg/m and its atomic mass is 63.5 a.m.u. (1 a.m.u. = 1.66×10^{-27} kg, $h = 6.63 \times 10^{-34}$ J-sec, $m_e = 9.11 \times 10^{-31}$ kg).

146. Show that for a system of particles under most probable distribution, the distribution modulus $\beta = 1/KT$

147. (*a*) Distinguish between spontaneous emission and stimulated emission of radiation.
(*b*) What is population inversion? How is it achieved?

148. Derive the expression for relativistic velocity addition or subtraction of a body when S′ frame moving with velocity v along X-axis with respect to S frame.
Show $m = m_0/(1 - v^2/c^2)^{1/2}$, where the symbols have their usual significance.

EXAMINATION PAPERS

WBUT Examination Questions

Engineering & Technology Examinations, December – 2005

ENGINEERING PHYSICS

SEMESTER – I

Time : 3 Hours] [*Full Marks : 70*

The figures in the margin indicate full marks.
Candidates are required to give their answers in their own words as far as practicable.

Note : *(i) Question No. 1 is compulsory.*
(ii) Answer any five full questions from the remaining.

1. (A) Select the correct answers from the given alternatives in each case : $10 \times 1 = 10$

(*i*) Degree of freedom of a hydrogen molecule is

(*a*) 1 (*b*) 2 (*c*) 5.

(*ii*) The angle between $\hat{i}$ and $\left(2\hat{i} + \hat{j}\right)$ is

(*a*) $\cos^{-1}\frac{2}{5}$ (*b*) $\cos^{-1}\frac{2}{\sqrt{5}}$ (*c*) $\cos^{-1}\frac{2}{3}$.

(*iii*) In a free space Poisson's equation is

(*a*) $\nabla^2 V = 0$ (*b*) $\nabla^2 V = \frac{\epsilon_0}{\rho}$ (*c*) $\nabla^2 V = \infty$.

(*iv*) If $\vec{B} = \vec{\nabla} \times \vec{A}$, $\vec{B}$ and $\vec{A}$ are any vectors then

(*a*) $\vec{\nabla} . \vec{B} = 0$ (*b*) $\vec{\nabla} . \vec{B} = 1$ (*c*) $\vec{\nabla} . \vec{B} = -1$

(*v*) Maxwell's electromagnetic wave equations in terms of electric field vector $\vec{E}$ in free space is

(*a*) $\nabla^2 \vec{E} = -\mu_0 \epsilon_0 \frac{\partial^2 \vec{E}}{\partial t^2}$ (*b*) $\nabla^2 E = \mu_0 \epsilon_0 \frac{\partial^2 E}{\partial t^2}$ (*c*) $\nabla^2 \vec{E} = \mu_0 \epsilon_0 \frac{\partial^2 \vec{E}}{\partial t^2}$,

(*vi*) On superimposing two mutually perpendicular simple harmonic motion, we get circular Lissajous figures when their phase difference (ϕ) and amplitudes (P, Q) are as follows :

(*a*) $\phi = O$. $P = Q$ (*b*) $\phi = \frac{\pi}{2}$, $P = Q$ (*c*) $\phi = 0$, $P \neq Q$.

(*vii*) Polarisation of light establishes that light is essentially

(*a*) corpuscular (*b*) longitudinal wave (*c*) transverse wave.

(*viii*) The central fringe in Newton's ring experiment with a monochromatic light is

(*a*) bright (*b*) dark (*c*) white.

(*ix*) The intensity of principal maxima in the spectrum of a grating with N number of lines is proportional to

(*a*) $\frac{1}{N}$ (*b*) N (*c*) N^2

(*x*) The relation between the Hamiltonian H and the Lagrangian L of a system of particles is given by

(*a*) H = 2T + L (*b*) H = 2T – L (*c*) H = T + L

(where T is the kinetic energy).

(B) Answer any five from the following questions : $5 \times 2 = 10$

(*i*) Write down the Lagrangian for a particle moving freely in one dimension.

(*ii*) Show that the derivative of a vector of constant direction is parallel to that vector.

(*iii*) State Stoke's theorem.

(*iv*) What is Biot-Savart law in electromagnetism?

(*v*) What is meant by the relaxation time of a disturbed system?

(*vi*) What is the difference between the extraordinary and the ordinary ray propagating through an uniaxial crystal?

(*vii*) Can X-rays be diffracted from a single slit of width 0.1 mm? Explain.

(*viii*) Define resolving power of a grating.

2. (*a*) Define Hamiltonian of a system. When is it equal to the total energy of the system? 2 + 3 = 5

(*b*) A particle is falling vertically under the influence of gravity. Obtain the Hamiltonian and the equations of motion corresponding to that Hamiltonian. 3 + 2 = 5

3. (*a*) In an electric field the electric potential is given by

$$V(x, y, z) = (2x^2 + 4y^2 + 3z^2)^{-1/2}.$$

Find the electric field at a point (1, 1, 1). 4

(*b*) Distinguish between scalar and vector fields with examples. 4

(*c*) Define line integral of a vector. 2

4. (*a*) Show that the equation of continuity is given by $\vec{\nabla} . \vec{J} + \frac{\partial \rho}{\partial t} = 0$ where $\vec{J}$ and ρ have their usual meaning. 6

(*b*) Find the potential of a uniformly charged sphere of radius R having a constant charge density ρ at a distance r from the centre of the sphere where $r > R$. 4

5. (*a*) Define curl of a vector field. 2

(*b*) If $\phi = x^2 - y^2 + 2z$, find $\vec{\nabla} . \left(\vec{\nabla} \phi\right)$. 4

(*c*) Show that $\vec{B}(x, y, z) = 2xyz\,\hat{i} + (x^2 z + 2y)\,\hat{j} + x^2 y\,\hat{k}$ is irrotational. 4

6. (*a*) Write down the Maxwell's field equations. 5

(*b*) From those equations identify Gauss' law, Ampere's law and Faraday's law. 3

(*c*) How does velocity of light depend on the properties of the medium? 2

7. (*a*) Define damped vibration.

(*b*) Write down the differential equation for damped vibratory motion explaining the physical significance of each term in the equation.

(*c*) Obtain solution for the damped oscillatory motion.

8. (*a*) What is meant by polarization of light?
   (*b*) State Brewster's law and hence prove that the angle between the reflected and the refracted rays is 90°.
   (*c*) The refractive index of a glass plate is 1.6. Calculate the polarization angle and corresponding angle of the refracted ray.
9. (*a*) What are coherent sources?
   (*b*) Show that the laws of conservation of energy is not violated in the interference process.
   (*c*) In a Young's double slit experiment the distance between the two coherent sources is 1.15 mm. Calculate the fringe width that would be observed on screen placed at a distance of 85 cm from the source. The wavelength of light used is 5893 Å.
   (*d*) Will the radius of the Newton's ring change if air is replaced by oil in between the lens and the glass plate?
10. (*a*) Distinguish between Fraunhoffer and Fresnel diffractions.
    (*b*) Obtain an expression for the resultant intensity in a single slit Fraunhoffer diffraction process and show the intensity pattern graphically. 6
11. Write short notes on any two of the following : $2 \times 5 = 1$
    (*a*) Fresnel's biprism
    (*b*) Nicol prism as polarizer and analyzer
    (*c*) Displacement current
    (*d*) Generalized potential in classical mechanics.

WBUT Examination Questions

Engineering & Management Examinations, December – 2006

ENGINEERING PHYSICS

SEMESTER – I

Time : 3 Hours] [*Full Marks : 70*

Group – A
(Multiple Choice Type Questions)

1. Choose the correct alternatives for any *ten* of the following questions : $10 \times 1 = 10$
   (*i*) A system consists of three point masses. If the mutual distances between the masses do not vary with time then the degree of freedom of such system is
   (*a*) zero (*b*) three (*c*) six (*d*) nine.
   (*ii*) Condition of co-planarity of three vectors $\left(\vec{\alpha}, \vec{\beta}, \vec{\gamma}\right)$ is
   (*a*) $\vec{\alpha} \cdot \left(\vec{\beta} + \vec{\gamma}\right) = 0$ (*b*) $\vec{\alpha} \cdot \left(\vec{\beta} \times \vec{\gamma}\right) = 0$

(c) $\vec{\alpha} \times \left(\vec{\beta} \times \vec{\gamma}\right) = 0$ (d) $\vec{\alpha} \cdot \left(\vec{\beta} - \vec{\gamma}\right) = 0$.

(iii) The angle between $\nabla\varphi$ and the surface φ = constant is

(a) $\frac{\pi}{2}$ (b) $\frac{\pi}{4}$ (c) π (d) zero.

(iv) Flux of the electric field for a point charge (q) at origin through a spherical surface centered at the origin is

(a) $\frac{2q}{\varepsilon_0}$ (b) $\frac{q}{\varepsilon_0}$ (c) $\frac{q}{4\pi\varepsilon_0}$ (d) zero.

(v) Which of the following statements is not characteristic of a static magnetic field ?

(a) It is solenoidal (b) It is conservative

(c) Magnetic flux lines are always closed (d) It has no sink or source.

(vi) If $\vec{E}$ and $\vec{B}$ are the electric field and the magnetic field of a electromagnetic wave travelling in vacuum with propagation vector $\vec{K}$ then

(a) $\vec{K} \cdot \vec{E} = 0$ (b) $\vec{K} \times \vec{E} = 0$ (c) $\vec{B} \times \vec{E} = 0$ (d) $\vec{K} \times \vec{E} = -\vec{B}$.

(vii) To get a circular Lissajous figure, the phase difference (ϕ) and the amplitudes (a & b) of two superimposing mutually perpendicular simple harmonic motions respectively are

(a) $\phi = 0$; $a = b$ (b) $\phi = \frac{\pi}{2}$; $a = b$

(c) $\phi = \frac{\pi}{2}$; $a \neq b$ (d) $\phi = 0$; $a \neq b$.

(viii) The refractive indices of E-ray and O-ray are respectively 1.65 and 1.45. Then the thickness of the material required for a quarter-wave plate for light of wavelength 5000 Å is

(a) 2500 Å (b) 1250 Å (c) 6250 Å (d) none of these.

(ix) Two waves having the intensities in the ratio of 9 : 1 produce interference. The ratio of maximum to minimum intensity is equal to

(a) 10 : 8 (b) 9 : 1 (c) 4 : 1 (d) 2 : 1.

(x) Newton's rings are example of interference fringes of

(a) equal inclination, coherence is obtained by amplitude division

(b) equal thickness, coherence is obtained by amplitude division

(c) equal thickness, coherence is obtained by wave-front division

(d) equal inclination, coherence is obtained by wave-front division.

(xi) Angle of diffraction of the first order maximum in a diffraction grating

(a) increases if the separation between the rulings is increased

(b) decreases if the separation between the rulings is increased

(c) does not change if the separation between the rulings is increased

(d) change depends on the wavelength of the incident light.

(*xii*) The plane of vibration and the plane of polarization of a beam of plane polarized light

(*a*) are identical to each other

(*b*) are orthogonal to each other

(*c*) make an angle, which depends on the colour of the light

(*d*) rotate with respect to each other along the path of the beam.

Group – B
(Short Answer Type Questions)

Answer any *three* questions. $3 \times 5 = 15$

2. (*a*) Obtain the differential equation of motion of a simple pendulum from its Lagrangian representation. 1 + 2

   (*b*) What is meant by "ignorable" or cyclic co-ordinate ? Explain with an example. 1 + 1

3. (*a*) Show that $\vec{F} = (2xy + z^3)\,\hat{i} + x^2\hat{j} + 3xz^2\,\hat{k}$ is a conservative force field. 2

   (*b*) Show that $\oint_S \vec{B}.d\hat{s} = 0$, when $\vec{B}$ is the magnetic field and S is a closed surface. State the theorem that you have used. 2 + 1

4. Write down the Faraday's law of electromagnetic induction. Express it in differential form. 1 + 3

   Give an example of an electrical circuit carrying non-steady current where Ampere's circuital law is not applicable. 1

5. State the algebraic relation how the displacement is related to time in case of a damped harmonic motion. Derive the relation between the damping constant and logarithmic decrement. Derive the relation between the first throw of a ballistic galvanometer with the amplitude of undamped oscillation in terms of the logarithmic decrement. 1 + 2 + 2

6. (*a*) A polarizer and an analyzer are oriented so that the maximum intensity is achieved. To what fraction of its maximum value is the intensity reduced when the analyzer is rotated through 60° ? 2

   (*b*) The Intensity distribution for single slit diffraction is

   $$I = I_o (\sin^2 \beta / \beta^2),$$

   where $$\beta = \pi b(\sin\theta) / \lambda$$

   b is the width of the slit, and λ is the wavelength of light.

   Show that secondary maxima are given by the equation

   $$\tan\beta = \beta.$$ 3

Group – C
(Long Answer Type Questions)

Answer any *three* of the following. $3 \times 15 = 45$

7. (*a*) Distinguish between holonomic and non-holonomic constraints. 2

   (*b*) Write down the equation of constraint in each of the following cases :

   (*i*) a particle constrained to move on the surface of a sphere

   (*ii*) a simple pendulum with a fixed support.

Specify the nature of constraint and calculate the degrees of freedom in each case. 3 + 3

(*c*) Obtain the equation of motion of a simple pendulum and its time period, using Hamiltonian method. 5

(*d*) When does the Hamiltonian of a system represent the total energy of the system? 2

8. (*a*) Write down the differential equation of a forced vibration, explaining each term. 3

(*b*) Starting from the solution of the equation for forced vibration, explain the phenomenon of amplitude and velocity resonance deriving the value of the driving frequency in each case. 4 + 4

(*c*) Starting from the equation of a plane progressive wave in one dimension derive the differential equation for the wave. 4

9. (*a*) Explain the difference between interference and diffraction. 1

(*b*) Derive the expression of intensity at a point for Fraunhoffer diffraction due to a double slit. Draw the intensity distribution curve (diffraction pattern) and explain it. 6 + 5

(*c*) A parallel beam of light of wavelength (5890 Å) falls normally on a plane transmission grating having 4250 lines/cm. Find the angle of diffraction for maximum intensity in first order. 3

10. (*a*) State Gauss' law of electrostatics. 3

(*b*) Use this law to calculate the electric field between two infinite extent parallel plate capacitor carrying charge density σ and mutual separation d. Draw the necessary diagrams. 6

(*c*) Write the expression of the magnetic field due to a current carrying conductor. Draw a diagram necessary to explain the symbols. Show that this field is solenoidal. 3 + 3

11. (*a*) Write down Maxwell's field equations, explaining the terms used. Show that in vacuum both electric and magnetic vectors obey wave equation. Assuming a plane wave solution show that magnetic field is always orthogonal to the electric field. 5 + 5

(*b*) Find the displacement current within a parallel plate capacitor in series with a resistor which carries current I. Area of the capacitor plates are A and the dielectric is vacuum. 5

WBUT Examination Questions

Engineering & Management Examinations, June – 2006

ENGINEERING PHYSICS

SEMESTER – II

Time : 3 Hours] [*Full Marks : 70*

Group – A
(Multiple Choice Type Questions)

1. Choose the correct alternative for any *ten* of the following : $10 \times 1 = 10$

(*i*) 1 milligram of matter after getting completely converted to energy will yield

(*a*) 90 joules (*b*) 9×10^{10} joules

(*c*) 900 joules (*d*) 9 joules.

(*ii*) Lorentz Transformation is LT, Galilean Transformation is GT, velocity between two frames is v and velocity of light is given by c.

Which of the following statements is correct ?

(*a*) LT can never be reduced to GT under any approximation

(*b*) LT is reduced to GT when $v >> c$

(*c*) LT is reduced to GT when $v = c$

(*d*) LT is reduced to GT when $v << c$.

(*iii*) Mass of a photon of frequency v is given by (Planck's constant = h and velocity of light = c)

(*a*) hv (*b*) hv/c (*c*) hv/c^2 (*d*) hv^2/c.

(*iv*) An α-particle is 4 times heavier than a proton. If a proton and an α-particle are moving with the same velocity, how do their de Broglie wavelengths compare ?

(*a*) $\lambda_p = \lambda_\alpha$ (*b*) $\lambda_p = 4\lambda_\alpha$ (*c*) $\lambda_p = \lambda_\alpha /2$ (*d*) $\lambda_p = \lambda_\alpha /4$.

(*v*) Uncertainty principle tells that

(*a*) A particle can have only position but no momentum

(*b*) A particle can have only momentum but no position

(*c*) One can determine simultaneously the position and momentum of a particle

(*d*) One cannot determine simultaneously the position and momentum of a particle.

(*vi*) Which of the following statements is wrong ?

(*a*) Maxwell-Boltzmann statistics deals with distinguishable particles

(*b*) Fermi-Dirac statistics deals with indistinguishable particles

(*c*) Bose-Einstein statistics deals with indistinguishable particles

(*d*) Bosons are subjected to Pauli exclusion principle.

(*vii*) Miller indices of a plane which cut intercepts of 2,3 and 4 units along the three axes are

(*a*) (6, 4, 3) (*b*) (2, 3, 4) (*c*) 3, 2, 1) (*d*) (2, 3, 2).

(*viii*) Which of the following wavelengths falls in the X-ray range ?

(*a*) 1 nm (*b*) 100 nm (*c*) 0.001 nm (*d*) 1000 nm.

(*ix*) The population inversion in preparing laser beam can be achieved

(*a*) when one of the excited states is more populated than the ground state

(*b*) when one of the excited states is less populated than the ground state

(*c*) when the population of one excited state and the ground state are equal

(*d*) irrespective of any relationship between the population in the excited an the ground states.

(*x*) Which of the following statements is correct ?

(*a*) Optical fibres can bend light using the principle of Poiseuille's law

(*b*) Optical fibres cannot bend the light beam

(*c*) Optical fibres bend the light beam using the principles of total internal reflection

(*d*) Optical fibres use a liquid which helps in bending the light beam.

(*xi*) Nuclear binding energy of an isotope is $c^2 \Delta M$, where c is the velocity of light and ΔM is

(*a*) difference between the mass of the particles forming the nucleus and the mass of the nucleus itself.

(*b*) difference between the mass of the protons and the neutrons present

(*c*) difference between the mass of the isotope and its most abundant variety
(*d*) difference between the mass of the protons and the electrons.

(*xii*) Emissive power of a black body kept at an absolute temperature T is proportional to
(*a*) T^3 (*b*) T^4 (*c*) T^5 (*d*) T^{-1}

(*xiii*) When two parallel rays of X-rays, of wavelength λ, are incident at an angle θ on a crystal with lattice separation d, constructive interference would be observed when (n is an integer)
(*a*) $n\lambda = 2d \sin\theta$ (*b*) $n\lambda = d \sin\theta$
(*c*) $n\lambda = d \sin 2\theta$ (*d*) $n\lambda = 2d \sin 2\theta$.

(*xiv*) Which of the following statements is wrong ?
(*a*) Strong forces are of short range
(*b*) Electromagnetic forces can be both attractive and repulsive
(*c*) Weak forces are of longer range than electromagnetic forces
(*d*) Electromagnetic forces are weaker than the strong force.

Group – B
(Short Answer Type Questions)

Note. *(i) Each answer should not exceed 50 words.*
(ii) Answer any three questions. $3 \times 5 = 15$

2. Write down the Schrodinger equation for a particle confined in an one-dimensional box.
3. Consider a two-state system each of which can exist in states ε_1, ε_2 and ε_3. What are the possible states if the particles are (*a*) bosons (*b*) fermions?
4. What are the postulates of special theory of relativity?
5. What is the ratio of Compton shift for scattering at O° and 90°?
6. The distance between (100) planes in a body-centred cubic structure is 0.232 nm. What is the size of the unit cell ? What is the radius of the atom?
7. Why are slow neutrons used in nuclear reactor?
8. Describe briefly the principle of operation of a laser.

Group – C
(Long Answer Type Questions)

Note : *Answer any three of the following questions.* $3 \times 15 = 45$

9. Explain the terms packing fraction, mass defect and binding energy of a nucleus. Draw a curve showing the variation of binding energy per nucleon against the mass number. Use this curve to explain the release of energy in fusion of light nucleus and in fission of heavy nucleus.
10. State clearly, explaining all the terms, the Planck's law, Rayleigh-Jeans law and Wien's displacement law for radiation. Find out the two limits at which the Planck's formula reduces to the other two.
11. Describe briefly the Davisson-Germer experiment. What inferences may you draw from this experiment? Explain what you mean by "wave packet".
12. Describe briefly the Michelson-Morley experiment and the significance of the experimental results.
13. What is a crystal band ? How to differentiate between metal, insulator and semiconductor on the basis of energy band diagram ?

14. What is an optical resonator ? The r.i.s. of the core and cladding of an optical fibre are 1.523 and 1.472 respectively. Find numerical aperture and the angle of acceptance of the optical fibre.

WBUT Examination Questions

Engineering & Management Examinations, December – 2007

ENGINEERING PHYSICS

SEMESTER – I

Time : 3 Hours] [*Full Marks : 70*

Group – A
(Multiple Choice Type Questions)

1. Choose the correct alternatives for any *ten* of the following: $10 \times 1 = 10$

(*i*) Earth moves round the sun due to gravitational force which is central in nature. Sun-earth distance (R) changes periodically round the year. So which of the following in correct?

(*a*) Angular velocity of the earth round the sun is proportional to R

(*b*) Angular velocity is constant

(*c*) Angular velocity is proportional to $\frac{1}{R}$

(*d*) Angular velocity is proportional to $\frac{1}{R^2}$.

(*ii*) A body was at rest when it explodes and breaks up into three pieces of masses in the ratio 1 : 1 : 3. The two pieces of equal mass fly off perpendicular to each other with a speed of 30 m/s each. If the mass of the body be 1 kg then the velocity of the heavier piece is

(*a*) 40 m/s (*b*) 60 m/s (*c*) $10\sqrt{2}$ m/s (*d*) $6\sqrt{2}$ m/s.

(*iii*) Generalized co-ordinate can be of any dimension. However,

(*a*) the product of generalized co-ordinate and generalized momentum is always of dimension ml^2/t

(*b*) dimension of the work done depends on the dimension of the generalized co-ordinate

(*c*) the generalized momentum is always of the dimension ml/t

(*d*) the generalized force is always of the dimension ml/t^2.

(Given, mass = m, length = l, time = t)

(*iv*) Rigid body has constraints classified in which of the following group?

(*a*) Rheonomic and Holonomic (*b*) Rheonomic and Non-holonomic

(*c*) Scleronomic and Holonomic (*d*) Scleronomic and Non-holonomic.

(*v*) Two mutually perpendicular oscillations with same frequency, amplitude but phase difference δ will produce closed curve with non-zero area enclosed

(*a*) for all values of δ except $\delta = 0$

(*b*) only for $\delta = \frac{\pi}{2}$

(*c*) for all values of δ except $\delta = 0$ and $\delta = \pi$

(*d*) for all values of $\delta \geq \frac{\pi}{2}$.

(*vi*) Waves originating from a point source and travelling in an isotropic medium is described as

(*a*) $\phi = \phi_0 \exp i\,(kr - \omega t)$ (*b*) $\phi = \phi_0 \exp i\,(kr - \omega t)/r$

(*c*) $\phi = \phi_0 \exp i\,(kr - \omega t)/r^2$ (*d*) $\phi = \phi_0 \exp i\,(kr + \omega t)/r$

(*vii*) Example of weakly damped harmonic oscillator is

(*a*) Dead-bead galvanometer

(*b*) Tangent galvanometer

(*c*) Ballistic galvanometer

(*d*) Discharge of a charged capacitor through a resistance.

(*viii*) If the current density $\vec{j} = k\hat{r}$, where $\hat{r}$ is a unit vector along $x\hat{i} + y\hat{j}$, then the current through the surface $x^2 + y^2 = a^2$, bounded by $z = 0$ and $z = h$, is

(*a*) $\pi a^2 hk$ (*b*) zero (*c*) $2\pi\, ahk$ (*d*) $\frac{a^3 k}{\pi h}$.

(*ix*) Let (r, θ, ϕ) represent the spherical polar co-ordinate of a point in a region where the electrostatic potential V is given by $V = k\phi^2$. Then the charge density in that region

(*a*) is also a function of ϕ only

(*b*) is constant in that region

(*c*) is a function of all the co-ordinates (r, θ, ϕ)

(*d*) is a function of (r, θ) only.

(*x*) At a point just outside a current carrying conducting wire

(*a*) electric field is zero while the magnetic field is non-zero

(*b*) magnetic field is zero while the electric field is not

(*c*) both electric and magnetic fields are zero

(*d*) neither the electric field is zero nor the magnetic field is zero.

(*xi*) Electromagnetic wave is propagated through a region of vacuum, which does not contain any charge or current. If the electric vector is given by

$\vec{E} = E_0 \exp i(kx - wt)\ \hat{j}$ then the magnetic vector is

(*a*) in x direction (*b*) in y direction

(*c*) in z direction (d) rotating uniformly in the $x - y$ plane.

(*xii*) The electric flux through each of the faces of a cube of side 1 m if a charge q coulomb is placed at its centre is

(*a*) $\frac{q}{4\epsilon_0}$ (*b*) $4\epsilon_0\, q$ (*c*) $\frac{q}{6\epsilon_0}$ (*d*) $\frac{\epsilon_0}{6q}$

(*xiii*) A conducting loop carrying a current I is placed in a uniform magnetic field pointing in to the plane of the paper as shown. The loop will have a tendency to

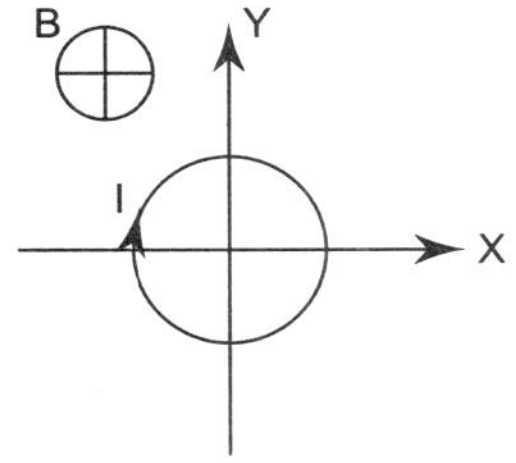

(*a*) contract (*b*) expand

(*c*) move towards positive X-axis (*d*) move towards negative X-axis.

(*xiv*) A copper wire is bent in the form of a sine wave of wavelength λ and peak to peak value *a* as shown in figure. A magnetic field of flux density *B* tesla acts perpendicular to the plane of the figure in the entire region. If the wire carries a steady current I ampere, the magnetic force on the wire is

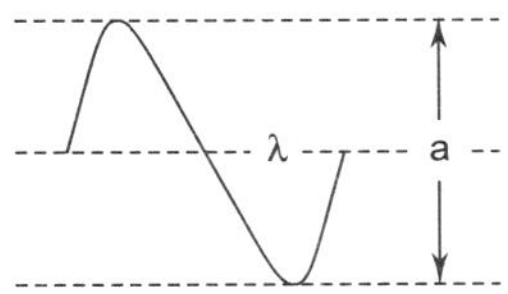

(*a*) $I\sqrt{(a^2 + \lambda^2)}B$ (*b*) I*a*B

(*c*) I (*a* + λ) B (*d*) IλB

(*xv*) Displacement current arises due to

(*a*) negative charges only (*b*) time varying electric field

(*c*) positive charges only (*d*) time varying magnetic field.

Group – B
(Short Answer Type Questions)

Answer any *three* of the following. $3 \times 5 = 15$

2. A sphere impinges directly on an equal sphere which is rest. Show that a fraction $\frac{1}{2}(1 - e^2)$ of the original K.E. is lost during the impact. 5

3. (*a*) What is the form of Lagrange's equation of motion for a non-conservative force ?

   (*b*) Starting from the definition of current and current density derive the equation of continuity in current electricity. 1 + (1 + 1 + 2)

4. (*a*) State Ampere's law in magnetostatics in integral form and from that deduce its differential form.

   (*b*) Write down the condition of steady state current. Show that Ampere's law implies that the current is in the steady state. (1 + 1½) + (1 + 1½)

5. A very long cylindrical object carries charge distribution proportional to the distance from the axis (r). If the cylinder is of radius a, then find the electric field both at $r > a$ and $r < a$ by the application of Gauss's law in electrostatics. 5

6. If a harmonic oscillator of mass m and natural frequency w_0 is driven by a force $F_0 \sin wt$ and the damping is proportional to 2μ times the velocity of the oscillator then the displacement is given by

$$x = \frac{F_0/m}{\sqrt{(w_0^2 - w^2) + 4\mu^2 w^2}} \sin(wt - \theta)$$

where $$\tan\theta = \frac{2\mu w}{w_0^2 - w^2}.$$

Using the above relation (you need not deduce the relation), show that at velocity resonance the velocity is in phase with the driving force. 5

Group – C
(Long Answer Type Questions)

Answer any three questions. $3 \times 15 = 45$

7. (a) If a proton is 1800 times heavier than an electron then show that the centre of man of hydrogen atom is located practically at proton. Find the reduced man of this system and write the Lagrangian of the reduced man. 1 + 1 + 2

(b) Show that if a co-ordinate is cyclic in Lagrangian it will be cyclic in Hamiltonian also. 3

(c) Two objects of mass 5 kg and 10 kg moving at an angle of 60° with respect to each other, symmetric to an axis on a frictionless surface with velocities 20 m/s and 10 m/s respectively collide and diverge out at angles symmetric to the above axis the velocity of the first body is 8 m/s after the collision, find the velocity of the second body after collision in the lab frame as well as centre of mass frame. Is the collision elastic ? 5+ 1

(d) Give example of system with non-holonomic and rheonomic (one each) constraints. 2

8. (a) Find out Hamiltonn's equations of motion for a system comprising of masses m_l and m_2 connected by a massless string of length L through a frictionless pulley such that $m_1 > m_2$. 5

(b) Find out the degree of freedom of a rigid body constrained to move along the equator and remain on the surface of the earth. 2

(c) If $y = f_1(x - vt) + f_2(x + vt)$ where f_1 and f_2 are the two functions then show that y satisfies wave equation. 3

(d) Define ampere from the consideration of force between two parallel currents. A test charge having charge 0.4C is moving with a velocity of $4\hat{i} - \hat{j} + 2\hat{k}$ m/s through an electric field of intensity $10\hat{i} + 10\hat{k}$ of and a magnetic field $2\hat{i} - 6\hat{j} - 6\hat{k}$. Determine the magnitude and direction of the Lorentz force acting on the test charge. 2 + 3

9. (a) Calculate the time period of the liquid column of length l in a U-tube, if it is depressed in one arm by x, d is the density of liquid and A is the cross-sectional area of each arm of U-tube. 4

(*b*) In damped harmonic motion, calculate the time in which the energy of the system falls to *l*/*e* times the initial value. 3

(c) Write down the differential equation of a series L-C-R circuit driven by a sinusoidal voltage. Identify the natural frequency of this circuit. Find out the condition that this circuit will show an oscillatory decay and find out the relaxation time. 1 + 1+2+1

(*d*) Find out the value of the driving frequency at which the voltage across the capacitor is maximum (you may use the expression provided in question no 6). 3

10. (*a*) Find if the work done in moving an object in the field

$$\vec{F} = (2xy + z^3)\hat{i} + x^2\hat{j} + 3xz^2\hat{k}$$

from point (1, – 2, 1) to (3, 1, 4) is independent of the path chosen. 3

(*b*) If the vector potential A = $(x^2 + y^2 - z^2)\ \hat{j}$ at position (x, y, z). Find the magnetic field at (1, 1, 1). 2

(*c*) Find a unit vector perpendicular to $x^2 + y^2 - z^2 = 100$ at the point (1, 2, 3). 2

(*d*) Find the magnetic induction $\vec{B}$, at a point on the axis of an infinitely long solenoid carrying current I, no. of turns per unit length being *n*. 5

(*e*) Derive Coulomb's Theorem using Gauss's law of electrostatics. 3

11. (*a*) Two concentric spheres of radii *a* and *b* are kept it *x* potentials V_a and V_b. If the intervening space is vacuum then write the appropriate differential equation that the electrostatic potential satisfies. Solve this equation to find out potential in any point between the spheres and also for a point outside the sphere. Calculate the total charge on the outer sphere. 2 + 3 + 2 + 3

(*b*) Find the magnetic field of a circular loop carrying current I on a point on the axis of the loop. 5

12. (*a*) Use Faraday's law of electromagnetic induction and the fact that magnetic induction $\vec{B}$ can be derived from a vector potential $\vec{A}$, show that the electric field can be expressed as

$$\vec{E} = -\bar{\nabla}\phi - \frac{\partial \vec{A}}{\partial t}$$, where ϕ is a scalar potential. 5

(*b*) If *r* is the magnitude of the position vector $\vec{r}$ then prove that

$$\nabla^2 (\log_e r) = \frac{1}{r^2}$$ 5

(*c*) If in a region electric field is always in the *x*-direction then prove that the potential is independent of *y* and *z* co-ordinates. If the field is constant there is no free charge in that region. 3 + 2

WBUT Examination Questions

Engineering & Management Examinations, December – 2007

ENGINEERING PHYSICS

SEMESTER – I

Time : 3 Hours] [*Full Marks : 70*

Group – A
(Multiple Choice Type Questions)

1. Choose the correct alternatives for any *ten* of the following: $10 \times 1 = 10$

(*i*) The centrifugal force, due to rotation of the earth about the axis passing through the poles

(*a*) does not influence the gravitational acceleration

(*b*) reduces gravitational acceleration and the effect is most prominent at the poles

(*c*) reduces gravitational acceleration and the effect is most prominent at the equator

(*d*) increases gravitational acceleration in the southern hemisphere.

(*ii*) When a thick lens is surrounded on both sides by the same medium then

(*a*) the nodal points are always co-incident with the corresponding principal points

(*b*) two nodal points coincide but the principal points do not

(*c*) two principal points coincide but the nodal points do not

(*d*) there are no nodal points.

(*iii*) Two photons are moving in the same direction in vacuum. The velocity of one photon with respect to the other is

(*a*) zero (*b*) C (*c*) $\frac{C}{\sqrt{2}}$ (*d*) $\sqrt{2}\,C$.

(*iv*) A ball *b* falling freely through a liquid terminal velocity of the ball

(*a*) will decrease if radius of the ball increases

(*b*) will increase if viscosity of the liquid is decreased

(*c*) will increase if the ball has some initial velocity while entering the liquid column

(*d*) will not depend on the radius of the ball at all.

(*v*) In a damped harmonic motion, velocity of the particle

(*a*) is constant with time (*b*) decays exponentially with time

(*c*) shows damping oscillation (*d*) shows undamped oscillation.

(*vi*) Which of the following phenomenon proves the transverse nature of light?

(*a*) Interference (*b*) Reflection (*c*) Refraction (*d*) Polarization.

(*vii*) Sound wave is

(*a*) an electromagnetic wave (*b*) a transverse wave

(*c*) an elastic wave (*d*) a radio wave.

(*viii*) In Fresnel's Bi-prism experiment, the co-herent sources are achieved by

(*a*) two different coherent sources (*b*) division of wave front

(*c*) division of amplitude (*d*) none of these methods.

(*ix*) E-rays and O-rays have

(*a*) equal velocities along optical axis

(*b*) equal velocities perpendicular to the optical axis

(*c*) equal refractive indices along optical axis

(*d*) refractive indices less than one.

(*x*) x is the displacement of a particle executing S.H.M., its acceleration is proportional to

(*a*) x^2 (*b*) x (*c*) $\frac{1}{x}$ (*d*) $\frac{1}{x^2}$

(*xi*) Nicol prism is based on the action of

(*a*) double refraction (*b*) diffraction

(*c*) scattering (*d*) absorbtion

(*xii*) The sources of light are said to be coherent of the waves produced by them have the same

(*a*) wavelength

(*b*) amplitude

(*c*) wavelength and constant phase difference

(*d*) amplitude and the same wavelength.

(*xiii*) In Fraunhoffer diffraction, the incident wavefront is

(*a*) planer (*b*) cylindrical (*c*) spherical (*d*) hyperbolic.

(*xiv*) The relation between Hamiltonian and Lagrangian is

(*a*) H = 2T + L (*b*) H = 2T − L

(*c*) $H = \sum p_i \dot{q}_i - L$ (d) $H = \sum q_i \dot{p}_i - L$.

(*xv*) Number of degrees of freedom for a particle moving on a plane is

(*a*) one (*b*) two (*c*) three (*d*) six.

(*xvi*) In a solenoidal field.

(*a*) $\overline{\nabla} . \overline{A} = 0$ (*b*) $\oint_C \overline{A} . d\overline{r} = 0$ (*c*) $\overline{\nabla} \times \overline{A} = 0$ (*d*) $\nabla^2 \vec{A} = 0$.

(*xvii*) Maxwell's electromagnetic wave equation in terms of electric field vector $\vec{E}$ free space is

(*a*) $\nabla^2 \vec{E} = -\mu_0 \epsilon_0 \frac{\partial^2 \vec{E}}{\partial t^2}$ (*b*) $\nabla^2 \vec{E} = \mu_0 \epsilon_0 \frac{\partial^2 \vec{E}}{\partial t^2}$

(*c*) $\nabla^2 \vec{E} = \mu_0 \epsilon_0 \frac{\partial \vec{E}}{\partial t}$ (*d*) $\nabla^2 \vec{E} = \frac{1}{\mu_0 \epsilon_0} \frac{\partial^2 \vec{E}}{\partial t^2}$

(*xviii*) In a charge free region, electrostatic potential depends only on x coordinat then its general form is (a, b, c, s are constants) (*a*) $V = ax + b$ (*b*) $V = ax^2 + bx + c$

(*c*) $V = a \sin (bx)$ (*d*) $V = ae^{ibx}$.

(*xix*) In a region where magnetic field is changing with time, electric field $\left(\vec{E}\right)$

(*a*) does not exist (b) is irrotational

(*c*) is not irrotational (*d*) is given by $-\bar{\nabla}\phi$.

where ϕ is a scalar field.

(*xx*) The force between two antiparallel current carrying conductor separated at small distance is

(*a*) attractive

(*b*) repulsive

(*c*) zero

(*d*) sinusoidally varying between attractive and repulsive phases.

Group – B
(Short Answer Type Questions)

Answer any *three* of the following. $3 \times 5 = 15$

2. A particle is moving with a velocity *u* in *x* direction while another particle is moving with a velocity *v* in *y* direction with respect to some frame of reference. Starting from the Lorentz transformation of space and time, calculate the relative velocity of the second particle with respect to the first.
3. Write down the matrices corresponding to translation and refraction in geometrical optics.
4. Describe how the population inversion is achieved in Ruby laser or in He-Ne laser.
5. State Bernoulli's theorem and derive Torricelli's theorem from that.
6. Derive the expression for the generalized kinetic energy.
7. Show that $\vec{\nabla} \times \left(\bar{\nabla} \times \bar{A}\right) = \vec{\nabla} \times \left(\vec{\nabla} \times \vec{A}\right) - \nabla^2 \vec{A}$.
8. Calculate the electric field at the centre of a circle of radius *a* whose circumference contains a total charge Q uniformly distributed.
9. Assuming a plane wave solution to the electromagnetic wave equation, show that the wave is transverse in nature.
10. (*a*) In Newton's Ring experiment, the diameter of the mth dark ring is 12 mm and that of $(m + 10)^{th}$ ring is 13.11 mm. If the radius of curvature of the lower surface of the lens is 10 m, calculate the wavelength of the light used. 3

    (*b*) Show graphically the intensity distribution in a single slit diffraction pattern. Label the location of the minima. 2
11. A vibrator of mass 1 gm is acted upon by a restoring force of 10^7 dynes/cm of displacement, a retarding force of 4×10^3 dynes /velocity and a driving force of 10^5 cost dynes. Find the value of the maximum possible amplitude and the quality factor.

Group – C
(Long Answer Type Questions)

Answer any *three* questions. $3 \times 15 = 45$

12. (*a*) If a particle is dropped from a height h from rest and it falls freely under gravity towards the surface of the earth, then calculate the direction and amount of deviation of it assuming the latitude to be λ. 7

(*b*) Lagrangian of a system is given by

$$L = \frac{1}{2}\dot{x}^2 - \frac{1}{2}w^2\dot{x}^2 - \alpha x^3 + \beta x\,\dot{x}^2$$

obtain the Hamiltonian and hence the cannonical equations. 8

13. (*a*) Construct the system matrix for refraction of paraxial rays through a thick lens. Find the matrix for a thin lens. Hence prove, for a thin convex lens, that

$$\frac{1}{f} = (n-1)\left[\frac{1}{r_1} - \frac{1}{r_2}\right]$$

where symbols have usual meanings. 10

(*b*) A biconvex thin lens of *r.i*, 1. 5 have focal length 50 cm in air. Show that when this lens is immersed in liquid of *r.i*, 1.36, the focal length is increased five times. 5

14. (*a*) State and explain Brewster's law. 4

(*b*) What is meant by double refraction ? Explain how this property may be utilised in polarizing light. 5

(*c*) Show that when unpolarized light is incident on an ideal polarizer, the intensity of the polarized light transmitted through the polarizer is half the incident intensity, no matter what the orientation of the polarizer is. 6

15. (*a*) What is log-decrement of a decaying oscillation? Derive the relation between log-decrement and damping constant. If log-decrement is λ and the time period of this oscillation is T, then calculate the time period in absence of damping. 2 + 3 + 3

(*b*) Write down the differential equation of a forced harmonic oscillator in presence of damping, solve this equation to find out the displacement as a function of time. 7

16. (*a*) What is generalized co-ordinates ? What is the advantage of Lagrangian formulation over Newtonian formulation in classical mechanics ? What is meant by a cyclic co-ordinate? Show that the momentum corresponding to a cyclic co-ordinate is a constant of motion. 8

(*b*) Write down the Lagrangian of a simple pendulum and hence obtain the equation of motion and find out the general solution. 7

17. (*a*) Prove that if $\bar{B} = \bar{\nabla} \times \vec{A}$, where $\vec{A}$ is a vector field then $\vec{B}$ is always solenoidal. 5

(*b*) Write down Maxwell's equation in differential form in free space. Show that electric field $\vec{E}$ satisfies wave equation. What is the velocity of this wave? 10

18. (*a*) What is the relation between electrostatic field and ϕ electrostatic potential ? Show that the lines of forces one always perpendicular to the equipotential surfaces. 1 + 3

(*b*) Find the potential and also the electric field due to a charge Q uniformly distributed within sphere of radius a at points inside and outside the sphere. 6

(*c*) State Faraday's law of electromagnetic induction and derive the differential form. 5

19. Write short notes on any *two* of the following : $2 \times 7\frac{1}{2}$
    (*a*) Fresnel's biprism
    (*b*) Spatial and temporal co-herence
    (*c*) Missing order spectra.

20. (*a*) Deduce the expression of intensity distribution in Young's double slit interference and show that the total energy is conserved. 8 + 3

    (*b*) In a diffraction grating width of each ruling is 0.0001 cm and the seperation between the rulings are 0.0002 cm of the wavelength of the light is 5.5×10^{-5} cm then what is the grating element and angle of first order spectrum? If the width of the grating is 5 cm, then how many orders would be observable? 1 + 2 + 1

WBUT Examination Questions

Engineering & Management Examinations, June – 2007

ENGINEERING PHYSICS

SEMESTER – 2

Time : 3 Hours] [*Full Marks : 70*

Group – A
(Multiple Choice Type Questions)

1. Choose the correct alternatives for any *ten* of the following $10 \times 1 = 10$

   (*i*) Time interval between two events measured in inertial frames S1, S2, S3 and S4 are 1*s*, 1.1*s*, 1.5*s* and 2*s* respectively. The events occurred in the same spatial location in one those frames of reference, which is
   (*a*) S1 (*b*) S2 (*c*) S3 (*d*) S4.

   (*ii*) How fast a particle must travels so that its mass becomes twice its rest mass?
   (*a*) 0.5 *c* (*b*) 2 *c* (*c*) $\frac{\sqrt{3}}{2}c$ (*d*) 0.25*c*.

   (*iii*) Davisson and Germer studied electron diffraction with Nickel crystal and found a first order peak at 65° with electron beam of 54 eV. If, instead a 216 eV beam were used then the peak would have been at
   (*a*) 27° (*b*) 54° (*c*) 130° (*d*) 260°.

   (*iv*) If visible light is used to study the Compton scattering then the Compton shift will be
   (*a*) negative (wave length of the scattered light will be lesser)
   (*b*) more positive than what is observed with *x*-ray
   (*c*) zero
   (*d*) positive but not detectable in the visible window.

   (*v*) What is the value of $\left[\hat{x}, \frac{\hat{\partial}}{\partial x}\right]$?
   (*a*) – 1 (*b*) 0 (*c*) – 2 (*d*) 1.

(*vi*) Which of the following wave function is the solution of Schrödinger Equation?

(*a*) A sec (*x*) (*b*) A tan (*x*) (*c*) A exp $(-x^2)$ (*d*) A exp (x^2).

(*vii*) Fermi-Dirac distribution approaches Maxwell-Boltzmann distribution at

(*a*) low temperature and high pressure

(*b*) low temperature and low particle mass

(*c*) high temperature and high particle mass

(*d*) high density and low energy range.

(*viii*) A nucleus is more stable if both atomic number (Z) and the mass number (A) are even. This indicates that

(*a*) inter-nuclear force is charge dependent

(*b*) inter-nuclear force is spin dependent

(*c*) inter-nuclear force saturates over a small distance

(*d*) nucleons are Fermions.

(*ix*) In the nuclear reaction ${}^{23}_{11}Na + {}^{4}_{2}He \rightarrow {}^{26}_{13}Al + \ldots$, the missing particle is

(*a*) electron (*b*) proton (*c*) neutron (*d*) positron.

(*x*) Assuming that the atoms in a crystal are spheres of equal size and touching each other, it can be shown that atomic radius of *bcc* is equal to

(*a*) $\frac{a}{2}$ (b) $\sqrt{3}\,\frac{a}{4}$ (c) $\sqrt{3}\,\frac{a}{2}$ (d) $a/2\sqrt{2}$.

(*xi*) Average energy of an electron in a metal at T = OK is

(*a*) $\in_F$ (*b*) $\frac{1}{2}\in_F$ (*c*) $\frac{3}{5}\in_F$ (*d*) $\frac{5}{3}\in_F$

(*xii*) Origin of continuous X-ray is due to the process of

(*a*) ionization (*b*) inner orbital transition

(*c*) bremsstrahlung (*d*) none of these.

(*xiii*) In an optical fibre the intensity of light

(*a*) attenuates linearly with the length of the fibre

(*b*) attenuates exponentially with the length of the fibre

(*c*) attenuates logarithmically with the length of the fibre

(*d*) does not attenuate with the length of the fibre.

(*xiv*) In a Ruby laser, population inversion is achieved by

(*a*) optical pumping (*b*) inelastic atom-atom collisions

(*c*) chemical reactions (*d*) applying strong electric field.

Group – B
(Short Answer Type Questions)

Answer any *three* of the following. $3 \times 5 = 15$

2. (*a*) When a particle moves with a velocity much less compared to the velocity of light in free space show that the relativistic expression of kinetic energy approaches the classical result. 3

(*b*) If two photons are approaching each other in free space then find the relative velocity of one photon with respect to the other. 2

3. (*a*) What are the features of photoelectric effect which cannot be explained from wave theory of light ? 3

(*b*) Show graphically how the energy density *vs* frequency plot of blackbody radiation is changed if the temperature is increased. 2

4. (*a*) If the wave function, $\psi(x)$, of a quantum mechanical particle is given by

$$\psi(x) = a \sin \frac{\pi x}{L}, \text{ for } 0 \le x \le L$$

$$= 0 \text{ otherwise}$$

then determine the value of a. 1

Also determine the value of x where the probability of finding the particle is maximum. 2

(*b*) Define, with an example, the Q-value of a nuclear reaction. 2

5. What are energy bands? What is the difference between valency band and conduction band? What is the relation between the energy bands and fermi level of a conductor? 2 + 2 + 1

6. (*a*) 3 distinguishable particles, each of which can be in one of the $\in$, $2\in$, $3\in$, $4\in$ energy states, have total energy $6\in$.

Find all possible distributions of particles in the energy states. 2

(*b*) Find the number of microstates in each case. 3

Group – C
(Long Answer Type Questions)

Answer any *three* questions. $3 \times 15 = 45$

7. (*a*) Two rods of proper length l_0 move length-wise towards each other parallel to their common axis with the same velocity v relative to laboratory frame. Show that the length of each rod in the reference frame fixed to the other rod is $l = l_o \frac{1-\beta^2}{1+\beta^2}$ where $\beta = \frac{v}{C}$. 4

(*b*) Starting from the Lorentz transformation show that if (x_1, y_1, z_1, t_1) and (x_2, y_2, z_2, t_2) represent two events then the quantity $(x_1 - x_2)^2 + (y_1 - y_2)^2 + (z_1 - z_2)^2 - c^2 (t_1 - t_2)^2$ is invariant in all inertial frames. 4

(*c*) Show that $\hat{x} = i\hbar \frac{\partial}{\partial p}$ and $\hat{p} = p$ is consistent with the commutation relation $[\hat{x}, \hat{p}] = i\hbar$. 3

(*d*) Starting from De-Broglie's hypothesis show that the group velocity associated with a particle is same as the particle velocity. 4

8. (*a*) Describe briefly Michelson-Morley's experiment to measure the velocity of light. What was the observation? 7

(*b*) A system has two energy eigenstates $\in_0$ and $3\in_0$. ψ_1 and ψ_2 are the corresponding normalized wave functions. At an instant the system is in a superposed state $\psi = c_1 \psi_1 + c_2 \psi_2$ and $c_1 = \dfrac{1}{\sqrt{2}}$.

(*i*) Find the value of c_2 if ψ is normalized. 2

(*ii*) What is the probability that an energy measurement would yield a value $3\in_0$? 2

(*iii*) Find out the expectation value of energy. 2

(*c*) Given that for a particle in a rigid box (spanning from $x = 0$ to a) the eigen functions are given by $\psi_n = \sqrt{\dfrac{2}{a}} \sin \dfrac{n\pi x}{a}$. What are the energy eigen values in these states? 2

9. (*a*) Assume that an electron is inside a nucleus of radius 10^{-15} m. Calculate from the uncertainty principle the minimum kinetic energy of the election.
[Given, $h = 6.63 \times 10^{-34}$ J-s, electron mass $= 9.1 \times 10^{-31}$ kg]. 4

(*b*) If a light of wavelength 6000Å falls on a metal surface & emits photoelectrons with velocity of 4×10^5 m/s, what is the threshold wavelength ?
[given $c = 3 \times 10^8$ m/sec] 3

(*c*) Explain the working principle of He-Ne laser with energy level diagram. 5

(*d*) Determine the numerical aperture of a step index fibre when the core *r.i.* $n_1 = 1.50$ and that of cladding $n_2 = 1.48$. Find the maximum angle for entrance of light if the fibre is placed at air. 3

10. (*a*) Calculate the total number of particles in a Fermionic gas in terms of the Fermi level at absolute zero temperature. 8

(*b*) What is meant by Miller indices? Explain with example. 2

(*c*) Calculate the interplanar spacing '*d*' of planes (1, 1, 1) in a simple cubic lattice of side a. Deduce the formula, that you use. 5

11. (*a*) Show how binding energy per nucleon ($\bar{B}$) varies with mass, number A. From that curve explain the possibility of fusion and fission energy sources. 2 + 2 + 2

(*b*) Can a ${}^{238}_{92}U$ nucleus decay into a ${}^{206}_{82}Pb$ nucleus?

Given $\bar{B}$ of ${}^{238}_{92}U$ is 7.57 MeV and $\bar{B}$ for ${}^{206}_{82}Pb$ is 7.83 MeV. 2

(*c*) A beam of X-rays of wavelength 0.842 Å is incident on a crystal at a glancing angle of 8°.35′, when first order Bragg's reflection occurs. Calculate the glancing angle for third order reflection. 4

(*d*) The energy gap in Ge is 0.72 eV. Find the maximum wavelength of e.m. radiation which can cause transition from the valence bond to the conduction band. 3

WBUT Examination Questions

Engineering & Management Examinations, June – 2008

ENGINEERING PHYSICS

SEMESTER – 2

Time : 3 Hours] [*Full Marks : 70*

Group – A
(Multiple Choice Type Questions)

1. Choose the correct alternatives for any *ten* of the following : 10 x 1 = 10

(*i*) Relativistically correct expression of kinetic energy of a particle with rest mass m_0 and velocity v is (velocity of light in vacuum is c)

(*a*) $\frac{1}{2} m_o v^2$ (*b*) $\frac{1}{2} m_o c^2$

(*c*) $\frac{1}{2} \frac{1}{\sqrt{1-\frac{v^2}{c^2}}} m_o c^2$ (*d*) $\left(\frac{1}{\sqrt{1-\frac{v^2}{c^2}}} - 1\right) m_o c^2$

(*ii*) With respect to the earth, one rocket is moving in x direction with velocity v and another rocket is moving in y direction with a velocity u. Then the x component of the velocity of the second rocket with respect to the first is given by

(*a*) v (*b*) $-v$ (c) $-v/\sqrt{1-\frac{v^2}{c^2}}$ (*d*) $\frac{u-v}{1-\frac{uv}{c^2}}$

(*iii*) In the limit of large wavelengths, Planck's law of radiation reduces to

(*a*) Wien's law (*b*) Kirchhoff's law

(*c*) Rayleigh-Jeans law (*d*) Stefan's law.

(*iv*) Operator corresponding to kinetic energy is

(*a*) $\frac{\hbar^2}{2m} \nabla^2$ (*b*) $-\frac{\hbar^2}{2m} \nabla^2$ (*c*) $\frac{2\hbar}{m} \nabla^2$ (*d*) $\frac{-2m}{\hbar^2} \nabla^2$

(*v*) Which of the pairs of variable cannot be simultaneously measured?

(*a*) z, p_z (*b*) p_x, p_y (*c*) x, p_y (*d*) p_x^2, p_x

(where symbols represent the usual variables)

(*vi*) The relation between group velocity and particle velocity is

(*a*) $V_g = \sqrt{v}$ (*b*) $V_g = 2v$ (*c*) $V_g = v/2$ (*d*) $V_g = v$

where V_g = group velocity and v = particle velocity.

(*vii*) The spin of a photon is

(*a*) $\frac{1}{2}\hbar$ (*b*) $\frac{3}{2}\hbar$ (*c*) $\hbar$ (*d*) 0.

(*viii*) The Fermi-Dirac distribution function is expressed as

(*a*) $F(E) = \dfrac{1}{e^{(E-E_f)/kT}+1}$ (*b*) $f(E) = \dfrac{1}{e^{(E-E_f)/kT}}+1$

(*c*) $f(E) = \dfrac{1}{e^{(E-E_f)/kT}}-1$ (*d*) $f(E) = \dfrac{1}{e^{(E-E_f)/kT}-1}$.

(*ix*) Particles with half integral spin obey which of the following statistics ?

(*a*) M – B (*b*) B – E (*c*) F – D (*d*) None of these.

(*x*) Three fermions can be distributed in 4 states in which of the following number of ways ?

(*a*) 1 (*b*) 3 (*c*) 6 (*d*) 12.

(*xi*) In case of simple cube, effective number of lattice point (or atoms) per unit cell is

(*a*) 2 (*b*) 3 (*c*) 1 (*d*) none of these.

(*xii*) The Miller indices of a plane parallel to $x - y$ plane is

(*a*) (100) (*b*) (010) (*c*) (110) (*d*) (001).

(*xiii*) X-rays are suitable for crystal diffraction studies because

(*a*) the wavelength of X-ray is of the same order of interatomic separation spacing of the crystal

(*b*) X-rays are not visible

(*c*) X-rays are electromagnetic wave

(*d*) none of these.

(*xiv*) In the nuclear reaction:

$${}_{11}Na^{23} + {}_{2}He^{4} \rightarrow {}_{13}Al^{26} + \ldots\ldots\ldots\ldots$$

the missing particle is

(*a*) proton (*b*) neutron (*c*) α-particle (*d*) electron.

(*xv*) Origin of nuclear force is due to which of the following interaction ?

(*a*) Strong (*b*) Electromagnetic

(*c*) Weak (*d*) Gravitational.

(*xvi*) Isobars are nuclei which have

(*a*) same number of protons but different atomic numbers

(*b*) same atomic number but different number of protons

(*c*) equal number of protons and neutrons

(*d*) equal mass number but different atomic numbers.

(*xvii*) If ϕ is the electrostatic potential and $\vec{E}$ is the electrostatic field, then

(*a*) $\phi = \vec{\nabla}\cdot\vec{E}$ (*b*) $\vec{E} = \vec{\nabla}\phi$ (*c*) $\vec{E} = -\vec{\nabla}\phi$ (*d*) $|\vec{E}| = \phi$.

(*xviii*) If $\vec{\nabla}\times\vec{B} = \mu_0\vec{j}$ ($\vec{B}$ = magnetic induction, $\vec{j}$ = current density, μ_0 = permeability of vacuum), then (*a*) $\vec{\nabla} \cdot \vec{j} = 0$ (*b*) $\vec{\nabla} \times \vec{j} = 0$ (*c*) $\vec{\nabla} \cdot \vec{j} = 1 - \frac{\partial p}{\partial t}$ (*d*) $\vec{\nabla} \cdot \vec{j} = 1 + \frac{\partial p}{\partial t}$.

(*xix*) If three vectors $\vec{A}, \vec{B}, \vec{C}$ are coplaner then

(*a*) $\left(\vec{A} + \vec{B}\right) \cdot \vec{C} = 0$ (*b*) $\left(\vec{A} - \vec{B}\right) \cdot \vec{C} = 0$

(*c*) $\left(\vec{A} \times \vec{B}\right) \cdot \vec{C} = 0$ (*d*) $\vec{A} \times \vec{B} \times \vec{C} = 0$.

(*xx*) For an irrotational field $\vec{v}$ (x, y, z)

(*a*) $\vec{\nabla} \times \vec{v} = 0$ (*b*) $\vec{\nabla} \cdot \vec{v} = 0$ (*c*) $\vec{\nabla} \times \vec{v} \neq 0$ (*d*) $\vec{\nabla} \cdot \vec{v} \neq 0$.

Group – B
(Short Answer Type Questions)

Answer any *three* of the following. $3 \times 5 = 15$

2. What is an irrotational field ? What is a conservative field ? Show that if a vector field is irrotational it is also conservative by using Stokes law. 1 + 1 + 3
3. Find out the capacitance of a capacitor consisting of two concentric spherical conductors of radii a and b. Dielectric constant of the substance between the plates is k and the outer sphere is grounded. 5
4. What is Seebeck effect ? What is neutral temperature and why is it called so ? State the law of intermediate metals. 1 + 2 + 2
5. Discuss briefly the failure of classical physics to explain the experimental observations of photoelectric effect. 5
6. (*a*) An electron is confined to a box of length 10^{-10} m. Calculate the minimum uncertainty in the measurement of its velocity.
   ($h = 6.626 \times 10^{-34}$ Js, electron mass $= 9.1 \times 10^{-31}$ kg)

   (*b*) If the wave function of a particle is given by $\psi = \sqrt{\frac{2}{L}} \sin \frac{\pi x}{L}$ then find the probability of finding the particle from $x = 0$ to $x = L$. 2 + 3.
7. Derive the formula of length contraction and time dilation from Lorentz transformation.
8. (*a*) State Bragg's law of X-ray diffraction. 5

   (*b*) The spacing between successive (100) planes in NaCl is 2.82 Å. X-rays incident on the surface of the crystal is found to give rise to first order Bragg reflection at glancing angle 8-8°. Calculate the wavelength of the X-rays. 2 + 3
9. Define Q-value of a nuclear reaction and threshold energy in endothermic reaction. Establish the relation between them. 2 + 3
10. (*a*) Prove that for a system at T > OK obeying Fermi-Dirac statistics, the probability of occupation at a level lying Δε above the Fermi level is the same as the probability that a level lying Δε below the Fermi level is unoccupied.

    (*b*) Draw the occupation-probability curve at a temperature T = OK according to F-D statistics. 3 + 2

Group – C
(Long Answer Type Questions)

Answer any *three* questions. 3 × 15 = 45

11. (*a*) Prove that

$$\vec{\nabla} \cdot \left(\vec{\nabla} \phi \right) = \nabla^2 \phi.$$ 4

(*b*) Use Gauss Divergence theorem to prove that

$$\vec{\nabla} \cdot \vec{j} + \frac{\partial \rho}{\partial t} = 0.$$

where $\vec{j}$ = current density, ρ = charge density. 4

(*c*) State Faraday's law of electromagnetic induction. Express the equation in differential form. 2 + 2

(*d*) Write three Maxwell's equations other than Faraday's law. 3

12. (*a*) State Heisenberg's uncertainty principle. 2

(*b*) Applying uncertainty principle, prove that a free electron cannot exist in an atomic nucleus. 4

(*c*) Write down the uncertainty relation for energy and time. 1

(*d*) Find the wavelength spread of a 1 mano-second pulse from a ruby laser with a wavelength of 630 nm. 3

(*e*) Calculate the de Broglie wavelength of a baseball of mass 1 kg, moving at a speed of 10 m/sec. Discuss the reason why its wave length cannot be observed. 3

(*f*) Derive Rayleigh-Jean's law from Planck's law of radiation. 2

13. (*a*) An incident X-ray photon of frequency v_0 is scattered by a free electron at *r* through an angle ϕ. Using relativistic expression of electron energy, show that, the change in wavelength of the photon is given by 4

$$\Delta\lambda = \frac{h}{m_0 c}(1 - \cos\phi)$$

where m_0 = the rest mass of the electron.

(*b*) Write time-dependent Schrödinger equation and show that the wave function $\psi(x, y, z, t)$ can be written as $\psi(x, y, z, t) = \psi(x, y, z) \frac{-iEt}{\hbar}$, write down the equation $\psi(x, y, z)$ satisfy. 2 + 3 + 3

(*c*) Write down the Schrödinger equation for a particle confined in a three dimensional box. [You do not have to solve the equation]. 3

14. (*a*) Derive Coulomb's law of electrostatics from Gauss' law. 3

(*b*) Calculate the magnetic field due to a infinitely long straight wire carrying current I using Ampere's circuital law. 4

(c) A particle with mass m and charge e is moving along y-axis with a velocity v_y in uniform magnetic field B_z along z-axis. Write down the x-y-z components of force on this particle. 3

(d) What is Peltier effect ? How is it different from Joule's heating effect ? 1 + 2

(e) Write the relation among B, H and M (symbols represent their usual variables in magnetism). 2

15. (a) Derive a relation between lattice constant and density of the material of a cubic crystal. 4

(b) If the lattice constants are a, b and c then find the Miller indices of the plane which intersects at $\left(a, \frac{b}{2}, 3c\right)$. 3

(c) Find the number of quantum state of photon in the energy range ε to ε + $d6$. 4

(d) Obtain Planck's law of radiation by making use of B-E statistics. 4

16. (a) State the conditions the particles must satisfy in order to be decribed by Maxwell-Boltzmann, Fermi-Dirac or Bose-Einstein statistics. 3

(b) Give examples of Fermions and Bosons, three each. 3

(c) What is Bravais lattice? Give three examples with figures. 1 + 3

(d) What do you mean by conduction bond and valence bond of a solid? How conductors and semiconductors can be distinguished by their Fermi levels ? Explain with the help of a diagram. 2 + 3

17. (a) Describe the experiment of Michelson-Morby. Also discuss the significance of the inferences drawn from this experiment. 5 + 2

(b) Starting from the formula of velocity dependence of mass

$$m(v) = \frac{m_0}{\sqrt{1 - v^2/c^2}}$$

establish the relation, $E^2 = c^2p^2 + m_0^2 c^4$

where, E is the total energy, other symbols represent usual variables. 4

(c) Show that the rest mass of photon is zero. Also show that if the frequency of a photon is ν then its momentum is given by $\frac{h\nu}{c}$. 2 + 2

18. (a) Why do you need population inversion in a laser ? Discuss with the help of a diagram, how the population inversion is achieved in a He-Ne laser ? 2 + 4

(b) What are the application of optical fibre ? 4

(c) What is mass defect ? The atomic man of ^{16}O is 15.994915 u, calculate the mass defect in u and binding energy per nucleon in MeV. 3

(d) Draw a rough sketch of the curve representing the variation of binding energy per nucleon with mass number. 2

WBUT Examination Questions

Engineering & Management Examinations, December – 2008

ENGINEERING PHYSICS

SEMESTER – I

Time : 3 Hours] [*Full Marks : 70*

Group – A
(Multiple Choice Type Questions)

1. Choose the correct alternatives for any *ten* of the following : $10 \times 1 = 10$

(*i*) When a body collides elastically with another of equal mass at rest

(*a*) the first particle comes to rest and the second particle moves with the velocity of the first particle

(*b*) two bodies move with equal velocities after impact

(*c*) two bodies move with equal accelerations

(*d*) the first particle comes to rest and the second particle moves with twice the velocity of the first particle.

(*ii*) A system consists of three point masses. If the mutual distance between the point masses is fixed with time then the degree of freedom of such system is

(*a*) zero (*b*) three (*c*) nine (*d*) six.

(*iii*) The Lagrangian of a system is given in terms of generalized coordinates r, θ. ϕ and associated velocities as

$$L = \frac{1}{2}\theta^2 + \frac{1}{2}r^2\,\phi^2 + \frac{1}{2}\alpha\,\theta^2$$

The cyclic co-ordinate/s is/are

(*a*) ϕ (*b*) θ and ϕ (*c*) r and θ (*d*) ϕ.

(*iv*) The Lagrangian of a harmonic oscillator is given by $L = \frac{1}{2}m\dot{x}^2 - \frac{1}{2}kx^2$. The Hamiltonian is given by

(*a*) $\frac{1}{2}m\dot{x}^2 + \frac{1}{2}kx^2$ (*b*) $\frac{p^2}{2m} - \frac{1}{2}kx^2$, where $p = \frac{\partial L}{\partial x}$

(*c*) $\frac{1}{2}m\dot{x}^2$ (*d*) $\frac{1}{2}kx^2$.

(*v*) Two mutually perpendicular S.H.Ms with equal time periods but different amplitudes are superposed. If the phase difference between these oscillations is 45°, then they form a

(*a*) circle (*b*) straight line (*c*) ellipse (*d*) parabola.

(*vi*) Motion of a system in critical damped condition is

(*a*) oscillatory (*b*) damped oscillatory

(*c*) harmonic (*d*) non-oscillatory.

(*vii*) A plane wave travels in a medium in the positive x-direction. The displacements of the particles are given by

$$y(x, t) = 10 \sin 2\pi (t - 0.2x)$$

where x and y are measured in metre and t in seconds. The wavelength of the wave is

(*a*) 0.2 m (*b*) 5 m (*c*) 2π m (*d*) $0.4\ \pi$m.

(*viii*) For arbitrary scalar and vector fields ϕ and $\vec{A}$ which of the following is always correct ?

(*a*) $\vec{\nabla} \times \left(\vec{\nabla} \times \vec{A}\right) = 0$ and $\vec{\nabla} \times \left(\vec{\nabla} \phi\right) = 0$ (*b*) $\vec{\nabla}\left(\vec{\nabla} \cdot \vec{A}\right) = 0$ and $\vec{\nabla} \cdot \left(\vec{\nabla} \phi\right) = 0$

(*c*) $\vec{\nabla} \cdot \left(\vec{\nabla} \times \vec{A}\right) = 0$ and $\vec{\nabla} \times \left(\vec{\nabla} \phi\right) = 0$ (*d*) $\vec{\nabla} \cdot \left(\vec{\nabla} \times \vec{A}\right) = 0$ and $\vec{\nabla} \cdot \left(\vec{\nabla} \phi\right) = 0$

(*ix*) When the magnitude of $\vec{A}$ is constant which one of the following is true ?

(*a*) $\dfrac{d\vec{A}}{dt} = 0$ (*b*) $\vec{A} \cdot \dfrac{d\vec{A}}{dt} = 0$ (*c*) $\vec{A} \times \dfrac{d\vec{A}}{dt} = 0$ (*d*) $\left|\dfrac{d\vec{A}}{dt}\right| = 0.$

(*x*) In an electromagnetic wave in free space the electric and magnetic fields are

(*a*) parallel to each other (*b*) perpendicular to each other

(*c*) inclined at acute angle (*d*) inclined at obtuse angle.

(*xi*) In the following figure, *AB* is an infinite line of charge distribution. *P* and *Q* are points as shown. The ratio of electric field at *P* and *Q* is

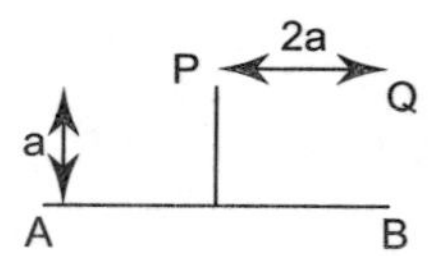

(*a*) 1 : 1 (*b*) 1 : 2 (*c*) 2 : 1 (*d*) 1 : 4.

(*xii*) The direction of magnetic induction due to a straight infinitely long current carrying wire is

(*a*) perpendicular to the wire (*b*) parallel to the wire

(*c*) at an inclination of 30° to the wire (*d*) none of these.

(*xiii*) Which of the following statements is not correct regarding electrostatic field vector $\vec{E}$?

(*a*) $\oint_C \vec{E} \cdot d\vec{r} = 0$, where C is a simple closed curve.

(*b*) $\int_a^b \vec{E} \cdot d\vec{r}$ is independent of the path for given end points a and b.

(*c*) $\vec{E} = \vec{\nabla} \times \vec{A}$ for some vector potential $\vec{A}$.

(*d*) $\vec{E} = \vec{\nabla} \varphi$, for some scalar field ψ.

(*xiv*) A current carrying straight wire cannot move, but a current carrying square loop adjacent to it can move under the influence of magnetic force.

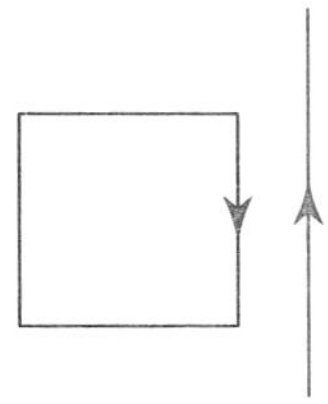

The square loop will

(*a*) remain stationary (*b*) move towards the wire
(*c*) move away from the wire (*d*) none of these.

(*xv*) The electromagnetic wave is called transverse because

(*a*) the electric field and magnetic field are perpendicular to each other
(*b*) the electric field is perpendicular to the direction of propagation
(*c*) the magnetic field is perpendicular to the direction of propagation
(*d*) both the electric field and magnetic field are perpendicular to the direction of propagation.

(*xvi*) Optic axis is a direction along which

(*a*) the extra-ordinary ray travels faster than the ordinary ray
(*b*) the ordinary ray travels faster than the extra-ordinary ray
(*c*) both the extra-ordinary ray and the ordinary ray travel with same velocity
(*d*) none of these.

(*xvii*) A plane polarized light of intensity I_0 is incident on a polarizer whose axis of polarization is 30° with respect to the direction of propagation of the incident light. The intensity of light coming out of the polarizer is

(*a*) I_0 (*b*) $\frac{\sqrt{3}\, I_0}{2}$ (*c*) $\frac{3\, I_0}{4}$ (*d*) $\frac{I_0}{4}$.

(*xviii*) The radii of the Newton's rings (R_n) depend on the wavelength of the incident light (λ) and the radius of curvature of the convex surface of the plano-convex lens (r) as follows :

(*a*) R_n increases with λ and decreases with r
(*b*) R_n increases with r and decreases with λ
(*c*) R_n increases both with λ and r
(*d*) R_n decreases both with λ and r.

(*xix*) The relation between the path difference (Δl) and phase difference ($\Delta\phi$) is

(*a*) $\Delta\phi = \frac{2\pi}{\lambda}\Delta l$ (*b*) $\Delta\phi = \frac{\pi}{\lambda}\Delta l$ (*c*) $\Delta l = \frac{2\pi}{\lambda}\Delta\phi$ (*d*) $\Delta l = \frac{\lambda}{\pi}\Delta\phi$.

(*xx*) In Fraunhoffer diffraction, the incident wavefront is

(*a*) plane (*b*) spherical (*c*) cylindrical (*d*) of arbitrary shape.

(*xxi*) The separation between diffraction lines (Δ) for the pattern formed by a transmission grating depends on the number of rulings per inch (n) and the wavelength of the incident light (λ) as

(*a*) Δ increases with n but decreases with λ

(*b*) Δ decreases with *n* but increases with λ

(*c*) Δ increases both with *n* and λ

(*d*) Δ decreases both with *n* and λ.

Group – B
(Short Answer Type Questions)

Answer any *three* of the following. $3 \times 5 = 15$

2. A particle is executing one dimensional Simple Harmonic Motion under the action of a potential $V = \frac{1}{2} Kx^2$. Write down the Lagrangian. Derive the Hamiltonian and Hamilton's equations. 1 + 2 + 2

3. A particle is subjected to a harmonic restoring force and a damping. Its equation of motion is given by

$$m \frac{d^2 x}{dt^2} = - sx - K \frac{dx}{dt}$$

Under the condition of critical damping, find the expression for displacement as an function of time. Can you justify why this condition (critical damping) corresponds to the fastest non-oscillatory decay ? 4 + 1

4. Given $\phi = \phi(r)$ and $\vec{\nabla} = \hat{i}\frac{\partial}{\partial x} + \hat{j}\frac{\partial}{\partial y} + \hat{k}\frac{\partial}{\partial z}$, find $\nabla^2 \phi = \vec{\nabla} \cdot \left(\vec{\nabla} \phi\right)$ without using directly the explicit form, of ∇^2 in the spherical polar coordinate. 5

5. If ϕ is the electrostatic potential in a region V, free of charge, then show that $\int_V \left|\vec{\nabla} \phi\right|^2 dV = \oint_\Sigma \phi \vec{\nabla} \phi \cdot d\vec{S}$, where Σ is the surface enclosing V. Hence show that if either the potential or the electric field vanishes identically on the surface Σ, the potential has to be constant inside V. 3 + 2

6. A metal bar slides frictionlessly on two parallel conducting rails at distance *l* apart. A resistor *R* is connected across the rails and a uniform magnetic field *B*, pointing into this plane fills the entire region. If the bar moves to the right a constant speed *v*, then what is the current in the resistor ?

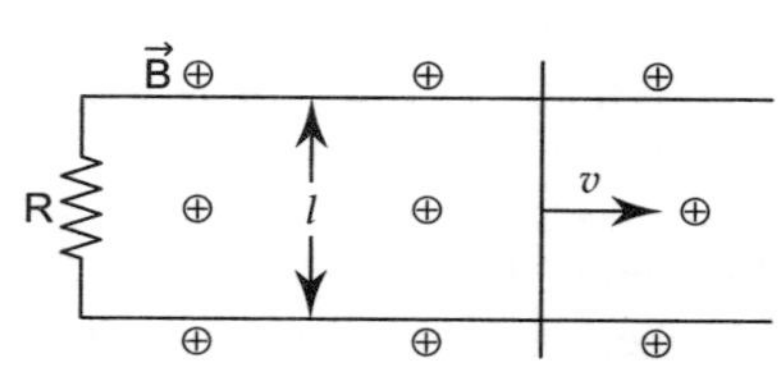

5

7. Sodium light of wavelength 589 nm and 589.6 nm are made incident normally on a grating having 500 lines per mm. Calculate the angular dispersion of these lines in the spectrum of first order. By a simple calculation can you Justify whether these lines are resolved or not ? 3 + 2

8. The width of each slit of a double slit is 0.15 mm and they are separated by a distance of 0.45 mm. If the double slit produces Fraunhoffer diffraction, find the angular position of the first minima and the missing orders. 2 + 3

Group – C
(Long Answer Type Questions)

Answer any *three* of the following questions. $3 \times 15 = 45$

9. (*a*) What are the advantages of Lagrangian formulation over Newtonian formulation ?

   (*b*) A particle of mass m moves in a force field of potential V. The kinetic energy and the potential energy of the system is

$$T = \frac{1}{2} m \left(\dot{r}^2 + r^2 \dot{\theta}^2 + r^2 \dot{\phi}^2 \sin^2 \theta\right)$$
$$V = V(r, \theta, \phi)$$

   Write the equation of motion using the Lagrangian method and the Hamiltonian method and compare them. 3 + (5 + 7)

10. Establish the equation of a forced harmonic oscillator. Solve this equation to obtain the amplitude of steady state oscillations. What is Q-factor of a forced oscillator ? The amplitude of an oscillator of frequency 200 cps falls to $1/10^{th}$ of its initial value after 2000 cycles. Calculate its relaxation time, quality factor, time in which its energy falls to $1/10^{th}$ of its initial value. 2 + 5 + 2 + (2 + 2 + 2)

11. (*a*) State Gauss law in electrostatics and hence obtain Poisson's equation.

    (*b*) Derive Coulomb's law from Gauss law.

    (*c*) The potential in a medium is given by $\phi(r) = \dfrac{qe^{-r/\lambda}}{4\pi\varepsilon_0 r}$.

    Obtain corresponding electric field.

    (*d*) Find the charge density that may produce the potential mentioned above. (2 + 3) + 4 + 3 + 3

12. (*a*) State Biot-Savart law in Magnetostatics. Find the magnetic field of an infinitely long straight wire at a transverse distance of d from the expression of $\vec{B}$ found in Biot-Savart law. 2 + 4

    (*b*) Express Biot-Savart law in terms of current density and hence show that the magnetic field is solenoidal. 2 + 4

    (*c*) Express Ampere's circuital law in terms of vector potential.

    (You may use $\vec{\nabla} . \vec{A} = 0$, where $\vec{A}$ is the vector potential.) 3

13. (*a*) State Maxwell's equations. From these equations derive the wave equations for an electromagnetic wave. What is the velocity of this wave ? 4 + 5+1

    (*b*) Assuming a plane, wave solution, establish the relation between the propagation vector $\left(\vec{K}\right)$, electric field $\left(\vec{E}\right)$ and. magnetic field $\left(\vec{B}\right)$. 5

14. (*a*) State the principle of conservation of linear momentum. Show that in the absence of external forces acting on the system of particles, the total linear momentum is conserved. 2 + 5

(*b*) Given $\vec{F} = f(r)\,\vec{r}$, show that $\vec{\nabla} \times \vec{F} = 0$ and hence show that $\oint_C \vec{F} \cdot d\vec{r} = 0$, where C is a simple closed curve. 5

(*c*) A series circuit is formed using an inductor (L), a capacitor (C) and a resistance (R). Obtain the differential equation describing the charge on the capacitor as function of time. Identify the natural frequency of the circuit. 3

15. (*a*) Find the intensity distribution for the Fraunhoffer diffraction pattern through a single slit.

(*b*) Hence find the positions of maxima and minima.

(*c*) A parallel beam of light of wavelength 500 nm is incident normally on a narrow single slit of width 0.2 mm. For a Fraunhofer diffraction pattern, find the angular positions of the first and the second maxima. 7 + (2 + 2) + (2 + 2)

16. (*a*) What is Brewster's law ?

(*b*) How is Brewster's angle connected to the refractive index of the medium ?

(*c*) Discuss the nature of polarization of the reflected and refracted rays.

(*d*) Describe a Nicol prism and discuss its use as a polarizer. 1 + 3 + 2 + 9

WBUT Examination Questions

Engineering & Management Examinations, June – 2008

ENGINEERING PHYSICS

SEMESTER – 2

Time : 3 Hours] [*Full Marks : 70*

Group – A
(Multiple Choice Type Questions)

1. Choose the correct alternatives for any *ten* of the following : $10 \times 1 = 10$

(*i*) Double slit interference pattern is the limiting case of Double slit diffraction pattern when the

(*a*) distance between the slits tends to zero

(*b*) distance of the source and the slits tends to infinity

(*c*) slit widths tend to zero

(*d*) distance between the slits tends to infinity.

(*ii*) A light ray is passing through a calcite crystal. If the plane of vibration of the light ray is perpendicular to the optical axis then the ray

(*a*) is an O-ray

(*b*) is an E-ray

(*c*) has both non-zero O and E components

(*d*) cannot propagate through the crystal.

(*iii*) A source of light emits lights of frequencies between v to $v + \Delta v$. Coherence time of the emergent light beam is τ_c. then

(*a*) $\tau_c \propto \Delta v$ (*b*) $\tau_c \propto 1/\Delta v$ (*c*) $\tau_c \propto v$ (*d*) $\tau_c \propto 1/v$.

(*iv*) If the quantum mechanical state of a particle is described by

$$\psi(x) = \psi_0 e^{ik(x-a)}, \text{ for } |x| \le L$$

and $\psi(x) = 0$, otherwise

then

(*a*) the particle has a definite position but uncertainty in momentum

(*b*) the particle has a definite momentum but uncertainty in position

(*c*) the particle has uncertainties both in position and momentum

(*d*) the particle has no uncertainties in either momentum or in position.

(*v*) Which of the facts about optical fibre is not true ?

(*a*) In multimode graded index fibre the distortion due to difference in path length is lesser than in multimode step index fibre

(*b*) The diameter of the core in single mode fibre is more than 10 times the maximum wavelength to be conducted through

(*c*) Distortion of the signal due to dispersion is less in single mode fibre than in multimode

(*d*) Single mode fibre is preferred over multimode to long distance communication.

(*vi*) The intensity of principal maxima in the spectrum of a grating with N number of lines is proportional to

(*a*) 1/N (*b*) N (*c*) N^2 (*d*) $1/N^2$.

(*vii*) A stone is dropped from the top of a building. What happens to the de-Broglie wavelength of the stone as it falls ?

(*a*) It increases (*b*) It decreases

(*c*) Remains constant (*d*) de-Broglie wavelength cannot be defined.

(*viii*) Number of oscillation modes for the electromagnetic standing waves of frequency v to $v + \Delta v$ or the cavity radiation is proportional to

(*a*) v (*b*) v^2 (*c*) v^4 (*d*) $\dfrac{hv}{e^{\frac{hv}{kT}-1}}$

(*ix*) If $\psi(x, t)$ is a normalized wave function we must have

(*a*) $\int \psi^* \psi \, dx = 1$ (*b*) $\int \psi^* \psi \, dx = 0$

(*c*) $\int \psi^* \psi \, dx = 1/2$ (*d*) $\int \psi \psi \, dx = 1$.

(*x*) Which one of the following functions is an eigen function of the operator $\dfrac{d^2}{dx^2}$?

(*a*) x (*b*) x^2 (*c*) $e-x^2$ (*d*) $\cos x$.

(*xi*) The number of possible arrangement of two fermions in 3 cells is

(*a*) 9 (*b*) 6 (*c*) 3 (*d*) 1.

(*xii*) If n_i is the number of identical and indistinguishable particles in the ith energy state with degeneracy g_i then classical statistics can be applied if

(*a*) $\frac{n_i}{g_i} = 1$ (*b*) $\frac{n_i}{g_i} << 1$ (*c*) $\frac{n_i}{g_i} >> 1$ (*d*) $g_i = 0$.

(*xiii*) The nearest neighbour distance in the case of *bcc* structure is

(*a*) $\frac{a\sqrt{3}}{2}$ (*b*) $\frac{a\sqrt{2}}{2}$ (*c*) $\frac{2a}{\sqrt{3}}$ (*d*) $\frac{2a}{\sqrt{2}}$

(*xiv*) The primitives of a crystal are 1.2Å, 1.8Å, 2Å, 2Å along the three axes where a plane with Miller indices (2 3 1) cut intercept 1.2Å along *x*-axis. What will be the lengths of intercept along *y* and *z* axis ?

(*a*) 1.5Å and 5Å (*b*) 1.9Å and 2.5Å (*c*) 1.2Å and 4Å (*d*) 1.5Å and 4Å.

(*xv*) In He-Ne Laser, Neon atoms obtain energy

(*a*) on collision with He atoms (*b*) from chemical reactions

(*c*) from electrical pumping (*d*) optical pumping.

Group – B
(Short Answer Type Questions)

Answer any *three* of the following. $3 \times 5 = 15$

2. Two polarizers are placed at crossed position. (angle between the polarizing planes are 90°), a third polarizer with angle θ with the first one is placed between them. An unpolarizerd light of intensity *I* is incident on the first one and passes through all three polarizers. Find the intensity of the light that comes out. 5

3. (*a*) Calculate the polarizing angle for light travelling from water of refractive index 1.33 to glass of refractive index 1.53. 2

(*b*) The spacing of the planes in a crystal is 1.2Å and the angle for the first order Bragg reflection is 30°. Determine the energy of the X-ray in eV. 3

4. (*a*) An electron and proton has same de Broglie wavelength. Prove that the energy of electron is greater. 2

(*b*) Why Compton effect cannot be observed with visible light ? 1

(*c*) Show that the temperature dependence in Stephan's law can be derived from Plancks radiation law. 2

5. (*a*) When electromagnetic radiations of wavelength λ is incident on a photosensitive surface, the kinetic energy of the photoelectrons emitted from the surface is 2eV. When the wavelength of the incident radiation is 2λ the kinetic energy of the photoelectrons emitted from the surface is 0.5 eV. Then find the threshold wavelength (maximum wavelength) for photoelectric emission from the surface in terms of λ. 3

(*b*) In a He-Ne laser transition from 3*s* to 2*p* level gives a laser emission of wavelength 6328 Å. If the energy of the 2*p* level is 15.2×10^{-19}J, how much pumping energy is required, if there is no loss in He-Neon laser ? 2

6. (*a*) A step index optical fibre has refractive indices of the core and the cladding as 1.44 and 1.40 respectively and is surrounded by air. Calculate the numerical aperture. 1

(*b*) A system has non-degenerate single-particle states with 0, 1, 2, 3 energy units. Three particles are to be distributed in these states such that the total energy of the system is 3 units. Find the number of micro states if the particles obey

(*i*) MB statistics (*ii*) BE statistics (*iii*) FD statistics. 2 + 1 + 1

Group – C
(Long Answer Type Questions)

Answer any *three* questions. 3 × 15 = 45

7. (*a*) Explain briefly the action of a dichroic Polaroid. 3

(*b*) In a Young's double slit interference experiment distance between the coherent sources are 1.15 mm. Calculate the fringe width that would be observed on a screen placed at a distance of 85 cm from the source. The wavelength of light used is 5893Å. 3

(*c*) Interference phenomenon does not violate the principle of conservation of energy. Justify it. 3

(*d*) A film of oil of r.i. 1.70 is placed between a plane glass plate and an equi-convex lens. The focal length of the lens is 1 metre. Determine the radius of the 10th dark ring when light of wavelength is 6000 Å. 4

(*e*) A convex lens of focal length 40 cm is employed to focus the Fraunhoffer diffraction pattern of a single slit of 0.3 mm width. Calculate the linear distance of the first order dark band from the central band. The wavelength of the light is 589 mm. 2

8. (*a*) Calculate the least width that a grating must have to resolve two components of the sodium-D line in the second order, the grating having 800 lines/cm. The wavelength for D_1 and D_2 lines of sodium are 5893 Å and 5896 Å respectively. 3

(*b*) Deduce the missing orders for a double slit Fraunhoffer diffraction pattern if the slit width are 0.16 mm and they are 0.8 mm apart. 3

(*c*) Two stars situated at a distance of 9.5×10^{12} km from a telescope of diameter 20 cm are sending light of wavelength 600 nm. Find the distance of the separation of the stars for which, they are just resolved. 2

(*d*) Find out the packing fraction for F.C.D. structure. Find out the expression for the separation between the planes with Miller indices h, k, l and lattice constant a for a simple cubic structure. 3 + 4

9. (*a*) Derive the expression for Compton shift when a photon interacts with electron. Show that it is impossible to transfer all the energy of the photon to the electron. 5 + 3

(*b*) Describe an experiments which verifies de Broglie's hypothesis about the wave nature of particles. 4

(*c*) Show that the product of group velocity and phase velocity of the de-Broglie wave are constant. 3

10. (*a*) A particle of mass m is confined within $x = 0$ to $x = L$.

(*i*) Write down the Schrödinger equation for such a system. 1

(*ii*) Solve the equation to find out the normalized eigenfunctions. 3

(*iii*) Show that the eigenfunctions corresponding to two different eigenvalues are orthogonal. 2

(*iv*) If p denotes momentum then find $< p >$ as well as $< p^2 >$ in the ground state. 4

(*b*) Using the principle of operator correspondence write down the operator $\hat{L}_x, \hat{L}_y, \hat{L}_z$ (components of angular momentum) in terms of position and momentum operators. Then show that $[L_x, L_y] = i\hbar L_z$. 2 + 3

11. (*a*) Write down the energy eigenfunction of a particle in a three dimensional box of length L. (You need not derive this). Hence find out the energy as well as degeneracy's in the ground and the first excited state. 1 + 1 + 1

(*b*) Sketch the Fermi distribution for T = 0 and T > 0 and explain. 3

(*c*) What is the occupation probability at $E = E_F$? 1

(*d*) Calculate how the degeneracy function *g*(E) depends on E for a Fermionic gas. 4

(*e*) Express the Fermi level in a metal in terms of free electron density. 4

12. (*a*) Give two examples of bosons. 2

(*b*) Define Einstein's *A*, *B* coefficients and deduce their mutual relation. 4

(*c*) Show that the ratio of spontaneous and stimulated emission is proportional to the cube of the frequency. 3

(*d*) Explain the construction and action of the optical resonator in a Ruby laser. 3

(*e*) If the energy levels of K, L, M state of Pt are 78 keV, 12 keV and 3 keV respectively, then calculate the wavelengths of K_α and K_β lines emitted from platinum. 3

($h = 6.626 \times 10^{-34}$ JS; $c = 3 \times 10^3$ m/s)

WBUT Examination Questions

Engineering & Management Examinations, June – 2009

ENGINEERING PHYSICS

SEMESTER – 2

Time : 3 Hours] [*Full Marks : 70*

Group – A
(Multiple Choice Type Questions)

1. Choose the correct alternatives for any *ten* of the following 10 × 1 = 10

(*i*) Newton's ring experiment is based on

(*a*) division of amplitude (*b*) division of wave-front (*c*) none of these.

(*ii*) de Broglie wavelength of a particle of mass m and kinetic energy E is

(*a*) $\lambda = \dfrac{h}{2mE}$ (*b*) $\dfrac{h}{\sqrt{2mE}}$ (*c*) $\dfrac{\sqrt{2mE}}{h}$

(*iii*) Mass of a photon of frequencies ν is given by

(*a*) $\dfrac{h\nu}{c}$ (*b*) $\dfrac{h\nu}{c^2}$ (*c*) $\dfrac{h\nu^2}{c}$

(*iv*) The eigenvalue of the eigenfunction e^{ix} for the operator $\frac{d^2}{dx^2}$ is

(*a*) 1 (*b*) 0 (*c*) – 1

(*v*) Miller indices of a plane which cut intercepts of 2, 3 and 4 units along the three axes are

(*a*) (2, 3, 2) (*b*) (2, 3, 4) (*c*) (6, 4, 3).

(*vi*) In a plane transmission grating, light

(*a*) diffracts to produce the resultant pattern

(*b*) diffracts and interfares to produce the resultant pattern

(*c*) interfares to produce the resultant pattern,

(*vii*) The atomic radius of a face centred cubic crystal of lattice constant a is

(*a*) $\frac{a}{2}$ (*b*) $\frac{\sqrt{3}\,a}{4}$ (*c*) $\frac{\sqrt{2}\,a}{4}$.

(*viii*) An X-ray tube is subjected to a potential difference of 50 kV with the corresponding current of 8 mA through it. The number of electrons striking per second on the target material is

(*a*) 5×10^{16} (*b*) 6×10^{11} (*c*) none of these.

(*ix*) In He-Ne laser, the laser light emits due to the transition from

(*a*) $3s \rightarrow 2p$ (*b*) $3s \rightarrow 3p$ (*c*) $2s \rightarrow 2p$.

(*x*) For an optical fibre, if n_0, n_1 and n_2 are the refractive index of air, core and cladding region respectively, then

(*a*) $n_0 > n_2 > n_1$ (*b*) $n_1 > n_2 > n_0$ (*c*) $n_2 > n_1 > n_0$.

(*xi*) If the speed of an electron increases, the specific charge

(*a*) increases (*b*) decreases (*c*) remains constant.

(*xii*) One milligram of matter converted into energy will give

(*a*) 90 joule (*b*) 9×10^{10} joule (*c*) None of these.

(*xiii*) Relative velocity of two particles moving with velocity (C) of light in opposite direction is

(*a*) C (*b*) 2C (*c*) O.

Group – B
(Short Answer Type Questions)

Answer any *three* of the following. $3 \times 5 = 15$

2. (*a*) What is the difference between temporal coherence and spatial coherence ? 2

(*b*) If the amplitudes of two coherent light waves are in the ratio 1 : 4, find the ratio of maximum and minimum intensity in the interference pattern. 3

3. (*a*) What is Compton effect ? Calculate the Compton wavelength for an electron. $1\frac{1}{2} + 1\frac{1}{2}$

(*b*) Why does the unmodified line appear in Compton scattering ? 2

4. (*a*) Deduce the formulae for interplaner spacing of a simple cubic crystal. 4

(*b*) Why X-ray diffraction is used for crystal structure analysis ? 1

5. Find the possible arrangements of two particles in three cells for

(*i*) Bose-Einstein Statistics & (*ii*) Fermi-Dirac Statistics. $2\frac{1}{2} + 2\frac{1}{2}$

6. (*a*) Describe briefly the working principle of laser action. 3
   (*b*) Mention at least three main advantages of optical fibre over wire or cable. 2
7. (*a*) What are the basic postulates of special theory of relativity ? 2
   (*b*) Show that for V << C, Lorentz transformation reduces to the Galilean transformation. 3
8. (*a*) Discuss the important characteristics of nuclear forces. 2
   (*b*) Why nuclear fission reaction must be done first before nuclear fusion ? 3

Group – C
(Long Answer Type Questions)

Answer any *three* questions. $3 \times 15 = 45$

9. (*a*) What is double refracting crystal ? 2
   (*b*) Discuss Nicol prism as polarizer and analyzer. 4
   (*c*) Determine the Brewster's angle for glass of refractive index 1.5 immersed in water of refractive index 1.33. 3
   (*d*) Prove that the intensity of secondary maxima formed for Fraunhoffer diffraction at a single slit are of decreasing order. 3
   (*e*) In a plane transmission grating the angle of diffraction for 2nd order maxima for wavelength 5×10^{-5} cm is 30°. Calculate the number of lines in one centimeter of the grating surface. 3
10. (*a*) State and explain de Broglie hypothesis. 2
    (*b*) Prove that the product of phase velocity and group velocity for a de Broglie wave is equal to the square of the velocity of light. 5
    (*c*) Compute the smallest possible uncertainty in the position of an electron moving with velocity 3×10^7 m/s. The rest mass of electron is 9.3×10^{-31} kg. 3
    (*d*) Derive the Wein's displacement law from Planck's radiation law. 5
11. (*a*) Write down the postulates of Fermi-Dirac statistics. 3
    (*b*) Plot electron distribution function governed by Fermi-Dirac statistics in metal at T = OK and T > OK. Explain their physical significance 2 + (3 + 3)
    (*c*) Why Compton effect cannot be observed with visible light but can be observed due to X-rays ? 4
12. (*a*) Discuss the operation of a Ruby laser with the help of energy level diagram. 5 + 1
    (*b*) Find the numerical aperture and angle of acceptance of a given optical fibre. [Given, the refractive indices of core and cladding are 1.562 and 1.497 respectively]. 3
    (*c*) In a He-Ne laser transition from 3*s* to 2*p* level gives a laser beam of wavelength 632.8 nm. If the 2*p* level has energy equal to 15.2×10^{-19} J, calculate the required pumping energy (assuming no loss of energy). 4
    (*d*) What is the role of optical resonator in laser production ? 2
13. (*a*) Write down Schrödinger equation for-one dimensional motion of a free particle in a one dimensional potential box. Find its eigenfunction and eigenenergy. 1 + 3 + 3
    (*b*) Prove that the first excited energy state of a free particle in a cubical box has three fold degeneracy. 4
    (*c*) Copper has FCC structure and the atomic radius is 0.1278 nm. Find its density and the interplaner spacing for (321) planes. The atomic weight of copper is 63.5. 4

14. (*a*) What are the differences between inertial and non-inertial frame of references ? 2
(*b*) Deduce an expression of time dilation on the basis of Lorentz transformation equation. 5
(*c*) Find the mass and speed of 2 MeV electron. 2 + 2
(*d*) If the total energy of a particle is thrice of its rest energy, find the velocity of the particle. 4

15. (*a*) What is nuclear binding energy ? 2
(*b*) Draw a binding energy curve. What informations do you set from such a curve ? 2 + 5
(*c*) Find the energy released for the following nuclear fussion reaction : 3
$4\,_1H^1 \rightarrow {}_2He^4 + 2 + {}_1e^0 + \text{energy}$
Given that
mass of $_1H^1$ atom = 1.007825 amu
mass of $_2He^4$ atom = 4.0026 amu
mass of + $_1e^0$ (positron) = 0.00055 amu.
(*d*) Differentiate conductor, insulator and semiconductor on the basis of energy band diagram. 3

WBUT Examination Questions

CS/B.Tech/SEM-I/PH-101/2009-10

2009

ENGINEERING PHYSICS

Time : 3 Hours] [*Full Marks : 70*

The figures in the margin indicate full marks.
Candidates are required to give their answers in their own words as far as practicable.

Group – A
(Multiple Choice Type Questions)

1. Choose the correct alternatives of the following : $10 \times 1 = 10$
(*i*) The centre of mass of a body is a point at which
(*a*) only translational motion occurs
(*b*) both translational and rotational motion occurs
(*c*) none of these.
(*ii*) If a system has *f* degrees of freedom the number of Hamilton's equations for the system is
(*a*) 2 (*b*) *f* (*c*) 2*f*.
(*iii*) The value of $\bar{\nabla} \times \bar{r}$ is equal to
(*a*) 1 (*b*) 0 (*c*) – 1.
(*iv*) The electric flux through the surface vector $\vec{S}$ equal to $6\hat{j}$ in a region of electric field $3\hat{i} + \hat{j}$ is

(*a*) 10 $Nm^2 C^{-1}$ (*b*) 6 N $m^2 C^{-1}$ (*c*) none of these.

(*v*) In free space Poisson's equation reduces to

(*a*) $\nabla^2 V = 0$ (*b*) $\nabla^2 V = \rho/\varepsilon_0$. (*c*) $\nabla^2 V = -\rho/\varepsilon_0$.

(*vi*) The divergence of a magnetic flux density $(\bar{B})$ is

(*a*) 0 (*b*) 1 (*c*) – 1.

(*vii*) The differential form of Faraday's laws of electromagnetic induction is

(*a*) $\bar{\nabla} \times \bar{E} = \partial\bar{B}/\partial t$ (*b*) $\bar{\nabla} . \bar{E} = 2\bar{B}$ (*c*) $\bar{\nabla} \times \bar{E} = -\partial\bar{B}/\partial t$.

(*viii*) When a spring with spring constant *K* is cut into three equal parts, the force constant of each of the part would be

(*a*) K/3 (*b*) 3K (*c*) K.

(*ix*) If *a* is the force constant of an oscillating body of mass m, the Q-factor is

(*a*) $Q = C\sqrt{am}$ (*b*) $Q = C\sqrt{m/a}$ (*c*) $Q = C\sqrt{a/m}$.

where C = relaxation time.

(*x*) Superposition of two S.H.M.s of equal time period and equal amplitude with phase difference $\phi = \pi/2$ forms

(*a*) circle (*b*) ellipse (*c*) parabola.

Group – B
(Short Answer Type Questions)

Answer any *three* of the following. $3 \times 5 = 15$

2. (*a*) Show that, $y = ae^{i(\omega t - kx)}$. Do the solution of wave equation.

(*b*) The potential energy of a particle with mass 10 g is given by, $V(x) = 32x^2 + 0.2$, where *x* is in metre and V is in joule. Write down the equation of motion and solve it. $2\frac{1}{2} + 2\frac{1}{2}$

3. (*a*) Prove that $\bar{E} = \cos(y - t)\,\hat{k}$ and $\bar{B} = \cos(y - t)\,\hat{i}$ constitute possible electromagnetic wave.

(*b*) Define displacement current. 4 + 1

4. (*a*) Calculate the magnetic field intensity just outside and inside of a hollow cylinder of radius 4 cm carrying 50 A current.

(*b*) Differentiate between electrostatic field and magnetic field. 3 + 2

5. Prove that $\bar{A} \times \bar{B} \times \bar{C} = \bar{B} \times (\bar{A} . \bar{C}) - \bar{C} \times (\bar{A} . \bar{B})$.

6. (*a*) What is cyclic co-ordinate ? Explain with an example. 5

(*b*) Derive the Lagrangian for a particle falling freely under the influence of gravity. (1 + 2) + 2

Group – C
(Long Answer Type Questions)

Answer any *three* of the following. $3 \times 15 = 45$

7. (*a*) Define Hamiltonian function and explain its general significance. 2 + 3

(*b*) If Lagrangian of a system is given as $L = \frac{1}{2}\dot{x}^2 + \dot{x} - \frac{x^2}{2}$, find the Hamiltonian and equation of motion. 2 + 3

(*c*) At what angle must a body be incident on a perfectly hard plane, so that the angle between the directions before and after may be a right angle ?

[Coefficient of restitution, $e = \frac{1}{3}$] 5

8. (*a*) Distinguish between scalar and vector fields with example. 3

(*b*) If the potential of a field is given by
$V(x, y, z) = (4x^2 + 2y^2 + z^2)^{1/2}$, find the field intensity at the point (1, 1, 1). 3

(*c*) Prove that curl grad $\phi = 0$. 3

(*d*) Show that the potential function $x^2 - y^2 + z$ satisfies Laplace's equation. 3

(*e*) Find the potential of a uniformly charged sphere of radius *R* having a constant charge density ρ at a distance *r* from the centre of the sphere where $r > R$. 3

9. (*a*) State and prove Ampere's circuital law. 2 + 3

(*b*) Define magnetic vector potential.

If the vector potential $\bar{A} = (10x^2 + y^2 - z^2)\,\hat{j}$ at any position, then find the magnetic field at the point (1, 1, 2). 2 + 3

(*c*) Write down the condition of steady state current. Show that Ampere's law implies that the current is in the steady state. 1 + 2

(*d*) Calculate the magnetic field intensity just outside a hollow cylinder of radius 4 cm. 2

10. (*a*) Write down Maxwell's equations for free space. 4

(*b*) Show that for free space the electromagnetic wave equation for $\bar{E}$ is $\nabla^2 \bar{E} = \mu_0 t_0 \frac{\partial^2 \bar{E}}{\partial t^2}$, where symbols have their usual significance. Prove that electromagnetic wave moves with the velocity of light in free space. 3 + 2

(*c*) Prove that $\bar{E} = \sin(y - t)\,\hat{k}$ and $\bar{B} = \sin(y - t)\,\hat{i}$ constitute a possible electromagnetic wave. 3

(*d*) Establish the integral form of Faraday's law of electromagnetic induction. 3

11. (*a*) Establish the differentiation equation of damped harmonic motion. 3

(*b*) Solve the equation for light damping and prove that the amplitude of vibration decreases exponentially with time. 4 + 2

(*c*) A cubical block of side L *cm* and density *d* is floating in a water of density ρ $(\rho > d)$. The block is slightly depressed and released. Show that it will execute simple harmonic motion and hence determine the frequency of oscillation. 4 + 2

WBUT Examination Questions

CS/B.Tech/SEM-2/PH-101/2010

2010

ENGINEERING PHYSICS

Time : 3 Hours] [*Full Marks : 70*

The figures in the margin indicate full marks.
Candidates are required to give their answers in their own words as far as practicable.

Group – A
(Multiple Choice Type Questions)

1. Choose the correct alternatives for any *ten* of the following : $10 \times 1 = 10$

(*i*) The average energy of a free electron in a metal at absolute zero is

(*a*) zero (*b*) more than $\frac{1}{2}E_f$

(*c*) less than or equal to $\frac{1}{2}E_f$ (*d*) equal to E_f when E_f is the Fermi level.

(*ii*) In an interference pattern produced by the identical coherent sources of monochromatic light, the intensity at the site of central maximum is I. If the intensity at the same spot when either of the two slits is closed is I_0 we must have

(*a*) $1 = I_0$ (*b*) $I = 2I_0$,

(*c*) $I = 4I_0$ (*d*) I and I_0 are not related.

(*iii*) Optic axis in a crystal is a

(*a*) particular direction perpendicular to which ordinary and extraordinary rays travel with the same velocity

(*b*) particular plane along which ordinary ray travels faster than extraordinary ray

(*c*) particular direction along which ordinary and extraordinary rays travel with the same velocity

(*d*) particular direction along which extraordinary rays cannot travel.

(*iv*) The resolving power of a grating having number of rulings exposed in the *n*th order is

(*a*) $\frac{n}{N}$ (*b*) nN (*c*) $\frac{N}{n}$ (*d*) $\frac{N}{n^2}$

(*v*) In a single slit Fraunhoffer diffraction, when the slit width is increased the width of the central maximum

(*a*) increases

(*b*) decreases

(*c*) remains unchanged

(*d*) may increase or decrease and that depends upon the initial value of the slit width.

(*vi*) A black body is something which

(*a*) does not emit any radiation in any temperature

(*b*) in room temperature absorbs more energy than it radiates

(c) it absorbs incident radiations of all frequencies completely
(d) absorption coefficient varies with the third power of frequency.

(vii) Work function of a metal may be obtained by irradiating the metal surface with light and from
(a) the x-intercept of the stopping potential vs frequency plot
(b) the slope of the stopping potential vs frequency plot
(c) the y-intercept of the intensity vs photocurrent plot
(d) the slope of the photocurrent vs frequency plot.

(viii) Group velocity of the wave packet associated with a particle of energy E and momentum P is

(a) $\frac{d^2 p}{d\text{E}^2}$ (b) $\frac{dp}{d\text{E}}$ (c) $\frac{d\text{E}}{dp}$ (d) $\frac{d\text{E}^2}{dp}$.

(ix) If $f(x)$ denotes the wave function of a particle in a one-dimensional box then the dimension $f(x)$ is of

(a) length (b) $\frac{1}{\sqrt{\text{length}}}$
(c) none, $f(x)$ is dimensionless (d) $\frac{1}{\text{length}}$.

(x) If x-component of the momentum operator, $\hat{p}_x$ is represented by $-i\hbar\frac{\partial}{\partial x}$ then the commultator of $\hat{x}$ and $\hat{p}_x^2, \left[\hat{x}, \hat{p}_x^2\right]$ is represented by (here x is the position operator)

(a) $-\hbar^2\frac{\partial^2}{\partial x^2}$ (b) $i\hbar$ (c) $2\hbar^2\frac{\partial}{\partial x}$ (d) zero.

(xi) Example of a boson is
(a) neutron (b) deuterium nucleus
(c) Proton (d) Positon.

(xii) The coordination number for BCC structure is
(a) 6 (b) 8 (c) 12 (d) 4.

(xiii) The activator action in Ruby laser is
(a) Aluminium (b) Oxygen (c) Gron (d) Chromium.

(xiv) The angle between the crystal planes (1, 0, 0) and (0, 0, 1) in SC structure is
(a) zero (b) 90° (c) 45° (d) 135°.

(xv) If the power output of an optical fibre is $\frac{1}{10}$th of the power launched into it then the attenuation loss is
(a) 10 dB (b) 1 dB (c) 0.1 dB (d) 100 dB.

Group – B
(Short Answer Type Questions)

Answer any *three* of the following. $3 \times 5 = 15$

2. (a) A film of oil of refractive index 1.70 is placed between plane glass plate and a plan convex lens. The radius curvature of the lens is 1 metre. Determine the radius of the 10th dark ring formed in the Newton ring set-up given that the wavelength of light is 6000 Å 3

(*b*) State and explain Malus's law of polarization. 2

3. (*a*) A light of wavelength 500 nm is incident normally on a slit of variable width and large length. A detector is placed at a large distance on the other side at an angle of 20° with the normal. For that width of the slit would the receiver show zero intensity ? 2

(*b*) How are continuous X-rays produced ? Find out the relation between the accelerating voltage of the *x*-ray tube and the minimum wavelength of X-ray produced. 1 + 2

4. (*a*) If m_0 is the rest mass of electron and C is the velocity of light and the kinetic energy of an electron is equal to its rest energy then find its de Broglie wavelength in terms of Planck's constant m_0 and C. 3

(*b*) If a system has two eigenstates ψ_1 and ψ_2 with eigenvalues E_1 and E_2, under what condition, will linear combination ($\psi = a\psi_1 + b\psi_2$) be also eigenstate? 2

5. (*a*) If the wave function (*x*) of quantum mechanical particle is given by

$$\psi(x) = a \sin(\pi x/L) \quad \text{for } 0 \le x \le L$$
$$= 0 \quad \text{otherwise,}$$

then determine the value of *a*.

Also determine the value of *x* where probability of finding the particle is maximum. 2 + 2

(*b*) Mention one application of graded index optical fibre. 1

6. (*a*) A system has non-degenerate single particle states with energy levels 0, E, 2E, 3E. Three particles are to be distributed in these states such that the total energy is 6E. Write down in each case all possible microstates assuming that the particles obey M-B statistics or- F-D statistics. 4

(*b*) What is acceptance angle of an optical fibre ? Explain with a diagram. 1

Group – C
(Long Answer Type Questions)

Answer any *three* of the following. 3 × 15 = 45

7. (*a*) Fraunhoffer double slit diffraction pattern is observed in the focal plane of a lens of focal length 50 cm. The distance between two maxima adjacent to principal maxima is 5 mm and the fourth order is missing. Find the width of each slit and the distance between their centres. Given, λ = 5000 Å. 5

(*b*) Write down the expression for intensity distribution of single slit diffraction pattern and explain the symbol used. Sketch how the intensity varies with angle of diffraction, θ. Find the values of the angles at which minima are located. Also find the condition that θ satisfies in order to be the position of a secondary maximum. 2 + 2 + 4

(*c*) In a Young's double slit interference experiment distance between the coherent sources is 1.15 mm. Calculate the fringe width that would be observed on a screen placed at a distance of 85 cm from the source. The wavelength of light used is 5893 Å. 2

8. (*a*) Calculate the minimum thickness of a calcite plate which would convert plane polarized light into circularly polarized light. μ_e = 1.486 and μ_0 = 1.658 and λ = 5890 Å. 2

(*b*) What is Brewster's angle ? What is the value of this angle for glass of refractive index 1.5 immersed in water of refractive index 1.33? 1 + 2

(*c*) What do you mean by temporal and spatial coherence? Define coherent length. 3

(*d*) Find the relation between lattice constant and density of the crystalline material for a cubic cell. 3

(*e*) In a certain crystal, the crystal plane of Miller indices (110) has a separation 1.12 Å. If these planes show X-ray diffraction with wavelength 1.54 Å, calculate up to which order, Bragg's reflection can be observed. 3

(*f*) What is a Bravais lattice? 1

9. (*a*) What is Compton shift? 2

(*b*) Find the change in wavelength of an X-ray photon where it is scattered through an angle of 135° C by a free electron. Use approximate values of the constants. 2

(*c*) Why in case of macroscopic bodies uncertainty principle is not relevant ? 1

(*d*) Describe photoelectric effect. How does classical theory fail to explain this effect? 2 + 2

(*e*) The ground state and the excited state normalized wave functions of an atom are ψ_0 and ψ_1 respectively, the corresponding energy being E_0 and E_1. If the probability of finding the atom in the ground state is 90% and that for the excited state is 10%, then find the average energy of the atom. Also determine the normalized wave function of the atom. 2 + 2

(*f*) Show that the momentum of a photon of frequency ν is given by $p = \frac{h\nu}{c}$, where h is Planck's constant and C is the velocity of light in vacuum. 2

10. (*a*) A particle, of mass m, moving in three dimensions is confined within a box $0 < x < a$, $0 < y < b$, $0 < z < c$. (The potential is zero inside and infinite outside.) Write down Schrödinger equation for the particle. By considering a stationary state wave function of the form $\chi(x, y, z) = f_1(x)\, f_2(y)\, f_3(z)$, show that the allowed energies are $E = \frac{\pi^2 \hbar^2}{2m}\left(\frac{n_1^2}{a^2} + \frac{n_2^2}{b^2} + \frac{n_3^2}{c^2}\right)$. What is the degeneracy of the first excited energy level when $a = b = c$? Explain. What is the ground state energy ? 1 + 4 + 2 + 1

(*b*) Show that the group velocity of a wave packet associated with a particle is same as the particle velocity. 3

(*c*) Write down Planck's law of frequency distribution of Black body radiation. Convert the expression into the wavelength distribution and find out the limiting form of this distribution at very small wavelengths. How does this result compare with the predictions from Rayleigh-Jeans formula ? 1 + 2 + 1

11. (*a*) Write down the time-dependent Schrödinger equation explaining the symbols used. Derive time-independent Schrödinger equation assuming that the potential is time independent. 1 + 2

(*b*) What are Fermions ? Give two examples. 1 + 1

(*c*) Write the expression of B-E distribution law and derive Planck's back body radiation law from B-E statistics. 1 + 6

(*d*) Sketch the Fermi distribution for T = 0 and T > 0 and explain. 3

12. (*a*) Define the following: 3

(*i*) Microstate. (*ii*) Macrostate.

(*iii*) Thermodynamic probability.

(*b*) What do you mean by 'spontaneous', and 'stimulated' emission of radiation ? Why is population inversion necessary in a laser ? 2 + 2

(*c*) Describe He-Ne laser source with relevant energy level diagram. 6

(*d*) Find the atomic packing factor for body centred cubic and face centred cubic crystals. 2

APPENDIX

SI BASED UNITS AND SI DERIVED UNITS WITH SPECIAL NAMES

Basic Units

Quantity	*Name of Units*	*Symbol*
length	meter	m
mass	kilogram	kg
time	second	s
electric current	ampere	A
thermodynamic temperature	kelvin	K
amount of substance	mole	mol
luminous intensity	candela	cd

Derived Units

Quantity	*Name of Units*	*Symbol*	*Equivalent Units*
frequency	hertz	Hz	s^{-1}
mass density (density)	kilogram per cubic meter	kg/m^3	
angular velocity	radian per second	rad/s	
acceleration	meter per second squared	m/s^2	
angular acceleration	radian per second squared	rad/s^2	
force	newton	N	$kg . m/s^2$
pressure (mechanical stress)	pascal	Pa	N/m^2
kinematic viscosity	square meter per second	m^2/s	
dynamic viscosity	newton-second per square meter	$N . s/m^2$	
work, energy, quantity of heat	joule	J	N . m
power	watt	W	J/s
quantity of electricity	coulomb	C	A . s
potential difference, electromotive force	volt	V	J/C, W/A
electric field strength	volt per meter	V/m	N/C
electric resistance	ohm	Ω	V/A
capacitance	farad	F	A . s/V
magnetic flux	weber	Wb	V . s
entropy	joule per kelvin	J/K	
specific heat capacity	joule per kilogram-kelvin	J/kg . K	
thermal conductivity	watt per meter-kelvin	W/m . K	
inductance	henry	H	V . s/A
magnetic flux density	tesla	T	Wb/m^2
magnetic field strength	ampere per meter	A/m	
magnetomotive force	ampere	A	
luminous flux	lumen	lm	cd . sr
luminance	candela per square meter	cd/m^2	
illuminance	lux	lx	lm/m^2
wave number	1 per meter	m^{-1}	
radiant intensity	watt per steradian	W/sr	

PERIODIC TABLE OF THE ELEMENTS

Period \ Group	1	2	3	4	5	6	7	8	9	10	11	12	13	14	15	16	17	18
1	1 **H** 1.008																	2 **He** 4.003
2	3 **Li** 6.941	4 **Bi** 9.012											5 **B** 10.811	6 **C** 12.011	7 **N** 14.007	8 **O** 15.999	9 **F** 18.998	10 **Ne** 20.180
3	11 **Na** 22.990	12 **Mg** 24.305											13 **Al** 26.982	14 **Si** 28.086	15 **P** 30.974	16 **S** 32.065	17 **Cl** 35.453	18 **Ar** 39.948
4	19 **K** 39.098	20 **Ca** 40.078	21 **Sc** 44.956	22 **Ti** 47.867	23 **V** 50.942	24 **Cr** 51.996	25 **Mn** 54.938	26 **Fe** 55.845	27 **Co** 58.933	28 **Ni** 58.693	29 **Cu** 63.546	30 **Zn** 65.409	31 **Ga** 69.723	32 **Ge** 72.64	33 **As** 74.922	34 **Se** 78.96	35 **Br** 79.904	36 **Kr** 83.80
5	37 **Rb** 85.468	38 **Sr** 87.62	39 **Y** 88.906	40 **Zr** 91.224	41 **Nb** 92.904	42 **Mo** 95.94	43 **Tc** (98)	44 **Ru** 101.07	45 **Rh** 102.905	46 **Pd** 106.42	47 **Ag** 107.868	48 **Cd** 112.411	49 **In** 114.818	50 **Sn** 118.710	51 **Sb** 121.760	52 **Te** 127.60	53 **I** 126.904	54 **Xe** 131.293
6	55 **Cs** 132.905	56 **Ba** 137.327	71 **Lu** 174.967	72 **Hf** 178.49	73 **Ta** 180.948	74 **W** 183.84	75 **Re** 186.207	76 **Os** 190.23	77 **Ir** 192.217	78 **Pt** 195.078	79 **Au** 196.967	80 **Hg** 200.59	81 **Ti** 204.383	82 **Pb** 207.2	83 **Bi** 208.980	84 **Po** (209)	85 **At** (210)	86 **Rn** (222)
7	87 **Fr** (223)	88 **Ra** (226)	103 **Li** (262)	104 **Rf** (261)	105 **Db** (262)	106 **Sg** (266)	107 **Bh** (264)	108 **Hs** (269)	109 **Mt** (268)	110 **Ds** (271)	111 **Uuu** (272)	112 **Uub** (285)	113 **Uut**	114 **Uuq** (289)	115 **Uup**	116 **Uuh**	117 **Uus**	118 **Uuo**

Lanthanoids	57 **La** 138.905	58 **Ce** 14.116	59 **Pr** 140.908	60 **Nd** 144.24	61 **Pm** (145)	62 **Sm** 150.36	63 **Eu** 151.964	64 **Gd** 157.25	65 **Tb** 158.925	66 **Dy** 162.500	67 **Ho** 164.930	68 **Er** 167.259	69 **Tm** 168.934	70 **Yb** 173.04
Actinoids	89 **Ac** (227)	90 **Th** (232)	91 **Pa** (231)	92 **U** (238)	93 **Np** (237)	94 **Pu** (244)	95 **Am** (243)	96 **Cm** (247)	97 **Bk** (247)	98 **Cf** (251)	99 **Es** (252)	100 **Fm** (257)	101 **Md** (258)	102 **No** (259)

SI Supplementary Units

Quantity	*Name of Units*	*Symbol*
plane angle	radian	rad
solid angle	steradian	sr

Fundamental Physical Constants

Gravitational Constant (G)	$G = 6.672 \times 10^{-11}$ N.m^2. kg^{-2}
Normal acceleration due to gravity (g)	$g = 9.80665$ m/s$^2 \approx 9.81$ m/s^2
Normal atmospheric pressure (P)	$P = 1.0129 \times 10^5$ N/m^2
Avagadro's number (N)	$N = 6.022045 \times 10^{23}$ molecules/mol
Volume of one mole of ideal gas at NTP (V)	$V = 22.41383 \times 10^{-3}$ m^3/mole
Universal gas constant	$R = 8.3144$ J/mole/K
Boltzmann's constant	$k = R/N = 1.380662 \times 10^{-23}$ J/K
Velocity of light in vacuum	$C = 2.9979258 \times 10^8$ m/s $\approx 3 \times 10^8$ m/s
Magnetic Permeability	$\mu_0 = 4\pi \times 10^{-7}$ H/m – $12.5663706 \times 10^{-7}$ Wb/A.m.
Electric permittivity of evacuated free space	$\varepsilon_0 = 8.85418782 \times 10^{-12}$ F/m
Rest mass of electron	$m_e = 9.109534 \times 10^{-31}$ kg
Rest mass of proton	$m_p = 1.6726485 \times 10^{-27}$ kg
Rest mass of neutron	$m_n = 1.6749543 \times 10^{-27}$ kg
One Atomic mass unit	1 a.m.u. $= 1.6605655 \times 10^{-27}$ kg
Elementary charge	$e = 1.6021892 \times 10^{-19}$ C
Electron charge mass ratio	$e/m_e = 1.7588047 \times 10^{-11}$ C/kg
Faraday constant	$F = 9.648456 \times 10^4$ C/mole
Plank's constant	$h = 6.622176 \times 10^{-34}$ J.s
Solar constant	$= 1370$ W/m^2
Stefan-Boltzman constant	$\sigma = 5.67032 \times 10^{-8}$ W/(m^2.k^4)
Wien constant	$= 0.00289782$ m.k.
Rydberg constant	$R_\alpha = 10{,}973{,}731$ m^{-1}

Density of Substances (SOLIDS)

Substance	*Density in kg/m³*	*Substance*	*Density in kg/m³*
Aluminium	2.7×10^3		
Aluminium chloride	1.5×10^3	Iron, steel	7.8×10^3
Brass	8.5×10^3	Lead	1.14×10^4
Brick	1.8×10^3	Magnanin	8.5×10^3
Coal	1.4×10^3	Mica	2.8×10^3
Constantan	8.9×10^3	Nicrome	8.3×10^3
Copper	8.9×10^3	Nickel	8.9×10^3
Copper sulphate	2.2×10^3		
Diamond	3.5×10^3	Platinum	9.28×10^3
Ebonite	1.2×10^3	Porcelain	2.3×10^3
Germanium	5.32×10^3	Salt common	2.1×10^3
Glass	2.5×10^3	Silver	1.05×10^4
Gold	1.93×10^4	Tin	7.3×10^3
Graphite	2.1×10^3	Tungsten	1.93×10^3
Ice (0°C)	0.9×10^3	Uranium	1.87×10^4
Iridium	2.24×10^4	Zinc	7.1×10^3

Density of Substances (LIQUIDS at 20°C)

Substance	*Density in kg/m^3*	*Substance*	*Density in kg/m^3*
Alcohol, ethyl	7.9×10^2		
Benzene	9×10^2		
Glycerin	1.20×10^3	Petrol	7.0×10^2
Kerosene	8.0×10^2	Turpentine	8.7×10^2
Mercury at 0°C	1.36×10^4	Water at 227 K	1.0×10^3
Nitrobenzene	1.2×10^3	Water at 373 K	0.958×10^3

Density of Substances (GASES at N.T.P)

Substance	*Density in kg/m^3*	*Substance*	*Density in kg/m^3*
1. Hydrogen	0.09	8. Nitrogen	1.25
2. Helium	0.18	9. Air	1.29
3. Methane	0.72	10. Oxygen	1.43
4. Krypton	3.74	11. Argon	1.78
5. Ammonia	0.77	12. Carbon dioxide	1.98
6. Neon	0.90	13. Chlorine	3.21
7. Acetylene	1.17	14. Xenon	5.85

Elastic Constants

Substance	*Young's modulus 10^{10} N/m^2*	*Rigidity – n 10^{10} N/m^2*	*Bulk modulus k 10^{10} N/m^2*	*Poisson's Ratio*
Aluminium	7.2 — 7.5	2.5 — 3.4	7.46	0.84
Brass	9.7 — 10.1	3.5	10.6	0.34 — 0.40
Copper	10.5 — 12.5	3.5 — 4.5	14.3	0.26
Glass	5.1 — 7.1	3.1	3.75	0.13 — 0.32
Iron (cast)	10 — 13	3.5 — 5.3	9.6	0.23 — 0.31
Iron (wrought)	19 — 20	7.7 — 8.3	14.6	0.27
Lead	1.6	5.6	5	—
Rubber	.00015 — .0005	.00005	—	0.46 — 0.40
Silver	7.1	2.5	10.9	—
Steel	20.9	8.12	16.4	0.29

Coefficient of Linear Expansion of Solids

Substance	*α in K^{-1}*
Aluminium	2.3×10^{-5}
Brass	1.9×10^{-5}
Cast iron	1.8×10^{-5}
Copper	1.7×10^{-5}
Ebonite	7.0×10^{-5}
Glass	9.0×10^{-6}
Gold	1.4×10^{-5}
Invar	6×10^{-7}
Iron Steel	1.2×10^{-5}
Lead	2.9×10^{-5}
Nickel	1.28×10^{-5}
Platinum	9.0×10^{-5}
Tin	2.1×10^{-5}
Tungsten	4.0×10^{-6}
Zinc	2.9×10^{-5}

Coefficient of Volume Expansion of Liquids

Substance	*β in K^{-1}*
Aceton	1.2×10^{-3}
Alcohol-ethyal	1.1×10^{-3}
Ethyl ether	1.6×10^{-3}
Glycerin	5.0×10^{-4}
Kerosene	1.0×10^{-3}
Mercury	1.8×10^{-4}
Petrol	1.0×10^{-3}
Petroleum	1.0×10^{-3}
Suphuric acid	5.7×10^{-4}
Transformer oil	6.0×10^{-4}
Water (10–20°C)	1.5×10^{-4}

Surface Tension (at 293 K)

Substance	*T in N/m*
Acetone	0.024
Alcohol ethyl	0.022
Castor oil	0.033
Ethyal ether	0.017
Glycerin	0.059
Kerosene	0.024
Mercury	0.470
Petrol	0.029
Turpentine	0.027
Water	0.072

Specific Heats

Substance	*C in (J/Kg – K)*
SOLIDS	
Aluminium	880
Brass	380
Brick	750
Cement	800
Concrete	880
Copper	380
Glass	840
Gold	125
Ice	2090
Iron (cast)	550

Iron steel	460
Lead	120
Naphthalene	1300
Paraffin	3200
Platinum	125
Sand	970
Silver	250
Sulphur	712
Tin	250
Zinc	400
Wood	2700
LIQUIDS	
Alcohol, ethyl	2430
Glycerin	2430
Iron	830
Kerosene	2140
Mercury	125
Oil machine	2100
Oil transformer	2093
Sulphate	2330
Water	4187
GASES (at Constant Pressure)	
Air	1000
Ammonia	2100
Carbon dioxide	830
Helium	5200
Hydrogen	14300
Nitrogen	1000
Oxygen	920
Water vapour	220

Dielectric Constants

Substance	*Relative permittivity -no units*
Air (at 1 atm)	1.0006
Air (at 100 atm)	1.055
Amber	2.8
Aniline	84.0
Ebonite	2.7
Glycerin	39
Hydrogen	1.0003
Ice (at–18°C)	3.2
Kerosene	2.0
Paraffin	2.2
Paraffined paper	2.0
Petrol	2.3
Porcelain	4.7
Vacuum	1
Wax	5.8
Water	81.0

Specific Resistance

Substance	*S, in Ω-m*
Aluminium	2.7×10^{-8}
Brass	6.3×10^{-8}
Constantan	4.7×10^{-7}
Copper	1.68×10^{-8}
Gold	2.2×10^{-8}
Iron	9.9×10^{-8}
Lead	2.07×10^{-7}
Magnanin	3.9×10^{-7}
Mercury	9.54×10^{-7}
Nichrome	1.05×10^{-6}
Nickel	7.3×10^{-8}
Platinum	1.05×10^{-7}
Silver	1.58×10^{-8}
Tin	1.13×10^{-7}
Tungsten	5.3×10^{-8}
Zinc	5.95×10^{-8}

Temperature Coefficients of Resistance (α)

Substance	*a in K^{-1}*
Constantan	0.000005
Magnanin	0.000008
Nichrome	0.0002
Nickeline	0.0001
Rheotan	0.0004
Tungsten	0.0050

Electrochemical Equivalent

Substance	*Z in kg/C*
Aluminium	9.32×10^{-8}
Calcium	2.077×10^{-7}
Chlorine	3.67×10^{-7}
Copper	3.294×10^{-7}
Gold	6.81×10^{-7}
Hydrogen	1.044×10^{-8}
Lead	1.074×10^{-6}
Mercury	2.072×10^{-6}
Magnesium	1.26×10^{-7}
Nickel	3.04×10^{-7}
Oxygen	8.29×10^{-8}
Potassium	4.052×10^{-7}
Silver	1.118×10^{-6}
Sodium	2.383×10^{-7}
Zinc	3.388×10^{-7}

Refractive Indices

Substance	μ
Alcohol ethyl	1.36
Alcohol methyl	1.33
Acetone	1.36
Air	1.0003
Aniline	1.59
Carbon tetrachloride	1.46
Carbon bisulphide	1.63
Diamond	2.42
Glass (crown glass)	1.5
Glycerin	1.47
Ice	1.31
Petrol	1.5
Quartz	1.54
Salt (Rock)	1.54
Sugar	1.56
Turpentine	1.51
Water	1.33

CONVERSION FACTORS

Length

1 km = 0.6215 m
1 mi = 1.609 km
1 m = 1.0936 yd = 3.281 ft = 39.37 in
1 in = 2.54 cm
1 ft = 12 in = 30.48 cm
1 yd = 3ft = 91.44 cm
1 lightyear = 1 ly = 9.461×10^{15}m
1 Å = 0.1 nm

Area

1 $m^2 = 10^4\ cm^2$
1 $km^2 = 0.3861\ mi^2$ = 247.1 acres
1 $in^2 = 6.4516\ cm^2$
1$ft^2 = 9.29 \times 10^{-2}\ m^2$
1 $m^2 = 10.76\ ft^2$
1 acre = 43.560 ft^2
1 mi^2 = 460 acres = 2.590 km^2

Volume

1$m^3 = 10^6\ cm^3$
1 L = 1000 $cm^3 = 10^{-3}\ m^3$
1 gal = 3.786 L
1 gal = 4 qt = 8 pt = 128 oz = 231 in^3
1 $in^3 = 16.39\ cm^3$
1 $ft^3 = 1728\ in^3$ = 28.32 L = $2.832 \times 10^4\ cm^3$

Speed

1 km h^{-1} = 0.2778 m s^{-1} = 0.6215 mi h^{-1}
1mi h^{-1} = 0.4470 m s^{-1} = 1.609 km h^{-1}
1mi h^{-1} = 1.467 ft s^{-1}

Angle and Angular Speed

π rad = 180°
1 rad = 57.30°
1° = 1.745×10^{-2} rad
rev min^{-1} = 0 1047 rad s^{-1}
1 rad s^{-1} = 9.549 rev min^{-1}

Mass

1 kg = 1000 g
1 tonne = 1000 kg = 1 Mg
1 u = 1.6606×10^{-27} kg
1 kg = 6.022×10^{26} u
1 slug = 14.59 kg
1 kg = 6.852×10^{-2} slug
1 u = 931.50 MeV/c^2

Density

1 g cm^{-3} = 1000 kg m^{-3} = 1 kg L^{-1}

Force

1 N = 0.2248 lbf = 10^5 dyn
1 lbf = 4.4482 N
1 kgf = 2.2046 lbf

Time

1 h = 60 min = 3.6 ks
1 d = 24 h = 1440 min = 86.4 ks
1y = 365.24 d = 31.56 Ms

Pressure

1 Pa = 1 N m^{-2}
1 bar = 100 kPa
1 atm = 101.325 kPa = 1.01325 bar

Magnetic Field
1 G = 10^{-4} T
1 T = 1 Wb m^{-2} = 10^4 G

Energy
1 kW h = 3.6 MJ
1 cal = 4.186 J
1 ft 1bf = 1.356 J = 1.286×10^{-3} Btu
1 L atm = 101.325 J
1 L atm = 24.217 cal
1 Btu = 778 ft 1b = 252 cal = 1054.35 J
1 eV = 1.602×10^{-19} J
1 u c^2 = 931.50 MeV
1 erg = 10^{-7} J

1atm = 14.7 1bf/in^2 = 760 mm Hg
= 29.9 in Hg = 33.8 ft H_2O
1 1bf in^{-2} = 6.895 kPa
1 torr = 1 mm Hg = 133.32 Pa

Power
1 horsepower (hp) = 550 ft 1bf/s
= 745.7 W
1 Btu min^{-1} = 17.58 W
1 W = 1.341×10^{-3} hp
= 0.7376 ft 1bf/s

Thermal Conductivity
1 W m^{-1} K^{-1} = 6.938 Btu in/hft^2 °F
1 Btu in/hft^2 °F = 0.1441 W/m K

MATHEMATICAL FORMULAE

Quadratic Formula
If $ax^2 + bx + c = 0$.

then $$x = \frac{-b \pm \sqrt{b^2 - 4ac}}{2a}$$

Trigonometric Functions of Angle θ

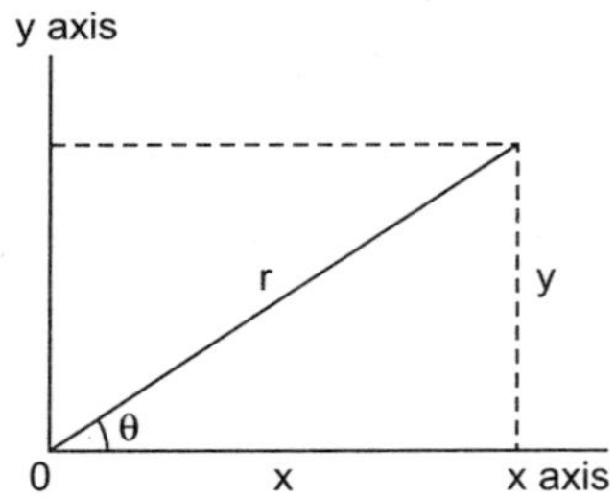

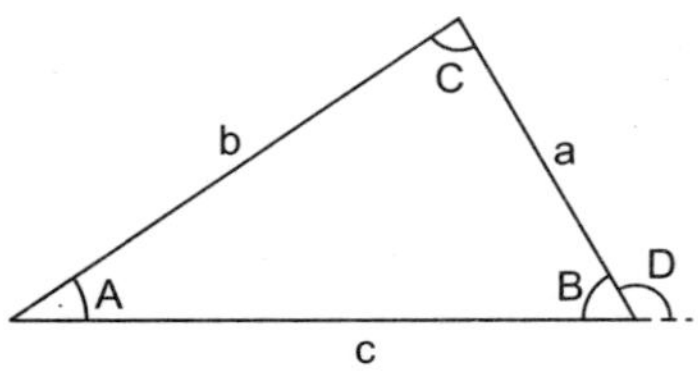

Mathematical Signs and Symbols

= equals

≅ equals approximately

~ is the order of magnitude of

Triangles
Angles are A, B, C
Opposite sides are a, b, c
Angles A + B + C = 180°

$$\frac{\sin A}{a} = \frac{\sin B}{b} = \frac{\sin C}{c}$$

$$c^2 = a^2 + b^2 - 2ab \cos C$$

Exterior angle D = A + C

$\cos \alpha + \cos \beta$

$$= 2 \cos \frac{1}{2}(\alpha + \beta) \cos \frac{1}{2}(\alpha - \beta)$$

$\cos \alpha - \cos \beta$

$$= -2 \sin \frac{1}{2}(\alpha + \beta) \sin \frac{1}{2}(\alpha - \beta)$$

Binomial Theorem

$$(a + b)^n = a^n + na^{n-1}b + \frac{n(n-1)a^{n-2}b^2}{2!} + \frac{n(n-1)(n-2)a^{n-3}b^3}{3!} + \ldots..$$

$$(1 \pm x)^n = 1 \pm \frac{nx}{1!} + \frac{n(n-1)x^2}{2!} + \ldots.. \quad (x^2 < 1)$$

$\neq$ is not equal to

$\equiv$ is identical to, is defined as

$>$ is greater than ($>>$ is much greater than)

$<$ is less than ($<<$ is much less than)

$\geq$ is greater than or equal to (or, is no less than)

$\leq$ is less than or equal to (or, is no more than)

$\pm$ plus or minus

$\propto$ is proportional to

Σ the sum of

$\bar{x}$ or $< x >$ or x_{av} the average value of x

Trigonometric Identities

$$\sin(90° - \theta) = \cos\theta$$

$$\cos(90° - \theta) = \sin\theta$$

$$\sin\theta / \cos\theta = \tan\theta$$

$$\sin^2\theta + \cos^2\theta = 1$$

$$\sec^2\theta - \tan^2\theta = 1$$

$$\text{cosec}^2\theta - \cot^2\theta = 1$$

$$\sin 2\theta = 2\sin\theta\cos\theta$$

$$\cos 2\theta = \cos^2\theta - \sin^2\theta = 2\cos^2\theta - 1$$

$$= 1 - 2\sin^2\theta$$

$$\sin(\alpha \pm \beta) = \sin\alpha\cos\beta \pm \cos\alpha\sin\beta$$

$$\cos(\alpha \pm \beta) = \cos\alpha\cos\beta \mp \sin\alpha\sin\beta$$

$$\tan(\alpha \pm \beta) = \frac{\tan\alpha \pm \tan\beta}{1 \mp \tan\alpha\tan\beta}$$

$$\sin\alpha \pm \sin\beta = 2\sin\frac{1}{2}(\alpha \pm \beta)\cos\frac{1}{2}(\alpha \mp \beta)$$

$$(1 \pm x)^{-n} = 1 \mp \frac{nx}{1!} + \frac{n(n+1)x^2}{2!} + \ldots\ldots (x^2 < 1)$$

$$(1 \pm x)^{-n} = 1$$

Exponential Expansion

$$e^x = 1 + x + \frac{x^2}{2!} + \frac{x^3}{3!} + \ldots\ldots$$

Logarithmic Expansion

If $\log a = x$,

then $a = 10^x$, $\log(ab)$

$= \log a + \log b$ $\log(a/b)$

$= \log a - \log b$

If $\ln a = x$

then $a = e^x$ $\ln(ab) = \ln a + \ln b$ $\ln(a/b)$

$= \ln a - \ln b$

$$\log a^n = n\log a, \ \log 10 = 1$$

$$\ln a^n = n\ln a \quad \ln e = 1$$

$$\ln(1 + x) = x - \frac{1}{2}x^2 + \frac{1}{2}x^3 - \ldots\ldots (|x| < 1)$$

Trigonometric Expansion

(θ in radians)

$$\sin\theta = \theta - \frac{\theta^3}{3!} + \frac{\theta^5}{5!} - \ldots\ldots$$

$$\cos\theta = 1 - \frac{\theta^2}{2!} + \frac{\theta^4}{4!} - \ldots\ldots$$

$$\tan\theta = \theta + \frac{\theta^3}{3} + \frac{2\theta^5}{15} - \ldots\ldots$$

Science Input to Important Technologies

Scientific Principle(s)	*Technology*
Laws of thermodynamics	Steam engine
Controlled nuclear fission	Nuclear reactor
Generation, propagation and detection of electromagnetic waves	Radio and Television
Digital logic	Computers
Light amplification by stimulated emission of radiation	Lasers
Superconductivity	Production of ultra high magnetic fields
Newton's laws of motion	Rocket propulsion
Faraday's laws of electromagnetic induction	Electric generator
Conversion of gravitational potential energy into electrical energy	Hydroelectric power
Bernoulli's principle in fluid dynamics	Aeroplane
Motion of charged particles in electromagnetic fields	Particle accelerators
Reflection of ultrasonic waves	Sonar
Total internal reflection of light	Optical fibers
Thin film optical interference	Non-reflecting coatings
Wave nature of electrons	Electron microscope
Photoelectric effect	Photocell
Magnetic confinement of plasma	Fusion test reactor (Tekamah)
Detection of cosmic radio waves	Giant Metrewave Radio Telescope (GMRT)
Trapping and cooling of atoms by laser beams and magnetic fields.	Bose-Einstein condensate.
Bell's theorem	Quantum computation

Range and Order of Time Intervals

Event	*Time Intervals*
Life-span of most unstable particle	10^{-24}
Time required for light to cross a nuclear distance	10^{-22}
Period of x-rays	10^{-19}
Period of atomic vibrations	10^{-15}
Period of light wave	10^{-15}
Life time of an excited state of an atom	10^{-8}
Period of radio wave	10^{-6}
Period of a sound wave	10^{-3}
Wink of eye	10^{-1}
Time between successive human heart beats	10^{0}
Travel time for light from moon to the Earth	10^{0}
Travel time for light from the Sun to the Earth	10^{2}
Time period of a satellite	10^{4}
Rotation period of the Earth	10^{5}

Rotation and revolution periods of the moon	10^6
Revolution period of the Earth	10^7
Travel time for light from nearest star	10^8
Average human life-span	10^9
Age of Egyptian pyramids	10^{11}
Time since dinosaurs became extinct	10^{15}
Age of the universe	10^{17}

$$\frac{\textit{Maximum Mass}}{\textit{Minimum Mass}} = 10^{41}$$

Range and Order of Lengths

Size of object or distance	*Length (m)*
Size of a proton	10^{-15}
Size of atomic nucleus	10^{-14}
Size of hydrogen atom	10^{-10}
Length of typical virus	10^{-8}
Wavelength of light	10^{-7}
Size of red blood corpuscle	10^{-5}
Thickness of a paper	10^{-4}
Height of the Mount Everest above sea level	10^4
Radius of the Earth	10^7
Distance of moon from the Earth	10^8
Distance of the Sun from the Earth	10^{11}
Distance of Pluto from the Sun	10^{13}
Size of our galaxy	10^{21}
Distance to Andromeda galaxy	10^{22}
Distance to the boundary of observable universe	10^{26}

$$\frac{\textit{Maximum Length}}{\textit{Minimum Length}} = 10^{41}$$

Object	*Mass (kg)*
Electron	10^{-30}
Proton	10^{-27}
Uranium atom	10^{-25}
Red blood cell	10^{-13}
Dust particle	10^{-9}
Rain drop	10^{-6}
Mosquito	10^{-5}
Grape	10^{-3}
Human	10^2
Automobile	10^3
Boeing 747 aircraft	10^8
Moon	10^{23}
Earth	10^{25}
Sun	10^{30}
Milky way galaxy	10^{41}
Observable Universe	10^{55}

$$\frac{\textit{Maximum Mass}}{\textit{Minimum Mass}} = \left(10^{41}\right)^2$$

Negative Powers of 10

Prefix	*Symbol*	*Multiplier Factor*	*Power of 10*
Yocto	*y*	10^{-24}	-24
Zepto	*z*	10^{-21}	-21
Atto	*a*	10^{-18}	-18
Femto	*f*	10^{-15}	-15
Pico	*p*	10^{-12}	-12
Nano	*n*	10^{-9}	-9
Micro	µ	10^{-6}	-6
Milli	*m*	10^{-3}	-3
Centi	*c*	10^{-2}	-2
Deci	*d*	10^{-1}	-1

Positive Power of 10

Prefix	*Symbol*	*Multiplier Factor*	*Power of 10*
Yotta	Y	10^{24}	24
Zetta	Z	10^{21}	21
Exa	E	10^{18}	18
Peta	P	10^{15}	15
Tera	T	10^{12}	12
Giga	G	10^{9}	9
Mega	M	10^{6}	6
Kilo	K	10^{3}	3
Hecto	H	10^{2}	2
Deca	da	10^{1}	1

Bibliography

Following is the list of some popular books, which the students can consult for further understanding of the subject.

Classical Mechanics

1. J.W. LEECH, *Classical Mechanics*, Methuen and Co., 1965.
2. T.W.B. KIBBLE, *Classical Mechanics*, McGraw Hill, 1973
3. Herbert Goldstein, *Classical Mechanics*, 1998
4. A.P. French & M.G. EBISON, Introduction to Classical Mechanics, Chapman & Hall, 1986.
5. Gupta, Kumar, Sharma, *Classical Mechanics*, Pragati Prakashan.
6. N.C. Rana and P.S. Joag, *Classical Mechanics*, Tata McGraw Hill,
7. D.P. Roychowdhury, *Classical Mechanics*
8. J.C. Upadhya, *Classical Mechanics*

Field Theory

1. Resnick Halliday/Walker, *Fundamental of Physics*, 2001
2. Mahajan & Rangwala, *Electricity and Magnetism*, 2000
3. Rajput, *Mathematical Physics*, 1999
4. P.K. Chakraborti, *Advanced Textbook of Physics*
5. Gaur Gupta, *Engineering Physics*, 2000
6. A. K. Mukhopadhyay, *Mathematical Methods for Engineers and Physics*
7. D. Chattopadhyay & P.C. Rakshit, *Electricity & Magnetism*

Vibration & Waves

1. D.P. Raychowdhury, *Advanced Accoustics*, 2001
2. L.E. Kinsler *et al.*, *Fundamental of Accoustics*, 4th edition, John Wiley, 2000
3. H.J. PAIN, *Applied Accoustics*, Peninsula Publishing, 1977
4. N.K. Bajaj, *The Physics of Waves and Oscillations*
5. D.V. Chopra, *Waves and Oscillation*
6. D.P. Roychowdhury, *Advanced Accoustics*

Optics

1. P.K. Chakraborty, *Optics*, 2000
2. Ajoy Ghatak, *Optics*
3. White, *Fundamental of Physics*
4. E. HECHT, Optics, 3rd edition, Addison – Wesley, 1998
5. B. Ghosh & K.G. Mazumdar, *A Textbook on Light*, Sreedhar Publishers
6. C. R. Dasgupta, *Optics (Degree Physics)*

Special Theory of Relativity

1. Robert Resnick, *Introduction to Special Relativity*
2. Arthur Beiser, *Concept of Modern Physics*, 1999
3. W. Rindler, *Essential Relativity*, Springer – Verlag, 1977
4. H.S. Mani and G.K. Mehta, *Modern Physics*

Quantum Physics

1. L. I. Schiff, *Quantum Mechanics*, 3rd Edition, McGraw – Hill
2. A.I.M. RAE, *Quantum Mechanics*, McGraw – Hill, 1986
3. S.M. Mcmurry, *Quantum Mechanics*, Addison Wesley, 1994
4. L. Pauling and E. B. Wilson, *Introduction to Quantum Mechanics*, McGraw – Hill, 1935
5. R. L. Liboff, *Introductory Quantum Mechanics*, 3rd edition,, 1986 Addison Wesley, 1935

Quantum Statistics

1. R. Bowley and M. Sanchez, *Introductory Statistical Mechanics*, Oxford Science,1996
2. L.D. Landau and E.M. Lifshitz, *Statistical Mechanics*, 3rd edition, pergamon, 1980
3. M. R. Spiegel, *Statistics*, *Schaum Outline Series*. McGraw – Hill, 1998
4. E. Schrodinger, *Statistical Thermodynamics*, Dover Publication, 1989
5. N.G. Das, *Statistical Methods*

Crystallography and Solid State Physics

1. C. Kittle, *Introduction to Solid State Physics*, Wiley, 7th edition, 1996.
2. H.E. Hall and J.R. Hook, *Solid State Physics*, 2nd edition Wiley, 1991
3. G. Burns, *Solid State Physics*, Academic Press, 1985
4. N.W. Ashcroft and Mermin, *Solid State Physics*, Hold Saunders, 1976
5. S. O. Pillai, *Solid State Physics*, New Age
6. A. J. Dekker, *Solid State Physics*

Laser and Fibre Optics

1. O. Svelto, *Principles of Lasers*, 4th edition, Plenum, 1998
2. M. I. Beesley, *Laser and their Applications*, Taylor Francis.
3. S. D. Smith, *Optoelectronic Devices*, Prentice Hall International, 1995
4. K. Thyagarajan, A.K.Ghatak, *LASER*
5. B.B. Laud, *Laser and Non-Linear Optics*

Nuclear Physics

1. A.B. Gupta and D. Ghosh, *Atomic and Nuclear Physics*
2. R. J. Blin-Stoyle, *Nuclear and Particle Physics*, Chapman and Hall, 1991
3. N.A. Jelley, *Fundamentals of Nuclear Physics*, Cambridge University press, 1990
4. S. N. Ghoshal, *Atomic and Nuclear Physics*, S.Chand
5. J.B. Rajam, *Atomic & Nuclear Physics*

Index

A

B

C

D

Q

R

S

T

U

V